A Half-Century of the

JOURNAL OF

Polymer Science

A Half-Century of the

JOURNAL OF

Polymer Science

Edited by

DAVID A. TIRRELL
ELI M. PEARCE
MITSUO SAWAMOTO
ERIC J. AMIS

A Wiley-Interscience Publication

John Wiley & Sons, Inc.

New York • Chichester • Weinheim • Brisbane • Singapore • Toronto

PREFACE

A Half-Century of the *Journal of Polymer Science*

In December 1945, P. M. Doty, H. Mark, and C. C. Price announced that "in response to the rapidly increasing literature in the field of polymers" they would transform *Polymer Bulletin,* which during 1945 published summaries of seminars and research at the Polytechnic Institute of Brooklyn, into a new international journal. Their efforts were immediately successful, and thus in 1995 the *Journal of Polymer Science* concluded its first half-century of continuous publication. This 50-year period has witnessed enormous changes in the science and technology of polymeric substances, and in the impact of polymeric materials on the quality of life for literally billions of people. In no small measure, the *Journal of Polymer Science* has played a role in these developments, providing a reliable forum for dissemination of new ideas and new information regarding the preparation and properties of macromolecules.

To highlight this role, the Editors selected a representative set of significant papers published in the *Journal* over the last five decades, which were reprinted in their original format, one per issue, throughout 1996. In addition, the Editors solicited commentary on each paper, from the original authors when possible, but in every case from experts in the field who described the role of the paper in the development of that particular area of polymer science. We hope that this has provided a new window on the history of our discipline, while at the same time to recognizing the contributions of our colleagues and of the *Journal* in establishing the foundations of the field. All of the papers and commentaries have been collected in the special volume. We hope that this collection will serve an educational purpose for students entering the field of polymer science as well as our other stated purposes.

We have been asked many times how this particular set of papers was chosen. As in all of the editorial processes of the *Journal,* we solicited advice from many colleagues and especially from our Editorial Advisory Boards. We considered a large set of candidate papers, and applied where possible quantitative measures of significance such as citation frequency. Nevertheless, the final selections were our own, and while we would never assert that we have chosen uniquely well, we are confident that our readers will be impressed by the quality, scope, and impact of the work that has appeared in these pages over the years.

David A. Tirrell, Mitsuo Sawamoto, and Eli M. Pearce
Editors, *Polymer Chemistry*

Eric J. Amis
Editor, *Polymer Physics*

CONTENTS
Part A: Polymer Chemistry

CONTENTS
Part B: Polymer Physics

A Half-Century of the

JOURNAL OF
Polymer Science

Part A
POLYMER CHEMISTRY

Edited by

DAVID A. TIRRELL
ELI M. PEARCE
MITSUO SAWAMOTO

PERSPECTIVE

Comments on "Intramolecular Reactions in Vinyl Polymers as a Means of Investigation of the Propagation Step" by E. Merz, T. Alfrey, and G. Goldfinger, *J. Polym. Sci.*, 1, 75 (1946)

H. JAMES HARWOOD

The University of Akron, Akron, Ohio 44325-3909

At the time the following article appeared, copolymerization theory was under vigorous development. Several research groups, including Alfrey's, had already analyzed the terminal copolymerization model, had developed the copolymer equation, and had shown how some features of copolymer structure could be calculated from copolymerization information. It is interesting historically that present conventions for monomer reactivity ratios were not even agreed upon until the following summer at a Gibson Island Conference on High Polymers (later to become the Gordon Conference on Polymers).[1]

The Merz–Alfrey–Goldfinger paper was important for the development of copolymerization theory because it introduced the penultimate model for copolymerization, derived the equivalent of the copolymer equation for the penultimate model, and showed how conditional probabilities for monomer enchainments in copolymers could be calculated from monomer concentration ratios and monomer reactivity ratios for both terminal and penultimate copolymerization models. It furthermore showed how the conditional probabilities could be used to calculate such features of copolymer structure as monomer unit sequence distributions, relative monomer pair concentrations (dyad distributions), and the yields that could be expected for cyclization reactions that involved adjacent pairs of like monomer units in the copolymers. The article points out that information related to copolymer structure can

be used to distinguish between the terminal and penultimate copolymerization models and that monomer reactivity ratios could be obtained from copolymer structure and monomer feed ratio information.

The concepts and approaches presented in this article and others in its series were basic to the further development of copolymerization theory and the statistical characterization of copolymer structure. An extensive literature now exists concerning copolymerization systems that are thought to obey the penultimate copolymerization model; other more elaborate copolymerization models such as the antipenultimate model, the monomer complex participation model, and the monomer complex dissociation model have also appeared and many efforts have been made to use copolymer composition and structure information to distinguish among these models.

The Merz–Alfrey–Goldfinger article was the first to propose that copolymer structure information could be used to distinguish among copolymerization models. At the time the article was written, the powerful tools that are available today for characterizing polymer structure, such as IR and NMR spectroscopy, were either not available or were just beginning to be used, and polymer chemists had to rely on the yields obtained in polymer modification or degradation reactions to obtain structure information. Such reactions included ozonolysis, epoxidation, destructive hydrogenation, and cyclization reactions.

Cyclization reactions involving adjacent pairs of monomer units in polymers were particularly attractive to theoreticians at that time because of the interesting challenge of treating them statistically and because information concerning the yields obtained in several such reactions was becoming available, mostly through the efforts of C. S. Marvel's group. Flory had already outlined the general approach to such problems and had shown that the fraction of monomer units remaining uncyclized after random cyclization pairs of adjacent monomer units in a homopolymer was e^{-2} or 0.1356. This was in reasonable accord with results obtained by Marvel and coworkers who treated poly (vinyl chloride) with zinc, expecting that reactions between pairs of head-to-tail enchained monomer units would yield 3-membered rings. (We now know that 5-membered rings also form.) Wall had applied Flory's approach to the random cyclization of adjacent pairs of like monomer units in *random* copolymers and showed that the fraction of such units that would be uncyclized was e^{-2x}, where x is the mole fraction of cyclizable units, provided that the monomers were enchained in a head-to-tail manner.

The Merz–Alfrey–Goldfinger article extended the Flory–Wall treatment to cover intrasequence cyclization reactions of copolymers prepared by either the terminal or penultimate copolymerization models. In the case of the terminal model, they showed that x in the Wall equation should be replaced by $P(A/A)$, the conditional probability that an A unit is attached to another A unit in the copolymer. The article showed that cyclization yield results could be used to distinguish between the two models and probably represents the first published realization of the importance of copolymer structure information for copolymerization mechanism studies.

Also in 1946, but in another journal, Alfrey, Lewis, and Nagel extended the analysis to cover cyclization reactions occurring in copolymers between unlike repeating units (intersequence cyclization). In these two articles, Alfrey and coworkers provided a complete solution for the treatment of cyclization reactions that involve terminal model copolymers. Although the cyclization reactions investigated at the time were rather unsuitable for careful quantitative study, the equations developed by Alfrey and coworkers have been very valuable for subsequent studies where the cyclization reactions occurred cleanly, yielding lactam, lactone, and cyclic imide rings.[2,3]

This article, the ninth to be published in *Journal of Polymer Science*, was probably solicited to help launch the new journal because it is part of a series of papers that appear mostly in the *Journal of the American Chemical Society*. It helped give the new journal a good start because it provided major lasting contributions to several areas of polymer science and established a high standard for subsequent contributors to emulate.

REFERENCES AND NOTES

1. T. Alfrey, F. R. Mayo, and F. T. Wall, *J. Polym. Sci.*, **1,** 581 (1946).
2. H. J. Harwood, in *Reactions on Polymers*, J. A. Moore, Ed., D. Reidel Publishing Co., Boston, 1973, pp. 188–229.
3. N. A. Plate, A. D. Litmanovich, and O. Noah, *Macromolecular Reactions*, Wiley, New York, 1995, pp. 396–398.

Intramolecular Reactions in Vinyl Polymers as a Means of Investigation of the Propagation Step

E. MERZ,* T. ALFREY, and G. GOLDFINGER,† *Polymer Research Institute, Polytechnic Institute of Brooklyn, Brooklyn 2, New York*

Received October 1, 1945

Synopsis — The Flory-Wall expressions for the fraction of substituents remaining on a vinyl copolymer, when substituents are removed at random from adjacent 1,3 positions, are extended to the case of nonrandom copolymerization. A copolymer composition equation is derived for the case in which the type of monomer preceding the active free-radical chain end affects the propagation reaction. The distribution of lengths of sequences in copolymers is discussed.

INTRODUCTION

EXPERIMENTAL COPOLYMERIZATION studies have been shown to afford a unique tool for the investigation of the propagation step in vinyl polymerization. In the polymerization of a single component, the observable variables (over-all rate, molecular weight, etc.) are not determined by the rate constant for any single reaction step, but by rather complicated combinations of rate constants. A low over-all rate of polymerization, for example, could be ascribed to a low rate of initiation, a low rate of propagation, or a high rate of termination.

The *complete* kinetic analysis or a copolymerization reaction is still more complicated, since at least eight distinct unit reactions are involved. However, if one ignores such questions as over-all rate and degree of polymerization and considers only the more limited problems of copolymer composition, simplification can be achieved. It has been shown that the chemical composition of a copolymer, both in the crude sense of over-all composition and in the more detailed sense of the distribution of the units along the chain, is controlled almost completely by the nature of the propagation steps.

Considering four distinct propagation steps—as originally advanced by Norrish and Brookman (3)—Alfrey and Goldfinger (1) and Lewis and Mayo (2) independently derived the following equation regarding the over-all copolymer composition with the composition of the monomer mixture:

$$\frac{dA}{dB} = \frac{a}{b} = \frac{A}{\alpha B} \cdot \frac{\alpha B + A}{\beta B + A} \tag{1}$$

where a and b represent the molar concentrations of the two monomers in the initial copolymer, A and B the molar concentrations in the monomer mixture, and α and β the

---

* Monsanto Chemical Co. Research Fellow.
† Present address: Department of Chemistry, University of Buffalo.

MERZ, ALFREY, AND GOLDFINGER

two significant ratios of propagation rate constants. Wall (12) has emphasized the analogy between copolymerization and the distillation of a binary liquid mixture.

On the same basis, the former authors (1) derived equations for the distribution of lengths of sequences of monomer units of the same kind in the copolymer chain. In certain reactions such as the dechlorination of polyvinyl chloride with zinc, reaction takes place very predominantly with two adjacent monomer units at a time. If this reaction proceeds in a random fashion, it is clear that some monomer units will remain unreacted because of isolation of some chlorines between two reacted pairs. In the case of a copolymer, the number of isolated unreacted groups will be higher. Wall has calculated the percentage of residual isolated groups which remain after the completion of such a reaction in the special case of a "random copolymer" where α and β equal 1. In this paper, it is attempted to generalize these theoretical developments as follows:

(1) The Flory-Wall equations for the percentage of unreacted units in a reaction of the polyvinyl chloride–zinc type will be extended to the general case of a nonrandom copolymer where α and β may have any values.

(2) The copolymer composition equation will be extended to take account of the effect of the group preceding the active chain end upon the reactivity of the chain end.

(3) The Flory-Wall expressions for fraction of monomer units isolated in a reaction of the polyvinyl chloride–zinc type will be further extended in the light of the more detailed picture of the propagation steps.

(4) The distribution of lengths of sequences of monomeric units of a given kind will be calculated for the case where the group preceding the active chain end has an influence on the rate of the addition reactions.

(5) The number of a-b and b-a linkages in a copolymer will be given for both the simple and more general treatment of copolymerization.

(6) The number average size of sequences of monomers in the molecule will be calculated for both treatments of copolymerization.

(7) Equations will be developed for the calculation of some of the copolymerization ratios of rate constants from a limited number of experiments.

GENERALIZATION OF THE FLORY-WALL CALCULATIONS

Wall (4) has shown that, in a random "1,3" oriented copolymer the number of groups, n_{ai},[*] containing i members can be expressed as a function of the total number, N_a, of monomers in the polymer chain which are a's, and the mole fraction, x, of the a monomers in the copolymer:

$$n_{at} = N_a(1-x)^2 x^{i-1} \qquad (2)$$

If the average number of a's remaining untouched in the copolymer after random removal of 1,3 pairs of substituents from a type monomer sequences is S_i for a sequence of i members, then the total remaining number of these substituents, N_r, in the infinite polymer is:

$$N_r = \sum_i S_i n_{at} \qquad (3)$$

The fraction of the substituents left on the a type monomers will be:

$$f = \frac{N_r}{N_a} \qquad (4)$$

[*] In Wall's paper (4), n_{ai} is called P_i and N_a is called N_x.

INTRAMOLECULAR REACTIONS IN VINYL POLYMERS

Recent data (2,5,6,13) on copolymerizations show that the monomers rarely enter the growing chain at random. To replace equation (2) with an equation embodying the condition of nonrandomness of entry of the monomers into the growing chain, we can use equation (5), which gives the number distribution of sequence lengths, $N_{(n)}$, of monomer a as a function of the probability, P_{aa}, that an A monomer will add to an a free-radical chain end:

$$N_{(n)} = (1-P_{aa})P_{aa}^{n-1} \tag{5}$$

By multiplying equation (5) by $N_a(1-P_{aa})$ where N_a is the total number of monomers of type a in the polymer, we obtain equation (6), which states that the total number of sequences of a monomers of lengths i, n_{ai}, in the copolymer is the total number of a sequences times the fraction of these sequences which are i-mers:

$$n_{ai} = N_a(1 - P_{aa})^2 P_{aa}^{i-1} \tag{6}$$

Substituting equation (6) into equation (4) and solving in the manner of Wall (4), we obtain finally:

$$f = 1/e^{2P_{aa}} \tag{7}$$

which states that the fraction of substituents left on the a monomers in the copolymer after the copolymer has been treated so as to remove substituents two at a time from 1–3 positions is equal to $1/e^{2P_{aa}}$. If $P_{aa} = 1$, that is, if only a monomer is present, then the fraction of the substituents left will be $1/e^2$ or 0.1353, in agreement with the calculations of Flory (8) and Wall (4), and the experiments of Marvel and his co-workers (9). It has been show that:

$$P_{aa} = A/(A + \alpha B) \tag{8}$$

where A and B are molar concentrations of the monomers in the polymerizing monomer mixture and α is defined as the rate constant of addition of B monomer to an A free-radical chain end with respect to the rate constant of addition of an A monomer to an A free-radical chain end. Alpha can be determined in principle from equation (1) using the chemical analysis of two copolymers prepared from different initial monomer concentrations. If the appropriate monomers are used, from the chemical analysis of only one copolymer before and after a reaction of the dechlorination-with-zinc type, α can be determined from the following equation:

$$\alpha = \frac{A}{B}\left(1 - \frac{2}{\ln f}\right) = \frac{A}{B}\left(1 - \frac{0.868}{\log f}\right) \tag{9}$$

which is obtained by substituting equation (8) into equation (7) and solving explicitly for α.

EXTENSION OF THE COPOLYMER COMPOSITION EQUATIONS

The average number of substituents left over after an n-mer is subjected to the treatment described above is substantially different from the average number of substituents left on an $(n-1)$- or $(n + 1)$-mer when n is small (10). When the sequences become long (n becomes large), this effect diminishes. It is in the length of the sequences that the effect of the monomer in the chain preceding the free-radical chain end would become noticeable, and perhaps capable of experimental study. We must first derive the copolymer composition equation and set up probabilities of types of addition in order to

MERZ, ALFREY, AND GOLDFINGER

reach an expression for n_{at} as in equation (6). Table I lists the eight possible addition reactions in a two-component copolymerization and also gives the reaction rate constant for each step. The rate of disappearance of monomer A divided by the rate of disappearance of monomer B is then:

$$\frac{dA}{dB} = \frac{a}{b} = \frac{k_2^{aaa}(AA^*)A + k_2^{bba}(BB^*)A + k_2^{aba}(AB^*)A + k_2^{baa}(BA^*)A}{k_2^{aab}(AA^*)B + k_2^{bbb}(BB^*)B + k_2^{abb}(AB^*)B + k_2^{bab}(BA^*)B} \quad (10)$$

TABLE I
TABULATION OF PROPAGATION REACTIONS

Growing chain	Adding monomer	Rate constant (11)	Product
.....AA*	A	k_2^{aaa}	.....AA*
	B	k_2^{aab}	.....AB*
.....BA*	A	k_2^{baa}	.....AA*
	B	k_2^{bab}	.....AB*
.....AB*	A	k_2^{aba}	.....BA*
	B	k_2^{abb}	.....BB*
.....BB*	A	k_2^{abb}	.....BA*
	B	k_2^{bbl}	.....BB*

Equation (10) also expresses the composition, a/b, of the initial polymer. Under steady-state conditions, the following may readily be seen to be true:

$$(AA^*) = \frac{k_2^{baa}}{k_2^{aab}} \frac{(BA^*)A}{B} \quad (11)$$

$$(BB^*) = \frac{k_2^{abb}(AB^*)B}{k_2^{bba}A} \quad (12)$$

$$(AB^*) = \frac{(BA^*)(k_2^{baa}A + k_2^{bab}B)}{k_2^{aba}A + k_2^{abb}B} \quad (13)$$

$$(BA^*) = \frac{(AB^*)(k_2^{abb}B + k_2^{aba}A)}{k_2^{baa}A + k_2^{bab}B} \quad (14)$$

Substituting equations (11), (12), (13), and (14) into equation (10), we obtain:

$$\frac{dA}{dB} = \frac{a}{b} = \frac{k_2^{abb}B + k_2^{aba}A + \left(k_2^{baa}A + \frac{k_2^{aaa}k_2^{baa}}{k_2^{aab}} \frac{A^2}{B}\right)\left(\frac{k_2^{abb}B + k_2^{aba}A}{k_2^{baa}A + k_2^{bab}B}\right)}{\frac{k_2^{bbb}k_2^{abb}}{k_2^{baa}} \frac{B^2}{A} + k_2^{abb}B + (k_2^{baa}A + k_2^{bab}B)\left(\frac{k_2^{abb}B + k_2^{aba}A}{k_2^{baa}A + k_2^{bab}B}\right)} \quad (15)$$

Introducing the following ratios of rate constants into equation (15):

$$\alpha_1 = \frac{k_2^{aab}}{k_2^{aaa}} \qquad\qquad \alpha_2 = \frac{k_2^{bab}}{k_2^{baa}}$$

$$\beta_1 = \frac{k_2^{abb}}{k_2^{aba}} \qquad\qquad \beta_2 = \frac{k_2^{bbb}}{k_2^{bba}}$$

INTRAMOLECULAR REACTIONS IN VINYL POLYMERS

we obtain for the equation linking initial copolymer composition with initial monomer concentrations:

$$\frac{dA}{dB} = \frac{a}{b} = \frac{1 + \dfrac{A}{\alpha_1 B}\left(\dfrac{\alpha_1 B + A}{\alpha_2 B + A}\right)}{1 + \dfrac{\beta_1 B}{A}\left(\dfrac{\beta_2 B + A}{\beta_1 B + A}\right)} \tag{16}$$

If the influence of the monomer in the chain preceding the free-radical chain end is negligible, then $\alpha_1 = \alpha_2 = \alpha$ and $\beta_1 = \beta_2 = \beta$, where α and β are the constants used by Alfrey and Goldfinger (1) and equation (16) reduces to:

$$\frac{dA}{dB} = \frac{a}{b} = \frac{1 + \dfrac{A}{\alpha B}}{1 + \beta \dfrac{B}{A}} = \frac{A}{\alpha B}\frac{\alpha B + A}{\beta B + A} \tag{17}$$

FURTHER GENERALIZATION OF THE FLORY-WALL CALCULATIONS

We now define two probabilities, P_{baa} and P_{aaa}. P_{baa} is the probability that a chain ending in . . . . . BA* will add another A monomer:

$$P_{baa} = \frac{A}{A + \alpha_2 B} \tag{18}$$

and P_{aaa} is the probability that a chain ending in . . . . . AA* will add another A monomer:

$$P_{aaa} = \frac{A}{A + \alpha_1 B} \tag{19}$$

where A and B are molar concentrations of the A and B monomers, respectively, in the polymerizing monomer mixture.

The number distribution $N_{(n)}$ of lengths of sequences of A monomers is then:

$$N_{(n)} = P_{baa}P_{aaa}^{n-2}(1 - P_{aaa}) \tag{20}$$

Equation (20) is not entirely adequate since for the fraction of 1-mers the following holds:

$$N_{(1)} = 1 - P_{baa} \tag{21}$$

The total number of a monomer sequences, n_{ai}, of length i in the copolymer is then the total number of a sequences, $N_a(1 - P_{aaa})$ times equation (20), the fraction of a sequences which are i-mers:

$$n_{ai} = P_{baa}P_{aaa}^{i-2}(1 - P_{aaa})^2 N_a \tag{22}$$

Substituting equation (21) into equation (4), we obtain:

$$f = \sum_i P_{baa}P_{aaa}^{i-2}(1 - P_{aaa})^2 S_i \tag{23}$$

Correcting equation (23) for the fact that the distribution function for lengths of sequences is incorrect when $i = 1$, we subtract the fraction of 1-mers according to (20) from equation (23) and add the correct fraction of 1-mers from equation (21):

MERZ, ALFREY, AND GOLDFINGER

$$f = [\sum_i P_{baa} P_{aaa}^{i-2}(1 - P_{aaa})^2 S_i] - \frac{P_{baa}}{P_{aaa}}(1 - P_{aaa}) + (1 - P_{baa}) \quad (24)$$

The fraction of 1-mers is of course equal to the fraction of substituents left on 1-mers after the prescribed reaction since substituents must come off in pairs from 1,3 positions. Solving equation (24) as before, we obtain:

$$f = 1 + \frac{P_{baa}}{P_{aaa}}(e^{-2P_{aaa}} - 1) \quad (25)$$

which, if the influence of the monomer preceding the free-radical chain end is negligible, reduces to equation (7), since $P_{baa} = P_{aaa}$ and both reduce to P_{aa} (Eq. 8).

From equation (16) α_1, α_2, β_1, and β_2 can be obtained from analyzing the copolymers obtained from four copolymerizations with different monomer concentrations. However, the constants can be obtained from only two different copolymerizations if α_1 and α_2 are calculated from equation (25) (which must be solved by the method of successive approximations) and substituted into equation (16).

POSSIBILITY OF EXPERIMENTAL DETECTION OF INFLUENCE OF MONOMER PRECEDING THE ACTIVE CHAIN END

The fraction of substituents on a monomers which remain after the prescribed reaction is calculated according to equation (25) for assigned values of the various probabilities, and is plotted in Figure 1. As can be readily seen, when the arbitrary parameter, X, in Figure 1 is chosen equal to 1, the plot is of equation (7). From a few rough calculations we have found that if α_1 and α_2 differ by about 10% from α, the calculation of the fraction of substituents remaining according to equation (24) is different by about 2% from the fraction of substituents remaining as calculated from equation (7). Dechlorination as worked out by Marvel and his co-workers (9) would not be sufficiently accurate to determine the magnitude of the variation of α_1 and α_2 from α, but would be sensitive enough to determine whether such a difference existed.

CALCULATION OF AVERAGE SEQUENCE LENGTH

The number of a-b linkages is given by the probability of an a-b linkage being formed, P_{ab}, multiplied by the number of a monomers in the polymer, a. Similarly, the number of b-a linkages is given by bP_{ba}. The number of a-b linkages must of course be equal to the number of b-a linkages, since each sequence of a's must be bounded by at least one b at each end. The total number of a-b and b-a linkages is then:

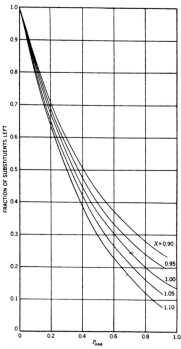

Fig. 1—Fraction of substituents left as a function of the probability, P_{aaa}. $X = P_{baa}/P_{aaa}$.

$$N_{(ab)} = aB\left(\frac{\alpha_1}{A + \alpha_1 B} + \frac{\alpha_2}{A + \alpha_2 B}\right) \quad (26)$$

INTRAMOLECULAR REACTIONS IN VINYL POLYMERS

Equation (26) reduces to:

$$N_{(ab)} = 2\,aP_{ab} = \frac{2\alpha_a B}{A + \alpha B} \tag{27}$$

if we set $\alpha_1 = \alpha_2 = \alpha$. These equations, incidentally, provide another way of determining copolymerization constants, namely, by counting the number of a-b and b-a linkages.

When the average length of sequences of the monomers in the copolymer is short as compared with the total length of the molecule, then the total number of a's divided by the number of a-b linkages gives the number average length of a sequences:

$$\bar{N}_{(a)} = \frac{2}{B}\frac{(A + \alpha_1 B)(A + \alpha_2 B)}{\alpha_1(A + \alpha_2 B) + \alpha_2(A + \alpha_1 B)} \tag{28}$$

which, if α_1 and α_2 are set equal to α, reduces to:

$$\bar{N}_{(a)} = \frac{1}{B}\frac{A + \alpha B}{\alpha} = \frac{A}{\alpha B} + 1 \tag{29}$$

and substituting α from equation (9):

$$\bar{N}_{(a)} = \frac{\ln f}{\ln f - 2} + 1 \tag{30}$$

Thus, from determining f, $N_{(a)}$ can also be easily obtained.

The special cases that come to mind in which the amount of a-b linkages can be readily determined are:

1. Heteropolymers: Since one component, B, cannot add to itself, the total number of B monomers in the chain times two gives the total amount of $(a$-$b)+(b$-$a)$ linkages.

2. Copolymers of vinyl acetate and acrylic esters*: Should one of these copolymers be subjected to hydrolyzing conditions (step 1, figure 2), the formation of a γ-lactone

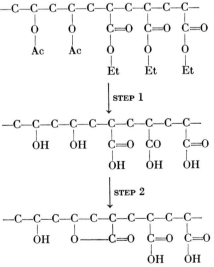

Figure 2

* We are indebted to Dr. C. C. Price for suggesting this to us.

MERZ,   ALFREY,   AND   GOLDFINGER

would proceed readily (Step 2). The nonlactone OH groups could be determined, and hence the number of *a-b* linkages (lactones) could be calculated easily. As can be seen from Figure 2, any sequence of acrylic esters longer than two will have left carboxyl groups not tied up in lactone formation and hence titratable.

   3. Dehalogenation of copolymers of vinyl halide and a nonhalide-bearing monomer: Equations (26) and (27) are directly applicable to these cases.

CONCLUSIONS

   Wall (4) has shown that, in the special case of random 1,3 oriented vinyl copolymers, chemical removal of substituents in pairs yields information about the internal structure of the copolymer. In this paper, Wall's treatment is generalized for the case of nonrandom copolymers (Eq. 7) and refined to account for the influence of the monomer in the chain preceding the free-radical chain end on the addition reactions in nonrandom copolymerizations (Eq. 25). For this latter case, it was necessary first to develop an equation which would relate the copolymer composition to the initial monomer concentrations under this assumption (Eq. 16). The investigation of the possibility of verifying this assumption experimentally led to the conclusion that the influence of the monomer preceding the active chain end, if present, could be detected but not measured quantitatively at present.

   For both treatments of copolymerization, equations were developed which describe the distribution of monomer sequence lengths (Eqs. 20 and 21), give the number average length of the monomer sequences (Eqs. 28 to 30), and count the number of intersequence linkages (Eqs. 26 and 27). Several methods are given for the calculation of some of the copolymerization ratios of rate constants from a limited number of experiments (Eqs. 9, 26, 27, and 29 with 30).

References

(1) T. Alfrey and G. Goldfinger, *J. Chem. Phys.*, **12**, 205, 322 (1944).
(2) F. Lewis and F. Mayo, *J. Am. Chem. Soc.*, **66**, 1594 (1944).
(3) R. Norrish and E. Brookman, *Proc. Roy. Soc. London*, **A171**, 147 (1939).
(4) F. Wall, *J. Am. Chem. Soc.*, **62**, 803 (1940).
(5) T. Alfrey and E. Lavin, *ibid.*, **67**, 2044 (1945).
(6) T. Alfrey, E. Merz, and H. Mark, *J. Polymer Science*, **1**, 37 (1946).
(7) T. Alfrey and E. Merz, *Polymer Bull.*, **1**, 86 (1945).
(8) P. Flory, *J. Am. Chem. Soc.*, **61**, 1518 (1939); **64**, 177 (1942).
(9) C. S. Marvel, G. D. Jones, T. W. Mastin, and G. L. Schertz, *ibid.*, **64**, 2356 (1942).
(10) W. Stockmayer, *J. Chem. Phys.*, **13**, 199 (1945).
(11) J. Abere, G. Goldfinger, H. Mark, and H. Naidus, *Ann. N. Y. Acad. Sci.*, **44**, Art. 4, 267 (1943).
(12) F. Wall, *J. Am. Chem. Soc.*, **66**, 2050 (1944).
(13) F. Lewis, F. Mayo, and W. Hulse, *ibid.*, **67**, 1701 (1945).

COMMENTARY

Reflections on "Relative Reactivities in Vinyl Copolymerization," Turner Alfrey, Jr. and Charles C. Price, *J. Polym. Sci.,* 2, 101 (1947)

CHARLES C. PRICE

The Quadrangle, Haverford, Pennsylvania

When I was a young faculty member in the University of Illinois Chemistry Department, with an active background and interest in reaction mechanisms, "Speed" Marvel suggested that I ought at least study *important* reactions, such as vinyl polymerization. At that time (ca. 1939), Marvel and his students were studying the polymerization of D- & DL-*sec*-butyl α-chloroacrylate, with an interest in the effect of optical activity on the polymer properties. He had found that the optical activity of the monomer and the polymer were quite different, suggesting to me that the kinetics of this reaction could be followed simply by following the change in optical rotation.

We found very simple and straight-forward kinetics, with the rate proportional to the square root of the peroxide initiator concentration and the first power of the monomer concentration, much simpler than earlier kinetics reported for styene polymerization. These observed kinetics were exactly in accord with a slow first-order decomposition to form free radicals (*initiation*), rapid repeated addition of monomer to the radicals (*propagation*), and rapid bimolecular *termination* by radical recombination, confirming a mechanism proposed by Paul Flory.[1]

This rapidly led to speculation that initiation should attach a radical from the initiator to the polymer chain, which was repeatedly confirmed by experiments using labeled initiators at Illinois, Harvard, and the Polytechnic Institute of Brooklyn.

This early research led to two invitations from Herman Mark. One was to join him and Charles Overberger as co-editors of the newly founded Journal of Polymer Science. The second was to come to Brooklyn Poly during the summer of 1945 to give weekly evening lectures. These lectures were subsequently published as a book by Interscience, *Reactions at the Carbon–Carbon Double Bond* (1946).

The opportunity to spend most of the summer of 1945 sharing an office with Turner Alfrey and with opportunities for many stimulating discussions with Herman Mark, Paul Doty, Bruno Zimm, and many frequent visitors such as Frank Mayo and Cheves Walling was a remarkable experience!

Much data was becoming available on the relative rates of the steps in copolymerization of many vinyl monomers. This had led to some of my early[2] qualitative suggestions. Through a review of substitution in the benzene ring[3] I was very familiar with the Hammett σ, ρ scheme.

With considerable stimulus and interaction with Turner Alfrey, this led finally to exploring somewhat similar considerations for vinyl polymerization and thence in 1945 to our proposal of the Q, e scheme.[4,5]

It seemed of some interest to me that the σ-factor of Hammett (assessing an electrical charge effect on ionic reactions in benzene derivatives) is linearly related to the e-factor (assessing an electrical charge effect on free radical reactions in vinyl compounds).

There have been a few added published interpretations of the Q, e scheme.[6]

REFERENCES AND NOTES

1. C. C. Price and R. Kell, *J. Am. Chem. Soc.,* **63,** 2798 (1941).
2. C. C. Price, *Ann. N.Y. Acad. Sci.,* **XLIV,** 351 (1943).
3. C. C. Price, *Chem. Revs.,* **29,** 37 (1941).
4. (a) C. C. Price, *J. Polym. Res.,* **1,** 83 (1946). (b) C. C. Price, *Record of Chemical Progress,* Jan–April 1947, p. 5. (c) C. C. Price, *Faraday Soc. Disc.,* 1947, No. 2.
5. T. Alfrey and C. C. Price, *J. Polym. Sci.,* **2,** 101 (1947).
6. (a) T. C. Schwan and C. C. Price, *J. Polym. Sci.,* **40,** 457 (1959). (b) C. C. Price, *J. Polym. Sci. Part B, Polym. Lett.,* **1,** 433 (1963).

PERSPECTIVE

Comments on "Relative Reactivities in Vinyl Copolymerization," Turner Alfrey, Jr. and Charles C. Price, *J. Polym. Sci.*, 2, 101 (1947)

KENNETH F. O'DRISCOLL

Institute for Polymer Research, University of Waterloo, Waterloo, Ontario, Canada N2L 5Y9

To place the Alfrey–Price Q-e scheme in a current perspective, it is important to read carefully the second paragraph of the original article. There the authors clearly set out their objective: they wanted to characterize each individual monomer's copolymerization behavior by a set of numerical constants and thus correlate reactivity with molecular structure. The success of their work in achieving this objective is attested by (e.g.) the current listing of Q-e values for over 200 monomers in the Polymer Handbook.[1]

As time passed after the appearance of the original article, it became apparent that the Q-e scheme has some flaws, and it is now common to regard it as an empirical scheme which works rather well in a qualitative fashion. As one of the authors has written[2]:

In fact, if we were to look for the simplest *possible* empirical method of correlating copolymerization reactivity with monomer structure, through a linear expression for the free energy of activation, we would probably be led to [the Q-e equations] or their equivalent. Although several investigators have proposed theoretical interpretations of the significance of the parameters, the Q-e scheme is generally regarded as an empirical method of correlation.

Besides the quantum chemical approaches discussed in Alfrey and Young,[2] other theoretical approaches have included Kawabata's inclusion of a surplus conjugation stabilization energy[3] and the "Patterns" scheme of Bamford and Jenkins.[4,5] Both of the latter schemes introduce a third parameter, thus increasing the likelihood of better describing the observed monomer behaviors while necessarily increasing the complexity of the scheme. Despite many other, well-founded or well-intentioned efforts to improve on the Q-e scheme, it remains in the original form.

So we can today regard the Q-e scheme as a minimalist description of monomer reactivity, which is quite good in a qualitative fashion, but which must, because of its empiricism and simplicity, fail for some monomers and monomer combinations. As an example of some failure, we can note that in the most recent tabulation,[1] the correlation coefficients for monomers as common as acrylamide, vinyl chloride, and vinylidene chloride are very low. The low values of these coefficients are a measure of the poor "fit" of the Q and e values to the experimentally obtained reactivity ratios. That these coefficients are so low results from the possible compounding of at least four effects: (1) the empiricism of the Q-e scheme, (2) the incorrectness of the basic Mayo–Lewis model of copolymerization, (3) the experimental inaccuracy of the reactivity ratios used to determine Q and e, and (4) the statistical inaccuracy of the method used to calculate Q and e from the data. In recent times each of the last three has been brought into question: With regard to the second point, it is by no means certain that the Mayo–Lewis model is correct for many systems, but "Bootstrap"[6] or penultimate unit effects may be operative,[7] thus rendering the reactivity ratios inaccurate when they are based on the simple Mayo–Lewis model. With regard to the third point, many experimental reactivity ratio determinations, as recognized by Greenley,[1] are suspect and those parameters should properly be determined only in well-designed experiments using nonlinear least squares analyses.[8] With regard to the fourth point, the Greenley method[1] of computing Q and e values is less than statistically optimal. An alternate approach to computing Q and e from reactivity ratio data has been presented.[9]

The Alfrey–Price Q-e scheme has followed the path of most original and creative science: in spite of its imperfections, it has left us with a strong qualitative sense of the effect of structure on monomer reactivity, and with a hope that future improvements in our understanding of copolymerization may enable us to refine the quantitative use of the scheme.

REFERENCES AND NOTES

1. R. Z. Greenley, *Polymer Handbook,* J. Brandrup and E. Immergut (eds.), Wiley-Interscience, New York, 1989, p. II/267.

2. T. Alfrey, Jr. and L. J. Young, *Copolymerization,* G. E. Ham (ed.), Wiley-Interscience, New York, 1964, p. 67.

3. N. Kawabata, T. Tsuruta, and J. Furukawa, *Makromol. Chem.,* **51,** 80 (1962).

4. C. H. Bamford and A. D. Jenkins, *J. Polym. Sci.,* **53,** 149 (1961).

5. A. D. Jenkins, *Eur. Polym. J.,* **25,** 721 (1989).

6. H. J. Harwood, *Makromol. Chem., Macromol. Symp.,* **10/11,** 331 (1987).

7. T. Fukuda, K. Kubo, and Y.-D. Ma, *Prog. Polym. Sci.,* **17,** 875 (1992).

8. G. Laurier, K. F. O'Driscoll, and P. M. Reilly, *J. Polym. Sci., Polym. Symp.,* **72,** 17 (1985).

9. R. C. McFarlane, P. M. Reilly, and K. F. O'Driscoll, *J. Polym. Sci.,* **18,** 251 (1980).

Relative Reactivities in Vinyl Copolymerization

TURNER ALFREY, JR., *Institute of Polymer Research, Polytechnic Institute of Brooklyn, Brooklyn, New York*

CHARLES C. PRICE, *Department of Chemistry, University of Notre Dame, Notre Dame, Indiana*

Received September 17, 1946

Synopsis. *From data in the literature on relative rates of copolymerization it has been possible to evaluate two constants, Q and e, characteristic of an individual monomer, which appear to account satisfactorily for its behavior in copolymerization. The constant Q describes the "general monomer reactivity" and is apparently related to possibilities for stabilization in a radical adduct. The constant e takes account of polar factors influencing copolymerization.*

THE STUDY of copolymerization reactions is rapidly providing a large body of data relating to the relative reactivities of unsaturated compounds with free radicals. This paper is concerned with the interpretation of such data in the case of vinyl compounds and 1,1-disubstituted ethylenes. The most complete experimental study reported in the literature is that of Lewis, Mayo, and Hulse[1] who investigated the copolymerization of all possible pairs among the four monomers: styrene, methyl methacrylate, acrylonitrile, and vinylidene chloride. Their results with these systems, summarized in Table I, illustrate the well-recognized fact that there exists no unique order of monomer reactivities; the relative rates of reaction of a series of monomers follow different orders, depending upon the character of the free radical with which they are reacting.

It would, of course, be extremely useful if a simple pattern of copolymerization behavior could be found which would allow each individual monomer to be described by characteristic constants. In the first place, this would simplify the experimental task of determining the behavior, in copolymerization, of large numbers of monomer pairs. The present theory of copolymerization permits the quantitative description of any *pair* of monomers in terms of two numbers—the two characteristic relative reactivity ratios. No quantitative method has yet been devised for expressing such experimental results in terms of constants characteristic of individual monomers. The relative reactivity ratios for a given pair of monomers must be determined from experiments involving that particular pair. In order to determine the relative reactivity ratios for all possible pairs among a large number of monomers, experimental studies must be made on every one of the tremendous number of such pairs. If it were possible to copolymerize a given monomer with a limited number of "reference monomers," to compute values of constants characteristic of the individual monomer from these results, and then to predict the behavior of this monomer in other (unstudied) combinations, the practical advantages of

[1] F. M. Lewis, F. R. Mayo, and W. F. Hulse, *J. Am. Chem. Soc.*, **67**, 1701 (1945).

T. ALFREY, JR., AND C. C. PRICE

such a procedure would be considerable.   Furthermore, if each *monomer*, rather than each *pair* of monomers, can be characterized by a set of numerical constants, the correlation of reactivity with molecular structure becomes a more attainable goal than if one must consider the structures of two monomers at a time.   It may even become possible to predict relative reactivities.

Lewis, Mayo, and Hulse[1] have discussed at some length the qualitative aspects of *monomer reactivity* as an important factor influencing the course of copolymerization. Another factor which has been considered of importance is the polarity induced by substituents in the monomer.[1,2]   We may summarize the conclusions of the above authors more or less in the following manner.   A definite order of general monomer reactivity with free radicals does exist.   The position of a monomer in the general reactivity series depends upon the amount of conjugation of the double bond with the substituent group or groups and the associated degree of resonance stabilization of the free radical which is formed by the addition of the monomer to a growing chain.   This general order of monomer reactivity is complicated by a tendency of monomers to *alternate* in copolymerization. This alternation tendency varies from one monomer pair to another.   At least one important source of such alternation is a difference in polarity between the double bonds of the two monomers concerned.   The present paper presents an approach to the quantitative evaluation of these two independent factors affecting copolymerization.

The basic data for the calculations are the relative reactivities of different monomers for different radical end groups as determined by Lewis, Mayo, and Hulse.   Their data, summarized in Table I, indicate the reactivity of each monomer for a particular end group, assigning a value of unity for the reactivity of a monomer for its own end group. Since the unknown absolute values for the rate constants for the individual propagation steps are undoubtedly related to the differing reactivity of the radical end groups, it has been arbitrarily assumed that this factor might be expressed as an unknown constant, γ. Thus, the numbers in the first column in Table I may be considered to be the actual chain-propagation rate constants multiplied by a factor, γ_1, related to the reactivity of the styrene radical end group.   The numbers in the second column would similarly be the actual chain-propagation constants multiplied by a different factor, γ_2.

TABLE I

RELATIVE RATES OF MONOMER ADDITION[1]

Monomer	Radical			
	Styrene	Acrylo-nitrile	Methyl methacrylate	Vinylidene chloride
Styrene	(1.0)	25.0	2.0	7.0
Acrylonitrile	2.5	(1.0)	0.8	2.7
Methyl methacrylate	2.0	7.0	(1.0)	4.1
Vinylidene chloride	0.5	1.1	0.4	(1.0)

Table II represents a generalization of the treatment of relative reactivities in this manner.   The constants k_{11}, k_{23}, etc., represent the actual rate constants for propagation. The four numbers, γ_1, γ_2, γ_3, and γ_4, are four independent unknown constants.   The products, such as $\gamma_2 k_{23}$, correspond to the values found in Table I.   It follows that no direct comparison can be made of individual values from different columns of such a

[2] C. C. Price, *J. Polymer Sci.*, **1**, 83 (1946).

REACTIVITIES IN VINYL COPOLYMERIZATION

table. We can, however, compare the *geometric mean of the reactivities of a given monomer with all four radical types* with the corresponding geometric mean for a different monomer. The geometric mean of the four values in any horizontal row of Table I (or Table II) is

TABLE II

ANALYSIS OF RELATIVE RATES OF MONOMER ADDITION

Monomer	Radical			
	Styrene 1	Acrylo-nitrile 2	Methyl methacrylate 3	Vinylidine chloride 4
1 Styrene	$\gamma_1 k_{11}$	$\gamma_2 k_{21}$	$\gamma_3 k_{31}$	$\gamma_4 k_{41}$
2 Acrylonitrile	$\gamma_1 k_{12}$	$\gamma_2 k_{22}$	$\gamma_3 k_{32}$	$\gamma_4 k_{42}$
3 Methyl methacrylate	$\gamma_1 k_{13}$	$\gamma_2 k_{23}$	$\gamma_3 k_{33}$	$\gamma_4 k_{43}$
4 Vinylidene chloride	$\gamma_1 k_{14}$	$\gamma_2 k_{24}$	$\gamma_3 k_{34}$	$\gamma_4 k_{44}$

equal to the geometric mean of the actual propagation rate constants of the monomer in question, multiplied by the fourth root of the product of the unknown constants. The geometric mean of the values in a different horizontal row involves exactly the same unknown factor. Thus, the geometric means for the four monomers bear the same relation to each other as do the geometric means of the actual propagation rate constants. In the case of the data in Table I, the values of these geometric mean reactivities for the four monomers in question are summarized in Table III.

TABLE III

GEOMETRIC MEAN REACTIVITIES

Monomer	General mean reactivity	Q	e
Styrene	4.3	1.00	−1
Methyl methacrylate	2.75	0.64	0
Acrylonitrile	1.53	0.34	+1
Vinylidene chloride	0.68	0.16	0

This order of the four monomers fits in very well with the theoretical expectations based upon resonance stabilization as the principal factor influencing general monomer reactivity. Styrene, the most "reactive" of the set, exhibits the greatest number of resonance structures for the radical which is formed by addition to the chain. Methyl methacrylate comes second, acrylonitrile next, and vinylidene chloride last. In fact, no ordinary "conjugation" at all exists in the vinylidene chloride radical. A *small* amount of resonance stabilization might be expected from weakly contributing structures in which the carbon-chlorine bond has a certain double-bond character analogous to that postulated in chlorobenzene.

If the numbers in Table I are now divided by the geometric mean reactivities of the monomers concerned and then each vertical column is multiplied by the proper arbitrary factor, the set of numbers found in Table IV can be obtained. These new numbers can be taken to represent the relative rates of addition of the various monomers to each type of free radical *after correction is made for differences in general monomer reactivity.* These values fit a simple pattern. All numbers corresponding to addition processes in which either methyl methacrylate or vinylidene chloride is involved, as monomer or as radical, are approximately equal to unity. The four numbers involving only styrene and acrylonitrile differ considerably from unity. The numbers for styrene adding to styrene and for acrylonitrile adding to acrylonitrile are both considerably *less* than unity, while the numbers corresponding to styrene adding to acrylonitrile, and vice versa, are decidedly *greater* than unity. These values are consistent with the hypothesis that the

TABLE IV

RELATIVE RATES OF MONOMER ADDITION (TABLE I) CORRECTED FOR GENERAL MONOMER REACTIVITY (Q)

Monomer	Radical			
	Styrene	Acrylo-nitrile	Methyl methacrylate	Vinylidene chloride
Styrene	0.32	2.71	1.03	1.1
Acrylonitrile	2.24	0.33	1.16	1.2
Methyl methacrylate	0.997	1.27	0.81	1.01
Vinylidene chloride	1.01	0.81	1.3	0.99

double bond in styrene—and therefore the carbon atom with the odd electron in the styrene radical—is electron-rich (negative), the double bond in acrylonitrile is electron-poor (positive), while the double bonds in methyl methacrylate and vinylidene chloride are approximately "neutral." If this is true, one would expect the relative rate of addition of each neutral monomer to *all* radicals to be determined primarily by the over-all reactivity of the monomer, and also for the competition of *all monomers* for the neutral radicals to be determined by the over-all reactivities of the monomers. Furthermore, the rate of addition of styrene to styrene radical and acrylonitrile to acrylonitrile radical should be reduced by electrostatic repulsion to values smaller than those expected on the basis of the mean reactivities of the monomers; likewise, the rate of styrene adding to acrylonitrile, and vice versa, should be increased by electrostatic attraction beyond the values expected on the basis of mean monomer reactivities alone.

Just as the empirically determined mean reactivities fit in reasonably well with theoretical deductions from electronic structure, so do the polarities. This general pattern of behavior is consistent with the hypothesis that the rate constant for the addition of a monomer of type 2 to a free radical of type 1 (see Table II) can be given to a good approximation by an expression such as the following (at constant temperature):

$$k_{12} = A_{12}e^{-(p_1 + q_2 + e_1e_2)}$$

In this expression for the propagation step, A_{12} represents the probability factor, p_1 is an activation factor related to the general reactivity of the polymer end group, q_2 is a similar factor related to the general monomer reactivity, and e_1 and e_2 are the two electrical factors. Since the monomers considered have all been those with one free methylene end, it has been assumed that the factor A will be essentially constant. One may then modify the expression as follows:

$$k_{12} = P_1Q_2e^{-e_1e_2}$$

P_1 is characteristic of radical 1 (see Table II); Q_2 is the mean reactivity of monomer 2; e_1 is proportional to the charge on the end group of radical 1; and e_2 is proportional to the charge on the double bond of monomer 2.

REACTIVITIES IN VINYL COPOLYMERIZATION

It would be a reasonable speculation, for example, to write:

$$k_{12} = P_1 Q_2 e^{-(c_1 c_2 / rDkT)}$$

where c_1 and c_2 are the actual charges on radical 1 and monomer 2, D is the effective dielectric constant, and r the distance of separation in the activated complex. The characteristic constant, e, would then be related to the actual charge, c, by the proportionality:

$$e_1 = c_1 / \sqrt{rDkT}$$

The relative rates at which monomers 1 and 2 compete for radicals of type 1 is thus given by an expression such as the following:

$$k_{11}/k_{12} = (Q_1/Q_2) e^{-e_1(e_1 - e_2)}$$

$$k_{21}/k_{22} = (Q_1/Q_2) e^{-e_2(e_1 - e_2)}$$

The constant P, connected with the reactivity of the radical (and related to the γ constants discussed above), cancels out.

This leads to the possibility of calculating relative copolymerization ratios if the two characteristic numbers, Q and e, are known for both monomers. The above analysis of the data of Mayo, Lewis, and Hulse[3] leads to the values of these two constants included in Table III for each of the four monomers involved. If the characteristic constants of Table III are used to calculate a set of relative reactivities for the various propagation steps, the calculated values agree fairly well with the experimental values from which the constants were derived. Calculated and experimental values of these relative reactivities are shown in Table V.

As a suggested experimental program for determining the characteristic constants Q and e for a given monomer, one might first copolymerize the monomer with one or two "neutral" monomers in order to determine the general reactivity of the monomer in question and then follow this by copolymerization with one positive and one negative monomer in order to fix the polarity value. Actually, if one calculates Q values for the four monomers in question from their behavior with the two neutral monomers alone, the results

TABLE V

COMPARISON OF OBSERVED AND CALCULATED RELATIVE RATES OF MONOMER ADDITION

Radical	Monomer	Relative rate of addition	
		Calculated	Observed
Styrene	Styrene	1.0	(1.0)
Styrene	Acrylonitrile	2.5$_2$	2.5 (2.0–3.0)
Styrene	Methyl methacrylate	1.74	2.0 (1.9–2.1)
Styrene	Vinylidene chloride	0.43	0.5 (0.47–0.53)
Acrylonitrile	Styrene	22.0	25.0 (12–100)
Acrylonitrile	Acrylonitrile	1.0	(1.0)
Acrylonitrile	Methyl methacrylate	5.1	7 (5–12)
Acrylonitrile	Vinylidene chloride	1.28	1.1 (1.0–1.2)
Methyl methacrylate	Styrene	1.6	2.0 (1.9–2.1)
Methyl methacrylate	Acrylonitrile	0.53	0.83 (0.75–0.95)
Methyl methacrylate	Methyl methacrylate	1.0	(1.0)
Methyl methacrylate	Vinylidene chloride	0.25	0.40 (0.38–0.41)
Vinylidene chloride	Styrene	6.3	7 (5–11)
Vinylidene chloride	Acrylonitrile	2.13	2.7 (2.1–3.7)
Vinylidene chloride	Methyl methacrylate	4.0	4.1 (3.7–4.7)
Vinylidene chloride	Vinylidene chloride	1.0	(1.0)

[3] F. M. Lewis, F. R. Mayo, and W. F. Hulse, *J. Am. Chem. Soc.,* **67,** 1701 (1945).

T. ALFREY, JR., AND C. C. PRICE

obtained are reasonably similar to those deduced from the behavior with all four monomers.

Alternatively, it will be noted that division of the two relative reactivities leads to the following simple expression involving only the difference in polarity.[4]   If the reference

$$(k_{11}/k_{12})/(k_{21}/k_{22}) = e^{-(e_1-e_2)^2}$$

monomer of the pair has had the value of e established, the other value is thus readily estimated.   It is then a simple matter to establish the value of Q for the new monomer.

NOTE ADDED IN PROOF: Cheves Walling, of the U. S. Rubber Laboratories, has pointed out that only the *differences in polarity* among a group of monomers can be unambiguously determined from copolymerization data, using our equation.   The location of the reference zero of polarity is entirely arbitrary.   Once the e value of one monomer has been set, however, those of the other monomers are fixed if our equation holds.

The specific procedure used in this paper *arbitrarily* sets the scale of e values so that the average for the four monomers studied by Lewis, Mayo, and Hulse[3] is zero.   Fortunately, for interpretation in terms of electronic structure, this appears to be very nearly the correct location for the scale, although *any* arbitrary location would, with suitable readjustment of Q values, lead to an equally satisfactory calculation of copolymerization ratios.

Résumé

À partir de données prises dans la littérature, concernant des vitesses relatives de copolymérisation, il a été possible de déterminer deux constantes, Q et e, caractéristiques d'un monomère individuel et qui semblent rendre compte de manière satisfaisante de son comportement dans la copolymérisation.   La constante Q exprime la "réactivité générale du monomère" et est apparemment reliée aux possibilités de stabilisation du radical.   La constante e prend en considération les facteurs polaires influençant la copolymérisation.

[4] The authors are indebted to F. T. Wall of the University of Illinois for pointing out this simplification.

PERSPECTIVE

Comments on "Une Nouvelle Classe de Polymeres d'α-Olefines ayant une Régularité de Structure Exceptionnelle," by G. Natta, *J. Polym. Sci.*, XVI, 143 (1955)

PAOLO CORRADINI

Dipartimento di Chimica, Università Federico II, Via Mezzocannone 4, I-80134, Napoli, Italy

The discovery of how to synthesize stereoregular polymers of olefins was made by Natta and his co-workers in 1954[1,2] (soon after that of Ziegler, for the preparation of linear ethylene polymers), using as catalysts combinations of aluminum alkyls and transition metal chlorides. In a few years after the discovery, an entire new chapter had been disclosed in the field of macromolecular chemistry. It was possible to obtain polymers from olefinic and various other hydrocarbon monomers, with an extraordinary regularity of structure from both their chemical constitution and the configuration of the successive monomeric units along the chain of each macromolecule ("stereoregular" polymers).

At the end of the 1960s, new catalytic systems for the polymerization of ethylene were implemented, in which the titanium chloride is supported on a matrix, as, for instance, magnesium oxide or chloride. These new catalysts show a very high activity in the polymerization of ethylene, with yields of the order of 10^6 instead of 10^4 grams of polymer per gram of titanium.

The new supported catalysts, however, were unsatisfactory for the polymerization of propylene, where control of the succession of the m versus the r configurations along the polymer chain is necessary.

Research to design a high-yield supported catalyst for the isotactic polymerization of propylene has led, with the help of the previous experience in the field, to the implementation in the industrial research laboratories of new catalytic systems, which are capable of such high yields (thousands of kilograms of polymer per gram of titanium) that the purification process can be eliminated in the production plants; on the other side, the isotacticity is so high that the heavy cost of the extraction of the amorphous fraction can be considerably reduced. The research efforts, started already by Professor Natta soon after the discovery, for a better comprehension of the action mechanism of the polymerization catalysts, of the nature of the catalyst surface, and of the influence of various chemical agents are now paying back, in this way, also in terms of a simpler industrial process. The production of isotactic polypropylene, in the whole world, is presently of the order of 15 million tons per year!

The discovery of homogeneous stereospecific catalysts for the polymerization of α-olefins—a further big breakthrough—was achieved 10 years ago by Ewen,[3,4] on the basis of earlier research on metallocenes in combination with alkyl-Al-oxanes by Kaminsky and Sinn.[5] It has opened up new prospects for research on stereospecific polymerization and on stereoregular polyolefins.

Depending on the specific metallocene π-ligands used, these systems present completely different stereospecific behaviors. For example, catalytic systems containing the metallocene stereorigid ligand ethylene-*bis*(1-indenyl) or ethylene-*bis*(4,5,6,7-tetrahydro-1-indenyl) polymerize α-olefins to isotactic polymers while catalytic systems comprising the metallocene stereorigid ligand isopropyl(cyclopentadienyl-1-fluorenyl) instead polymerize α-olefins to syndiotactic polymers (Fig. 1). In the case of propylene, it is possible to obtain at will syndiotactic and isotactic (Fig. 2), hemitactic and atactic polymers, tuning the catalyst structure to the desired polymer properties.

One of the most exciting features of these homogeneous catalysts is that the structure of the catalyst precursors can be accurately determined and the influence of the ligand geometry on the stereospecificity of the polymerization reactions can be studied in detail.

Correlations of the polymer microstructure with the structure of the catalyst precursor have revealed an extraordinary amount of information on the po-

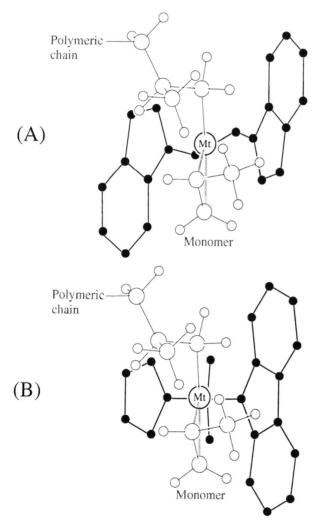

Figure 1. Models of homogeneous catalytic sites for the polymerization of propylene. The structures of the ligands π-coordinated to the metal (Mt) in the catalytic intermediates, which precede the insertion of a monomer unit, are evidenced with a marked line (and black small circles for the carbon atoms). (A) *rac*-ethylene-*bis*-(1-indenyl); (B) isopropyl(cyclopentadienyl-1-fluorenyl).

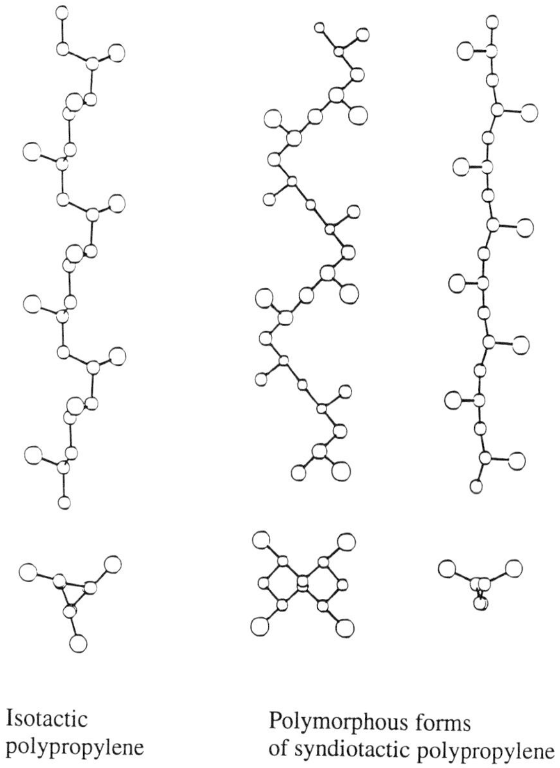

Isotactic polypropylene

Polymorphous forms of syndiotactic polypropylene

Figure 2. Models of the chain of stereoregular propylene polymers in the crystalline state. Projections along and perpendicular to the chain axis.

lymerization mechanism and the origin of stereodifferentiation for the polymerization reactions. It has been possible to show that, for the homogeneous as well as for the heterogeneous Ziegler–Natta catalysts of polymerization of olefins, nonbonded interactions, enforcing a chiral orientation of the first C—C bond of the growing chain, play the most important role in determining their surprising and tunable specificities.[6]

A main feature of the new homogeneous catalytic systems is that they can be single site, that is, they comprise all identical catalytic sites, thus allowing better control of the molecular mass distribution as well as, for copolymers, better control of the comonomer composition and distribution. This can be a great advantage with respect to the heterogeneous catalytic systems, for which several sites with different reactivities and regio- and stereospecificities are present. Relevant industrial applications may be foreseen in the future.

Hence, it has been possible to tune the structure of these catalysts to the preparation of a series of new stereoregular polymers, in particular of a series of new crystalline syndiotactic polymers.

The discovery of soluble catalysts has also allowed the preparation of new types of stereoregular hydrocarbon polymers, giving rise to a reflourishing of structural studies in our and other laboratories, in particular on the newly synthesized syndiotactic polystyrene and on many new syndiotactic polyolefins.[7] For the hydrocarbon polymers, which can be obtained with Ziegler–Natta heterogeneous and homogeneous catalysts, a new flowering of fundamental studies and industrial utilizations is to be anticipated in the near future.

REFERENCES AND NOTES

1. (a) G. Natta, *Acc. Naz. Lincei. Mem.*, **4**(sez. 2), 61 (1955); (b) G. Natta, *J. Polym. Sci.*, **16**, 143 (1955).

2. (a) G. Natta and P. Corradini, *Acc. Naz. Lincei. Mem.*, **4**(sez. 2), 73 (1955); (b) G. Natta, P. Pino, P. Corradini, F. Danusso, E. Mantica, G. Mazzanti, and G. Moraglio, *J. Am. Chem. Soc.*, **77**, 1708 (1955).

3. J. A. Ewen, *J. Am. Chem. Soc.*, **104**, 6355 (1984).

4. J. A. Ewen, R. L. Jones, A. Razavi, and J. D. Ferrara, *J. Am. Chem. Soc.*, **110**, 6255 (1988).

5. H. Sinn and W. Kaminsky, *Adv. Organomet. Chem.*, **18**, 99 (1980).

6. (a) P. Corradini, V. Barone, R. Fusco, and G. Guerra, *Eur. Polym. J.*, **15**, 133 (1979); (b) P. Corradini, V. Barone, R. Fusco, and G. Guerra, *J. Catal.*, **77**, 32 (1982); (c) G. Guerra, L. Cavallo, G. Moscardi, M. Vacatello, and P. Corradini, *J. Am. Chem. Soc.*, **116**, 2988 (1994).

7. P. Corradini and G. Guerra, *Adv. Polym. Sci.*, **100**, 183 (1992).

JOURNAL OF POLYMER SCIENCE   VOL. XVI, PAGES 143–154 (1955)

Une Nouvelle Classe de Polymeres d'α-Olefines ayant une Régularité de Structure Exceptionnelle*†

G. NATTA, *Istituto di Chimica Industriale del Politecnico, Milano*

Les polymères vinyliques, dérivant de la polymérisation tête-queue de monomères du type $CH_2=CHR$, sont caractérisés par la possession de chaînes principales dans lesquelles les atomes de carbone asymétriques alternent avec des groupes méthyléniques.

L'incapacité de cristalliser généralement reconnue des polymères contenant une succession de groupes $-CH_2-CHR-$ lorsque le groupe R a des dimensions bien plus grandes que l'hydrogène lié au carbone avait été justement attribuée à la configuration différente des atomes de carbone asymétriques. C'est pour cela que l'on n'obtient pas dans la chaîne de tels polymères une répétition régulière d'éléments de substitution spatialement identiques, mais une succession, la plupart du temps désordonnée statistiquement, de deux unités structurales, différentes, qui ne diffèrent entre elles que par la configuration stérique des atomes de carbone asymétriques.

Tous les polymères, décrits dans la littérature scientifique, dérivant d'oléfines du type sus-mentionné, dans lesquels R est un groupe alkylique, sont en effet des liquides ou des solides amorphes.

La présence de groupes méthyliques latéraux (ayant un rayon effectif de van der Waals d'à peu près 2 A.), ou d'autres groupes de dimensions encore plus grandes dans ce type de polymères était en outre considérée comme un empêchement sûr de la formation de chaînes complètement étendues ayant la structure en zig-zag des paraffines et se trouvant sur un plan.

Selon Flory,[1] il y aurait une situation plus favorable, pour obtenir une

* Le présent mémoire, qui a un caractère d'introduction sur l'identification et sur quelques propriétés générales d'une nouvelle catégorie très intéressante de polymères d'α-oléfines, préparés par nous, ayant une régularité extraordinaire de structure, une haute cristallinité et des propriétés physiques et mécaniques exceptionnelles, ne représente qu'une partie d'un travail complexe effectué en 1954 en collaboration avec la Société Montecatini et de nombreux chercheurs, auprès de l' "Istituto di Chimica Industriale del Politecnico di Milano" et avec la collaboration de M. Piero Pino pour l'organisation de recherches de chimie organique et pour la discussion des résultats, de Giorgio Mazzanti, Ettore Giachetti et, en un premier temps, de Paolo Chini pour la préparation et purification des nouveaux polymères, de Paolo Corradini pour les recherches aux rayons X, de Ferdinando Danusso et Giovanni Moraglio pour les mesures de viscosité intrinsèque et dilatométriques, d'Enrico Mantica, Mario Peraldo et Luisa Bicelli pour la spectrographie I.R. et de Giuseppe Lutzu pour les essais mécaniques.

† Presentée à l' "Accademia Nazionale dei Lincei" in Rome, le 11.12.1954.

chaîne à structure régulière plane: ce serait celle où alternativement se succèderaient le long de la chaîne les atomes de carbone asymétriques avec une configuration stérique opposée.

Une structure semblable avait été proposée pour les polymères cristallins du vinylisobutyléther, obtenus par C. E. Schildknecht, S. T. Gross, H. R. Davidson, I. M. Lambert, et A. O. Zoss[2] à température très basse avec une vitesse de polymérisation extrêmement lente.

Les empêchements stériques s'atténuent seulement si les groupes R présent (comme le F, OH) un rayon d'action inférieur à 1,4 A., c'est-à-dire des dimensions pas beaucoup plus grandes que celles de l'hydrogène lié au carbone, et, dans ce cas, la formation de la chaîne rigide en zig-zag plane, typique de la structure rombique des paraffines, est encore possible.

C'est pour cela que nous avons été bien surpris, lorsque dans les premiers jours de mars 1954 dans le fractionnement de hauts polymères solides du propylène de haut poids moléculaire, que nous avions préparé par des procédés et dans des conditions permettant d'obtenir des polymères avec des chaînes essentiellement linéaires (c'est-à-dire sans ramification plus longues que celles correspondant au groupe R), nous avons réussi à séparer une proportion considérable de polymères qui à l'examen aux rayons X apparaissaient comme nettement cristallins.

Par la suite, nous avons préparé des polymères cristallins d'autres α-oléfines appartenant à la série aliphatique et du styrène et nous avons approfondi l'étude de leur structure pour nous rendre compte de la constitution de ces produits nouveaux.

Nous parelerons ailleurs des méthodes, de caractère général, qui nous ont permis d'obtenir avec facilité, des quantités importantes de ces polymères cristallins que nous pouvoir rendre, à volonté, cristallisés ou amorphes. Ce dernier point fait encore de l'objet de recherches systématiques.

Nous nous bornons à la description des méthodes que nous avons employées pour la séparation des polymères cristallins, sur leurs propriétés vraiment singulières et sur les relations entre les propriétés et la structure des chaînes.

Séparation des Polymères Cristallins des Polymères Amorphes et Leurs Propriétés

Le fractionnement, par extraction par des solvants, de certains polymères bruts du propylène, nous avait donné déjà en mars 1954 une série de fractions, dont les propriétés n'apparaissaient pas une fonction régulière de la viscosité intrinsèque des solutions. Par exemple, des fractions ayant une viscosité intrinsèque de 1,20, insolubles dans l'éther, étaient nettement cristallines jusqu'à la température d'au moins 130°, alors que des fractions insolubles dans l'acétone et solubles dans l'éther, ayant une viscosité intrinsèque de 1,0, étaient complètement amorphes, même à basse température. Les fractions intermédiaires ne semblaient pas constituées par un polymère de propriétés intermédiaires, mais par un mélange hétérogène de deux parties, l'une cristalline et l'autre amorphe.

Ces observations nous ont amenés à étudier la cause de ces anomalies que l'on n'observe pas, par ex., lorsque l'on fractionne des polymères obtenus dans des conditions de polymérisation ordinaires.

C'est pour cela que nous avons effectué des extractions successives de polymères bruts par des solvants choisis parmi les suivants: acétone, acétate d'éthyle, éther, n-heptane, benzène, toluène, qui nous enumerons ici dans l'ordre croissant de leur pouvoir de dissolution à leur température d'ébullition.

Le produit extrait par l'éther du polypropylène est complètement amorphe, celui insoluble dans le n-heptane bouillant est pratiquement tout à fait cristallin. Pour le poly-α-butylène, une séparation des deux types produits peut être effectuée moyennant si l'éther est bouillant. Le polystyrène cristallin est facilement séparable de celui amorphe qui l'accompagne à cause de sa complète insolubilité dans le n-heptane bouillant.

Dans la Tableau I sont comparées certaines propriétés caractéristiques, communes aux polypropylènes, poly-α-butylènes, polystyrènes, pour les deux types de polymères que nous avons ainsi séparés.

TABLEAU I

Polymères	Amorphes non cristallisables	Cristallins
Aspect	Caoutchouteux à une température supérieure à la température de transition de 2ème ordre, vitreux aux températures inférieures	Poudre blanche qui donne par moulage, à des températures proches de la température de transition de 1er ordre, des lamelles tenaces et flexibles.
Examen roentgeno-graphique	Bandes typiques des polymères amorphes, linéaires	Cristallinité jusqu'aux températures d'au moins 120°.
Spectres I.R. à lumière polarisée	Absence de dichroïsme	Dichroïsme marqué
Analyse thermique	Fusion ou ramollissement aux températures inférieures à 100°	Température de transition de 1er ordre > 120°
Densité	Plus basse	Plus forte
Comportement aux déformations mécaniques	Visqueux-élastique aux températures supérieures à celle de transition de 2ème ordre	Orientation des cristaux et anisotropies correspondantes. elasticité reversible jusqu'à des deformations d'environ 20%
Résistance à la traction	Basse	Très élevée pour les échantillons orientés de polymères à poids moléculaire très élevé (30–80 kg./mm.²)

Bien que les mesures de viscosité intrinsèque effectuées sur une centaine d'échantillons aient fourni des valeurs généralement plus hautes (>1,5) pour les produits cristallins que pour ceux amorphes, séparés du même produit brut, on ne doit pas attribuer cependant à un poids moléculaire différent la cause des différentes propriétés des deux types de produits.

146   G. NATTA

Nous avons en effet pu séparer des fractions, les unes cristallines, les autres amorphes, ayant dans les solutions la même viscosité intrinsèque.

En outre, une solution en tétraline de 1,7% de polypropylènes fortement cristallins, ayant une viscosité intrinsèque de 1,85 a fourni par dépolyméri-

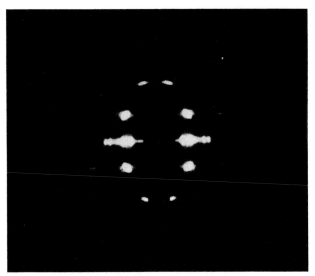

Plate I. Polypropylène cristallin.

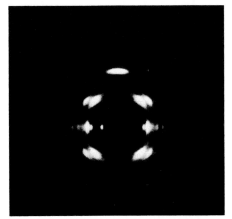

Plate II. Polystyrène cristallin.

sation thermique après 37 h à 335° en atmosphère d'azote très pur (0,002% O_2) des produits encore cristallins ayant une viscosité intrinsèque de 0,17.

Par contre, nous avons séparé des polypropylènes complètement amorphes, insolubles dans l'éther, ayant une viscosité intrinsèque supérieure à 1,5.

Dans le Tableau II, on compare les propriétés de quelques fractions de polypropylènes et polybutylènes des deux types. Les viscosités intrin-

TABLEAU II

Polymère	Viscosité intrinsèque, 100 cm.³/g.	Densité expérimentale	Température de transition de 1er ordre, °C.	Temp. de fusion initiale	Solubilité dans		
					éther	n-heptane	toluène
Polypropylène cristallin de haut poids moléculaire	2,40	0,92	158–160	—	ins.	ins.	sol.
Polypropylène cristallin de bas poids moléculaire	0,17	0,91	149	—	ins.	sol.	sol.
Polypropylène amorphe	0,55	0,85	—	75	sol.	sol.	très sol.
Poly-α-butylène cristallin	1,02	0,91	126–128	—	ins.	sol.	très sol.
Poly-α-butylène amorphe	0,35	0,87	—	65	sol.	très sol.	très sol.

sèques ont été déterminées en employant comme solvant la tétraline à 135°. La température de transition indiquée représente la température à laquelle les diffractions avec les rayons X, typiques de la phase cristalline, disparaissent complètement. Les solubilités dans les différents solvants à la température d'ébullition sont indiquées qualitativement (ins. = insoluble; sol. = soluble, très sol. = très soluble).

On observe des différences encore plus marquées entre le polystyrène cristallin et le polystyrène amorphe. Un polystyrène insoluble dans l'acétone, dans l'éther et dans le n-heptane, soluble dans le toluène, ayant une viscosité intrinsèque à 25°: $[\eta] = 3,8$ (10^2 cm.³/g. en benzène), qui se maintient cristallin presqu'au dela de 200°, et qui présente une température de transition de 1er ordre déterminée par voie dilatométrique de ∼220°C, est comparé ci-dessous avec les polystyrènes amorphes du commerce:

	Densité	Temp. de transition
Polystyrène cristallin	1,08	∼220 °C (1er ordre)
Polystyrènes amorphes	1,04–1,065	70–100 °C (2ème ordre)

Les différences considérables dans les propriétés physiques des deux types de polymères doivent être attribuées à une cause commune, qui doit être cherchée dans la structure des chaînes principales des macromolécules.

Forme Linéaire de la Chaîne Principale

Il résulte de la littérature sur la chimie macromoléculaire que des différences considérables dans les propriétés des hauts polymères peuvent dériver de la présence ou de l'absence de ramifications fréquentes de chaînes principales. Par exemple, le polyméthylène, exempt de ramifications, obtenu par décomposition avec du BF₃, du diazométhane, est trè cristallin, tandis que le polyéthylène, obtenu par amorçage de radicaux

148   G. NATTA

libres à haute température, est fortement ramifié et il contient normalement 50% environ de partie amorphe; on a trouvé qu'un polyéthylène du type précédent, auquel nous avions greffé des nombreuses ramifications éthyliques, était complètement amorphe.[4]

La haute cristallinité trouvée dans certains de nos polypropylènes et polystyrènes (voir figure hors texte) ne peut pas être compatible avec la présence de ramifications fréquentes dans la chaîne principale.

Même le polypropylène amorphe (sous-produit de la préparation du polypropylène cristallin), soluble dans l'éther, préparé par nous, ne devrait pas être très ramifié, car pour des poids moléculaires supérieurs à 5.000 il a la consistance d'un solide, tandis qu'il résulte de la littérature que les polypropylènes ramifiés, obtenus par polymérisation avec du $AlBr_3$,[5] sont des liquides visqueux, même pour des poids moléculaires de plusieurs dizaines de milliers.

Le spectre aux rayons X d'un polypropylène amorphe, vulcanisé après sulfochloration avec introduction d'environ 1% de soufre, maintenu sous tension, présente la bande principale de l'amorphe correspondant à 5,3 A., sur l'équateur du spectre, et cela démontre l'orientation parallèle de bouts de chaînes linéaires, distancés normalement par des groupes de dimensions non supérieures à celles du groupe méthylique.

TABLEAU III

SPECTRE INFRAROUGE DU POLYPROPYLÈNE ENTRE 7 ET 15 μ

	Polypropylène cristallisable			
	Produit cristallin orienté direction d'allongement ⟶			
Produit cristallin	Vecteur électrique ⟵⟶	Vecteur électrique ↕	Produit fondu	Polypropylène amorphe non cristallisable
7,53 f	—	7,53 f		
7,67 m	7,67 m	—	—	7,67 f
7,71 e	—	7,72 f	—	—
7,97 m	7,96 m	—	7,97 f	7,97 m
				8,13 m
8,20 f	—	8,20 f	—	—
8,57 F	8,56 F	e		
8,66 e	e	8,67 m	8,70 F, l	8,66 F
9,06 f	9,06 f	9,06 f	9,06 f	9,06 e
9,57 f	9,57 f	—	—	—
9,65 e	—	—	—	—
10,02 F	10,02 F	10,02 f	10,02 f, m	10,02 f, m
10,28 F	10,28 F	10,28 m	10,28 F	10,28 F
10,64 f	—	10,63 f	—	
11,12 m	11,12 ff	11,12 m	Large zone d'absorption entre 11 et 12,7 μ avec des minima à 12,3 et 11,2 μ	Large zone d'absorption entre 11 et 12,7 μ avec des minima à 11,12–12,33 et épaulement 11,90
11,89 F	11,89 F	11,89 f		
12,36 m	12,36 ff	12,36 m		

ff = très faible.   m = medium.   f = faible.   F = forte.   e = épaulement.   l = large.

La grande ressemblance entre les spectres I.R. du polypropylène soluble et de celui insoluble dans l'éther, examiné à l'état fondu, indique que le degré de ramification ne peut pas être très différent dans les deux cas.

On doit par conséquent conclure que le polymère solide soluble dans l'éther ne peut pas être très ramifié. Les spectres I.R. sont en effet différents des spectres des polypropylènes très ramifiés, obtenus avec le AlBr₃.

Par contre on observe des différences remarquables dans le spectre I.R. du produit cristallin, lequel, à l'état orienté, présente un dichroïsme élevé (Tab. III).

On sait que les spectres I.R. de composés rendus optiquement actifs par la présence d'atomes de carbone asymétriques diffèrent des spectres des composés racémiques, seulement si on examine de tels composés à l'état solide tandis qu'ils ne diffèrent pas à l'état de solution ou de fusion. De l'examen des spectres I.R., il paraît logique d'attribuer la différence de structure entre les deux types de produits à la différente distribution, le long de la chaîne, d'atomes de carbone asymétriques, ayant une configuration différente. Dans le cas des polymères amorphes cette distribution devrait être statistique, dans les cas des polymères cristallins elle devrait être ordonnée.

Evidence Roentgenographique des Régularités Exceptionnelles de Structure de la Chaîne des Polymères Cristallins (Figs. 1, 2, et 3)

Nous nous bornons ici à donner quelques uns des résultats qui ont contribué à résoudre le problème de la structure de nos polymères.

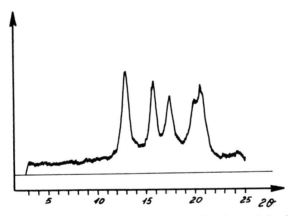

Fig. 1. Polypropylène cristallin.   Ordonnée: intensité relative.

On a obtenu des spectres de fibre sur des produits extrudés à des températures légèrement supérieures à celles de transition de 1er ordre et étirés par la suite à des températures plus basses (voir Tab. I).

Nous donnons quelques résultats de l'examen de ces photogrammes et justement la période d'identité le long de l'axe de la fibre *c* et les projec-

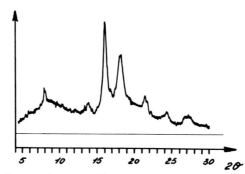

Fig. 2. Polystyrène cristallin.   Ordonnée: intensité relative.

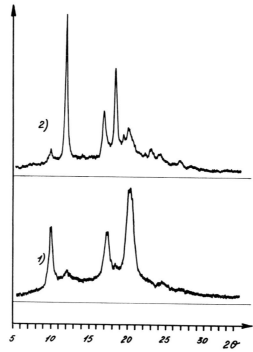

**Fig. 3. Poly-α-butylène.   (1) Prévalantement forme 1er.   (2) Prévalantement
forme 2ème.   Ordonnée: intensité relative.**

tions a' et b' sur un plan normal à l'axe c des constantes minima a et b de la cellule élémentaire, qui interprètent les réflets hko observés.

Polypropylène:

$$c = 6,50 \pm 0,05 \text{ A.}$$
$$a' = 6,56 \pm 0,05 \text{ A.}$$
$$b' = 5,46 \pm 0,05 \text{ A.}$$
$$\gamma' = 106° 30'$$

Le volume d'une cellule avec ces dimensions est de 223,4 A.   La densité, en supposant que cette cellule contient 3 unités monomères, est 0,936, en accord avec la densité expérimentale: 0,922.

Polystyrène et poly-α-butylène:

	Polystyrène	Poly-α-butylène
$a' = b' =$	$12,64 \pm 0,05$ A.	$10,0 \pm 0,1$ A.
$c =$	$6,65 \pm 0,05$ A.	$6,7 \pm 0,1$ A.
$\gamma' =$	$120°$	$120°$
Volume des cellules de ces dimensions	$921,5$ A.^{3}	$580,2$ A.^{3}
Densité calculée pour des cellules conte-		
nant 6 unités monomères	$1,124$	$0,96$
Densité expérimentale	$1,085$	$0,915$

Les réflets équatoriaux et tous les réflets d'ordre supérieur du polysty-rène peuvent être interprétés en considérant une cellule hexagonale ayant les dimensions suivantes: $a = 21,9$ A., $c = 6,65$ A. (groupe spatial R3c ou R$\bar{3}$c).

Au sujet du dimorphisme du polybutylène (voir fig. 3), qui par recristal-lisation de solutions peut donner des cristaux ayant une structure diffé-rente, nous examinerons la question ailleurs.

La grande clarté et le nombre considérable de réflets, appartenant à des couches différentes, des photogrammes aux rayons X obtenus par nous sur les fibres orientées ne sont compatibles qu'avec une grande régularité de structure, et ils doivent par conséquent être attribués à une succession, le long de l'axe des fibres, coincidant avec l'axe des chaînes, d'unités struc-turales parfaitement identiques.

Structure des Chaînes des Polymères Isotaxiques

La cause de l'exceptionnelle régularité de structure des polymères cris-tallisés examinés ici pourrait être attribuée à une des hypothèses structu-rales suivantes:

(*1*) Tous les atomes asymétriques adjacents présentent, au moins pour des longs bouts d'une chaîne, la même configuration stérique.

(*2*) Chaque chaîne contient plusieurs atomes de carbone asymétriques de configuration différente, mais la succession des atomes asymétriques, dans l'une aussi bien que dans l'autre configuration, se produit avec une régularité déterminée.

De ces deux hypothèses, nous croyons que l'on peut exclure le deu-xième, car la période d'identité trouvée le long de l'axe (environ 6,5 A.) correspond à 3 unités monomères —CH_2—CHR— disposées probablement à forme de spirale le long de l'axe du cristal, tandis qu'une succession régu-lière d'atomes de carbone asymétriques, de différente configuration stérique et présents en nombre égal, exigerait un nombre pair de ces unités dans la période d'identité.

Le cas d'une succession régulière de deux unités d'un type et d'une unité de l'autre n'a pas été considéré, car il paraît extrêmement improbable.

Comme nous attribuons au phénomène observé une importance fonda-mentale pour la connaissance d'une vaste catégorie de macromolécules, dont on a commencé une production destinée à de grands développements, nous avons voulu, pour en rendre plus facile la description, assigner aux atomes de carbone asymétriques ayant une configuration égale, le terme "isotaxiques," du grec ἴσος = égal et τάττω = disposer en ordre.

En outre, nous appelerons "isotaxiques" les chaînes et les molécules contenant des atomes de carbone isotaxiques, "isotaxiques" les polymères contenant des molécules isotaxiques, et "isotaxie" le phénomène décrit.

Pour rendre mieux compréhensible la structure des chaînes "isotaxiques," nous pouvons imaginer de disposer arbitrairement dans un plan la chaîne principale paraffinique en zig-zag du polymère. Dans ce cas les groupes R peuvent être (I) tous au dessous ou tous au dessus du plan de la chaîne (laquelle dans le schéma ci-dessous correspond au plan de la feuille) ou alors il peuvent avoir une distribution statistique (II):

(I) Structure isotaxique

(II) Exemple de structure
    non isotaxique

Evidemment une molécule isotaxique peut présenter deux configurations différentes comme il résulte du schème 2, où on indique à titre d'exemple deux types de molécules isotaxiques saturées, non superposable (énantiomorphes), l'une étant image spéculaire de l'autre:

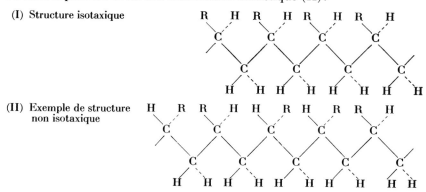

Si on mettait à l'envers une des deux molécules (en mettant la tête à la place de la queue, et en la faisait tourner de 180° autour de son axe) les deux molécules apparaitraient superposables sur presque toute leur longueur à l'exception des groupes finals.

En ce qui concerne l'activité optique, à la lumière polarisée, de chaque molécules dissoute, on peut s'attendre à ce qu'il se produise des phénomènes marqués de compensation interne entre les atomes de carbone asymétriques se trouvant à la même distance du centre de chaque chaîne isotaxique, tandis que dans chaque chaîne non isotaxique on a une compensation statistique. En outre, la présence de mélanges statistiques de deux catégories de molécules enantiomorphes est une cause ultérieure de disparition de toute activité optique dans de tels polymères synthétiques dissous.

Les polymères isotaxiques, comme nous l'avons vu, sont considérablement différents des polymères anisotaxiques à l'état solide, non seulement dans le comportement à la lumière polarisée, mais aussi dans beaucoup d'autre propriétés non vectorielles (température de fusion, solubilité, densité, etc.). Les différences de propriétés remarquées par nous sont beaucoup plus grandes que celles que l'on observe en général entre des composés de bas poids moléculaire respectivement avec des configurations différentes. Cela est dû, pour nos composés macromoléculaires, à l'existence dans chaque molécule d'un très grand nombre (souvent des milliers) d'atomes asymétriques, et à leur collaboration pour exalter un effet, qui serait petit s'il était limité à un nombre très petit d'atomes différenciés stériquement.

Les poly-α-oléfines, que nous avons décrites dans cette note, représentent les premier exemples de polyhydrocarbures, pour lesquels on ait démontré la présence d'ordonnements structuraux isotaxiques.

Nous sommes pourtant d'avis que la présence de chaînes isotaxiques, probablement du même type, soit la cause de la cristallinité observée dans certains polymères synthétiques, produits normalement avec des structures non isotaxiques, comme par exemple dans le polyvinylisobutyléther, qui présente une période d'identité le long de l'axe de la chaîne de 6,2 A.[2]

Le phénomène de l'isotaxie se vérifie automatiquement lorsque des macromolécules sont préparées par condensation de molécules asymétriques de configuration égale, contenant deux différents groupes réactifs liés à l'atome de carbone asymétrique, par exemple dans la condensation de α-aminoacides lévogyres. On doit par conséquent considérer comme isotaxiques les chaînes de beaucoup de protéines fibreuses cristallisables.

Nous sommes d'avis que la phénomène peut se produire dans d'autres polymères synthétiques et qu'il offre la possibilité de construire des nouvelles espèces de macromolécules, ayant des propriétés particulières, très différentes de celles des polymères correspondants non isotaxiques connus à ce jour.

Referénces

(1) P. J. Flory, *Principles of Polymer Chemistry*, Cornell University Press, New York (1953).

(2) *Ind. Eng. Chem.*, **40**, 2104 (1948).

(3) Procédés dont les demandes de brevets ont été déposées.

(4) G. Natta et P. Corradini, Travail en cours de publication dans la revue *Ricerca Scientifica* (présenté au Symposium de Chimie Macromoléculaire, Milan-Turin, Septembre–Octobre 1954).

(5) C. Fontana, R. I. Herold, E. I. Kimegaud, et R. C. Miller, *Ind. Eng. Chem.*, **44**, 2955 (1952).

154   G. NATTA

Résumé

Les propriétés exceptionnelles d'une nouvelle classe d'hydrocarbures macromoléculaires à structure linéaire obtenus par polymérisation d'α-oléfines sont décrites.   La haute température de fusion, la cristallinité élevée, la faible solubilité, les propriétés mécaniques spéciales de ces polymères sont attribuées à une régularité particulière de structure due à l'existence, dans chaque macromolécule, de longues successions ordonnées d'atomes de carbone asymétriques ayant la même configuration stérique.   Pour ce type particulier d'ordonnement d'atomes de carbone asymétriques dans des macromolécules linéaires, on propose le terme "isotaxique."

Synopsis

The exceptional properties of a new class of linear polymeric hydrocarbons, obtained by the polymerization of α-olefins, are described.   The high melting points, the high degree of crystallinity, the low solubility, and the special mechanical properties of these polymers are attributed to a particular regularity of structure, due to the existence, in each macromolecule, of a long sequence of asymetric carbon atoms, all having the same steric configuration.   It is suggested that this special type of order of asymetric carbon atoms in linear macromolecules be called "isotactical."

Zusammenfassung

Die aussergewöhnlichen Eigenschaften einer neuen Klasse von makromolekularen Kohlenwasserstoffen von linearer Struktur, die durch Polymerisation von α-Olefinen hergestellt wurden, werden beschrieben.   Die hohen Schmelztemperaturen, die erhöhte Kristallinität, die schwache Löslichkeit und die speziellen mechanischen Eigenschaften dieser Polymere werden einer besonderen Regelmässigkeit der Struktur zugeschrieben, die durch das Bestehen von langen und geordneten Folgen von asymmetrischen Kohlenstoffatomen, die die gleiche sterische Konfiguration haben, in jedem Makromolekül bedingt ist.   Für diesen besonderen Typus der Anordnung von asymmetrischen Kohlenstoffatomen in linearen Makromolekülen wird der Ausdruck "isotaxisch" vorgeschlagen.

Received February 17, 1955

COMMENTARY

Reflections on "Interfacial Polycondensation. I.," by Emerson L. Wittbecker and Paul W. Morgan, *J. Polym. Sci.,* XL, 289 (1959)

EMERSON L. WITTBECKER

Sarasota, Florida

DuPont started a revolution in synthetic textiles when nylon, the first wholly man-made fiber, was commercialized in 1939. This was followed by Dacron® polyester fiber, produced in pilot plant quantities and sold to the trade for evaluation by the end of 1951. Both polyamides and polyesters are made using melt polymerization, very useful as a laboratory technique for the synthesis of new polymers, and as an economical, attractive commercial process. The very high temperatures required, however, make this method unsuitable for the preparation of unstable and infusible polymers.

While engaged in broad scouting research on new polymers in the early 1950s, Emerson L. Wittbecker, in the Pioneering Research Laboratory of the Textile Fibers Department, discovered that a low temperature polymerization, which became known as interfacial polycondensation, was a powerful technique for preparing new polymers that cannot be made by melt polymerization. His initial work was directed toward synthesizing a polyurethane with a melting point high enough to make it potentially useful as a new fiber.

The book, *Science and Corporate Strategy: DuPont R&D, 1902–1980,* by David A. Hounshell and John Kelly Smith, Jr., and published by Cambridge University Press (1988) was used as a source for some of the information in this perspective of polymer research before and during the exploration of interfacial polycondensation.

Dacron® and Lycra® are trademarks of the E. I. DuPont Company for its polyester and spandex fibers. Nomex® and Kevlar® are trademarks for its aramid fibers.

There were postwar reports from Germany that polyurethanes had been synthesized from diamines and bischloroformates made from glycols and phosgene. The polymers that were prepared, however, were too low melting for use in fibers. Wittbecker predicted that a polyurethane with a structure analogous to polyethylene terephthalate (Dacron®) would have a melting point approaching that of the polyester, and that fabrics would be more like cotton with respect to wicking and absorption of water.

Polyethylene piperazine-1,4-dicarboxylate was synthesized at room temperature, from piperazine and the bischloroformate from ethylene glycol, using a home blender. The polymerization was run in an immiscible two-phase system, using an organic solvent for the bischloroformate, and water for the diamine and acid acceptor. Fibers were prepared using solvent spinning. As predicted, the melting point was somewhat below that of its polyester analog, but high enough for use in fibers. Fabrics were more hydrophylic, but unfortunately they did not have the excellent wash–wear performance of Dacron® polyester garments. The early phrase used for nylon "made from coal, air and water" was mimicked by saying that this polyurethane fiber was made from antifreeze, an obsolete war gas, and a chicken deworming agent.

The potential of interfacial polycondensation for preparing polyamides was confirmed by making a high molecular weight polymer from 2,5-dimethylpiperazine and terephthaloyl chloride. This infusible polymer was dry spun into fibers using a solvent, and the heat resistance was demonstrated dramatically by holding a glowing cigarette against the fabric and subliming the dye without burning a hole.

The rapid, easy preparation of these polymers, and the great potential of this polymerization process for preparing thousands of new polymer structures at room temperature so impressed W. Hale Charch, Director of the Pioneering Research, that he immediately assigned all of the organic chemists in his organization to explore this technique. They soon found that interfacial polycondensation could be used to synthesize not only polyurethanes and polyamides, but also polysulfonamides, polyureas, and certain polyesters.

After preparing and characterizing hundreds of new polymers, and spinning fibers from many of them, it became evident that it would be difficult, if not impossible, to find and commercialize another general purpose fiber. At this point, the direction of the research on low temperature polymerization was turned to finding specialty fibers. Modified and improved polymerization processes were developed, that ultimately resulted in the commercialization of new fibers with outstanding properties, notably Nomex® and Kevlar® aramid fibers and Lycra® spandex fiber.

COMMENTARY

Reflections on "Interfacial Polycondensation. II. Fundamentals of Polymer Formation at Liquid Interfaces," by Paul W. Morgan and Stephanie L. Kwolek, *J. Polym. Sci.,* XL, 299 (1959)

S. L. KWOLEK

Wilmington, Delaware

It was in the early 1950s that a group of chemists in the Textile Fibers Department, Pioneering Research Laboratory of the DuPont Company began to explore interfacial polycondensation as a route to polymers that could not be prepared by the melt polymerization technique. These polymers and/or intermediates did not melt or were unstable to the conditions normally required by the melt process.

Interfacial polycondensation, on the other hand, requires simple equipment, atmospheric conditions, and two fast-reacting intermediates dissolved in a pair of immiscible liquids, one of which is preferably water. This process for the rapid preparation of condensation polymers at about room temperature (about 0 to 40°C) resulted in the synthesis of a wide variety of polymers: polyamides, polyurethanes, polysulfonamides, polyureas, polyphenyl esters, and copolymers. This work is described in a collection of 11 papers published in the *Journal of Polymer Science,* **XL,** 289–418 (1959).

Interfacial polycondensation can be carried out in either a stirred or unstirred system. The stirred method usually produces in high yield a powder or granular polymer product that is easily isolated and purified. In the case of highly reactive intermediates that are easily hydrolyzed by water, lowering the temperature of the reaction to about 5 to 10°C, can improve both yield and polymer molecular weight.

Because of its ease of operation, the low-temperature stirred interfacial polycondensation method was used to prepare hundreds of polymers, to evaluate procedures, elucidate mechanisms, and search for polymers with new property combinations.

In the unstirred interfacial polycondensation system, a film of high molecular weight polymer can form at the interface of the two immiscible liquids, provided the organic liquid is not a solvent for the polymer. When this film is removed without stirring, it is instantly and continuously replaced by more film until one or the other of the two polymer intermediates is depleted. This unstirred interfacial polycondensation system not only provides a demonstration of polymer formation, but has made it possible to observe polymerization behavior and to study the effects of many variables upon the process.

The unstirred interfacial polycondensation method is particularly well-suited for dramatic classroom demonstrations of polymer formation. A polyamide film that forms at the interface of the two immiscible liquids can be pulled away as a continuous collapsed tube, the size of which is determined by the diameter of the interphase. This experiment is described in the article entitled "The Nylon Rope Trick," in *J. Chem. Ed.,* **36,** 182, 530 (1959). Thousands of these reprints have been distributed at the request of teachers, students, and various exhibitors.

In retrospect, interfacial polycondensation provided a much needed alternative route to polymers that could not be prepared by the melt polycondensation process. In addition, it was a simpler and faster procedure that was especially useful in a laboratory with an objective to search for new polymers and fibers with new property combinations. With time, even interfacial polycondensation had its limitations, however, new avenues of thinking had been opened. Other new low-temperature polycondensation procedures were developed to synthesize the more intractable aromatic polyamides.

Although requests for reprints have been greatly reduced, the Nylon Rope Trick experiment was part of the Material World exhibit at the Smithsonian Institute in 1989 and 1990.

Both stirred and unstirred interfacial polycondensation have found use in research and in industry. Some of these applications include paper, fiber coatings, and encapsulation of dyes, catalysts, and insecticides. Polycarbonates are manufactured by interfacial polycondensation in Europe, Japan, and the United States.[1]

REFERENCES AND NOTES

1. P. W. Morgan, in *Encyclopedia of Polymer Science and Engineering,* Vol. 8, J. I. Kroschwitz (ed.), Wiley-Interscience, New York, 1987, pp. 224–225, 233–234.

PERSPECTIVE

Comments on "Interfacial Polycondensation. I.," by Emerson L. Wittbecker and Paul W. Morgan, *J. Polym. Sci.,* XL, 289(1959) and "Interfacial Polycondensation. II. Fundamentals of Polymer Formation at Liquid Interfaces," by Paul W. Morgan and Stephanie L. Kwolek, *J. Polym. Sci.,* XL, 299(1959)

VIRGIL PERCEC

Macromolecular Science Department, Case Western Reserve University, 10900 Euclid Avenue, Cleveland, Ohio 44106

Interfacial polycondensations carried out by the Schotten–Baumann reaction of an acid chloride with a compound containing an active hydrogen atom were only sporadically reported in the literature before 1955. Most probably, the first interfacial polycondensation was reported by Einhorn[1] in 1898. He prepared a polycarbonate by reacting an alkali-aqueous solution of hydroquinone with a solution of phosgene in toluene. Obviously, this experiment was well ahead of its time because no polymer community existed, and this work was forgotten. In the mid-1950s, Schnell[2] of Bayer, Germany, reported the interfacial synthesis of polycarbonates, which were soon to become commercially available. Almost simultaneously, Conix[3] of Gevaert, Belgium, announced the interfacial synthesis of thermoplastic polyesters. Thus, interfacial polycondensation, which originally was considered to be unsuitable for the synthesis of polymers, started to receive increasing interest. However, none of these publications influenced the polymer community in the way the publications of Wittbecker and Morgan and Morgan and Kwolek did. Their work was first presented in part at the 134th ACS Meeting in Chicago,

September 9, 1958, and at the Gordon Conference on Polymers, New London, NH, July 3, 1958.

These two publications discussed for the first time and in a comprehensive way the conventional melt phase condensation polycondensation and compared it with interfacial polycondensation. The preparation of a high molecular weight polymer by the first technique requires a perfect stoichiometric ratio between two extremely pure monomers, very high conversion, no side reactions, very long reaction times, a higher reaction temperature than the melting or softening point of the resulting polymer, and sophisticated laboratory equipment including vacuum–inert gas capabilities. Such reversible polycondensation can be used only for the synthesis of polymers which melt without decomposition. These two papers demonstrated in a very elegant way that all of these restrictions are lifted by the interfacial polycondensation technique. Irreversible interfacial polycondensation does not require stoichiometric balance between the two monomers, and polymers of extremely high molecular weight are obtained even at very low conversion at room temperature and in only a few minutes reaction time. Interfacial polycondensation was demonstrated to be the most easily accessible, simple and rapid technique for the preparation of polyamides, polyesters, polycarbonates, polyurethanes, polysulphonates, polyphosphonates, organometallic polymers, etc. This method allows the direct fabrication of fibers and films and is extensively used both in the laboratory and on a commercial scale for the preparation of membranes with ion-exchange properties and for the encapsulation of inks, catalysts, insecticides, etc. Several polymers such as aromatic polyamides, polycarbonates, aromatic polyesters, etc., may not have become commercially available without this technique. These two publications also demonstrated the role of all reaction parameters and reagents, the locus of reaction, and the limitations of the interfacial polycondensation.

Much of the excitement and general interest generated by these two publications is also due to the fascinating Nylon Rope Trick experiment[4] which was described for the first time in the second of these two papers. Most probably, this experiment has been done and viewed by more people than any other in the area of polymer chemistry. After all these years it continues to be part of all polymer preparation courses around the world.

Another important (and perhaps neglected) contribution to the field of preparative organic chemistry was also generated by these two publications. The use of an onium salt detergent to enhance the rate of an organic reaction via the transfer of a phenolate from an aqueous to an organic phase was clearly demonstrated here. To my knowledge, the discovery of phase transfer catalysis[5] and the sub-

sequent developments in this field[6,7] have never recognized this pioneering contribution.

Soon after these two and the other publications in this series, Morgan published a book[8] which, in spite of the extremely large volume of literature generated since then on this topic, remains a classic. Interfacial polycondensation was extended from water-immiscible solvents to water-miscible solvents and to gas–liquid interfacial systems.[9] Its role in the rapid development of condensation polymers, many of them inaccessible by any other preparative method, remains invaluable.

REFERENCES AND NOTES

1. A. Einhorn, *Liebigs Ann. Chem.*, **300,** 135 (1898).
2. H. Schnell, *Angew. Chem.*, **68,** 633 (1956); H. Schnell, *Ind. Eng. Chem.*, **51,** 157 (1959).
3. A. J. Conix, *Ind. Chim. Belge*, **22,** 1457 (1957); A. J. Conix, *Ind. Eng. Chem.*, **51,** 147 (1959).
4. P. W. Morgan and S. L. Kwolek, *J. Chem. Ed.*, **36,** 182, 530 (1959).
5. C. M. Starks, *J. Am. Chem. Soc.*, **93,** 195 (1971).
6. C. M. Starks, C. L. Liotta, and M. Halpern, *Phase Transfer Catalysis. Fundamentals, Applications and Industrial Perspective,* Chapman and Hall, New York, 1994.
7. E. V. Dehmlov and S. S. Dehmlov, *Phase Transfer Catalysis,* 3rd ed., VCH, Weinberg, 1993.
8. P. W. Morgan, *Condensation Polymers by Interfacial and Solution Methods,* Interscience, New York, 1965.
9. P. W. Morgan, *J. Macromol. Sci.-Chem.*, **A15,** 683 (1981).

JOURNAL OF POLYMER SCIENCE   VOL. XL, PAGES 289–297 (1959)

Interfacial Polycondensation.   I.*

EMERSON L. WITTBECKER and PAUL W. MORGAN, *Pioneering Research Division, Textile Fibers Department, E. I. du Pont de Nemours & Company, Inc., Wilmington, Delaware*

INTRODUCTION

Polycondensations, in general, have been based on slow, reversible organic reactions which require elevated temperature and reduced pressure for the formation of high polymer.   Polyamidation and polyesterification, for example, are carried out in the melt at temperatures well above 200°C. in order to remove the low molecular weight byproducts, water and alcohol. While many examples of polyamides and polyesters have been prepared successfully by melt polymerization, they have been limited to intermediates and polymers that are stable under the severe conditions normally required.   Since the melt polymerization technique is also a time-consuming and difficult laboratory procedure, there has been a need for a rapid polycondensation reaction which can be run in common equipment under atmospheric conditions.

We have found the Schotten-Baumann reaction of an acid chloride with a compound containing an active hydrogen atom (—OH, —NH, —SH) to be the basis for a simple, versatile laboratory process which we have designated *interfacial polycondensation*.   In this process the irreversible polymerization of two fast-reacting intermediates occurs near the interface of the two phases of a heterogeneous liquid system.   The objective of this paper is to outline the general nature of interfacial polycondensation, list the major variables, and compare the principles and laboratory techniques with those pertaining to melt polymerization.   Detailed studies of the preparation of various classes of polymers and the mechanism of the polymerization are reported in the following papers.

POLYCONDENSATION WITH DIACID CHLORIDES

Throughout the literature it is possible to find repeated reference to the use of diacid chlorides in polymer preparations.   In at least one patent[1] it is stated that polyamides may be prepared from diamines and diacid chlorides with the use of inert organic diluents and acid acceptors.   The products, however, are reported to be "frequently of relatively low molec-

* Presented at the Symposium on Polyethers and Condensation Polymers at the 134th Meeting of the American Chemical Society, Chicago, Ill., September 1958.

ular weight.'' Heating at elevated temperatures is recommended for the purpose of increasing the molecular weight. Polyesters have been prepared[2] successfully from acid chlorides and certain glycols with the liberation of hydrogen chloride, but this was done at elevated temperature with the same limitations as those inherent in melt polymerizations.

At first glance, the Schotten-Baumann reaction which is run in an aqueous system does not appear to fulfill the requirements for a polymerization method. Hydrolysis of acid chloride groups would be expected to upset the balance of functional groups, and the chain growth would be terminated. However, a polymerization based on similar chemistry was reported[3] as early as 1946. Polyurethanes were prepared by reaction of bischloroformates with diamines in a two-phase system, but application of this route was primarily to the same polymers that were prepared by the diisocyanate–glycol method. For example, hexamethylenediamine and tetramethylene bis(chloroformate) were used as an alternate route to the polyurethane prepared from hexamethylene diisocyanate and tetramethylene glycol. While some advantages were noted for the former process (greater latitude and elimination of the need for costly diisocyanates), the latter was preferred because of the narrower molecular weight distribution obtained.

Our research on polyurethanes has led us to conclude that the interfacial polycondensation of diamines and bischloroformates is a much more general route to linear high molecular weight polymers than the diisocyanate–glycol method. Moreover, interfacial polycondensation is not limited to polyurethanes. Polyamides, polysulfonamides, and polyureas have been synthesized from diamines with dicarboxylic acid chlorides, disulfonyl chlorides, and phosgene; polyphenyl esters and polythiol esters have been prepared from diacid chlorides with diphenols and dithiols.

LABORATORY PROCEDURES

The preparation of poly(hexamethylene sebacamide) 610, a polyamide normally prepared by melt polymerization, will be used to illustrate interfacial polycondensation. Sebacoyl chloride in a water-immiscible organic solvent such as benzene reacts with hexamethylenediamine in water to yield a polyamide:

$$H_2N(CH_2)_6NH_2 \underset{\substack{H_2O \\ NaOH}}{} + \underset{C_6H_6}{Cl\overset{\overset{O}{\|}}{C}(CH_2)_8\overset{\overset{O}{\|}}{C}Cl} \longrightarrow \left[-NH(CH_2)_6NH\overset{\overset{O}{\|}}{C}(CH_2)_8\overset{\overset{O}{\|}}{C}- \right]_n$$
$$NaCl + H_2O + C_6H_6$$

An inorganic base, such as sodium hydroxide, in the aqueous phase accepts the hydrogen chloride formed in the condensation.

There are two different procedures suitable for running this type of polycondensation reaction, neither of which requires special equipment or precautions.

Unstirred Interfacial Procedure

When an aqueous phase containing diamine and acid acceptor and an organic phase containing diacid chloride are placed carefully in a beaker as distinct layers, a film of high polymer is formed near the interface. When this is removed without stirring it is instantly and continuously replaced. This method is particularly well-suited for demonstration purposes. A more detailed discussion is given in a paper that follows.[4] The use of this experiment as a lecture demonstration has been described.[5,6]

Stirred Interfacial Procedure

While stirring is not essential for the interfacial preparation, it is desirable in order to obtain high yields of powdered or granular products which are easily isolated, washed, and dried. The two phases are combined rapidly with vigorous agitation. Since the diacid chloride is the more unstable of the two starting materials, it is generally advisable to add the solution of it in the organic solvent to the stirred aqueous diamine solution. Polymerization started at room temperature will warm up to 50–60°C. from the heat of the reaction. With diacid chlorides that are particularly sensitive to hydrolysis, it is better to cool the solutions to 5–10°C. After a few minutes stirring the high polymer will probably have precipitated and may be separated by filtration of the mixture. Polymers that are soluble in the organic solvent may be isolated by evaporation of the solvent or by addition of a precipitant with agitation.

VARIABLES IN INTERFACIAL POLYCONDENSATIONS

Superficially, this polymerization appears unusually simple and in fact, from a preparative standpoint, it is. However, from a theoretical viewpoint, it is subject to multiple variables that are difficult to separate. Some of the more important variables are (1) organic solvent, (2) reactant concentration, and (3) detergents. Others will be treated under the comparison with melt polycondensation.

Organic Solvent

The choice of the organic solvent is most important since it will affect several other polymerization factors such as the potential partition of the reactants between the two phases, the diffusion of the reactants, reaction rate, and the solubility, swelling, or permeability of the growing polymer. While the solvent must be capable of preventing precipitation before high molecular weight has been attained, it need not dissolve the final polymer. It will influence the physical character of the product isolated from the interfacial polycondensation. The common organic solvents immiscible with water and inert to the reactive intermediate are quite satisfactory for interfacial polycondensation. Chlorinated hydrocarbons and aromatic hydrocarbons are particularly useful.

Reactant Concentration

For many polymerizations it has been found that there is an optimum ratio of concentration of reactant in the organic phase to the concentration of reactant in the water phase for the production of the highest polymer. This may result from favorable changes in the diffusion rates and location of the polymerization zone. The optimum ratio is affected by the properties of the reactants, the organic solvent, agitation, additives in the aqueous phase such as salt and alkali, and the overall concentration of the reactants in both phases.

For smooth operation in a stirred interfacial polycondensation, the polymer should be 5% or less on the basis of combined weights of water and organic solvent. At concentrations of 10% or higher, all of the liquid may be absorbed by the swollen polymer so that the mass can not be stirred. This can lead to incomplete polymerization and low molecular weight. On the other hand, hydrolysis of the acid chloride can become an important factor if polymerization is conducted in systems which are too dilute.

Detergents

Detergents may be added with benefit to some polymerizations, but consideration should be given to possible reaction with the intermediates. Of a large number of detergents which have been tested, sodium lauryl sulfate detergent (Duponol ME, Du Pont trademark) has been found to be most satisfactory. Small amounts (ca. 0.2%) are beneficial in the stirred preparation as an aid to better mixing. Larger amounts are sometimes harmful even though no chemical side reactions have been recognized. For polyurethane preparation, however, the presence of a detergent (up to 1%) is an important factor, and the highest molecular weights were not readily obtained without it.

COMPARISON OF INTERFACIAL AND MELT POLYCONDENSATION

Procedure

Polyamides, one of the more important classes of condensation polymers, have been prepared in the laboratory by the following procedure. A balanced salt is prepared from the diamine and dibasic acid and heated (above 200°C.) in a glass tube sealed under vacuum. Condensation proceeds to equilibrium, thus "fixing" the diamine through the formation of polymer with molecular weight sufficiently high so that no loss occurs on subsequent heating. The polymer is allowed to solidify, and after the tube is opened and the contents remelted, heating is continued at atmospheric pressure in order to remove the water formed. Oxygen is excluded in order to prevent degradation of the melt. Frequently, heating at 215–285°C. for several hours under vacuum with a bleed of nitrogen is required to force the condensation to proceed to high molecular weight. After cooling, the glass tube is broken away, if it has not already shattered violently when the polymer, which adheres well to glass, solidified and contracted. The remainder of the glass is filed from the plug of polymer which is then cut into pieces to be used in subsequent operations such as dissolution.

Interfacial polycondensation is a considerably more satisfactory laboratory procedure than melt polycondensation. It is carried out rapidly at room temperature in simple equipment, and a larger number of polymers (heat-sensitive and infusible as well as stable and meltable), representing more polymer classes, can be synthesized. Polymers are obtained as useful granular or powdered products rather than unwieldy plugs. The two methods are contrasted further in Table I.

TABLE I
Comparison of Interfacial and Melt Polymerization

	Interfacial	Melt
Intermediates		
Purity	Moderate to high	High
Balance	Unnecessary	Necessary
Stability to heat	Unnecessary	Necessary
Polymerization conditions		
Time	Several minutes	Several hours
Temperature	0–40°C.	>200°C.
Pressure	Atmospheric	High and low
Equipment	Simple, open	Special, sealed
Products		
Yield	Low to high	High
Structure	Unlimited	Limited by stability to heat and fusibility

Principles

The principles used to guide research in the condensation polymer field were derived from melt polymerization and have remained essentially unchanged for over twenty years. For the polycondensation of two complementary bifunctional molecular species (for example, diamine plus dibasic acid), stringently purified intermediates must be brought together in stoichiometric amounts in a reaction that is essentially quantitative— a procedure which stresses the principles of purity, balance, and high yield (no side reactions). In interfacial polycondensation, all three of these appear to have been violated.

Purity

It is not surprising that most impurities cannot be tolerated in a homogeneous melt polymerization with vigorous heating at, for example, 285°C. However, because of the mild reaction conditions of interfacial polycondensation a greater number of the impurities frequently present in organic intermediates are unreactive. In other words, functional groups which are much less reactive than the two upon which the polycondensation is based can be present without interfering with the formation of polymer of high molecular weight. In addition, the heterogeneity of this polymerization keeps some of the impurities—especially those that are very soluble in water—from ever reaching the zone of reaction. In some cases, relatively impure intermediates have yielded polymers with molecular weights of 10,000 to 15,000 (normal for many condensation polymers). White polymers frequently can be prepared from colored intermediates. As is the case with melt polymerization, reactive monofunctional intermediates will limit the molecular weight by chain termination, and reactive intermediates with a functionality greater than two will branch and finally crosslink the polymer.

While stringently purified intermediates are not required for the preparation of some polymers with satisfactory molecular weights, they are necessary for the attainment of optimum results. In addition, purity can be a very important variable in cases where, for other reasons, the polymers are difficult to prepare in high molecular weight.

Balance

In a homogeneous polycondensation the two different functional groups must be present in equivalent amounts or the polymer chains stop growing when they are all terminated by the same group. Deliberate excesses of glycol can be used in the preparation of polyesters from diesters, since the excess is lost (providing that the glycol is volatile) by the same alcoholysis reaction used for the initial exchange reaction. A chemical balance is attained in polyamides by the use of a balanced diamine-dibasic acid salt.[7] When a diamine is reacted with a diester or a diurethane the balance must be secured by careful weighing.

Because interfacial polycondensation is a heterogeneous reaction and the reactive intermediates must diffuse to the interface, large excesses of one of the intermediates can be tolerated without limiting the molecular weight. In other words, reactant equivalence is attained in the polymerization zone even though the system taken as a whole is unbalanced. Thus two moles of diamine and one mole of diacid chloride will produce high molecular weight polymer, although the stoichiometric reactant ratio is 1 to 1. Small molar excesses (5–10%) of the diamine frequently help to produce higher molecular weights.

Yield

In the literature, much emphasis has been placed on the necessity for any successful polycondensation reaction to proceed very nearly to completion, without side reactions that consume the functional groups that are needed for the production of large linear molecules. This is indeed true for melt-polycondensation, and as a result, not too many organic reactions are suitable for polymer formation at high temperatures. In contrast, it is not necessary for interfacial polycondensation to proceed quantitatively— high molecular weight polymers have been prepared in yields as low as 5%. As might be expected, hydrolysis of acid chloride groups can be an important side reaction. Since this occurs primarily in the aqueous phase, it becomes a serious competitive reaction only when the polymerization reaction is relatively slow or when the diacid chloride is quite water-soluble. Hydrolysis of both ends of a diacid chloride merely changes the reactant ratio, and although the polymerization may not take place under optimum conditions, the major result is only a loss in yield of polymer.

Molecular Weight and Molecular Weight Distribution

A priori it might be expected that interfacial polycondensation would fail to yield high polymers, since either hydrolysis of one end of diacid chloride just before the other end reacts with an amine group or hydrolysis of an acid chloride group on the end of a growing polymer chain would lead to early termination of the polymerization. One must conclude, however, that actually these side reactions are negligible, since high molecular weights are obtained from properly run interfacial polycondensations. In fact, in certain cases where intermediates have been especially pure and the reaction conditions for optimum results have been determined, we have prepared condensation polymers with weight-average molecular weights as high as 500,000! In the synthesis of polyamides, polyesters, and polyurethanes by melt methods, number-average molecular weights above 25,000 (weight-average, 50,000) are seldom obtained even when purity and equivalence requirements are met rigidly.

Since interfacial polycondensation is an irreversible random coupling of complementary components with the elimination of a byproduct such as hydrogen chloride, there is no interchange between linkages once they are formed. Thus a major difference between this polymerization and conventional melt polycondensation is that there is no equilibration among polymer species of varying molecular weights.[8] The distribution of chain lengths resulting from melt polycondensation is known from theoretical considerations[9] and actual fractionation data.[10] Interfacial polycondensation, however, leads to various distributions, depending on the reaction conditions. Although polymers with molecular weight distributions broader than normal have been observed, distributions similar to or even narrower than that resulting from melt polymerizations have also been obtained.

References

1. W. H. Carothers, U. S. Pat. 2,130,523 (9/20/38), assigned to Du Pont Co.

2. P. J. Flory and F. S. Leutner, U. S. Pat. 2,594,144 (4/22/52), assigned to Wingfoot Corp.

3. L. H. Smith, *Synthetic Fiber Developments in Germany*, Textile Research Institute,

New York, 1946; W. Hechelhammer and M. Coenen, German Pat. 818,580 (1951). assigned to Farbenfabriken A.-G.; L. Orthner, H. Wagner, and A. Siebert, German Pat. 904,471 (2/18/54), assigned to Farbwerke Hoechst A.-G.; L. Orthner, G. Wagner, and P. Schlack, German Pat. 912,863 (6/3/54), assigned to Farbwerke Hoechst A.-G.; M. Coenen and P. Schlack, German Pat. 915,868 (7/29/54), assigned to Farbwerke Hoechst A.-G.; H. G. Trieschmann and G. Wenner, German Pat. 919,610 (10/28/54); assigned to Badische Anilin- u. Soda-Fabrik A.-G.; L. Orthner, H. Wagner, G. Wagner, M. Coenen, and W. Kimpel, German Pat. 925,612 (3/24/55), assigned to Farbwerke Hoechst A.-G.; M. Coenen and G. Wagner, German Pat. 929,214 (6/23/55), assigned to Farbwerke Hoechst A.-G.

4. P. W. Morgan and S. L. Kwolek, *J. Polymer Sci.*, **40,** 299 (1959).

5. P. W. Morgan and S. L. Kwolek, *J. Chem. Educ.*, **36,** 182 (1959).

6. P. W. Morgan, *SPE Journal*, **15,** 485 (1959).

7. D. D. Coffman, G. J. Berchet, W. R. Peterson, and E. W. Spanagel, *J. Polymer Sci.*, **2,** 306 (1947).

8. L. F. Beste and R. C. Houtz, *J. Polymer Sci.*, **8,** 395 (1952).

9. P. J. Flory, *J. Am. Chem. Soc.*, **58,** 1877 (1936).

10. G. B. Taylor, *J. Am. Chem. Soc.*, **69,** 635 (1947).

Synopsis

Interfacial polycondensation is a rapid, irreversible polymerization at the interface between water containing one difunctional intermediate and an inert immiscible organic solvent containing a complementary difunctional reactant. It is based on the Schotten-Baumann reaction in which acid chlorides are reacted with compounds containing active hydrogen atoms (—OH, —NH and —SH). A large number of polymers (heat-sensitive and infusible as well as stable and meltable) can be prepared. The method has been applied to the preparation of polyurethanes, polyamides, polyureas, polysulfonamides, and polyphenyl esters. Interfacial polycondensations are run in simple, open laboratory equipment with or without stirring. With suitable agitation granular or powdered polymers with high molecular weight are prepared at room temperature and isolated within a few minutes. The intermediates need not be absolutely pure or in balance nor is a quantitative yield needed in order to obtain high polymer. The major variables in the interfacial polycondensation process are discussed and the laboratory techniques and principles are contrasted with melt polycondensation.

Résumé

La polycondensation interfaciale consiste en une polymérisation rapide et irréversible à l'interface de l'eau contenant un des réactifs de départ et d'un solvant organique inerte immiscible à l'eau et contenant l'autre réactif. Elle est basée sur la réaction de Schotten-Baumann dans laquelle les chlorures acides réagissent sur des comporés à hydrogéne réactionnel (—OH, —NH et —SH). Un grand nombre de polymère (thermolabiles et infusibles aussi bien que les polymères stables et fusibles) peut être préparé ainsi. La méthode a été appliquée à la préparation de polyuréthannes, polyamides, polyurées, polysulfonamides et esters polyphénylés. Ces polycondensations interfaciales sont effectuées dans un équipement de laboratoire simple et ouvert avec ou sans agitation. Sous agitation convenable, on obtient à température de chambre et endéans quelques minutes des polymères granulaires ou poudreux de poids moléculaires élevés. Les réactifs ne doivent pas être absolument purs ni en proportions exactes; il n'est pas nécessaire d'obtenir un rendement quantitatif de haut polymère. Les variables principales dans le processus de polycondensation interfaciale sont discutées et les techniques expérimentales et principes sont mis en contraste avec la polycondensation à l'état fondu.

Zusammenfassung

Als Grenzflächenpolykondensation wird eine rasche, irreversible Polymerisation bezeichnet, die an der Grenzfläche zwischen Wasser, das einen reaktionsfähigen Stoff enthält, und einem inerten, nicht mischbaren organischen Lösungsmittel abläuft, das die andere komplementäre Komponente enthält. Sie beruht auf der Schotten-Baumann-Reaktion, bei welcher Säurechloride mit Verbindungen, die aktive Wasserstoffatome (—OH, —NH und —SH) enthalten, zur Reaktion gebracht werden. Eine grosse Zahl von Polymeren (sowohl hitzeempfindliche und unschmelzbare als auch stabile und schmelzbare) können dargestellt werden. Das Verfahren wurde zur Darstellung von Polyurethanen, Polyamiden, Polyharnstoffen, Polysulfonamiden und Polyphenylestern angewendet. Grenzflächenpolykondensationen werden in einer einfachen, offenen Laboratoriumsapparatus, mit oder ohne Rührung, durchgeführt. Bei geeignet durchgeführter Bewegung werden bei Raumtemperatur körnige oder pulverförmige Polymerisate mit hohem Molekulargewicht gebildet und innerhalb einiger Minuten isoliert. Die reagierenden Stoffe müssen nicht absolut rein sein oder in bestimmtem Mengenverhältnis zu einander stehen; es ist nicht erforderlich eine quantitative Ausbeute en Hochpolymerem zu erhalten. Die wichtigsten Versuchsvariabeln beim Grenzflächenpolykondensationsprozess werden diskutiert und die Laboratoriumsmethoden und -prinzipien werden denen der Schmelzpolykondensation gegenübergestellt.

Received June 22, 1959

JOURNAL OF POLYMER SCIENCE   VOL. XL, PAGES 299–327 (1959)

Interfacial Polycondensation. II. Fundamentals of Polymer Formation at Liquid Interfaces*

PAUL W. MORGAN and STEPHANIE L. KWOLEK, *Pioneering Research Division, Textile Fibers Department, E. I. du Pont de Nemours & Company, Inc., Wilmington, Delaware*

INTRODUCTION

In the preceding article in this series,[1] a general laboratory process for the rapid preparation of condensation polymers at room temperature is described. This process consists of bringing a solution of a diacid halide in an inert, water-immiscible, organic solvent, into contact with an aqueous solution of a diamine or a bisphenol and alkali, stirring the mixture vigorously, and isolating the product. Stirred interfacial polycondensation may produce polymer dissolved in a separated or emulsified organic phase or a polymer precipitate which may be dispersed, pasty, powdery, or granular. Such differences depend upon the polymer and solvent combination.

If the complementary phases for an interfacial polycondensation are brought together *without stirring* and if the organic liquid is a nonsolvent for the polymer, a thin film of polymer will be formed at once at the interface. Under the right conditions, this polymer gel (a swollen, pliable, noncrosslinked, noncrystalline polymer precipitate) is tough and has high molecular weight. When the film is grasped and pulled from the area of the interface, more polymer forms at once and a collapsed sheet or tube of polymer may be withdrawn continuously[2,3] (Figs. 1 and 11). This type of polymerization provides a fascinating lecture or laboratory demonstration. Furthermore, the omission of stirring decreases the number of variables and permits observation of the rate and manner of film growth not possible with the stirred polymerization.

This paper will outline the effects of variables on unstirred interfacial polycondensations, interpret the effects in terms of the polymerization mechanism, and present some properties of the films and polymers. The conclusions drawn here will apply as well to stirred polymerizations. A subsequent paper will integrate these findings with those pertaining to the added complexities of a stirred system. A study of endgroups and molecular weights will also be presented later.

* Presented at the Symposium on Polyethers and Condensation Polymers at the 134th Meeting of the American Chemical Society, Chicago, Ill., September 1958.

299

Fig. 1. Polyamide film forming at and being withdrawn from a liquid interface.

DISCUSSION

Chemistry

The chemistry of interfacial polycondensation is very simple. A typical reaction employs a diacid chloride and a diamine. Useful intermediates of this type are believed to react by an S_N2 (nucleophilic) mechanism[4-6] to form a protonated amide from which a proton is rapidly eliminated in the presence of more base. The proton acceptor is presumably an amine group on a diamine molecule or the end of an oligomer chain. Water also could act as the proton carrier. The amine can be regenerated in the aqueous phase by the use of inorganic base.

$$
\text{ClC(CH}_2)_8\text{CCl} + \text{H}_2\text{N(CH}_2)_6\text{NH}_2 \xrightarrow{k_2}
$$

$$
\left[\text{ClC(CH}_2)_8\text{C—N(CH}_2)_6\text{NH}_2} \right] \xrightarrow[k_{H^+}]{\text{Diamine}}
$$

$$
\text{ClC(CH}_2)_8\text{C—N(CH}_2)_6\text{NH}_2} + \text{Diamine salt}
$$

$$
k_{H^+} > k_2 > \text{``}k_{\text{reactant mixing}}\text{''}
$$

Mechanism of Polymer Formation

The experiments show that when the solution of the diacid chloride is brought into contact with an aqueous diamine solution, high polymer forms in a fraction of a second near the interface.   In this short period of time, equivalents of the intermediates must combine nearly quantitatively, eliminate two equivalents of hydrogen chloride, and form a precipitate.

The reaction rates must be high, of the order of magnitude discussed in a following section.   The rate requirement is met by any reaction which is faster than mixing and considerably faster than any important side reaction at the polymerization site.   There must be sufficient time to essentially complete the polymerization before the polymer precipitates or gels.

Interface Function

Direct and indirect evidence to be cited points to the formation of most condensation polymers in the organic phase.   The primary functions of the aqueous phase are to serve as a solvent medium for the diamine or bisphenoxide and the acid acceptor and to remove byproduct acid from the polymerization zone.   We do not believe that the interface has any special orienting or aligning effect on the reactants, but it does provide, through solubility differences, a controlled introduction of the aqueous reactant into an excess of diacid halide in the organic phase next to the interface.

Amine Acylation

Upon phase contact, both reactants and solvents tend to become partitioned with the opposing phase.   The diamine nearly always has an appreciable potential partition toward the organic phase, whereas the acid chloride has very little solubility in water.   Measured equilibrium partition coefficients for diamines in useful solvent systems have varied from 400 to less than 1 $(C_{\mathrm{H_2O}}/C_{\mathrm{solvent}})$.   The values have been used to estimate the relative tendency of diamines to transfer to the organic phase under polymerization conditions.

Partition equilibria are never achieved during polymerization because acylation takes place in the organic phase as rapidly as diamine is transferred.   The mass transfer of the diamine is the rate-controlling step at all concentrations.   Were this not so, the diamine would have time to penetrate more deeply into the acid chloride layer, and reaction and polymer precipitation would take place more diffusely.

During polycondensation, the first diamine meets a high concentration of acid chloride and presumably is acylated to a large extent at both ends. The following diamine finds a layer of acid chloride-terminated oligomers plus diacid chloride (see Fig. 5).   The reaction proceeds by an irreversible coupling of the oligomers by the diamine.   The concentration and size of oligomers increases until a layer of high polymer is obtained.   Thus, high polymer forms because of the high reaction rate and the increasing probability that diamine reacts with an oligomer rather than with new diacid chloride.

Fig. 2. Electron micrograph of a polyamide 610 film.   Face toward the aqueous phase.
10,000×.

During this period of rapid reaction, there is presumably some diffusion of oligomers toward the organic phase along with incoming diamine, while from the side of the organic phase there is an entry and reaction of some diacid chloride.   The diffusion of polymer chains is relatively slow.   To get the highest polymer, the balance of transfer rates must be such that equivalents of reactants enter the polymerization zone just within the solution period.   For this reason, the system is sensitive to those factors which change the transfer rate of diamine to the organic phase as well as to changes in reaction rate and solvent swelling of the polymer.

Since the reaction is appreciably exothermic, local movement may be increased by the thermal energy imparted to the reactants and solvent molecules.   Much of this energy is absorbed by the solvents.   Some inequality of movement in the polymerization area or of later penetration of the initial polymer layer is indicated by the roughness of the organic solvent face of the films shown in Figures 3 and 4.

As an extension of the preceding mechanism of polymerization, one may consider that the reaction takes place in a series of incremental layers through which the diamine advances from the liquid interface.   As before, the reaction rate exceeds diamine flow.   A first layer of polymer forms as the amount of diamine becomes equal to the acid chloride.   The polymer in this layer becomes capped with amine groups and diamine passes on to the next.   In successive layers, the acid chloride concentration is initially higher than that of the diamine.   At the same time, the diamine flow drops off because of the increased distance from the interface and the need to

Fig. 3. Electron micrograph of a polyamide 610 film. Face toward the organic solvent phase. 10,000×.

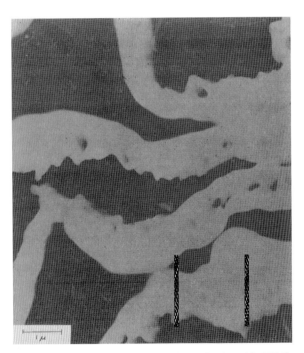

Fig. 4. Electron micrograph of the cut cross-sections of polyamide 610 film prepared by interfacial polycondensation. 10,000×.

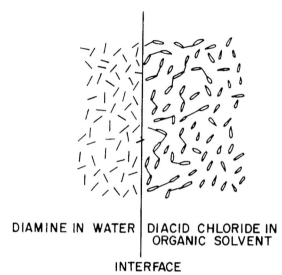

DIAMINE IN WATER | DIACID CHLORIDE IN
ORGANIC SOLVENT

INTERFACE

Fig. 5. Diagram of early relationship of reactants in interfacial polyamidation.

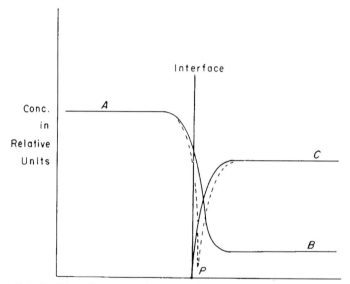

Fig. 6. Relationships of concentrations in an interfacial polyamidation (schematic). Amine groups (A) in water phase.   Acid chloride (C) and amine (B, at partition equilibrium) groups in organic phase. P and dotted lines are gradients to an equivalence in polymerization.

penetrate the first layers of polymer.   There may exist in the final film a gradation of average molecular weights from face to face which would vary with the original concentration of the reactants.

Figure 6 shows diagramatically the relation of the concentration of amine and acid chloride groups about the interfaces of water and an organic solvent, such as carbon tetrachloride.   A is the level of amine in water and

B is an equilibrium concentration in the organic phase. C is the initial concentration of acid chloride. The dotted lines show the relation of the reactive groups when the first incremental layer of polymer has formed.

Acid Elimination

Following the initial fast addition of amine to acid chloride, hydrogen chloride is eliminated in a still faster step and transferred to the aqueous phase. Elimination and transfer of the proton and chloride ion are presumed to occur with the assistance of water or amine. The amine hydrochloride is not an acylatable species. It is usually very insoluble in the organic phase but is soluble in the aqueous phase. The hydrogen chloride may be neutralized in the aqueous phase by inorganic bases.

Acid Chloride Hydrolysis

Experiments have shown that acid chloride hydrolysis is exclusively in the aqueous phase. Hydrolysis may be by water alone or by the faster reaction with hydroxyl ion. Most diacid chlorides have low solubility in water and so are protected from hydrolysis. The shorter chain aliphatic diacid chlorides are the most water-soluble and are hydrolyzed to the greatest extent, as shown by lowered yield of polymer and difficulty in maintaining continuous film formation. Hydrolysis also becomes a more important competitive reaction when the polycondensation rate is low. Since the hydrolysis products are water-soluble salts, they are held out of the polymerization by the aqueous phase.

Precipitation

As the concentration of polymer species increases, interchain contacts increase until a contact network is formed and precipitation results. We presume that there is a finite and rather reproducible solution period which increases with the strength of the interaction of the polymer and solvent. Further, the initial gel is more open or less compact as the polymer-solvent interaction increases. These two factors provide a variation in the polymerization time with the nature of the solvent.

The moment a coherent film is formed, polymerization does not entirely stop but is greatly decreased in rate because of the lower mobility of the polymer chains and the decreased diffusion rate of the intermediates. While the initial polymer reaches the maximum molecular weight permitted by its degree of freedom in the gel and the availability of new reactants, there is a secondary growth of low polymer in the spaces of the gel and upon the face of the film in the organic phase. This leads to a lowering of the average molecular weight and a broadening of molecular weight distribution (Table VI).

The completeness of polymer precipitation may be variable in another sense. That is, xylene and carbon tetrachloride cause the complete precipitation of polyamide 610 at all molecular weights. (The notation used in this paper is described in a footnote in Table V.) Chloroform precipitates high polymer but holds some low polymer in solution. This effect enhances the mobile lifetime of the polymer chain in the stronger solvents.

Location of the Polycondensation

The condensation reaction should occur wherever the reactive intermediates meet. This could be exactly at the liquid interface, in the aqueous phase, or in the organic phase. The evidence shows that most water-insensitive polymers form and grow on the organic solvent side of the interface. Some reactions occur primarily at the interface. Even here the polymer may accumulate in the organic phase. No examples of polymerization in the aqueous phase are known.

Experiments and polymerization behavior which point to the formation of polymer in the organic phase are as follows.

1. Monofunctional compounds can be shown to react in the organic phase. For example, when one equivalent of aniline and one equivalent of ammonia in the aqueous phase compete for one equivalent of benzoyl chloride in benzene, only benzanilide and ammonium chloride are obtained. Ammonia is not only more basic than aniline (pK_a 9.26 vs. 4.58) but reacts several thousand times as fast with benzoyl chloride[4] in homogeneous solution. The competitive reaction goes wholly to benzanilide because the site is in the organic phase and because the partition coefficient

TABLE I
Monofunctional Reactants Added to Interfacial Polycondensations of Polyamide 610[a]

Additive	Mole ratio, additive/diamine or diacid	η_{inh}	Film quality
None	0	1.48	tough
Benzoyl chloride	0.063	0.56	fairly strong
Benzoyl chloride	0.118	0.28	discontinuous
Propionyl chloride	0.032	0.81	tough
	0.063	0.39	weak; stirred
n-Butylamine	0.032	0.92	weak; stirred
	0.063	0.36	weak; discontinuous
N-(3-Aminopropyl)morpholine	0.063	1.04	tough
N-(3-Aminopropyl)morpholine	0.118	0.81	tough

[a] 0.8M diamine and 1.6M NaOH in water; 0.117M sebacoyl chloride in CCl_4.

of aniline is 0.1 ($C_{H_2O}/C_{benzene}$) compared to 75 for ammonia. Ammonia as the stronger base is the acid acceptor.

2. Monofunctional reactants dissolved in the aqueous phase do not always decrease the molecular weight in proportion to the amount used, whereas benzoyl chloride and propionyl chloride in the organic phase are effective in proportion to the amount used (Table I). The more water-soluble the aqueous monofunctional reactant, the less effect it has in reducing molecular weight.

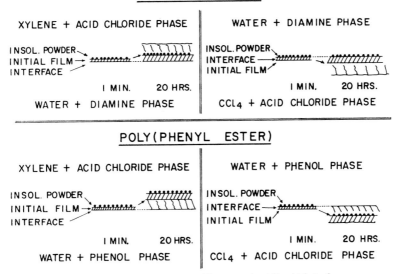

Fig. 7. Diagrams of polymer film growth at liquid interfaces.

Relative reactivity must also be considered in competitive reactions. The mono- and diprimary amines should have reaction rates of the same order of magnitude. Benzoyl chloride reacts less fast than sebacoyl chloride. Therefore, the comparison of the chain terminating effect of propionyl chloride with the monoprimary amines is more appropriate.

Butylamine depressed the molecular weight of 610 more than did aminopropylmorpholine. The reactivities are similar, but the partition coefficient for n-butylamine is 0.7. For aminopropylmorpholine and hexamethylenediamine, K ($C_{H_2O}/C_{benzene}$) values are 80 and 58 (at 25°C.; C_{H_2O} about 0.3M). These relative values would be roughly maintained in carbon tetrachloride with alkali present.

3. Observation of the manner of polymer film and solution formation shows that the polymer accumulates in the organic phase at the interface. The formation and manner of film growth can be checked for some polymers by the introduction of colored powders upon the interface or the initial film (Fig. 7). No polymer has been observed to form on the aqueous side of the interface. Some polymers which are water-sensitive yield a film in the organic phase and then this is slowly penetrated by water which shows as small drops or domes of newly-forming polymer film.

4. A fourth consideration in assigning the location of the polymerization is the response of the system to changes in polymerization conditions, as discussed in a following section. All of these responses are best interpreted in terms of polymerization in the organic phase. A polymerization exactly at the interface would be best satisfied by a 1:1 concentration ratio of the reactants. Such is usually not the case.

Time as a Polymerization Factor

The absolute reaction rates of unhindered aliphatic acid chlorides with primary diamines have so far been immeasurably fast. Rates of such reactions are of the order of 10^2–10^4 l. mole^{-1} sec.$^{-1}$ in homogeneous solution—as fast as the propagation step of the fast vinyl polymerizations.[9] This means that if reactants could be brought together at a concentration of $1M$, one could make several tons of polymer per minute.

Interfacial polycondensations can be carried out successfully with reactants for which the rate constants are considerably less than those above; see, for instance, the preparation of piperazine polyurethanes described by Wittbecker and Katz.[10] The principal requirements are that the polymerization rate be appreciably faster than all side reactions and that the polymer should be in solution or "mobile" for a sufficient period to attain a high degree of polymerization.

The rate of production of a specific polymer in the unstirred system is dependent upon the choice of solvent, the reactant concentrations, the area of the interface, and the rate at which the film is wound up. The wind-up speed is limited by the toughness of the gel film. It has not been possible to pull a film from an interface before it reached the range of maximum molecular weight. Therefore, by the time cohesive film precipitates and is isolated, the maximum molecular weight permitted by the system has been attained.

Poly(hexamethylene sebacamide) of high molecular weight has been made in less than 0.02 sec., if only the average time required to form a film fold is counted. This is an average half-life of 0.003 sec., if eight half-lives are required to reach high polymer. Some experimental rates of polymer formation are given in Tables VIII and IX. These are discussed in the experimental section.

Optimum Polymerization Conditions for Polyamide 610

Concentration Ratios

For every pair of reactants in a given solvent, there are optimum polymerization conditions which are best expressed as a ratio of the concentra-

TABLE II
Variables Affecting Interfacial Polycondensation

1. Chemical reaction rate.
2. Precipitation rate of polymer (solubility).
3. Impurities in reactants and solvents.
4. Transfer rate of diamine.
5. Partition potential of reactants.
6. Concentration ratio of phases.
7. Stirring.
8. Concentration level.
9. Hydrolysis of acid halides.
10. Transfer rate of salts.
11. Fluid and polymer solution viscosities.
12. Polarity of solvent.
13. Interfacial energy barriers.

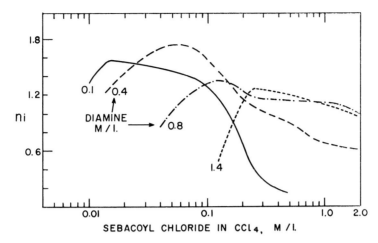

Fig. 8. Polyamide 610: product solution viscosity vs. reactant concentrations in an unstirred polymerization.

tion of the reactants in their respective phases. These optimum conditions must represent the best balance of the factors which affect the polymerization (Table II). It is usually possible to interpret and even partially predict changes in optimum conditions in terms of major variables.

The curves in Figure 8 show the variation in viscosity number of polyamide 610 as the diamine concentration is held constant and acid chloride concentration is varied. The log plot is used for convenience in shortening the right-hand side of the scale. For each dilution, the highest polymer is obtained at a diamine:acid chloride ratio of about 6.5. At the left, where diamine is present in a higher ratio, the inherent viscosity is down and the films are weak and poorly coherent. This is interpreted as due to the greater diffusion of the diamine into the organic phase and the consequent formation of a less compact polymer network. Toward the right, increasing acid chloride concentration causes a decrease in viscosity number of the product but the effect is not as marked as for excess diamine. This effect is believed to result from the fact that diffusion of acid chloride can only occur up to the interface, owing to its extremely low solubility in water. An increase in concentration increases the concentration gradient and restricts the depth or thickness of the polycondensation zone. Diamine, on the other hand, knows no restriction in the organic phase below partition equilibrium, although in many polymerizations this is appreciably below the acid chloride concentration in the bulk phase.

When the solvent is changed, the diamine partition coefficient changes (Table III) and as a consequence there is a corresponding shift in the reactant concentration ratios at which the highest polymer is obtained. A solvent which extracts diamine strongly, such as chloroform, must have a correspondingly higher concentration of acid chloride to give a balanced reaction.

TABLE III
Effect of Diamine Partition on Optimum Polymerization Conditions for Polyamide 610

Solvent	Diamine partition coefficient[a]		Optimum reactant concentration ratio, diamine/acid chloride[b]	Peak η_{inh}
	K	C_{H_2O} moles/l.		
Cyclohexane	182	0.40	17	0.86
Xylene	50	0.392	8	1.47
CCl$_4$	35	0.40	6.5	1.75
C$_6$H$_5$NO$_2$	13.8	0.391	3.4	(1.11)
CHCl$_3$/CCl$_4$ (30/70 (v/v)	6.4	0.39	3	1.90
ClCH$_2$CH$_2$Cl	5.6	0.432	2.3	(1.76)
CHCl$_3$	0.70	0.457	1.7	2.75

[a] $K = C_{H_2O}/C_{solvent}$ at 25°C. at equilibrium with 2 moles sodium hydroxide per mole of diamine in the aqueous phase.

[b] Polymerizations were carried out with the diamine at $0.40M$.

Concentration Level

Measurement of the diamine partition coefficients at different concentrations frequently shows appreciable changes. Generally, the partition favors water upon dilution as is the case for hexamethylenediamine plus sodium hydroxide in water–carbon tetrachloride. In chloroform–water the opposite effect was obtained. In spite of this, the optimum reactant concentration ratios for polyamide 610 was little affected by dilution. The lack of an exact correlation is probably due to inability to assess exactly the partition conditions during the actual reaction and a need for a refined film-forming technique which would be more sensitive to partition and concentration variables.

Effects of Polymer-Solvent Interaction

One might infer from the important role of the potential diamine partition that precipitation must occur in all solvents in about the same period of time. This is not necessary because polycondensation may occur as an advancing face or throughout an expanding zone during the "mobile" period, so that films of increasing thickness are obtained as the time for precipitation increases. The order of polymer-solvent interaction for polyamide 610 is believed to increase in the order given in Table III. Therefore, the thickest films would be formed in chloroform and the thinnest in cyclohexane. This is true for samples made at the same concentration levels. Higher yields of polymer are also obtained in chloroform for a given rate and time of film removal (Tables VIII and IX). We do not know what differences in gel density, if any, exist at the precipitation point or in the final film. The polymer-solvent interaction has another

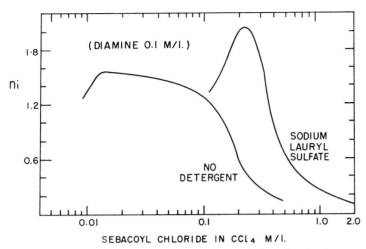

Fig. 9. Polyamide 610: effect of detergent on product solution viscosity in an unstirred polymerization.

important effect. For comparable polymerization conditions, the stronger the solvent, the higher the maximum inherent viscosity. This is illustrated by comparing the values for 610 in chloroform, chloroform–carbon tetrachloride, carbon tetrachloride, xylene, and cyclohexane. The values obtained in the other solvents cannot be compared because the polymerizations were carried out with less pure reactants.

Detergents and Salts

The first paper in this series has shown that small amounts of detergents are sometimes helpful to stirred polycondensations. The benefit appears to be derived from improved mechanical mixing and contact of the reactants, resulting, possibly, in the formation of smaller droplets of organic phase. Sodium lauryl sulfate is an acceptable detergent, whereas many others show great interference.

The effect of detergent on the unstirred polymerization of polyamide 610 is shown in Figure 9. The best polymerization conditions were shifted toward higher acid chloride concentrations. This effect is equivalent to raising the diamine concentration. Detergent may assist the transfer of both diamine and salt across the interface.

Although a higher maximum viscosity number was obtained, continuous film formation failed at the lower acid chloride concentrations and weak, dispersed polymer was obtained. The breakdown was accompanied by a large increase in acid ends in the polymer.

The break could not be associated with changes in surface tension, but the polymer was observed to be dispersed in and by the aqueous phase. Solvent was still occluded in the polymer. This may mean that under certain conditions the polymerization is interrupted by a transfer of the polymer from the organic to aqueous phases.

TABLE IV

Effect of Some Bases, Salts, and Detergents on Partition Coefficients of Hexamethylene-diamine (25°C.)

Additive[a]	Solvent	C_{H_2O}	pH	K ($C_{H_2O}/C_{solvent}$)
None	CHCl$_3$	0.43	12.1	1.39
NaCl (0.8M)	CHCl$_3$	0.49	11.7	1.61
NaCl + HCl	CHCl$_3$	0.36	10.3	10.2
NaCl + HCl	CHCl$_3$	0.39	9.5	550
None	CCl$_4$	0.098		76
NaOH (0.19M)	CCl$_4$	0.095		56
NaCl (0.19M)	CCl$_4$	0.096		69
NaCl (1.59M)	CCl$_4$	0.079		41
None	CCl$_4$	0.32		76
NaOH (0.67M)	CCl$_4$	0.28		29
NaHCO$_3$ (0.8M)	CCl$_4$	0.42		1250
Sodium lauryl sulfate (1%)	CCl$_4$	0.32		75

[a] Concentration in aqueous phase.

When a detergent, such as trihydroxytitanium stearate, was added to the organic phase, a similar failure in film continuity occurred, but this time the polymer film fragments were wetted by the organic phase

Salt is formed during interfacial polycondensations and salts may be added to the aqueous phase prior to the reaction. Salts, particularly neutral salts, shift the partition of diamine toward the organic phase (Table IV). This is probably a change in the amount of unionized diamine as well as salting-out effects. Large amounts (>10%) of sodium chloride change the diamine partition coefficient to such an extent that a shift in the peak of the polymerization curve can be recognized.

Salts in the aqueous phase also reduce the solubility of the acid chloride in the water. Any hydrolysis is thereby reduced. The addition of salts improves polymerizations employing the more water-soluble and -sensitive aliphatic acid halides.

Optimum Conditions for Some Other Polyamides

In comparing optimum conditions for one polymer with those of another, limited relationships are apparent. Dominant factors, as before, must be reaction rate, diamine partition, and polymer precipitation rate. For the continuous removal of films the reaction rates must be high and so all the polymer intermediates described have this common characteristic. Figure 9 and Table V compare the preparation of polyamides 6-T, Pip-T, and 610 in chloroform. 6-T is the least soluble and high polymer is obtained by reducing the acid chloride concentration. Pip-T is the most soluble, but the partition coefficient of piperazine favors the water, and so the polymerization maximum falls between the maxima for polyamides 6-T and 610.

The peak viscosity numbers do not fall in the order of solubilities, but the molecular weights do. Polyamide Pip-T has a higher molecular weight than polyamide 610 at the same viscosity number. Of course, purity of intermediates and side reactions can also limit the maximum molecular weight and their importance must be assessed for each polymerization.

Table V also shows conditions at which the peak viscosities are obtained in the preparation of several other polyamides. Of these, the viscosity data for Pip-10 in carbon tetrachloride compared to polyamide

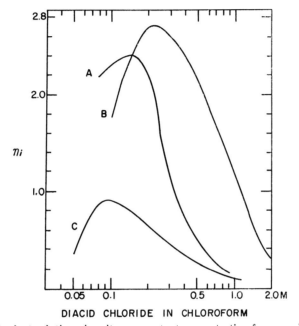

Fig. 10. Product solution viscosity vs. reactant concentration for several polyamides prepared by unstirred polycondensation. A, Poly(terephthaloyl piperazine); B, poly(hexamethylene sebacamide); C, poly(hexamethylene terephthalamide); diamines, 0.4M.

610 fits the preceding interpretation. Piperazine has a high partition coefficient which calls for a higher relative diamine concentration in the polymerization. However, Pip-10 is much more swollen by carbon tetrachloride than 610; therefore, diamine should be somewhat withheld in order to retain the polymerization near the interface. The result is that the same reactant concentration ratio is best for both polymers.

The data for polyamides 66 and 410 do not fit this picture. The curve for 410 showed a poorly defined peak. This was probably due to insufficiently pure diamine. Polymerizations of 66 often developed considerable turbidity in the liquid phases. Hydrolysis of the acid chloride is no doubt an interfering factor here.

Polycondensations Other Than Polyamidation

Thus far, the discussion has been concerned with polyamidation. The principles which apply to it apply with little modification to all other fast condensations employing diamines in the aqueous phase, such as the preparation of polyurethanes and polysulfonamides.

Reactions which do not eliminate a byproduct, such as the formation of a polyurea from a diisocyanate and a diamine, proceed on the organic solvent side of the interface. This has been demonstrated in an examination of the film formed with hexamethylene diisocyanate in benzene and aqueous hexamethylenediamine. Another point of interest in this reaction is that the benzene solution of diisocyanate does not show any hydrolysis and formation of polymer even after many hours of contact with a water phase which contains no diamine.

The formation of polyphenyl esters differs from polyamidation in several respects. First, the reaction of bisphenols with a diacid halide is a slow reaction at room temperature unless the phenol is in the form of the phenoxide ion. Furthermore, the formation of the bisphenoxide ion is usually necessary in practice in order to attain sufficient solubility for the phenolic intermediate in the aqueous phase.

Polymer

The illustrated reaction is believed to proceed by a nucleophilic attack of phenoxide on acid chloride. Though no data are known to the authors on rates and mechanisms for the phenoxide-acid halide reaction, the related reaction of epoxides with phenoxides has been shown to be of this type.[7]

The sodium bisphenoxide has high solubility in the aqueous phase and very little solubility in the organic phase. Therefore, in this polymerization the diacid chloride must be free to approach the liquid interface for a reaction to take place. Because of this requirement, solvents or near-solvents for the polymer are necessary in order to provide continued mobility for the oligomers and permeability for the diacid chloride. Fox, Conix, and Schnell[8] each report that quaternary ammonium salts and bases assist the polymerization. The improvement is presumed to result from the formation of organic solvent-soluble phenoxides and subsequent transport of the chloride ion to the aqueous phase as the quaternary ammonium salt.

TABLE V

Diamine Partition and Optimum Polymerization Conditions for Several Polyamides[a]

| Polymer | Solvent | Diamine partition coefficient[b] | | Optimum reactant concentration ratio, diamine/acid chloride[c] | Peak η_{inh} |
		K	C_{H_2O} moles/ liter		
410	CCl₄	405	0.40	3	1.00
66	CCl₄	35	0.40	3.5	1.70
Pip-10	CCl₄	460	0.40	7	1.68
Pip-T	CHCl₃	25.5	0.385	3	2.44
6-T	CHCl₃	0.70	0.457	4	0.93

[a] The notation in this paper uses numbers to refer to the number of carbon atoms in the amide or acid portion of the polyamide; the amine is designated first. T represents terephthaloyl and Pip stands for piperazine.

[b] K is the equilibrium ratio at 25°C. of $C_{H_2O}/C_{solvent}$ with 2 moles of NaOH per mole of diamine in the aqueous phase.

[c] Polymerizations were carried out with the diamine at 0.40M except for Pip-10, where the piperazine concentration was 0.817M.

An examination of the unstirred interfacial polycondensation of 2,2-bis-(hydroxyphenyl)propane and alkali with terephthaloyl chloride (Fig. 7) showed that the polymer film formed in the organic phase. The initial film continued to expand some within its original surface boundaries. Much of the additional growth was on the face adjacent to the liquid interface and there was no growth on the surface in the organic phase. No change in the manner of film growth was noted when quaternary ammonium salts were added to the aqueous phase. Tetraethylammonium chloride, in particular, greatly accelerated the rate of film formation.

This picture of the mechanism is in accord with the findings of Eareckson[11] that polyphenyl esters are best prepared from sodium phenoxides in solvents or near-solvents in rapidly stirred systems with detergents. Such a method provides small droplets and large surface areas about which the reaction may occur. Thereby, as much forming polymer as possible is kept in contact with the interface and the obstruction to the passage of acid chloride is kept at a minimum. Since the acid chloride must react near the interface and be exposed to a high concentration of hydroxyl ions, the less readily hydrolyzed aromatic acid chlorides yielded higher polymers than the aliphatic acid chlorides.

Yield

From the interfacial film-forming experiments it is clear that the polycondensation is dependent only on the conditions near the interface. Since diffusion is quite slow, the polymerization is uninfluenced by the distant reserves of reactants. High polymer is obtained at only a fraction of a per

316 P. W. MORGAN AND S. L. KWOLEK

cent yield, based on the total charge. The yield and freedom from terminating reaction in the polymerization zone is high. In unstirred interfacial polycondensations, when film is pulled away, unreacted portions of both aqueous and organic phases are carried off and the yield, when one of the phases is consumed or film fails to form continuously, is 25–55% from equivalents of reactants.

Film Properties

The continuous formation of film is not a test of polymer quality, for coherent polyamide films have been obtained with inherent viscosity numbers as low as 0.10, while high polymer precipitates are sometimes not coherent.

The thickness of films varies with the polymer, solvent, concentrations, and reaction times. Short reaction times may yield wet films from 0.1 to 200 μ in thickness. Commonly the thickness of polyamide 610 film which was continuously removed was about 3 μ, after washing and drying. The density of the gel films was 0.10–0.25 g./cm.^{3} polymer. In some experiments the apparent gel density decreased with reaction time. The density of dried film varied with the final washing solvent but was often below 0.3 g./cm.^{3} Polyphenyl ester films could be isolated from the xylene system in a tough, translucent form having normal density.

Figure 4 shows an electron micrograph of the cut cross sections of films of 610. The thickness is about 1 μ. There are pairs of smooth faces and rough faces together. The rough side also contains large holes. Electron micrographs (Figs. 2 and 3) of the two faces of a film show a fairly smooth and a rough side. The smooth side is always toward the water. In the first few seconds of reaction the films are nearly clear. Roughness and opacity increase rapidly and the films quickly turn white if left in the reaction mixture. The increased roughness can be observed with a light microscope. The opacity is not crystallization, for the films of 610 polyamide were not crystallized until the organic solvent was displaced by alcohol or water. Likewise, poly(hexamethylene terephthalamide) film was amorphous if the fresh film was kept frozen in liquid nitrogen.

There are several conditions under which films of greater clarity are obtained. Slow reacting intermediates produce clearer films because film growth is retarded. A similar effect applies to faster reactions when a poor solvent for the diamine, such as cyclohexane, is employed. The more soluble or solvent-sensitive polymers, such as from piperazine and sebacoyl or isophthaloyl chlorides, produce quite clear films, because of the greater plasticity of the polymer gel.

Washed and undried films of 610 polyamides were readily permeable to inorganic salts and to small dye molecules; however, permeability to large dye molecules, such as common direct dyes, was restricted.

Air-dried, undrawn polyamide films resembled tissue paper in general character. Films which had been pulled from an interface as a rope and then drawn some and dried under tension after washing with aqueous alcohol had tenacities of about 1.5 g./den.   The relaxed tube or straw dried from water had a tenacity of 0.8 g./den.

Polymer Properties

The question naturally arises as to what differences, if any, exist between melt condensation polymers and polymers from interfacial polycondensation.   The broad answer is that interfacial polymers are usually normal, linear condensation polymers.   They may differ in endgroup relationships and they may have unusual or nonrandom molecular weight distributions.[8c]

Melting Points and Crystallinity

Wherever melting points and crystallinity (by x-ray pattern) of interfacial polymers and those made by other methods can be compared, no differences are found.

Branching

The possibility of branching and crosslinking exists in reactions employing primary diamines.   This could arise from a second acylation taking place at the amide link.   One might expect this to be common since the polyamide is in the organic phase soaking in acid chloride solution.   Unintentional branching or crosslinking has been recognized in the case of the polysulfonamides and is promoted by excess alkali.   The aliphatic polyamides also are often branched when prepared with concentrated acid chloride solutions.

Endgroups and Molecular Weight

The polymers have been described only in terms of their inherent viscosities.   A forthcoming paper will report on the endgroup determinations and molecular weight relationship of 610 polyamide prepared by melt and interfacial methods.   Interfacial polyamides often have a high ratio of carboxyl to amine ends.   This arises from the formation of the polymer in the organic phase where it may be finally saturated and capped by acid chloride.   Hydrolysis of the acid chloride ends of the polymer or hydrolysis of one end of a diacid chloride molecule during polymerization could contribute to this effect.

EXPERIMENTAL

Range of Applicability

Interfacial polycondensation is widely applicable as a preparative method as is indicated by other papers in this series.   More often than not, the polymers form as precipitates.   Many polymers yield coherent films with the proper choice of solvents and other conditions.   In particular, the polyamides and polyurethanes will form self-supporting films under a wide variety of conditions.   Some of the best of these are the polyamides from aliphatic or aromatic diacid halides with aliphatic and alicyclic diamines and piperazines.   They have been employed in this work.

Purity of Materials and Control Experiments

The best test of purity of a material is its ability to form a high polymer.   Although interfacial polycondensation can proceed when many types of impurities are present, intermediates and solvents should be purified further if high polymer is not obtained.   In seeking new polymers one can often proceed by proving the purity of some of the materials first in known, successful polymerization recipes.

The evaluation of polymerization variables presupposes that some combination of conditions yields high polymer and that purity is not the restricting factor.   Throughout this study control experiments on reagent quality have been rerun many times.   Some of the variations which appear in the maximum viscosities for different groups of experiments are due to small changes in reagent purity.

Materials

Solvents in general were reagent or A.C.S.-grade materials.   Baker's Analyzed grade of chloroform was used after it was freed from ethanol by washing three times with water, rough drying with calcium chloride, and then storing in brown bottles over a mixture of calcium chloride and anhydrous potassium carbonate.   Under these conditions, neither hydrogen chloride nor phosgene form in the chloroform for as long as three weeks.

Hexamethylenediamine was nylon-production grade of proven quality. It may also be purchased from chemical supply houses as 70% aqueous solutions or a 97% solid.   Piperazine hydrate was obtained from the Polychemical Co., New York.   For general use, the amines and sodium hydroxide were made up in 0.2 g./ml. solutions which were dispensed from automatic burets equipped with stopcocks of Teflon (Du Pont trademark) tetrafluoroethylene resin.   Carbon dioxide was excluded by nitrogen or a soda-lime drying tube.

High-grade adipoyl and sebacoyl chlorides were prepared by the reaction of excess thionyl chloride on the acids or by distillation of commercially available products.   Such distillations must be carried out at pressures below 2 mm. in order to reduce the pot temperature.   Otherwise, general and vigorous decomposition will occur.   There must be a series of traps or tubes containing alkali pellets to absorb evolved gases.   Late in the work, high-grade acid chlorides were obtained directly from City Chemical Co., New York, N.Y.   Acid chloride solutions were prepared just prior to their use.

General Procedure

The general procedure is illustrated as follows by the preparation of polyamide 610 with a high inherent viscosity under preferred conditions of isolation:

An organic phase consisting of 8.00 ml. (0.0374 mole) sebacoyl chloride in 632 ml. carbon tetrachloride was prepared and placed in an 800 ml. beaker. In this, a 5-in. section of glass tubing with a 2.34-in. diameter was mounted vertically so that about half the tube was above the liquid. Into this tube was poured 374 ml. of a water solution containing 4.34 g. (0.0374 mole) hexamethylenediamine and 3.00 g. (0.075 mole) sodium hydroxide. A film of polymer formed at once at the liquid interface. This was grasped with tweezers and raised as a rope of continuously forming polymer film, which was passed over a variable-speed, driven roller and guided quickly into a stirred bath of 1% aqueous hydrochloric acid. The acid neutralized the diamine and assured the termination of polymerization.

The arrangement of apparatus and phases described provides a means for conveniently bringing together equivalents of reactants at an interface of constant size when the phase concentrations differed widely. When the phase concentrations were more nearly equal, simple tubes or shortened graduated cylinders were used. As is brought out in the discussion, there is no need, in an unstirred system, for reserves of equivalents of the two reactants. However, unless otherwise indicated (as in Table VII), the experiments were always started with equivalents of reactants and with each phase at least $1/2$ in. deep. This was done to obviate the possibility that reserve reactant was a variable. Two moles of sodium hydroxide were always added as the acid acceptor per mole of diamine.

The polymer was shredded in the dilute acid, washed well with distilled water, shredded in 1% aqueous sodium hydroxide, washed again, and finally titrated as a stirred suspension in water to a pH of 7. The latter step was carried out with a Beckman automatic titrator. Such careful neutralization at the end was considered necessary only in order to prepare the polymer for a correct determination of chain endgroups. This procedure and other purification methods will be described in a subsequent paper on endgroups and molecular weight relationships. The damp sample was partly air-dried and then dried in a vacuum oven at 80–100°C.

Viscosity determinations on the polymer solutions are reported here as inherent viscosities ($\eta_{inh} = [\ln \eta_{rel}]/c$) determined at 30°C. in m-cresol, with $c = 0.5$ g. polymer/100 ml. of solution.

Variations in Procedure

For demonstrations and many other experiments, small jars, drinking glasses, or beakers make excellent reaction vessels. A simple recipe for a demonstration experiment, which has been described elsewhere,[2,12] calls for 1.5 ml. sebacoyl chloride in 50 ml. perchloroethylene as the lower phase and

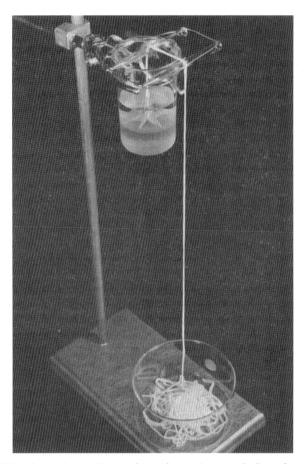

Fig. 11. Guiding device for continuous formation and removal of a polyamide film by gravity.

2.2 g. hexamethylenediamine and 4.0 g. sodium carbonate in 50 ml. water as the upper phase.

In the study of the polymerization mechanism the withdrawal rate of polymer film was measured and kept constant by rolls turning at a controlled rate. The polymerization can be made self-propelled by passing the collapsed film from the middle area of the vessel over a glass rod or tube and then out and down over a second rod as shown in Figure 11. At a height of about 3 to 4 ft. the weight of the wet film will provide the force to continue the formation of film until one phase is exhausted. Polymerization time may be extended by continuously replenishing the phases through inlet tubes. The arrangement of the phases in an open tube and beaker, described under General Procedure, has been used in this way.

Aqueous acidic alcohol was used as a reaction quenching medium in some early experiments. Two difficulties can arise. Some polymer films, such as poly(sebacoyl piperazine), are disintegrated in mixtures of chlorinated solvents and alcohols; under some conditions the acid chloride ends of the polymers become esterified, making normal endgroup determinations impossible.

It is obvious that if the organic layer is lighter than the aqueous layer, then the former should be entered at the top. Film toughness and continuity of film formation are not dependent on the position of the phases. Difficulty does arise when the specific gravities of the two phases are nearly equal, as are those of chlorobenzene and water.

Organic and aqueous phases can be alternated so as to produce two or more tubes of polymer film within one another. For example, 20 ml. CCl₄ with 1 ml. sebacoyl chloride was placed in a beaker, then 30 ml. 2.5% hexamethylenediamine solution with two equivalents of sodium hydroxide was added; the polymer film was pulled up and held as a rope while the third layer of 20 ml. xylene with 1 ml. adipoyl chloride was added. The two-layered tube of film was pulled out at 7 ft./min. It appeared to be one product, and could be drawn out to tough, smooth filaments. Careful examination of the fiber showed that the layers were not fused. The average inherent viscosity was 1.0. More complex systems were formed by adjusting the gravity of various layers by mixing solvents and adding salts to the aqueous phases.

Flat uncollapsed films can be collected in long strips by pulling the forming polymer off over the edge of the vessel, preferably a rectangular one. By this technique, it is almost impossible to avoid the formation of a thinner second film on the underside where the upper phase wets the vessel. Film samples were also isolated by placing plastic rings in the lower phase, adding the second phase and clearing the surface to get a smooth film, and finally collecting the sample on the raised ring. For this purpose, the ring should have a vertical wall at least 0.25 in. high to prevent film from folding around to the underside. The films were washed on the ring and then dried on glass plates or on screens for electron micrographs. The thinnest ones can be used in electron microscope work without sectioning.

Thicknesses of thin, wet films were measured by draping them over the square edge of a microscope cover glass of known thickness which was held vertically on the microscope stage. Illumination from top or bottom clearly showed the vertical layer of wet film.

Variations in Polymerization Time and Phase Dimensions

Table VI shows the effect of changing the speed of removal of polyamide 610 film from polymerizations at different concentrations. The interface areas were constant and equivalents of reactants were used. The viscosity number drops when the rate of removal is low. This can be interpreted as follows. When polymerization conditions are properly chosen, the gel film which forms initially is fairly uniform high polymer. Following this, polymerization proceeds more slowly, owing to the presence of the gel separating the reactants. The secondary polymer has low molecular weight; it fills the spaces in the gel and builds up on the organic solvent side of the film. When left at the interface, the films show a rapid increase in opacity and surface roughness (on the side of the organic solvent) and the average molecular weight drops. Films of polyamide 610 were split parallel to the interface after the reaction had proceeded several hours. The polymer on the organic solvent side had an inherent viscosity of 0.25 while the layer including the original film dropped in inherent viscosity from 1.4 to 0.8. The effect is not due to polymer degradation, because isolated film of high molecular weight can be soaked for many hours in portions of either phase without molecular weight change.

TABLE VI

Effect of Rate of Film Removal on Inherent Viscosity of Polyamide 610

Reactant concentrations, moles/l.			Initial height of H$_2$O phase, in.[a]	Speed, ft./min.	η_{inh}	Yield, g./1000 ft.
Diamine	NaOH	Sebacoyl chloride				
1.36	2.72	0.234	0.53	50	1.43	
				20	1.47	
				8	1.33	
				1.5	1.01	
0.40	0.80	0.059	1.73	40	1.54	
				20	1.54	
				8	1.47	
				1.5	0.98	
0.40	0.80	0.059	1.73	40	1.78[b]	4.6
				20	1.74	6.7
				10	1.79	7.5
				5	1.66	8.9

[a] Interface radius, 1.17 in.

[b] Better quality sebacoyl chloride than in first experiments.

Changes in the period of exposure of the polymer to the reactants are also produced by dimensional changes in the polymerization system. No change in the viscosity number of polyamide 610 (prepared in carbon tetrachloride–water) was noted when the distance between the interface and the quenching bath was varied from 10 to 90 in., while the upper, aqueous phase was 1 in. deep and the film speed was 20 ft./min. We believe that the initial high polymer has formed as soon as the film rises and folds just above the interface (Figs. 1 and 12). The folding action and some additional drainage down the tube excludes the lower phase and cuts down the formation of secondary polymer. Prolonged standing of the film rope (> ca. 1 min.) without a neutralization step often causes a decrease in average viscosity number. Similarly, lengthened exposure to one phase may lower the viscosity number (Table VII).

Table VII shows experiments on other dimensional changes in which the reactant concentrations and film removal speed were held at 20 ft./min. Although the polymerization was not overly sensitive to the size of the in-

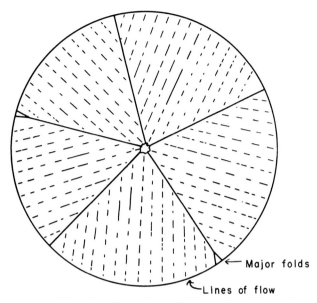

Fig. 12. Movement of forming film when withdrawn from the center of a circular interface (top view).

terface or the height of the uppermost phase, a decrease in average inherent viscosity occurred when the interface area was large or the aqueous phase was excessively high. Experiments in which xylene–acid chloride was the uppermost phase were inconclusive because sticking of the film to the glass interfered with rapid and steady removal of the film tube.

TABLE VII
Effect of Phase Dimensions on 610 Polymer Formation[a]

Interface		Height of H₂O phase, in.	Mole ratio,[b] diamine/acid chloride	η_{inh}
Radius, in.	Area, in.²			
0.94	2.77	1.0	1	1.35
1.44	6.52	1.0	1	1.35
2.0	12.5	1.0	1	1.25
0.94		0.5	1	1.39
0.94		1.5	3	1.41
0.94		3.0	6	1.45
0.94		9.5	19	1.25

[a] The molar reactant concentrations were: sebacoyl chloride, 0.117; diamine, 0.863; NaOH, 1.726. The wind-up speed was 20 ft./min.
[b] Ratio of quantities of reactants in entire system.

In subsequent experiments the interface area was kept small and constant, film removal speed was 20 ft./min. wherever possible, and the first gram of film was used for viscosity number and molecular weight determinations.

TABLE VIII
Yield of Polyamide 610 with Changes in the Organic Solvents[a]

Solvent	Time interval, min.	Yield, g./min.	η_{inh}
Carbon tetrachloride[b]	0–6	0.13	1.74
	6–12	0.13	1.78
	12–18	0.13	1.74
	18–24	0.13	1.83
Chloroform/carbon tetrachloride, 30/70 (v/v)[c]	0–3	0.53	1.95
	3–6	0.47	1.83
	6–7.4	0.29	1.71
Chloroform[d]	0–1	1.5	1.87[e]
	1–2	1.5	1.85
	2–3	1.3	1.73
	3–4	1.2	1.59
	4–5	0.8	1.47
	5–8	0.8	1.24
	8–10.6	0.3	0.94

[a] Interface radius, 1.17 in.; rate of film removal, 20 ft./min.

[b] Diamine, 0.40M; sebacoyl chloride, 0.06M.

[c] Diamine, 0.40M; sebacoyl chloride, 0.133M.

[d] Diamine, 0.40M; sebacoyl chloride, 0.20M.

[e] Higher molecular weights obtainable under these conditions with purer reactants or solvent.

Table VIII shows some actual rates of formation of polyamide 610 in carbon tetrachloride, chloroform, and a mixture of the two solvents. The rate was highest in chloroform. This was also recognizable by the thickness and toughness of the film. The polymerizations were carried out at the reactant concentration ratios which yield the highest polymer in the several solvents.

The data in Table IX show that, even when the concentrations in the carbon tetrachloride and chloroform systems are identical, the rate in the latter is much higher. Increasing the acid chloride concentration alone increases the rate of polymer formation in both solvents. The high rate of polymer formation in chloroform means that because of the more favorable diamine partition coefficient the reactants were brought together at a faster

TABLE IX
Yield Variation with Reactant Concentration Ratio[a]

Molar conc. diamine/acid chloride	Solvent	Time interval, min.	η_{inh}	Yield, g./min.
0.4/1.2	CHCl₃	0–1	1.06	2.6
0.4/0.8	CHCl₃	0–1	1.71	1.8
0.4/0.4	CHCl₃	0–1	1.90	1.8
0.4/0.2	CHCl₃	0–1	1.83	1.5
0.4/0.2	CCl₄	0–6	1.49	0.18
0.4/0.06	CCl₄	0–6	1.75	0.13

[a] Interface radius, 1.17 in.; rate of film removal, 20 ft./min.

rate. The postulated longer period of solution in chloroform enhances the yield further. Increasing the acid chloride concentration could increase the rate by a mass effect and by providing more intermediates in the same volume. In chloroform, the rate is sufficient to produce a marked temperature rise, and this must affect all the variables.

Technique for Locating Film Formation

Finely divided, colored powders, such as copper metal, pigments, and dyes, were introduced at the interface to show where and how the polymeric films were forming. The substance should not be soluble in the phase which it contacts.

Copper powder and inorganic pigments were floated upon the interface of a water-carbon tetrachloride system containing sebacoyl chloride in the lower phase. Hexamethylenediamine was carefully added and polymer film allowed to form. For the most part, the particles remained free at the film surface and were not trapped by continued reaction. Particles remained free after 20 hours when added to a similar 610 system following the initial formation of film. When a system of xylene–water was used, so that the organic phase was uppermost, particles added at the interface were caught in the film but remained at the lower film surface. Particles added just after the initial film formation were buried by a continuing growth of film in the organic phase above the initial film (see Fig. 7).

Some polymers which were water-sensitive appeared to start forming in the organic phase, but were then penetrated by the aqueous phase. This was shown by the dissolution of water-soluble dyes or powdered potassium dichromate.

References

1. E. L. Wittbecker and P. W. Morgan, *J. Polymer Sci.*, **40**, 289 (1959).
2. P. W. Morgan and S. L. Kwolek, *J. Chem. Ed.*, **36**, 182, 530 (1959).
3. E. E. Magat and D. R. Strachan, U.S. Pat. 2,708,617 (5/17/1955), assigned to E. I. du Pont de Nemours & Co., Inc.
4. G. H. Grant and C. N. Hinshelwood, *J. Chem. Soc.*, **1933,** 1351.
5. H. K. Hall, Jr., and P. W. Morgan, *J. Org. Chem.*, **21**, 249 (1956).
6. H. K. Hall, Jr., *J. Am. Chem. Soc.*, **77**, 5993 (1955).
7. H. H. Jaffé, *Chem. Revs.*, **53**, 191 (1953).
8. (a) D. W. Fox and E. P. Goldberg, lecture at Brooklyn Polytechnic Institute, Brooklyn, N. Y. (1958); (b) A. Conix, *Ind. chim. belge*, **22**, 1457 (1957); *Ind. Eng. Chem.*, **51**, 147 (1959); (c) H. Schnell, *Angew. Chem.*, **68**, 633 (1956); *Ind. Eng. Chem.*, **51**, 157 (1959).
9. H. W. Melville and G. M. Burnett, *J. Polymer Sci.*, **13**, 417 (1954).
10. E. L. Wittbecker and M. Katz, *J. Polymer Sci.*, **40**, 367 (1959).
11. W. M. Eareckson, *J. Polymer Sci.*, **40**, 399 (1959).
12. P. W. Morgan, *SPE Journal*, **15**, 485 (1959).

Synopsis

If a solution of a fast-reacting diacid halide in a water-immiscible solvent is brought together with an aqueous solution of a diamine without stirring, a thin film of high polymer is formed at once at the interface. Polyurethanes and polyamides in particu-

326   P. W. MORGAN AND S. L. KWOLEK

lar form tough films which can be grasped and pulled continuously from the interface as a folded rope of film. This unstirred interfacial polycondensation not only provides a dramatic demonstration of polymer formation but has permitted the observation of polymerization behavior and the study of the effects of many variables upon the process. The polymer-forming reactions proceed by nucleophilic displacement, and many have rate constants of at least 10^2–10^4 l. mole^{-1} sec.$^{-1}$. Polymers derived from diamines were found to form in the organic solvent phase. Therefore the rate of polymer formation is controlled by the transfer of diamine from the aqueous phase. It is believed that the liquid interface does not have any beneficial orienting effect on the reactants but that it provides for the regulated flow of one reactant into an excess of the other. Furthermore, the aqueous phase provides for the removal of the interfering byproduct, hydrogen halide, from the polymerization site. Some of the interrelated variables which have been studied and which are discussed in relation to the physical mechanism are the solvent sensitivity of the polymer, partition coefficients of the reactants, reactant concentration, duration of the polymerization, and the addition of monofunctional reactants, detergents, and salts. The formation of polyphenyl esters is discussed briefly.

Résumé

Quand une solution d'halogénure de diacyle très réactionnel dans un solvant non-miscible à l'eau est mise en présence d'une solution aqueuse de diamine sans agitation, il se développe un fin film de haut polymère à l'interface. Les polyuréthannes et polyamides en particulier forment des films durs qui peuvent être pris et tirés continuellement de l'interface comme une corde de film plissé. Cette polycondensation interfaciale en absence d'agitation ne démontre pas seulement de façon spectaculaire la formation de polymère, mais permet de suivre le processus de polymérisation et d'étudier l'influence des nombreuses variables. Les réactions de formation de polymère consistent en un déplacement nucléophile; elles ont des constantes de vitesse d'au moins 10^2 à 10^4 litre mole^{-1} sec^{-1}. Les polymères dérivés de diamines so forment dans la phase du solvant organique; il en résulte que la vitesse de formation du polymère est controlée par le transfert à la diamine au départ de la phase aqueuse. On croit que l'interface liquide n'a aucun effet d'orientation sur les réactifs, mais elle rend possible le passage régulier d'un réactif donné vers l'excès de l'autre. En outre, la phase aqueuse écarte de l'endroit de la polymérisation les produits secondaires gènants, tel l'hydracide halogéné. Quelques unes des variables qui ont été étudiées et discutées en rapport avec le mécanisme physique sont: la sensibilité du polymère à l'égard du solvant, les coefficients de pontage des réactifs, la concentration en réactif, la durée de polymérisation, l'addition de réactif mono-fonctionnel, de détergents et de sels. La formation d'esters polyphénylés est discutée brièvement.

Zusammenfassung

Beim Zusammenbringen einer Lösung eines rasch reagierenden Dikarbonsäurehalogenids in einem mit Wasser nicht mischbaren Lösungsmittel mit einer wässrigen Lösung eines Diamins ohne Rühren bildet sich an der Berührungsfläche sofort ein dünner Film eines Hochpolymeren. Besonders Polyurethane und Polyamide bilden zähe Filme, welche angefasst und kontinuierlich von der Grenzfläche als gefalteter Strang abgezogen werden können. Diese Grenzflächenpolykondensation ohne Rühren liefert nicht nur eine anschauliche Demonstration der Polymerbildung sondern erlaubt auch die Beobachtung des Polymerisationsverlaufs und die Untersuchung des Einflusses vieler Variabler auf den Vorgang. Die Polymerbildungsvorgänge verlaufen über eine nukleophile Substitution und viele davon haben Geschwindigkeitskonstanten von mindestens 10^2 bis 10^4 lit. Mol^{-1} sec^{-1}. Es wurde gefunden, dass sich die von Diaminen abgeleiteten Polymeren in der organischen Lösungsmittelphase bilden. Die Geschwindigkeit der Polymerbildung wird daher durch den Übergang des Diamins aus der Wasserphase bestimmt. Man kommt zu der Ansicht, dass die Flüssigkeitsgrenzfläche zwar keinen günstigen, orientierenden Einfluss auf die reagierenden Stoff hat, dass sie jedoch für einen geregelten Fluss des einen Reaktionsteilnehmers in einen Überschuss des anderen sorgt. Ausserdem bewirkt die Wasserphase die Entfernung des Halogenwasserstoffes, der als Nebenprodukt auf die Reaktion Einfluss hat, vom Polymerisationsort. Einige den Reaktionsverlauf beeinflussenden Variabeln, die untersucht wurden und die in Bezug auf den physikalischen Mechanismus diskutiert werden, sind die folgenden: die Empfindlichkeit des Polymeren gegen das Lösungsmittel, Verteilungskoeffizienten der reagierenden Stoffe, ihre Konzentration, Polymerisationsdauer und der Zusatz von monofunktionellen Stoffen, Detergents und Salzen. Die Bildung von Polyphenylestern wird kurz diskutiert.

Received June 29, 1959

PERSPECTIVE

Comments on "Polymer NSR Spectroscopy.* II. The High Resolution Spectra of Methyl Methacrylate Polymers Prepared with Free Radical and Anionic Initiators," F. A. Bovey and G. V. D. Tiers, *J. Polym. Sci.,* XLIV, 173 (1960)

KOICHI HATADA

Faculty of Engineering Science, Osaka University, Toyonaka, Osaka 560, Japan

In writing the Perspective on this article, I cannot help mentioning the preceding paper, "Polymer NSR Spectroscopy. I. The Motion and Configuration of Polymer Chains in Solution," [*J. Polym. Sci.,* **XXXVIII,** 73 (1959)]. This first report on high resolution NMR of polymers in solution covers almost all the fundamental ideas regarding how we can use NMR spectroscopy for polymers, such as: line broadening and macroscopic or local viscosity; spin-lattice relaxation and segmental mobility; chemical shift and diamagnetic shielding; deuterium substitution technique for peak assignment; effect of tacticity or comonomer sequence on NMR spectra; and chain conformation and NMR spectra. As to the tacticity dependence of NMR spectra, the authors measured the spectra of polystyrenes prepared with several kinds of initiators, including isotactic polystyrene, but they found no difference in the positions and appearance of the peaks for all polymers. However, I suppose that Dr. Bovey believed in the sensitivity of NMR spectra to polymer tacticity because he discussed the effect of conformation of polymer molecules on chemical shift; conformation is affected by tacticity. This belief resulted in their second paper, in which the tacticity dependence of NMR signals was first demonstrated.

At the time when their second paper was published, I was a graduate student and studying stereospecific polymerization of methyl methacrylate. Before this publication, only qualitative data on

polymer tacticity could be obtained by X-ray fiber analyses or empirical infrared absorption correlations. Moreover, as most of the poly(methyl methacrylate)s prepared with anionic initiators at that time were amorphous, X-ray analysis told us nothing. Publication of Bovey's paper was more than 30 years ago, but it is still very fresh in my memory how I was excited to read it; it enabled us to measure the polymer tacticity quantitatively. According to their NMR method, triad tacticity could be determined from α-methyl proton signals and diad tacticity from backbone methylene proton signals. Appearance of the methylene signals gives an absolute and independent confirmation of the type of tacticity; AB quartet for an isotactic polymer and singlet for a syndiotactic polymer. For the designation of chain tacticity it was proposed to name the methylene group located in *dd* or *ll* diads as *"meso"* (*m*) and that located in *dl* or *ld* diads as *"racemic"* (*r*) (the latter was renamed as *"racemo"* by IUPAC Macromolecular Nomenclature Commission to avoid confusion with "racemic" used in low molecular weight chemistry). The proposed terms are now used conveniently for the description of polymer tacticity of long sequences as well as of diads.

In addition to the spectroscopic achievements mentioned above, Bovey and Tiers also studied the effects of initiator, solvent, and temperature on the stereoregulation of the polymerization of methyl methacrylate. From the results they concluded that the free radical propagation can be described by a single value of σ, which is the probability that a growing polymer chain will add a monomer unit to

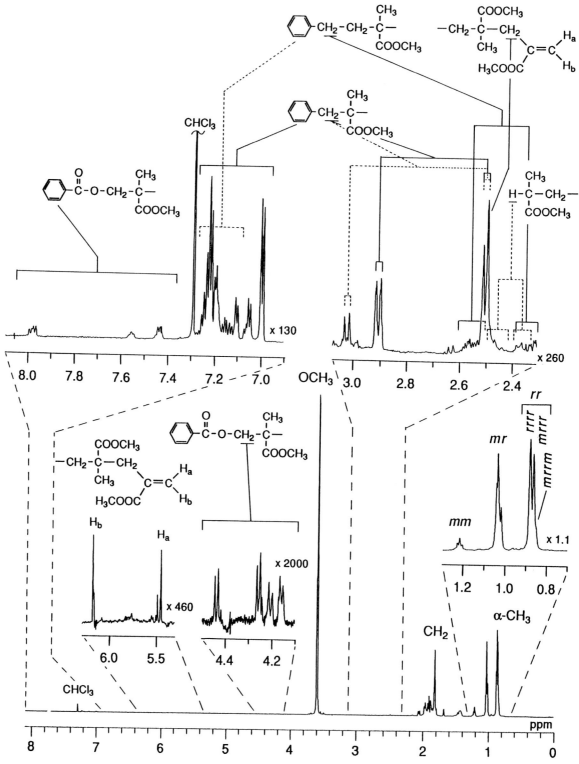

Figure 1.   750 MHz ¹H-NMR spectrum of poly(methyl methacrylate) prepared with benzoyl peroxide at 100°C in toluene (measured in CDCl₃ at 55°C).

give the same configuration as that of the last unit at its growing end. They also indicated that the propagation in anionic polymerization is not a single-σ process. These were the first statistical treatments of polymer tacticity and established the way of studying the mechanism of stereoregulation in polymerization. For these analyses the plot of the probabilities of formation of isotactic, heterotactic and syndiotactic triads as a function of σ value was used. This plot was later called "Bovey's plot" and is cited in most polymer textbooks.

Figure 1 shows a 750 MHz ^{1}H-NMR spectrum of poly(methyl methacrylate) prepared with benzoyl peroxide at 100°C in toluene using the latest Varian instrument; the conditions of polymer preparation are very similar to those for the polymer described in Figure 1(a) of the Bovey's paper. Compared with Bovey's spectrum, the resolution and signal-to-noise ratio are greatly enhanced, which provides us new information on longer tactic sequences and chain-end structures as indicated in the figure. The great improvement in the quality of NMR spectra is due to the increasing development of NMR technology, including the introduction of superconducting magnets and Fourier transform methods. NMR spectroscopy has played an increasingly important role in a variety of fields of polymer science, including studies on structure, molecular motion, and reaction mechanism. Its importance in polymer research and the development during the last three decades are within Dr. Bovey's vision as his pioneering works published during 1959 and 1960 indicate clearly. Dr. Bovey is surely the pioneer who opened up the development of NMR spectroscopy in polymer science.

JOURNAL OF POLYMER SCIENCE   VOL. XLIV, PAGES 173–182 (1960)

Polymer NSR Spectroscopy.* II. The High Resolution Spectra of Methyl Methacrylate Polymers Prepared with Free Radical and Anionic Initiators

F. A. BOVEY and G. V. D. TIERS, *Central Research Laboratories, Minnesota Mining and Manufacturing Company, St. Paul, Minnesota*

The development of methods for producing stereoregulated vinyl polymers has brought with it the necessity for the development of methods for determining the structure (i.e., stereochemical configuration) and conformation of their chains with the highest possible discrimination. Up to the present time, the polymer chemist has had x-ray fiber diagrams and empirical infrared absorption correlations[1] at his disposal for this purpose. In this paper we describe a new method, based on the high resolution nuclear spin resonance (NSR) spectra of polymers in solution. This method has been employed only on polymers of methyl methacrylate but should have a broader applicability.

EXPERIMENTAL

Free radical polymerization of methyl methacrylate was carried out in bulk at 0° by exposure to approximately 10 Mrep of gamma radiation; the monomer was sealed *in vacuo* in an ampule which was held in ice in a Dewar flask placed in a waste fission-product source for 96 hr. at a dose rate of approximately 0.1 Mrep/hr. Bulk polymerization was also carried out in the presence of 0.5% benzoyl peroxide. Solution polymerizations were carried out on 10% monomer solutions with 0.5% initiator for periods of 4–16 hr. (See Table I for details concerning initiator type and polymerization temperature.)

n-Butyllithium was obtained as a 2.12*M* solution in heptane from the Lithium Corp. of America; a total of 1–2 ml. of this solution was incrementally fed under dry nitrogen pressure into a stirred reaction vessel of approximately 150 ml. capacity containing 100 ml. of a 10% solution of methyl methacrylate in the chosen solvent.[2] Table I may be consulted for solvent composition and reaction temperatures. Reaction times of 2–6 hr. were employed. The polymer was recovered by precipitation in a large volume

---

* The first paper of this series is F. A. Bovey and G. V. D. Tiers, *J. Polymer Sci.*, **38**, 73 (1959).

173

174   F. A. BOVEY AND G. V. D. TIERS

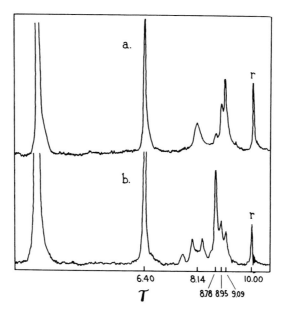

Fig. 1. NSR spectra of methyl methacrylate polymers prepared with (*a*) benzoyl peroxide in toluene at 100° and (*b*) *n*-butyllithium in toluene at −62°. The spectra are run on 15% solutions of the polymers in chloroform with 1% tetramethylsilane reference (*r*).

of methanol and was dried *in vacuo* at about 50°. In case of severe contamination with lithium salts (a grayish appearance being noted), the polymer was dissolved in butanone and reprecipitated in methanol. Ethylene glycol dimethyl ether (glyme) was obtained from Ansul Chemical Co.; interfering impurities initially present, such as water, were removed by treatment with P_2O_5 followed by simple distillation.

Sodium naphthalenide was prepared according to the directions of Scott et al.[3] Polymerization of methyl methacrylate[4] was carried out in a test tube immersed in a Dry Ice–acetone bath by rapid addition of ca. 0.5 ml. of sodium naphthalenide solution to 10 ml. of a 10% monomer solution in glyme. Polymerization appeared to be instantaneous.

Polymethyl methacrylate prepared by means of phenylmagnesium bromide initiator was supplied through the kindness of Dr. C. Agre, Internal Chemicals Department, and Dr. Paul Trott, Tape Research Laboratory, Minnesota Mining and Manufacturing Company. Polymerization was carried out in toluene solution at 0°. The sample supplied by Dr. Trott was prepared exactly according to directions provided in the patent literature.[5] It is felt that some degree of fractionation may occur during the isolation.

Spectra were obtained on 15% (w/v) solutions of the polymers in chloroform, 0.5 ml. of solution being placed in 5 mm. o.d. Pyrex tubes. A Varian V-4300-2 40.00 Mcycles/sec. spectrograph was employed, together with a Varian heated probe, sample spinner, audio-oscillator, Hewlett-Packard

522-B frequency counter, and Varian recorder. In order to achieve the required spectral resolution, all samples were maintained at approximately 90°. The peak at the extreme right in Figure 1 is that of the internal reference standard, tetramethylsilane; the scale is expressed in parts per million, referred to this peak as 10.00. Values on this scale, termed τ-values,[6] were reproducible to ca. ±0.02 p.p.m. standard deviation in the present work. Peak areas in the α-methyl region of the spectrum were determined by sweeping through this region at a rate of approximately 10 cm./ppm. The individual peak outlines were then constructed and their areas estimated with a planimeter. The sum of the individual areas equalled the total area within ±5%.

EXPERIMENTAL RESULTS AND INTERPRETATION

In Figure 1 are shown typical spectra for 15% solutions of two methyl methacrylate polymers in chloroform. The polymers were prepared with (a) benzoyl peroxide in toluene at 100°; and (b) n-butyllithium at −62°. The large peak at the left is that of the chloroform solvent. The ester methyl group appears at 6.40τ in both spectra, and is not affected by the chain conformation. There are three α-methyl peaks, at 8.78τ, 8.95τ, and 9.09τ, whose relative heights vary greatly with the method of polymer preparation. Polymers prepared with n-butyllithium show a very prominent peak at 8.78τ, the others being much smaller. Polymers prepared with benzoyl peroxide initiator show the same three peaks, but now the peak at 9.09τ is the most prominent.

We interpret these very marked differences as follows. Polymethyl methacrylate prepared with n-butyllithium initiator in hydrocarbon solvents is believed to be predominantly isotactic,[2] as is also the polymer prepared in hydrocarbon solvents in the presence of 9-fluorenyllithium.[7] The peak at 8.78τ must therefore be due to the α-methyl groups of monomer units which are flanked on both sides by units of the same configuration, i.e., *ddd* or *lll*. This we term an *isotactic configuration*, the central unit being termed an *i* unit. The most prominent peak in free radical polymers (at 9.09τ), is attributed to α-methyl groups of central monomer units in *syndiotactic configurations* (*s* units), i.e., *ldl* or *dld*, since in free radical polymers, at least when prepared at low temperatures, this structure tends to predominate.[7] The peak at 8.95τ is believed to be due to α-methyl groups of central monomer units in *heterotactic configurations* (*h* units), i.e., *ldd*, *dll*, *ddl*, or *lld*. The 8.78τ, 8.95τ, and 9.09τ peaks will be proportional to the numbers of *i*, *h*, and *s* units, respectively.

Let us designate by σ the probability that a polymer chain will add a monomer unit to give the same configuration as that of the last unit at its growing end. (Here, σ is essentially the same as Coleman's α.[8])

As a matter of fact, the placement of the end unit is not decided until the next unit has been added, since the end unit itself, whether a radical, anion, or cation, is probably unable to maintain optical asymmetry. This does not affect the statistical argument.

Let us assume that σ is controlled only by the configuration of this end unit and not by that of the penultimate unit, and that the propagation can be described by a single value of σ. We then find that:

$$P_i = \sigma^2$$

$$P_s = (1 - \sigma)^2$$

and

$$P_h = 1 - P_i - P_s = 2(\sigma - \sigma^2)$$

where P_i, P_s, and P_h represent the probabilities of forming i, s, and h units, respectively. In Figure 2, these relationships are plotted. It will be noted that the proportion of h units rises to a maximum at $\sigma = 0.5$, which corresponds to random propagation. For a random polymer, the proportion $i{:}h{:}s$ will be $1{:}2{:}1$.

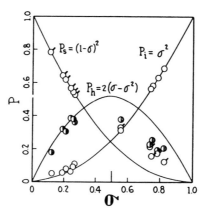

Fig. 2. The probabilities P_i, P_s, and P_h of formation of isotactic, syndiotactic, and heterotactic triads, respectively, as a function of σ, the probability of isotactic placement of monomer units during chain propagation. Experimental points at the left are for methyl methacrylate polymers prepared with free radical initiators; those at the right for polymers prepared with anionic initiators: i (O) peaks; (◌) s peaks; and (◑) h peaks.

In Table I are shown the analyses of the spectra of a number of free radical polymethyl methacrylates prepared under conditions varying as indicated; the spectra are interpreted in terms of the proportions of i, h, and s units. In Figure 2, these results have been fitted to the probability curves by placing the s points on the curve and letting the other points fall where they may. It will be observed that for the free radical polymers, the proportions of h and i units fall satisfactorily close to the calculated curves. We interpret this as indicating that the free radical propagation can be described by a single value of σ. There is a noticeable effect of temperature on σ (compare polymers 1 and 2), but the syndiotactic tendency is always predominant.[1] We shall treat the effect of temperature in more detail in a later paper.

TABLE I

Structure of Polymers of Methyl Methacrylate Prepared with Free Radical Initiators

Polymer no.	Polymerization conditions	$P \times 100$			
		i	h	s	σ
1	Irradiation in bulk, 0°	7.5	30.0	62.5	0.21
2	Benzoyl peroxide in bulk, 100°	8.9	37.5	53.9	0.27
3	Lauroyl peroxide in 10% hexane soln., 50°	10.5	35.8	53.8	0.26
4	Benzoyl peroxide in 10% nitromethane soln., 100°	4.9	30.8	64.2	0.20
5	Azoisobutyronitrile in 10% toluene soln., 50°	6.3	37.6	56.0	0.25

Polymers 3 and 5 represent products prepared in solvents of low dielectric constant which are poor and good solvents, respectively, for the polymer It has been suggested by Szwarc[9] that polymerization in a poor solvent may encourage regular growth because of the necessity of conforming to a helical chain growth pattern.   Our present data do not encourage this idea, at least for this system, since the spectra of polymers 3 and 5 are very similar.   Polymerization in a solvent of high dielectric constant (polymer 4) appears to encourage syndiotactic placement somewhat.

In Table II are shown results obtained for polymers prepared with anionic initiators.   These polymers appear to be predominantly isotactic, provided effective complexing solvents are not employed.   (An interpretation in terms of σ cannot be offered for reasons to be discussed below.)   This result is in general agreement with the finding of Fox et al.[7]   Glyme (polymer 9) is apparently much more effective in this respect than dioxane (polymer 8).   Fox et al.[7] have reported that in glyme alone, with 9-fluorenyllithium as initiator, syndiotactic polymer is obtained.   We agree that the tendency is in this direction in the presence of this ether, but we were not able to obtain polymer using it alone as solvent.   Fox also reported that in the presence of dioxane, block polymers consisting of alternating syndiotactic and isotactic regions were obtained.   This point is discussed in the next section.

TABLE II

Structure of Polymers of Methyl Methacrylate Prepared with Anionic Initiators

Polymer no.	Polymerization conditions	$P \times 100$		
		i	h	s
6	n-BuLi in toluene, −62°	0.63	0.19	0.18
7	n-BuLi in toluene, 25°	0.68	0.20	0.12
8	n-BuLi in 60:40 toluene:dioxane, −62°	0.56	0.23	0.21
9	n-BuLi in 50:50 toluene:glyme, −62°	0.31	0.37	0.32
10a	C_6H_5MgBr in toluene, 0°	0.58	0.26	0.16
10b	C_6H_5MgBr in toluene, 0°	1.00	0.0	0.0
11	Sodium naphthalenide in glyme, −70°	$i + h \simeq 0.10$		0.90

Phenylmagnesium bromide (polymer 10a) appears to function in essentially the same manner as *n*-butyllithium. Polymer 10b is discussed in the next section. Sodium naphthalenide, which is stable only in an ether solvent and decomposes in toluene, gives a highly syndiotactic polymer, with a σ of approximately 0.05.

A comparison of polymers 6 and 7 in Table II shows that the effect of temperature on metal alkyl-initiated polymerization is not large.

The magnetic fields experienced by α-methyl groups in *i*, *h*, and *s* configurations may be expected to be different because of the strong magnetic anisotropy of the ester carbonyl groups. These differences will not be averaged to zero by the segmental motion of the polymer chains, since the various conformations are not equally populated. The methyl resonance of pivalic acid, $(CH_3)_3CCOOH$, is at 8.77τ, corresponding closely to that of *i* units; in *h* and *s* configurations the α-methyl group is somewhat more shielded than this.

The "backbone" methylene resonance would be expected to be a single peak in a syndiotactic polymer, as observed, because from simple geometrical considerations both protons must on a time average experience the same magnetic environment. In an isotactic polymer, however, this is not true; the two protons will be differently shielded and may therefore be expected to show electron-coupled spin-spin splitting to give a fourfold resonance. (A somewhat parallel situation has been studied by Wiberg.[6f]) This will consist of two "nonequivalent" doublets having nearly equal peaks if the difference in shielding considerably exceeds the coupling constant J, whereas if the shielding difference is approximately equal to J, the peaks will be more closely spaced and the outer peaks will be weaker.[10,11] In Figure 1a the methylene resonance is a single peak at 8.14τ. In Figure 1b three of the expected four peaks can be seen; the fourth is assumed to be under the α-methyl resonance. The center of the quartet is in the same position, 8.14τ, as the single peak of *a*. The coupling constant J is found to be 15.5 cycles-sec.$^{-1}$ and the shielding difference 0.60 ppm. *The observation of the methylene resonances provides absolute, independent confirmation of the structures deduced from x-ray fiber diagrams.*

The designation of chain configuration in terms of *d* and *l* units we consider to be unsatisfactory for two reasons: (a) it is not operational, i.e., it draws a distinction between enantiomers which is not reflected in optical properties, terminal group effects being essentially nil in medium to high polymers; and (b) it is not convenient because it fails to make evident the experimentally observable classes of polymer structure. One system of nomenclature which we prefer for NSR studies has already been indicated. In considering the methylene resonances, however, another system is more convenient. The fraction of methylenes giving a quartet resonance will be equal to the fraction of *dd* or *ll* units, i.e., we now must consider the chain structure in terms of dyads of monomer units rather than in terms of triads as above. The fraction of methylene groups which are located between *dl* or *ld* configurations we shall designate as the fraction of racemic or r units; the fraction lying between dd or ll units we shall designate as meso or m units. If P_r designates the probability of forming r units, and P_m the probability of forming m units, then:

$$P_m = \sigma = P_i^{1/2}$$

and

$$P_r = 1 - \sigma = P_s^{1/2}$$

the restrictions on σ being maintained as described earlier. If either of these restrictions is removed, then these simple relationships do not hold. For a random polymer, we find that the fraction of r units will be 0.50.

DISCUSSION

While free radical polymerization of methyl methacrylate appears to be describable by a single value of σ, there are clear indications that the anionic polymerizations, particularly in the presence of complexing solvents, are not. In the presence of 40% of dioxane (polymer 8 in Table II), the s peak is comparable to the h peak in area, which is not in accord with expectation for a "single-σ" propagation. In the presence of 50% of glyme (polymer 9), a large deviation from expectation occurs, as can be seen from the conspicuous failure of the peak areas for this polymer to fit the theoretical curves (see the points near about 0.55 on the abscissa in Fig. 2.) It has been suggested by Fox et al.[7] that in the presence of complexing solvents, isotactic–syndiotactic block polymers may be formed. We may imagine that in the presence of such solvents, the metal counterion is sometimes associated with the growing chainend (isotactic propagation) and sometimes with the solvent (syndiotactic propagation) and that the rate of complex formation and decomposition is slower than the rate of propagation. Such block chains would in general be formed if any given chain stereoregularity, once formed, tended to propagate itself preferentially but not exclusively. It can readily be shown that if we assign lengths to the blocks in the manner illustrated by this example:

$$(l) \ldots \underbrace{dddd \ldots dddd}_{\bar{n}_i} \underbrace{ldldl \ldots dldl}_{\bar{n}_s} \ldots (d)$$

then

$$(\bar{n}_i - 1)/(\bar{n}_s - 1) = i/s$$

$$\bar{n}_i/\bar{n}_s = m/r$$

$$\bar{n}_i/(\bar{n}_i + \bar{n}_s) = m/(r + m)$$

$$(\bar{n}_i + \bar{n}_s)/2 = 1/h = \bar{n}$$

where $\bar{n}_i$ and $\bar{n}_s$ are number-average lengths of the isotactic and syndiotactic blocks, respectively, and $\bar{n}$ is the number-average length of all blocks. (If the propagation *can* be described by a single value of σ, then: $\bar{n}_i/\bar{n}_s =$

$m/r = \sigma/(1 - \sigma)$, $\bar{n}_s = 1/\sigma$, and $\bar{n}_i = 1/(1 - \sigma)$.) Thus, a prominent h peak means that block lengths must be relatively short. It can also readily be seen that $h/2 \leq s$, the limiting equality $h = 2s$ applying to a structure consisting of single h units between isotactic blocks. From these relationships, we find that for polymer 6, $\bar{n}_i = 7.6$ and $\bar{n}_s = 3.2$; 40% of dioxane decreases the apparent $\bar{n}_i$ to 6.0, while 50% of glyme decreases it to 2.6.

Consideration of polymer 10b, however, makes it necessary to modify these conclusions. This polymer was prepared in the same manner as 10a, but was fractionated by precipitation in petroleum ether; it constitutes about 90% of the polymer prepared in this run and contains no h or s sequences, being purely isotactic within the limits of detection of the present method. For polymer 10a, an average isotactic sequence of only 5.8 is calculated. These results strongly suggest that the apparent h and s peaks in the spectrum of polymer 10a (and, by implication, in the spectra of polymers 6 and 7 as well) must be due to a contaminating product, possibly of much lower molecular weight and formed by a reaction entirely separate from that producing the isotactic polymer.[12] When dioxane or glyme is present, h and s sequences may actually occur in the high polymer chains, but again the presence of this contaminating product probably leads to a gross underestimate of $\bar{n}_i$. In order to proceed further it will be necessary to fractionate polymers prepared in complexing and noncomplexing solvents, and to examine the fractions separately by NSR. This should permit a correct estimate of $\bar{n}_i$ and $\bar{n}_s$.

Another imaginable type of "non-sigma" propagation is that in which isotactic blocks of opposite configuration alternate, i.e.,

$$\ldots dddddlllll \ldots$$

If such a structure occurred exclusively, the following relations would hold:

$$s = 0$$

$$\bar{n} = 2/h = 1/r$$

No polymer showing these characteristics has yet come under our observation, an h peak being always found to be accompanied by an appreciable s peak.

The authors thank Miss Lucy Stifer for preparation of polymer samples and Mr. Donald Hotchkiss for excellent maintenance and operation of the nuclear magnetic resonance spectrometer.

References

1. Fox, T. G., W. E. Goode, S. Gratch, C. M. Huggett, J. F. Kincaid, A. Spell, and J. D. Strange, *J. Polymer Sci.*, **31**, 173 (1958).

2. We are indebted to Prof. H. F. Mark for information concerning this method of polymerization.

3. Scott, N. D., J. F. Walker, and V. L. Hansley, *J. Am. Chem. Soc.*, **58**, 2442 (1936).

4. Szwarc, M., and A. Rembaum, *J. Polymer Sci.*, **22**, 189 (1956).

5. Australian Pat. 36,684, assigned to Rohm and Haas, March 4, 1958.

6. (*a*) Tiers, G. V. D., *J. Phys. Chem.*, **62**, 1151 (1958); (*b*) G. V. D. Tiers, *J. Chem. Phys.*, **29**, 963 (1958); (*c*) G. V. D. Tiers and F. A. Bovey, *J. Phys. Chem.*, **63**, 302 (1959); (*d*) J. A. Eldridge and L. M. Jackman, *Proc. Chem. Soc.*, **1959**, 89; (*e*) M. S. Barber, L. M. Jackman, and B. C. L. Weedon, *Proc. Chem. Soc.*, **1959**, 96; (*f*) K. B. Wiberg and H. W. Holmquist, *J. Org. Chem.*, **24**, 578 (1959); (*g*) F. A. Bovey and G. V. D. Tiers, *J. Am. Chem. Soc.*, **81**, 2870 (1959).

7. Fox, T. G., B. S. Garrett, W. E. Goode, S. Gratch, J. F. Kincaid, A. Spell, and J. D. Stroupe, *J. Am. Chem. Soc.*, **80**, 1768 (1958); J. D. Stroupe and R. E. Hughes, *J. Am. Chem. Soc.*, **80**, 2341 (1958).

8. Coleman, B., *J. Polymer Sci.*, **21**, 155 (1958).

9. Szwarc, M., *Chem. and Ind. (London)*, **1958**, 1589.

10. Hahn, E. L., and D. E. Maxwell, *Phys. Rev.*, **88**, 1070 (1952).

11. Pople, J. A., W. G. Schneider, and H. J. Bernstein, *High Resolution Nuclear Magnetic Research*, McGraw-Hill, New York, 1959, pp. 119–123, p. 292.

12. Glusker, D. L., I. Lysloff, and E. Stiles, *Abstracts of Papers 136th Am. Chem. Soc. Meet., Sept 13–18, 1959, Atlantic City*, p. 8T.

Synopsis

Spectra of methyl methacrylate polymers in chloroform solution at 90° show three α-methyl proton peaks.   Measurements of their areas in spectra of polymers prepared with free radical and anionic initiators indicate that these may be attributed to (*1*) isotactic sequences (*ddd* or *lll*), (*2*) syndiotactic sequences (*ldl* or *dld*), and (*3*) heterotactic sequences (*ldd, dll, ddl,* or *lld*).   The conformation and stereochemical configuration of the chains may thus be examined with considerable discrimination; degrees of regularity, block sizes, etc. appear to be determinable.   Free radical polymers of methyl methacrylate are found to be predominantly syndiotactic, whereas those produced with anionic initiators, such as *n*-butyllithium, are shown to be predominantly isotactic, in agreement with the findings of others.   The backbone methylene resonance also shows striking changes with chain stereochemical configuration, such as to provide absolute, independent confirmation of the assigned structures.

Résumé

Les spectres des polymères de méthacrylate de méthyle dans le chloroforme à 90° montrent trois pics correspondants aux protons α-méthyliques.   Les mesures de leurs surfaces dans les spectres des polymères préparés par radicaux libres et par initiateurs anioniques indiquent que ces derniers peuvent être attribués à (*1*) des séquences isotactiques (*ddd* ou *lll*), (*2*) syndiotactiques (*ldl* ou *dld*) et (*3*) des séquences hétérotactiques (*ldd, dll, ddl* ou *lld*).   La conformation des chaînes doit donc être examinée avec un discrimination considérable; les degrés de régularité, les dimensions des blocs, etc. peuvent être déterminés.   Les polymères par radicaux libres des méthacrylate de méthyle ont été trouvés être principalement syndiotactiques tandis que ceux produits par initiateurs anioniques tel que le *n*-butyllithium ont été prouvés être principalement isotactiques; ces résultats sont en accord avec les résultats des autres auteurs.   La résonance du méthylène de la chaîne montre également des changements marqués avec la conformation de la chaîne, de telle sorte que cela donne une confirmation indépendante des structures assignées.

Zusammenfassung

Die Spektren von Methylmethacrylatpolymeren in Chloroformlösung bei 90° zeigen drei α-Methylprotonenmaxima.   Die Messung ihrer Flächen in Spektren von Polymeren, die mit radikalischen und anionischen Startern dargestellt wurden, zeigt, dass diese (*1*)

182   F. A. BOVEY AND G. V. D. TIERS

isotaktischen (*ddd* oder *lll*), (*2*) syndotaktischen Sequenzen (*ldl* odor *dld*) und (*3*) hetero-
taktischen Sequenzen (*ldd, dll, ddl* oder *lld*) zugeordnet werden können.   Die Kon-
formation der Ketten kann so mit grosser Schärfe untersucht werden; der Grad der
Regelmässigkeit, die Blockgrösse etc. scheint damit einer Bestimmung zugänglich zu
sein.   Es wird gefunden, dass radikalische Polymere von Methylmethacrylate vorwiegend syndyotaktisch sind, während gezeigt wird, dass die mit anionischen Startern,
wie *n*-Butyllithium, erzeugten, in Übereinstimmung mit den Befunden anderer Autoren,
vorwiegend isotaktisch sind.   Die Resonanz der Methylengruppen der Hauptkette zeigt
ebenfalls auffällige Änderungen mit der Kettenkonformation und liefert damit eine
absolute, unabhängige Bestätigung der Strukturzuordnung.

Received August 5, 1959
Revised September 22, 1959
Revised January 15, 1960

COMMENTARY

Reflections on "Recent Developments in Polymerization by an Alternating Intra-Intermolecular Mechanism," by George B. Butler, *J. Polym. Sci.,* XLVIII, 279 (1960)

GEORGE B. BUTLER

Department of Chemistry, University of Florida, Gainesville, Florida 32611

This paper deals with a summary of two separate experimental programs which failed in their original objectives. Fortunately, some unusual results were obtained in both cases, which prompted us to continue the programs rather than abandon them. Having been schooled in the classical observations reported by Hermann Staudinger, our reaction was a natural one of surprise and disbelief when the dienes failed to crosslink, yet the olefinic bonds were consumed. The afternoon the experiment was conducted was exciting because quaternary ammonium salts had not previously been polymerized and there was no assurance that there would not be an inhibitory action with the generated radicals. Little did we know that the most interesting aspect of the afternoon was that crosslinking did not occur. In due time, crosslinked quaternary ammonium polymers were obtained and they were shown to be strongly basic ion exchangers, the original objective of the program. As described in the paper, the "alternating intra-intermolecular" mechanism was proposed and considerable evidence gathered in its support. However, in discussions with "Speed" Marvel before we published our papers, he commented: "Why get more evidence? It can't be anything else!"

A later thorough examination of Staudinger's work showed that he did not report on *o*-divinylbenzene, but did specifically mention the *m*- and *p*-isomers as crosslinking. The *o*-isomer has now been shown to undergo predominantly cyclization. Consequently, it is interesting to speculate why the experiment did not include the *o*-isomer.

Another aspect of vinyl polymerization had been discussed by Paul Flory in a classical paper. He proposed that propagation should proceed through the more stable radical intermediate which led us to propose the six-membered ring. Recent work, however, has shown that in the case of the most widely studied monomer, diallyldimethylammonium chloride, that the five-membered ring is formed almost exclusively.

It is interesting to note that this work was submitted for presentation before the Organic Division, ACS, in 1955 but was turned down. The Secretary of the Division, a friend of mine, in his rejection letter offered his own explanation of what was happening. The paper was ultimately accepted for presentation at a 1956 ACS meeting.

The second section of this paper deals with copolymerization of maleic anhydride with divinyl ether, which represents the second experiment which failed. It is tempting to describe this experiment as a careful deliberation of this monomer pair in terms of the proposed mechanism mentioned above, and that the structure arrived at was an obvious one. However, this was not the case. This experiment was done on Thanksgiving Day 1951, with no apparent connection with the earlier work. During a research consultants meeting at Smith, Kline & French Laboratories, it was pointed out that in treating sodium edema in human patients suffering from high blood pressure, it would be advantageous to have a crosslinked polycarboxylic acid of low equivalent weight. A likely comonomer pair to accomplish this, maleic anhydride and divinyl ether, was selected for study. The literature has ample references to the use of divinyl ether as a crosslinker. In fact, there is a patent which describes a highly crosslinked copolymer with maleic anhydride, but included no evidence that the material was actually insoluble in water and aqueous base. In our case, the product was completely solubilized in aqueous base, much

to our surprise and disbelief. The results lay dormant for some time but ultimately our thinking process brought these apparently isolated bits of information together, and led to presentation of a paper before a 1958 ACS meeting entitled "A Bimolecular Alternating Intra-intermolecular Chain Propagation." This copolymer has been investigated extensively for its biological properties.

Although the terms "cyclopolymerization" and "cyclocopolymerization" have evolved into popular usage to describe these processes, they are less specific and scientifically accurate than the original ones.

PERSPECTIVE

Comments on "Recent Developments in Polymerization by an Alternating Intra-Intermolecular Mechanism," by George B. Butler, *J. Polym. Sci.*, XLVIII, 279 (1960)

CHARLES L. McCORMICK

Department of Polymer Science, The University of Southern Mississippi, Hattiesburg, Mississippi 39406-0076

During the course of a career, a scientist rarely makes a discovery that has a profound impact on both academic research and technological development. The discovery of cyclopolymerization by George B. Butler at the University of Florida in the 1950s represents such an event. The classic manuscript published in the *Journal of Polymer Science* in 1960 and the subject of this commentary, articulated key mechanistic principles for controlled, alternating, intra-intermolecular steps involved in forming linear polymers with cyclic units as part of the repeating structure. The resulting architecture has been classified as the eighth major structural feature of synthetic high polymers.

Since the initial discovery, over 3000 publications have appeared in international scientific journals, books, monographs, and patents.[1] Butler and his students have contributed some 150 research papers and 45 research theses directed toward understanding and exploiting the fundamental chemistry of cyclopolymerization and related processes. Notably, this chemistry has had dramatic impact on commercial production with millions of pounds of cyclopolymers produced annually throughout the world. Over 100 patents have appeared on this subject per year for the past two decades.

The initial observation[2] of non-crosslinked, water-soluble polymers from the free-radical polymerization of diallyl quaternary ammonium salts was quite surprising based on classical dogma of the era that nonconjugated dienes produced only crosslinked polymers.[3] Less careful investigators might have dismissed this failed experiment to form cross-linked ion-exchange gels without further concern. However, Butler and students, supported by funds from the Office of Naval Research, immediately began studies to explain this curious discovery. Utilizing the best analytical techniques available at that time, the researchers found evidence of virtually complete cyclization from infrared spectroscopy studies and by ring cleavage reactions.

From 1955 through 1960, Butler[4,5] proposed a reaction pathway in which diene monomers polymerized through an alternating intramolecular-intermolecular chain propagation. The key alternating steps were intramolecular cyclization ($R_c = k_c[\mathbf{I}]$) and intermolecular propagation ($R_p = k_p[\mathbf{II}][M]$) where R and k represent the respective rates and rate constants, M the monomer concentration, and $\mathbf{I}$ and $\mathbf{II}$ the concentrations of respective intermediate radicals shown on the first page of Butler's *Journal of Polymer Science* article. A similar pathway was proposed on the sixth page for cyclocopolymerization of divinyl ether-maleic anhydride (structures $\mathbf{XI}$ to $\mathbf{XIII}$) to yield $\mathbf{XIV}$ (the latter polymer has been studied extensively for its biological activity including anti-tumor properties[6]).

The proposed cyclopolymerization pathway sparked considerable discourse in the academic community at research group meetings, ACS meetings, Gordon Conferences, etc. As NMR methods became routinely available in the 1960s, additional curiosity arose with the discovery that cyclization of diene monomers also yielded (often predominantly) the less thermodynamically stable primary radical via closure to the five-membered ring. Over the period from 1960 to 1970, at least three plausible mechanisms were proposed which addressed the issues of

101

virtually complete alternation, lack of crosslinking side reactions, and ring size from the cyclization step.[7] An early proposal was based on electronic interactions between nonconjugated double bonds in the ground state, predisposing the monomer to cyclization. No satisfactory spectroscopic evidence has been reported to support this mechanism although indirect kinetic evidence is consistent with the model.

The second mechanism involves a spatially-favored intramolecular interaction of a radical, cation, or anion (depending on initiator type) with the double bond.

I

Both ground state and transition state interactions have been proposed in the free energy of activation of this step leading to the cyclized product. A strong intramolecular complexation would be reflected in a high frequency factor. The substituents R and X influence both electronic and steric interactions.

The third mechanism invokes a simple kinetic argument that the proximity of the intramolecular double bond and its steric effect on limiting the approach of an unreacted diene favors cyclization.

Several important studies have utilized the Arrhenius relationship $\ln R_c/R_p$ or $\ln k_c/k_p$ vs $\frac{1}{T}$ to determine the activation energy differences and pre-exponential ratios, A_c/A_p, for the competing steps with a variety of unconjugated diene monomers.[7] In many instances, for example, the free radical cyclopolymerization of methacrylic anhydride,[8] the activation energy of cyclization was found to be higher than that for intermolecular propagation; however, the A_c/A_p ratio was very high resulting in a faster rate of cyclization. Similar results were reported for free radical cyclopolymerization of o-divinyl benzene.[9] Interestingly for the latter monomer, the A_c/A_p ratios changed from 50 for free radical initiation in benzene to 2.2×10^4 for cationic initiation in toluene.

The above kinetic studies, as well as other studies on anionically initiated cyclopolymerizations,[10] strongly support mechanism 2. However, the steric exclusion principle proposed in mechanism 3 is also likely to be operative in each case. The values of A_c (orientation and collision frequency factors) may also explain preference for five- or six-membered

rings based on favorable bond overlap in the transition state. Ring size has been shown to depend upon the nature of X in Structure **I**.[7] Recently, for example, the ratio of six- to five-membered rings was shown to be 2 : 1 for free radically cyclopolymerized diphenyl diallyl phosphonium salts as determined from P^{31} NMR studies.[11] Butler et al. had previously prepared cyclopolymers (and proposed six-membered rings) from a number of diallyl phosphonium salts.[12] Ratios of ring sizes have also been shown to be dependent on the temperature and solvent polarity during cyclopolymerization.[7]

Forty years after discovery, we evidence the impact of water-soluble cyclopolymers on environmental stewardship—the major challenge of the 1990s. Those polymers synthesized in, processed from, and utilized in water circumvent problems associated with volatile organic compounds. Is it not appropriate that Butler's first cyclopolymer (dimethyldiallylammonium chloride) is a major polymer utilized commercially today for water remediation? Over 33 million pounds alone are sold annually for water purification in mining and in industrial and municipal waste treatment. Another 1.5 to 2.0 million pounds are used in personal care formulation. From serendipitous discovery, intellectual curiosity, directed research, and finally industrial development, the pathway of cyclopolymerization has been remarkable.

REFERENCES AND NOTES

1. G. B. Butler, *Cyclopolymerization and Cyclocopolymerization,* Marcel Dekker, New York, 1992.
2. G. B. Butler and F. L. Ingley, *J. Am. Chem. Soc.,* **73,** 894 (1951).
3. H. Staudinger and W. Heuer, *Ber.,* **67,** 1159 (1934).
4. G. B. Butler, Gordon Research Conference June 15, 1955, *Program in Science,* **121,** 574 (1955).
5. G. B. Butler and R. J. Angelo, *J. Am. Chem. Soc.,* **79,** 3128 (1957).
6. D. S. Breslow, *Pure Appl. Chem.,* **45,** 103 (1976).
7. G. B. Butler, *Acc. Chem. Res.,* **15,** 370 (1982).
8. T. F. Gray and G. B. Butler, *J. Macromol. Sci.-Chem.,* **A9**(1), 45 (1975).
9. G. B. Butler, *Proceedings of the International Symposium in Macromolecules, Rio de Janeiro, 1974,* E. B. Mano, Ed., Elsevier, Amsterdam, 1975, p. 57.
10. C. L. McCormick and G. B. Butler, *J. Macromol. Sci., Revs. Macromol. Chem.,* **C8**(2), 201 (1972).
11. D. Seyferth and T. C. Masterman, *Macromolecules,* **28,** 3055 (1995).
12. G. B. Butler, D. L. Skinner, W. C. Bond, and C. L. Roger, *J. Macromol. Sci.-Chem.,* **A**(4), 1437 (1970).

JOURNAL OF POLYMER SCIENCE   VOL. XLVIII, PAGES 279–289 (1960)

Recent Developments in Polymerization by an Alternating Intra-Intermolecular Mechanism

GEORGE B. BUTLER, *Department of Chemistry, University of Florida, Gainesville, Florida*

In 1951, Butler and Ingley[1] reported that diallyl quaternary ammonium salts polymerized in the presence of catalytic quantities of *tert*-butylhydroperoxide to form water-soluble, presumably noncrosslinked polymers. The polymerization was observed to occur exothermally at room temperatures in certain cases. The monoallyl derivatives did not polymerize, while monomers containing three or more allyl groups resulted in crosslinked, water-insoluble polymers which were found to be useful as strongly basic ion exchangers. Subsequent quantitative hydrogenation experiments on the purified polymer in conjunction with infrared data indicated that these soluble polymers contained no unsaturation. These results could be interpreted on the assumption that both double bonds of the original diallyl compound had been consumed in the polymerization process without resulting in crosslinking.

In view of the fact that these results were not in agreement with the widely accepted hypothesis advanced by Staudinger[2] in 1934 that nonconjugated dienes produce exclusively crosslinked polymers when polymerized, a polymerization mechanism to account for these results was proposed.[3] This mechanism can be defined most accurately as an alternating intramolecular-intermolecular chain propagation. According to this mechanism, diene monomers capable of forming six-membered ring systems by an intramolecular attack of a free radical on the terminal methylene of the second double bond within the same molecule can result in linear saturated polymers:

$$
\begin{array}{cc}
\text{—CH}_2\text{—CH}\cdot \quad \overset{\text{CH}_2}{\underset{\text{CH}_2}{\overset{\|}{\text{CH}}}} & \rightarrow \quad \text{—CH}_2\text{—CH} \quad \overset{\text{CH}_2}{\underset{\text{CH}_2}{\text{CH}\cdot}} \\
\text{CH}_2 \quad \text{CH}_2 & \text{CH}_2 \quad \text{CH}_2 \\
\diagdown \text{X} \diagup & \diagdown \text{X} \diagup \\
\text{I} & \text{II}
\end{array}
$$

The process consumes two double bonds per molecule, each molecule being converted to a six-membered ring.

While it does not appear too probable that such a process could occur exclusive of crosslinking, considerable support for such a mechanism can be found in the literature. For example, Walling[4] commented on the

279

fact that observed gel points in polymerizing systems capable of cross-linking should occur slightly later than calculated by use of the gel-point equation of Stockmayer, because no account had been taken of the occasional formation of cyclic structures in deriving the equation. Systems studied were methyl methacrylate/ethylene dimethacrylate and vinyl acetate/divinyl adipate, and the ratio of observed to calculated gel point was often as high as 15–20. Simpson, Holt, and Zeite[5] studied the polymer of diallyl phthalate at the gel point and found that about 40% of the reacted monomer had been used up in the formation of cyclic structures. Haward[6] calculated from probability considerations that 31% of the reacted monomer in polymerizing diallyl phthalate should be used up in cyclization of some kind. The experimental results of Simpson[5] are in good agreement with these calculations. Simpson and Holt[7] and Haward and Simpson[8] extended their work along these lines. More recently, Oiwa and Ogata[9] have shown that as high as 81% cyclization occurs when diallyl phthalate is polymerized in solution.

Since, in the case of dimethyldiallylammonium bromide and similar compounds, polymerization apparently occurred exclusively by the proposed mechanism, steric aid by the methyl groups was considered to be a possible important factor in favoring the intramolecular attack over an intermolecular propagation at this point. Consequently, compounds were synthesized, for example, diallyl *tert*-butylamine, which would lend greater steric aid to cyclization. However, when diallylamine hydrochloride resulted in formation of a linear, saturated polymer, it became apparent that steric assistance was not necessary for cyclization to occur exclusive of crosslinking.

As a means of proving the proposed mechanism, representative polymers were degraded to show conclusively the presence of the cyclic structure in the polymer chain.[10] The polymer of diallylamine hydrobromide was degraded as follows:

The structure of V has shown to be correct by the following: (a) The analysis was that required for this structure. (b) Potentiometric titrations in dimethylformamide solution afforded typical titration curves for carboxylic acids under these conditions. (c) The infrared spectrum showed absorption bands that correlate with such a structure. (d) On heating slightly above its melting point, the product yields a sublimate of benzoic acid, and the residual polymer was clearly crosslinked, being insoluble and forming gels with several solvents. This product has thus resulted from the cleavage of rings present in the original poly(diallylammonium bromide), III.

ALTERNATING INTRA-INTERMOLECULAR POLYMERIZATION **281**

Poly(diallyldimethylammonium bromide), VI, was degraded as follows:

This degradation also has given products that can only reasonably be explained by assuming a structure for poly(diallyldimethylammonium bromide), VI, as shown.

Polymerization by this mechanism has now been extended by Marvel and Vest,[11] who polymerized α,α'-dimethylenepimelate derivatives to essentially saturated linear polymers. Marvel and Stille[12] described essentially saturated linear polymers from 1,5-hexadiene and 1,6-heptadiene by use of the trialkyl aluminum–titanium tetrachloride catalyst, although some crosslinking occurred in the case of 1,5-hexadiene. 1,6-Heptadiene gave no crosslinking, although the soluble polymer contained 4–10% of monomer units which had not cyclized, and each contained one double bond. The polymer had a melting point of 210–230°, an inherent viscosity of 0.38, and conversion of monomer to polymer was 80%. 1,5-Hexadiene gave 87% conversion; however, only 46% of the polymer was soluble. This polymer had a melting point of 85–90°, and an inherent viscosity of 0.23. Crawshaw and Butler[13] and Jones[14,15] independently described polymerization of acrylic anhydride under a variety of conditions to a soluble polymer. Molecular weight of these polymers is as high as 95,000. In an effort to synthesize diacrylylmethane by a Claisen condensation of methyl vinyl ketone with ethyl acrylate, Jones[16] isolated poly(diacrylylmethane) which he attributed to an anionic polymerization of the monomeric compound as it formed. Jones[17] has also described a cyclic polymer from alloocimene. Marvel, Kiener, and Vessel,[18] however, have described a linear unsaturated polymer from this monomer by Ziegler catalysis. Marvel and Vest[19] have described soluble polymers by polymerization of α,α'-dimethylenepimelonitrile as well as copolymers of this monomer with other well known monomers. Marvel and Gall[20] have described the preparation of 2,6-diphenyl-1,6-heptadiene, and Field[21] has described polymerization of this monomer to a polymer having cyclic structures and having a melting point of 300°C. The inherent viscosity was 0.49. All the known general types of initiation, free radical, Ziegler-type, cationic, and conventional anionic, have been found to polymerize this monomer and all resulted only in intermolecular-intramolecular polymerization.

Price[22] has described soluble copolymers of diallylalkylamine oxides with acrylonitrile and other monomers. These copolymers were assumed to possess one unreacted double bond for each molecule of diallylamine oxide entering the chain; however, in view of the polymerization mechanism under consideration, copolymerization has probably occurred according to this mechanism to produce saturated copolymers. Schuller, Price, Moore, and Thomas[23] have described a number of soluble copolymers involving several diallyl monomers and several conventional monomers and have postulated that these copolymerizations follow the mechanism under discussion. Friedlander[24] described cyclization of allyl ether and allyl sulfide to substituted tetrahydropyrans and tetrahydrothiopyrans, respectively, when attempts were made to add various reagents under free radical conditions. The activity of the reagent as a chain transfer reagent was found to be important in determining the degree to which addition is accompanied by cyclization. Berlin and Butler[25] have described soluble polymers from diallylarylphosphine oxides, and Butler, Skinner, and Stackman[26] have described soluble polymers from diallyldialkylphosphonium salts and diallyldialkylsilanes, the former by free radical initiation, and the latter by Ziegler-type catalysis. Marvel and Garrison[27] obtained varying ratios of soluble polymers from higher α-diolefins in which the terminal double bonds are separated by from four to eighteen methylene groups. The results obtained in this study are summarized in Table I.

These percentages of cyclization correspond roughly to those which have been realized in other cyclization reactions. Seven-membered rings form fairly readily, the intermediate ring sizes with greater difficulty, the larger rings (14–15) are easier to obtain, and finally the yields drop again with the higher ring sizes.

The authors found internal *trans*-unsaturation in the soluble polymers as well as terminal olefin. Terminal olefin decreased with higher dilution; however, the *trans*-unsaturation remained constant.

Stille[28] has described polymerization of 1,6-diynes by use of Ziegler catalysts to linear, cyclic polymers in which the polymer backbone consists of a continuous conjugated system of double bonds.

TABLE I
Polymerization of Higher α-Diolefins with Metal Alkyl Coordination Catalysts

Monomer (0.2M) CH_2=$CH(CH_2)_n$ CH=CH_2	Smallest possible ring size	Estimated % cyclic units in soluble polymer	Conversion, %	
			Soluble	Insoluble
$n = 4$	7	25	54	2
$n = 5$	8	9	20	40
$n = 6$	9	6	18	38
$n = 7$	10	10	25	41
$n = 8$	11	11	27[a]	40
$n = 9$	12	11	25[a]	36
$n = 11$	14	15	38[a]	32
$n = 12$	15	15	17	57
$n = 14$	17	4	40[b]	22
$n = 18$	21	8	38[a]	37

[a] 0.05M monomer concn.
[b] 0.6M monomer concn.

An examination of the literature concerning certain unsymmetrical dienes indicated that cyclization may have occurred during polymerization. Blout and Ostberg[29] and Cohen, Ostberg, Sparrow, and Blout[30] studied the polymerization of allyl methacrylate at elevated temperatures using benzoyl peroxide as catalyst and at low temperatures using biacetyl and ultraviolet radiation to initiate polymerization. At the elevated temperature, gelation occurrred at 6% conversion, while at 1°C. 39% conversion was obtained before gelation occurred. Some evidence presented by these authors indicated that cyclization may have occurred, although the results were not attributed to this factor. A brief investigation of allyl acrylate[31] indicated that this monomer resulted in gelation at 4% conversion, and for this reason the soluble polymer was not examined beyond the extent of establishing the presence of unsaturation.

Because of the copolymerizing tendencies of maleate and fumarate esters, unsaturated esters of these acids were prepared and studied.[32] The allyl, crotyl, and 3-butenyl esters of monomethylmaleic and fumaric acids were found to produce soluble polymers containing from 23 to 63% cyclization. These results are summarized in Tables II–IV.

TABLE II

Physical Properties of Polymers of
$$\begin{matrix} R_1 & & H \\ & C{=}C & \\ R_2 & & COOCH_3 \end{matrix}$$

Compound	R_1	R_2	Soften. point, °C.	Inherent viscosity (at conc., g./100 ml.)
Ia	H	$-COOCH_2CH{=}CH_2$	95	0.061 (0.306)
Ib	H	$-COOCH_2CH{=}CHCH_3$	110	0.036 (0.297)
Ic	H	$-COOCH_2CH_2CH{=}CH_2$	105	—
IIa	$-COOCH_2CH{=}CH_2$	H	100	—
IIb	$-COOCH_2CH{=}CHCH_3$	H	75	0.044 (0.243)
IIc	$-COOCH_2CH_2CH{=}CH_2$	H	100	0.072 (0.262)

TABLE III

Gel Times, Percentage Conversion at Gelation, and Catalyst-to-Monomer Ratio

Polymer	Gel time, hr.	Conversion at gelation, %	Benzoyl peroxide as % of monomer
Poly-Ia	9	18	2
Poly-Ib	a	>28[a]	2
Poly-Ic	8.25	12	2
Poly-IIa	11.5 (1)[b]	9	0.1
Poly-IIb	56	17	2
Poly-IIc	12 (1)[b]	14	0.1

[a] No gelation after 243 hr. The per cent conversion at gelation represents the yield of soluble polymer after this period of heating at 62°C.

[b] Using 2% benzoyl peroxide.

TABLE IV
Analytical Data Derived from Bromine Titrations

Polymer	Residual alcohol bond, %	Residual acid bond, %	Approximate cyclization, %
Poly-Ia	35.8 ± 0.9	15.3 ± 2.7	49
Poly-Ib	21.1 ± 0.6	18.6 ± 2.1	60
Poly-Ic	32.8 ± 0.1	29.2 ± 0.2	38
Poly-IIa	29.3 ± 0.4	7.76 ± 0.22	63
Poly-IIb	69.7 ± 0.6	6.52 ± 0.20	23
Poly-IIc	43.4 ± 0.2	14.4 ± 0.3	43

From interpretation of infrared data, it was shown that formation of five-, six- and seven-membered lactone rings were formed in this series of polymerizations.

Recent work has shown that certain 1,4-dienes will undergo copolymerization with certain olefins to produce linear polymers containing a six-membered ring formed during the process.[33] Divinyl ether and maleic anhydride undergo free radical-catalyzed copolymerization to produce a polymeric anhydride containing the tetrahydropyran ring according to the following scheme:

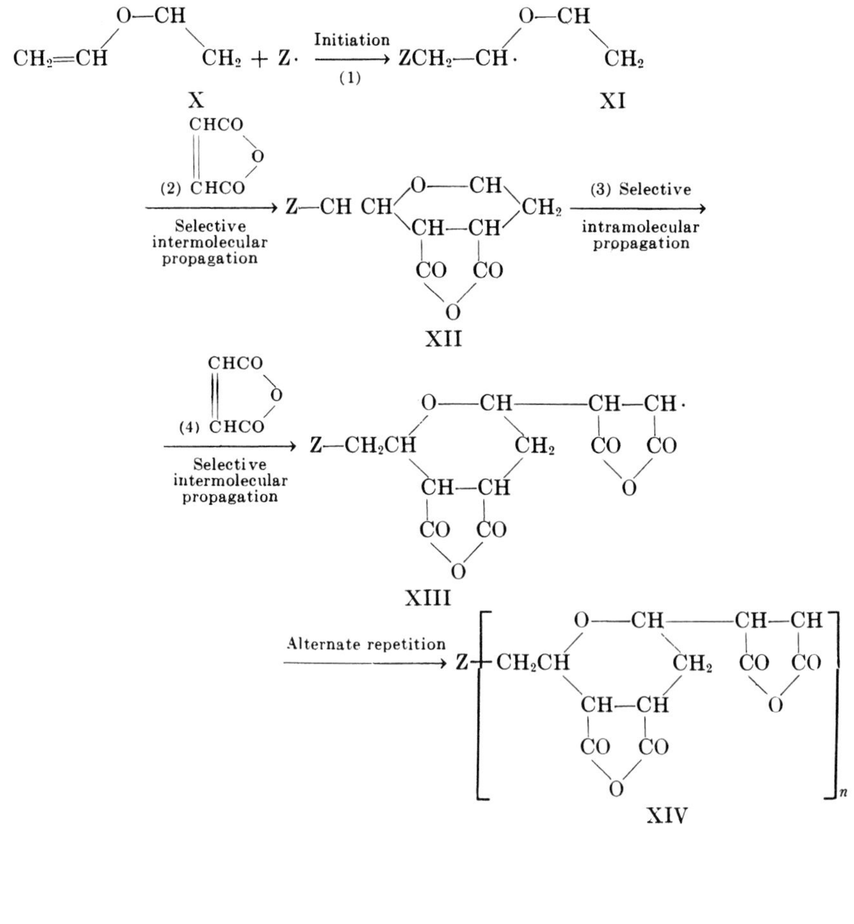

The copolymer has a melting point of 350°C. with decomposition. It is soluble in acetone, dimethylformamide, and aqueous sodium hydroxide and undergoes the usual reactions of anhydrides. The intrinsic viscosity was determined to be 0.175 in both dimethylformamide (DMF) and 2N NaOH. Infrared data confirm the proposed structure of the copolymer. Table V lists several additional copolymers of this type which have been studied.

TABLE V
Copolymers of 1,4-Diolefins and Olefins

		Intrinsic viscosity	
1,4-Diene	Olefin	DMF	2N NaOH
Divinyl ether	Fumaronitrile	—	—
Divinyldimethylsilane	Maleic anhydride	0.121	0.082
Divinyl sulfone	Maleic anhydride	0.064	0.052
Divinylcyclopentamethylenesilane	Maleic anhydride	0.106	—
Divinyl ether	Dimethyl fumarate	0.123	—
Divinyl ether	Diethyl maleate	0.235	—
Divinyldimethylsilane	Vinyl acetate	0.063	—
Divinyldimethylsilane	Fumaryl chloride	0.142	—

A copolymerization which has been studied briefly involves 1,5-dienes and molecules which are capable of inserting one atom into the propagating chain. An example of such a system is the 1,5-hexadiene–sulfur dioxide comonomer pair. Copolymerization probably occurs according to the following scheme:

$$n CH_2{=}CH{\diagdown}{\overset{CH_2}{\underset{CH_2-CH_2}{\diagup}}}CH + 2nSO_2 \xrightarrow{\text{Free radical}}$$

$$\left[CH_2{-}CH{\diagdown}{\overset{SO_2-CH_2}{\underset{CH_2-CH_2}{\diagup}}}CH{-}SO_2\right]_n$$
XVI

Other comonomer pairs which can reasonably be expected to produce similar copolymers are allyl vinyl ether and sulfur dioxide, allylvinyldimethylsilane and sulfur dioxide, 1,5-dienes and carbon monoxide, and possibly alkyl isocyanides and 1,5-dienes.

Delmonte and Hays[34] have described a very interesting example of cyclization of a diolefinic monomer which is also accompanied by chain transfer. These authors studied copolymerizations of ethyl acrylate, methyl methacrylate, and styrene with 1–3% of vinyl acrylate. The reduced specific viscosities of the copolymers were only about 5% of the values for the corresponding homopolymers. There was no apparent decrease in rate in the copolymerizations as compared with the homopolymerizations. Conversions were on the order of 90%. It was shown that these results were not the effect of hydrolysis of the vinyl ester. An infrared spectral analysis of the styrene–vinyl acrylate copolymer (97:3) showed the presence of about 2% of a $\Delta^{\beta,\gamma}$-butenolide structure. Also, about 1% of pendant vinyl ester groups were present. In order to account for this unusual phenomenon, these authors have postulated the following reaction scheme:

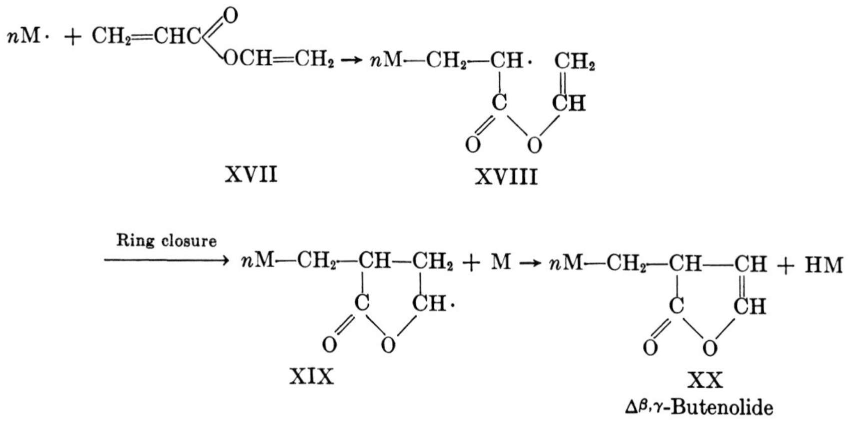

Infrared analysis confirmed the presence of the terminal butenolide unit, and the exceptionally low specific viscosities confirm the postulated chain transfer.

These authors pointed out that chain transfer is not observed in polymerization of acrylic anhydride, which forms a six-membered ring.

In order to justify the chain transfer tendency in this monomer which forms a five-membered ring, it was pointed out that the saturated five-membered lactone has four more bond oppositions between hydrogens than the unsaturated butenolide structure. Each such bond opposition will increase the conformational strain by about 1 kcal., for a total of 4 kcal. The unsaturated structure has more angular strain than the saturated structure, but evidence indicates that conformational strain far outweighs the angular strain involved in the change from sp^3 to sp^2 bonds. Consequently, the unsaturated structure would be expected to possess greater stability than the saturated lactone which would be required to form if propagation had occurred without chain transfer. Bond opposition is practically nonexistent in the six-membered polyacrylic anhydride rings.

In an effort to explain the strong polymerization tendency of 1,6-heptadienes relative to their monoolefinic counterparts, a structure which involves some sort of intramolecular electronic interaction between the olefinic linkages has been proposed.[33] This interaction can be represented in its simplest form by the structures shown on following page.

This proposed interaction is supported by the work of Marvel and Stille,[12] Schuller et al.,[23] Price,[22] and Mikulasova and Hvirik.[35] The latter authors

ALTERNATING INTRA-INTERMOLECULAR POLYMERIZATION   **287**

have shown that the total activation energy of radical polymerization for diallyldimethylsilane is about 9 kcal./mole double bond less than that for allyltrimethylsilane.   Such an interaction is also supported by numerous published articles, several of which are listed.[35-51]

References

1. Butler, G. B., and F. L. Ingley, *J. Am. Chem. Soc.*, **73,** 894 (1951).

2. Staudinger, H., and W. Heuer, *Ber.*, **67,** 1159 (1934).

3. Butler, G. B., and R. J. Angelo, *J. Am. Chem. Soc.*, **79,** 3128 (1957); Ion Exchange, Gordon Research Conference, June 15, 1955, Program in Science, **121,** 574 (1955).

4. Walling, C., *J. Am. Chem. Soc.*, **67,** 441 (1945).

5. Simpson, W., T. Holt, and R. Zeite, *J. Polymer Sci.*, **10,** 489 (1953).

6. Haward, R. N., *J. Polymer Sci.*, **14,** 535 (1954).

7. Simpson, W., and T. Holt, *J. Polymer Sci.*, **18,** 440 (1955).

8. Haward, R. N., and W. Simpson, *J. Polymer Sci.*, **18,** 440 (1955).

9. Oiwa, M., and J. Ogata, *J. Chem. Soc. Japan*, **79,** 1506 (1958).

10. Butler, G. B., A. Crawshaw, and W. L. Miller, *J. Am. Chem. Soc.*, **80,** 3615 (1958).

11. Marvel, C. S., and R. D. Vest, *J. Am. Chem. Soc.*, **79,** 5771 (1957).

12. Marvel, C. S., and J. K. Stille, *J. Am. Chem. Soc.*, **80,** 1740 (1958).

13. Crawshaw, A., and B. B. Butler, *J. Am. Chem. Soc.*, **80,** 5464 (1958).

14. Jones, J. F., *J. Polymer Sci.*, **33,** 15 (1958).

15. Italian Patent 563,941, to B. F. Goodrich Company, June 7, 1957.

16. Jones, J. F., *J. Polymer Sci.*, **33,** 7 (1958).

17. Jones, J. F., *J. Polymer Sci.*, **33,** 513 (1958).

18. Marvel, C. S., P. E. Kiener, and E. D. Vessel, *J. Am. Chem. Soc.*, **81,** 4694 (1959).

19. Marvel, C. S., and R. D. Vest, *J. Am. Chem. Soc.*, **81,** 984 (1959).

20. Marvel, C. S., and E. J. Gall, *J. Org. Chem.*, **24,** 1494 (1959).

21. Field, N. D., *J. Org. Chem.*, **25,** 1006 (1960).

22. Price, J. A., U.S. Pat. 2,871,229, to American Cyanamid Company, January 27, 1959.

23. Schuller, W. H., J. A. Price, S. T. Moore, and W. M. Thomas, *J. Chem. Eng. Data*, **4,** 273 (1959).

288   G B BUTLER

24. Friedlander, W. S., paper presented to Division of Organic Chemistry, American Chemical Society Meeting, San Francisco, April 1958, paper No. 29.

25. Berlin K. D., and G. B. Butler, *J. Am. Chem. Soc.*, **82,** 2712 (1960).

26. Butler, G. B., D. L. Skinner, and R. W. Stackman, paper presented at Conference on High Temperature Polymer and Fluid Research, WADC, Dayton, Ohio, May 1959.

27. Marvel, C. S., and W. E. Garrison, Jr., *J. Am. Chem. Soc.*, **81,** 4737 (1959).

28. Stille, J. K., paper presented to Division of Polymer Chemistry, American Chemical Society Meeting, Boston, April 1959, paper No. 43.

29. Blout, E. R., and B. E. Ostberg, *J. Polymer Sci.*, **1,** 230 (1946).

30. Cohen, S. G., B. E. Ostberg, D. B. Sparrow, and E. R. Blout, *J. Polymer Sci.*, **3,** 264 (1948).

31. Crawshaw, A., and G. B. Butler, unpublished work.

32. Barnett, M. D., A. Crawshaw, and G. B. Butler, *J. Am. Chem. Soc.*, **81,** 5946 (1959).

33. Butler, G. B., unpublished work.

34. Delmonte, D. W., and J. T. Hays, paper presented at Delaware Science Symposium, Newark, Delaware, February 1959.

35. Mikulasova, O., and A. Hvirik, *Chem. zvesti*, **11,** 641 (1957).

36. Braude, E. A., E. R. H. Jones, F. Sondheimer, and J. B. Toogood, *J. Chem. Soc.*, **1949,** 607.

37. Bartlett, P. D., and E. S. Lewis, *J. Am. Chem. Soc.*, **72,** 1005 (1950).

38. Kumler, W. B., L. A. Strait, and E. L. Alpen, *J. Am. Chem. Soc.*, **72,** 1463, 4558 (1950).

39. Bennett, W. B., and A. Burger, *J. Am. Chem. Soc.*, **75,** 84 (1953).

40. Cram, D. J., and H. Steinberg, *J. Am. Chem. Soc.*, **73,** 5691 (1951).

41. Cram, D. J., and J. D. Knight, *J. Am. Chem. Soc.*, **74,** 5839 (1952).

42. Fleischacker, H., and G. F. Woods, *J. Am. Chem. Soc.*, **78,** 3436 (1951).

43. Parker, E. D., and L. A. Goldblatt, *J. Am. Chem. Soc.*, **72,** 2151 (1950).

44. Pines, H., and J. Ryer, *J. Am. Chem. Soc.*, **77,** 4372 (1955).

45. Simonetta, M., and S. Winstein, *J. Am. Chem. Soc.*, **76,** 18 (1954).

46. Cristol, S. J., and R. L. Snell, *J. Am. Chem. Soc.*, **76,** 5000 (1954); *ibid.*, **80,** 1950 (1958).

47. Winstein, S., and M. Shatavsky, *Chem. & Ind. (London)*, **1956,** 59; *J. Am. Chem. Soc.*, **78,** 592 (1956).

48. Turner, R. B., W. R. Meador, and R. E. Winkler, *J. Am. Chem. Soc.*, **79,** 4116 (1957).

49. Turner, R. B., and R. H. Garner, *J. Am. Chem. Soc.*, **80,** 1424 (1958).

50. Turner, R. B., D. E. Nettleton, Jr., and M. Perlman, *J. Am. Chem. Soc.*, **80,** 1432 (1958).

51. Huntsman, W. D., V. C. Soloman, and D. Eros, *J. Am. Chem. Soc.*, **80,** 5455 (1958).

Synopsis

Recent developments in polymerization by an alternating intra-intermolecular mechanism are summarized and discussed briefly. The methods used to prove the structure of typical polymers, which in turn established the mechanism of polymerization, are reviewed. A study of larger ring formation by Marvel and Garrison is also summarized. Ring of sizes from 7 to 21 members were formed from α-diolefins in yields of from 4 to 25%. A study of certain unsymmetrical dienes has shown that cyclization occurs in yields up to 60%, but that gelation eventually occurs because of the difference in relative reactivities of the two olefinic double bonds. Copolymerization of certain 1,4-dienes with certain olefins to produce linear copolymers containing six-membered rings is discussed. In an effort to explain the strong driving force to close a six-membered ring during polymerization of 1,6-diolefins, a pronounced interaction between the double bonds, which can be represented by several resonance forms of the molecule, is proposed.

Résumé

Les nouveaux développements en polymérisation par un mécanisme intra- et inter-moléculaire alternatif sont résumés et discutés brièvement. Les méthodes utilisées pour déterminer la structure des polymères typiques, et qui établissent le mécanisme de poly-merisation sont revus. Une étude de Marvel et Garrison sur la formation de larges cycles est également résumée. Des cycles ayant de 7 à 21 membres sont formés à partir d'α-diolefines les pourcentages variant de 4 à 25%. Une étude sur certains diènes asymétriques a montré qu'une cyclisation a lieu jusqu'à 60%, mais qu'il apparaît éven-tuellement un gel à cause de la différence des réactivités relatives des deux doubles liaisons oléfiniques. La copolymérisation de certains diènes oléfines pour produire des copolymères linéaires contenant des cycles de 6 membres est discutée. En vue d'ex-pliquer la force nécessaire pour fournir un cycle à 6 membres pendant la polymérisation de dioléfines 1,6, on suppose qu'il existe une forte interaction entre les doubles liaison, la molécule pouvant être représentée par plusieurs formes de résonance.

Zusammenfassung

Neuere Entwicklungen bei der Polymerisation über einen abwechselnd intra-inter-molekularen Mechanismus werden zusammengefasst und kurz diskutiert. Es wird ein Uberblick über die Methoden gegeben, mittels welcher die Struktur typischer Polymerer bewiesen wurde, welche dann ihrerseits die Aufstellung des Polymerisationsmechanismus erlaubten. Auch eine Untersuchung über die Bildung grösserer Ringe von Marvel und Garrison wird besprochen. Ringe mit sieben bis einundzwanzig Gliedern wurden aus α Diolefinen mit Ausbeuten von 4 bis 25% gebildet. Eine Untersuchung gewisser un-symmetrischer Diene Zeigte, dass bis zu 60% Ausbeute an Cyklisierung auftritt, dass aber Gelbildung unter Umständen infolge der Reaktivitätsunterschiede der beiden olefinischen Doppelbindungen stattfindet. Die Copolymerisation gewisser 1,4-Diene mit gewissen Olefinen unter Bildung linearer Copolymerer mit sechsgliedrigen Ringen wird diskutiert. Zur Erklärung der starken Triebkraft zur Schliessung eines sechs-gliedrigen Ringes während der Polymerisation von 1,6-Diolefinen wird eine bevorzugte Wechselwirkung zwischen den Doppelbindungen angenommen, welche durch mehrere Resonanzformen des Moleküls dargestellt werden kann.

Received September 15, 1960

PERSPECTIVE

Comments on "Polybenzimidazoles, New Thermally Stable Polymers," by Herward Vogel and C. S. Marvel, *J. Polym. Sci.,* L, 511 (1961)

GEORGE A. SERAD

Hoechst Celanese Corporation, P.O. Box 32414, Charlotte, North Carolina 28232-6085

The research of Vogel and Marvel truly represents a pioneering effort in the synthesis of high performance polymers. In this landmark article, they extended the knowledge of the thermal stability of imidazole derivatives into the realm of high molecular weight, fully aromatic polybenzimidazole (PBI) polymers. They demonstrated the ability to produce high molecular weight polymers having potentially useful properties. One key to achieving a high molecular weight was the use of the phenyl ester derivative of the aromatic dioic acids. This provided superior polymer to that produced by using the free acid or the methyl ester. The commercial PBI products of today still rely on that finding.

Their exploratory studies were extended through funding by the Air Force Materials Laboratory and NASA, who addressed a specific set of aerospace and defense needs. This subsequently led to some of the earlier developments of a prepolymer that could be used as a high temperature adhesive, a carbon foam made possible by the high carbon yield of PBI, and the development of PBI fibers.

Essential to the practical utility of Vogel and Marvel's polymers was that several PBI compositions formed fully soluble, high molecular weight polymer solutions. This allowed subsequent processing into useful forms avoiding the nonprocessable, proverbial "brick-dust," encountered with some other high performance polymers.

From a commercial viewpoint, the balance of processability and polymer performance led to the selection of poly[2,2'-(*m*-phenylene)-5,5'-bibenzimidazole] as the principal product of commerce. Monomers needed to produce this PBI composition were a major concern. Diaminobenzidine (now more commonly referred to as tetraaminobiphenyl) was not commercially available and the diphenyl isophthalate was a limited-supply specialty chemical. Hence, dedicated facilities for manufacturing tetraaminobiphenyl (Hoechst AG) and PBI polymer and fiber (Celanese Corporation, now Hoechst Celanese) were constructed and began operation in 1982/1983.

Barring a significant discovery to reduce monomer costs, PBI remains a specialty polymer targeted at high performance niches. Even so, its unique performance attributes (e.g., has a glass transition temperature of 435°C; does not burn in air, contribute fuel to flames or produce much smoke; forms a tough flexible char in high yield; is hydrophilic, having high moisture regain; forms very comfortable fabrics for garments; has good chemical resistance) permits PBI to be used in a cost-effective manner in particular situations.

Fire blocking of aircraft seats was among the earliest large-volume uses for PBI fiber. Currently, fire service applications are the predominant commercial area. Vogel and Marvel, in the course of their early investigations, may not have envisioned the future when people who had used PBI products in these commercial applications would have acknowledged that "PBI saved my life." They would have had great satisfaction in the knowledge that many lives were to be saved because of their scientific contributions.

PBI still remains a material of interest for both scientific and commercial investigation. PBI forms, molded from powder by high temperature and pressure techniques, have compressive strengths greater than any other commercial polymer. A possible contributor to this property is that water forms strong hydrogen bonds between the PBI chains. Additionally, recent NMR investigations indicate that PBI forms sheet-like layers in the manner of a glassy solid. Its thermal stability is also evident in that molecular motion above the glass transition temperature is similar to that below it.

Primary commercial applications continue to be with fibers, predominantly in thermal protection markets. However, PBI's unique properties seem to continually spark new investigations. Currently, there is growing interest in the use of PBI for battery and fuel cell applications. PBI can absorb three to five times its weight in strong acids (e.g., sulfuric and phosphoric) without degradation. In film form, these materials act as solid polymer electrolytes. Because water is not the predominant transport mechanism, they can operate above 100°C under atmospheric pressure which is a potential advantage.

As PBI enters its second decade as a commercial product it remains indebted to the keystone scientific foundation developed by Vogel and Marvel.

JOURNAL OF POLYMER SCIENCE   VOL. L, PAGES 511–539 (1961)

Polybenzimidazoles, New Thermally Stable Polymers *

HERWARD VOGEL and C. S. MARVEL, *Noyes Chemical Laboratory, University of Illinois, Urbana, Illinois*

In connection with the search for new high molecular weight materials with superior properties such as stability, retention of stiffness, toughness at elevated temperatures, etc., a study of the preparation of fully aromatic polybenzimidazoles has been started.

Imidazole derivatives are known to be remarkably stable compounds.[1-3] Many of them are resistant to the most drastic treatments with acids and bases and are not readily attacked by oxidizing reagents. They have high melting points and are stable at elevated temperatures (430–645°C.).[3] These properties suggest that a condensation polymer characterized by the recurrence of benzimidazole nuclei would also exhibit outstanding heat stability. We were able to synthesize a variety of polybenzimidazoles by melt polycondensation of suitable aromatic tetraamines and aromatic dicarboxylic acids.

DISCUSSION

A most versatile and practical synthesis of benzimidazoles is provided by the interaction of an *o*-phenylenediamine with a carboxylic acid.[4] Its applicability in polymer chemistry has recently been shown by Brinker and Robinson,[5] who discovered that the reaction of bis-*o*-diaminophenyl compounds with aliphatic dicarboxylic acids could be employed to form linear condensation polymers.

The preparation of a polybenzimidazole containing only aromatic recurring units required the condensation of an aromatic tetraamine with an aromatic dicarboxylic acid or derivative thereof. The interaction of 3,3'-diaminobenzidine and 1,2,4,5-tetraaminobenzene with aromatic dibasic acids, eq. (1) and (2), and the polycondensation of the trifunctional 3,4-diaminobenzoic acid, eq. (3), were investigated in order to accomplish this objective.

* The early part of this work was supported by grants (NSF-G-2626 and NSF-G-5906) from the National Science Foundation and the latter part was carried out under the sponsorship of Contract No. AF-33(616)-5486 with the Nonmetallic Materials Laboratory of Wright Air Development Division, Wright-Patterson Air Force Base, Ohio. This paper may be reproduced for any purpose of the United States Government.

$$n \; \text{H}_2\text{N} \underset{\text{H}_2\text{N}}{\longleftrightarrow} \text{NH}_2 \quad + \quad n \; \overset{O}{\underset{X}{>}} \text{C} - \text{R} - \text{C} \overset{O}{\underset{X}{<}} \quad \xrightarrow[-2n\text{H}_2\text{O}]{-2n\text{HX}}$$

$$\left[\cdots \right]_n \quad (1)$$

$$n \; \text{H}_2\text{N} \longleftrightarrow \text{NH}_2 \quad + \quad \overset{O}{\underset{X}{>}} \text{C} - \text{R} - \text{C} \overset{O}{\underset{X}{<}} \quad \xrightarrow[-2n\text{H}_2\text{O}]{-2n\text{HX}}$$

$$(2)$$

where R is the aromatic nucleus and X is OH, Cl, OCH$_3$, or OC$_6$H$_5$.

$$n \; \overset{O}{\underset{X}{>}} \text{C} \longleftrightarrow \underset{\text{NH}_2}{\overset{\text{NH}_2}{}} \quad \xrightarrow[-n\text{H}_2\text{O}]{-n\text{HX}}$$

$$(3)$$

where X is OH, OCH$_3$, or OC$_6$H$_5$.

1. Poly 2,5(6)-Benzimidazole

The polycondensation of 3,4-diaminobenzoic acid, a readily obtainable and stable compound, was investigated in particular. All attempts to employ the acid in a melt condensation polymerization failed. At temperatures around 180°C., with p-toluenesulfonic acid and diaminobenzoic acid dihydrochloride as catalysts, and phenol and diphenyl ether as fluxes, the diaminobenzoic acid underwent partial decarboxylation which prevented the formation of the expected benzimidazole polymer:

$$\underset{\text{HO}}{\overset{O}{>}} \text{C} \longleftrightarrow \underset{\text{NH}_2}{\overset{\text{NH}_2}{}} \quad \begin{array}{c} \xrightarrow{-\text{CO}_2} \\ \\ \xrightarrow{-2\text{H}_2\text{O}} \end{array}$$

In a modification of the Phillips[6] method for the preparation of benzimidazoles, several experiments were run with 3,4-diaminobenzoic acid in refluxing dilute or concentrated mineral acid solutions. The isolated polymeric materials were soluble in concentrated sulfuric acid only, showing inherent viscosities around 0.2. Elemental analyses and ultraviolet absorption spectra indicated low molecular weight polybenzimidazoles.

Melt condensation of methyl 3,4-diaminobenzoate proved unsatisfactory. Decomposition of the monomer occurred under the conditions required to effect reaction by catalysts and/or heat.

Phenyl diaminobenzoate, however, could be polymerized by melt condensation to yield a high molecular weight polybenzimidazole which was soluble in formic acid and showed outstanding heat stability.

2. Polybenzimidazoles with Mixed Aromatic Units

The phenyl ester grouping was found to be generally suitable for the polycondensation of aromatic dioic acids and aromatic tetraamines to give polybenzimidazoles with all aromatic units. The use of the free acids and the corresponding dimethyl esters gave inferior results. The phenol liberated from the diphenyl esters during the condensation reaction seemed to exert an advantageous plasticizing effect which the low-boiling water and methanol did not achieve.

High molecular weight materials were obtained by heating together the aromatic tetraamino compounds and various diphenyl esters of aromatic dibasic acids in an inert atmosphere at high temperatures. The polycondensations of 3,3'-diaminobenzidine and 1,2,4,5-tetraaminobenzene with diphenyl isophthalate are as follows:

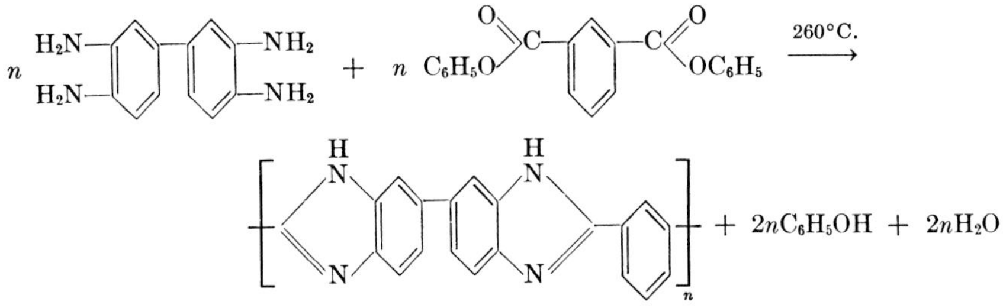

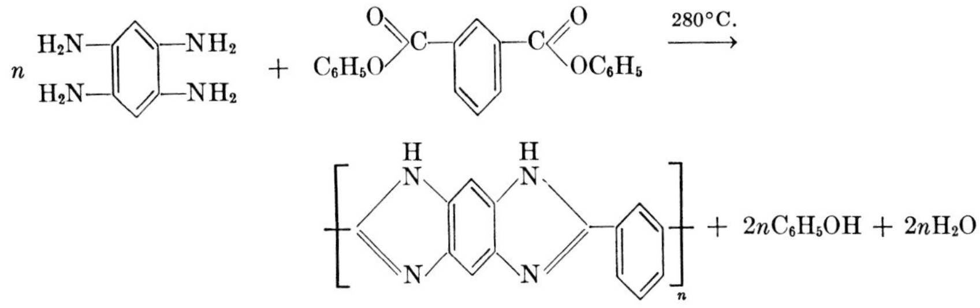

The condensations were initiated at temperatures around 250°C. In some cases, temperatures as high as 290°C. were necessary. Evolution of phenol and water indicated that the reaction had commenced. Depending on the rigidity and symmetry of the polymer structure, the viscous melt which was formed first changed very quickly to a solid cake or gradually to

TABLE I

Data on Preparation, Melting Points, Inherent Viscosities,
and Ultraviolet Absorption of Polybenzimidazoles and Some Model Compounds

Compound or repeating unit	Reaction temp., °C.	Reaction time, hr.	Melting point, °C.[a]	η_{inh}[b]	λ_{max},[c] mμ	$E_{1\ cm}^{1\%}$
	d	d	300		300 243	1000 680
	280–400	4		1.27	365 257	2200 1140
			355		327 254	1330 1100
	250–270	0.75	450 (dec.)	3.19	293 228	800 1590

POLYBENZIMIDAZOLES **515**

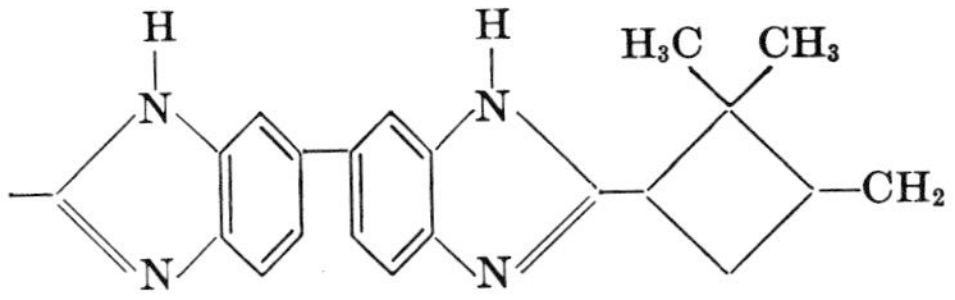 240–280	1.5	400 (dec.)	1.03	296 / 232	800 / 1360
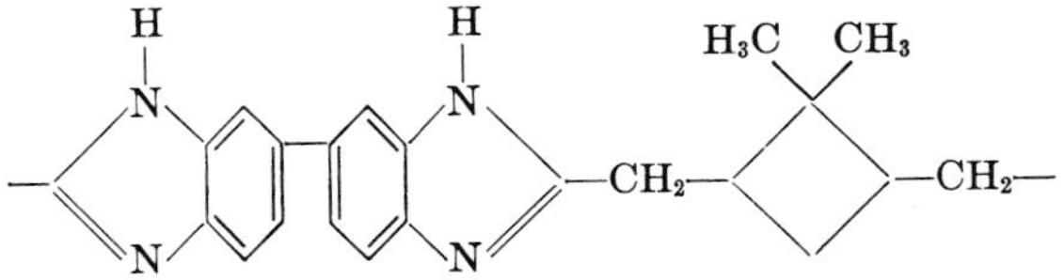 240–280	1.5	400 (dec.)	1.38	293 / 230	700 / 1400
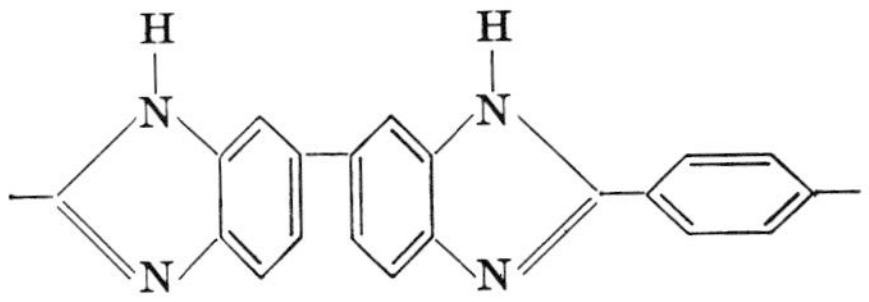 240–400	5		1.00	379 / 257	1400 / 660
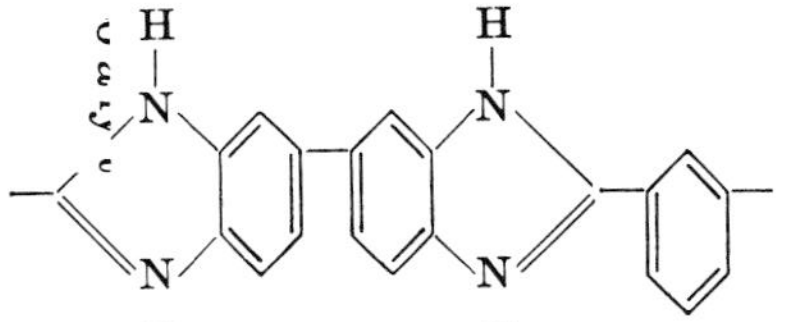 260–400	10		3.34	336 / 252	1250 / 920
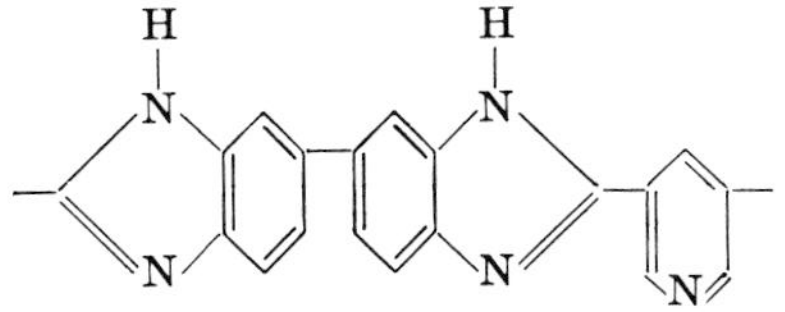 260–400	5		1.48	370 / 247	840 / 790

(continued)

516   H. VOGEL AND C. S. MARVEL

TABLE I (continued)

Compound or repeating unit	Reaction temp., °C.	Reaction time, hr.	Melting point, °C.[a]	η_{inh}[b]	λ_{max},[c] mμ	$E_{1\ cm.}^{1\%}$
(structure)	260–400	5	480 (dec.)	0.74	405 (428) (shoulder) 295	1530 (1440) 370
(structure)	240–400	5		2.70	334 235	980 670
(structure)	250–400	6		0.86 (H₂SO₄)	368 275	1700 340
(structure)	285–400	6	430	2.99	328 238	770 910

POLYBENZIMIDAZOLES

Polymer structure					
(biphenyl-linked PBI)	270–300	1.5	550		339 224	1970 1010
$(CH_2)_4$-linked PBI			490 (dec.)	2.51	295 (284) (shoulder)	1420 (920)
p-phenylene-linked PBI	280–360	4.5		0.80	385 248	1340 570
phenylene-linked PBI	280–400	5		1.10	347 240	1600 1030

(CH₂)₄

a Where data are not given, the melt temperatures are in excess of 600°C.

b Measured at a concentration of 0.2% in formic acid or, in one case, in sulfuric acid. The values shown were, however, influenced by polyelectrolyte effects differing in extent with different polymer structures. It was possible to eliminate these effects by use of dimethyl sulfoxide solutions. The inherent viscosities of various polymers in dimethyl sulfoxide were found to be roughly independent of the concentration, and the values ranged from approximately 0.4 to 1.1.

c The ultraviolet absorption spectra were measured on concentrated sulfuric acid solutions of the polymers.

d Data of Walther and Von Tulawski.[7]

a glassy foam. These initial low molecular weight materials were powdered and reheated under high vacuum for several hours at temperatures gradually rising to 400°C. Throughout this treatment a solid-state reaction was in progress. A list of the polymers which have been prepared is given in Table I in order to illustrate this novel class of condensation polymers possessing heteroaromatic nuclei as recurring units.

The elemental analyses of the polymers indicated material corresponding to the expected polybenzimidazole. Also, the infrared spectra of the polymers agreed well with those of model compounds. The ultraviolet spectra revealed an absorption peak of high intensity in the 330–420 mµ region. These curves were very similar in shape to those of model compounds. However, a bathochromic shift of 50 to 100 mµ was observed. A correlation was observed in a given polymer structure between the viscosity as a measure of the length of the aromatic chain system, and optical density and wave length of maximum absorption. Increased viscosity caused a bathocromic and/or hyperchromic displacement of the absorption band. The polymers were obtained as yellowish to orange-brown powders. They were characterized by a high degree of stability, showing great resistance to treatment with hydrolytic media. They were soluble in concentrated sulfuric acid and in formic acid, producing stable solutions. Many of the polymers were soluble also in dimethyl sulfoxide and some also in dimethylformamide. Since most of the polymers were infusible, their ability to form concentrated solutions was a feature of great importance. A list of solubilities is shown in Table II. The table also includes information on the crystallinity of some of the polybenzimidazoles, which was obtained from x-ray diffraction studies.

It can be seen from these data that the solubility and crystallinity are strongly dependent on the polymer chain rigidity, symmetry, and intermolecular attraction. The x-ray diffraction studies were made as qualitative tests on the polymer powders or unstretched films; as a rule distinctive ring patterns were present in the more symmetrical polymer structures, indicating some degree of crystallinity. Figure 1 shows three typical diffraction patterns.

The higher molecular weight polymers could be cast into stiff and tough films from formic acid and dimethyl sulfoxide solutions. A film made from poly-2,2'-(m-phenylene)-5,5'-bibenzimidazole was evaluated for typical properties, and yielded the results listed in Table III.

The zero-strength temperature was above 770°C. A determination by light scattering of the molecular weight of the same sample, which showed an inherent viscosity of 0.8 in 0.5% concentrated dimethyl sulfoxide solution, indicated a weight-average molecular weight of 54,000.

3. Thermal Stability of Polybenzimidazoles

The polybenzimidazoles were characterized by an ability to withstand continued exposure to extremely high temperatures without degradation. The thermal degradation studies of the polymers were made by measuring

POLYBENZIMIDAZOLES

519

TABLE II
Crystallinity and Solubility of Some Polybenzimidazoles[a]

Repeating unit	Crystallinity[b]	Solubility, g./100 ml. solvent		
		HCOOH	$(CH_3)_2SO$	DMF
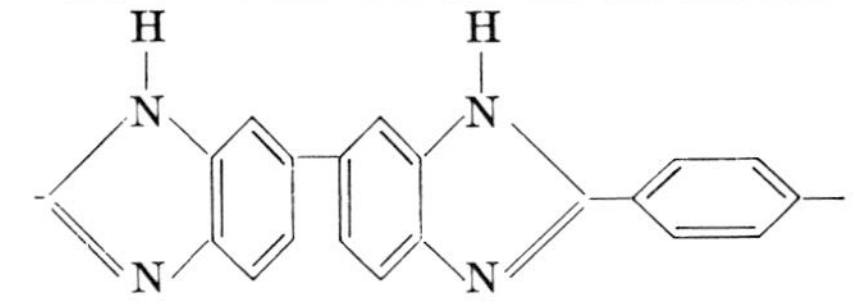	+	2–3	0.5–1	Insoluble
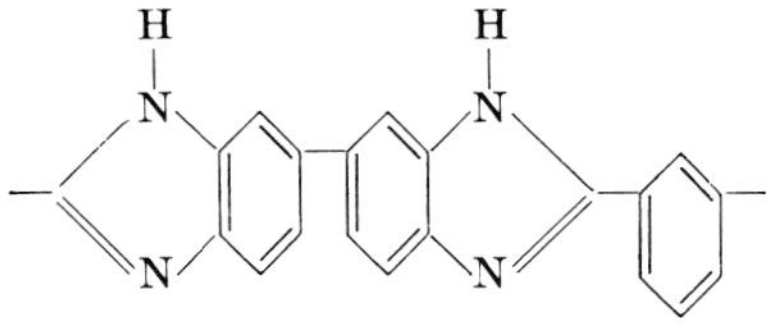	0	5–6	Soluble	Partially soluble
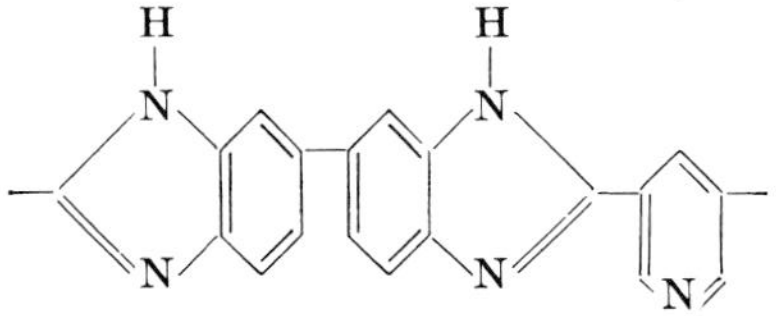	+	10–15	Soluble	Partially soluble
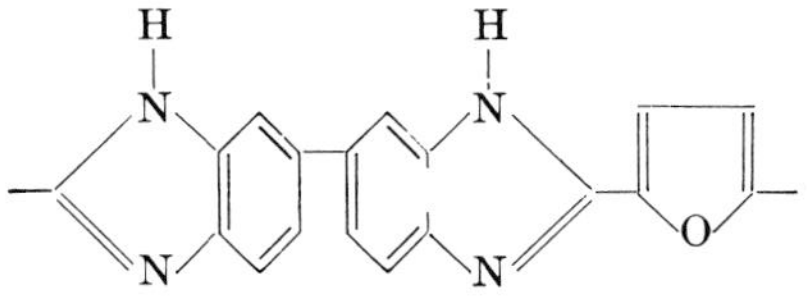	+	3–4	Soluble	Soluble

(continued)

520 H. VOGEL AND C. S. MARVEL

TABLE II (continued)

Repeating unit	Crystallinity[b]	Solubility, g./100 ml. solvent		
		HCOOH	(CH₃)₂SO	DMF
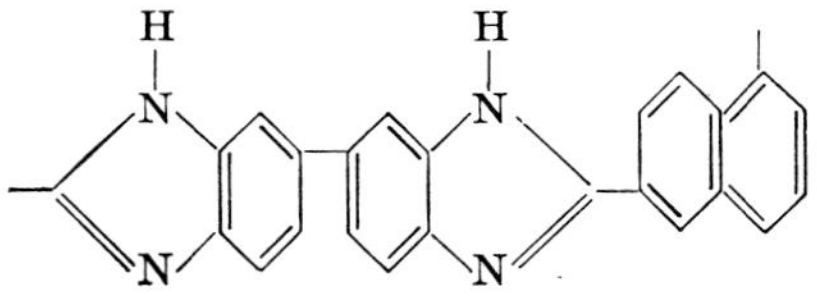	0	8-10	Soluble	Insoluble
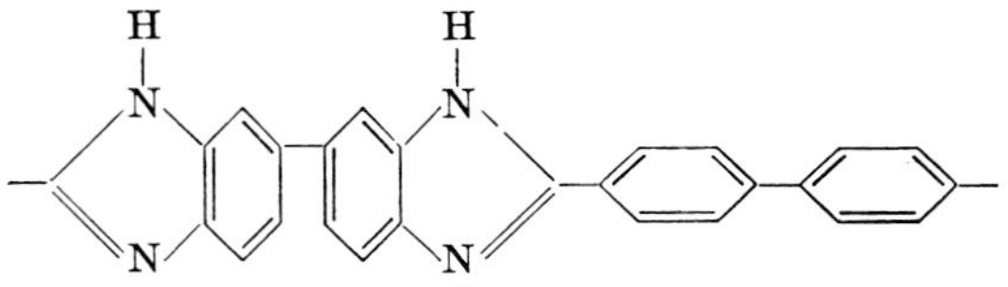	+	Partially soluble	Insoluble	Insoluble
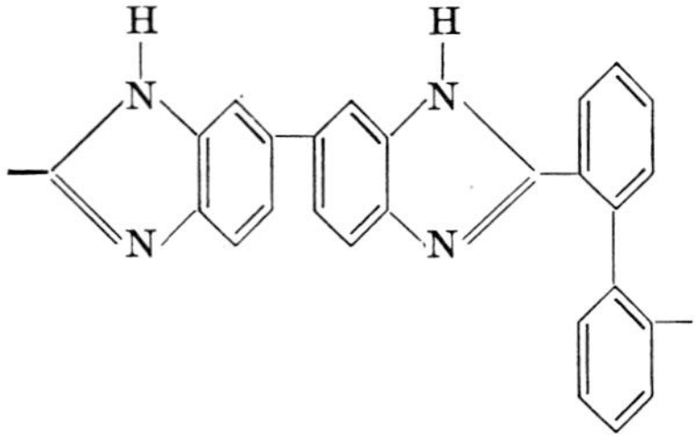	0	Soluble	Soluble	Soluble

POLYBENZIMIDAZOLES   521

Structure	Crystallinity[b]			
(structure with —(CH₂)₄—)	0	Soluble	Soluble	Partially soluble
(structure with para-phenylene)	+	2–3	Insoluble	Insoluble
(structure with meta-phenylene)	+	5–6	Insoluble	Insoluble
(structure with —(CH₂)₄—)	+	Soluble	Partially soluble	Insoluble

[a] Soluble indicates a 20% or greater solution.
[b] +, Crystallinity detected; 0, crystallinity not detected.

522   H. VOGEL AND C. S. MARVEL

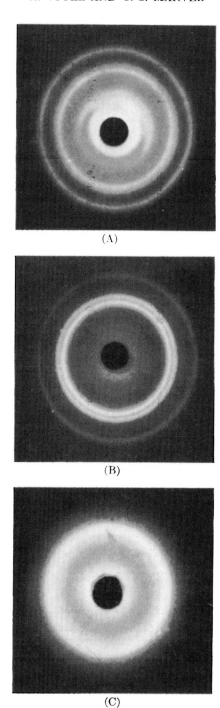

(A)

(B)

(C)

Fig. 1. X-ray patterns of three polybenzimidazoles synthesized from 3,3′-diamino-benzidine and (A) diphenyl pyridine-3,5-dicarboxylate, (B) diphenyl terephthalate, and (C) diphenyl isophthalate.   Flat-film diagrams, nickel-filtered CuKα radiation.

TABLE III
Properties of Poly-2,2'-(m-phenylene)-5,5'-bibenzimidazole Film

Temperature, °C.	Tensile strength, g./den.	Elongation, %	Modulus, g./den.
25	0.7	7	35
200	0.5	9	15

the rate of weight loss in a nitrogen atmosphere at 400, 450, 500, 550, and 600°C. consecutively for 1 hr. at each temperature. The results obtained are listed in Table IV.

The weight loss during heating at 400, 450, and 500°C. was believed to be due to endgroup reaction. In some cases the volatilization at 600°C. was still less than 5%. The superiority of the wholly aromatic polymers is further evidenced by the TGA curves[8] of four different polybenzimidazoles, presented in Figure 2.

These curves show that the aliphatic derivatives undergo complete decomposition around 470°C., due to the presence of aliphatic C—C linkages, whereas the slowly increasing volatilization of the aromatic polymers,

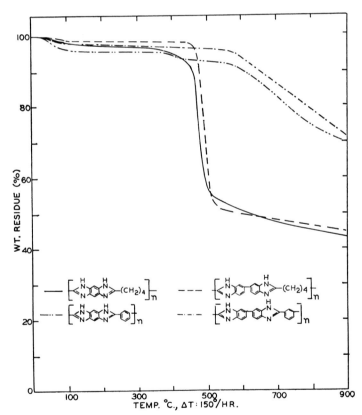

Fig. 2. Thermographic analyses of four polybenzimidazoles.

TABLE IV
Weight Loss in Polymers Heated at Various Temperatures Under Nitrogen[a]

Repeating unit	Weight lost during 1 hr. of heating, wt.-%					Total weight loss, wt.-%
	400°C.	450°C.	500°C.	550°C.	600°C.	
	1.1	0.4	0.4	0.4	5.0	7.3
	1.0	1.0	0	1.7	1.0	4.7
	0.6	0	0.4	1.3	2.2	4.5
	0	0	0.3	3.9	3.7	7.9

POLYBENZIMIDAZOLES   **525**

Structure						
(phenyl)	22.5	9.0	7.0	5.2	1.7	0
(pyridyl)	5.6	2.7	1.4	0.5	0.8	0.2
(furyl)	10.0	2.0	2.3	2.6	1.7	1.4
(naphthyl)	6.5	3.7	1.2	0.8	0.4	0.4
(biphenyl)	3.5	2.1	0.3	0.8	0	0.3

(continued)

526   H. VOGEL AND C. S. MARVEL

TABLE IV (continued)

Repeating unit	Weight lost during 1 hr. of heating, wt.-%					Total weight loss, wt.-%
	400°C.	450°C.	500°C.	550°C.	600°C.	
	0	0.5	8.0	8.5	0.5	17.5
	2.8	0.5	1.0	1.9	4.0	10.2
	0.7	1.4	0.3	1.4	1.4	5.2

[a] Each polymer sample was heated for 1 hr. at each of the given temperatures, consecutively.
[b] Heated under 0.1 mm. Hg pressure.
[c] Heated in air.

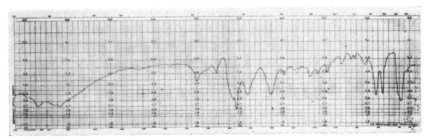

Fig. 3. Infrared spectrum of 2,2′-diphenyl-5,5′-bibenzimidazole (in KBR).

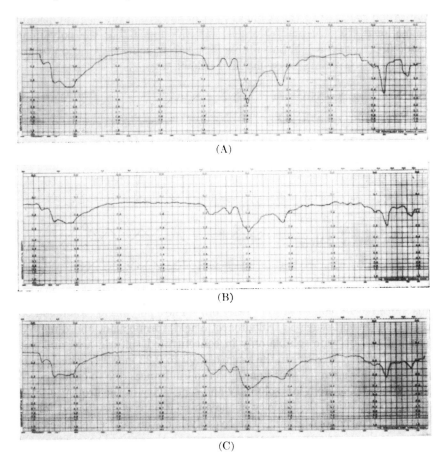

(A)

(B)

(C)

Fig. 4. Infrared spectra of a film of poly-2,2′-(*m*-phenylene)-5,5′-bibenzimidazole heated under nitrogen for consecutive hourly periods: (*A*) one hour at 400°C., (*B*) one additional hour at 500°C., (*C*) one additional hour at 600°C.

gen at 400, 500, and 600°C., consecutively, for 1 hr. at each temperature. The absorption peaks at 3450, 3100, 1620, 1530, 1445, 1285, 800, and 685 cm.$^{-1}$ remained unchanged except for a minor decrease in intensity. The spectrum of a model compound is given also, for comparison. Continued heating at about 500°C. caused the polymers to become gradually insoluble, perhaps owing to crosslinking via free radical reactions.

4. Interfacial Polycondensation of Aromatic Tetraamines and Aromatic Dicarboxylic Acid Chlorides

Experiments were undertaken with the aim of using the interfacial polymerization method[9] in the condensation of the aromatic tetraamines and bifunctional aromatic acid chlorides. It was hoped that the polyaminoamide first formed would be suitable for a subsequent dehydration to give the polybenzimidazole, as exemplified below:

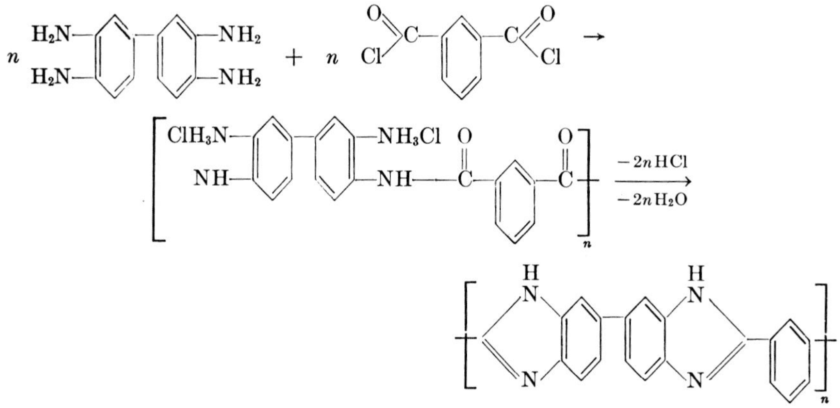

However, only low molecular weight materials have been obtained from this interfacial condensation reaction. The polymeric aminoamides swelled in solvents such as trifluoroacetic acid, formic acid, and dimethylformamide, but were soluble only in concentrated sulfuric acid and possessed inherent viscosities in the range of 0.09 to 0.22 in 0.2% concentrated solution. It is assumed that in addition to hindrances due to experimental difficulties arising from low solubility and sensitivity of the aromatic tetraamines to oxidation, the basicity of the amino groups is insufficient for the interfacial polycondensation. The hydrolysis of the acid chloride component may therefore come into competition·with the desired reaction.

EXPERIMENTAL*

Monomers

Phenyl 3,4-Diaminobenzoate. 3-Nitro-4-acetaminobenzoic acid was prepared by the procedures of Salkowsky[10] and Ritsert and Epstein[11] in 87% yield based on *p*-aminobenzoic acid.

---

* The melting points are uncorrected.

Two hundred and twenty-four grams (1 mole) of 3-nitro-4-acetamino-benzoic acid and 800 g. of thionyl chloride were heated on a water bath for one hour at gentle boiling. When all had dissolved, the excess of thionyl chloride was distilled and the residue dissolved in 800 ml. of hot benzene. On cooling, the 3-nitro-4-acetaminobenzoyl chloride separated in yellow needles. The yield was 198 g. (82%), m.p. 106–107°C.

ANAL. Calcd. for $C_9H_7N_2O_4Cl$: C, 44.53%; H, 2.85%; N, 11.54%. Found: C, 44.33%; H, 2.90%; N, 11.24%.

A mixture of 180 g. (0.75 mole) of 3-nitro-4-acetaminobenzoyl chloride and 100 g. of phenol was heated to 150–160°C. for about $^1/_2$ hr. until the evolution of hydrogen chloride had ceased. The melt was then dissolved and recrystallized from 3 l. of hot alcohol. The yield was 196 g. (87%) of phenyl 3-nitro-4-acetaminobenzoate.

Fifty grams of the phenyl nitroacetaminobenzoate was heated in 600 ml. of concentrated hydrochloric acid for $1^1/_2$ hr. with the temperature not exceeding 80°C. After cooling and dilution with water, the phenyl 3-nitro-4-aminobenzoate was recovered.

For reduction, the crude material was added in five portions to a mixture of 30 g. of tin metal, 30 g. of stannous chloride, and 600 ml. of concentrated hydrochloric acid while the temperature was maintained around 70°C. After cooling, the tin complex salt of the phenyl 3,4-diaminobenzoate was filtered off, dissolved in hot water, and treated with hydrogen sulfide in order to remove the tin salt. The addition of a saturated sodium acetate solution led to the precipitation of the phenyl 3,4-diaminobenzoate which was recrystallized from benzene/ether. The yield was 14 g., m.p. 120–122°C. The overall yield based on p-aminobenzoic acid was 26%.

ANAL. Calcd. for $C_{13}H_{12}N_2O_2$: C, 68.42%; H, 5.22%; N, 12.28%. Found: C, 68.10%; H, 5.30%; N, 12.18%.

3,3′-Diaminobenzidine. The diaminobenzidine tetrahydrochloride was prepared in 55% yield from benzidine according to a combination of the procedures given by Strakosch,[12] Hodgson,[13] and Hoste.[14]

Thirty-six grams (0.1 mole) of diaminobenzidine tetrahydrochloride was dissolved in 200 ml. of boiled, oxygen-free water and poured into 400 ml. of a 4% sodium hydroxide solution to effect neutralization. In all manipulations the materials were blanketed by nitrogen. The collected precipitate was immediately recrystallized twice from 1500 ml. of boiling methanol. The yield of the slightly pink diaminobenzidine was 12.5 g. (58%), m.p. 179–180°C. The overall yield based on benzidine was 32%.

ANAL. Calcd.: C, 67.19%; H, 6.54%; N, 26.17%. Found: C, 67.15%; H, 6.50%; N, 25.93%.

1,2,4,5-Tetraaminobenzene. The 1,2,4,5-tetraaminobenzene tetrahydrochloride was prepared in 66% yield by the procedure of Nietzki and Schedler[15] from m-dichlorobenzene by nitration, reaction with ammonia, and reduction.

A solution of 22 g. of the tetrahydrochloride in 100 ml. of boiled, oxygen-free water was poured with shaking into 100 ml. of an ice-cold 15% sodium hydroxide solution while nitrogen was bubbled through the solution. The precipitate was filtered with suction, washed with ice-cold methanol, and finally dried to constant weight at 50°C. at 0.1 mm. Hg pressure and sealed under nitrogen. The yield was 5.33 g. (55%), m.p. 274–276°C. The overall yield from *m*-dichlorobenzene was 36%.

The tetraaminobenzene could also be liberated from a suspension of the tetrahydrochloride in methanol containing the equivalent amount of lithium methoxide. The separating tetraaminobenzene was collected and used as such. The melting point was 276–277°C.

Diphenyl Pinate. A mixture of 50 g. of pinic acid and 150 ml. of thionyl chloride was heated to reflux on the water bath for $2^1/_2$ hr. The excess of thionyl chloride was then distilled off and the residue heated with 50 g. of phenol for $^1/_2$ hr. at 100–150°C. The reaction mixture was dissolved in ether and extracted three times with dilute sodium hydroxide solution. Fractional distillation under reduced pressure gave a forerun at 160–175°C./ 0.8 mm. Hg; the bulk of diphenyl pinate distilled at 222–223°C./0.8 mm. Hg. The yield was 34 g. It was obtained as white crystals from *n*-hexane/ ether, m.p. 68–69°C.

ANAL. Calcd. for $C_{21}H_{22}O_4$: C, 74.56%; H, 6.51%. Found: C, 74.13%; H, 6.73%.

Diphenyl Homopinate. The ester was prepared in the same way as the diphenyl pinate. From 50 g. of homopinic acid 38 g. of diphenyl homopinate was obtained, b.p. 242–243°C./0.8 mm. Hg, m.p. 56–57°C.

ANAL. Calcd. for $C_{22}H_{24}O_4$: C, 75.00%; H, 6.82%. Found: C, 74.94%; H, 7.30%.

Diphenyl Pyridine-3,5-dicarboxylate. Pyridine-3,5-dicarboxylic acid was converted into the diacid chloride by use of the procedure of Meyer and Tropsch.[16] Thirty grams of dinicotinic acid hydrochloride and 200 g. of thionyl chloride were refluxed on a water bath until all had dissolved (18 hr.). The excess of thionyl chloride was then distilled and the residue dissolved in 200 ml. of benzene. After the mixture had been heated under reflux for four hours, the benzene was removed and the residue recrystallized from *n*-hexane. The yield was 28 g., m.p. 67–69°C. Fifteen grams of the diacid chloride was then melted together with 15 g. of phenol and the resulting diphenyl pyridine-3,5-dicarboxylate purified by repeated recrystallization from methanol. The yield was 14.5 g., m.p. 128°C.

ANAL. Calcd. for $C_{19}H_{13}NO_4$: C, 71.47%; H, 4.07%. Found: C, 71.25%; H, 3.71%.

Diphenyl Dehydromucate. Mucic acid was dehydrated and cyclized to dehydromucic acid according to the procedure of Yoder and Tollins.[17] Ten grams of dehydromucic acid was then mixed with 80 ml. of phosphorus oxychloride and 25 g. of phosphorus pentachloride and heated at reflux for

1 hr. The phosphorus oxychloride was then distilled and the residue recrystallized from n-hexane. The yield was 8 g. of the acid chloride, m.p. 79–80°C.

Eight grams of the dehydromucyl chloride was then heated with 9 g. of phenol for 1 hr. at 160–230°C. The reaction mixture was twice recrystallized from methanol with application of the decolorizer Darco. The yield was 9.5 g. of diphenyl dehydromucate, m.p. 144–145°C.

ANAL. Calcd. for $C_{18}H_{12}O_5$: C, 70.13%; H, 3.90%. Found: C, 70.42%; H, 3.89%.

Diphenyl Diphenate. A mixture of 25 g. of biphenyl-2,2'-dicarboxylic acid (diphenic acid), 40 g. of phosphorus pentachloride, and 50 ml. of phosphorus oxychloride was heated at reflux for 2 hr. The phosphorus oxychloride was then stripped off and the residue was heated with 20 g. of phenol for ½ hr. at 150°C. The reaction mixture was then dissolved in ether and extracted with sodium hydroxide solution. The remainder of the ether phase was recrystallized twice from methanol with application of Darco. The diphenyl diphenate melted at 84–85°C. The yield was 16.2 g.

ANAL. Calcd. for $C_{26}H_{18}O_4$: C, 79.18%; H, 4.57%. Found: C, 79.10%; H, 4.46%.

Diphenyl Biphenyl-1,8-dicarboxylate. A mixture of 25 g. of biphenyl-1,8-dicarboxylic acid, 40 g. of phosphorus pentachloride, and 50 ml. of phosphorus oxychloride was heated at reflux for 3 hr. The phosphorus oxychloride was then stripped off and the residue recrystallized from toluene. The acid chloride melted at 186–188°C.

Ten grams of the acid chloride was melted together with 20 g. of phenol and heated for ½ hr. at 180–200°C. The reaction product was then recrystallized once from a toluene/methanol mixture and again from toluene with application of decolorizing Darco. The diphenyl biphenyl-1,8-dicarboxylate melted at 214–215°C. The yield was 6.4 g.

ANAL. Calcd. for $C_{26}H_{18}O_4$: C, 79.18%; H, 4.57%. Found: C, 79.13%; H, 4.82%.

Diphenyl Naphthalene-1,6-dicarboxylate. A mixture of 20 g. of 1,6-naphthalenedicarboxylic acid, 40 g. of phosphorus pentachloride, and 50 ml. of phosphorus oxychloride was heated at reflux for 1½ hr. After the phosphorus oxychloride had been stripped off, the residue was recrystallized from benzene. The yield of 1,6-naphthalenedicarboxylic acid chloride was 16 g., m.p. 128–129°C. Sixteen grams of the acid chloride and 15 g. of phenol were heated together for 30 min. at 150°C. The reaction product was recrystallized twice from methanol with application of decolorizing Darco. The yield of the diphenyl ester was 12 g., m.p. 134–135°C.

ANAL. Calcd.: C, 78.33%; H, 4.38%. Found: C, 78.18%; H, 4.51%.

2,2'-Diphenyl-5,5'-bibenzimidazole. A charge of 1.07 g. of diaminobenzidine and 2.0 g. of phenyl benzoate was heated under nitrogen for 1 hr. at 270–280°C. The reaction product was recrystallized from glacial acetic acid/ether. The recovered 2,2'-diphenyl-5,5'-bibenzimidazole was dried for 8 hr. under vacuum at 118°C. The yield was 1.6 g.; the melting point was 350–352°C.

Anal. Calcd. for $C_{26}H_{18}N_4$: C, 80.83%; H, 4.66%; N, 14.51%. Found: C, 79.77%; H, 4.75%; N, 13.90%.

2,6-Diphenyl-diimidazobenzene (1,2;4,5). A mixture of 1.20 g. of tetra-aminobenzene and 2.0 g. of phenyl benzoate under a nitrogen atmosphere was placed in a bath preheated to 280°C., whereupon a melt formed. This was then heated for 1 hr. at 290°C. The reaction product was re-crystallized from glacial acetic acid/ether and sublimed at 480°C. and 15 mm. Hg pressure. The yield was 1.0 g. The 2,6-diphenyl-diimidazoben-zene melted with subliming at about 550°C.

Anal. Calcd. for $C_{20}H_{14}N_4$: C, 77.42%; H, 4.52%; N, 18.06%. Found: C, 77.26%; H, 4.85%; N, 18.42%.

Melt Polycondensations

General Procedure. Equimolar amounts of the monomer components were placed in a round-bottomed flask having a capacity of about 10 times the volume of the reactants. The flask was connected to tubing leading to a receiver and outlet. The apparatus was purged with pure nitrogen by repeated evacuation and refilling. The flask was then placed in a silicone oil bath preheated to 220°C. in the case of the diaminobenzidine deriva-tives, and to 280°C. in the case of the tetraaminobenzene derivatives. The reaction mixture formed a melt and was then heated to the temperature necessary for reaction. The evolution of phenol and water indicated that the reaction had commenced. After 10 to 30 min., when the melt had changed to a more or less solid mass, the flask was carefully put under high vacuum and heating was continued for 30 min. at the particular tempera-ture. In many instances a sort of glassy foam was produced during this treatment. The apparatus was then filled with nitrogen and cooled. The materials were pulverized and reheated for several hours under high vac-uum at temperatures gradually rising to 400°C. The yields were quanti-tative. If not noted otherwise, the inherent viscosities were measured at 0.2% concentration in formic acid at 25°C.

Poly-2,5(6)-benzimidazole. A charge of 3 g. of phenyl 3,4-diaminoben-zoate was heated under nitrogen for 1 hr. at 290°C., producing a pale yellow solid mass with an inherent viscosity of 0.3 (0.2% concn. in formic acid). This mass was then finely pulverized and reheated under vacuum for $1^1/_2$ hr. at 350°C. and $1^1/_2$ hr. at 400°C. The polymer thus prepared had an inherent viscosity of 1.27.

Anal. Calcd. for $(C_7H_4N_2)_n$: C, 72.41%; H, 3.45%; N, 24.14%. Found: C, 72.50%; H, 4.30%; N, 23.31%.

Poly-2,2'-(tetramethylene)-5,5'-bibenzimidazole. Diphenyl adipate was prepared by treating adipyl chloride with phenol at 100°C. The ester was recrystallized from alcohol. The melting point was 108–109°C. A mixture of 3.0 g. (0.01 mole) of the ester and 2.14 g. (0.01 mole) of diaminobenzidine was heated at 250°C. for 10 min. Heating was continued for $1/2$ hr. at 0.1 mm. Hg pressure while the temperature was raised to 270°C. The inherent viscosity of the polymer was 3.19.

ANAL. Calcd. for $(C_{18}H_{18}N_4)_n$: C, 75.00%; H, 5.60%; N, 19.40%. Found: C, 74.87%; H, 5.51%; N, 19.33%.

Polybenzimidazoles from Diphenyl Pinate and Diphenyl Homopinate. A mixture of 3.21 g. (0.015 mole) of 3,3'-diaminobenzidine and 5.28 g. (0.015 mole) of diphenyl pinate was heated under nitrogen for 1 hr. at temperatures rising gradually from 240 to 280°C. The foamy, glassy mass obtained was then powdered and reheated for $1/2$ hr. at 280°C./0.1 mm. Hg. The inherent viscosity of the polymer was 1.03.

ANAL. Calcd. for $(C_{21}H_{20}N_4)_n$: C, 76.83%; H, 6.10%; N, 17.06%. Found: C, 76.35%; H, 6.25%; N, 16.16%.

The polybenzimidazole from diphenyl homopinate was prepared under the same conditions.

ANAL. Calcd. for $(C_{22}H_{22}N_4)_n$: C, 77.19%; H, 6.43%; N, 16.37%. Found: C, 77.80%; H, 6.68%; N, 15.27%.

Poly-2,2'-(m-phenylene)-5,5'-bibenzimidazole. A mixture of 4.28 g. (0.02 mole) of diaminobenzidine and 6.36 g. (0.02 mole) of diphenyl isophthalate[18] was melted at 220°C. The reaction was initiated at 260°C. and heating was continued for 30 min. Evacuation to 0.1 mm. Hg pressure produced a glassy foam and removed most of the liberated phenol and water. After 30 min. the material was pulverized and reheated at 0.1 mm. Hg for 9 hr. while the temperature was raised from 280 to 400°C. The inherent viscosity of the polymer was 3.34 in formic acid. A 0.5% solution of the polymer in dimethyl sulfoxide showed an inherent viscosity of 1.02.

ANAL. Calcd. for $(C_{20}H_{12}N_4)_n$: C, 77.92%; H, 3.90%; N, 18.18%. Found: C, 78.06%; H, 4.39%; N, 18.05%.

Poly-2,2'-(p-phenylene)-5,5'-bibenzimidazole. The melt of 2.14 g. (0.01 mole) of diaminobenzidine and 3.18 g. (0.01 mole) of diphenyl terephthalate[13] was heated for 10 min. at 250°C. A yellow, solid cake was formed. Heating was continued at 0.1 mm. Hg pressure for 30 min. at 260°C., and after the product had been powdered, for $4^1/2$ hr. at temperatures gradually rising to 400°C. The inherent viscosity was 1.00.

ANAL. Calcd. for $(C_{20}H_{12}N_4)_n$: C, 77.92%; H, 3.90%; N, 18.18%. Found: C, 76.65%; H, 4.14%; N, 18.10%.

Poly-2,2'-(pyridylene-3'',5'')-5,5'-bibenzimidazole. A mixture of 2.14 g. (0.01 mole) of diaminobenzidine and 3.19 g. (0.01 mole) of the diphenyl pyridine-3,5-dicarboxylate was heated for 30 min. at 260°C. Then high vacuum was employed for 30 min. to remove most of the liberated phenol and water. This material was finely ground and reheated at 0.1 mm. Hg for 4 hr. while the temperature was gradually raised to 400°C. The inherent viscosity of the polymer was 1.48 in formic acid and 0.38 in 0.2% solution in dimethyl sulfoxide.

ANAL. Calcd. for $(C_{19}H_{11}N_5)_n$: C, 73.78%; H, 3.56%; N, 22.65%. Found: C, 73.71%; H, 3.91%; N, 21.03%.

Poly-2,2'-(furylene-2'',5'')-5,5'-bibenzimidazole. A mixture of 3.21 g. (0.015 mole) of 3,3'-diaminobenzidine and 4.62 g. (0.015 mole) of the diphenyl dehydromucate was heated under nitrogen for 30 min. at 260°C. and then under high vacuum at this temperature for 30 min., which produced a foamed glassy product. The material was then finely powdered and reheated at 0.1 mm. Hg for 4 hr. while the temperature was gradually raised to 400°C. The inherent viscosity of the final polymer was 0.74 in formic acid and 0.3 in 0.2% solution in dimethyl sulfoxide.

ANAL. Calcd. for $(C_{18}H_{10}N_4O)_n$: C, 72.48%; H, 3.35%; N, 18.79%. Found: C, 72.94%; H, 4.02%; N, 18.59%.

Polybenzimidazole from Diphenyl Naphthalene-1,6-dicarboxylate. A mixture of 2.14 g. (0.01 mole) of diaminobenzidine and 3.68 g. (0.01 mole) of the ester was melted together at 220°C. The temperature was then raised to 240°C. where the condensation reaction started. After the mixture had been heated for 30 min. at about 270°C. the flask was evacuated, whereupon a solid foam was formed filling the flask. The flask was then heated for 30 min. at 0.1 mm. Hg and 270°C. Powdering and reheating of the polymer for 4 hr. at 0.1 mm. Hg and temperatures rising to 400°C. gave a product with an inherent viscosity of 2.7 in formic acid and 0.86 in 0.2% solution in dimethyl sulfoxide.

ANAL. Calcd. for $(C_{24}H_{14}N_4)_n$: C, 80.51%; H, 3.94%; N, 15.65%. Found: C, 80.36%; H, 4.33%; N, 15.21%.

Polybenzimidazole from Diphenyl Biphenyl-1,8-dicarboxylate. In a 50-ml. round-bottomed flask were placed 2.14 g. (0.01 mole) of diaminobenzidine and 3.94 g. (0.01 mole) of the ester. A melt was formed at 220°C. and after the temperature had been increased to 250°C. the reaction started. After about 10 min. the viscous melt became a solid mass. Careful evacuation to 0.1 mm. Hg for 45 min. at 260°C. led to a yellow solid cake which was then powdered and reheated at 0.1 mm. Hg and temperatures rising within 5 hr. to 400°C. The polymer had an inherent viscosity of 0.86 (0.2% concn. in sulfuric acid, 25°C.).

ANAL. Calcd. for $(C_{26}H_{16}N_4)_n$: C, 81.30%; H, 4.20%; N, 14.55%. Found: C, 78.69%; H, 4.50%; N, 14.12%.

Polybenzimidazole from Diphenyl Diphenate. A mixture of 2.14 g. (0.01 mole) of diaminobenzidine and 3.94 g. (0.01 mole) of the ester was melted together at 230°C. by means of a Wood's metal bath. The temperature was raised to 285°C. when evolution of phenol bubbles indicated that reaction had commenced. After heating for $2\frac{1}{2}$ hr. at temperatures around 300°C., the reddish-brown melt became very viscous. Vacuum was then applied and a solid foam was formed filling the flask. The polymer was heated for 30 min. at 310°C. at 0.1 mm. Hg pressure. After cooling, it was powdered and then reheated under 0.1 mm. Hg for 3 hr. at temperatures gradually rising from 300° to 400°C. The material obtained had an inherent viscosity of 2.99 in formic acid and 1.17 in 0.2% concentrated dimethyl sulfoxide solution.

ANAL. Calcd. for $(C_{26}H_{16}N_4)_n$: C, 81.30%; H, 4.20%; N, 14.55%. Found: C, 81.53%; H, 4.18%; N, 14.02%.

Poly-2,6-(tetramethylene)-diimidazobenzene. A mixture of 1.87 g. (0.0137 mole) of tetraaminobenzene and 4.11 g. (0.0137 mole) of diphenyl adipate was melted at 270°C. by means of a silicon oil bath preheated to this temperature. Reaction started immediately. The brown viscous melt changed after 20 min. to a dark blue solid. This mass was then powdered and reheated at 0.1 mm. Hg for 1 hr. at 300°C. The polymer had an inherent viscosity of 2.51; it melted at 490°C. with decomposition.

ANAL. Calcd. for $(C_{12}H_{12}N_4)_n$: C, 67.92%; H, 5.66%; N, 26.42%. Found: C, 66.69%; H, 5.73%; N, 23.85%.

Poly-2,6-(m-phenylene)-diimidazobenzene. A mixture of 1.665 g. (0.012 mole) of 1,2,4,5-tetraminobenzene and 3.82 g. (0.012 mole) of diphenyl isophthalate was placed in a 50-ml. still. The flask was put in a Wood's metal bath preheated to 280°C. A reddish-brown melt was first formed which solidified after about 10 min., during which time phenol and water had been distilling. After 30 min., vacuum was employed for $^1/_2$ hr. The material was then powdered and reheated at 0.1 mm. Hg pressure for 3 hr., the temperature being allowed to rise gradually to 400°C. The polymer had an inherent viscosity of 1.1.

ANAL. Calcd. for $(C_{14}H_8N_4)_n$: C, 72.41%; H, 3.45%; N, 24.14%. Found: C, 71.11%; H, 3.82%; N, 21.82%.

Polycondensation of Dimethyl Isophthalate and Diaminobenzidine. Dimethyl isophthalate from Eastman Kodak was recrystallized twice from methanol, m.p. 65–66°C. Then 0.97 g. (0.005 mole) of the ester and 1.07 g. (0.005 mole) of diaminobenzidine were placed in a high-pressure tube. The tube was filled with nitrogen, sealed, and heated for 1 hr. at 250°C. A red-brown glassy mass was formed which was pulverized and reheated at 0.1 mm. Hg pressure for 3 hr. at temperatures rising from 260° to 360°C. The polymer had an inherent viscosity of 0.46.

ANAL. Calcd. for $(C_{20}H_{12}N_4)_x$: C, 77.92%; H, 3.90%; N, 18.18%. Found: C, 75.08%; H, 4.97%; N, 17.43%.

Polycondensation of Isophthalic Acid and Diaminobenzidine. Isophthalic acid from Eastman Kodak was recrystallized twice from methanol, m.p. > 330°C. A mixture of 1.36 g. (0.01 mole) of the acid and 2.14 g. (0.01 mole) of diaminobenzidine was treated in a manner similar to the general procedure described earlier in this paper. A melt was formed at 250°C. while the reaction was starting. Heating was continued for 30 min. at 270°C., during which time the melt became a solid red-brown mass. This material was pulverized and reheated under nitrogen for 2 hr. at 280°C. and finally at 0.1 mm. Hg for $2^{1}/_{2}$ hr. at temperatures rising from 300 to 350°C. The inherent viscosity of the polymer was 0.68.

ANAL. Calcd. for $(C_{20}H_{12}N_4)_n$: C, 77.92%; H, 3.90%; N, 18.18%. Found: C, 76.71%; H, 4.24%; N, 18.18%.

Interfacial Polycondensation of Diaminobenzidine and Isophthalyl Chloride. To a vigorously stirred solution of 3.21 g. (0.015 mole) of diaminobenzidine in 1.5 l. of water was added, as fast as possible, a solution of 3.045 g. (0.015 mole) of isophthalyl chloride in 300 ml. of methylene chloride. Stirring was continued for 40 min. The precipitate was collected, placed in 2 l. of boiling 1% sodium hydroxide solution, and finally collected on a filter, washed, and dried. The yield was 3.8 g. The inherent viscosity was 0.22 in 0.2% concentrated sulfuric acid solution.

ANAL. Calcd. for $(C_{20}H_{16}N_4O_2)_n$: C, 69.75%; H, 4.68%; N, 16.27%. Found: C, 70.61%; H, 4.41%; N, 12.54%.

Interfacial Polycondensation of Tetraaminobenzene Tetrahydrochloride and Tetrephthalyl Chloride. In a 3-l., three-necked, round-bottomed flask which was purged with nitrogen was placed 800 ml. of boiled, oxygen-free water, 4.26 g. (0.015 mole) of tetraminobenzene tetrahydrochloride, and 3.2 g. (0.03 mole) of sodium carbonate. To the vigorously stirred solution was added a solution of 3.0 g. (0.015 mole) of terephthalyl chloride in 500 ml. of chloroform. Stirring was continued for 15 min. The precipitated polymer was filtered off and placed in 1 l. of boiling water containing 10 g. of sodium carbonate. The polymer was then collected, washed with hot water, and dried under vacuum. The yield was 3.7 g.; the inherent viscosity in 0.2% concentrated sulfuric acid solution was 0.08.

ANAL. Calcd. for $(C_{14}H_{12}N_4O_2)_n$: C, 62.74%; H, 4.51%; N, 20.91%. Found: C, 61.95%; H, 4.56%; N, 16.10%.

Thermal Stability Measurements

Samples of approximately 300 mg. of the polymers were placed in 5-ml. vials and at first dried under vacuum at 200°C. for about 30 min. to remove adsorbed moisture. The vials were then placed in a larger flask which was purged with nitrogen and heated by means of a Wood's metal bath consecutively at 400, 450, 500, 550, and 600°C. The weight loss was measured after 1 hr. of heating at each of these temperatures (see Table IV).

Hydrolytic Stability

Two samples of 0.5 g. of poly-2,2'-(m-phenylene)-5,5'-bibenzimidazole having an inherent viscosity of 3.34 (0.2% concn. in formic acid, 25°C.)

were heated to reflux in 50 ml. of 70% sulfuric acid and 50 ml. of 25% potassium hydroxide solution, respectively, each for 10 hr. The acidic reaction mixture was then poured in dilute sodium hydroxide solution. The basic reaction mixture was diluted with water. The polymeric materials were collected on a filter, washed, and dried. The polymer now showed an inherent viscosity of 3.25 in the case of the acidic treatment (0.48 g. recovered) and 3.58 in the case of the basic treatment (0.38 g. recovered).

The authors wish to express their thanks to Dr. M. Morton, University of Akron, for the determination of the molecular weight by the light-scattering technique and to Dr. A. F. Smith, Textile Fibers Department, E. I. du Pont de Nemours & Company, for the determination of the physical properties of the polymer listed in Table III. For the TGA curves, we wish to thank Dr. G. Ehlers, Wright Air Development Division, Wright-Patterson Air Force Base, Ohio. The analyses and infrared and ultraviolet spectra were performed by Mr. J. Nemeth, Mr. P. E. McMahon, and Miss C. Juan at the Microanalytical Laboratory and the Spectroscopic Laboratories at the University of Illinois. We are grateful to Dr. C. C. Pfluger, University of Illinois, for help in obtaining and interpreting the x-ray diffraction patterns. The pinic acid and homopinic acid were furnished by Dr. Glen W. Hedrick, Naval Stores Station, Southern Utilization Research and Development Division, Agricultural Research Service, U. S. Department of Agriculture, Olustee, Florida.

Dr. Herward Vogel is indebted to the International Educational Exchange Service (Fulbright Commission), in cooperation with the Exchange Visitors Program, for a travel grant.

References

1. Hofmann, K., *Imidazole and Its Derivatives*, Interscience, New York, 1953.

2. Steck, E. A., F. C. Nachold, G. W. Ewing, and N. H. Gorman, *J. Am. Chem. Soc.*, **70**, 3406 (1948).

3. Dale, J. W., I. B. Johns, E. A. McElhill, and J. O. Smith, WADC Technical Report 59-95, January, 1959.

4. Wagner, E. C., and W. H. Millett, *Org. Syntheses*, Coll. Vol. II, Wiley, 1946, p. 65.

5. Brinker, K. C., and J. M. Robinson, U. S. Pat. 2,895,948 (1959).

6. Phillips, M. A., *J. Chem. Soc.*, **1930**, 1415.

7. Walther, R., and Th. von Tulawski, *J. prakt. Chem.*, **59**, 249 (1899).

8. Doyle, C. D., General Electric Company, WADC Technical Report 59-136, ASTIA Document No. 216453, June, 1959.

9. Wittbecker, E. L., and P. W. Morgan, *J. Polymer Sci.*, **40**, 289 (1959), and following papers.

10. Salkowsky, H., *Ann.*, **173**, 39 (1874).

11. Ritsert, E., and W. Epstein, *Zentralblatt*, **1904**, I, 1587.

12 Strakosch, J., *Ber.*, **5**, 236 (1872).

13. Hodgson, H. H., *J. Chem. Soc (London)*, **1926**, 1757.

14. Hoste, J., *Anal. Chim. Acta*, **2**, 402 (1948).

15. Nietzki, R., and A. Schedler, *Ber.*, **30**, 1666 (1892).

16. Meyer, H., and H. Tropsch, *Monatsh. Chem.*, **35**, 782 (1914).

17. Yoder, P. A., and B. Tollins, *Ber.*, **34**, 3447 (1901).

18. Schreder, J., *Ber.*, **7**, 708 (1874).

Synopsis

Wholly aromatic polybenzimidazoles were synthesized from aromatic tetraamines and difunctional aromatic acids and characterized as new thermally stable polymers. The melt polycondensation of aromatic tetraamines and the diphenyl esters of aromatic dicarboxylic acids was developed as a general procedure of wide applicability. Polybenzimidazoles containing mixed aromatic units in the chain backbone were prepared from 3,3'-diaminobenzidine, 1,2,4,5-tetraaminobenzene and a variety of aromatic diphenyl dicarboxylates. Phenyl 3,4-diaminobenzoate could also be polymerized by melt condensation to give poly-2,5(6)-benzimidazole. The polymers were characterized by a high degree of stability, showing great resistance to treatment with hydrolytic media and an ability to withstand continued exposure to elevated temperatures. Most of the polymers were infusible, but some had melting points above about 400°C. Many of the polymers exhibited no change in properties on being heated to 550°C. and showed a weight loss of less than 5% when heated under nitrogen for several hours to 600°C. The polymers were soluble in concentrated sulfuric acid and formic acid, producing stable solutions. Many of the polymers were soluble in dimethyl sulfoxide and some also in dimethylformamide. The inherent viscosities of a number of polymers in 0.5% dimethyl sulfoxide solution ranged from approximately 0.4 to 1.1. The higher polymers could be cast into stiff and tough films from formic acid and dimethyl sulfoxide solutions.

Résumé

Des polybenzimidazols aromatiques ont été synthétisés à partir de tétramines aromatiques et d'acides aromatiques bifonctionnels; ils sont caractérisés comme de nouveaux polymères thermiquement stables. La polymérisation en fusion des tétramines aromatiques et des esters diphénylés des acides aromatiques dicarboxyliques, a été développée comme un procédé général de large application. Des polybenzimidazols contenant des unités aromatiques dans la chaîne latérale, ont été préparés à partir des 3,3'-diaminobenzidine, 1,2,4,5-tétraminobenzène et une série de diphényldicarboxylates aromatiques. Le phényl 3,4-diaminobenzoate peut être aussi polymérisé par condensation en fusion pour donner le poly, 2,5(6)-benzimidazol. Les polymères sont caractérisés par un haut degré de stabilité montrant une grande résistance en traitement dans un milieu hydrolytique et une résistance à une exposition continue à des hautes températures. La plupart de ces polymères sont infusibles, certains cependant ont des points de fusion au dessus de 400°C environ. Beaucoup de polymères ne montrent pas de changement dans les propriétés, par chauffage à 550°C et montrent une perte de poids de moins de 5% par chauffage sous atmosphère d'azote pendant quelques heures à 600°C. Les polymères sont solubles dans l'acide sulfurique concentré et l'acide formique, donnant des solutions stables. Beaucoup de polymères sont solubles dans le diméthylsulfoxide et quelques uns aussi dans le diméthylformamide. La viscosité inhérente de la plupart des polymères dans une solution à 0,5% dans le diméthylsulfoxide varie entre les valeurs de 0.4 à 1.1. Les polymères les plus élevés, peuvent être coulés en film rigide à partir de solutions dans l'acide formique et le diméthylsulfoxide.

Zusammenfassung

Aus aromatischen Tetraaminen und difunktionellen aromatischen Carbonsäuren wurden rein aromatische Polybenzimidazole dargestellt und als neue, thermisch stabile Polymere charakterisiert. Die Schmelzpolykondensation von aromatischen Tetraaminen und den Diphenylestern aromatischer Dikarbonsäuren wurde zu einem allgemeinen Verfahren weiter Anwendbarkeit entwickelt. Polybenzimidazole mit gemischten, aromatischen Einheiten in der Hauptkette wurden aus 3,3'-Diaminobenzidin, 1,2,4,5-Tetraaminobenzol und verschiedenen aromatischen Diphenyldicarboxylaten dargestellt. Auch Phenyl-3,4-diaminobenzoat konnte durch Schmelzkondensation unter Bildung von Poly-2,5(6)-benzimidazol polymerisiert werden. Die Polymeren besitzen insofern einen hohen Grad von Stabilität, als sie eine gross Beständigkeit gegen Behandlung mit hydrolysierenden Medien und die Eignung, einer längeren Erwärmung auf höhere Temperaturen zu widerstehen, besitzen. Die meisten Polymeren waren unschmelzbar,

POLYBENZIMIDAZOLES   **539**

einige hatten jedoch Schmelzpunkte oberhalb etwa 400°C. Viele Polymere zeigten beim Erhitzen auf 550°C keine Änderung der Eigenschaften und beim mehrstündigen Erhitzen unter Stickstoff auf 600°C einen Gewichtsverlust kleiner als 5%. Die Polymeren waren unter Bildung stabiler Lösungen in konzentrierter Schwefelsäure und Ameisensäure löslich. Viele Polymere waren in Dimethylsulfoxyd und manche auch in Dimethylformamid löslich. Die Viskositätszahlen einer Anzahl von Polymeren in 0,5% Dimethylsulfoxydlösung lagen bei angenähert 0,4 bis 1,1. Die höheren Polymeren konnten aus Ameisensäure- und Dimethylsulfoxydlösungen zu steifen und zähen Filmen gegossen werden.

Received November 27, 1960

COMMENTARY

Reflections on "Polymerization by Oxidative Coupling. II. Oxidation of 2,6-Disubstituted Phenols," by Allan S. Hay, *J. Polym. Sci.,* 58, 581 (1962)

ALLAN S. HAY

Department of Chemistry, McGill University, Montreal, Quebec, Canada H3A 2K6

The discovery, in August 1956,[1,2,3] that a high molecular weight poly(phenylene) oxide (**2**: R=CH$_3$) could be produced by the oxidative polymerization of 2,6-dimethylphenol (**1**: R=CH$_3$) was followed by an intensive study of the scope and the mechanism of this novel polymerization reaction.[4]

Many different phenols were studied. The reaction was found to be general for 2,6-disubstituted phenols with the exceptions that if the substituents are large the carbon–carbon coupled product, a diphenoquinone **3**, predominated and if the substituents were too electronegative no reaction took place because of the high oxidation potential of the phenol.[5] By the use of bulky ligands attached to the copper only limited success was achieved in the oxidative polymerization of phenol and mono-*ortho*-substituted phenols.[6]

A preliminary study of the preparation and properties of the polymer obtained from 2,6-dimethylphenol, which was called PPO® resin, was carried out in parallel. The polymer has a very low degree of crystallinity and a very high glass transition temperature, 210°C, and is readily soluble in aromatic and chlorinated hydrocarbons. The completely aromatic poly(phenylene oxide) (**2**; R=Ph) obtained from 2,6-diphenylphenol (**1**; R=Ph) was also synthesized and it has a T_g of 235°C and crystallizes readily with a T_m of 490°C.[7,8] The monomer is readily available in two steps from cyclohexanone.[9] Attempts were made to commercialize this material by AKZO (TENAX® resin).[10]

Commercial development of PPO began in 1960, which culminated in a decision by GE in 1965 to build a commercial plant to introduce PPO to the market as an engineering thermoplastic. A facile synthesis of the monomer, 2,6-dimethylphenol, based on a vapor phase reaction of phenol and methanol made the material economically attractive.[11]

Catalysts that were developed for the oxidative coupling of phenols were also found to be very effective for the oxidative coupling of acetylenes,[12,13] thiols,[14] and aromatic amines.[15] Oxidative coupling also provided an attractive route to the synthesis of biphenols by the reduction of diphenoquinones obtained by an oxidative carbon–carbon coupling reaction from 2,6-disubstituted phenols.[16]

In parallel, studies were in progress preparing blends of PPO with various other polymers, for example, polycarbonate,[17] polyethylene,[18] and nylon.[19] Surprisingly, it was found that PPO was completely miscible with polystyrene.[20,21] The blends with polystyrene are considered to be a classical example of miscible blends and have been extensively studied. PPO, because of its high glass transition temperature, was found to be difficult to process because careful control of the processing conditions was reprocess. Oxidative coupling polymerization is only one of many of his highly innovative contributions to synthetic polymer chemistry. After moving from General Electric to McGill University, he has continued to find novel approaches to polymerization of high performance materials ranging from polyphenylene sulfide to polyheterocyclics.[8]

REFERENCES AND NOTES

1. A. S. Hay, *Polym. Eng. Sci.,* **16,** 1 (1976).
2. A. S. Hay, *Adv. Polymer Sci.,* **4,** 496 (1967).
3. E. P. Cizek, U.S. Patent 3,383,435 (May 14, 1968).
4. A. S. Hay, H. S. Blanchard, G. F. Endres, and J. W. Eustance, *J. Am. Chem. Soc.,* **81,** 6335 (1959).
5. G. D. Staffin and C. C. Price, *J. Am. Chem. Soc.,* **82,** 3632 (1960).
6. D. M. White and H. J. Klopfer, ACS Symposium Series, Number 6, *Polyethers,* American Chemical Society, Washington, DC, 1975, p. 169.
7. H. A. P. de Jongh et al., *J. Polym. Sci.,* **11,** 345 (1973).
8. For example, M. Strukelji, J. Harnier, E. Elce, and A. S. Hay, *J. Polym. Sci. Part A.: Polym. Chem.,* **32,** 193 (1994).

PERSPECTIVE

Comments on "Polymerization by Oxidative Coupling. II. Oxidation of 2,6-Disubstituted Phenols," by Allan S. Hay, *J. Polym. Sci.,* **58,** 581 (1962)

T. TAKEKOSHI and D. M. WHITE

GE Corporate Research and Development, Schenectady, New York 12301

The discovery that high molecular weight polymers could be prepared by oxidative coupling reactions has opened up many routes to new classes of polymers and led to new major industrial processes. The initial work by Allan S. Hay grew from work at the General Electric Research Laboratory on metal catalyzed oxidation of simple organic materials, such as xylenes, to produce phthalic acids for alkyd resins. Among other substrates examined were phenols, thiophenols, and amines. The observation that a highly selective polymerization reaction could be achieved by blocking reactive sites in the *ortho* positions of phenol with methyl groups spurred an extensive research program on oxidative coupling.[1] From this work came knowledge of the poly-1,4-phenylene ether structure of the polymers from 2,6-disubstituted phenols, the scope of the reaction, the mechanism of the polymerization, and the determination of the physical properties of the polymeric products.[2]

The polymerization process for the conversion of 2,6-dimethylphenol to poly(2,6-dimethyl-1,4-phenylene oxide) (PPO® resin) was sufficiently straightforward to be commercially attractive, and efforts were made to scale up and optimize the process. However, an unusual characteristic of the polymer was its high glass transition temperature of ∼ 208°C, which meant high processing temperatures would be required for commercial utilization. Under these conditions, polymer degradation via oxidation and thermal side reactions became an issue. The problem was overcome by the discovery that PPO resin formed homogeneous blends with polystyrene[3] and that the blends could be extruded and molded with normal polymer processing conditions. As a result, the blends under the Noryl® resin trademark have become a major family of engineering thermoplastics and have found many applications in the electrical, plumbing, appliance, and automotive industries.

Shortly after the initial communication on oxidative coupling polymerization of phenols,[4] a different method for preparation of PPO was described.[5] This was a free-radical initiated halide displacement reaction on 4-bromo-2,6-dimethylphenol. Although not as commercially attractive as the oxidative coupling route because of the presence of halogen, this procedure has been useful for the preparation of brominated PPO resins.[6]

Hay found that a large number of phenols could be oxidatively polymerized analogously to 2,6-dimethylphenol. 2,6-Diphenylphenol was converted to the totally aromatic polyether and was found to have utility in applications such as an organic phase for gas chromatography and in gas absorption equipment. Oxidative coupling of diacetylenes such as 1,3-diethynylbenzene and its isomers led to soluble all aromatic polyacetylenes with high carbon contents that could be converted to high performance carbon fibers in high yield. Diaminoaromatics were found to be readily converted to polymeric *azo* compounds and di-mercapto compounds produced polymeric arylene disulfides. The work spurred others to oxidize a variety of activated hydrogen compounds (e.g., bifunctional arylcyanoacetic esters)[7] to form novel polymers.

Oxidative coupling polymerization serves as an example of how Hay's leadership allowed a unique type of polymerization to become a major industrial process. Oxidative coupling polymerization is only one of many of his highly innovative contributions to synthetic polymer chemistry. After moving from General Electric to McGill University, he has continued to find novel approaches to polymerization of high performance materials ranging from polyphenylene sulfide to polyheterocyclics.[8]

REFERENCES AND NOTES

1. A. S. Hay, *Polym. Eng. Sci.,* **16,** 1 (1976).
2. A. S. Hay, *Adv. Polymer Sci.,* **4,** 496 (1967).
3. E. P. Cizek, U.S. Patent 3,383,435 (May 14, 1968).
4. A. S. Hay, H. S. Blanchard, G. F. Endres, and J. W. Eustance, *J. Am. Chem. Soc.,* **81,** 6335 (1959).
5. G. D. Staffin and C. C. Price, *J. Am. Chem. Soc.,* **82,** 3632 (1960).
6. D. M. White and H. J. Klopfer, ACS Symposium Series, Number 6, *Polyethers,* American Chemical Society, Washington, DC, 1975, p. 169.
7. H. A. P. de Jongh et al., *J. Polym. Sci.,* **11,** 345 (1973).
8. For example, M. Strukelji, J. Harnier, E. Elce, and A. S. Hay, *J. Polym. Sci. Part A.: Polym. Chem.,* **32,** 193 (1994).

JOURNAL OF POLYMER SCIENCE   VOL. 58, PAGES 581–591 (1962)

Polymerization by Oxidative Coupling.
II. Oxidation of 2,6-Disubstituted Phenols

ALLAN S. HAY, *General Electric Research Laboratory,
Schenectady, New York*

INTRODUCTION

In a preliminary communication[1] we described the catalytic oxidation of 2,6-disubstituted phenols with oxygen which gave as products high molecular weight polyphenylene ethers or diphenoquinones.

This is the first paper in a series and will be concerned with the scope of the reaction with regard to the oxidation of 2,6-disubstituted phenols.

EXPERIMENTAL

The general procedure, as carried out for the synthesis of poly(2,6-dimethyl-1,4-phenylene ether) was as follows. To a 250-ml. wide-mouthed Erlenmeyer flask equipped with a Vibromixer stirrer, oxygen inlet tube, and thermometer was added 135 ml. of pyridine and 1 g. (0.01 mole) of copper(I) chloride. Oxygen was passed through the vigorously stirred solution which becomes dark green in color, then 5 g. (0.041 mole) of 2,6-dimethylphenol was added. The reaction mixture became dark orange-brown and in 7 min. the temperature rose from 28°C. to 46°C., at which point the solution became viscous and the color of the solution returned to the original dark green. The reaction mixture was slowly added with stirring to 500 ml. of methanol, then filtered and washed thoroughly with methanol containing a small amount of hydrochloric acid. There remained an almost colorless, bulky solid which was dissolved in chloroform, filtered,

TABLE I

Oxidation of 2,6-Disubstituted Phenols Yielding Polymers

R_1	R_2	Solvent[a]	Yield, %[b]	M.W.[c]	$[\eta]$, dl./g.[d]	Formula	C, % Calc.	C, % Found	H, % Calc.	H, % Found
Methyl	Methyl	A	85	31,000	0.72	C_8H_8O	80.0	80.1	6.7	6.8
Methyl	Ethyl	B	82	25,400	0.40	$C_9H_{10}O$	80.6	80.8	7.5	8.0
Methyl	Isopropyl	A	62	15,350	0.24	$C_{10}H_{12}O$	81.0	81.4	8.2	8.3
Ethyl	Ethyl	B	81	32,000	0.53	$C_{10}H_{12}O$	81.0	80.7	8.2	8.2
Methyl	Chloro[e]	A	83	71,000	0.47	C_7H_5ClO	59.8	61.9	3.6	3.9
Methyl	Bromo[f]	A	18	—	0.03	C_7H_5BrO	45.5	58.0	2.7	2.5
Methyl	Methoxy	A	60	13,100	0.27	$C_8H_8O_2$	70.6	69.5	5.9	6.0
Methyl	Phenyl	A	60	—	—	$C_{13}H_{10}O$	84.8	82.6	6.6	5.1
Phenyl	Phenyl	B	46	—	0.05	$C_{18}H_{12}O$	88.5	88.4	5.0	5.2

[a] Solvents: A = pyridine, B = 22% pyridine + 78% nitrobenzene.
[b] Reprecipitated polymer.
[c] Osmotic MW in $CHCl_3$ at 25°C.
[d] In $CHCl_3$ at 25°C.
[e] Anal., Cl: calc., 25.2%; found, 21.7%.
[f] Anal., Br: calc., 45.4%; found, 26.0%.

and reprecipitated in methanol. After drying at 110°C. at 3 mm. Hg for 3 hr. the polymer weighed 4.2 g. (0.035 mole; 85% yield). It begins to soften at approximately 240°C. but does not melt up to 300°C.

ANAL. Calcd. for C_8H_8O: C, 80.0%; H, 6.7%. Found: C, 80.1%; H, 6.8%.

The polymer had an intrinsic viscosity of 0.72 dl./g. ($CHCl_3$, 25°C.).

Table I lists the other 2,6-disubstituted phenols that were oxidized in a similar manner. Table II lists those 2,6-disubstituted phenols which gave diphenoquinones as products. Table III contains the 2,6-disubstituted phenols which gave low molecular weight oils as products or did not oxidize.

TABLE II
2,6-Disubstituted Phenols Yielding Diphenoquinones

R_1	R_2	Yield, %	M.P., °C.
Methyl	*tert*-Butyl	45	217
Isopropyl	Isopropyl	53	225
tert-Butyl	*tert*-Butyl	97	246
Methoxy	Methoxy	74	>300

TABLE III
2,6-Disubstituted Phenols Giving Miscellaneous or No Reaction

R_1	R_2	Result
Methyl	Allyl	Low M.W. oil
Allyl	Allyl	Low M.W. oil
Chloro	Chloro	No reaction
Nitro	Nitro	No reaction

DISCUSSION AND RESULTS

Previous work on oxidation of simple monohydric phenols has generally yielded very complex products,[2] unless the phenol is substituted in both *ortho* positions. Thus Waters[3] found that 2,6-dimethylphenol on oxidation with alkaline ferricyanide gave only 3,3′,5,5′-tetramethyldiphenoquinone (50%) and a very low molecular weight yellow resin as products. Oxidation with Fenton's reagent gave the diphenoquinone and the corresponding dihydroxybiphenyl as well as a small amount of 2,6-dimethyl-1,4-dihydroxybenzene.[2] When the substituents are large and bulky, as in 2,6-di-*tert*-butylphenol, then the diphenoquinone can be obtained by using oxygen[4] in an alkaline medium as oxidizing agent.

Very little work has appeared on the catalytic oxidation of phenols with oxygen. Shibata[5] studied the oxidaselike action of some sixty complex salts of cobalt, nickel, copper, zinc, cadmium, chromium, iron, and silver using an alcoholic solution of myricetin (3,3′,4′,5,5′,7-hexahydroxyflavone) as indicator. The most active catalysts were those coordination compounds that were least stable hydrolytically, i.e., those that by hydrolysis would form aquo complexes and hydroxo complexes.[6] Acids were found to be inhibitors, as would be expected, since they reduce the concentration of the hydroxo complex.   Further work on pyrogallol,[7] and ascorbic acid[8] as substrates and on the decomposition of hydrogen peroxide[9] substantiated these conclusions.

Recently Brackman and Havinga[10] found the phenols are readily oxidized with copper(II) salts and oxygen in methanol containing a primary or secondary amine.   The product obtained from phenol with morpholine as the amine is IV (where R_2 is $[—(CH_2)_2]_2O—$).

They found that no reaction occurred in the presence of tertiary amines and demonstrated that hydrogen peroxide was a necessary intermediate in the reaction which proceeds via an initial hydroxylation of phenol to catechol. They postulated that the *ortho*-hydroxylation of phenol was effected by means of the copper (II) complex V (where X = Y = morpholine).

In 1955 Terent'ev and Mogilyanski[11] reported on the oxidation of primary aromatic amines to azo compounds with oxygen using as catalyst copper(I) chloride in pyridine.   The reaction occurs readily at room temperature and probably proceeds as follows:

$$2\ ArNH_2 \rightarrow 2\ ArNH\cdot \rightarrow ArNHNHAr \rightarrow ArN{=}NAr$$

The function of the oxygen is simply to raise the copper(I) salt to the divalent state, and they found that the catalyst absorbed one mole of oxygen per four moles of copper.

Kinoshita[12] has also studied this reaction and postulates that the actual oxidant is a hydroxy cupric chloride pyridine complex formed by autoxidation of copper(I) chloride in pyridine.   He had previously found[13,14] that benzoin was oxidized to benzil and thence to benzoic acid with this catalyst system.   However, with copper(II) chloride as catalyst, benzoin was converted almost quantitatively to benzil; no benzoic acid being obtained.

Recently Havinga[15] found the air oxidation of analine in the presence of copper(II) acetate to be a complex reaction in contrast to the results obtained with copper(I) chloride.   As in the oxidations of phenol that he had previously studied, the first step in the reaction appeared to be an *ortho*-hydroxylation.

Phenols are very similar in reactivity to aromatic amines and the first step in the oxidation of a phenol is generally considered to be formation of an aryloxy radical.[2,3]   Thus, it appeared probable that catalysts such as these would be effective for the oxidation of phenols.   Indeed, when phenol itself was treated with oxygen in pyridine solution in the presence of copper-(I) chloride a reaction occurred; however, only a complex tarry residue was formed.

The phenoxy radical is a resonance hybrid of the following structures:[16]

Carbon–carbon coupling of radicals can occur at the *ortho* and *para* positions as well as carbon–oxygen coupling.   Furthermore, the initial products formed will react further on oxidation.   Therefore, it is not surprising that oxidation of phenol yields only a complex mixture.

When we oxidized 2,6-dimethylphenol (I; $R_1 = R_2 = CH_3$) with oxygen in pyridine solution in the presence of copper(I) chloride we obtained a high molecular weight polyphenylene ether (II; $R_1 = R_2 = CH_3$) and a small amount of a diphenoquinone (III; $R_1 = R_2 = CH_3$) as sole products.

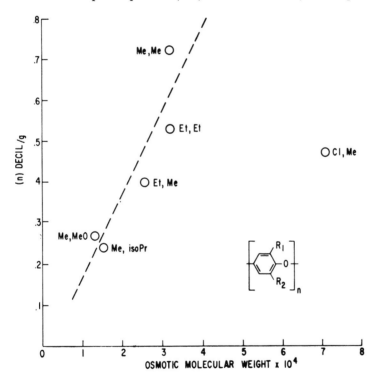

Fig. 1. Intrinsic viscosity–molecular weight relationships for 2,6-disubstituted-1,4-phenylene ethers.

586   A. S. HAY

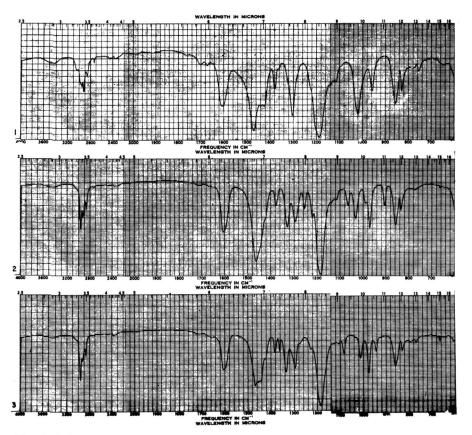

Fig. 2. Infrared spectra of 2,6-disubstituted-1,4-phenylene ethers (II): (1) $R_1 = R_2 = CH_3$; (2) $R_1 = CH_3$, $R_2 = CH_2CH_3$; (3) $R_1 = CH_3$, $R_2 = CH(CH_3)_2$; (4) $R_1 = R_2 = CH_3$; (5) $R_1 = CH_3$, $R_2 = Cl$; (6) $R_1 = CH_3$, $R_2 = OCH_3$.

The reaction is very exothermic and extraordinarily rapid, being over in a few minutes at room temperature if the reaction mixture is vigorously stirred. Under certain conditions, to be described later, colorless polymers with intrinsic viscosities as high as 3.4 dl./g. (chloroform, 25°C.) which is equivalent to a molecular weight (osmotic) of approximately 150,000 have been obtained. The polymer is soluble in chlorinated hydrocarbons such as chloroform and s-tetrachloroethane and aromatic solvents such as nitrobenzene and toluene, and evaporation of a solution of the polymer leaves a transparent, tough, flexible film. The polymer has a softening temperature of about 250°C.

This was an extraordinary result because of the extreme facility of the reaction and because it represents not only the first synthesis of a high molecular weight polyphenylene ether but the first example of a polymerization that occurs by an oxidative coupling reaction utilizing oxygen as the oxidizing agent.

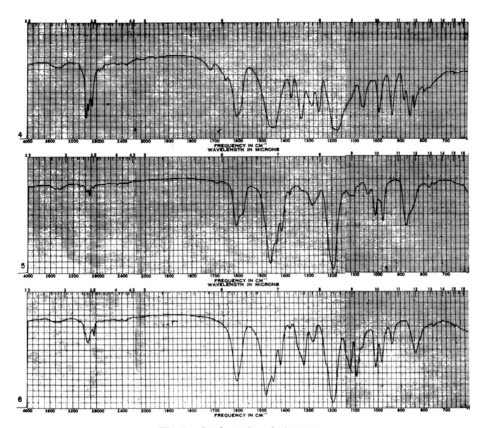

Fig. 2.   See legend on facing page.

Süs[17] attempted to make the parent polymer (II; R = H) by photolysis of benzene-1,4-diazooxide and obtained an insoluble and infusible product. Dewar[18] attempted the thermal decomposition of 2,6-dibromo-1,4-diazooxide but did not obtain the polyether (II; R = Br).   In an extension of early work reported by Hunter and co-workers,[19] Staffin and Price[20] obtained polymers with molecular weights in the range 2,000–10,000 by treatment of 4-bromo-2,6-dimethylphenol with various oxidizing agents.   The infrared spectrum of this polymer is almost identical to that of the polymer we have obtained by the oxidative polymerization of 2,6-dimethylphenol.[20] Very recently, Price has succeeded in making higher molecular weight polymers by a modification of the method he earlier described.[21]

Lindgren has very recently synthesized a low molecular weight polymer by oxidation of 2,6-dimethylphenol with silver(I) oxide.[22]

A large number of other 2,6-disubstituted phenols have been oxidized in a similar fashion, and these results are summarized in Tables I, II, and III.

It can be seen (Tables I–III) that polymer formation readily occurs only if the substituent groups are relatively small and not too electronegative.

When the substituents are large and bulky (Table II), as in 2,6-di-*tert*-butylphenol, diphenoquinones are formed by a tail-to-tail coupling. The head-to-tail coupling, which would give polymer, is precluded on the basis of steric hindrance. One *tert*-butyl group is enough to force the reaction toward diphenoquinone formation, whereas two isopropyl groups are necessary. Phenols with substituents of intermediate size, such as 2-ethyl-6-methylphenol, 2,6-diethylphenol, and 2-methyl-6-isopropylphenol, are readily oxidized to high molecular weight polymers.

Oxidation of 2-chloro-6-methylphenol gives a polymer; however, the reaction is considerably slower and higher temperatures are necessary. No appreciable reaction occurs when 2,6-dichlorophenol or 2,6-dinitrophenol is oxidized, even at 100°C. This is not surprising, since the redox potential of these phenols is considerably higher because of the electronegative substituents; thus, they would be expected to be considerably more difficult to oxidize. Polymers obtained by oxidation of 2-chloro-6-methylphenol are apparently somewhat branched. Elemental analyses show that up to 15% of the chlorine in the original phenol is lost during the oxidation.

Table I gives the intrinsic viscosity and molecular weight data for six polymers from different 2,6-disubstituted phenols. (Detailed data on osmotic molecular weight-viscosity relationships will be presented in a subsequent paper.) A plot of intrinsic viscosity versus molecular weight (Fig. 1) for these polymers shows that data for all of these polymers, with the exception of those for 2-chloro-6-methylphenol, fall on a straight line. The conclusion is that the polymer from 2-chloro-6-methylphenol is branched, whereas the others are essentially linear. Thus, 2-chloro-6-methylphenol is acting partially as a trifunctional monomer by reacting through the 2-position by elimination of chlorine. A structure such as VI would then account for branching in the polymer. The oxidation of

VI

2-bromo-6-methylphenol is much more complex. Only a very small amount of material insoluble in methanol was obtained. This means that a greater proportion of the reaction took place at the 2-position, and this agrees with the expected greater activity of bromine compared to that of chlorine in a reaction of this type. Dewar[18] and Price[20] have both shown that bromine is very readily displaced by a phenoxy radical.

Oxidation of 2-allyl-6-methylphenol and 2,6-diallylphenol also gave only low molecular weight oils as products. Fujisaki[23] has shown that oxidation of 2,6-di-*tert*-butyl-4-methylphenol with potassium ferricyanide gives 3,3″,5,5′-tetra-*tert*-butylstilbenequinone and indicates that the hydrogen atom is first removed from the hydroxyl and that dimerization probably proceeds through 2,6-di-*tert*-butylmethylenequinone. In the present cases, a reaction might also be expected to occur through the methylene groups, and this could explain the formation of complex products.

The infrared spectra (KBr pellets) of polymers from six different 2,6-disubstituted phenols (2,6-dimethyl-, 2-ethyl-6-methyl-, 2-isopropyl-6-methyl-, 2,6-diethyl-, 2-chloro-6-methyl-, and 2-methoxy-6-methylphenol) are recorded in Figure 2. There are striking similarities in these spectra, as would be expected for polymers of such similar structure. (The absorption at ca. 3450 cm.$^{-1}$ is due to water.)

In the four polymers from 2,6-dialkylphenols there are weak doublets at 1663 ± 10 cm.$^{-1}$ and 1710 ± 10 cm.$^{-1}$ which are apparently overtones of out-of-plane CH frequencies characteristic of a tetrasubstituted aromatic.[24] The relatively strong absorption in all these polymers at 855 ± 20 cm.$^{-1}$ and for the first four at 835 ± 5 cm.$^{-1}$ confirms this tetrasubstitution.[24]

All six spectra show the characteristic C=C stretching vibrations at 1607 ± 4 cm.$^{-1}$ and 1470 ± 20 cm.$^{-1}$ and the very strong absorption at 1188 ± 8 cm.$^{-1}$ due to the C—O absorption.

All of the polymers show absorption at 984 ± 9 cm.$^{-1}$ (C—H in-plane deformation?), 1292 ± 13 cm.$^{-1}$, and 1379 ± 4 cm.$^{-1}$ (aliphatic C—H bending).

The absence of carbonyl absorption and, in general, the relative simplicity of the spectra points to a 1,4-polyether structure. Further structure proof for the linear 1,4 structure of the polymer from 2,6-dimethylphenol will be presented in a later paper. This will include chemical evidence and physical evidence such as viscosity–molecular weight data and comparison of infrared spectra with model compounds.

It is a pleasure to acknowledge the very capable assistance of Mrs. B. M. Boulette and Mr. R. J. Flatley. Many of the phenols were prepared by Dr. J. R. Ladd.

References

1. Hay, A. S., H. S. Blanchard, G. F. Endres, and J. W. Eustance, *J. Am. Chem. Soc.*, **81**, 6335 (1959). See also A. S. Hay, French Pat. 1,234,336 (May 16, 1960).

2. For a bibliography see S. L. Cosgrove and W. A. Waters, *J. Chem. Soc.*, **1951**, 1726.

3. Haynes, C. G., A. H. Turner, and W. A. Waters, *J. Chem. Soc.*, **1956**, 2823.

4. Kharasch, M. S., and B. S. Joshi, *J. Org. Chem.*, **22**, 1439 (1957).

5. Shibata, Y., and K. Shibata, *J. Tokyo Chem. Soc.*, **41**, 35 (1920); *Chem. Absts.*, **14**, 2590 (1920).

6. For a discussion on hydrolysis of complex salts, see F. Basolo in *Chemistry of the Coordination Compounds*, Reinhold, New York, 1956, p. 425.

7. Shibata, Y., and H. Kaneko, *J. Chem. Soc. Japan*, **43**, 833 (1922).

8. Wasaki, T. I., *J. Chem. Soc. Japan*, **63**, 820 (1942).

9. Shibata, Y., and H. Kaneko, *J. Chem. Soc. Japan*, **44**, 166 (1923).

10. Brackman, W., and E. Havinga, *Rec. trav. chim.*, **74**, 1107 (1955) and preceding articles.

11. Terent'ev, A., and R. D. Mogilyanskiǐ, *Doklady Akad. Nauk. S.S.S.R.*, **103**, 91 (1955).

12. Kinoshita, K., *Bull. Chem. Soc. Japan*, **32**, 777 (1959).

13. Kinoshita, K., *J. Chem. Soc. Japan, Pure Chem. Sect.*, **75**, 48 (1954).

14. Kinoshita, K., *J. Chem. Soc. Japan, Pure Chem. Sect.* **75**, 173 (1954).

15. Engelsma, G., and E. Havinga, *Tetrahedron*, **2**, 289 (1958).

16. Muller, E., K. Ley, and W. Kiedaisch, *Ber.*, **87**, 1605 (1954).

17. Süs, O., K. Muller, and H. Heiss, *Ann.*, **598**, 123 (1956).

18. Dewar, M. J. S., and A. N. James, *J. Chem. Soc.*, **1958**, 917.

19. Hunter, W. H., and M. A. Dahlen, *J. Am. Chem. Soc.*, **54**, 2459 (1932) and preceding papers.

20. Staffin, G. D., and C. C. Price, *J. Am. Chem. Soc.*, **82**, 3632 (1960).

21. Price, C. C., private communication.

22. Lindgren, B. O., *Acta Chem. Scand.*, **14**, 1203 (1960).

23. Fujisaki, T., *Nippon Kagaku Zasshi*, **77**, 869 (1956).

24. See L. J. Bellamy, *The Infrared Spectra of Complex Molecules*, Wiley, New York, 1958, pp. 68–79.

Synopsis

Many 2,6-disubstituted phenols are readily oxidized with oxygen, a tertiary amine and a copper (I) salt being used as catalyst. The products are diphenoquinones (I) or 1,4-polyphenylene ethers (II):

The effect of variation of the substituents on the phenol and the infrared spectra of the polymers is discussed. Amine complexes of copper salts are the only effective catalysts found for the polymer-forming reaction. The effect on the catalyst of varying the copper salt and the ligand is discussed.

Résumé

Nous avons trouvé que beaucoup de phénols-2.6-disubstitués sont facilement oxydés par l'oxygène en utilisant comme catalyseur une amine tertiaire et un sel de cuivre (I). Les produits de réaction sont des diphénoquinones (I) ou des éthers de 1.4-polyphénylène (II) (v. le résumé anglais). L'influence de la nature des substituants du phénol est étudiée, et les spectres infra-rouges des polymères discutés. Les complexes aminés des sels de cuivre sont les seuls catalyseurs efficients pour la formation de polymères. L'effet sur le catalyseur de la nature du sel de cuivre et du liant sera discuté ultérieurement.

Zusammenfassung

Wir haben gefunden, dass viele 2,6-disubstituierte Phenole durch Sauerstoff unter der katalytischen Einwirkung eines tertiären Amins und eines Kupfer-I-salzes leicht oxydiert werden. Die Oxydationsprodukte sind Diphenochinone (I) oder 1,4-Polyphenylenäther (II) (siehe englische Zusammenfassung). Der Einfluss einer Variierung der Substituenten am Phenol sowie die Infrarorspektren der Polymeren werden diskutiert. Amikomplexe von Kupfersalzen sind die einzigen, bis jetzt aufgefundenen, wirksamen

POLYMERIZATION BY OXIDATIVE COUPLING. II   591

Katalysatoren für die Bildungsreaktion des Polymeren.   Der Einfluss einer Variierung des Kupfersalzes und des Linganden auf den Katalysator wird diskutiert.

Discussion

H. Z. Lecher (*Plainfield, N. J.*): Is the oxidizing-polymerizing agent formed in pyridine from CuCl and O_2 cuprous oxychloride or is it an intermediate peroxide?

A. S. Hay: In pyridine solution the CuCl absorbs approximately 0.25 moles of oxygen, and this solution, purged of oxygen, will oxidize 0.5 moles of 2,6-dimethylphenol to high polymer.   We do not know the exact structure of the oxidized species.

J. Nazy (*Central Research Laboratory, General Mills, Inc., Minneapolis, Minn.*): Would you give state of the physical and chemical properties of these polymers?

A. S. Hay: A polymer from 2,6-dimethylphenol with intrinsic viscosity 0.97 (dec./g./ $CHCl_3$ at 25°C.; osmotic molecular weight, 29,000) has a density of 1.06, a softening temperature of about 250°, a tensile strength of 9000 psi, and an elongation at break of 80%.

R. W. Lenz (*Dow Chemical Company, Eastern Research Laboratory, Framingham, Mass.*): *1.* What are the thermal stabilities of these polymers? *2.* Do you obtain essentially all head-to-tail addition? *3.* Are the polymers crystalline?

A. S. Hay: *1.* In a Chevenard thermobalance under nitrogen and at a heating rate of 150°/hr. a polymer from 2,6-dimethylphenol showed a sharp loss in weight around 400°. *2.* Minor amounts of tail-to-tail coupling, giving the diphenoquinone, usually occur. *3.* The polymers obtained from the reaction mixture are generally amorphous but can be crystallized.

O. D. Trapp (*Westinghouse Electric Corporation, Pittsburgh, Pa.*): What conditions are required to vary the molecular weight from 30,000 to above 100,000?.

I would expect that it is essential to match the ionization potentials of the phenol and the metal–amine catalyst.   Further, by careful choice of the metal–amine complex it should be possible to oxygen-couple any desired phenol.   Do you find this to be the case?

A. S. Hay: In pyridine solution, polymers with molecular weights in the range 20,000–30,000 are obtained; however, in nitrobenzene–pyridine, polymers with molecular weights of about 100,000 are obtained.

J. E. Franz (*Monsanto Chemical Company, St. Louis, Mo.*): *1.* In general, do 2,6-disubstituted phenols, other than 2,6-dimethylphenol, form polymers in high conversion? *2.* What catalysts other than Cu_2Cl_2 are effective for this type of polymerization? *3.* Have copolymers been prepared by your elegant method?

A. S. Hay: *1.* Yes. *2.* Other copper(I) salts, such as the bromide and acetate. *3.* Yes.

COMMENTARY

Reflections on "General Theory of Stationary Random Sequences with Applications to the Tacticity of Polymers," by Bernard D. Coleman and Thomas G Fox, *J. Polym. Sci. A*, 1, 3183 (1963)

BERNARD D. COLEMAN

Department of Mechanics and Materials Science, Rutgers University, Piscataway, New Jersey 08855-0909

I met Thomas G Fox for the first time in early July 1957, on the day of my arrival in Pittsburgh to start as a new appointee of the Mellon Institute. Tom arrived that same day. He was a Staff Fellow, which meant that he reported directly to Paul Flory, the Scientific Director of the Institute. I was a Senior Fellow (but not in age) and was a member of the new Polymer Group which Tom administered. For the next 20 years I received, from a wise and magnanimous friend, valuable advice and unstinting support in all my scholarly efforts.

I recall, as if it were yesterday, sitting with Tom in the coffee shop of the old Webster Hall Hotel a few days later and discussing his research interests. With his colleagues at the Rohm and Haas Company in Philadelphia, Tom had been working on the problem of controlling the stereoregularity of polymers resulting from polymerizations of α-olefins in homogeneous media. [The then available methods[1,2] of preparing stereoregular poly-(α-olefins) used heterogeneous catalysis.] Tom told me at that time that there was good reason to believe that not all "atactic" polymers prepared from a given monomer by free-radical polymerization were the same in their diastereosequence distributions. He mentioned that in 1944 Huggins[3] had suggested the temperature of polymerization should affect the diastereosequence distribution and that, in experiments on the free-radical polymerization of methyl methacrylate, Tom and colleagues at Rohm and Haas had observed that the lower the polymerization temperature, the greater is the tendency of the resulting polymer to be syndiotactic. (A paper[4] on the experiments came out 1958).

The things I learned in that discussion made a strong impression on me. I became immersed in the problem of finding mathematical concepts with which one could describe the diastereosequence distributions of atactic polymers. It was clear from the beginning that what mattered was not the absolute chirality (*d* or *l*) of an asymmetric carbon, but rather the relation of its chirality to that of the previous asymmetric carbon: did they form a repeating pair (*dd* or *ll*), that is, were their side groups in *isotactic placement,* or did they form an alternating pair (*dl* or *ld*) with their side groups in *syndiotactic placement?* I concluded that the diastereosequence distributions formed in free-radical polymerizations were governed by simple Markov chains or Bernoulli trials (terms from probability theory that are discussed below). Tom agreed with me and encouraged me to publish my thoughts. Paul Flory suggested that in my article I concentrate on the relation between the freezing point of a polymer and an easily visualized measure of the polymer's tacticity. The result was a paper[5] that appeared in the same issue of this journal as the important, although brief, paper[4] on the free-radical polymerization of methyl methacrylate that Tom wrote with his Rohm and Haas colleagues, and which I have mentioned above.

In our 1963 paper here reprinted, $p_1\{I\}$ is the probability that a placement selected at random along the macromolecule is isotactic, and $p_1\{S\}$ the probability that it is syndiotactic. We write $p_{1|1}\{I|S\}$ for the probability that a placement is isotactic given that the previously formed placement is syndiotactic,

$p_{1|1}\{I|I\}$ for the conditional probability that it is isotactic given that the previously formed placement is isotactic, $p_{1|2}\{I|SS\}$ for that probability given that the two previously formed placements are syndiotactic, etc. If no matter how much information we are given about the previous placements, the probability of a placement being isotactic is equal to $p_1\{I\}$ (and hence, in particular, $p_1\{I\} = p_{1|1}\{I|I\}$ = $p_{1|1}\{I|S\} = p_{1|2}\{I|SS\}$ etc.), then one says that the diastereosequence distribution is governed by *Bernoulli* trials. If information about the N immediately preceding placements affects the probability that a given placement is isotactic, but information about placements that occurred before those N placements has no influence on that probability, then the diastereosequence distribution is said to be governed by a *Markov chain of order* N. A Markov chain of order 1 is called a *simple Markov chain*. A diastereosequence distribution not governed by a Markov chain of any finite order is said to be *non-Markovian*.

Methyl methacrylate is the first monomer for which it was shown that homogeneous anionic polymerization initiated by organic derivatives of Group I metals can be stereospecific and can, under appropriate circumstances, yield "stereoblock copolymers."[6,7] A year or two after Refs. 4 and 5 appeared, Tom brought it to my attention that work, done by a group at Rohm and Haas which was the direct descendent of his old group and after Tom's departure was directed by Donald Glusker, strongly indicated that, in contrast to the (also homogeneous) free radical case, homogeneous anionic polymerization of methyl methacrylate can be governed by a two-state mechanism.[8,9] We decided to investigate the mathematical implications of the assumption that in homogeneous anionic polymerizations "the reactive end of a growing polymer can have several, say N, states $\{1\}, \{2\}, \ldots, \{N\}$ which are in dynamic equilibrium and that each such state is capable of adding monomer with its own rate and stereospecificity." [The quote is from Ref. 10; the case of principal interest is that in which $N = 2$.] The details of our analysis of the relation of diastereosequence distributions to the various rate constants appearing in such multistate mechanisms may be found in Ref. 11. That analysis showed that when N is 2 or greater the resulting diastereosequence distribution is, in general, non-Markovian. (Although one cannot determine from NMR spectra alone whether a diastereosequence distribution is non-Markovian, it was later[12] shown that such spectra can be employed to determine whether such a distribution is governed by a *simple* Markov chain, and experimental confirmation, by NMR, that diastereosequences arising from homogeneous anionic polymerizations need not obey the statistics of simple Markov chains was then obtained.[13])

Once it was clear that diastereosequence distributions need not be Markovian of any order, there was a need for a theory of sufficient generality to permit one to determine which procedures used to check the self-consistency of NMR data and to obtain from such data the mean length of closed diastereosequences are valid for both Markovian and non-Markovian distributions. The development of such a theory was the goal of the paper here reprinted.

REFERENCES AND NOTES

1. C. E. Schildknecht, S. T. Gross, H. R. Davidson, T. M. Lambert, and A. D. Zoss, *Ind. Eng. Chem.*, **41**, 2104 (1948).
2. G. Natta, P. Pino, P. Corradini, F. Danusso, E. Mantica, G. Mazzanti, and G. Moraglio, *J. Am. Chem. Soc.*, **77**, 1708 (1955).
3. M. L. Huggins, *J. Am. Chem. Soc.*, **66**, 1991 (1944).
4. T. G. Fox, W. E. Goode, S. Gratch, C. M. Huggett, J. F. Kincaid, A. Spell, and J. D. Stroupe, *J. Polym. Sci.*, **31**, 173 (1958).
5. B. D. Coleman, *J. Polym. Sci.*, **31**, 155 (1958).
6. T. G Fox, W. E. Goode, S. Gratch, C. M. Huggett, J. F. Kincaid, A. Spell, and J. D. Stroupe, *J. Am. Chem. Soc.*, **77**, 1768 (1958).
7. F. J. Glavis, *J. Polym. Sci.*, **36**, 547 (1959).
8. D. L. Glusker, E. Stiles, and B. Yoncoskie, *J. Polym. Sci.*, **49**, 297 (1961).
9. D. L. Glusker, I. Lysloff, and E. Stiles, *J. Polym. Sci.*, **49**, 315 (1961).
10. B. D. Coleman and T. G. Fox, *J. Polym. Sci. C*, **4**, 345 (1963).
11. B. D. Coleman and T. G Fox, *J. Chem. Phys.*, **38**, 1065 (1963).
12. M. Reinmöller and T. G Fox, *Polymer Division Preprints*, Vol. 7, No. 2, 152nd National Meeting of the American Chemical Society, September 1966, pp. 987–999.
13. B. D. Coleman, T. G Fox, and M. Reinmöller, *J. Polym. Sci., Polym. Lett.*, **4**, 1029 (1966).

PERSPECTIVE

Comments on "General Theory of Stationary Random Sequences with Applications to the Tacticity of Polymers," Bernard D. Coleman and Thomas G Fox, *J. Polym. Sci. A,* 1, 3183 (1963)

ROBERT M. WAYMOUTH

Stanford University, Stanford, California 94305

The subtleties of macromolecular stereochemistry have intrigued experimentalists and theorists ever since the pioneering work of Schildknecht[1] and Natta[2] on the synthesis of stereoregular polymers. The allure of macromolecular stereochemistry derives from the fact that a polymer containing n stereogenic centers contains 2^{n-1} possible diastereosequences. As with many problems in polymer science, advances in macromolecular stereochemistry required an appropriate statistical theory. The development of statistical analyses of diastereosequence distribution in polymer chains played a critical role in advancing the field of stereospecific polymerization chemistry. The analysis of diastereosequence distribution (or polymer microstructure) provides information that can be correlated to physical properties. Analysis of the microstructure also provides a wealth of information on the polymerization mechanism because the microstructure of a polymer is a stereochemical record of the process of enchainment. In a series of papers in the late 1950s and early 1960s, Coleman and Fox provided the theoretical foundation for the development of modern statistical analysis of polymer microstructure.

Coleman was among the first to appreciate and prove that mathematical treatments of statistical Bernoullian trial processes and Markov chains could be applied to the problem of diastereosequence distribution in polymer chains.[3] Prior to this, atactic polymers were just atactic; there was no means of assessing the degree of randomness in a polymer that was neither highly isotactic or syndiotactic. The

significance of Coleman and Fox's approach in applying these statistical treatments was that the degree or type of randomness provided information about the polymerization mechanism. These concepts were critical in the evolution of our understanding of anionic and radical polymerization processes. For example, Coleman predicted in 1958[3] that the stereochemistry of enchainment in radical polymerization processes should obey Bernoullian trial statistics; this was elegantly confirmed by Bovey in 1960 by NMR spectroscopy.[4]

The application of Bernoullian and Markovian statistical treatments to polymer diastereosequence distribution is an extraordinarily powerful theoretical approach. However, Coleman and Fox appreciated that Bernoullian trial processes or Markovian chains might not explain all polymerization processes or all types of diastereosequence distributions in polymer chains.[5] In their paper in the *Journal of Polymer Science* in 1963, they developed a statistical theory of diastereosequence distributions that encompasses situations under which Bernoullian and Markovian statistics are applicable, but which is more general. This theory provided equations for describing the average isotactic (or syndiotactic) sequence length for *any* polymer chain (here presented using both Coleman's[6] and Bovey's[4] nomenclature):

mean length of isotactic sequences = $\mu\{I\}$

$$= \frac{p_1\{I\}}{p_1\{S\} - p_2\{SS\}} = \frac{[m]}{[r] - [mm]}$$

$$\text{or } \mu\{I\} = \frac{p_1\{I\}}{p_2\{IS\}} = \frac{[m]}{[mr]}$$

These developments anticipated Randall's derivations[7] of equations for diastereosequence distributions in polymer chains. Coleman and Fox's theoretical treatment also provided a means of assessing whether a particular diastereosequence distribution conformed to Bernoullian and Markovian statistics. This was important because there are many polymerization systems which do not obey Bernoullian or Markovian statistics, in particular the stereospecific Ziegler–Natta catalysts.[8] Coleman and Fox's multistate mechanism for the synthesis of block polymers, described in the same year,[5] is another fascinating situation under which neither Bernoullian or Markovian statistics are applicable.

A major contribution of their theoretical approach[3,6] was the choice of relative stereochemistry (isotactic, syndiotactic, and heterotactic placements of adjacent stereocenters) rather than absolute stereochemistry as the basis for the theoretical development. Coleman was among the first to appreciate the limited applicability of absolute stereochemistry (*d* or *l* nomenclature) in describing polymer microstructure; relative, not absolute stereochemistry is the most important determinant of physical properties (such as melting point).[3] The propitious choice of relative stereochemistry considerably simplified the mathematics (compare Price's treatment of Markovian statistics of polymer chains using absolute stereochemistry)[9] and provided a theoretical treatment that could be directly tested against experimental data that was just emerging from studies of polymers by high resolution NMR spectroscopy.[4] The concept of using relative stereochemistry was most elegantly manifested in Bovey's simple but powerful *m/r* nomenclature to describe the microstructure of vinyl polymers.

The theoretical treatment by Coleman and Fox[3,5,6] and others,[7-10] coupled with the experimental application of high resolution NMR spectroscopy by Bovey,[4,10] revolutionized our understanding of polymer microstructure. These developments provided critical tools to study the relationship of polymer structure to polymer properties and the mechanism of polymerization through analysis of diastereosequence distribution in vinyl polymers.

REFERENCES AND NOTES

1. C. E. Schildknecht, S. T. Gross, H. R. Davidson, T. M. Lambert, and A. D. Zoss, *Ind. Eng. Chem.,* **41,** 2104 (1948).
2. G. Natta, P. Pino, P. Corradini, F. Danusso, E. Mantica, G. Mazzanti, and G. Moraglio, *J. Am. Chem. Soc.,* **77,** 1708 (1955).
3. B. D. Coleman, *J. Polym. Sci.,* **31,** 155 (1958).
4. F. A. Bovey and G. V. D. Tiers, *J. Polym. Sci.,* **44,** 173 (1960).
5. B. D. Coleman and T. G Fox, *J. Chem. Phys.,* **38,** 1065 (1963).
6. B. D. Coleman and T. G Fox, *J. Polym. Sci., A,* **1,** 3183 (1963).
7. J. C. Randall, *Polymer Sequence Determination, Carbon-13 NMR Method,* Academic Press, New York, 1977, p. 138.
8. R. A. Sheldon, T. Fueno, T. Tsunetsuga, and J. Furukawa, *J. Polym. Sci.,* **B3,** 23 (1965).
9. F. P. Price, *J. Chem. Phys.,* **36,** 209 (1962).
10. H. L. Frisch, C. L. Mallows, and F. A. Bovey, *J. Chem. Phys.,* **45,** 1565 (1966).

JOURNAL OF POLYMER SCIENCE: PART A   VOL. 1, PP. 3183–3197 (1963)

General Theory of Stationary Random Sequences with Applications to the Tacticity of Polymers*

BERNARD D. COLEMAN and THOMAS G FOX, *Mellon Institute, Pittsburgh, Pennsylvania*

Synopsis

When, in a poly-α-olefin, the probability that a given placement be isotactic depends upon the tacticity of only a finite number of immediate predecessors, the resulting diastereosequence distribution obeys the theory of Markoff chains. When this is not the case, one says that the resulting diastereosequence distribution is non-Markoffian. A special case of a Markoffian distribution is given by a *simple* Markoff chain in which the tacticity of a given placement is assumed to be affected by only the tacticity of the immediately preceding placement. Another special case is, of course, the Bernoulli trial distribution in which the probability that a given placement be isotactic is independent of the tacticity of all other placements. A high resolution NMR spectrum can sometimes yield a quantitative determination of the concentrations of isotactic and syndiotactic placements and the concentrations of the three types of possible adjacent pairs of such placements (i.e., isotactic, syndiotactic, and heterotactic pairs). When this is the case, the spectrum can be used to determine whether or not a given diastereosequence distribution is Bernoullian. However, because the longest diastereosequences whose concentration can be measured by NMR spectroscopy involve only two placements, an NMR spectrum cannot check whether a given non-Bernoullian distribution be simple Markoffian or Markoffian in general. In fact, non-Markoffian distributions are compatible with existing NMR spectra on polymers prepared by anionic polymerizations. In this paper we work within the framework of Kac's theory of stationary statistical processes and point out some general results which are valid for both Markoffian and non-Markoffian processes. The results are applied to NMR spectroscopy and it is pointed out which calculations used to check the self-consistency of NMR data and to obtain the mean length of closed diastereosequences are valid for both Markoffian and non-Markoffian distributions.

Introduction

Let us consider a polymer molecule formed by head-to-tail addition polymerization of an unsymmetrical α-olefin. Assuming that the end-groups of this molecule are distinguishable, every second carbon atom in its principal (backbone) chain is an asymmetric carbon atom. We number these asymmetric carbon atoms in the order in which they were added during polymerization and use the following notation:[1] if the mth and $(m + 1)$th asymmetric chain atoms have the same stereoconfiguration then we say that the mth *placement* of our polymer molecule is *isotactic;* if these

* Paper presented at the 142nd Meeting of the American Chemical Society, Atlantic City, New Jersey, September 1962.

167

two asymmetric atoms have opposite stereoconfigurations we say that the *m*th *placement* is *syndiotactic*. The property of being isotactic or syndiotactic is called "tacticity."

We define a *diastereosequence* of length *n* to be an ordered set of *n* consecutive placements, say, the *m*th, (*m* + 1)th, . . . , (*m* + *n* − 1)th placements which express, respectively, the stereorelationship between the *m*th and the (*m* + 1)th, the (*m* + 1)th and (*m* + 2)th, . . . , the (*m* + *n* − 1)th and the (*m* + *n*)th asymmetric chain atoms. Thus, a diastereosequence of length *n* involves *n* + 1 monomer units and a poly-α-olefin of degree of polymerization *D* constitutes a diastereosequence of length *D* − 1.

There are 2^{D-1} possible distinct diastereosequences of length *D* − 1: The first placement may be isotactic or syndiotactic, in either case the second placement may be isotactic or syndiotactic, and so on, for *D* − 1 times. Usually, each of these 2^{D-1} stereosequences constitutes a distinct diastereoisomer of our polymer. If the end groups are not distinguishable, the number of distinct diastereoisomers may be less than 2^{D-1}, but it is always of this order of magnitude.

Since *D* is often greater than a thousand or even ten thousand, if there is the slightest amount of randomness involved in the formation of successive placements, there is little hope of isolating in pure form every diastereoisomer resulting from a given polymerization, even when the molecular weight distribution is perfectly sharp. Furthermore, when the molecular weight is high and formation of successive placements is governed by processes which are strongly random (for example, by a process which can be represented by a game in which the tacticity of each placement is determined by an independent toss of a not-too-biased coin, i.e., by Bernoulli trials) so many distinct diastereoisomers of nearly equal probability are possible that most of them must be absent from any sample of reasonable size and the concentration of those present can hardly exceed one molecule per sample.

The situation just described is the rule rather than the exception for free radical polymerizations;[1-4] for anionic polymerizations the coin-tossing analogy breaks down, but there is usually still sufficient randomness to make the general conclusion valid.

Thus we see that even if we ignore dispersion in molecular weight, the stereochemistry of addition polymers must differ markedly in concepts and techniques from classical stereochemistry.

The sort of questions which are meaningful here are questions of the following types. What is the probability that the diastereosequence of length *n* which runs from the *m*th through the (*m* + *n* − 1)th placement has a particular pattern (say, contains only isotactic placements or consists of an alternating sequence of isotactic and syndiotactic placements)? What is the mean number of isotactic placements between successive syndiotactic placements? This second question is related to questions of "mean recurrence times."[5,6]

There is a large mathematical literature dealing with questions of these types, but, with one notable exception,[5,6] (unfortunately, unknown to us until the present investigation was completed) it is restricted to cases in which the sequences under consideration are generated by either Bernoulli trials or by Markoff chains. True, for *free radical polymerizations* it has been proposed,[1,2] on the basis of experience with the magnitude of penultimate effects in copolymer polymerization kinetics, that to a high approximation diastereosequences should be generated by Bernoulli trials, and this has recently been confirmed, for methyl methacrylate, by high resolution nuclear magnetic resonance spectroscopy.[3] However, NMR spectra indicate that polymers prepared by anionic polymerizations do not obey Bernoulli trial statistics,[3,7] and there is no evidence supporting Markoff chain statistics for those polymers. A recent theoretical study of plausible mechanisms indicates that non-Markoffian distributions may occur frequently in anionic polymerizations.[8] Thus, if we are to have a theory in which we can discuss the questions of interest in sufficient generality to insure applicability of our results to all polymerizations of α-olefins, including anionic polymerizations, then we must leave behind Bernoullian and Markoffian simplifications and examine afresh the theory of random sequences.

The terms "Bernoullian," "Markoffian," and "non-Markoffian" will be precisely defined below. For the present we note that since an NMR spectrum determines the concentrations of only the diastereosequences which have length less than three it cannot be used to check for statistical interactions extending beyond two placements. NMR data alone can rule out Bernoullian statistics but cannot establish Markoffian statistics.

Fortunately, a diastereosequence is a binary sequence; there are only two possibilities for each component element: a given element can be either an isotactic placement or a syndiotactic placement. When this fact is combined with but one statistical assumption, that of stationarity, a rich general theory is obtained in which much can be said without further assumptions.

To explain the chemical significance of statistical stationarity, let us note that since we are here interested in polymers of high molecular weight it is reasonable to neglect end effects. One part of such neglect is the assumption that the finite diastereosequences mentioned in our questions are always finite pieces of random diastereosequences of infinite length: this is the assumption that every placement has both successors and predecessors. Related to the physical notion that end effects should be unimportant is the assertion that the probability that the mth placement of a polymer is isotactic (given no information about the tacticity of other placements) should be independent of m and should be equal to the probability that a placement selected at random is isotactic. This assertion can be strengthened by assuming that, for each n, the probability that the diastereosequence of length n running from the mth to the $(m + n - 1)$th placement has a particular pattern is independent of m (if no information is given about the tacticity of the placements preceding and succeeding the sequence): this is the essence of our assumption of the statistical stationarity. It should be emphasized that to have stationarity one need not assume that the probability that the mth placement is isotactic be unaffected by a knowledge of the tacticity of the $(m-1)$th placement; nor need it be assumed that there exists a fixed integer N, such that the probability that the mth placement is isotactic be affected by knowledge of the tacticity of placements $m-1, m-2, \ldots, m-N$ but given such knowledge, be then independent of the tacticity of placement $m-N-1$. In the former case the present theory would reduce to that of Bernoulli trials while in the latter to that of stationary Markoff processes.

Definitions and Postulates

Let us replace the polymer molecule by an infinite sequence $\mathcal{S}$ of letters S and I, where S plays the role of a syndiotactic placement and I an isotactic placement. The observations made in the previous section about diastereosequence distributions are equivalent to the assertion that the occurrence of the letters S and I in $\mathcal{S}$ are governed by statistical laws which, though stationary along $\mathcal{S}$, are not necessarily given by independent Bernoulli trials or even finite Markoff chains.

One way of developing our present theory would be to consider a sample space in which the entire sequence $\mathcal{S}$ is taken as the basic random variable. Such a procedure involves some subtle concepts, because in it the basic sample space, i.e., the space of all sequences $\mathcal{S}$ of binary digits, is an infinite and nondenumerable sample space. We shall here avoid mentioning these subtle concepts by axiomatizing our theory in such a way that we refer only to probability distributions p_n over finite sample spaces. Although we shall have an infinite (yet countable) number of these finite distributions p_n the procedures we use here do not require any sophisticated mathematical concepts.

We denote finite sequences of letters (to be interpreted as consecutive subsequences of $\mathcal{S}$) as follows: S represents, of course, the single letter S; SI represents an I followed by an S; etc. (This convention of building up sequences from right to left is motivated by applications of the theory to the example of Fox and Coleman[8] in which the basic probabilities are given by algorithms involving matrix products. This convention also has certain advantages when one manipulates conditional probabilities.) We use the symbols $U^{(n)}$, $V^{(n)}$, $W^{(n)}$ to denote unspecified consecutive sequences of length n. For example, $U^{(1)}$ may be either S or I; $U^{(2)}$ may be II, SI, IS, or SS. etc. Since all sequences which we consider are consecutive, for short we call $U^{(n)}$ simply a *sequence* of length n. We can combine the symbols S and I with the symbols $U^{(n)}$ and $V^{(n)}$. For example $IU^{(2)}S$ represents a sequence of length four about which we specify only that it begin with an S and end with an I; i.e., $IU^{(2)}S$ may be any one of the following: $IIIS$, $IISS$, $ISIS$, $ISSS$. We use the following abbreviation: I^k represents k I's and S^k, k S's; hence SI^3S^2 is short for $SIIISS$. In order to state some of our propositions concisely, it is convenient to assign a meaning to the superscript k in $U^{(k)}$, I^k, and S^k when $k=0$. The symbols

$U^{(0)}$, $I^{(0)}$, and $S^{(0)}$ always occur in sequences involving other symbols and always have the same meaning, namely,

$$U^{(0)}V^{(m)} = V^{(m)}U^{(0)} = I^0V^{(m)} = V^{(m)}I^0 = S^0V^{(m)} = V^{(m)}S^0 = V^{(m)}$$

The symbol $p_n\{U^{(n)}\}$ denotes the probability of occurrence of the sequence $U^{(n)}$; e.g., $p_3\{IIS\}$ is the probability that a given sequence of length three consists of an S followed by an I which is, in turn, followed by an I. (It is implicit in our notation that $p_n\{U^{(n)}\}$ is independent of the position of $U^{(n)}$ along $\mathcal{S}$: the random sequences $\mathcal{S}$ covered by our theory are stationary.)

We assume that for every n the function p_n exists and is indeed a probability distribution. It follows from this that

$$0 \leq p_n\{U^{(n)}\} \leq 1 \tag{1}$$

and that the summation of $p_n\{U^{(n)}\}$, over the 2^n different ways of forming $U^{(n)}$ from sequences of S's and I's, is unity:

$$p_1\{I\} + p_1\{S\} = 1 \tag{2a}$$

$$p_2\{II\} + p_2\{IS\} + p_2\{SI\} + p_2\{SS\} = 1 \tag{2b}$$

$$\vdots$$

etc.

We assume that for each n the probability distribution p_n is related to the distribution p_{n+1} by means of the formulae

$$p_{n+1}\{U^{(n)}I\} + p_{n+1}\{U^{(n)}S\} = p_n\{U^{(n)}\} \tag{3a}$$

$$p_{n+1}\{IU^{(n)}\} + p_{n+1}\{SU^{(n)}\} = p_n\{U^{(n)}\} \tag{3b}$$

The identities (3) hold for every fixed sequence $U^{(n)}$. They say that no matter what the length of a sequence $U^{(n)}$, it may be regarded as a contraction of a sequence one letter longer. In particular, eq. (3a) says that $U^{(n)}I$ (the sequence $U^{(n)}$ preceded by an I) and $U^{(n)}S$ (the sequence $U^{(n)}$ preceded by an S) give two mutually exclusive (and exhaustive) ways of preceding $U^{(n)}$ with another letter; while eq. (3b) says that $SU^{(n)}$ and $IU^{(n)}$ are two mutually exclusive (and exhaustive) ways of following $U^{(n)}$ with another letter.

Note that eqs. (3) tell us that if the function p_n is known for all $U^{(n)}$ then p_{n-1} is determined for all $U^{(n-1)}$, but not (in general) conversely.

In order to avoid vacuous theorems, we assume further that

$$p_1\{I\} \neq 0 \quad \text{and} \quad p_1\{S\} \neq 0 \tag{4}$$

Our postulates have now been stated: everything that we do henceforth shall follow from the equations and inequalities (1)–(4). We note that these axioms are symmetric in the symbols S and I. Hence, every proposition we prove will remain valid if in it one replaces each S by an I and each I by an S.

Application of a theory based on stationarity assumptions to a real polymer sample always involves an idealization. Indeed, if we interpret $p_n\{U^{(n)}\}$ as the probability that a partial diastereosequence, of known distance m from the initiating end, is of type $U^{(n)}$, then we must face the fact that every imaginable mechanism of synthesis yields a dependence of $p_n\{U^{(n)}\}$ on m. Usually this dependence on m falls off rapidly with increasing m and becomes unimportant for samples of high molecular weight.

It seems possible to increase the scope of applicability of the present theory by interpreting $p_n\{U^n\}$ as the probability that a diastereosequence *selected at random* is of type $U^{(n)}$. Such an interpretation suggests that the present theory can be applied even in circumstances in which end effects are not negligible, i.e., in circumstances in which $p_n\{U^{(n)}\}$, when given the interpretation of our previous paragraph, is strongly dependent on m even for large m. Yet an idealization is involved even here. Under our present interpretation, for polymer chains of finite length, eq. (3) neglects the probability that a randomly selected placement involves an endgroup and is thus lacking a successor or a predecessor. For chains which are really of infinite length the expression "randomly selected" becomes devoid of precise meaning.

We now give definitions of conditional probabilities and then Markoffian, Bernoullian, and non-Markoffian sets of distributions, p_n.

Now $p_{q+r}\{V^{(r)}U^{(q)}\}$ is the probability that a particular sequence $U^{(q)}$ of length q occurs and is immediately followed by a certain sequence $V^{(r)}$ of length r. The conditional probability $p_{r|q}\{V^{(r)}|U^{(q)}\}$ that the sequence $V^{(r)}$ occupies the positions $q + 1, q + 2, ..., q + r$, given that $U^{(q)}$ occupies the positions $1,2, ..., q$, is defined by

$$p_{r|q}\{V^{(r)}|U^{(q)}\}p_q\{U^{(q)}\} = p_{q+r}\{V^{(r)}U^{(q)}\} \tag{5}$$

where $p_q(U^q)$ is, of course, the probability of occurrence of the sequence $U^{(q)}$.

We say that our set of probability distributions p_n is Markoffian if there exists an integer N such that for each sequence $V^{(N)}$ of length N

$$p_{n|N}\{U^{(q)}|V^{(N)}\} = p_{n|N+m}\{U^{(n)}|V^{(N)}W^{(m)}\}; \tag{6}$$

this equation is to be interpreted as an identity holding for all integers n and m and all sequences $U^{(n)}$ and $W^{(m)}$.

If the identity (6) holds for N then it holds for all $m > N$. If N is the smallest integer for which eq. (6) holds, then we say that the set of distributions p_n is Markoffian of order N. In other words, if our collection of distributions is Markoffian of order N, then the probability of occurrence of a sequence $U^{(n)}$ is, in general, affected by a knowledge of the first preceding N letters, but if these first preceding N letters are known, then a knowledge of any other preceding letters does not further affect the probability of $U^{(n)}$. If eq. (6) holds for $N = 1$ then the p_n are said to form a simple Markoff chain.

If, for all pairs of sequences $W^{(m)}$ and $U^{(n)}$ we have

$$p_{n|m}\{U^{(n)}|W^{(m)}\} = p_n\{U^{(n)}\} \tag{7}$$

then we say that our set of distributions p_n is Bernoullian. Of course, if eq. (7) holds, then eq. (6) holds for every N.

If the identity (7) does not hold, and if there does not exist a finite integer N for which the identity (6) holds, then we say that our set of distributions is non-Markoffian. The theory which we present here, because it rests on only our postulates (1)–(4), applies whether the set of distributions p_n is Bernoullian, Markoffian, or non-Markoffian.

Elementary Propositions

On putting $n = 1$ and $U^{(1)} = S$ in the eqs. (3) we obtain

$$p_2\{SI\} + p_2\{SS\} = p_1\{S\} \tag{8a}$$

$$p_2\{IS\} + p_2\{SS\} = p_1\{S\} \tag{8b}$$

from which we read off Proposition 1.

Proposition 1:

$$p_2\{IS\} = p_2\{SI\} \tag{9}$$

i.e., the probability that a letter selected at random is an S preceded by an I is equal to the probability that it is an S followed by an I. This observation has the following generalization which was pointed out by Professor Morris De Groot.

Proposition 2: For all $m \geq 1$,

$$p_{m+1}\{I^m S\} = p_{m+1}\{SI^m\} \tag{10}$$

Proof: On putting $U^{(n)} = SI^{n-1}$ in eq. (3a), and then $U^{(n)} = I^{n-1}S$ in eq. (3b), we get

$$p_n\{SI^{n-1}\} = p_{n+1}\{SI^n\} + p_{n+1}\{SI^{n-1}S\} \tag{11a}$$

$$p_n\{I^{n-1}S\} = p_{n+1}\{I^nS\} + p_{n+1}\{SI^{n-1}S\} \tag{11b}$$

from which it is obvious that if eq. (10) holds for $m = n - 1$ then it holds for $m = n$; hence by induction it holds for all m; q.e.d.

Proposition 3: The two limits

$$\lim_{N\to\infty} p_N\{I^N\} = q \tag{12a}$$

and

$$\lim_{N\to\infty} \sum_{n=1}^{N} p_{n+1}\{I^nS\} \tag{12b}$$

exist and are related by

$$\sum_{n=1}^{\infty} p_{n+1}\{I^nS\} = p_1\{I\} - q \tag{13}$$

Proof: Equation (3a) tells us that

$$p_1\{I\} = p_2\{IS\} + p_2\{I^2\}$$

$$p_2\{I^2\} = p_3\{I^2S\} + p_3\{I^3\}$$

$$\cdot$$
$$\cdot$$
$$\cdot$$

$$p_n\{I^n\} = p_{n+1}\{I^nS\} + p_{n+1}\{I^{n+1}\}$$

and on adding these equations we get the intuitively evident formula

$$p_1\{I\} = \sum_{n=1}^{N} p_{n+1}\{I^nS\} + p_{N+1}\{I^{N+1}\} \tag{14}$$

From eqs. (14) and (1) we have

$$1 \geq \sum_{n=1}^{N} p_{n+1}\{I^nS\} \geq 0 \tag{15}$$

In other words, the series $\sum p_{n+1}\{I^nS\}$, consisting of positive terms, has a sum which is bounded. Hence, this series converges, i.e., the limit (12b) exists. It then follows from eq. (14) that the limit (12a) exists and that eq. (13) holds; q.e.d.

One might expect that our assumption $p_1\{S\} \neq 0$ gives us $q = 0$. Although this is true when the distributions are Bernoullian, it is not true in general. The following example, suggested by Professor De Groot, is compatible with postulates (1)–(4) yet does not yield $q = 0$. Suppose we toss a coin, and whenever we get heads we put all the letters in s equal to I and whenever we get tails we put all these letters equal to S. Then, $p_N\{I^N\} = p_1\{I\}$ for all N and thus $q = p_1\{I\} \neq 0$.

It is expected, however, that in all the applications of the present theory to real polymerization problems, we shall have $q = 0$.

The existence of the limit (12b) has the corollary

$$\lim_{n \to \infty} p_{n+1}\{I^nS\} = 0 \tag{16}$$

which is generalized in Proposition 4.

Proposition 4: Let $U^{(r)}$ and $V^{(q)}$ be two arbitrary fixed sequences such that the letter S occurs at least once in either $U^{(r)}$ or $V^{(r)}$, then

$$\lim_{n \to \infty} p_{q+n+r}\{V^{(q)}I^nU^{(r)}\} = 0 \tag{17}$$

Proof: If $V^{(q)}$ or $U^{(r)}$ has an S then $p_{q+n+r}\{V^{(q)}I^nU^{(r)}\}$ can be rewritten in either the form $p_{q+n+r}\{V^{(q)}I^{n+t}SW^{(r-t-1)}\}$ or the form $p_{q+n+r}\{W^{(q-t-1)}SI^{(n+t)}U^{(r)}\}$. Also, it follows from eqs. (1) and (3) that for any three sequences $U^{(x)}$, $V^{(y)}$, $W^{(z)}$, with $x \geq 0$, $y \geq 1$, $z \geq 0$,

$$p_{x+y+z}\{U^{(x)}V^{(y)}W^{(z)}\} \leq p_y\{V^{(y)}\} \tag{18}$$

On noting eq. (18), we see that eq. (17) now follows immediately from eq. (16).

We shall now use Proposition 4 to obtain a useful lemma.

Proposition 5. For any sequence $V^{(n)}$ which contains the letter S at least once, and for every integer $M \geq 0$,

$$\sum_{k=M}^{\infty} p_{n+k+1}\{SI^k V^{(n)}\} = p_{n+M}\{I^M V^{(n)}\} \tag{19}$$

Proof: From eq. (3b) we get

$$p_{n+k+1}\{SI^k V^{(n)}\} + p_{n+k+1}\{II^k V^{(n)}\} = p_{n+k}\{I^k V^{(n)}\} \tag{20}$$

Solving this for the first term on the left and then summing over k we find

$$\sum_{k=M}^{N} p_{n+k+1}\{SI^k V^{(n)}\} = \sum_{k=M}^{N} [p_{n+k}\{I^k V^{(n)}\} - p_{n+k+1}\{I^{k+1} V^{(n)}\}] \tag{21}$$

$$= p_{n+M}\{I^M V^{(n)}\} - p_{n+N+1}\{I^{N+1} V^{(n)}\}$$

If we now let $N \to \infty$ and note that, by Proposition 4,

$$\lim_{N \to \infty} p_{n+N+1}\{I^{N+1} V^{(n)}\} = 0 \tag{22}$$

we see that the proposition is proved.

When $V^{(n)}$ is just the letter S this lemma yields the formulae

$$\sum_{k=M}^{\infty} p_{k+2}\{SI^k S\} = p_{M+1}\{I^M S\} \tag{23}$$

$$\sum_{k=0}^{\infty} p_{k+2}\{SI^k S\} = p_1\{S\} \tag{24}$$

Theorems

We can now use the mathematical apparatus assembled in the previous section to write short proofs for some apparently nontrivial theorems on recurrence times and sequence lengths. In their present generality, these theorems were first proved by Kac.[5,6]

Let us consider the numbers f_n, $n \geq 1$, defined by

$$f_n = p_{n|1}\{SI^{n-1}|S\} \tag{25}$$

If, as we move along the sequence S, we observe that an S has occurred at a particular position, say position m, and are given no other information, then f_n tells us the probability that the positions $m+1, m+2, \ldots m+n-1$ are occupied by I's and that the next S to occur occupies the position $m+n$, i.e., f_n is the probability that the recurrence time for S has length n.

Theorem 1:

$$\sum_{n=1}^{\infty} f_n = 1 \tag{26}$$

Proof: By eqs. (25), (5), and the inequalities (4),

$$f_n = p_{n+1}\{SI^{n-1}S\}/p_1\{S\} \tag{27}$$

On summing we get

$$\sum_{n=1}^{\infty} f_n = \sum_{n=1}^{\infty} p_{n+1}\{SI^{n-1}S\} \Big/ p_1\{S\} = \sum_{k=0}^{\infty} p_{k+2}\{SI^kS\} \Big/ p_1\{S\} \tag{28}$$

The theorem now follows immediately from eq. (24).

The quantity $\chi\{S\}$ defined as

$$\chi\{S\} = \sum_{n=1}^{\infty} nf_n \tag{29}$$

may be called the *mean recurrence time* for the letter S.

Theorem 2:

$$\chi\{S\} = \frac{1-q}{p_1\{S\}} \tag{30}$$

Proof: By eqs. (29) and (28),

$$p_1\{S\}\chi\{S\} = \sum_{n=1}^{\infty} np_{n+1}\{SI^{n-1}S\} = \sum_{m=0}^{\infty} (m+1)p_{m+2}\{SI^mS\}$$

$$= \sum_{m=0}^{\infty} mp_{m+2}\{SI^mS\} + \sum_{m=0}^{\infty} p_{m+2}\{SI^mS\} \tag{31}$$

For the first summation in eq. (31) we write

$$\sum_{m=0}^{\infty} mp_{m+2}\{SI^mS\} = \sum_{M=1}^{\infty} \sum_{k=M}^{\infty} p_{k+2}\{SI^kS\} \tag{32}$$

and, on combining eqs. (23) and (13) we find that

$$\sum_{m=0}^{\infty} mp_{m+2}\{SI^mS\} = \sum_{M=1}^{\infty} p_{M+1}\{I^MS\} \tag{33}$$

$$= p_1\{I\} - q \tag{34}$$

Since the two summations on the right in eq. (32) are now both seen to be convergent (and, of course, absolutely convergent), eq. (32) is justified.

On putting eq. (34) into eq. (31) and noting that, by eq. (24), the second summation in eq. (31) is $p_1\{S\}$, we obtain

$$p_1\{S\}\chi\{S\} = p_1\{I\} - q + p_1\{S\} = 1 - q \tag{35}$$

which completes the proof.

When $q = 0$, eq. (30) reduces to the formula

$$\chi\{S\} = \frac{1}{p_1\{S\}} \tag{36}$$

Equation (36) states that the mean recurrence time for the letter S is equal to the reciprocal of the probability of an S, a plausible result.

For the coin-tossing same of Professor De Groot, mentioned above, eq. (30) yields $\chi\{S\} = 1$, which is, for that special case, also a plausible result.

A closed sequence of I's is a sequence of $n + 2$ letters, $n > 1$, which has an S at each end and n I's in the middle. Let us denote by u_n the probability that a given closed sequence of I's contains exactly n I's. In other words u_n is the probability, given the letter S and given that S is immediately followed by I, that exactly n I's occur before another S occurs. Now, the probability that a given letter S is followed by exactly n I's before the next S occurs, given no other information, is just f_{n+1}. The probability that a given letter S is immediately followed by an I is $1 - f_1$. Hence

$$(1 - f_1)u_n = f_{n+1} \tag{37}$$

and when $f_1 \neq 1$

$$u_n = \frac{f_{n+1}}{1 - f_1} \tag{38}$$

Of course, when $f_1 = 1$ there are no closed sequences of I's. When the quantity $\mu\{I\}$, defined as

$$\mu\{I\} = \frac{\sum\limits_{n=1}^{\infty} nf_{n+1}}{1 - f_1} = \sum\limits_{n=1}^{\infty} nu_n \tag{39}$$

exists, it may be interpreted as the *mean length of closed sequences of the letter I*.

On noting that

$$\sum_{n=1}^{\infty} nf_{n+1} = \sum_{m=2}^{\infty} (m-1)f_m = \sum_{m=1}^{\infty} (m-1)f_m = \sum_{m=1}^{\infty} mf_m - \sum_{m=1}^{\infty} f_m \tag{40}$$

and using eqs. (26), (27), and (29), we find

$$\mu\{I\} = \frac{\chi\{S\} - 1}{1 - p_2\{SS\}/p_1\{S\}} \tag{41}$$

This formula and Theorem 2 yield

Theorem 3:

$$\mu\{I\} = \frac{p_1\{I\} - q}{p_1\{S\} - p_2\{SS\}} \tag{42}$$

In the case of De Groot's coin tossing game, i.e., when $q = p_1\{I\}$ and $p_1\{S\} = p_2\{SS\}$, eq. (42) is indeterminate, and it should be, for in that game there are no closed sequences of I's.

Diastereosequences

In considering diastereosequences in an actual sample of a poly-α-olefin, the quantity $p_1\{I\}$ (or $p_1\{S\}$) may be interpreted as the concentration of asymmetric chain atoms which are in isotactic (or syndiotactic) placement with respect to their immediate predecessors, i.e., "the concentration of isotactic (or syndiotactic) placements."   Similarly, $p_2\{I^2\}$ (or $p_2\{S^2\}$) may be interpreted as the concentration of asymmetric chain atoms which are in isotactic (or syndiotactic) placement with respect to both their immediate predecessor and their immediate successor, i.e., "the concentration of isotactic (or syndiotactic) pairs."   The quantity

$$p\{IS_VSI\} = p_2\{IS\} + p_2\{SI\} \tag{43}$$

is, in the mathematical theory, the probability that a pair of adjacent letters selected at random is either the sequence SI or the sequence IS. In the application under consideration, $p\{IS_VSI\}$ is to be interpreted as the concentration of asymmetric chain atoms in isotactic placement with one nearest neighbor and in syndiotactic with the other, i.e., "the concentration of heterotactic pairs."   We call $p_1\{I\}$, $p_1\{S\}$ *placement concentrations* and $p_2\{I^2\}$, $p_2\{S^2\}$, $p\{IS_VSI\}$, *pair concentrations*.

The experiments of Bovey and Tiers on poly(methyl methacrylate) show that for this polymer placement concentrations and pair concentrations can be measured by high resolution NMR.[3]

Now $p_1\{I\}$ and $p_1\{S\}$ are, of course, related by eq. (2a) and it follows from eqs. (2b) and (43) that

$$p_2\{II\} + p_2\{SS\} + p\{IS_VSI\} = 1 \tag{2b'}$$

The equations (2a) and (2b') are not verified by NMR measurements, but are used instead to normalize the data before it is presented.   Thus, although the spectroscopist can report values of both of the placement concentrations $p_1\{I\}$, $p_1\{S\}$, and all three of the pair concentrations $p_2\{II\}$, $p_2\{SS\}$, $p\{IS_VSI\}$, we cannot regard all this data as being independent. The independent data obtained from spectroscopy consists of one placement concentration and two pair concentrations.

It follows from eqs. (3) that the experimentally independent data must obey some simple algebraic relations.

If one takes the independently measured quantities to be $p_1\{I\}$, $p_2\{I^2\}$, and $p\{IS_VSI\}$, then eqs. (3) yield

$$p\{IS_VSI\} = 2(p_1\{I\} - p_2\{I^2\}) \tag{44}$$

If $p_1\{I\}$, $p_2\{I^2\}$, and $p_2\{S^2\}$ are used, then

$$p_1\{I\} = {}^1\!/_2[1 + p_2\{I^2\} - p_2\{S^2\}] \tag{45}$$

and if $p_1\{I\}$, $p_2\{S^2\}$ and $p\{IS_VSI\}$ are used, then

$$p_1\{I\} = 1 - {}^1\!/_2\, p\{IS_VSI\} - p_2\{S^2\} \tag{46}$$

The other consequences of eqs. (3) for $n = 1$ can be obtained from eqs. (44)–(46), through interchange of the symbols S and I. Since for given spectroscopic measurement one set of independent data is equivalent to another, the experimenter need check only one of the equations (44)–(46):

if the equation he chooses is found to be consistent with his data, so then must all the others, and he can conclude that his measurements are consistent with eqs. (3). Under no circumstances should he regard an experimental confirmation of eqs. (44)–(46) as evidence for Markoffian statistics.

Let us now turn to Theorem 3. Clearly, in any stereochemical applications, we shall have $q = \lim\limits_{n \to \infty} p_n\{I^n\} = 0$. Hence, Theorem 3 tells us that a knowledge of the syndiotactic placement concentration $p_1\{S\}$ and the syndiotactic pair concentration $p_2\{SS\}$ determines the mean (i.e., number-average) length $\mu\{I\}$ of closed sequences of isotactic placements through the formula

$$\mu\{I\} = \frac{p_1\{I\}}{p_1\{S\} - p_2\{S^2\}} = \frac{1 - p_1\{S\}}{p_1\{S\} - p_2\{S^2\}} \tag{47}$$

Thus, NMR spectra, although they measure only placement concentrations and pair concentrations, can yet be used to calculate mean lengths of closed sequences of isotactic (and, of course, also syndiotactic) placements. The important point here is that this observation is true even when the diastereosequence distribution is non-Markoffian.

In the Bernoullian case, eq. (47) reduces to $\mu\{I\} = (p_1\{S\})^{-1}$ and is almost obvious. A result equivalent to eq. (47) has been derived by Johnsen[7] for simple Markoff chains. (Of course, to have $q = 0$ for simple Markoff chains one must suppose not only that $p_1\{I\} \neq 1$, but also that $p_{1|1}\{S|I\} \neq 0$.) Equations equivalent to eq. (47) have also been **proposed by** Bovey and Tiers[3] on the basis of a special model.

We remark that the placement and pair concentrations, by themselves, do not, in general, suffice to determine higher moments of the distribution of sequence lengths. Therefore, one cannot calculate weight-average lengths of closed sequences from NMR data unless one introduces special assumptions, such as the assumption that one is dealing with a simple Markoff chain. (This assumption was used for calculations reported by Johnsen;[7] as we remarked in the introduction, we do not believe such an assumption holds for homogeneous anionic polymerizations.)

Equations (8b), (9), and (43) may be used to rewrite eq. (47) in the perhaps more suggestive forms

$$\mu\{I\} = \frac{p_1\{I\}}{p_2\{IS\}} = \frac{2p_1\{I\}}{p\{IS_vSI\}} \tag{48}$$

Of course, in the case of Bernoulli trials, eq. (48) reduces to $\mu\{I\} = p_1\{S\}^{-1}$.

Let us consider the quantity ρ defined by

$$\rho = \mu\{I\} p_1\{S\} \tag{49}$$

From eq. (48) we have

$$\rho = \frac{2p_1\{I\} p_1\{S\}}{p\{IS_vSI\}} \tag{50}$$

3196   B. D. COLEMAN AND T. G FOX

It is easy to see that this same quantity ρ could also be defined by

$$\rho = \mu\{S\}\,p_1\{I\} \tag{51}$$

and that ρ is simply the ratio of the actual mean length of closed sequences of isotactic (or syndiotactic) placements to the mean length which one would calculate assuming that the diastereosequence distribution were Bernoullian with the same value of $p_1\{S\}$ (or $p_1\{I\}$). We call ρ the *persistence ratio*, for $\rho - 1$ measures the apparent "statistical after-effect," or departure from Bernoulli statistics. It follows from eq. (50) that ρ can be calculated from NMR data.

We are indebted to Professor Morris De Groot of the Mathematics Department, Carnegie Institute of Technology, for valuable discussions.

The research reported here was supported in part by the Air Force Office of Scientific Research under Contract AF 49(638)541.

References

1. Coleman, B. D., *J. Polymer Sci.*, **31**, 155 (1958).

2. Fox, T. G., W. E. Goode, S. Gratch, C. M. Huggett, J. F. Kincaid, A. Spell, and J. O. Stroupe, *J. Polymer Sci.*, **31**, 173 (1958).

3. Bovey, F. A., and G. B. D. Tiers, *J. Polymer Sci.*, **44**, 173 (1960).

4. Fox, T. G, and H. W. Schnecko, *Polymer*, **3**, 575 (1962).

5. Kac, M., *Probability and Related Topics in Physical Sciences*, Interscience, New York–London, 1959.

6. Kac, M., *Bull. Am. Math. Soc.*, **53**, 1012 (1947).

7. Johnsen, U., *Kolloid Z.*, **178**, 161 (1961).

8. Fox, T. G, and B. D. Coleman, paper presented to the Division of Polymer Chemistry, 142nd National Meeting of the American Chemical Society, September 1962; see also B. D. Coleman and T. G Fox, *J. Chem. Phys.*, **38**, 1065 (1963).

Résumé

Lorsque, dans une poly-α-oléfine, la probabilité d'une séquence isotactique dépend seulement d'un nombre limité de séquences précédentes, alors la distribution des segments qui en résulte suit la théorie des chaînes de Markoff. Dans l'autre cas, on dit que la distribution résultante est non-Markoffienne. Un cas spécial de cette distribution de Markoff est donné par une chaîne simple de Markoff dont la tacticité d'un placement donné est considérée comme étant influencée seulement par la séquence qui précède immédiatement. Un autre cas spécial est représenté naturellement par la distribution de Bernoulli dans laquelle la probabilté qu'un placement donné soit isotactique est indépendant de la tacticité de tout autre séquence. Un spectre NMR à haute résolution peut parfois fournir une détermination quantitative des concentrations des séquences isotactiques et syndiotactiques et des concentrations de trois types des paires adjacentes possibles de telles séquences, notamment des paires isotactique, syndiotactique et hétérotactique. Dans ce cas le spectre peut être employé pour déterminer si la distribution d'une séquence arbitraire est une distribution de Bernoulli ou pas. Pourtant, la plus longue séquence dont la concentration peut être mesurée par spectroscopie NMR consiste uniquement en deux unités; un spectre NMR ne peut pas définir dés lors si une distribution donnée non-Bernoullienne correspond à une distribution Markoffienne ou non-Markoffienne. En effet, les distributions non-Markoff sont compatibles avec des spectres de polyméres préparés par voie anionique. Dans cet article nous travaillons dans un cadre de la théorie de Kac qui traite des processus statistiques stationnaires et nous indiquons quelques résultats sui sont valables pour les deux processus: suivant

Markoff et non-Markoff. Les résultats sont appliqués à la spectroscopie NMR; on a indiqué quels calculs sont valables pour les distributions Markoffienne et non-Markoffienne descalculs employés à tester la validité des résultats NMR et à obtenir la longuer moyenne des diastérioséquences fermées.

Zusammenfassung

Falls bei einem Poly-α-Olefin die Wahrscheinlichkeit für die Isotaktizität einer gegebenen Plazierung nur von der Taktizität einer endlichen Zahl unmittelbarer Vorgänger abhängt, so gehorcht die resultierende Diastereosequenzverteilung der Theorie der Markoffketten. Wenn das nicht der Fall ist, bezeichnet man die resultierende Diastereosequenzverteilung als nicht-Markoffisch. Ein Spezialfall einer Markoffverteilung wird durch eine einfache Markoffkette gebildet, bei welcher die Taktizität einer gegebenen Plazierung als nur von der Taktizität der unmittelbar vorhergehenden Plazierung beeinflusst betrachtet wird. Einenweiteren Spezialfall bildet natürlich die Verteilung nach dem Bernoullischem Theorem, bei welcher die Wahrscheinlichkeit für die Isotaktizität einer gegebenen Plazierung unabhängig von der Taktizität aller übrigen Plazierungen ist. Ein Hochauflösungs-NMR-spektrum erlaubt manchmal eine quantitative Bestimmung der Konzentration isotaktischer und syndiotaktischer Plazierungen sowie der Konzentration der drei möglichen Typen benachbarter Paare solcher Plazierungen (nämlich isotaktische, syndiotaktische und heterotaktische Paare). In diesem Fall kann das Spektrum zur Entscheidung defür benutzt werden, ob eine gegebene Diastereosequenzverteilung vom Bernoulli-Typ ist. Da jedoch die längste Diastereosequenz, deren Konzzentration durch NMR-Spektroskopie bestimmt werden kann, nur zwei Plazierungen enthält, kann ein NMR-Spektrum nicht darüber entscheiden ob eine gegebene Nicht-Bernoulli-Verteilung vom einfachen oder allgemeinen Markoff-Typ ist. Tatsächlich sind Nicht-Markoff-Verteilungen mit den bekannten NMR-Spektren von anionisch hergestellten Polymeren verträglich. Die-vorliegende Mitteilung bewegt sich im Rahmen der Theorie von Kac und bringt einige allgemeine Ergebnisse, die sowohl für Markoff- als auch Nicht-Markoff-Prozesse gültig sind. Die Ergebnisse werden auf die NMR-Spektroskopie angewendet und es wird gezeigt, welche, bei der Überprüfung der Konsistenz der NMR-Daten und bei der Ermittlung der mittleren Länge geschlossener Diastereosequenzen verwendeten Berechnungen, sowohl für Markoff- als auch Nicht-Markoffverteilungen gültig sind.

Received August 23, 1962

PERSPECTIVE

Comments on "Gel Permeation Chromatography. I. A New Method for Molecular Weight Distribution of High Polymers," by J. C. Moore, *J. Polym. Sci. A*, 2, 835 (1964)

HENRI BENOÎT

CNRS, Institute Charles Sadron, Strasbourg F-67083, France

I remember very well, I was working in the library of our laboratory when Prof. Banderet (from the neighboring Ecole d'Application des Hauts Polymères) interrupted me and showed me the paper he was reading: the famous paper from J. C. Moore on Gel Permeation Chromatography.[1] At first sight I was so astonished by the results I did not believe it was possible to obtain so rapidly and so precisely the molar mass and polydispersity of polymeric samples. Immediately, I took the paper, read it carefully and began to dream about the new possibilities offered by this method.

Over thirty years ago, the problem of polymer characterization was still one of the major difficulties of polymer science. Measurement of molar masses was not straightforward and a big part of the research activity was devoted to this kind of problem. The determination of laws, like the viscosity law, to take a simple example, involved preparing a set of fractions, as monodisperse as possible and measuring their molar masses, and viscosities. The most time-consuming part of the work was the preparation of the fractions. The method was well established by G. V. Schulz[2] and P. J. Flory[3] but was extremely bothersome. I did make a fractionation one time, working as a PostDoc. in the laboratory of Prof. P. Doty[4] and I swore that I would never try to do it again (which, in fact, is the case).

The very important contribution of Moore was to extrapolate to organic solvents what was known about water solutions by people using Sephadex resins. This could not be done with silica beads or glass beads because of the strong interactions (mainly adsorption) between the polymer and the inorganic support and was successful only through the use of porous beads of polystyrene gels. The main contribution of Moore was to succeed in this preparation and to be able to prepare gels with the variable porosity, diameter of the holes and size. This was only possible for someone knowing perfectly the problems of polymerization even under unconventional conditions. It took many years before the know-how acquired by Moore could be repeated by other producers of columns.

When we read Moore's paper, we realize that the design of his instrument is rigorously copied in today's machines: the pump, the set of columns filled with porous beads, the detector, and the standardization curve. Of course a lot of progress has been made in the efficiency of the columns and the detectors, but all the principles have been kept constant. We regret that for his standardization curve Moore plotted ln M_w versus elution volume and not the reverse which would be more standard (for instance, nobody plots M_w or ln M_w versus intrinsic viscosity). It would still be much better to change but this is refused by all the users of this technique, who are prisoners of the tradition.

Moore did study carefully the effect of different parameters on the results : concentration of the injected solutions, amount of polymer, elution velocity, showing that with a minimum of precautions one could obtain reproducible results. This means, as it was shown later,[5] that a static explanation is much more appropriate than a dynamic one.

In conclusion, I would say that Moore's paper brought an important improvement in the techniques for the characterization of polymers, and opened a new area in this domain. Moore gave us a much faster technique and also much more precise. Before Gel Permeation Chromatography the expression *narrow fraction* had no meaning because no method was available to define precisely what it meant. Now Size Exclusion Chromatography (to use this new term) allows us to speak, for instance, about a ratio of weight to number average molar mass of 1.005.[6]

When the scientific community began to master this technique I was asking the question: should we not reconsider all the work done on polymer solutions because the improved accuracy could change the conclusions? Looking at the research recently done using GPC, one can answer negatively to this question: the work done by our predecessors with fractionation, membrane osmometers, viscometers, ultracentrifuges, and light scattering instruments is still valid showing that, even with simple techniques, one can obtain precise results.

REFERENCES AND NOTES

1. J. C. Moore, *J. Polym. Sci. A,* **2,** 835 (1964).
2. G. V. Schulz, *Makromol. Chem.,* **5,** 93 (1950).
3. P. J. Flory, *Principles of Polymer Chemistry,* Cornell University Press, Ithaca, NY, 1953.
4. H. Benoît, A. Holtzer, and P. Doty, *J. Phys. Chem.,* **58,** 635 (1954).
5. E. F. Casassa and Y. Tagami, *Macromolecules,* **2,** 14 (1969).
6. L. H. Tung, *J. Appl. Polym. Sci.,* **16,** 375 (1966); see also Z. Grubisic-Gallot, L. Marais, and H. Benoît, *J. Polym. Sci., Polym. Phys. Ed.,* **14,** 959 (1976).

JOURNAL OF POLYMER SCIENCE: PART A   VOL. 2, PP. 835–843 (1964)

Gel Permeation Chromatography. I. A New Method for Molecular Weight Distribution of High Polymers

J. C. MOORE, *Texas Basic Research Department, The Dow Chemical Company, Freeport, Texas*

Synopsis

Polystyrene gels crosslinked in the presence of diluents have been made in fine-mesh bead form suitable for packing into chromatographic columns. A series of narrow molecular weight range polymer fractions was eluted through such columns with aromatic and chlorinated solvents. Effluent concentrations were detected and recorded by a continuous differential refractometer. The fractions were shown to be efficiently separated. Columns capable of separating adjacent polymeric samples of high molecular weight were prepared from gels crosslinked in the presence of large amounts of diluents having little or no solvent action on polystyrene. Smaller proportions of diluents and those with more solvent action yielded columns with lower molecular weight permeability limits. Such studies provided a unique quantitative view of the topology of the gels. They also demonstrated that rapid repetitive molecular weight distribution data can be obtained in this way on polymers for which solvents compatible with the gels are available.

Introduction

Early in their work on ion-exclusion, Wheaton and Bauman[1] found that many nonionic substances of low molecular weight were separated by elution with water through a column packed with ion-exchange resin particles. Since that time the techniques of column chromatography on crosslinked gels have become broadly applicable to separations of large from small molecules in aqueous solutions. Lathe and Ruthven[2] showed that the separating range could be greatly extended by using swollen starch grains as column packing, differentiating between a globulin and hemoglobin, for example, of molecular weights 150,000 and 67,000, respectively. Porath and Flodin[3,4] have made available a series of hydrophilic gel column packings and have introduced the term "gel filtration" for the process, which has achieved considerable usefulness. Lea and Sehon[5] have recently described other hydrophilic gels of this nature. However, being swellable only in aqueous media, their use is limited to the separation of water-soluble substances.

Hydrophobic gels of high permeability have not been available. Several interesting separations were reported by Cortis-Jones[6] on columns packed with crosslinked polystyrene. His investigation did not, however, extend to large molecules. Vaughan[7] showed that some separation of low molecular weight polystyrenes did occur when they were eluted with benzene through polystyrene beads so lightly crosslinked that they had swollen in benzene to 51 times the volume of the dry polymer. However, all these gels were crosslinked in the absence of diluents, while the hydrophilic gels referred to were crosslinked in aqueous solution. This is a significant difference. Lloyd and Alfrey[8,9] have pointed out that gel networks of altered structure are produced by crosslinking in the presence of a diluent which is a solvent for the monomer. If the diluent is a nonsolvent for the resulting polymer, the gel may be still further altered by precipitation to give a rugged internal structure, and outstanding properties of stability and permeability are claimed for ion exchange resins based on such polymers.[10-12] The properties of gels prepared in the presence of various diluents have been investigated by Lloyd and Alfrey,[13] and diluents of intermediate solvent power were shown to give gels of intermediate properties.

Samples and Experimental Method

From the foregoing considerations, it appeared likely that we could prepare polystyrene beads with sufficient crosslinking to confer a desirable amount of rigidity, and still regulate the permeability of the network over a wide range by varying the amount and nature of the diluent present at the time of crosslinking. The permeabilities of the resulting gel networks were revealed by eluting a series of very similar compounds covering a wide range of molecular weights through a column packed with the gel. The properties of these marker materials are shown in Table I. In the low to medium molecular weight range the commercially available polypropylene glycols served as standards. A series of anionically polymerized polystyrenes covered the range from medium to high molecular weights.[14] In both cases the samples, while not monodisperse, gave well defined peaks. Aromatic or chlorinated eluting solvents chemically similar to the polystyrene gel were used to suppress adsorption and partition effects as much as pos-

TABLE I
Sample Materials

Polymer	Mol. Wt., $\bar{M}_w$	$\bar{M}_w/\bar{M}_n$
Anionic polystyrenes:		
S1	13,850	1.50
S102	82,000	1.05
S105	154,000	1.04
S108	267,000	1.08
S1159	570,000	1.05
S12	1,197,000	1.19
S114	3,500,000	1.24
Polypropylene glycols: The Dow Chemical Co.[a]		

[a] Average molecular weight by endgroup analysis as indicated in the identifying number, as: P-2000.

sible, and tailing was not evident with these samples. Eluent composition was followed by a Waters continuous differential refractometer, with a Rinco fraction collector adapted to mark the recorder chart at each volume increment.

Experimental Results and Discussion

A series of polystyrene gels with toluene as diluent in increasing proportions showed several significant changes. To avoid a soft compressible gel with resulting poor flow properties in the packed column, an increase in the proportion of diluent required a corresponding increase in the degree of crosslinking. With diluent and crosslinker increasing together, the permeability limit rose gradually at first, then rapidly at high dilution. With a rigidly crosslinked gel, the packed column became easier to pack and to use, less affected in volume by changes of eluting solvent. The compositions of these gels are shown in Table II. The permeability of each gel was then visualized by plotting the elution volume of each sample peak against the logarithm of the average molecular weight of the sample. These plots showed slanting lines, often quite straight for a considerable range, from an upper limit of permeability sharply defined at the interstitial volume, down to a nebulous lower limit around the total liquid

TABLE II
Permeabilities of Styrene Gels with Varying Proportions of Toluene

Gel	Gel composition			M.W. permeability limit	Notes
	Styrene, wt.-%	Divinyl-benzene, wt.-%	Toluene, wt.-%		
PSX8	92	8	0	1,000	
PSX4	96	4	0	1,700	
PSX1	99	1	0	3,500	Rubbery
PSXO.1	99.9	0.1	0		Too soft to pack
A	79.1	4.2	16.7	2,500	
B	65.7	5.7	28.6	c.a. 7,000	
C	30	10	60	7,000	
D	24.8	2.5	72.7		Too soft to pack
E	9	11	80	250,000	

volume of the column, in which region the elution volume was determined more by polarity factors than by molecular size. Examples of the elution tracings obtained and the resulting permeability curves are shown in Figures 1 and 2.

Great increases in permeability were found when the nature of the diluent was changed. In this series the gels were made by polymerizing a mixture of 21.8% styrene, 18.2% commercial (55%) divinylbenzene, and 60% diluent by weight. Varying the diluent by progressively replacing toluene with n-dodecane caused regular increases in permeability, but a limit was

838   J. C. MOORE

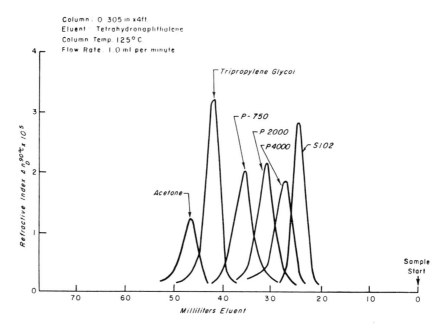

Column: 0.305 in x 4 ft
Eluent: Tetrahydronaphthalene
Column Temp. 125° C.
Flow Rate: 1.0 ml per minute

Fig. 1. Superimposed traces of elution peaks obtained with gel C.

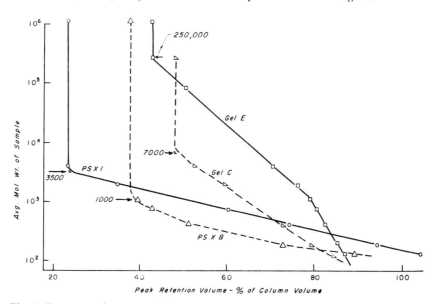

Fig. 2. Representative permeability curves for gels in Table II and determination of permeability limits.

reached when the diluent was 75% dodecane. Above that point the porous-structured beads were coated with a thin but impermeable skin. A non-solvent on the more polar side, isoamyl alcohol, progressively replacing diethylbenzene as diluent, showed very little increase in permeability at

TABLE III
Permeabilities of Gels with Various Diluents, All Made from 30% Styrene, 10% Divinylbenzene, 60% Diluent[a]

Gel	Diluents, parts/100 parts of gel	M.W. permeability limit
C	60 Toluene	7×10^3
F	30 Toluene, 30 diethylbenzene	1.5×10^4
G	60 Diethylbenzene	1.2×10^4
H	45 Toluene, 15 n-dodecane	1×10^5
I	30 Toluene, 30 n-dodecane	3×10^5
J	15 Toluene, 45 n-dodecane	2×10^6
K	10 Toluene, 50 n-dodecane	$<2 \times 10^3$
L	40 Diethylbenzene, 20 isoamyl alcohol	ca. 3.6×10^3
M	20 Diethylbenzene, 40 isoamyl alcohol	ca. 8×10^6
N	13.3 Diethylbenzene, 46.7 isoamyl alcohol	ca. 10^{10}
O	60 Isoamyl alcohol	Extremely high

[a] "Styrene" is a mixture of styrene and ethylvinylbenzene.

first, probably because of loss of some of the isoamyl alcohol to the aqueous continuous phase in which the beads were polymerized. With larger proportions of isoamyl alcohol in the diluent during polymerization, the permeability increased rapidly, far beyond the molecular sizes of the test materials used. This series of tests is summarized in Table III and Figures 3 and 4. The changes of gel structure indicated by the permeability curves were further shown in electron micrographs prepared by Dr. E. B. Bradford (Fig. 5). To preserve the swollen gel structure for electron microscopy, the porous polymer beads were soaked in a mixture of acrylate monomers, potted by polymerizing in a small gelatin capsule, and sectioned.

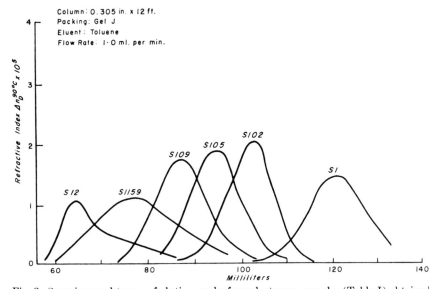

Fig. 3. Superimposed traces of elution peaks for polystyrene samples (Table I) obtained with gel J.

840 J. C. MOORE

Fig. 4. Representative permeability curves for gels from Table III.

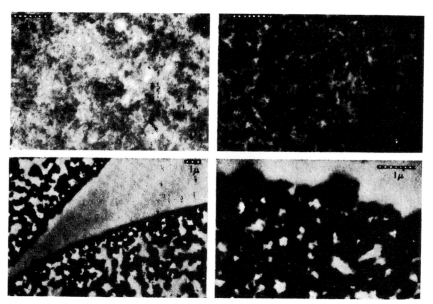

Fig. 5. Electron micrographs of thin sections of gels from Table III: (a) gel C; (b) ge G; (c) gel K; (d) gel M.

The chromatographic separation appears to be close to an equilibrium process, in which the solute molecules very rapidly diffuse into all parts of the gel network not mechanically barred to them. With small samples at slow flow rates, very sharp peaks can be obtained. With larger samples or with faster flow rates, the peaks are broader, but not earlier as would be expected if diffusion rates were the basis of separation. Even large mole-

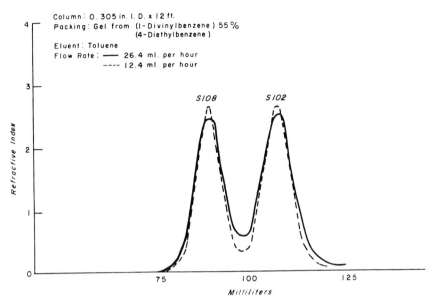

Fig. 6. Separation of two narrow-distribution styrenes at two flow rates.

cules do not appear to affect the permeability of the gel to small molecules, so that the elution pattern of a mixture can be expected to match the sum of the elution patterns of the components, as long as overloads are avoided. These effects were shown in the separation of two polystyrenes by elution with toluene through a column of 0.305 in. i.d. and 12 ft. long. The column was packed with a 200–325 mesh fraction of a bead polymer made from a mixture of one part commercial 55% divinylbenzene and four parts diethylbenzene. The 1.0 ml. sample contained 6.7 mg. each of styrene polymers S108 and S102 (Table I) in toluene. Figure 6 shows the tracings of effluent refractive index at two flow rates, 26.4 ml./hr. and again at 12.4 ml./hr. At the slower rate the resolution was improved, but the peaks were not shifted. Samples of the effluent were evaporated and examined by ultracentrifugation, and the separation was shown to be complete.

While the separation of the different polymeric species from the same monomer can be correlated quite well with molecular weight, nonhomologous compounds do not always correlate so simply. With larger molecules, the molecular size seems to be the prime factor with polarity becoming more important at lower molecular weights. These factors were demonstrated with a gel column of lower permeability, 1.3 cm. diameter and 109 cm. long. The packing was 100–200 mesh PSX4 polystyrene (Table II) swollen in methylene chloride, packed and eluted with the same solvent. Four compounds were sampled, using 0.25 ml. of 1% solution. The elution volumes of the resulting peaks show the expected inverse relationship to molecular model dimensions, with some deviation probably due to polarity, and no correlation with molecular weight, as shown in Table IV.

TABLE IV
Inverse Correlation of Peak Elution Volume with Molecular Size

Substance	Elution volume, ml.	Molecular size, A.	M.W.
Bis (glycidyl ether) of Bisphenol A $(CH_3)_2C(C_8H_4OCH_2CHCH_2O)_2$	75	18.5	340
Dibromotetrachloroxylene $m\text{-}(CH_2Br)_2C_6Cl_4$	108	13.5	402
Mixed xylenes, $(CH_3)_2C_6H_4$	109	8.9	106
Perchloroethylene, C_2Cl_4	122	7.7	166

It is apparent that polystyrene gels, crosslinked in the presence of appropriate diluents, can be used in column chromatography to make molecular size fractionations over an extremely wide range. The term "gel permeation chromatography" is proposed for this technique. It suggests and describes a mechanism of fractionation in which solute molecules are separated by their permeation into a gel which offers different internal volumes to molecules of different sizes over an extended range. This mechanism is prominent in the separation of small nonionic hydrophilic molecules by water elution through beds of ion-exchange resins, as noted much earlier by Wheaton and Bauman,[1] and in the "gel filtration" fractionation of hydrophilic molecules on Sephadex.[3,4] However, to us the term "gel filtration" has not seemed apt for a fractionation of extended range. It did seem apt for the short-column, sievelike separations with which the Sephadex gels were first reported.

This work has shown that structurally modified polystyrene gels, eluted with compatible solvents, have utility in making molecular size fractionations of hydrophobic macromolecules. It seems quite possible that suitable copolymer-diluent systems may be found to yield gels of adequate strength and rigidity with any desired combination of solvent compatibility and permeability. This would indeed open a broad field of usefulness to gel permeation chromatography.

Important contributions of many other persons to this study are gratefully acknowledged. Thanks are particularly due to W. G. Lloyd and T. Alfrey for consultations on the control of permeability in crosslinked gels, to D. R. Asher and D. B. Parrish for some of the fine-mesh bead preparations, to H. W. McCormick for the anionically polymerized styrenes and the ultracentrifuge work, to E. B. Bradford for the electron microscopy, and to M. C. Arrington for technical assistance.

References

1. Wheaton, R. M., and W. C. Bauman, *Ann. N. Y. Acad. Sci.*, **57**, 159 (1953).
2. Lathe, G. H., and C. R. Ruthven, *Biochem. J.*, **62**, 665 (1956).
3. Porath, J., and P. Flodin, *Nature*, **183**, 1657 (1959).
4. Flodin, J., and J. Porath, U. S. Pat. 3,002,823 (October 3, 1961).
5. Lea, D. J., and A. H. Sehon, *Can. J. Chem.*, **40**, 159 (1962).
6. Cortis-Jones, B., *Nature*, **191**, 272 (1961).
7. Vaughan, M. F., *Nature*, **188**, 55 (1960).

8. Lloyd, W. G., and T. Alfrey, papers presented at 139th Meeting, American Chemical Society, St. Louis, March 1961, Division of Polymer Chemistry, papers 6 and 7.

9. Lloyd, W. G., and T. Alfrey, *J. Polymer Sci.*, **62**, 301 (1962).

10. Meitzner, E. F., and J. A. Oline, Union S. Africa Pat. Appl. 59/2393 (May 19, 1959), to Rohm and Haas.

11. Bortnick, N. M., U. S. Pat. 3,037,052 (May 29, 1962).

12. Kressman, T., and J. Miller, Brit. Pat. 889,304 (to Permutit).

13. Lloyd, W. G., private communication.

14. McCormick, H. W., *J. Polymer Sci.*, **36**, 341 (1959).

Résumé

On the préparé des gels de polystyrène pontés en présence de diluants sous forme de fines perles convenant à l'entassement dans les colonnes chromatographiques. Une série de fractions polymériques de poids moléculaires voisins est éluée a travers de telles colonnes avec des solvants aromatiques et chlorés. On détermine et enrégistre les concentrations sortante à l'aide d'un réfractomètre continu différentiel. Les fractions sont ainsi efficacement séparées. On prépare des colonnes capables de séparer des échantillons de polymères de poids moléculaires voisins élevés à partir de gels pontés en présence de grandes quantités de diluants possédant une action solvatante faible ou nulle sur le polystyrene. Des proportions plus faibles de diluants et ceux possédant une plus forte action solvatante fournissent des colonnes présentant des limites de perméabilité des moléculaires plus faibles. De telles études donnent une vue quantitative unique de la topologie des gels. Elles démontrent également que des résultats rapides et reproductibles de distributions de poids moléculaires peuvent être obtenus de cette façon sur des polymères pour lesquels des solvants compatibles avec les gels sont existants.

Zusammenfassung

In Gegenwart von Verdünnungsmitteln vernetzte Polystyrolgele wurden in Feinsieb-Kugelform zur Verwendung in chromatographischen Säulen erzeugt. Eine Reihe eng-verteilter Polymerfraktionen wurde durch solche Säulen mit aromatischen und chlorier-ten Lösungsmitteln eluiert. Die Effluentkonzentrationen wurden mit einem kontin-uierlich arbeitenden Differentialrefraktometer bestimmt und registriert. Es trat eine wirksame Trennung der Fraktionen ein. Zur Trennung von hochmolekularen Poly-merproben mit benachbartem Molekulargewicht wurden Säulen mit Gelen gebaut, die in Gegenwart grosser Mengen von Verdünnungsmitteln mit geringer oder fehlender Lösungswirkung für Polystyrol vernetzt worden waren. Kleinerer Gehalt an Ver-dünnungsmittel und Verdünnungsmittel mit grösserer Lösungswirkung für Polystyrol lieferten Säulen mit niedrigeren Molekulargewichtspermeabilitätsgrenzen. Diese Un-tersuchungen ergaben einzigartige Aufschlüsse über die Topologie der Gele. Sie zeigter ferner, dass auf diese Weise für Polymere, für welche mit den Gelen verträgliche Lösung-smittel vorhanden sind, rasche und reproduzierbare Ergebnisse für die Molekular-gewichtsverteilung zugänglich sind.

Received December 14, 1962

COMMENTARY

Reflections on "Aromatic Polypyromellitimides from Aromatic Polyamic Acids," by C. E. Sroog, A. L. Endrey, S. V. Abramo, C. E. Berr, W. M. Edwards, and K. L. Olivier, *J. Polym. Sci. A.,* 3, 1373 (1965)

C. E. SROOG

Polymer Consultants Inc., Wilmington, Delaware

I received a Ph.D. in Organic Chemistry from the University of Buffalo in June 1950 and a month later joined the Research Division of the newly organized Film Department of the duPont Company. This Department was concerned with self-supporting films such as polyester film, polyethylene, polyvinyl fluoride film, and cellophane.

Film's only research lab was the Yerkes Research Lab located in Buffalo. It proved to be an excellent research site, combining scouting research with additional projects on applied objectives both product and process. I had the good fortune to report to Dr. Emmette Izard, an internationally recognized chemist who had brought polyesters to duPont. His support and very high standards helped greatly during my first program on synthesis of new polyesters and on a concurrently self-initiated project on synthesis of poly(etherketone)s in liquid SO_2. For the next three years, I reported to several different supervisors on polyester catalysis (some of my suggestions were used) and other projects. In April 1954, I received my first promotion to Research Supervisor. I now was responsible for a variety of scouting projects including poly(aramids), polyethylene, and corona-resistant polyesters. I was also given the important responsibility of developing and maintaining contact with key research people at duPont's Experimental Station in Wilmington.

During my first visit there in the summer of 1954, extensive contact with the Polychemicals Department resulted in my first acquaintance with polyimides. A group was involved with a polyimide derived from pyromellitic dianhydride (PMDA) and 4,4'-dimethylheptamethylene diamine; the polymer was prepared by heating at high temperature using procedures similar to aliphatic polyamides. The polyimide called Polymer E was touted as superior to poly(ethylene terephthalate), the future Mylar polyester film. I did not agree with this opinion. Regardless, I asked many questions regarding alternative diamines such as *m*-phenylene diamine (MPD), *p*-phenylene diamine (PPD) and hexamethylene diamine. I was told that these diamines resulted in useless "brick dust" and the very few molded objects prepared were very friable and useless. I came away with the feeling that research should be done on an intermediate stage but no specifics were yet in my head. On this same trip, I made the acquaintance of Dr. W. F. Gresham, who was Manager of the Exploratory Group in the same department. Frank Gresham, who was about 20 years my senior, was much respected. Our discussion went very well but I did not bring up polyimides at this meeting. Dr. Gresham proved to be enormously helpful later on in the development of polyimides.

Later in the year, I was told privately by the Yerkes Lab Director that Film was building a lab at the Experimental Station to conduct Film's scouting research. I let him know that I would be very interested in joining such a lab; early in 1955 I was transferred there, reporting to Dr. H. W. Gray, Director. A program review ensued wherein I brought up the polyimides, emphasizing the objective of a malleable intermediate as a route to polyimides; Hugh Gray endorsed this goal enthusiastically. Our lab opened late spring but with a skeleton staff; recruiting became our first objective. One of the first candidates was A. L. Endrey, who was offered a position because of his excellent impression on us. Andy Endrey accepted our employment offer and arrived in March 1956; he was assigned to the polyimide project which I had written three months earlier. In our planning discussion we discussed possible solvents and included DMF and DMAc in our group of solvents to assess.

Andy conducted his first experiments in late March with PMDA and MPD. Once we understood that the PMDA had to be added to the diamine solution, the reaction went very well in DMF and subsequently in DMAc. Cast dried film showed IR absorption for COOH and NH$_2$ and we realized that a poly(amic acid) had been synthesized. Thermal conversion to polyimide was the next step followed by testing of the polyimide properties; it was clear that thermal resistance was outstanding with an excellent balance of physical and electrical properties. One month later, chemical conversion was demonstrated using a mix of acetic anhydride/pyridine as the converting agent. Frank Gresham graciously noted Film's achievement and assigned Dr. S. V. Abramo the task of finding a moldable aromatic polyimide. Endrey had prepared a number of polyimides including the diamine, 4,4'-diaminodiphenyl ether (ODA) but had evaluated only the MPD polymer. It had shown excellent film properties but also had somewhat poor hydrolytic stability. Sam Abramo decided that he had the best chance of success in the search for a malleable molding resin if he concentrated on ODA-PMDA, due to the ether component. He not only laid the basis for a polyimide molding but he also determined that ODA-PMDA had very good hydrolytic stability. We switched to ODA immediately, and the rest was history.

Yerkes Laboratory was overwhelmed with other products at the final research stage. Therefore we took on the responsibility for early market development preparing small rolls of polyimide film and

Figure 1.   Photograph of the Zero Strength Test for the ODA-PMDA polyimide film which is the structure for Kapton. The test application to our earliest ODA-PMDA film was conducted by Dr. C. E. Berr; the polyimide film had a value of 750 to 800°C. The corresponding value for aluminum was in a range of 550 to 600°C. In discussing these results with duPont management, despite the major difference polyimide vs. aluminum, I tried to restrain my enthusiasm, because at first we had no specific indication of product significance. This developed very rapidly in our early visits to electrical companies in the area of heavy duty switch engines, where the combination of thermal stability and mechanical integrity at temperatures above 400°C gave the basis for industry acceptance. The ZST levels of polyimides were however very important in obtaining early attention at the highest levels of management; they were as excited as we were.

sampling electrical and aircraft companies as well as the Defense Department. The results were very positive; Yerkes assumed responsibility and did a superb job preparing for the first plant. This was started up in October 1965 and is still running at high productivity. By the end of 1965, in addition to polyimide film, duPont had a polyimide molding resin, and coating compositions.

Of the original authors of this paper, I am the only one still active in the polyimide field, writing, lecturing and consulting. It is still extremely interesting particularly with the amazing proliferation of polyimide structures serving unique industries.

PERSPECTIVE

Comments on "Aromatic Polypyromellitimides from Aromatic Polyamic Acids," by C. E. Sroog, A. L. Endrey, S. V. Abramo, C. E. Berr, W. M. Edwards, and K. L. Olivier, *J. Polym. Sci. A,* 3, 1373 (1965)

PAUL M. HERGENROTHER

NASA Langley Research Center, Hampton, Virginia

The paper entitled "Aromatic Polypyromellitimides from Aromatic Polyamic Acids" by C. E. Sroog, A. L. Endrey, S. V. Abramo, C. E. Berr, W. M. Edwards, and K. L. Olivier is a true classic in the field of high performance/high temperature polymers. The paper reports on the synthesis of poly(amic acids), their conversion to polyimides and the mechanical and physical properties of the polymers. The synthetic route to polyimides disclosed in this paper, namely the reaction of an aromatic dianhydride with an aromatic diamine, is still the most popular method for most applications. This article draws on several years of earlier work by this talented group of polymer scientists. The paper is the most frequently referenced paper in the field of polyimides. This premiere work is considered to be the seed that eventually produced the bud that blossomed into a worldwide market for polyimides that approaches one billion dollars today as reported in a survey on high temperature polymers in 1992 by Kline and Company, Fairfield, NJ.

The initial impetus to develop high temperature polymers was to satisfy the demands of the electrical and aerospace (defense or military) industries. This work started in the mid-1950s with polyimides and a few years later with the well known work of the late Prof. Carl Marvel in polybenzimidazoles. Since then, virtually every thermally stable heterocyclic unit has been incorporated in polymers. Although a myriad of polymers have been reported, only a few families of high performance/high temperature polymers are commercially available today. A favorable combination of price, processability, and performance and the strong support of a company are required to develop successful products. More work has been performed on polyimides than all the other high temperature polymers combined. This has been primarily due to the availability of monomers, the ease of polymer synthesis, and the excellent properties of the polymers.

The work described in this paper was initiated in the mid-1950s. At that time, it was difficult to convince management to invest research dollars in an esoteric project like polyimides. However, under the capable leadership of the principal author, Cyrus E. Sroog, a research budget was obtained at duPont and work was begun by a group of researchers who had the talent, insight, confidence, perseverance, and work ethics to accomplish much with little. As the research progressed and management began to understand and appreciate the potential of polyimides, support became more available and polyimide development flourished in the sixties and seventies. This pioneering work placed duPont in the enviable position of becoming the world leader in polyimides with well-known products such as Kapton® (films), Vespel® (moldings), Pyralin® (coatings), and Avimid® (composite matrices). Many other organizations have conducted research and development in polyimides but none have come close to approaching the success of duPont. It is unlikely polyimides could be developed in the tight industrial research environment of today.

Polyimide products such as those above, as well as others, find use in a variety of applications such as medical devices, separation systems, structural components in airplanes including jet engines, binders in abrasive and cutting discs, automotive components, liquid crystalline displays, household appliances, electronic/microelectronic components, wire insulation, and many other applications. Polyimides touch our lives in many different ways. For example, polyimides find use as insulation on electrical wiring in the large commercial airplanes in which we fly, traction motors in the trains we ride, electrical switches in automobiles we drive, household irons we use, pacemakers to regulate our heartbeat, eye lense implants for us to see, and computers that make us more efficient.

The future holds great potential for polyimides as present markets grow and new markets and products are developed. One particularly attractive potential market is in future high speed civil transports (HSCT) where adhesives and composite matrices represent enabling technology. Without these materials, economically viable HSCTs cannot be built.

Polyimides are currently the leading candidates for use as structural adhesives and composite matrices. The requirements are very demanding and the payoff is extremely attractive. For example, if 50% of the structural weight of the vehicle is composites and the cost of the prepreg is $100/pound, each airplane would require about $7.5 million of prepreg or a potential prepreg market of about $3.8 billion for 500 airplanes.

This paper published in 1965 opened a virgin area of polymer chemistry which has grown and matured into an attractive and prosperous business that has benefited many companies. Through the innovative work of these authors, a new area of polymers was developed that still thrives. This work illustrated the beneficial and unique performance that high temperature polymers could provide. The selection of this paper as one of the most influential publications during the first half-century of the *Journal of Polymer Science* is well-deserved and a tribute to the authors. The impact of this paper in forging the field of high performance/high temperature polymers is especially well-recognized by all who have worked in this area.

JOURNAL OF POLYMER SCIENCE: PART A   VOL. 3, PP. 1373–1390 (1965)

Aromatic Polypyromellitimides from Aromatic Polyamic Acids*

C. E. SROOG,† A. L. ENDREY,‡ S. V. ABRAMO, C. E. BERR, W. M. EDWARDS, and K. L. OLIVIER,§ *Film and Plastics Departments, E. I. du Pont de Nemours and Company, Inc., Experimental Station Laboratory, Wilmington, Delaware*

Synopsis

Aromatic diamines react with pyromellitic dianhydride in solvents such as dimethylacetamide to form solutions of high molecular weight polyamic acids. These soluble polymers can be converted to insoluble aromatic polyimides. The aromatic polyimides exhibit zero strength temperatures above 700°C. and possess excellent thermal and oxidative stability. The stability characteristics of films and the relation of chemical structure to thermal and hydrolytic stability and to polymer properties such as crystallinity are described.

INTRODUCTION

We have been interested for some time in the possibilities of preparing intractable condensation polymers by conversion of a high molecular weight, tractable precursor. Most studies of condensation polymerization have been concerned with rather simple elimination reactions to form polymers which are rheologically manageable, thereby permitting formation of variously shaped objects. We wish to report the preparation and properties of a new group of condensation polymers, the aromatic polyimides, which are unusually high melting, intractable, insoluble, and possess outstanding thermal stability.

Earlier workers in this field prepared polyimides by polyamide-salt techniques.[1-3] Thus, pyromellitic dianhydride was reacted with ethanol to form the diester-diacid, which was treated with diamine to precipitate the monomeric diester-diacid salt. The salt was then heated to form a polymerizing melt, which, after dehydration and dealcoholation in the melt, formed the polyimide. Recently several du Pont patents[4] and patent applications[5] disclosing other synthetic routes have appeared and have been

---

* Presented at the 147th meeting of the American Chemical Society, Philadelphia, April 1964, and at the Bad Nauheim meeting of the Fachgruppe *"Kunstoffe und Kautschuk," Gesellschaft Deutscher Chemiker*, April 1964.

† To whom inquiries should be addressed.

‡ Present address: Union Carbide Corp., Parma, Ohio.

§ Present address: Union Oil Company of California, Los Angeles.

1373

1374 SROOG, ENDREY, ABRAMO, BERR, EDWARDS, OLIVIER

cited in subsequent literature articles.[6-8] Published thermal stability data[7,8] have confirmed our observations.

This paper, which reviews activity in this laboratory over the last several years, describes the formation of polypyromellitimides by direct polymerization of pyromellitic dianhydride* and aromatic diamines in solvents such as dimethylacetamide or dimethylformamide to soluble, high molecular weight polyamic acids (also called polyamide acids). The polyamic acids are then dehydrated to polyimides of high molecular weight. The details of polymerization and the relationship of polyimide structure to properties as illustrated by specific diamines are included.

pyromellitic   dianhydride

Soluble polyamic acid   Insoluble polyimide

where R may be p-phenylene; m-phenylene; p,p'-biphenylene; 4,4'-diphenylene ether, sulfide, sulfone, methylene, and -isopropylidene.

EXPERIMENTAL

Reagents

Dimethylacetamide. du Pont technical grade material was distilled from phosphorus pentoxide at 30 mm. Hg.

Dimethylformamide. du Pont technical grade material was distilled from phosphorus pentoxide at 20 mm. Hg.

Diamines. Bis(4-aminophenyl) ether was purified by vacuum drying at 50°C. at ≤ 1 mm. Hg for 8 hr.

Bis(4-aminophenyl) sulfide was purified by decolorization of an aqueous solution of the dihydrochloride salt followed by precipitation with base and codistillation with dioctyl phthalate at ≤ 1 mm. Hg. White crystals of diamine which separated from the distillate on cooling were washed three times with petroleum ether and dried at ≤ 1 mm. Hg.

m-Phenylenediamine was purified by vacuum distillation.

Bis(4-aminophenyl) sulfone was best purified by recrystallization from an aqueous ethanol solution followed by vigorous vacuum drying.

* Polyimides derived from other dianhydrides will be described in future publications.

Pyromellitic Dianhydride. Technical grade material of the du Pont Explosives Department was sublimed through silica gel at 220–240°C. and 0.25–1 mm. mercury pressure.

Polymerizations

Typical Preparation of a Polyamic Acid. A 500-ml. flask fitted with mercury-seal stirrer, nitrogen inlet, drying tube, and stopper is carefully flamed to remove traces of water on the walls and is allowed to cool under a stream of dry nitrogen in a dry box. In the dry box 10.0 g. (0.05 mole) bis(4-aminophenyl) ether is added to the flask through a dried powder funnel, and residual traces are flushed in with 160 g. of dry dimethylacetamide. There is then added 10.90 g. (0.05 mole) pyromellitic dianhydride to the flask through a second dried powder funnel over a period of 2–3 min. with vigorous agitation. Residual dianhydride is washed in with 28 g. dry dimethylacetamide. The powder funnel is replaced by the stopper and the mixture is stirred for 1 hr. A small surge in temperature to 40°C. occurs as the dianhydride is first added, but the mixture rapidly returns to room temperature.

The above procedure yields a 10% solution which can be stirred without difficulty. In some cases the mixture becomes extremely viscous, and dilution to 5–7% solids may be required for efficient stirring. Polyamic acids so prepared exhibit inherent viscosities of 1.5–3.0 at 0.5% in dimethylacetamide at 30°C.

The polymer solutions may be stored in dry, sealed bottles at −15°C. until needed.

Preparation of Polyamic Acid Films. Thin layers (10–25 mils) of polyamic acid solutions are doctored onto dry glass plates and dried for 20 min. in a forced draft oven (with nitrogen bleed) at 80°C. The resulting films are only partly dry and after cooling can be peeled from the plates, clamped to frames, and dried further *in vacuo* at room temperature.

Polyimide Conversion. Films of the polyamic acids which have been dried to a solids level of 65–75% are clamped to frames, heated in a forced draft oven to 300°C. for an additional hour. Further heating at 300°C. causes essentially no increase in the imide infrared absorptions at 5.63 and 13.85 μ.

Property Measurement

Crease Test. A crease and bend test was used as a criterion of failure for films exposed in an oven with temperature control ±3°C. to 500°C. In the crease test, a narrow strip of the film was cut off and creased sharply between the fingers. In each case, several strips were cut from the sample and failure was recorded when more than half of the strips failed the test. The orders of stability as determined by this crease test correlated qualitatively with weight loss measurements.

Zero Strength Temperature. The du Pont Zero Strength Temperature Test was used to measure the temperature at which a polymer film (generally in the range of 20–75 μ thick) failed under a tensile load of 1.4 kg./ cm.^{2} (20 psi) 5 ± 0.4 sec. after contact with a preheated metal bar. This technique is relatively insensitive to film thickness within the range stated, and depending on sample uniformity, has the following precision: (1) about ±5°C. for Mylar polyester film (near 240°C.) (2) ±15–20°C. for poly-[$N,N'(p,p'$-oxydiphenylene)pyromellitimide] film (800–900°C.).

In practice, film strips 0.25 in. wide and 5 in. long are attached to a clamp at one end, and sufficient weight is attached to the other end to give a tensile load of 20 psi. The film strip is then contacted with a heated metal cylinder and a plot is made of the time required for failure (burning or melting through) versus the temperature of the cylinder. Measurements are made at several temperatures above and below that at which 5 sec. is required for failure, and a curve is drawn through the points. The zero strength temperature is then determined from the curve.

Van de Graaff Irradiation. The sample is moved under the electron beam 10 cm. below an aluminum window at a rate of 2.27 cm./sec. with a beam scanning width of 20 cm. For 2 M.e.v. electrons and a beam current of 250 μamp., one pass gives a flux at the surface of 12.5 watt-sec./ cm.^{2}. For a material with a density of one and an electron range of 1 cm., the average dose per pass is 1.25 Mrads (1.25 × 10^8 ergs/g.).

Brookhaven Pile Exposure. Samples of film placed in the Brookhaven Pile were exposed to thermal neutrons, γ-rays, and other radiation. The flux in the pile is expressed in neutrons/cm.^{2}/sec. ($\overline{nvt}$).

Viscosity Measurements. Solution viscosities were measured on dilute solutions utilizing the indicated anhydrous solvents, concentrations, and temperatures. Dissolution (or dilution in the case of polyamic acid 10–15% solutions) was generally achieved by shaking at the temperature of measurement (30 or 15°C.). The work with fuming nitric acid requires cooling to minimize molecular weight degradation. The glass viscometers used were conventional Ostwald-Fenske types with capillaries sized to give flow times in the 40–120 sec. range, and bath temperatures were held to within ±0.05°C.

Thermogravimetric Analyses. Measurements of weight loss were conducted in a continuous-recording thermobalance, (Aminco Thermo-Grav), in controlled atmospheres. Temperature-programmed curves were obtained at slow rates (3°C./min.), while isothermal runs were made by preheating the chamber, then inserting the sample and holding to within ±0.5°C. The rate of gas flow was 100 cc./min.

X-Ray Measurements. Diffraction patterns were obtained on a General Electric XRD-5 diffractometer, utilizing nickel-filtered copper Kα radiation (1.54 A.) at 50 kv. and 16 ma. and pinhole collimation. The flat-plate Laue photographs were obtained with a 2.5-cm. sample-to-film distance; a 1.25-mm (50 mils) specimen thickness and various appropriate exposure times were used.

Mechanical and Electrical Properties. These measurements were obtained by utilizing official ASTM methods (designation D-150-59 for the dielectric measurements and D-882-61T for the tensile properties). The electrical measurements utilizing an improved Schering Bridge and specimens were 25 μ (1-mil) films sprayed with silver paint electrodes to give precise areas in intimate contact. Tensile tests were conducted at strain rates in the range of 30–100%/min.

RESULTS AND DISCUSSION

Synthesis of Polyamic Acids

The preparation of polyamic acids is generally carried out by adding pyromellitic dianhydride to a solution or slurry of the diamine in a polar solvent such as dimethylacetamide or dimethylformamide. The polymerization may be conducted heterogeneously by addition of pyromellitic dianhydride as a solid or as a slurry in a relatively inert liquid such as acetone, γ-butyrolactone, or benzonitrile. The rate of reaction appears to be limited by rate of solution of pyromellitic dianhydride; normally 30–60

TABLE I
Effect of Temperature of Polymerization on Inherent Viscosity of Polyamic Acid
[Reaction of Bis(4-aminophenyl) Ether and Pyromellitic Dianhydride]

Solvent	Bis(4-amino-phenyl) ether		Pyromellitic dianhydride		Solids, %	Temp., °C.	Time at temp., min.	η_{inh}[a]
	g.	Moles	g.	Moles				
DMAc	10.00	0.05	10.90	0.05	10.0	25	120	4.05
DMAc	10.00	0.05	10.90	0.05	10.0	65	30	3.47
DMAc	20.00	0.10	21.80	0.10	10.6	85–88	30	2.44
DMAc	20.00	0.10	21.80	0.10	10.7	115–119	15	1.16
DMAc	20.00	0.10	21.80[b]	0.10	10.3	125–128	15	1.00
DMAc	20.00	0.10	21.80[c]	0.10	15.7	135–137	15	0.59
N-Me capro-lactam	10.00	0.05	10.90	0.05	14.2	150–160	2	0.51
N-Me capro-lactam	10.00	0.05	10.90	0.05	12.9	175–182	1–2	Only partly soluble
N-Me capro-lactam	20.00	0.10	21.80	0.10	15	200	1	Insol.

[a] Determined at 0.5% concentration in the particular solvent at 30°C.

[b] Increment of 0.35 g. pyromellitic dianhydride added before determination of η_{inh}.

[c] Increment of 0.25 g. pyromellitic dianhydride and then 0.21 g. of bis(4-aminophenyl)ether added before determination of η_{inh}.

min. is adequate for laboratory-scale polymerization. Total solutes average 10–15%, although it has been possible to attain high molecular weight polyamic acids at concentration levels of 35–40%. At such high concentrations, the slurry method of addition enables the best mixing, and efficient cooling is required to compensate for the heats of reaction and stirring. When the solid pyromellitic dianhydride is added, a red-orange color is observed at the solid–liquid interface; this rapidly lightens to a lemon-yellow as the pyromellitic dianhydride dissolves and reacts with the di-amine.

Best results are obtained from both types of procedure at temperatures of 15–75°C.; above 75°C. a decrease in the molecular weight of the poly-amic acid becomes marked. Above 100°C. cyclization to imide is ap-preciable, causing eventual precipitation of polyimide as well as a lowered molecular weight of the polyamic acid (Table I). Above 150°C., the cyclization is so rapid that polyimide sometimes precipitates before all of the dianhydride can be added.

Conversion to Polyimide

Conversion of Polyamic Acid to Polyimide. Cyclization to polyimide may be carried out by heating a film of polyamic acid from 25 to 300°C. at a carefully controlled rate. As cyclization proceeds, the color of the polyamic acid darkens from a pale yellow to a deep yellow or orange.

Complete infrared spectra and ultraviolet spectra for poly[$N,N'(p,p'$-oxydiphenylene)pyromellitimide] are shown in Figures 1–4. (The instru-

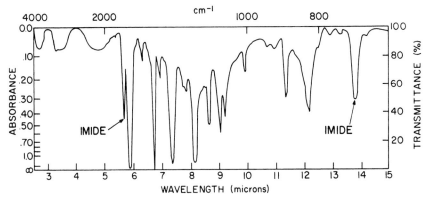

Fig. 1. Infrared spectrum of 0.1 mil poly[$N,N'(p,p'$-oxydiphenylene)pyromellitimide] (POP-PI) film. The film was heated to 300°C. over a period of 45 min., then held at 300°C. for 1 hr.

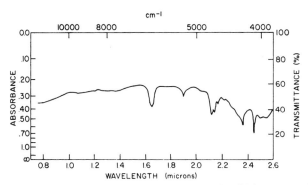

Fig. 2. Near infrared spectrum of POP-PI film 14 mils thick, scan 50 A./sec.

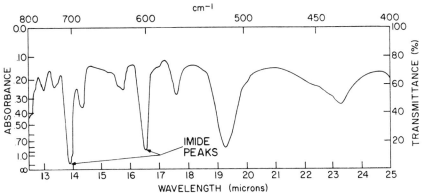

Fig. 3. Far infrared spectrum of POP-PI film 0.5 mil thick.

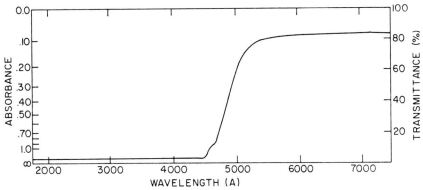

Fig. 4. Ultraviolet and visible spectrum of 0.1 mil POP-PI film

ments used were the Perkin-Elmer Model 21 with NaCl or KBr optics, and the Cary Model 14.) The cyclization can be followed spectrophoto-metrically[9] (Fig. 1) by disappearance of the N-H band (3.08 μ) and ap-

TABLE II
Nitrogen Analyses for Polypyromellitimides
(R = Diamine Component)

| | N, % | |
R	Calculated	Found
	9.7	9.8 9.5
	7.7	7.9 7.7
$-CH_2-$	7.4	7.5 7.6
$-O-$	7.3	7.3
$-SO_2-$	6.5	6.6 6.7

pearance of imide bands (5.63, 13.85 μ).   Nitrogen analyses correspond to the converted ring structure (Table II).

Properties

Polyamic Acids.   Polyamic acids are soluble in a variety of solvents, such as dimethylacetamide, dimethylformamide, dimethylsulfoxide, and

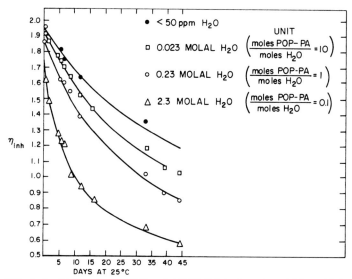

Fig. 5. Effect of added water on 0.23 molal POP-PA [polyamic acid from bis(4-amino-phenyl)ether and pyromellitic anhydride] solution in dimethylacetamide (DMAc).

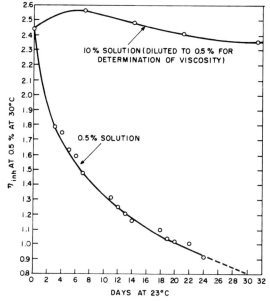

Fig. 6. Effect of concentration on stability of POP-PA in DMAc at 23°C.

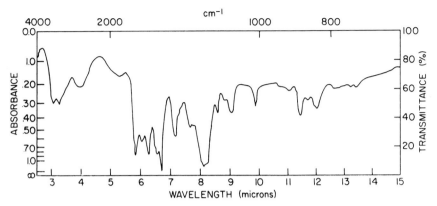

Fig. 7. Infrared spectrum of POP-PA film 0.1 mil thick, dried 2 hr. at 80°C.

tetramethyl urea. The amic acid polymer is sensitive to hydrolysis by water, similar to the monomeric phthalamic acids as previously noted by Bender[10] (Fig. 5). The stability at 23°C. of the polyamic acids in solution is dependent upon the polymer concentration, dilute polymer solutions decreasing in viscosity much more rapidly than concentrated solutions (Fig. 6).

Films of polyamic acids containing 25–35% of residual solvent can be prepared by careful drying of cast solutions. These films on dissolving exhibit considerably lower inherent viscosities than the original polyamic acids (Table III). Polyamic acids have no true softening points, since dehydration occurs with heating to form polyimide; as noted above this can occur in solution at temperatures as low as 100°C. The polymer, or

TABLE III
Decrease of Inherent Viscosity of Polypyromellitamic Acids During Film Casting and Drying (R = Diamine Component)

R	Solvent	η_{inh} solution	Drying conditions	η_{inh} film
⟨phenyl⟩–O–⟨phenyl⟩– [a]	Dimethylacetamide	2.1	30 min., 80°C.	1.3
		2.1	30 min., 80°C. 5 days, 23°C.	1.3
⟨phenyl⟩–	Dimethylacetamide	1.0	10 min., 120°C.	0.4
		1.2	30 min., 80°C.	0.33
	Benzonitrile/dimethylformamide (3/2)	1.0	15 hr., 25°C.	0.5
⟨phenyl⟩–CH₂–⟨phenyl⟩–	Dimethylacetamide	1.3	6 hr., 70°C.	0.7
	Dimethylacetamide	2.3	20 min., 110°C.	0.5
⟨phenyl⟩–S–⟨phenyl⟩–	Dimethylacetamide	1.2	30 min., 120°C.	0.5
⟨phenyl⟩–SO₂–⟨phenyl⟩–	Dimethylacetamide	0.7	30 min., 120°C.	0.4

[a] POP-PA film is polyamic acid from bis(4-aminophenyl) ether and pyromellitic dianhydride.

film cast from solution, will not cyclize extensively until temperatures of 120–150°C. are reached. A characteristic infrared spectrum of such a polyamic acid film from bis(4-aminophenyl) ether is shown in Figure 7.

Polyimides. A series of polypyromellitimides, some of which are summarized in Table IV, was prepared. Aromatic polypyromellitimides are extremely high melting, difficultly soluble polymers with outstanding resistance to heat, oxygen, and irradiation; their softening points, as determined by zero strength temperature of films, are normally well over 700°C. (in comparison, aluminum foil has a zero strength temperature of 550°C.). In contrast with polyamic acids, polyimides are colored compounds whose intensity of color depends upon the diamine component. The polyimide derived from bis(4-aminophenyl) sulfone is pale yellow; the polyimides from m-phenylenediamine, p-phenylenediamine, and bis(4-aminophenyl) ether are yellow to orange, while that based upon bis(4-aminophenyl)sulfide is deep red.

TABLE IV
Polypyromellitimide Films (R = Diamine Component)

R	η_{inh}[a]	Density
m-Phenylenediamine	>0.3	1.43
p-Phenylenediamine	0.5	1.41
p,p'-Biphenylenediamine	>0.3[b]	1.43
Bis(4-aminophenyl)methylene	1.7	1.36
Bis(4-aminophenyl) isopropylidene	0.5	1.30
Bis(4-aminophenyl) thioether	>0.3[b]	1.41
Bis(4-aminophenyl) ether	1.0[b]	1.42
Bis(4-aminophenyl) sulfone	>0.3	1.43
Bis(3-aminophenyl) sulfone	0.5	—

[a] 0.5% in H_2SO_4.
[b] Fuming HNO_3.

Thermal Stability. Thermal stability has been determined for polyimide films by weight loss measurements in air and helium atmospheres and by retention of film creasability upon heating in air. The data, summarized in Figures 8–10 and Table V, indicate that polyimides as a class possess outstanding thermal stability. When considered along with the retention of mechanical integrity as evidenced by zero strength temperatures in the metallic range, these stabilities become truly significant. Thus, the polypyromellitimide derived from bis(4-aminophenyl) ether showed isothermal weight loss in helium atmosphere of only 1.5% after 15 hr. at 400°C., 3.0% after 15 hr. at 450°C., and 7% after 15 hr. at 500°C. In other weight loss measurements during constant rate of temperature increase (3°C./min.) the polyimides derived from m-phenylenediamine, benzidine, and bis(4-aminophenyl) ether exhibited a 1–1.5% weight loss in the heating period to 500°C. The thermogravimetric analyses are lent additional significance by the retention of film toughness after thermal aging summarized in Table V. Films derived from wholly aromatic poly-

AROMATIC POLYPYROMELLITIMIDES 1383

Table V
Properties of Polypyromellitimides (R = Diamine Component)

R	Solubility	Crystallinity	Zero strength temperature, °C.	Thermal stability in air[a]	
				275°C.	300°C.
(phenylene ring)	Amorphous conc. H₂SO₄; crystalline insol.	Crystallizable	900	>1 yr.	>1 mo.
(phenylene ring)	Amorphous conc. H₂SO₄; crystalline insol.	Crystallizes readily	900	>1 yr.	
(biphenyl)	Fuming HNO₃	Highly crystalline	>900	—	1 mo.
(—CH₂— bridged rings)	Conc. H₂SO₄	Slightly crystalline	800	—	7–10 days
(—C(CH₃)₂— bridged rings)	Conc. H₂SO₄	Crystallizable with difficulty	580		15–20 days
(—S— bridged rings)	Fuming HNO₃	Crystallizable	800	10–12 mos. (estimated)	6 weeks
(—O— bridged rings)	Fuming HNO₃	Crystallizable	850	>1 yr.	>1 mo.
(—SO₂— bridged rings)	Conc. H₂SO₄	—	—	—	>1 mo.
(—SO₂— bridged rings)	Conc. H₂SO₄	—	—	—	>1 mo.

[a] As measured by retention of film creasability.

1384   SROOG, ENDREY, ABRAMO, BERR, EDWARDS, OLIVIER

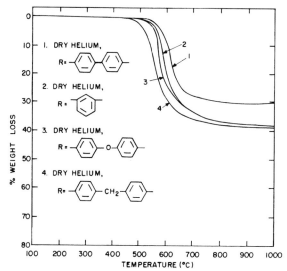

Fig. 8. Effect of diamine structure on weight loss of polypyromellitimides during constant rise in temperature (3°C./min.).

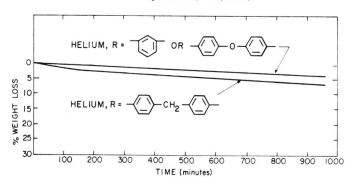

Fig. 9. Weight loss of $\left.\!-\!\!\left(\!R\!-\!N\!\!\left\langle\begin{smallmatrix}O\\\|\\C\\C\\\|\\O\end{smallmatrix}\right\rangle\!\!\right\rangle\!\!\left\langle\begin{smallmatrix}O\\\|\\C\\C\\\|\\O\end{smallmatrix}\right\rangle\!\!N\!\right)\!\!\right._{n}$ at 450°C.

pyromellitimides retain toughness for months at elevated temperatures in air.

Hydrolytic Stability. The hydrolytic stability of the polypyromellit-imides is, with certain exceptions, of a high order (Table VI) and may be considered quite unusual. The polypyromellitimides derived from bis(4-aminophenyl) ether and bis(4-aminophenyl) thioether, which already possess an outstanding level of thermal stability, retain toughness after one year and after three months respectively in boiling water. However, polyimides derived from diamines such as *m*-phenylenediamine and *p*-phenylenediamine exhibit poor hydrolytic stability, becoming embrittled after one week or less in boiling water.

TABLE VI
Hydrolytic Stability of Polypyromellitimides
(R = Diamine Component)

R	Retention of film flexibility in boiling H₂O
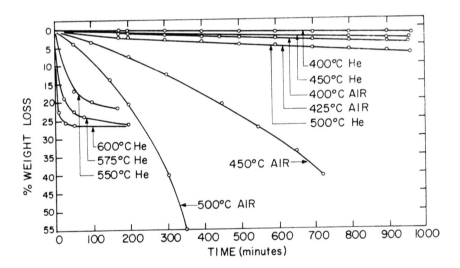	1 week
	1 week
	1 yr.
	>3 mo.

Radiation Resistance. The data in Table VII indicate that films of the aromatic polypyromellitimides are outstandingly resistant to irradiation from high energy electrons and from thermal neutrons. Thus films of the polypyromellitimide from bis(4-aminophenyl)methane exhibit good mechanical and electrical properties after high energy electron exposure of over 10,000 Mrad in the Van de Graaff generator while films of polystyrene and poly(ethylene terephthalate) become embrittled after 500–600 Mrad. Films of the polyromellitimide from bis(4-aminophenyl) ether remain creasable after 40 days exposure to thermal neutrons at 175°C. in the Brookhaven pile.

Fig. 10. Isothermal weight loss of

1386 SROOG, ENDREY, ABRAMO, BERR, EDWARDS, OLIVIER

TABLE VIIA
Effect of Radiation on Polypyromellitimide Films
(Exposure Van de Graaff 2 M.e.v. Electrons)

Polymer film	Thickness, mils	Number passes	Dose, Mrad	Remarks
DDM-PI[a]	2.0	8000	10,000	Retains toughness, good electrical properties
Mylar[b]	2.0–3.0	200	240	Creasable
	2.0–3.0	500	600	brittle, yellow
	3.0			yellow, brittle
Polystyrene	1.2	500	600	Yellow, extremely brittle
Polyethylene (branched)	6–10	200	240	Very weak
	6–10			sticky gum

[a] Polypyromellitimide from bis(4-aminophenyl)methane.
[b] Du Pont's registered trademark for its polyester film.

TABLE VIIB
Effect of Radiation on Polypyromellitimide Films
(Thermal Neutron Degradation[a])

Polymer film	Thickness, mils	Exposure, days	Temperature, °C.	Flux 10^{13} neutrons/ cm.²/sec.	Remarks
POP-PI[b]	2–2.7	40	50–75	0.4	Slightly darkened, brittle in spots
		40	175	0.5	Darkened, tough
		80	175	0.5	Darkened, brittle
Polystyrene	1.2	10	50–75	0.4	Yellow, very brittle
Polyethylene (branched)	3.0	10	50–75	0.4	Sticky, rubbery
	3.0	40	50–75	0.4	Brown varnish
Mylar[c]	3.0	10	50–75	0.4	Failed
	3.0	20	175	0.5	Yellow, brittle

[a] Courtesy of Brookhaven National Laboratory.
[b] Polyimide from bis(4-aminophenyl)ether and pyromellitic dianhydride.
[c] Du Pont registered trademark for its polyester film.

Solution Properties. The polyimides are insoluble in conventional solvents. Thus far, only fuming nitric acid has been found to be a general solvent for the species suitable for qualitative molecular weight evaluation at low temperature; concentrated sulfuric acid is effective for some polyimides. The stabilities of solutions of various polyimides in fuming HNO_3 and in concentrated H_2SO_4 are summarized in Figures 11 and 12.

Crystallinity. Several of the polyimides are crystalline as made, for example, the polypyromellitimide from p-phenylenediamine (Fig. 13). Others are more or less ordered, depending on their structure. Thus, polypyromellitimides derived from m-phenylenediamine, bis(4-aminophenyl)methane, bis(4-aminophenyl)isopropylidene, bis(4-aminophenyl)

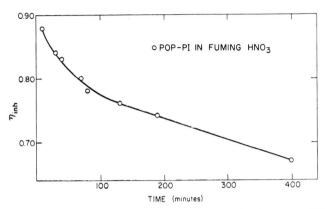

Fig. 11. Inherent viscosity vs. time for dilute solutions of polyimide; 15°C., 0.5 g. polymer/100 ml. solvent.

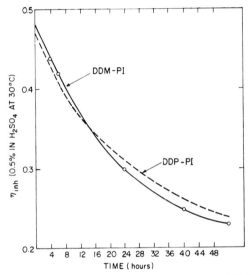

Fig. 12. Stability of polypyromellitimides in concentrated H_2SO_4; 30°C.

sulfide, and bis(4-aminophenyl) ether normally exhibit a low degree of order as prepared. High temperature treatment will cause crystallization of film derived from m-phenylenediamine and of films based upon bis(4-aminophenyl) ether (Fig. 14). Films, despite the slight weight loss upon heating noted previously, retain structural integrity for short exposures to 750–800°C. in air; up to this temperature there appears, for the polymers studied, to be no evidence of crystal melting. These include the polypyromellitimide derived from bis(4-aminophenyl) ether, m-phenylenediamine, and p-phenylenediamine. No true second-order transition temperatures have been determined.

Mechanical and Electrical Properties. Polyimides, in addition to the outstanding permanence characteristics noted above, possess an excellent

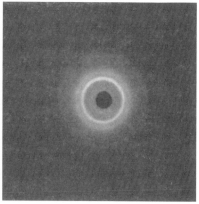

Fig. 13. Laue transmission x-ray diffraction pattern of unannealed poly(p-phenylene pyromellitimide) (filtered CuKα radiation).

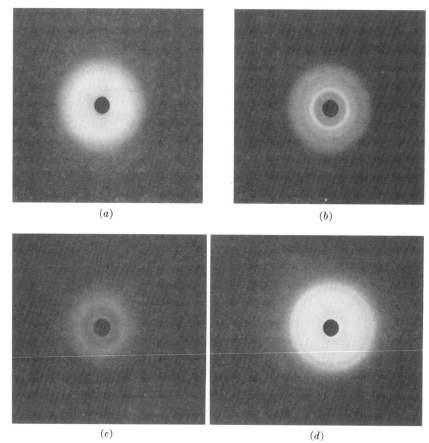

(a)

(b)

(c)

(d)

Fig. 14. Laue transmission x-ray diffraction patterns of (a) polypyromellitimide from m-phenylenediamine, not annealed; (b) same, annealed 2 min. at 400°C.; (c) polypyromellitimide from bis(4-aminophenyl) ether, not annealed; (d) same, annealed at 400°C.

balance of mechanical and electrical properties. The films are stiff and strong and retain mechanical strength to high temperatures. The electrical properties of the various polyimides are all excellent and are summarized as a range characteristic for polyimides in Table VIII. A great deal of information has been obtained for H film, which is derived from bis(4-aminophenyl) ether. The data are summarized in Table IX.

The aromatic polypyromellitimides exhibit outstanding thermal stability

TABLE VIII
Electrical Properties of Polypyromellitimide Films (1–2 mil)

	23°C.	150°C.	200°C.
Dielectric constant (1000 cycles/sec.)	3.1–3.7	2.9–3.1	2.8–3.2
Dissipation factor (1000 cycles/sec.)	0.0013–0.002	0.006–0.0014	0.0005–0.0010
Volume resistivity (23°C., 50% R.H.), ohm-cm.	10^{17}–5×10^{18}	$>10^{15}$	10^{14}–10^{15}
Dielectric strength, v./mil	4550–6900	—	4600–5900

TABLE IX
Properties of H Film[a]

Thermal stability in air[b]	
250°C.	10 yr. (extrapolated)
275°C.	1 yr.
300°C.	1 mo.
400°C.	1 day
Melting point, °C.	>900
Zero strength temperature, °C.	815
Solvents	None
Density	1.42
Glass transition temp., °C.	>500
Tensile strength, psi	
4°K.	Flexible
25°C.	25,000
200°C.	17,000
300°C.	10,000
500°C.	4,000
Elongation, %	
25°C.	70
200°C.	90
300°C.	120
500°C.	60
Tensile modulus	
25°C.	400,000
200°C.	260,000
300°C.	200,000
500°C.	40,000

[a] Data of Tatum et al.[11]
[b] Retention of film flexibility.

matched and supported by a unique property combination which has not been characteristic for other polymeric systems. In addition to thermal stability, the properties include outstanding radiation resistance, unusual resistance to solvent attack, film toughness and its retention under rigorous conditions of air aging, and a level of mechanical properties which make the films useful over a range from extremely low to extremely high temperatures.

Dr. Gordon D. Patterson, Jr. directed the analytical support for this work and contributed much valuable advice. Numerous helpful discussions were held with Professor George S. Hammond, California Institute of Technology.

References

1. Edwards, W. M., and I. M. Robinson, British Pat. 570,858; U.S. Pat. 2,710,853 (1955).
2. Gresham, W. F., and M. A. Naylor, U. S. Pat. 2,731,447 (1956)
3. Edwards, W. M., and I. M. Robinson, U. S. Pat. 2,900,369 (1959).
4. E. I. du Pont de Nemours & Co., French Pat. 1,239,491 (1960).
5. E. I. du Pont de Nemours & Co., Australian Application 58,424 (1960).
6. Anonymous, *Chem. Week*, **88**, 46, June 17, 1961.
7. Jones, J. I., F. W. Ochynski, and F. A. Rackley, *Chem. Ind. (London)*, **1962**, 1686 (Sept. 22, 1962).
8. Bower, G. M., and L. W. Frost, *J. Polymer Sci.*, **A1**, 3135 (1963).
9. Arcoria, A., and R. Passerini, *Bull. Sci. Fac. Chem. Ind. Bologna*, **15**, 121 (1957); *Chem. Abstr.*, **52**, 15244 (1958); S. P. Sadtler and Son, Inc., IR Spectrum #16607.
10. Bender, M. L., *J. Am. Chem. Soc.*, **79**, 1258 (1957); M. L. Bender, Y.-L. Chow, and F. Chloupek, *ibid.*, **80**, 5380 (1958).
11. Tatum, W. E., L. E. Amborski, C. W. Gerow, J. F. Heacock and R. S. Mallouk, paper presented at Electrical Insulation Conference, Chicago, Illinois, September 17, 1963.

Résumé

Des diamines aromatiques réagissent avec le dianhydride pyromellitique dans des solvants tels que le diméthylacétamide pour former des solutions d'acides polyamiques de poids moléculaires élevés. Ces polymères solubles peuvent être transformés en polyimides aromatiques insolubles. Lés polyimides aromatiques présentent une force nulle à des températures situées au-dessus de 700°C et possèdent une stabilité thermique et une stabilité à l'oxydation excellentes. Les caractéristiques de stabilité des films et la relation entre la structure chimique et la stabilité thermique et hydrolytique et les propriétés du polymère telles que la cristallinité sont décrites.

Zusammenfassung

Aromatische Diamine reagieren mit Pyromellitdianhydrid in Lösungsmitteln wie Dimethylacetamid unter Bildung von Lösungen hochmolekularer Polyamidsäuren. Diese löslichen Polymeren können in unlösliche aromatische Polyimide umgewandelt werden. Die aromatischen Polyimide verlieren ihre Festigkeit erst oberhalb 700°C und besitzen ausgezeichnete thermische und oxydative Stabilität. Die Stabilitätscharakteristik von Filmen sowie die Beziehung der chemischen Struktur zur thermischen und hydrolytischen Stabilität und zu Polymereigenschaften wie Kristallinität wird beschrieben.

Received April 12, 1964
(Prod. No. 4453A)

COMMENTARY

Reflections on "Alternating Copolymerization through the Complexes of Conjugated Vinyl Monomers-Alkylaluminum Halides," by Masaaki Hirooka, Hiroshi Yabuuchi, Jiro Iseki, and Yasuto Nakai, *J. Polym. Sci. A-1*, 6, 1381 (1968)

MASAAKI HIROOKA

Faculty of Information Science, Ryutsu Kagaku University, Kobe, Japan

It is a great honor for me that our article on alternating copolymerization has been selected as one of the most important papers in the *Journal of Polymer Science* during the past 50 years.

The new phenomenon of the alternating regulation of vinyl copolymerization was found in the course of trials to copolymerize propylene and acrylonitrile with coordinated catalysts. At that time I was working at the research laboratory of Sumitomo Chemical Company. After some research on polymerization of acrylonitrile for commercialization, I was studying the polymerization of olefins with Ziegler-type catalysts, and also working for nearly a year on the introduction of polypropylene technology from Montecatini. I was enthusiastic to develop proprietary technologies.

Although polar vinyl monomers had been reported not to be polymerized with Ziegler-type coordinated catalysts, I was challenged to polymerize acrylonitrile and found that the $AlEtCl_2$-$VOCl_3$ catalyst was effective. Further, I succeeded in copolymerizing acrylonitrile with propylene to obtain high molecular weight copolymers. It was, however, surprising that $AlEtCl_2$ itself initiated the copolymerization without $VOCl_3$ and that, unexpectedly, the compositions of copolymers were always equimolar irrespective of the monomer feed ratio. This was in 1963 and the first paper appeared in *Polymer Letters*[1] in 1967. The paper in 1968, reproduced in this issue, was the second report to prove that the copolymers certainly have alternating sequential structures. This was the first finding that alternating copolymers were synthesized from combinations of comonomers that gave random copolymers under conventional radical polymerization conditions. A summary of this area was reported in 1973[2] in which monomer combinations to give alternating copolymers in the presence of alkylaluminum halides were concluded to be chosen from an A group of monomers of electron-donating nature and a B group of monomers having a carbonyl or nitrile group conjugated with the double bond.

Three mechanisms for the alternating regulation were postulated. That is: (1) the three-component complex mechanism (N. G. Gaylord[3]) in which a complex formed among two monomers and an alkylaluminum halide causes alternating regulation; (2) the cross propagation mechanism (V. P. Zubov[4]) in which the rate of cross propagation becomes predominant in the presence of an alkylaluminum halide; and (3) the complexed radical mechanism (M. Hirooka[5]) in which alternating regulation is achieved by the charge-transfer interaction of a donor monomer A with the growing chain end of a B conjugated monomer unit complexed with the alkylaluminum halide. No definite conclusion for such debates has been reached so far, but I still believe my hypothesis of the complexed radical mechanism is most reliable.

In later studies, I found many abnormal behaviors of the alternating copolymerization with alkylaluminum halides. For example, although there was certain evidence for its radical mechanism, the copolymerization exhibits rather poor susceptibility to such chain transfer agents as carbon tetrachloride and such inhibitors as hydroquinone. The degradative chain transfer reactions of isobutylene and propylene through hyperconjugation do not take place but instead high molecular weight copolymers are produced. Internal olefins, such as butene-2, stilbene, and cyclopentene, are effectively copolymerized with B group monomers complexed with alkylaluminum halides. If one monomer in the alternating copolymerization is used in excess over the other, the excess monomer is not homopolymerized and remains unreacted after the completion of the copolymerization. Terpolymerization among monomers selected from A and B groups gives abnormal monomer reactivity ratios—different from those for conventional radical copolymerization.

Our analyses by UV, NMR, and cryoscopic measurements demonstrated that ethylaluminum dichloride quantitatively forms a complex with a B group monomer (e.g., methyl acrylate) but no complex with an A group monomer. Thus, it could be inferred that the polymer radical terminated with a B group monomer unit is stabilized by complexing with an alkylaluminum halide and then activated by a charge transfer interaction with an attacking A group monomer. This is the basis of my complexed radical mechanism. Whether this mechanism is accepted or not, it is important that the alternating copolymerization in the presence of alkylaluminum halides gives us much information to elucidate polymerization mechanism in general. Professor C. H. Bamford and I collaborated to determine the rate constants of elementary reactions of the alternating copolymerization and obtained informative results.[6]

The recent development of cationic and radical polymerizations indicates that stabilization of the growing chain end by additives is important to regulate reactivity and prevent side reactions (T. Higashimura, M. Sawamoto, and J. P. Kennedy). Also, the importance of coordinated metal complexes to regulate polymerization reactions has been recognized, and regulation of stereospecificity and molecular weight distribution of polymers have been achieved thereby; metallocene catalysts (W. Kaminsky and H. Yasuda), porphyrin complex catalysts (S. Inoue), acetylene polymerization (T. Masuda), and metathesis polymerization (R. Grubbs). These studies could lead us toward *precision polymerization* as an ideal process in which polymerization can be precisely controlled at the molecular and atomic level in terms of stereospecificity, sequential structures, and molecular weight distribution at the same time.

In the course of our studies on alternating copolymerization by alkylaluminum halides, we found[7] that the glass transition temperature (T_g) of alternating copolymers can be estimated from the monomer reactivity ratios and the T_g of the corresponding homopolymers and equimolar random copolymers. These findings suggest that the correlation between polymer properties and polymerization behavior is worth investigating. Further, precision polymerization will bring about enhanced properties of polymers and will produce more sophisticated polymer functions. In this connection, knowledge of the abnormal behaviors of alternating copolymerization could be informative for exploring precision polymerization.

At the end of my commentary, I recall that these results have been achieved by the enthusiastic help of my colleagues and by the support of Sumitomo Chemical Company and would like to express my sincere appreciation to them again. Also I was much indebted to Professor C. H. Bamford for his kind invitation to the University of Liverpool, U.K., and I thank him for his continued interest and fruitful discussion.

REFERENCES AND NOTES

1. M. Hirooka, H. Yabuuchi, S. Kawasumi, and K. Nakaguchi, *Polym. Lett.,* **5,** 47 (1967).
2. M. Hirooka, H. Yabuuchi, S. Kawasumi, and K. Nakaguchi, *J. Polym. Sci.,* **11,** 1281 (1973).
3. N. G. Gaylord and A. Takahashi, *Polym. Lett.,* **6,** 743 (1968).
4. V. P. Zubov et al., *J. Polym. Sci.,* **C23,** 147 (1968).
5. M. Hirooka, *Polym. Lett.,* **10,** 171 (1972).
6. C. H. Bamford and M. Hirooka, *Polymer,* **25,** 1791 (1984).
7. M. Hirooka and T. Kato, *Polym. Lett.,* **12,** 31 (1974).

PERSPECTIVE

Comments on "Alternating Copolymerization through the Complexes of Conjugated Vinyl Monomers-Alkylaluminum Halides," by Masaaki Hirooka, Hiroshi Yabuuchi, Jiro Iseki, and Yasuto Nakai, *J. Polym. Sci. A-1,* 6, 1381 (1968)

C. H. BAMFORD

Department of Clinical Engineering, University of Liverpool, Liverpool, U.K.

Studies of copolymerization have provided most of our knowledge about the reactivities of vinyl monomers and their derived radicals. The impressive investigations described in the following paper[1] are an excellent illustration of the truth of this statement.

Early workers interpreted copolymerization data in terms of an alternating tendency, connected with electrical polarity, according to which radical A· reacts preferentially with monomer B and vice versa when the monomers have equal reactivity. When the alternating tendency is sufficiently great a fully alternating copolymer may be formed; there are numerous examples of this when the monomer pair comprises a strong electron donor and a strong acceptor.

These thoughts naturally suggest that it should be possible to influence the degree of alternation of a given monomer pair if the polarity of the propagating species could be suitably changed by complexation. With this in mind, in 1957 we[2] studied the free-radical polymerization of acrylonitrile in the presence of lithium salts, determining the absolute rate coefficients of propagation and also chain transfer to CBr_4 and NEt_3. All three coefficients were changed by the presence of LiCl, k_p increasing by 88%, for example. Increase in the reactivity ratio of AN in copolymerization with vinyl acetate was also reported. Those re-

sults were interpreted in terms of formation of

$$\sim CH_2 - \dot{C}H - C = N^{\ominus}$$
$$|$$
$$Cl$$

the radical complexes and

$\sim CH_2 - \dot{C}H - C \equiv N^{\oplus}Li$, in which complexation changes the electron density at the carbon atom with the unpaired spin. In a related study of the polymerization of methyl methacrylate,[3] the presence of zinc chloride was found to increase k_p by 144% while k_t was virtually unaffected.

A few years earlier (1963) Imoto, Otsu, and Harada[4] had shown that the steady-state polymerization of methyl methacrylate is increased by the presence of zinc chloride and in the period 1968 through 1973 numerous observations were made of changes in monomer reactivities brought about the presence of salts acting as Lewis acids. Workers in the USSR were particularly active in this field.[5] A summary has been written by Kabanov.[6]

Reactivity ratios in copolymerization were also studied and large changes produced by salts found. Thus, Imoto et al.[7] reported an almost tenfold increase in the reactivity ratio of methyl methacrylate produced by the presence of zinc chloride in the copolymerization of the monomer with styrene and Yabumoto et al.[8] demonstrated an approach to strict alternation when the same salt is present in the copolymerization of acrylonitrile and styrene.

With hindsight, it is easy to see that in the late 1960s the field was ready for the remarkable generalization which was provided by the elegant investigations of Hirooka and his colleagues. In 1967,[9] they reported that the acrylonitrile—EtAlCl$_2$ complex (or a mixture of the components) copolymerizes spontaneously with olefinic hydrocarbon monomers such as propylene and styrene to give equimolar (alternating) copolymers. In the following paper,[1] this finding is generalized to include a wide range of the polar monomers, in which nitrile or carbonyl groups are conjugated to the double bond. With commendable caution, Hirooka identifies the equimolar polymers found as alternating. In 1971,[10] Hirooka et al. described the well-known classification of monomers into A and B groups, A-monomers (with low e-values) being donors and B-monomers (with higher e-values) acceptors.[11] The latter form complexes with Lewis acids through their polar groups which, by virtue of the existing conjugation, lower the (already low) electron density at the double bond. To obtain an alternating copolymer the reaction mixture must contain at least one monomer from each group together with a Lewis acid. The presence of a halogen atom in the molecule of the Lewis acid (alkyl aluminium halide) is important and no regulation occurs without it.

These developments greatly extended the range of alternating copolymers available and they allowed high-molecular-weight copolymers to be prepared from monomers not previously considered copolymerizable. Further, as stated in the paper,[1] they allowed comparison between the properties of alternating copolymers and random equimolar copolymers of the same monomers.

All the generalization we have described, which are either implicit or explicitly stated in the paper,[1] are striking in their simplicity. They undoubtedly constitute a significant advance in one of the most important areas of polymer science—the control of polymer structure.

REFERENCES AND NOTES

1. M. Hirooka, H. Yabuuchi, J. Iseki, and Y. Nakai, *J. Polymer. Sci. A-1,* **6,** 1381 (1968).
2. C. H. Bamford, A. D. Jenkins, and R. Johnston, *Proc. Roy. Soc. A.,* **241,** 364 (1957).
3. C. H. Bamford, S. Brumby, and R. P. Wayne, *Nature,* **209,** 292 (1966).
4. M. Imoto, T. Otsu, and Y. Harada, *Makromol. Chem.,* **65,** 174 (1963).
5. See C. H. Bamford, in *Alternating Copolymers,* J. M. Cowie, (ed.), Plenum Press, 1985, chapt. 3.
6. V. A. Kabanov, *J. Polym. Sci. C,* **67,** 17 (1980).
7. M. Imoto, T. Otsu, and Y. Harada, *Macromol. Chem.,* **65,** 180 (1963).
8. S. Yabumoto, K. Ishii, and K. Arite, *J. Polymer. Sci. A-1,* **7,** 1577 (1969).
9. M. Hirooka, H. Yabuuchi, S. Morita, S. Kawasumi, and K. Nakaguchi, *J. Polymer. Sci. B,* **5,** 47 (1967).
10. M. Hirooka, 23rd IUPAC Congress, Boston, 1971, *Macromol. Preprint* **1,** 311.
11. M. Hirooka, H. Yabuuchi, S. Kawasumi, and K. Nakaguchi, *J. Polymer. Sci. Chem. Ed.,* **11,** 1281 (1973).

JOURNAL OF POLYMER SCIENCE: PART A-1   VOL. 6, 1381–1396 (1968)

Alternating Copolymerization through the Complexes of Conjugated Vinyl Monomers–Alkylaluminum Halides

MASAAKI HIROOKA, HIROSHI YABUUCHI, JIRO ISEKI, and YASUTO NAKAI, *Central Research Laboratory, Sumitomo Chemical Co. Ltd., Takatsuki, Osaka, Japan*

Synopsis

A vinyl monomer that has the nitrile or carbonyl group conjugated to the C=C double bond, such as acrylonitrile, methyl acrylate, and methyl methacrylate, forms a complex with an alkylaluminum halide, and the complex reacts spontaneously with a hydrocarbon monomer such as styrene, propylene, or ethylene, giving a high molecular weight copolymer. The copolymers always contain the two monomer units in 1:1 ratio. Thus styrene, copolymerized with methyl acrylate or methyl methacrylate in the presence of ethylaluminum sesquichloride in homogeneous toluene solution, gives such an equimolar copolymer regardless of the initial monomer compositions. The NMR spectra of these copolymers are distinctly different from those of the equimolar copolymers obtained with azobisisobutyronitrile as initiator and have simpler and well separated patterns. The copolymers and the corresponding radical copolymers appear to be amorphous, judged by their x-ray diffraction patterns and their differential thermal analyses. Their infrared spectra resemble each other very closely. Hence, the difference in the NMR spectra may be ascribed to the matter of the sequence distribution. The infrared spectrum of ethylene–methyl acrylate copolymer shows no absorption near 720 cm.$^{-1}$ due to the methylene sequence arising from ethylene–ethylene linkage. These experimental data lead to the inference that the equimolar copolymers obtained in this work may have an alternating sequence.

INTRODUCTION

The authors[1] previously found that the acrylonitrile–ethylaluminum dichloride complex or a mixture of them *in situ* copolymerized spontaneously with olefinic hydrocarbon monomers, such as propylene, styrene, isobutylene and hexene-1, in high yield even at such low temperatures as −78°C. It was particularly noteworthy that an equimolar copolymer was obtained over a wide range of experimental conditions. Furthermore, some experimental results suggested that these equimolar copolymers have alternating monomer sequences. The crucial fact in this type of copolymerization appears to be that acrylonitrile is coordinated to an alkylaluminum halide. It was also interesting that copolymers of such novel monomer combinations as propylene and polar vinyl monomer were obtained, such copolymerization reactions having been scarcely known to take place.

1381

The authors have extended the studies of this type of reaction and of the properties of the copolymer. The present work relates to a similar copolymerization with methyl acrylate or methyl methacrylate in place of acrylonitrile. Emphasis has been placed on the clarification of the sequence distribution in the copolymers through the inferences obtained from the copolymerization process itself, the NMR and IR spectra, and other properties of the copolymers.

EXPERIMENTAL

Materials

Monomers. Methyl acrylate, methyl methacrylate, acrylonitrile, and styrene were purified, distilled, and dried in the usual way. Propylene and ethylene were polymerization grade, from the Sumitomo Chemical Co.

Solvents. n-Heptane was ASTM pure grade obtained from the Enjay Chemical Co. and further refined by being treated with concentrated sulfuric acid, washed with water, dried and distilled. Toluene, a commercial G.R. grade, was used after drying over silica gel.

Organoaluminums. Ethylaluminum sesquichloride and ethylaluminum dichloride, obtained from the Ethyl Corporation Ltd. were distilled under nitrogen atmosphere, diluted into purified n-heptane or toluene, and stored in glass ampules.

Polymerization

The polymerization procedure may be illustrated by an example of the copolymerization of methyl acrylate and styrene with ethylaluminum sesquichloride.

In a 200-ml. four-necked flask equipped with a stirrer, a thermometer, and a gas inlet tube, toluene and methyl acrylate were mixed under nitrogen atmosphere and cooled to $-78°C.$ by a mixture of solid carbon dioxide and methanol. When a toluene solution of ethylaluminum sesquichloride was added to this, the mixture became a homogeneous pale-yellow solution. Then the solution was brought to 25°C. No polymer was produced. Upon the addition of styrene a polymerization commenced. According to the progress of the copolymerization, the yellow, clear solution became viscous and yet remained homogeneous. At the end of the polymerization, the reaction mixture was poured into a large amount of methanol and white solid copolymer was coagulated. The copolymer was washed with methanol, dried *in vacuo*, and weighed. The purification of the copolymer was carried out by reprecipitation from acetone solution into methanol.

The copolymerizations with other monomer combinations were carried out by similar procedures. The copolymerization with propylene was conducted in a 300 ml. glass pressure vessel with a stirrer and that with ethylene in a 300 ml. stainless steel autoclave.

Measurements

Intrinsic Viscosity.   Viscosity measurements were carried out, in benzene solution in the case of methyl acrylate or methyl methacrylate copolymers and in dimethylformamide solution in the case of acrylonitrile copolymers, at 30°C. with an Ubbelohde viscometer.

NMR Measurements.   A 100 Mc./sec. high-resolution NMR spectrometer Model JNM4H-100, made by Japan Electron Optics Laboratory Inc., was used.   The NMR spectra of the copolymers were obtained at 70–80°C. in deuterated chloroform as about 10% solution.   Tetramethyl silane or hexamethyl disiloxane was used as the internal standard.

Infrared Measurements.   Infrared measurements of the copolymers were carried out with a Perkin-Elmer Model 125 infrared spectrophotometer.

RESULTS

Copolymerization

Various Monomer Combinations

Conjugated polar monomers that have the carbonyl or nitrile group, such as methyl acrylate, methyl methacrylate, and acrylonitrile, form complexes with alkylaluminum halides.   In the presence of the conjugated monomer and an alkylaluminum halide a hydrocarbon monomer such as propylene, styrene, or ethylene rapidly reacts and produces copolymer with high molecular weight.   Typical results of the copolymerization with *n*-heptane or toluene as solvent are given in Table I.   The copolymers were purified against the alkylaluminum component by reprecipitation from acetone, benzene, or dimethylformamide into methanol.   The compositions of the copolymers according to elementary analyses agreed well with the calculated value for the 1:1 copolymer.   The results are similar to those of the copolymerization of acrylonitrile and olefins with alkylaluminum halides, reported in a previous paper.[1]   Although the copolymerization of acrylonitrile proceeded heterogeneously in all cases, methyl acrylate or methyl methacrylate copolymerized in the homogeneous system if toluene was used as solvent.   Thus, in the copolymerization system of styrene and methyl acrylate (MA) or methyl methacrylate (MMA) with toluene as solvent both the complex of ethylaluminum sesquichloride and MA or MMA and the resulting copolymer were soluble in the solvent, and the copolymerization proceeded homogeneously at a moderate rate at about room temperature.   The authors have therefore chosen these homogeneous copolymerization systems for further detailed study.

Methyl Acrylate and Styrene

The results of copolymerizations effected over a wide range of monomer composition and given in Table II.   Ethylaluminum sesquichloride was used in one-half the molar amount of methyl acrylate, in order to prevent the

TABLE I
Copolymerization with Various Monomer Combinations

Expt. no.	Monomer,[a] g. A	Monomer,[a] g. B	Organoaluminum, mmoles	Solvent, ml.	Conditions Time, min.	Conditions Temp., °C.	Conditions Phase	Yield, g.	$[\eta]$, dl./g.	Elem. anal.,[b] % C	Elem. anal.,[b] % H	Elem. anal.,[b] % N
1	Pr, 100	AN, 2	AlEtCl₂, 50	n-heptane, 150	5	−78	heter.	2.48	0.63	—	—	14.46 (14.72)
2	" 100	" 2	" 50	toluene, 30	10	−78	"	2.87	1.84	—	—	14.39
3	" 20	MA, 4	" 50	—	10	−78	"	1.64	4.95	65.50 (65.60)	9.46 (9.44)	—
4	St, 10	AN, 2	AlEt₁.₅Cl₁.₅, 25	n-heptane, 20	10	−10	"	2.93	1.88	83.71 (84.04)	7.84 (7.05)	9.02 (8.91)
5	" 5	MA, 3	" 16	" 62	360	25	"	3.85	0.91	75.84 (75.76)	6.92 (7.42)	—
6	" 5	" 3	" 16	toluene, 62	240	25	homo.	3.46	3.18	75.76	7.85	—
7	" 5	MMA, 3	" 16	" 62	300	25	"	1.54	2.02	76.12 (76.40)	7.80 (7.90)	—
8	E, 100	MA, 4	" 50	n-heptane, 20	180	−78	heter.	0.82	1.53	63.22 (63.13)	9.11 (8.83)	—

[a] Pr, propylene; St, styrene; E, ethylene; AN, acrylonitrile; MA, methyl acrylate; MMA, methyl methacrylate.
[b] Figures in parentheses indicate value calculated for 1:1 copolymer.

ALTERNATING COPOLYMERIZATION 1385

TABLE II
Copolymerization of Methyl Acrylate and Styrene[a]

Expt. no.	Monomer concn., moles/l.		Polymn. time, min.	Yield, g.	Elem. anal., %		MA mole-% in polymer	Convsn., % of MA	$[\eta]$, dl./g.
	MA	St			C	H			
1	0.40	1.20	60	1.72	75.61	7.72	50.4	28.3	4.69
2	"	"	240	3.34	75.44	7.58	50.9	54.9	4.64
3	"	0.60	60	1.65	76.21	8.41	48.8	27.2	2.84
4	"	"	240	3.46	75.76	7.85	50.0	56.9	3.18
5	"	"	360	4.17	76.02	8.32	49.3	68.5	2.85
6	"	0.30	60	1.03	75.95	7.67	49.5	16.9	3.02
7	"	"	240	2.75	—	—	—	45.2	2.64
8	"	"	360	3.74	75.41	7.18	51.0	61.4	2.16
9	"	0.11	20	0.52	75.96	7.80	49.5	8.5	1.66
10	"	"	240	1.43	75.90	7.39	49.7	23.6	1.27

[a] Polymerization condition: 25°C.; AlEt$_{1.5}$Cl$_{1.5}$, 0.2 mole/l.; solvent, toluene; total liquid volume, 80 ml.

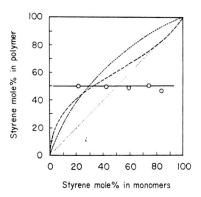

Fig. 1. Copolymerization of methyl acrylate and styrene: (O) with $AlEt_{1.5}Cl_{1.5}$; (— —) radical,[2,3] r_1 (St) = 0.75, r_2 (MA) = 0.18; (— - -) cationic,[4] r_1 (St) = 2.2, r_2 (MA) = 0.4.

homopolymerization of styrene by free ethylaluminum sesquichloride. In this condition no homopolymer of methyl acrylate was produced. The copolymerization proceeded homogeneously, and the viscosity of the clear, yellow solution gradually increased. Although the composition of the unreacted monomer varied according to the progress of the copolymerization, the resulting copolymer always contained each monomer component in equimolar proportions; this relationship held independently of initial monomer composition. All the compositions of the copolymers in elementary analyses represented extremely good agreement with the calculated value for the 1:1 copolymer. The molecular weights of the copolymers in general were high and increased in proportion to the increase of styrene concentration.

Figure 1 indicates the relationship between monomer feed and polymer composition in the present copolymerization compared with those of conventional cationic and radical copolymerizations. The polymer composition in either the cationic or the radical process varies according to the monomer composition, but the present copolymerization apparently differs from these conventional ones in monomer reactivity.

Methyl Methacrylate and Styrene

Table III gives the results of copolymerizing methyl methacrylate and styrene under conditions similar to those of methyl acrylate and styrene. In this case the effect of monomer composition was investigated under the condition of constant total monomer concentration and varying ratios of monomer components. Ethylaluminum sesquichloride was used in one-half the molar amount of methyl methacrylate. As the concentration of ethylaluminum sesquichloride increased, the polymer yield increased. The copolymerization was stopped at a relatively low conversion, 3–10%. The copolymer compositions may be seen to be equimolar over the wide range of monomer composition.

ALTERNATING COPOLYMERIZATION 1387

TABLE III
Copolymerization of Methyl Methacrylate and Styrene[a]

Expt. no.	Monomer concn.,[b] moles/l.			AlEt$_{1.5}$Cl$_{1.5}$ concn., moles/l.	Polymn. time, min.	Yield, g.	Elem. anal., %		MMA in polymer, %	Convsn., % of MMA	$[\eta]$, dl./g.
	St	MMA	Tot.				C	H			
11	2.0	0.5	2.5	0.25	90	1.08	75.82	7.85	51.9	10.6	2.72
12	1.5	1.0	"	0.50	60	1.23	75.60	8.07	52.6	6.1	3.81
13	1.0	1.5	"	0.75	15	1.61	74.82	7.95	55.0	5.3	3.58
14	0.5	2.0	"	1.00	7	1.14	74.79	7.96	55.1	2.8	3.30

[a] Polymerization conditions: 25°C.; MMA/AlEt$_{1.5}$Cl$_{1.5}$, molar ratio, 2.0; solvent, toluene; total liquid volume, 100 ml. Theoretical value of elementary analysis for alternating copolymer: C, 76.44%; H, 7.90%.
[b] St, Styrene; MMA, methyl methacrylate.

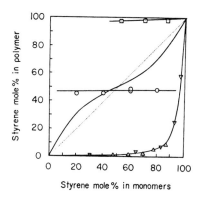

Fig. 2. Copolymerization of methyl methacrylate and styrene. (——) radical;[5] (□) cationic;[6] (△) anionic, Na catalyst;[7] (▽) anionic, BuLi catalyst;[7] (○) with AlEt$_{1.5}$Cl$_{1.5}$.

Figure 2 shows the relationship between monomer feed and copolymer composition compared with those in the conventional radical, cationic, and anionic copolymerizations. It is apparent that the results of the present copolymerization do not agree with any conventional relation.

Preparations and Properties for NMR and IR Analyses

The properties of the copolymers of the present copolymerization and of radical copolymerization, obtained for NMR and IR analyses, are summarized in Table IV. Random copolymers were prepared in bulk with azobisisobutyronitrile (AIBN) as initiator. To produce 1:1 copolymer, the initial monomer compositions were deduced from the monomer reactivity ratio. The reactions were stopped at a low conversion. The copolymers were purified by the reprecipitation procedure.

NMR Spectra of Copolymers

Styrene–Methyl Methacrylate Copolymers

The NMR spectra of the copolymers were obtained from the 10% deuterated chloroform solution at the resonance frequency of 100 Mc./sec. Figure 3 shows typical spectra of copolymers of 1:1 composition, prepared with AlEt$_{1.5}$Cl$_{1.5}$ and AIBN. Spectrum B is identical with that reported by Harwood and Ritchey,[8,9] but the spectrum A differs from them and also from the spectrum of block copolymer.[10] Spectrum A is comparatively simplified, and the resonance splits are sharp.

The authors[11] have analyzed the spectra of these copolymers and discussed the assignments of the resonances and sequence of the monomer units. By comparing spectra A and B the following facts may be summarized:

(1) The diamagnetic shift of the *ortho* proton in the phenyl group is seen in spectrum B but not A.

TABLE IV

Properties of Copolymers for NMR and IR Analyses[a]

Expt. no.	Monomer A	Monomer B	Catalyst	Elem. Anal.[b], % C	H	N	Polym. compn., mol.-%	$[\eta]$, dl./g.
1	St	MA	AlEt$_{1.5}$Cl$_{1.5}$	75.62 (75.76)	7.39 (7.42)	—	MA 50.4	5.37
2	"	"	AIBN	75.58	7.66	—	MA 50.5	1.11
3	St	MMA	AlEt$_{1.5}$Cl$_{1.5}$	75.93 (76.40)	7.50 (7.90)	—	MMA 51.6	3.83
4	"	"	AIBN	76.38	7.89	—	MMA 50.2	1.00
5	St	AN	AlEt$_{1.5}$Cl$_{1.5}$	83.88 (84.04)	7.59 (7.05)	8.71 (8.91)	AN 48.9	2.13
6	"	"	AIBN	82.48	7.83	9.71	AN 53.3	2.54
7	E	MA	AlEt$_{1.5}$Cl$_{1.5}$	63.22 (63.13)	9.11 (8.83)	—	MA 49.6	1.53
8	"	"	radical initiator	64.21	8.61	—	MA 45.4	2.05

[a] St, styrene; E, ethylene; MMA, methyl methacrylate; MA, methyl acrylate; AN, acrylonitrile; AIBN, azobisisobutyronitrile.

[b] Figures in parentheses indicate value calculated for 1:1 copolymer.

(2)   The resonance of the methoxy proton in the range of 6.4–7.7 τ is broad in spectrum B but splits sharply into three peaks at 6.6 (I), 7.0 (II), and 7.6 (III) τ in the case of spectrum A.

(3)   The methylene and methine resonances are also simplified in spectrum A.

(4)   A broad peak between 9 and 9.5 τ in spectrum B has been assigned to the α-methyl proton.   The corresponding resonance in spectrum A splits into three distinct peaks at 9.0 (IV), 9.3 (V), and 9.4 (VI) τ.

(5)   The resonance intensity ratio of phenyl proton to either methoxy or α-methyl proton is exactly 5:3 in both spectra A and B, as is to be expected from the 1:1 copolymer composition.   Furthermore, the corresponding peak intensity ratios of methoxy to α-methyl protons, i.e., (I) to (VI), (II) to (V), and (III) to (IV), are all equal and unity.

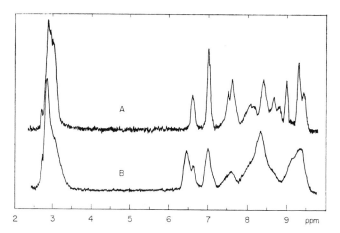

Fig. 3. NMR spectra of styrene–methyl methacrylate copolymers: (A) with $AlEt_{1.5}$-$Cl_{1.5}$; (B) with AIBN.

In spectrum A the configurational influence or the difference in the sequence of monomer units may be alternatively considered as the reason for the shift of the methoxy and α-methyl proton with the three well-separated peaks.   The relationship of the resonance intensities between the methoxy and α-methyl groups, however, can be explained only by the shifts' being due to the configurational effect.   This leads to the conclusion that the possible sequence is an alternating one.   This is also supported by the lack of the *ortho* proton shift in the phenyl group and the absence of an MMM methoxy triad at 6.4 τ, which appears in the spectrum of poly-(methyl methacrylate).

Three methoxy peaks are assigned to cosyndiotactic, coheterotactic, and coisotactic triads, respectively, from low to high field.   The existence of these three configurations means that the copolymer obtained with $AlEt_{1.5}$-$Cl_{1.5}$ is atactic.

Styrene–Methyl Acrylate Copolymers

Figure 4 shows the NMR spectra of styrene and methyl acrylate copolymers obtained with AlEt$_{1.5}$Cl$_{1.5}$ and AIBN. A distinct difference between them is observed. The split of the methoxy proton resonances into three peaks may be explained as analogous to that of styrene–methyl methacrylate copolymer. In this case, however, the diamagnetic shift of the *ortho* proton in the phenyl group is observed also in spectrum *A*.

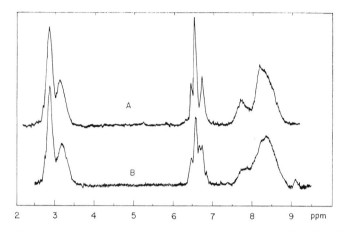

Fig. 4. NMR spectra of styrene–methyl acrylate copolymers: (*A*) with AlEt$_{1.5}$Cl$_{1.5}$; (*B*) with AIBN.

As to the split in the phenyl resonance, it is generally understood that it is due, not only to the successive styrene unit, but also to all environmental groups.[12] Therefore the possibility of an alternating sequence of the monomer unit is not excluded. In both spectra *A* and *B* the ratio of the resonance areas for the phenyl and methoxy protons is observed to be 5:3. Consequently, it is confirmed that each copolymer has a 1:1 composition.

IR Spectra of Copolymers

Copolymers Containing Styrene

The infrared spectra of the copolymers formed with AlEt$_{1.5}$Cl$_{1.5}$ and the 1:1 radical copolymers were compared over the range of 400–4000 cm.$^{-1}$. The spectra of styrene–methyl acrylate and styrene–acrylonitrile copolymers are shown in Figures 5 and 6. Both the spectra of the present copolymer and of the radical one are not clearly distinguished from each other except by a slight difference in the absorbance in a few bands. This is in contrast to the fact that there is a great difference between the NMR spectra of both copolymers.

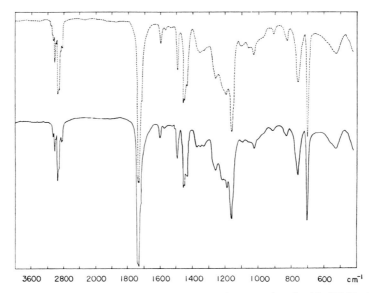

Fig. 5. IR spectra of styrene–methyl acrylate copolymers: (- - -) with AlBN; (——) with AlEt$_{1.5}$Cl$_{1.5}$.

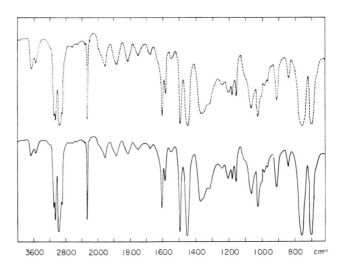

Fig. 6. IR spectra of styrene–acrylonitrile copolymers: (- - -) with AlBN; (——) with AlEt$_{1.5}$Cl$_{1.5}$.

IR Spectra of Ethylene–Methyl Acrylate Copolymers

Figure 7 shows the IR spectra of ethylene–methyl acrylate copolymers over the range of 700–900 cm.$^{-1}$ with a scale expander. Between the spectrum of the copolymer prepared from a monomer complex and that of the radical copolymer prepared by a high-pressure polymerization technique there is a clear difference. The absorption band at 750 cm.$^{-1}$

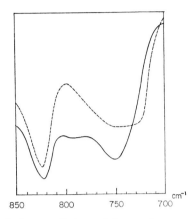

Fig. 7. IR spectra of ethylene–methyl acrylate copolymers: (- - -) with peroxide. (——) with AlEt$_{1.5}$Cl$_{1.5}$.

in the present copolymer was sharp and may be due to the skeletal vibration that can be seen also in the spectrum of poly(methyl acrylate). The corresponding absorption for the radical copolymer was broad and spread almost to 720 cm.$^{-1}$. This may be ascribed to the existence of a long chain of the methylene group; it is known that the rocking mode of at least four adjacent straight-chain methylene linkages appears near 720 cm.$^{-1}$. The lack of this kind of band in the spectrum of the present copolymer indicates that there is no substantial linkage of ethylene to ethylene. This may be considered in relation to the alternating sequence of the copolymer.

DISCUSSION

In a previous paper[1] attention was drawn to the fact that the copolymerization of acrylonitrile and an olefin in the presence of an alkylaluminum halide gave a copolymer of equimolar composition. Subsequent work, as shown in the present paper, has revealed that similar equimolar copolymers were obtained also when methyl acrylate or methyl methacrylate was used in place of acrylonitrile. When an alkylaluminum chloride alone was complexed with methyl acrylate or methyl methacrylate, polymerization did not occur and no homopolymer was produced. Unsaturated olefinic hydrocarbons, however, reacted instantaneously with the conjugated vinyl monomers in the presence of alkylaluminum halides, producing copolymers. The reaction gave 1:1 copolymer over a wide range of initial monomer composition and also independently of polymerization time. Consequently, each fact mentioned above supports the assumption that the present copolymer prepared through the monomer complex may have an alternating sequence.

Equimolar copolymers can be prepared also by conventional radical copolymerization. These copolymers, such as styrene–methyl acrylate and styrene–acrylonitrile copolymers, have not only 1:1 polymer composition but also patterns in their x-ray diagrams, differential thermal analyses, and infrared spectra that are indistinguishable from the corresponding copolymers formed with $AlEt_{1.5}Cl_{1.5}$. The data also indicate that both the radical and the present copolymers are amorphous. Nevertheless, the NMR spectra of styrene–methyl acrylate and styrene–methyl methacrylate copolymers formed in the present copolymerization were shown to be very different from those of conventional radical copolymers and to be simpler. We have analyzed these spectra and obtained confirmation that the styrene–methyl methacrylate copolymer prepared with $AlEt_{1.5}Cl_{1.5}$ has an alternating sequence. Besides, the lack of the absorption band near 720 cm.$^{-1}$ in the IR spectrum of ethylene–methyl acrylate copolymer from the present copolymerization means that a block sequence of ethylene–ethylene is not substantially included. On the basis of the results and considerations presented above it can be inferred that the 1:1 copolymer from the present copolymerization has an alternating sequence.

Although alternating copolymers from conventional radical polymerizations have been known to be obtained only from specific monomers such as maleic anhydride, vinylidene cyanide, and fumaric esters, which have little or no ability to homopolymerize, the authors have failed to find any report that an alternating copolymer was prepared from a monomer combination from which a random copolymer can be easily obtained. For this reason, the fact that alternating copolymers were obtained in the present copolymerization is noteworthy. If the conclusion concerning the alternating sequence of the copolymer is valid, it may be said that the present copolymerization is a peculiar process that enables us to compare an alternating copolymer with a random copolymer from the same monomer combination.

For an alternating copolymer a specific relationship between the monomer reactivity ratios, $r_1 = r_2 = 0$, is necessary. This means that the homopolymers do not appear and only copolymerization takes place. Such a phenomenon was observed in the copolymerization of acrylonitrile and propylene in the presence of ethylaluminum dichloride, as described in the previous paper. In the present work, too, it was confirmed that the complex of methyl acrylate or methyl methacrylate and ethylaluminum sesquichloride did not polymerize of itself. As shown in the present paper, the monomer reactivity in the present copolymerization cannot be explained by a conventional radical or ionic polymerization mechanism. The peculiarity of this copolymerization seems to consist in that the reaction proceeds through the complex of an alkylaluminum halide with a conjugated vinyl monomer. These complexes could be clearly recognized by means of IR spectroscopy in the cases of methyl acrylate, methyl methacrylate, and acrylonitrile. The complex is formed between aluminum and the carbonyl or nitrile group. It is considered that through the formation of the complex the electron density on the carbon-carbon double bond in the position conjugated to the polar group decreases and that the reactivity of the polymerization may be modified, as if it possessed the same characteristics as a monomer that is likely to give an alternating copolymer, such as maleic anhydride, fumaric ester, or vinylidene cyanide. It seems to be important that the C=C double bonding in this kind of monomer is located in the position conjugated to the polar group. Since this kind of complex does not polymerize of itself, it is believed that the activation of the complex may arise through the attack of a donor monomer possessing a polarity opposite of that of the complexed monomer. It is judged that in the present copolymerization the complex formation and the donor–accepter interaction play important roles in the activation of the copolymerization and at the same time induce a tendency to alternation.

The effect of metallic compounds on the polymerization of vinyl compounds has been studied in recent years. Imoto et al.[13–15] found that zinc chloride promotes the radical polymerization of acrylonitrile and methyl methacrylate and considered that the monomer reactivity varies because of the coordination of zinc chloride with the nitrile or carbonyl group. They further disclosed that when the copolymerization of acrylonitrile and vinylidene chloride was carried out in the presence of zinc chloride, the monomer reactivity ratios changed according to the increase of zinc chloride concentration.

A British patent applied for by Esso Research and Engineering Co.[16] outlines a method of producing copolymers that consists of copolymerizing a monomer having an electronegative group, such as acrylonitrile or methyl acrylate, and a second monomer that promotes a Friedel-Crafts polymerization but has no ability for radical polymerization, such as 2-methyl-pentene-1, isobutylene, or hexene-1, with a Friedel-Crafts halide, in the presence of a radical initiator. In this reaction system a complex formation between the first monomer and a metal halide is suggested. It is interesting that a copolymer can be obtained from the combination that includes a monomer known to be difficult to polymerize with a radical initiator. The compositions of the copolymers obtained, however, are random and differ from the present copolymers.

Although both the studies made by Imoto et al. and the Esso patent are characterized by the modification of a vinyl monomer through complex formation between the polar group of the vinyl monomer and a metal halide such as zinc chloride, other alternating copolymers have not been prepared. Consequently, it is evident that the tendency to an alternating sequence cannot be ascribed only to the coordination effect of the polar group in a conjugated vinyl monomer to a metal halide.

It is well known that the combination of trialkyl boron or trialkyl aluminum and oxygen or a peroxide acts as an initiator in radical polymerization. Hoechst's patent[17] indicates that the trialkyl aluminum–organic peroxide system initiates the copolymerization of a conjugated vinyl compound and an olefin, such as propylene, ethylene or isobutylene, which is known to polymerize with difficulty with a radical initiator. This means that these catalysts have different abilities to copolymerize with a conventional radical initiator. Further, Wexler and Manson[18] studied the polymerization reaction with trialkylaluminum and found that the copolymerization of methyl methacrylate and styrene could not be explained by a conventional radical mechanism. They concluded that the polymerization is neither clearly free-radical nor ionic, and some form of coordination may take place. From these reports it is deduced that the initiating action of trialkylaluminum cannot be identified with that of a conventional radical initiator but that an alkylaluminum compound itself does not cause alternating copolymerization.

It may be summarized that in the present copolymerization the coexistence of at least one alkyl group and one halogen atom attached to the aluminum metal is important, and an alternating copolymer can be obtained when the aluminum component is coordinated with a polar group of a conjugated monomer and the resulting complex is attacked by a donor monomer. Furthermore, it can be pointed out as a characteristic feature of the present copolymerization that it is not necessary to use deliberately a radical source and that the polymerization proceeds spontaneously.

The authors wish to express their gratitude to S. Kodama, K. Nakaguchi, and S. Kawasumi for their kind guidance and encouragement, Y. Kubota for his continued interest and helpful discussions, H. Oi, K. Shirayama, and T. Okada for determining and interpreting the infrared spectra and K. Sudo, T. Ishiyama, and T. Kawai for their assistance in the experimental work.

References

1. M. Hirooka, H. Yabuuchi, S. Morita, S. Kawasumi, and K. Nakaguchi, *J. Polymer Sci. B*, **5**, 47 (1967).
2. K. W. Doak, *J. Am. Chem. Soc.*, **72**, 4681 (1950).
3. F. M. Lewis, C. Walling, W. Cummings, E. R. Briggs, and F. R. Mayo, *J. Am. Chem. Soc.*, **70**, 1519 (1948).
4. Y. Lander, *J. Polymer Sci.*, **8**, 63 (1952).
5. G. E. Ham, *Copolymerization*, Interscience, New York, 1964, p. 791.
6. T. Higashimura and S. Okamura, *J. High Polymer (Japan)*, **17**, 635 (1960).
7. K. F. O'Driscoll and A. V. Tobolsky, *J. Polymer Sci.*, **37**, 363 (1959).
8. H. J. Harwood, *Chem. Eng. News*, **41**, 36 (1963).
9. H. J. Harwood and W. M. Ritchey, *J. Polymer Sci. B*, **3**, 419 (1965).
10. K. Ito and Y. Yamashita, *J. Polymer Sci. B*, **3**, 631 (1965).
11. J. Iseki, to be published.
12. F. A. Bovey and G. V. D. Tiers, *Fortschr. Hochpolymer-Forsch.*, **3**, S139 (1963).
13. M. Imoto, T. Otsu, and S. Shimizu, *Makromol. Chem.*, **65**, 174 (1963).
14. M. Imoto, T. Otsu, and Y. Harada, *Makromol. Chem.*, **65**, 180 (1963).
15. M. Imoto, T. Otsu, and M. Nakabayashi, *Makromol. Chem.*, **65**, 194 (1963).
16. Esso Research and Engineering Co., Brit. Pat. 946,052 (Jan. 8, 1964).
17. Farbewerke Hoechst A.-G., U.S. Pat. 3,156,675 (Nov. 10, 1964).
18. H. Wexler and J. A. Manson, *J. Polymer Sci. A*, **3**, 2903 (1965).

Received June 26, 1967
Revised October 2, 1967

COMMENTARY

Reflections on "Mesophasic Structures in Polymers. A Preliminary Account on the Mesophases of Some Poly-Alkanoates of *p,p'*-Di-Hydroxy-α,α'-Di-Methyl Benzalazine," by Antonio Roviello and Augusto Sirigu, *J. Polym. Sci., Polym. Lett. Ed.,* 13, 455 (1975)

AUGUSTO SIRIGU

Istituto Chimico, Universita degli Napoli, Napoli, Italia

Our paper, which the Editors and the Publisher have included among those chosen to celebrate the first half-century of *Journal of Polymer Science,* deserves probably the simple and modest credit for having shown to the scientific community an example, easy to follow, of how an entire class of liquid crystalline polymers could be conceived. For the authors, it has also a very personal meaning: it represents the first, happy result of a entirely new research activity undertaken by a 35-year-old crystallographer, rather dissatisfied with ordinary crystallography and in quest for "strange structures," and by his freshly graduated and scientifically voracious colleague. Their scientific partnership is still active.

Actually, the push to get a liquid crystalline polymer came from our desire to obtain a material possessing noncrystalline order in the solid state and to study its structure. The idea of having such a state by freezing a spontaneously ordered liquid was the second step, therefore: a liquid crystal to quench. However, what kind of liquid crystal? The literature on polymers, apart from some fascinating suggestions concerning poly-γ-benzyl-L-glutamate, was not very appealing to us. On the contrary, the harvest of low molecular weight liquid crystals known was already large. Then came the amusingly simple idea: let us take a typical liquid crystal, say, a nematic one, whose molecules are formed by a small rodlike segment and two flexible terminals; if, with an ideal experiment, we connect these terminals to build a linear rigid-flexible sequence haven't we got the desired structure? This was the start, what remained to do was to choose the specific polymers to synthesize, to get them made and characterized, and to compare results with expectations. A discussion with Paolo Corradini, leader, as he still now is, of the Polymer Group active at the Department of Chemistry of the Naples University, gave us a stimulating confirmation of the potential utility of our project.

The choice of what mesogen to use was preceded by that of what kind of polymerization reaction to perform. We chose polyesterification for its versatility. Polyamides were excluded under the assumption, not entirely reasonable, that interchain interactions would have brought exceedingly high melting temperatures. Some possible mesogens were selected on the basis of their recognised aptitude to be so, and on the assumed facility of their synthesis. Yet the very first results were not entirely satisfactory. Some polyalkanoates of 4,4'-dihydroxybenzalazine (not mentioned in the article) proved to be mesogenic but the isotropization (a crucial event, absolutely necessary to support our point) was not detectable because of chemical decomposition intervening. The isotropization temperature was too high! We had the first direct lesson about the role played by what eventually came to be called the spacer: it was too short.

However, an entirely satisfactory result did come not much later. At the beginning of 1974, a longer spacer and a somewhat thicker mesogen gave us nice and unmistakable evidence of a polymer exhibiting perfectly thermotropic liquid crystalline behavior. This result was soon confirmed with the synthesis of two homologous polymers having spacers of different length. We had made the point. Contemporary to this, ancillary research on the phase behavior of low molecular weight model compounds taught us a little bit about the utility and risks of using models.

We must say that we were not able to quench the liquid crystalline phase at room temperature without having it polluted with crystallinity (our desired solid polymer with intrinsic noncrystalline order had to wait for the liquid crystalline random copolymers to show up). Therefore, lacking at that time of any facility for performing X-ray diffraction experiments at high temperatures, our investigation of the me-sophase structure was seriously limited. However, funny enough, trials to quench the mesophase under shear produced a morphology, shown in the article as Figure 4, that we felt to be peculiar enough to deserve publication. This morphology is now recognisable as the crystalline evolution of a nematic band structure.

Finally, we had to have the article published. The answer of the reviewers was not exactly encouraging. Old doubts were rekindled: was the optically anisotropic liquid a true liquid crystal or, maybe, was it the intimate mixture of ordinary isotropic liquid polymer with a high-melting crystal fraction? We had no more doubts about this matter, however, no direct X-ray diffraction based proof of the absence of crystallinity in the anisotropic fluid phase was actually available. But, in the end, the belief of the Editor (H. F. Mark) was decisive: the article was accepted for publication in the *Journal of Polymer Science*.

PERSPECTIVE

Comments on "Mesophasic Structures in Polymers. A Preliminary Account on the Mesophases of Some Poly-Alkanoates of *p,p'*-Di-Hydroxy-*α,α'*-Di-Methyl Benzalazine," by Antonio Roviello and Augusto Sirigu, *J. Polym. Sci., Polym. Lett. Ed.,* 13, 455 (1975)

ROBERT W. LENZ

Polymer Science and Engineering Department, University of Massachusetts, Amherst, Massachusetts 01003

This publication by Roviello and Sirigu of the University of Naples, Italy, was the first description in journal literature of the ability of polymers with rodlike units in their backbones to self-organize into the liquid crystalline (LC) or mesomorphic state in the melt, above the crystalline melting transition temperature. But the paper is not only remarkable for its revelation that such polymers could melt into a stable, anisotropic liquid state without thermal decomposition, it is also notable for the sophistication of the analytical methods that the authors applied to characterizing the transition into the LC state and for their recognition of the type of LC state formed. Indeed, the group of analytical procedures they applied in this pioneering study became the standard method both for ascertaining that polymers with potentially mesogenic groups in their backbones (that is, rodlike repeating units) were LC in the melt (thermotropic LC polymers) and for determining the type of LC state formed. Of special importance in that regard was their use of hot-stage, polarized light microscopy in combination with differential scanning calorimetry (DSC) to assign the thermal transitions observed by the latter, in both the heating and cooling cycles, to the type of LC phase formed (nematic or smectic).

Until the publication by Roviello and Sirigu, it was generally assumed that polymers with rodlike backbones would have melting transitions that were too high to form a stable, observable LC state. The ability of such polymers to self-organize in solution into an LC state (lyotropic LC polymers) had been predicted by Flory in 1956[1] and had been put to practical use in the solution spinning of fibers based on all-aromatic polyamides in the Kevlar research program at du Pont.[2] Indeed, as a result of the success of the Kevlar program, researchers at du Pont, the Eastman Kodak Co., the Celanese Research Co., and in other industrial laboratories were searching for aromatic polyesters that could form a stable LC state in the melt in order to obtain fibers with similar mechanical properties to Kevlar by melt spinning.[3,4] Patents on such polymers began to issue in the same year as the Roviello–Sirigu paper,[5,6] but the first literature description and recognition of the LC properties of these all-aromatic, thermotropic LC polyesters in the melt was that by Jackson and Kuhfuss of the Eastman Kodak Co. That paper was also published in the *Journal of Polymer Science* and will be covered in this series.[7]

Prof. Augusto Sirigu and his student, Antonio Roviello, initiated a systematic study to determine if a polymer containing a "mesophase-originating atomic grouping" would "show the morphological characters of mesophases," meaning liquid crystalline phases, in the "solid" (i.e., melt) state rather than in solution. They predicted that "an extended mesophasic morphology should be a source of interesting optical and mechanical properties."

For their study they chose a bisphenol monomer containing a benzalazine group and reacted it with a series of aliphatic dicarboxylic acid chlorides. The aliphatic portion formed what was later referred to as a flexible spacer. The polyesters obtained had accessible melting transitions at about 200°C and formed stable anisotropic liquid phases. Furthermore, at higher temperatures, transitions into isotropic liquids (i.e., clearing transitions) were observed. Such isotropization transitions were unknown for the all-aromatic LC polyesters until much later.

Wide angle X-ray diffraction data were also obtained to confirm the presence of mesophasic order in the melt, and visual observations on a polarizing microscope revealed the formation of a Schieren texture indicating, without a doubt, that an LC phase was present (either nematic or smectic). Their studies also demonstrated the reversibility of the crystalline-LC and LC-isotropic transitions, and they demonstrated the effect of shearing the LC phase during cooling on the orientation of the crystalline phase formed. All of these properties became of great importance in the extensive, subsequent studies of the design, synthesis and characterization of main chain, thermotropic LC polymers, which were carried out later in many academic and industrial laboratories around the world. It can be fairly stated, therefore, that this paper formed the pattern and the laid the groundwork for many of those studies.

REFERENCES AND NOTES

1. P. J. Flory, *Proc. R. Soc. London, Ser. A,* **234,** 73 (1956).
2. S. L. Kwolek, U.S. Patent 3,600,350 (1971); 3,671,542 (1972).
3. G. W. Calundann and M. Jaffe, *Proc. Robert A. Welch Conf. on Chem. Res., XXVI, Synthetic Polymers,* November 1982, Houston, TX.
4. W. J. Jackson, Jr., *Mol. Cryst. Liq. Cryst.,* **169,** 23 (1989).
5. R. W. Lenz, *Polymer J.,* **17,** 105 (1985).
6. W. J. Jackson, Jr., *Br. Polym. J.,* **1980,** 154.
7. W. J. Jackson, Jr. and H. F. Kuhfuss, *J. Polym. Sci. A,* **14,** 2043 (1976).

POLYMER LETTERS EDITION   VOL. 13, PP. 455–463 (1975)

MESOPHASIC STRUCTURES IN POLYMERS. A PRELIMINARY ACCOUNT ON THE MESOPHASES OF SOME POLY-ALKANOATES OF p,p'-DI-HYDROXY-α,α'-DI-METHYL BENZALAZINE

The concepts of liquid crystallinity have been used in relation to polymers essentially in three different ways: (a) – To semiordered structures of mesophasic character have been ascribed some morphological aspects of amorphous polymers (1). (b) – A lyotropic liquid crystal phenomenology has been observed in solutions of polymers like polypeptides, typically poly-γ-benzyl-L-glutamate (2), or some polyesters of alkoxybenzoic acids (3) or some block copolymers like polystirene-polyoxyethylene (4). (3) – Some experimental work has been reported on polymerization in mesophasic media (5-10). Particular attention was paid to possible influences on molecular structure (tacticity) and molecular weight, the mesophasic medium being constituted either by the same monomers or by a substance not directly involved in the polymerization. Much of the experimental work focused on polymers containing the cholesteryl or cholestanyl group in the side chain.

The resulting picture of polymer formation is sometimes contradictory and by no means conclusive.

We have undertaken to systematically study the possibility of obtaining polymeric substances that in the "solid" state markedly show the morphological characters of mesophases.

Molecular aggregations of mesophasic type have been noted on solvent cast films of some synthetic polypeptides, particularly polyglutamates (11-12). A striking example of further induced superstructure is shown by films of poly-γ-benzyl-L-glutamate case in a magnetic field (13,14).

Many of the interesting properties of liquid crystals stem from their being liquid, nonetheless an extended mesophasic morphology should be a source of interesting optical and mechanical properties in materials even in the absence of fluidity.

An obvious approach to the problem of obtaining such polymers is to polymerize monomers that possesses the qualities for self-induced thermotropic mesophases. Alternatively the mesophase-originating atomic grouping could be built with the polymerization itself.

A second point arises concerning the position where the relevant atomic grouping has to be inserted relatively to the polymer chain. Most of the known experimental work deals with polymers having the relevant atomic grouping inserted in the side chain.

As to the kind of mesophase that one could expect to be developed from a particular chemical composition and molecular structure, the same criteria that hold for low molecular weight thermotropic liquid crystals could be tentatively

455

followed. In particular, for a polymer whose chain is formed by an ...ABAB... sequence of atomic groupings, which as to steric encumbrance and nonbonded interactions with neighboring chains are almost interchangeable, a nematic (if any) mesophase can presumably be expected. When the polymer chains are formed by different atomic grouping which are more strictly homophilic a smectic mesophase should be more probable.

On the basis of these hypotheses we have attempted to study and to synthetize polymers of the following formula:

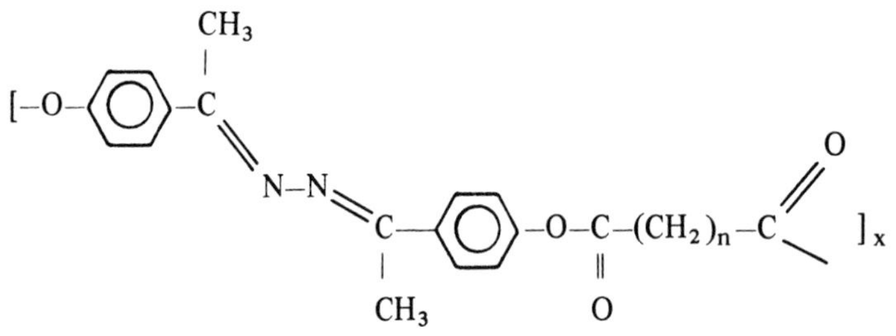

The present paper deals with the preparation and partial characterization of the unfractionated polymers.

Experimental

The p,p'-di-hydroxy-α,α'-di-methyl benzalazine:

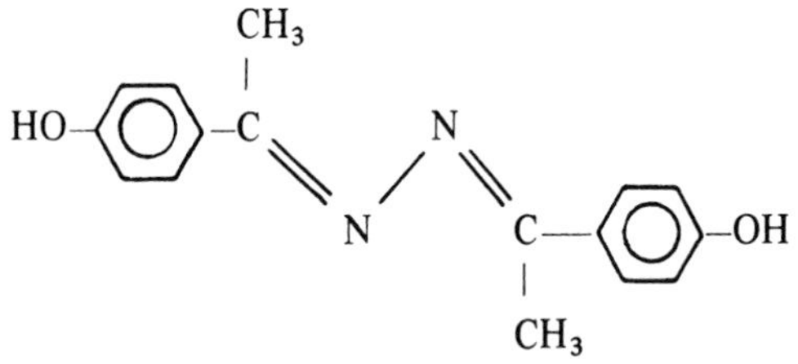

has been prepared from p-hydroxyacetophenone and hydrazine sulfate as reported in (15).

Melting point and NMR spectrum confirm the nature and purity of the prepared compound.

The acyl chlorides: $ClOC(CH_2)_nCOCl$ n = 6, 8, 10 have been prepared by standard procedures from the corresponding dicarboxylic acids (Fluka) and were vacuum distilled.

Polymerization. Polyesters of the given formula have been prepared in the following way:

A chloroformic solution of acyl chloride has been added, at room temperature, to a water solution of p,p'-di-hydroxy-α,α'-di-methyl benzalazine, sodium hydroxide, and benzyltriethylammonium chloride. After stirring six min in a

TABLE I

Phase Transition Temperatures and Enthalpy Changes (1)

	(2)	T(I)	ΔH(I)	T(II)	ΔH(II)
	h(a)	483	2.10		
	c	453	1.57		
P12	h(b)	{ 476 483	0.71 1.32	514	1.90
	c(c)	454	1.55	489	1.80
	h(d)	479	2.10	512	1.82
P10	(3)	476	1.90	529	2.33
P 8	(3)	511	2.53	568	2.68

(1)Temperatures are in K and are reproducible within 1%; enthalpies are given in Kcal mol^{-1} and are reproducible from sample to sample within 10%.

(2)h = heating run, c = cooling run, Letters in parentheses refer to Figure 1.

(3)From first heating run measurements.

(4)Substantial decomposition starts at $\simeq$ 580 K for P12 and P10 and at $\simeq$ 590 K for P8.

blender, n-heptane was added. The precipitate was filtered and washed with diethylether, ethylalcohol-diethylether, and water to finally eliminate Cl$^-$.

The quantitative elemental analysis gives the following results: (The polymers from here on in will be known as P12, P10, and P8—according to the number of carbon atoms in the aliphatic chain.)

P12— C% (72.70 calc.– 72.48 found); N% (6.06 calc.– 6.02 found);
    H% (7.36 calc.– 7.49 found)

P10— C% (71.89 calc.– 71.62 found); N% (6.45 calc.– 6.48 found);
    H% (6.91 calc.– 6.83 found)

P 8 C% (70.94 calc.– 70.89 found); N% (6.89 calc.– 6.77 found);
    H% (6.40 calc.– 6.49 found).

Solubility and viscosity. The polymers obtained are sparingly soluble in solvents like methyl- and ethylacetate, N,N dimethylformamide and pyridine. A good solubility has been observed in chloroform or in chloroform-phenol or 1,2 dichloroethane-phenol mixtures. Viscosity measurements in phenol containing mixtures show that the polymers are slowly degraded.

Reproducible viscosity measurements have been made with an Ubbelhode viscometer using chloroform as solvent.

TABLE II

X-ray Powder Diffraction Data for P12[a]

Original	Therm. Treated[b]
26.1 (w)	[c]
7.08 (mw)	5.37 (w)
4.80 (s)	4.80 (s)
4.13 (vs)	4.33 (vs)
	3.87 (s)
3.27 (m)	3.27 (m)
2.88 (w)	2.88 (w)
2.62 (w)	2.43 (w)
2.34 (w)	

[a]Spacings are given in Å, intensities are in relative arbitrary scale.

[b]The sample was previously brought to the liquid anisotropic phase: $483 \leqq T \leqq 500$ K.

[c]Not measurable by photographic means.

[d]The long spacing is 24.4 Å for P10 and 22.7 Å for P8.

At a temperature of $26.10 \pm 0.02°$C the following limiting viscosity numbers $[\eta]$ have been extrapolated:

$[\eta]$ P12 = 1.01 dl g^{-1}; $[\eta]$ P10 = 0.52 dl g^{-1}
$[\eta]$ P 8 = 0.79 dl g^{-1}.

No satisfactory reproducible molecular weight data could be obtained as yet from osmometric measurements on the unfractionated polymers.

Thermal analysis. A DSC analysis was performed on several samples of the polymers using a Perkin-Elmer DSC-1 apparatus. The relevant results are shown in Table I.

The samples were examined under dry nitrogen flow. The phase-transition temperatures refer to the endotherm maximum point. For the evaluation of the transition enthalpies an indium sample has been used as reference.

Thermal polarizing microscopy. For the microscopic analysis, a Leitz polarizing microscope equipped with heating stage and photographic camera was used. Samples were examined at different heating and cooling rates and for several cycles.

All the examined samples show a first transition to an anisotropic liquid

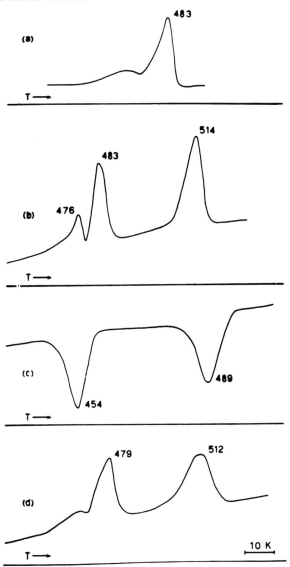

Fig. 1. DSC curves for P12: (a) first run, melting endotherm; (b) same sample, second run, melting endotherm and liquid crystal-isotropic liquid transition endotherm; (c) same sample, cooling run; (d) same sample, third heating run.

phase and a successive transition to an isotropic liquid. The phenomena observable on the microscope are very well reversible—provided that the decomposition temperature of the sample is not reached (580 K for P12 and P10, 590 K for P8).

X-ray analysis. X-ray powder spectra of the polymers with no previous

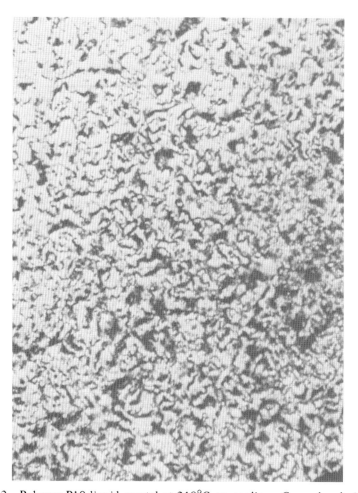

Fig. 2. Polymer P10 liquid crystal at 210°C on cooling. Crossed polarizers.

thermal manipulation were obtained photographically on the cylindrical camera of a Nonius Weissenberg apparatus. The CuK α radiation was used. The low angle portion of the spectrum was taken on a Philips powder goniometer and graphically recorded. The FeK α radiation was used. In both cases the samples were slightly cold-pressed.

The spectra (the wide angle portion) of the thermally treated samples were recorded by photographic means. The thermal treatment of the samples was made on the DSC apparatus.

Some X-ray data for P12 are given in Table II.

Discussion

The X-ray analysis shows the polymers to be at least partially crystalline.

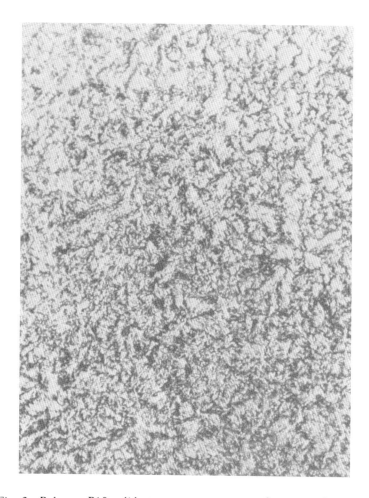

Fig. 3. Polymer P10 solid at room temperature. Same sample as in Fig. 2. Crossed polarizers.

The diffraction patterns depend upon the thermal history. The diffraction patterns obtained from the original samples and those obtained from samples previously melt to the anisotropic liquid phase are different. This is shown in Table II for P12; P10 and P8 show a similar behavior.

Rapid cooling from the isotropic liquid phase restores the original structure.

A parallel behavior is detectable from the DSC analysis (Table I and Figure 1). The melting endotherm of the original sample is broadly shouldered and it is resolved in two distinct endotherms in the successive runs provided that the transition to the isotropic liquid is not obtained. In the latter case, the melting endotherm of the following run is restored to its original form.

Polymers melt to give an anisotropic liquid with the morphological characters, as detectable from the microscopic observations, of a liquid crystalline nematic or a smectic C phase in the Schlieren texture (Figure 2).

Fig. 4. Polymer P8 solid at room temperature. A shear stress was applied during rapid cooling from the anisotropic liquid phase. Crossed polarizers.

The high enthalpic change involved in the liquid crystal-isotropic liquid transition as compared to the melting enthalpy should indicate in favor of a smectic phase.

By slow cooling ($\sim 1^\circ$C/min) of the anisotropic liquid trace of the threading is preserved down to room temperature in the solid phase (Figure 3).

It seems as if the linear singularities of the liquid crystal phase may act as nucleation lines for crystallization in such a way that the microcrystals ordering in the solid phase keeps some memory of the original semi-ordered liquid matrix.

A much more regular ordering of the microcrystals can be induced by shearing the liquid crystal phase on cooling (Figure 4).

Further work is in progress to characterize selected fractions of the original polymers.

We are grateful to Drs. S. di Martino and L. Pecoraro for helping in the preparative work and also to Drs. R. Palumbo and G. Maglio, and Prof. P. Corradini for their interest.

References

(1) G. S. Y. Yeh, Pure and Appl. Chem., 31, 65, (1972).
(2) C. Robinson, Mol. Crystals, 1, 467, (1966).

(3) V. N. Tsvetkow, E. I. Riumtsev, I. N. Shtennikova, E. V. Korneeva, B. A. Krentsel, and Y. B. Amerik, Europ. Polym. J., 9, 481 (1973).

(4) A. Douy, R. Mayer, J. Rossi, and B. Gallot, Mol. Crystals Liquid Crystals, 7, 103, (1969).

(5) Liquid Crystals, 3, part II, page 1041 f. Report on the 3rd Int. Liq. Cryst. Conf. (1970). G. H. Brown, M. M. Labes editors, Gordon and Breach Sc. Pub., (1972).

(6) W. J. Toth and A. V. Tobolsky, Polym. Lett., 8, 289, (1970).

(7) A. C. de Visser, K. de Groot, J. Feyen and A. Bantjes, J. Pol. Sc. A-1, 9, 1893, (1971).

(8) A. C. de Visser, K. de Groot, J. Feyen and A. Bantjes, Polym. Lett., 10, 851 (1972).

(9) H. Kamogawa, Polym. Lett., 10, 7, (1972).

(10) H. Saeki, K. Iimura, and M. Takeda, Polym. J., 3, 414, (1972).

(11) G. L. Wilkes, Mol. Crystals Liquid Crystals, 18, 165 (1972).

(12) G. L. Wilkes and B. The Vu, Polym. Sci. Technol., 1, 39, (1973).

(13) E. T. Samulski and A. V. Tobolsky, Macromolecules, 1, 555, (1968).

(14) G. L. Wilkes, Polym. Lett., 10, 935 (1972).

(15) G. Lock and K. Stach, Beric., 77/79, 293 (1944/46).

Antonio Roviello
Augusto Sirigu

Istituto Chimico della Università
Via Mezzocannone 4
Napoli, Italy

Received February 26, 1975
Revised March 20, 1975

COMMENTARY

Reflections on "Liquid Crystal Polymers. I. Preparation and Properties of *p*-Hydroxybenzoic Acid Copolymers," by W. J. Jackson, Jr. and H. F. Kuhfuss, *J. Polym. Sci., Polym. Chem. Ed.,* 14, 2043 (1976)

W. J. JACKSON, JR.

Research Laboratories, Eastman Chemicals Division, Eastman Kodak Company, Kingsport, Tennessee

With the objective of upgrading the physical properties of poly(ethylene terephthalate) (PET) in 1971, we attempted to react it with equimolar amounts of terephthalic acid (T) and hydroquinone diacetate (HQ) by a melt polymerization process, but the polymer melting points were too high for melt processability and the molecular weights were too low if we used more than about 20 mol % of the T(HQ) component. I then asked Herbert Kuhfuss to try an unsymmetrical monomer, an acyl ester of *p*-hydroxybenzoic acid (PHB), and our first attempt was successful (25 mol % PHB component and 75 mol % PET according to microfilm records of 1971-dated research notebooks of Tennessee Eastman Company, which was a division of Eastman Kodak Company at that time but is now Eastman Chemical Company). We subsequently were able to prepare copolyesters containing up to 90 mol % PHB.

This paper,[1] LCP I, describes the exciting and completely unexpected results which were obtained: several times the tensile strength and flexural (stiffness) properties in injection-molded test bars than ever before reported for an unreinforced thermoplastic and an unheard of zero coefficient of linear thermal expansion. Also, the melt viscosities under extrusion or molding conditions were a small fraction of those of high performance thermoplastics.

F. E. McFarlane thought that these properties might be due to the presence of thermotropic liquid crystallinity and, if so, the molten polymers should be orientable in a magnetic field. With the assistance of an NMR expert, V. A. Nicely, in 1973, he demonstrated the liquid crystalline nature of these copolyesters and presented their results in a lecture at the Gordon Research Conference on Polymers in July 1974. Because we had continued our study of the effect of polyester composition on thermotropic liquid crystallinity and polymer mechanical properties, I knew that we would be publishing additional papers, so I asked McFarlane to make their paper[2] the second in our Liquid Crystal Polymer (LCP) series.

We regret our concluding statement in LCP I: "these copolyesters appear to be the first thermotropic liquid crystal polymers to be recognized"; however, our manuscript was submitted to the journal in 1975, and it was much later that we became aware of a 1975 publication of Roviello and Sirigu,[3] who prepared several polyesters interfacially and demonstrated their thermotropic liquid crystallinity. Perhaps our statement is correct, but we certainly did not *publish* our work earlier.

Our initial PET/PHB copolyester discoveries led to 13 papers[1,2,4-14] in a LCP series and 49 U.S. Patents related to LCPs. Before publication, many of these papers were presented at international polymer conferences, and in conversations with scientists at these meetings, I was told that this initial paper had catalyzed much research at their universities or companies. Jim Economy, for instance, a pioneer in liquid-crystal type all-aromatic polyesters, told me at a Gordon Research Conference on Polymers in 1979 that he had not recognized his T/4,4'-biphenol/PHB copolyesters as being liquid crystalline, and in a 1984 review of his copolyesters he noted that our LCP I paper[1] "provided more definitive evidence of the thermotropic character" of their materials.[15]

We considered commercializing PET/80 mol % PHB as a high performance thermoplastic, but the aliphatic portion of the PET component prevented us from achieving the desired flammability resistance and heat resistance required for many electronic applications. We did license Unitika in Japan to produce these PET/PHB LCPs, however, and two of the copolyesters were commercialized there. Other companies in the United States, Europe, and Japan commercialized a number of all-aromatic LCPs, and these developments are included in an extensive history of aromatic LCPs which was written at the request of a journal editor.[12] These commercial copolyester thermoplastics have high tensile and flexural properties, of course, and our LCP I paper demonstrated that such high properties could be achieved!

REFERENCES AND NOTES

1. W. J. Jackson, Jr. and H. F. Kuhfuss, *J. Polym. Sci., A Polym. Chem.,* Ed., **14,** 2043 (1976).

2. F. E. McFarlane, V. A. Nicely, and T. G. Davis, *Contemporary Topics in Polymer Science,* Vol. 2, Plenum, New York, 1977, pp. 109–138.

3. A. Roviello and A. Sirigu, *J. Polym. Sci. C,* **13,** 455 (1975).

4. W. J. Jackson, Jr. and H. F. Kuhfuss, *J. Appl. Sci.,* **25,** 1685 (1980).

5. W. J. Jackson, Jr., *Br. Polym. J.,* **12,** 154 (1980).

6. W. J. Jackson, Jr., *Macromolecules,* **16,** 1027 (1983).

7. W. J. Jackson, Jr., *Contemporary Topics in Polymer Science,* Vol. 5, Plenum, New York, 1984, pp. 177–208.

8. W. J. Jackson, Jr. and J. C. Morris, *J. Appl. Polym. Sci., Appl. Polym. Symp.,* **41,** 307 (1985).

9. W. J. Jackson, Jr. and J. C. Morris, *J. Polym. Sci., Polym. Chem. Ed.,* **25,** 575 (1987).

10. W. J. Jackson, Jr. and J. C. Morris, *J. Polym. Sci., Polym. Chem. Ed.,* **26,** 835 (1988).

11. W. J. Jackson, Jr., in *Polymers for Advanced Technologies,* M. Lewis (ed.), VCH Publishers, New York, 1988, pp. 437–490.

12. W. J. Jackson, Jr., *Mol. Cryst. Liq. Cryst.,* **169,** 23 (1989).

13. W. J. Jackson, Jr. and J. C. Morris, in *Frontiers of Macromolecular Science,* T. Saegusa, T. Higashimura, and A. Abe (eds.), Blackwell Scientific Publications, London, 1989, pp. 405–410.

14. W. J. Jackson, Jr. and J. C. Morris, in ACS Symp. Series No. 435, *Liquid-Crystalline Polymers,* R. A. Weiss and C. K. Ober (eds.), American Chemical Society, 1990, pp. 16–32.

15. J. Economy, *J. Makromol. Sci.-Chem.,* **A21,** 1705 (1984).

PERSPECTIVE

Comments on "Liquid Crystal Polymers. I. Preparation and Properties of *p*-Hydroxybenzoic Acid Copolyesters," by W. J. Jackson, Jr. and H. F. Kuhfuss, *J. Polym. Sci., Polym. Chem. Ed.,* 14, 2043 (1976)

GORDON CALUNDANN

Hoechst Celanese Corporation, Summit, New Jersey

For decades people had speculated and heard rumors about the existence of a 400-lb primate living somewhere in Africa, but it was as late as the second half of the 19th century before the highland gorilla was actually observed by European naturalists. Albeit with much less drama, an analogous situation existed in some quarters of the polymer community in the 1960s and early 1970s. The Jackson–Kuhfuss paper of 1976 set many of us to wonder why the large and important world of thermotropic liquid crystal polymers was not discovered many years earlier.

This paper was widely anticipated after the summer of 1974, when F. E. McFarlane of Tennessee Eastman presented the essence of the research at the Gordon Research Conference on polymers at Colby-Sawyer College. An earlier (1973) patent[1] to Eastman describing and claiming key copolyesters of PET and *p*-hydroxybenzoic acid gave almost no hint of LC observation or the very thorough thermotropic liquid crystalline polymer characterization to come. By 1976, the race to patent and develop thermotropic LC polyesters with commercial import was well underway and clearly had been in progress for several years at the laboratories of Eastman, duPont, and Hoechst Celanese.

The 1972 Cottis–Economy patent[2] to the Carborundum Company claimed high melting wholly aromatic polyesters of *p*-hydroxybenzoic acid (HBA) and 4,4′-biphenol as key components. These mate-rials are, in fact, thermotropic LC polymers but no recognition of this property was indicated in the patent. The reasons for this are unclear. The high thermal transitions of the Carborundum polymers would have made LC characterization difficult at best; or, perhaps the behavior was noted and the authors thought it not relevant to the patent's purpose. Let me say a bit more on this point of missing a unique property. Over these many years since this period, more than several chemists (including one or two who were with ICI[3], Harrogate) have told me of polymers they had worked with in the 1960s or even the late 1950s that seemed curious because of odd morphology or cloudy (birefringent) melt phases. A few have mentioned that on rechecking, sometimes 30 years after the fact, that sure enough, the structure being studied at the time was a ther-motropic LC polymer. It's reasonable to assume that these kinds of materials sat unrecognized in several polymer laboratories, industrial and academic, years prior to the Jackson–Kuhfuss paper.

duPont's patents claiming wholly aromatic LC polyesters from aryl diacids and diols with ring-substituents began to appear in 1975.[4] Beyond recognizing the likelihood of thermotropic liquid crystallinity, and describing polymer melt light transmission between crossed polars as well as impressive fiber tensile properties, these patents made no attempt to rigorously demonstrate and characterize this new phenomenon in polymer chemistry. And, of course, given the immense commercial utility of certain polymers having the property of thermotropic liquid crystallinity, it would have been a very nice claim to own. The focus at Hoechst Celanese in the early 1970s was not to develop thermotropic LCPs, but rather to define polyester compositions with high performance fiber/film tensile properties which would also be melt processable at reasonable temperatures. By the summer of 1974, it began to be recognized in our laboratories that every polyester we prepared satisfying these two conditions was wholly aromatic, and a thermotropic, nematic liquid crystalline material. The first Hoechst Celanese patent issued in 1978.[5]

All of this aside, it was Tennessee Eastman that introduced the polymer community to well-characterized thermotropic LCPs within the closed circle of a Gordon Conference; and it was, in my view, the Jackson–Kuhfuss landmark paper that formalized this introduction and opened the floodgate of industrial/academic activity in LCP science and technology, research that continues to the present day.

The PET/HBA copolyester series studied in this paper was nearly the perfect system with which to illustrate the Jackson–Kuhfuss discoveries. Through the copolymer composition range, melt character evolved from isotropic to nematic, providing the authors with an excellent vehicle by which to compare key properties deriving from this sharply differing melt order: melt and film optical clarity; melt viscosity vs. composition; melt viscosity against shear rate. The copolyester system was also inexpensive, accessible, and relatively easily scaled. This afforded the authors a large quantity of polymer which in turn allowed a group of experiments of a pioneering and practical nature. Jackson–Kuhfuss taught the industry how thermotropic, nematic liquid crystalline copolyesters behave under typical injection molding conditions. Tensile properties as a function of melt order; tensile and flexural property anisotropy in a molded part; coefficient of linear thermal expansion anisotropy; skin-core effects. Behavior, now so well known and critical to LCP makers and users, was quite well described for the first time.

Because the article offered a readily available, easily synthesized LCP example, it was quickly duplicated in several industrial laboratories, including our own. This afforded each of us then active in the field to observe first-hand the physical texture, the wood-shavings morphology, of a shredded thermotropic LCP. From an industrial development viewpoint, more elaborate characterization was almost unnecessary. As one varied a polymer's structure, simple inspection of the solid gave in many instances the information needed to proceed.

Today, the thermotropic LCP industry is a $100 billion/year business with a double-digit growth rate anticipated to last well into the next decade. This paper was there at the start, and stands as a major contribution to the industrial development of a new class of useful polymeric materials.

REFERENCES AND NOTES

1. H. F. Kuhfuss and W. J. Jackson, Jr. (Tennessee Eastman), U.S. Patent 3,778,410 (1973).
2. S. G. Cottis, J. Economy, and B. E. Nowak (Carborundum), U.S. Patent 3,637,595 (1972); S. G. Cottis, J. Economy, and L. C. Woher (Carborundum), U.S. Patent 3,975,487 (1976).
3. Goodman et al. (ICI), British Patent 933,272 (1965).
4. J. J. Kleinschuster et al. (DuPont), Belgian Patent 828,935 (1975); J. J. Kleinschuster (DuPont), U.S. Patent 3,991,014 (1976).
5. G. W. Calundann (Celanese), U.S. Patent 4,067,852 (1978).

JOURNAL OF POLYMER SCIENCE: Polymer Chemistry Edition   VOL. 14, 2043-2058 (1976)

Liquid Crystal Polymers. I. Preparation and Properties of p-Hydroxybenzoic Acid Copolyesters*

W. J. JACKSON, JR. and H. F. KUHFUSS, *Research Laboratories, Tennessee Eastman Company, Division of Eastman Kodak Company, Kingsport, Tennessee 37662*

Synopsis

High molecular weight copolyesters were prepared by the acidolysis of poly(ethylene terephthalate) with p-acetoxybenzoic acid and polycondensation through the acetate and carboxyl groups. The mechanical properties of the injection-molded copolyesters containing 40–90 mole-% p-hydroxybenzoic acid (PHB) were highly anisotropic and dependent upon the PHB content, polyester molecular weight, injection-molding temperature, and specimen thickness. As the injection-molding temperature increased and the specimen thickness decreased, the tensile strength, stiffness, and Izod impact strength increased when measured along the direction of flow of the polymer melt, and the coefficient of thermal expansion was zero. In some compositions these properties were superior to those of commercial glass fiber reinforced polyesters. Maximum tensile strengths, flexural moduli, notched Izod impact strengths, and minimum melt viscosities were obtained with polyesters containing 60–70 mole-% PHB. Higher oxygen indicies (39–40) and heat deflection temperatures (150–220°C) were obtained with 80–90 mole-% PHB.

INTRODUCTION

Although poly(ethylene terephthalate) (PET) is widely used as a fiber and film, it has had only limited acceptance as a molding plastic. One important reason for this is the need for hot molds (140–150°C) to allow the polymer to crystallize. The upper use temperature of the amorphous polyester is limited by its T_g (69°C),[1] whereas the heat-deflection temperature of the unreinforced crystalline polyester is 85°C (264 psi load).[2] To meet certain specific flammability standards, fire-retardant additives, such as aromatic brominated components, must be incorporated in PET.

In an attempt to increase the T_g of the polyester, increase its flame resistance, and impart unusual physical properties, we increased the aromatic character of the polyester by modifying it with p-hydroxybenzoic acid. This was accomplished by a reaction similar to that developed by Hamb,[3,4] who found that copolyesters could be prepared by the reaction of poly(ethylene terephthalate) with 4,4′-isopropylidenediphenol diacetate and an equimolar amount of terephthalic acid. Our reaction of p-acetoxybenzoic acid (I) at 275°C with itself [eq. (1)] and with PET (III) [eq. (2)] gave short acetoxy-terminated and carboxyl-terminated segments, and these were condensed to-

---

* Paper presented in part at the Society of the Plastics Industry Annual Technical Conference, Washington, D.C., February 4–8, 1975.

2043

gether to give a high molecular weight polyester by heating under reduced pressure.

$$\text{I} \longrightarrow$$

$$\text{II} + CH_3C\text{—OH} \quad (1)$$

$$\text{III} + CH_3C\text{—O—}\bigcirc\text{—C—OH} \quad (2)$$

$$(3)$$

$$\text{IV}$$

The final copolyester then contained segments II and IV in addition to PET segments (III).

Polyesters with a somewhat similar structure prepared from methyl *p*-hydroxybenzoate, dimethyl terephthalate, and ethylene glycol have been reported.[5,6] These polymers are copoly(ether-esters) and contain *p*-ethyleneoxybenzoate units (V) and PET segments (III).

V

Several of these copoly(ether-esters) were prepared in order to compare their properties with those of our new copolyesters.

The preparation and properties of polymers from *p*-acetoxybenzoic acid and other types of polyesters in addition to PET are described in our patents.[7,8]

EXPERIMENTAL

Materials

Poly(ethylene terephthalate), with an inherent viscosity of 0.60, was prepared from dimethyl terephthalate and ethylene glycol with zinc acetate (65 ppm zinc) and antimony triacetate (230 ppm antimony) catalysts.

p-Acetoxybenzoic acid, mp 189–191°C, was prepared by a conventional acetylation of *p*-hydroxybenzoic acid with acetic anhydride in acetic acid at

120°C (sulfuric acid catalyst) followed by recrystallization from *n*-butyl acetate.

Methods

Inherent viscosities (IV) were measured at 25°C in 60:40 (by volume) phenol–tetrachloroethane at a concentration of 0.50 g/100 ml. The melt flow of an insoluble polyester was determined in accordance with ASTM D1238 with an 0.04-in. capillary and 2160 g load. Melting points were determined with a differential thermal analyzer (Du Pont 900). Glass transition temperatures T_g were determined with a differential scanning calorimeter (Perkin-Elmer DSC-1B). NMR spectra were determined on trifluoroacetic acid solutions of the polyesters with a Varian A-60 spectrometer at 60 MHz and a Varian HA-100 NMR spectrometer at 100 MHz. Melt viscosities were determined with an Instron Model 3211 capillary rheometer.

The polymers were injection-molded into unheated molds in a 1-oz Watson-Stillman injection-molding machine to give $2\frac{1}{2} \times \frac{3}{8} \times \frac{1}{16}$ in. tensile bars for tensile measurements and $5 \times \frac{1}{2} \times \frac{1}{8}$ in. flexure bars for determination of flexural modulus, Izod impact strength, and heat-deflection temperature. Several compositions were also injection-molded in a 6-oz New Britain 175-TP reciprocating screw machine to give $8\frac{1}{2} \times \frac{3}{4} \times \frac{1}{8}$ in. tensile bars and in a Newbury HV1-25T reciprocating screw machine to give $6\frac{1}{2} \times \frac{3}{4} \times \frac{1}{8}$ in. tensile bars. ASTM procedures were used for measuring tensile strength and elongation to break (ASTM D1708), flexural modulus (ASTM D790), flexural strength (ASTM D790), Izod impact strength (ASTM D256 Method A), Rockwell hardness (ASTM D785 Method A), heat-deflection temperature (determined at 264 psi, ASTM D648), mold shrinkage (ASTM D955), and coefficient of linear thermal expansion (ASTM D696). The oxygen index of $\frac{1}{8}$-in. bars was determined with a GE Flammability Index Tester, Model FL 101.

Polymer Preparation

Copolyesters from *p*-Acetoxybenzoic Acid and Poly(ethylene Terephthalate). A mixture of *p*-acetoxybenzoic acid (5–90 mole-%) and poly(ethylene terephthalate) particles (2-mm size, IV 0.60, 95–10 mole-%) was placed in a 500-ml flask equipped with a stainless steel stirrer and a short head with an inlet and an outlet for nitrogen. The nitrogen outlet was connected to a glass tube (distillation column), which led to a receiver with provision for applying vacuum. After the reaction flask was evacuated and purged with nitrogen three times to remove all air, it was heated under reduced pressure (0.5 mm) in a metal bath at 110°C to dry the reactants. The flask was removed from the bath, and the bath was heated to 275°C. The flask was then placed back in the bath; and while the contents were stirred in a nitrogen atmosphere, acetic acid slowly distilled out. After a low melt viscosity was obtained (about 30 min), a vacuum of about 0.5 mm was applied and stirring continued for about 4 hr at 275°C (300°C for the higher melting compositions, which solidified, prepared with 80 and 90 mole-% of *p*-acetoxybenzoic acid). High melt viscosities were obtained with polyesters prepared with 50 to 70 mole-% of p-acetoxybenzoic acid. The molten polymers were light tan and opaque, and IV values were about 0.6–0.8.

Lower melt viscosities and IV values were obtained with polyesters prepared with 5–40 mole-% p-acetoxybenzoic acid. These polymers were ground to pass a 20-mesh screen and then heated under reduced pressure (0.05–0.1 mm) at 210 to 220°C for 8–16 hr to increase the IV by solid-phase polymerization to about 0.4–0.6. The high-melting polyesters which had solidified during preparation were also built up further in the solid phase at 0.05–0.1 mm (4 hr at 260°C for the 80 mole-% PHB composition and 6 hr at 280°C for the 90 mole-% PHB copolyester).

Copoly(ether-esters) from Methyl p-Hydroxybenzoate, Dimethyl Terephthalate, and Ethylene Glycol. Copoly(ether-esters) prepared from 30, 60, and 80 mole-% of methyl p-hydroxybenzoate were made by the method of Example 1 in U.S. Patent 3,288,755.[6] The polymers were amber-colored and transparent; IV values were about 0.7–0.9.

RESULTS AND DISCUSSION

Polymer Preparation

A two-step process is involved in the preparation of the copolyesters: (1) cleavage of poly(ethylene terephthalate) by p-acetoxybenzoic acid at 275°C in an inert atmosphere, and (2) condensation under vacuum of the carboxyl-terminated and acetate-terminated segments (including the self-condensation of p-acetoxybenzoic acid) to form the high molecular weight copolyester. The highest IV values were obtained when strenuous precautions were taken to eliminate all air and moisture from the reactants just before polymerization. This was accomplished by evacuating and purging the flask containing the ingredients several times with nitrogen and then heating at 110°C and 0.5 mm for 30 min. When PET modified with 60 mole-% p-hydroxybenzoic acid (PHB) was prepared without these precautions, the polymer IV was 0.50, whereas IV values of 0.65–0.75 were obtained when the purging and drying procedures were followed.

Compositions modified with 40–90 mole-% PHB became hazy and then opaque during the polymerization step under vacuum, whereas the compositions modified with 5–30 mole-% PHB remained clear. Polymers modified with 50–75 mole-% PHB had IV values above 0.5 after 2–4 hr at 275°C under vacuum; the IV could not be determined with polymers containing 80 mole-% or more PHB because of insolubility in the 60:40 phenol–tetrachlorethane solvent used for measuring IV. The IV of polymers modified with 5–40 mole-% PHB was limited to about 0.35 in melt polymerization, but higher IV values were obtained by solid-phase polymerization under vacuum at 210–220°C.

The reaction of p-acetoxybenzoic acid with poly(ethylene terephthalate) appears to be primarily an acidolysis reaction [eq. (2)]; an appreciable amount of acetic acid is formed in this step of the reaction. Also, acidolysis of ethylene dibenzoate (PET model compound) occurred readily with 1-naphthoic acid in sealed tubes at 260°C whereas very little esterolysis occurred between ethylene dibenzoate and phenyl acetate at 270°C.[9] Hamb,[3] on the other hand, observed that acidolysis and esterolysis both occur readily when PET is heated at 275°C with terephthalic acid and 4,4'-isoproplyiden-ediphenol diacetate; very little acetic acid is formed.

The IV limitation with the lower levels of PHB in melt polymerization may be due to the formation of ethylene acetate end groups, -CH$_2$CH$_2$OAc. An NMR spectrum of the acidolysis product of PET and 25 mole-% p-acetoxy-benzoic acid indicated the presence of about 5 mole-%, based on the p-ace-toxybenzoic acid, of ethylene acetate end groups (peak at 2.1 ppm next to the peak for the methyl protons of acetic acid). After the melt-polymerization step under vacuum to an IV of 0.35, a similar amount of ethylene acetate end groups was observed by NMR analysis. The ethylene acetate groups will be produced if esterolysis occurs between PET an p-acetoxybenzoic acid or, per-haps more likely, if the acetic acid [produced in eqs. (1) and (3)] reacts with PET in an acidolysis reaction. The ethylene acetate groups may be responsi-ble for the low IV in melt polymerization because of the low reactivity of these groups. Few or no ethylene acetate groups were observed in the acidol-ysis product of PET and 60 mole-% p-acetoxybenzoic acid, and high IV values were obtained on melt polymerization.

Polymer Structure

The presence of segments II, III, and IV [eqs. (1)–(3)] has been confirmed by NMR spectra of compositions prepared from PET and 60 mole-% p-ace-toxybenzoic acid. No p-ethyleneoxybenzoate groups (V) were observed (no peak at 4.1 ppm due to ether methylene protons or doublet at 6.6 ppm due to the aromatic hydrogen atoms *ortho* to the ether linkage). The NMR spectra of these two types of polyesters [(1) PET reacted with p-acetoxybenzoic acid, and (2) PET copolyester prepared with methyl p-hydroxybenzoate] are given in our patent.[8]

NMR studies suggest that the PHB from p-acetoxybenzoic acid has a ran-dom distribution in the copolymer. Since the NMR chemical shift of the ar-omatic hydrogen atoms *ortho* to the hydroxyl group of PHB varies with the type of group attached to the oxygen atom, it was possible to measure the areas of the chemically shifted species and determine the dyad concentration of PHB-PHB units [II in eq. (1)] and terephthalate-PHB units [IV in eq. (3)]. The terephthalate-PHB concentration was found to be about 55% of the total PHB units in the samples, a value which is very close to the value of 56.5% obtained by calculation of the terephthalate-PHB concentration for a com-pletely random distribution, assuming equal reactivity of the PHB and ter-ephthalate carboxyl groups with the p-acetoxybenzoic groups.[10]

Gel-permeation chromatographic analysis of PET modified with 60 mole-% PHB has a normal molecular weight distribution.[10] The number-average molecular weight, determined ebulliometrically in hexafluoroisopropanol, was 14,800 for a sample having an IV of 0.57 and 20,600 for a sample having an IV of 0.70; PET has slightly higher molecular weights at these IV values.

X-ray diffraction curves were obtained with powders of PET modified with 5–90 mole-% PHB. The PET crystalline fraction disappeared continuously as the PHB concentration increased. At 60 mole-% PHB there was only a very slight indication of PET crystallinity or of PHB crystallinity. An injec-

JACKSON AND KUHFUSS

TABLE I
Properties of Poly(ethylene Terephthalate) Modified with
p-Hydroxybenzoic Acid (PHB)

PHB, mole-%	IV	T_g, °C	Mp, °C	Solubility[a]	Film appearance[b]
0	0.60	69	245	PTCE	Clear
5	0.41	74	241	PTCE	Clear
10	0.43	74	235	CHCl$_3$	Clear
15	0.46	74	231	CHCl$_3$	Clear
20	0.50	77	230	CHCl$_3$	Clear
25	0.62	83	235	CHCl$_3$	Clear
30	0.51	85, 166	227	CHCl$_3$	Clear
35	0.43	79, 158	226	CHCl$_3$	Opaque
40	0.55	81, 159	226	CHCl$_3$	Opaque
50	0.69	75, 159	228	PTCE	Opaque
55	0.59	—[c]	—[c]	PTCE	Opaque
60	0.74	—[c]	—[c]	PTCE	Opaque
65	0.68	—[c]	—[c]	PTCE	Opaque
70	0.86	—[c]	268[d]	PTCE	Opaque
80	Insol	—[c]	293[d]	Insol	Opaque
90	Insol	—[c]	—[c]	Insol	Opaque

[a] If insoluble in chloroform (CHCl$_3$), solubility was tested in 60:40 phenol—tetrachloroethane (PTCE).

[b] Films were pressed at about 300°C and quenched in water.

[c] None detected.

[d] Very weak endotherm.

tion-molded bar also showed very little crystallinity. The crystallinity due to PHB, a large single peak at 19.4° 2θ, CuKα, increased as the PHB concentration was increased further. The continuous manner in which the PET crystalline diffraction disappeared with the concomitant development of a diffraction pattern characteristic of the polyester of PHB suggests a continuously changing morphological system characteristic of random copolymers.

Polymer Properties

Table I lists a number of the properties of the copolyesters. The T_g increased as the PHB content increased up to 30 mole-%, and two T_g were observed for compositions containing 30–50 mole-% PHB. The upper T_g presumably is due to short PHB segments; a similar T_g has been reported for a copolyester of 45 mole-% p-hydroxybenzoic acid and 55 mole-% m-hydroxybenzoic acid.[11] A T_g was not observed when the PHB content of the PET copolyesters was increased above 50 mole-%. Crystalline melting points also were not observed except for very weak endotherms in two of the compositions. Copolyesters containing 10–40 mole-% PHB were soluble in chloroform; except for the highly crystalline compositions containing at least 80 mole-% PHB, the other copolyesters were soluble in 60:40 phenol–tetrachloroethane and not in chloroform. The molten polyesters and pressed films were opaque when at least 35 mole-% PHB was present.

One of the most unusual properties of these copolyesters is their melt viscosities. The effect of PHB content on the melt viscosity at 275°C is

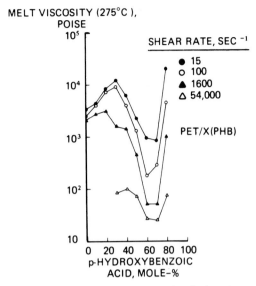

Fig. 1. Melt viscosity of PET modified with *p*-hydroxybenzoic acid.

shown in Figure 1.   Even though the inherent viscosities of the polymers containing 0–30 mole-% PHB (IV 0.54–0.57) were lower than those of the polymers containing 40–70 mole-% PHB (IV 0.61–0.64), they have higher melt viscosities.   The polyester containing 80 mole-% PHB in Figure 1 had a melt flow of about 5 g/10 min at 325°C as defined earlier.   It is noteworthy that the melt viscosities increased as the PHB content was increased to about 30 mole-% and then decreased as the PHB content was increased further to 60 mole-%; the molten copolymer also is clear at PHB contents up to 30 mole-% and opaque at higher PHB levels.   The effect of shear on the melt viscosities of these copolyesters at 275°C is shown in Figure 2.   As the PHB content increases the polymer becomes shear-sensitive at lower shear rates.   The poly-

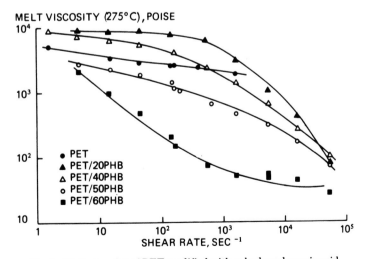

Fig. 2. Melt viscosity of PET modified with *p*-hydroxybenzoic acid.

TABLE II
Properties of Injection-Molded Poly(ethylene Terephthalate)
Modified with p-Hydroxybenzoic Acid (PHB)

Property	PHB content,							
	0	30 mole-%	40 mole-%	50 mole-%	60 mole-%	70 mole-%	80 mole-%	90 mole-%
Cylinder temperature, °C	275	250	250	260	260	280	340	400
Inherent viscosity								
Before molding	0.76	0.59	0.54	0.60	0.67	0.65	Insol	Insol
After molding	0.62	0.54	0.54	0.57	0.62	0.62	Insol[a]	Insol[b]
Tensile strength, 10^3 psi[c]	8.0	17.0	28.6	32.5	33.7	26.1	34.8	17.3
Elongation to break, %	240	12	10	26	20	10	24	18
Flexural modulus, 10^5 psi	3.3	5.8	11.1	14.1	18.1	14.5	14.0	10.3
Izod impact strength								
Notched, ft-lb/in.	0.3	1.0	1.2	1.6	7.8	2.3	2.2	0.4
Unnotched, ft-lb/in.	9.5	18.0	31.9	27.5	27.7	14.7	14.1	3.7
Rockwell hardness, L	73	82	65	51	42	53	65	78
Heat-deflection temperature, °C	66	73	71	65	64	74	154	221
Mold shrinkage, %	0.6	0.1	0.0	0.0	0.0	0.0	0.0	0.1
Oxygen index	21[d]	—	27	29	30	33	39	40

[a] Melt flow 28 g/10 min at 325°C (0.04-in. capillary).

[b] Melt flow 45 g/10 min at 380°C (0.04-in. capillary).

[c] Except for PET (break strength 7200 psi), none of the polymers exhibited a definite yield point.

[d] Determined on bar reinforced with 20 wt-% glass fibers to avoid abnormally high value because of dripping.

ester containing 60 mole-% PHB is particularly shear-sensitive; at a shear rate of 1000 sec^{-1} the melt viscosity is less than 5% that of PET.

As will be discussed in detail in the next paper of this series,[12] the opaque melts and melt viscosity behavior of the polymers are due to the presence of "liquid crystals."[13] These copolyesters appear to be the first thermotropic liquid crystal polymers to be recognized.

Molding Plastic Properties

Although the copolyesters are light tan in color, injection-molded specimens of compositions containing 40–90 mole-% PHB have a white, opaque surface, and bars of PET modified with 60 mole-% PET are also glossy. Table II shows the effect of PHB content on a number of the properties of the injection-molded copolyesters. Tensile strengths and flexural moduli are exceptionally high. Maximum flexural modulus (stiffness) and Izod impact strength and minimum Rockwell hardness were obtained with 60 mole-% PHB. The heat-deflection temperature of compositions containing up to 70 mole-% PHB appears to be limited by the T_g of the PET portion of the polymer. The copolyesters containing 80 and 90 mole-% PHB are more crystalline and, therefore, have higher heat-deflection temperatures. The absence of mold shrinkage in compositions containing 40–80 mole-% PHB is particularly noteworthy. The oxygen index increases with PHB content, as is expected, because of the increase in aromatic content of the copolyesters.

Even higher oxygen indices are attained with copolyesters prepared from

TABLE III
Effect of Molding Temperature on Properties of PET Modified with 60 Mole-% PHB

| Property | Cylinder temperature | | | | |
	210°C	240°C	250°C	260°C	280°C
Inherent viscosity[a]	0.62	0.61	0.63	0.62	0.60
Tensile strength, 10^3 psi	22.1	29.6	30.4	33.7	33.5
Elongation, %	8	9	20	20	12
Flexural modulus, 10^5 psi	15.1	15.9	19.1	18.1	16.5
Izod impact strength					
Notched, ft-lb/in.	2.9	4.5	5.6	7.8	5.5
Unnotched, ft-lb/in.	13	15	21	28	21
Rockwell hardness, L	50	36	45	42	39
Heat-deflection temperature, °C	—	—	66	64	64

[a] Value after molding.

p-acetoxybenzoic acid and poly(ethylene 2,6-naphthalenedicarboxylate) (PEN). The copolyester containing 60 mole-% PHB had an oxygen index of 42. Because of the high T_g of the PEN (111°C), the copolyester with 60 mole-% PHB had a heat-deflection temperature of 120°C (compared to 64°C for the similar PET/60 PHB composition in Table II). The PEN/PHB copolyesters also exhibit the low melt viscosity and the high tensile strength and stiffness of the PET/PHB copolyesters.

The temperature at which the PHB copolyesters are injection-molded affects the orientation of the "liquid crystal" polymer chains and, therefore, affects the mechanical properties. Table III shows the effect of cylinder temperature on the properties of PET modified with 60 mole-% PHB. As the temperature was increased, the melt viscosity decreased, the speed of the polymer melt injected into the mold increased, and the orientation of the polymer chains therefore increased.[14] The tensile strength, elongation, stiffness, and impact strength tended to increase as the cylinder temperature was increased from 210 to 260°C. The somewhat lower level of properties obtained when the polymer was molded at 280°C may be due to some loss of

TABLE IV
Effect of Inherent Viscosity on Properties of PET Modified with 60 Mole-% PHB[a]

| Property | Inherent viscosity[b] | | | | |
	IV-0.42	IV-0.50	IV-0.55	IV-0.62	IV-0.72
Tensile strength, 10^3 psi	14.7	24.7	31.0	33.7	37.9
Elongation, %	14	18	11	20	26
Flexural modulus, 10^5 psi	10.2	13.1	17.4	18.1	18.4
Notched Izod impact strength, ft-lb/in.	0.3	3.5	5.9	7.8	6.6
Heat-deflection temperature, °C	60	64	64	64	65

[a] Injection-molded at 260°C cylinder temperature.
[b] Value after molding.

TABLE V
Effect of Inherent Viscosity on Properties of PET Modified with 70 Mole-% PHB[a]

Property	Inherent viscosity [b]		
	IV-0.62	IV-0.72	IV-1.01
Tensile strength, 10^3 psi	26.1	44.4	38.2
Elongation, %	10	25	15
Flexural modulus, 10^5 psi	14.5	18.2	19.6
Izod impact strength			
Notched, ft-lb/in.	2.3	4.4	6.0
Unnotched, ft-lb/in.	14.7	18.5	27.9
Rockwell hardness, L	53	47	47
Heat-deflection temperature, °C	74	87	88

[a] Injection-molded at 280°C cylinder temperature.
[b] Value after molding.

orientation because of relaxation of the polymer chains in the melt before the polymer solidified.

The inherent viscosity affects the mechanical properties of the injection-molded copolyesters. Table IV shows the effect of viscosity on the properties of PET modified with 60 mole-% PHB and molded at the optimum cylinder temperature for this composition of about 260°C. The tensile strength, stiffness, and notched Izod impact strength increased as the IV was increased up to about 0.6–0.7 (number-average molecular weight about 16,000–20,000). In the case of PET modified with 70 mole-% PHB and molded at the optimum cylinder temperature for this composition of about 280°C (Table V), these properties increased as the IV was increased to 1.0.

Since the melt viscosity of PET modified with 80 mole-% PHB is appreciably higher than that of PET/70 PHB (Fig. 1), a higher injection-molding temperature is required in order to reduce the melt viscosity sufficiently so that high injection speed and polymer chain orientation can be achieved. The

TABLE VI
Effect of Molding Temperature on Properties of PET Modified with 80 Mole-% PHB[a]

Property	Cylinder temperature, °C					
	300	320	340	360	380	400
Tensile strength, 10^3 psi	12.0	21.2	26.9	33.2	33.7	33.7
Elongation, %	16	18	22	22	23	24
Flexural modulus, 10^5 psi	8.2	10.1	12.4	12.8	14.4	14.6
Izod impact strength						
Notched, ft-lb/in.	0.6	0.8	1.9	2.2	3.1	3.2
Unnotched, ft-lb/in.	2.9	7.6	10.2	8.7	12.1	—
Rockwell hardness, L	64	62	63	65	66	58
Heat-deflection temperature, °C	108	136	144	153	154	156

[a] Melt flow 5.1 g/10 min at 325°C (0.04-in. capillary); melt viscosity is shown in Fig. 1.

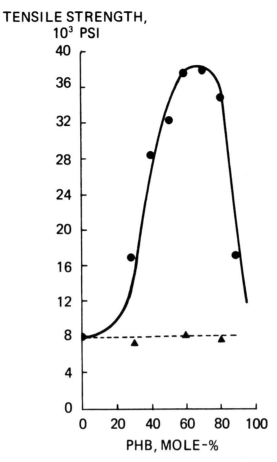

Fig. 3. Tensile strength of PET modified with p-hydrozybenzoic acid: (——) injection-molded copolyesters of Tables II–IV; (- -) injection-molded copolyesters containing segments of PET and p-ethyleneoxybenzoate (V).

dramatic effect of the cylinder temperature on the tensile strength, stiffness, impact strength, and heat-deflection temperature of PET/80 PHB is shown in Table VI. When the polyester melt flow rate was 28 g/10 min instead of 5 g/10 min as in Table VI, similar mechanical properties were obtained when bars were molded at a 20°C lower cylinder temperature. (Because of insolubility of the polymer, inherent viscosities could not be determined.)

The effect of PHB content on the maximum values obtained for tensile strength, flexural modulus, and notched Izod impact strength of the injection-molded copolyesters in Tables II–IV is shown graphically in the solid lines in Figures 3–5. As has been discussed, lower polymer IV values and lower molding temperatures give lower values of tensile strength, stiffness, and impact strength. When all compositions are injection-molded at 250–260°C, the maxima in these properties occur at about 60 mole-% PHT.[8] Significantly higher values of these properties were obtained when PET/70 PHT is injection-molded at 280°C (Table V) and PET/80 PHT is molded at 340 to 380°C (Table VI).

The dotted lines in Figures 3–5 show the lower level of properties obtained

2054   JACKSON AND KUHFUSS

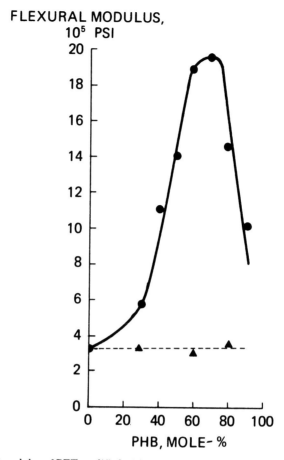

Fig. 4. Flexural modulus of PET modified with p-hydroxybenzoic acid.   Curves as in Fig. 3.

with the similar copolyesters containing segments of PET and p-ethylene-oxybenzoate (V) prepared from methyl p-hydroxybenzoate, dimethyl terephthalate, and ethylene glycol.   These compositions, injection-molded at 255–275°C, were amber-colored and transparent, and the IV values of the molded bars were 0.62–0.85.

The type of injection-molding machine which is used to mold the copolyesters affects the mechanical properties.   The polymers in Tables II–VI and Figures 3–5 were molded on a 1-oz plunger-type injection-molding machine.   Table VII lists the properties obtained when copolyesters of PET modified with 30–80 mole-% PHB were molded in reciprocating screw injection-molding machines.   The flexural moduli in Table VII are similar to the values in Figure 4, but notched Izod impact strengths are appreciably higher than those of Figure 5.   The tensile strengths of the copolyesters in Table VII, on the other hand, are lower than those of Table II and Figure 3, because of a lower degree of orientation of the polymer chains in the ⅛-in. thick tensile bars than in the thinner 1/16-in. thick tensile bars from the smaller machine.

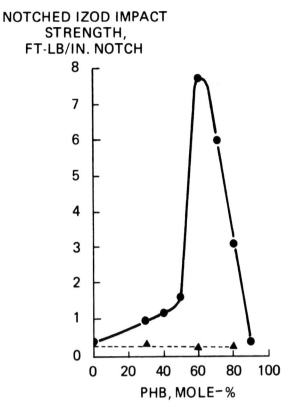

Fig. 5. Notched Izod impact strength of PET modified with *p*-hydroxybenzoic acid.   Curves as in Fig. 3.

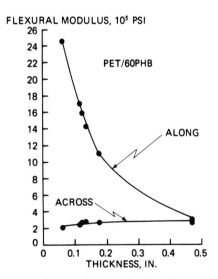

Fig. 6. Effect of thickness on along- and across-the-flow flexural modulus of PET modified with 60 mole-% *p*-hydroxybenzoic acid.

TABLE VII
Properties of PET/PHB Copolyesters Injection-Molded
in Reciprocating Screw Machines

| Property | PHB content | | | |
	30 mole-%[a]	40 mole-%[a]	60 mole-%[a]	80 mole-%
Inherent viscosity				
Before molding	0.61	0.59	0.65	Insol[c]
After molding	0.57	0.57	0.62	Insol
Tensile strength, 10^3 psi	9.8[d]	24.6	26.5	17.3
Elongation to break, %	21	13	11	6
Flexural modulus, 10^5 psi	6.6	11.9	19.4	14.8
Izod impact strength				
Notched, ft-lb/in.	3.2	7.1	13.6	4.6
Unnotched, ft-lb/in.	14[e]	19	No break	No break
Rockwell hardness, L	80	68	37	67
Heat-deflection temperature, °C	75	74	65	144

[a] 6-Oz New Britain 175-TP machine, cylinder temperature 260°C, mold temperature 23°C.

[b] 1-Oz Newbury HV1-25T machine, cylinder temperature 316°C, mold temperature 100°C.

[c] Melt flow 21 g/10 min at 325°C (0.04-in. capillary).

[d] Yield strength; break strength was 4700 psi; other compositions did not exhibit a yield point.

[e] Value for 60% of specimens which broke; 40% did not break.

The tensile strengths, elongations, flexural moduli, and Izod impact strengths in Tables II and VII of PET containing 60–80 mole-% PHB are similar to or higher than those properties of commercial glass fiber reinforced polyesters. PET reinforced with 30 wt-% of glass fibers, for instance, has a tensile strength of 19,500 psi, elongation at break of 3%, flexural modulus of 13.5×10^5 psi, and notched and unnotched Izod impact strengths of 1.5 and 7.0 ft-lb/in., respectively.[15]

Since the tensile strength, stiffness, and Izod impact strength increase as the degree of orientation of the copolyester increases, it was of interest to compare the "along-the-flow" properties with the "across-the-flow" properties. Plaques gated along one edge and ⅛ in. thick were injection-molded and then cut along and across the flow directions to give specimens of PET modified with 40, 60, and 80 mole-% PHB for testing. Table VIII shows that the mechanical properties are highly anisotropic. It is significant that the mold shrinkage and coefficient of linear thermal expansion are zero along the flow but not across the flow. Since the plaques have lower along-the-flow tensile strengths, flexural moduli, and Izod impact strengths then tensile bars of the same compositions and thickness molded similarly on the same machine (Table VII), the plaques apparently have a lower degree of orientation than the bars.

The effect of specimen thickness on along- and across-the-flow flexural modulus and flexural strength of PET/60 PHB is shown in Figures 6 and 7.

TABLE VIII
Anisotropic Properties of Injection-Molded PET/PHB Copolyesters

Property	40 PHB[a,b]		60 PHB[a,c]		80 PHB[d]	
	Along Flow	Across Flow	Along Flow	Across Flow	Along Flow	Across Flow
Tensile strength, 10^3 psi	17.6	7.6	15.5	4.2	14.9	5.4
Elongation, %	2	5	8	10	12	28
Flexural modulus, 10^5 psi	11.7	3.2	17.1	2.3	8.5	2.3
Flexural strength, 10^3 psi	19.4	8.7	15.9	4.9	15.9	6.2
Notched Izod impact strength, ft-lb/in.	1.0	1.7	6.1	0.6	4.0	2.3
Heat-deflection temperature, °C	77	62	66	55	—[e]	—[e]
Mold shrinkage, %	0.0	0.7	0.0	0.3	0.07	0.8
Coefficient of linear thermal expansion, 10^{-5} in./in./°C	0.0	5.1	0.0	4.5	0.0	4.3

[a] $4^1/_2 \times 4^1/_2 \times {}^1/_8$ in. plaques, gated along one edge and injection-molded in 6-oz New Britain 175-TP machine (cylinder temperature 260°C, mold temperature 23°C), were cut into 0.5-in. wide specimens; these specimens were milled into the standard tensile bar shape for tensile measurements.

[b] Inherent viscosity 0.59 before molding, 0.57 after molding.

[c] Inherent viscosity 0.64 before molding, 0.61 after molding.

[d] $3 \times 3 \times {}^1/_8$ in. plaques, gated along one edge and molded in 1-oz Newbury HV1-25T machine (cylinder temperature 340°C, mold temperature 23°C), were cut and formed into specimens as above. Melt flow 5.1 g/10 min at 325°C (0.04-in. capillary).

[e] Specimens not long enough to determine this property.

The thinnest (${}^1/_{16}$-in.) specimens had the most anisotropic properties (highest stiffness and along-the-flow strength and lowest stiffness and across-the-flow strength). These properties were almost isotropic when the thickness was approaching ½ in.

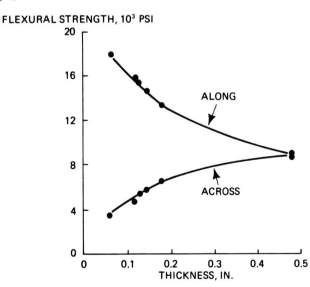

Fig. 7. Effect of thickness on along- and across-the-flow flexural strength of PET modified with 60 mole-% *p*-hydroxybenzoic acid.

2058 JACKSON AND KUHFUSS

CONCLUSIONS

High molecular weight copolyesters can be prepared by the reaction of p-acetoxybenzoic acid with poly(ethylene terephthalate). The reaction involves acidolysis of the pclyester and polycondensation of the acetate and carboxyl groups. Copolyesters with a p-hydroxybenzoic acid (PHB) content of at least 35 mole-% have opaque melts. The mechanical properties of the injection-molded copolyesters containing 40–90 mole-% PHB are highly anisotropic and dependent upon the PHB content, polyester molecular weight, injection-molding temperature, and specimen thickness. Maximum tensile strengths, flexural moduli, and notched Izod impact strengths and minimum melt viscosities are obtained with copolyesters containing 60–70 mole-% PHB. Higher oxygen indices and heat deflection temperature are obtained with 80–90 mole-% PHB or with copolyesters prepared from p-acetoxybenzoic acid and poly(ethylene 2,6-naphthalenedicarboxylate). The anisotropic properties of these polyesters are due to the orientation of the polymer chains during molding. The phenomena of the opaque melts, low melt viscosities, and anisotropic properties can be explained on the basis of liquid crystal formation;[12] these copolyesters appear to be the first thermotropic liquid crystal polymers to be recognized.

The authors acknowledge the excellent technical assistance of L. G. Holloway, who prepared many of the copolyesters. They are also indebted to R. H. Cox, who determined the melt viscosities; J. T. Dougherty, who obtained and interpreted the NMR spectra; C. A. Boye, Jr., who obtained and interpreted the x-ray diffractometric curves; and H. Gonzalez, who determined the anisotropic mechanical properties.

References

1. J. G. Smith, C. J. Kibler, and B. J. Sublett, *J. Polym. Sci. A-1*, **4**, 1851 (1966).

2. Anonymous, *Brit. Plastics*, **39**, 644 (1966).

3. F. L. Hamb, *J. Polym. Sci. A-1*, **10**, 3217 (1972).

4. F. L. Hamb, (to Eastman Kodak Co.), U. S. Pat. 3,772,405 (1973).

5. W. Griehl, *Chemiefasern*, **16** (10), 775 (1960).

6. W. Griehl and H. Lückert (to Inventa A.G. für Forschung und Patentverwertung), U. S. Pat. 3,288,755 (1966).

7. H. F. Kuhfuss and W. J. Jackson, Jr., (to Eastman Kodak Co.), U. S. Pat. 3,778,410 (1973).

8. H. F. Kuhfuss and W. J. Jackson, Jr., (to Eastman Kodak Co.), U. S. Pat. 3,804,805 (1974).

9. F. E. McFarlane (Tennessee Eastman Company), private communication.

10. J. R. Overton (Tennessee Eastman Company), private communication.

11. R. Gilkey and J. R. Caldwell, *J. Appl. Polym. Sci.*, **2**, 198 (1959).

12. F. E. McFarlane and V. A. Nicely, *J. Polym. Sci.*, in press.

13. J. A. Castellano and G. H. Brown, *Chem. Technol.*, **3**, 47, 229 (1973).

14. Z. Tadmor, *J. Appl. Polym. Sci.*, **18**, 1753 (1974).

15. J. R. Caldwell, W. J. Jackson, Jr., and T. F. Gray, Jr., in *Encyclopedia of Polymer Science and Technology Suppl.*, H. Mark and N. Bitiales, Eds., 1976, p. 444. Wiley, New York.

Received September 30, 1975

COMMENTARY

Reflections on "Photoinitiated Cationic Polymerization with Triarylsulfonium Salts," by J. V. Crivello and J. H. W. Lam, *J. Polym. Sci.: Polym. Chem. Ed.,* 17, 977 (1979)

JAMES V. CRIVELLO

Cogswell Laboratory, Department of Chemistry, Rensselaer Polytechnic Institute, Troy, New York 12180

In 1979 when this paper appeared, the field of photoinduced cationic polymerizations was just at its inception. The publication of this article disclosed the development of triarylsulfonium salts as the first new class of truly practical, latent and high quantum yield photoinitiators. In this paper we also presented evidence that the mechanism of initiation of cationic polymerization by triarylsulfonium salts involves the

Figure 1. The photograph shows a 13′6″ fiberglass-reinforced kyack fabricated using triarylsulfonium salt monomers together with an epoxidized vegetable oil as the resin. The kyack was constructed by ordinary wet layup techniques by impregnating glass cloth with the resin triarylsulfonium salt mixture in a mold and then cured by exposure to direct solar irradiation. The irradiation time was 30 min. In this fashion, the upper and lower portions of the kyack were made and then joined at the midseam using a resin impregnated tape and once again irradiating with sunlight. The photograph shows graduate student Ramesh Narayan (in the kyack) and his advisor, J. V. Crivello.

photochemical generation of a Brønsted acid followed by attack of the acid on the monomer. Lastly, it was demonstrated in this paper that triarylsulfonium salts are capable of mediating the photoinduced polymerization of virtually every known type of cationically polymerizable monomers.

In the intervening time, there has been considerable interest and research activity in these photoinitiators both by academic as well as industrial workers which has led to the rapid development of this area of polymer chemistry. Most significantly, triarylsulfonium salts have found applications in a number of commercial products. For example, currently, they are employed in photocurable can coatings, silicone release coatings, pressure sensitive adhesives and as printing inks. Photopolymerizable systems based on triarylsulfonium salts are used to generate three dimensional solid objects by stereolithography and the salts are also employed as photoacid generators in high resolution microelectronic photoresists to produce relief images with features smaller than 0.5 μm.

PERSPECTIVE

Comments on "Photoinitiated Cationic Polymerization with Triarylsulfonium Salts," by J. V. Crivello and J. H. W. Lam, *J. Polym. Sci.: Polym. Chem. Ed.,* 17, 977 (1979)

C. GRANT WILLSON

The Departments of Chemistry and Chemical Engineering, The University of Texas, Austin, Texas

This paper by Crivello and Lam is remarkably thorough and extremely well written. I often bring this paper to the attention of my students as a model of excellent technical writing. It includes general synthetic methods for the preparation of symmetrical and unsymmetrical analogs of the title compounds in which all of the products are fully and rigorously characterized. It includes a detailed and quantitative study of the photochemistry of these compounds including establishment of quantum yields and it reports the results of a series of quantitative experiments that demonstrate the utility of sulfonium salts as catalysts for the photoinitiated cationic polymerization of monomers from several classes of monomers. The authors could have chosen to separate the paper into several smaller publications and thereby added numbers to their lists of publications. I am pleased that they did not. They produced a single, comprehensive paper that succinctly but thoroughly tells the whole story.

This is not the first paper describing the photosensitivity of sulfonium salts. References to the earlier work are provided in the Crivello and Lam paper. Other authors had even noted the generation of acid upon photolysis of these compounds but mentioned it only in passing. Furthermore, none of the earlier publications report a quantum yield for the formation of any of the photoproducts and the irradiation times range from a few hours to more than two days. There is very little in these publications that would lead one to suspect the high photoefficiency for acid generation that was discovered by Crivello and Lam.

The photogeneration of acidic initiating species for cationic polymerization was not necessarily new at the time of this publication either. The photolysis of diazonium salts to produce Lewis acids had been known for a long time and Crivello and others had recently described the potential of diaryliodonium salts which they found to produce Brønsted acids as photoproducts. However, it must be noted that the thermal stability of the diazonium salts seriously limits their utility. Substances that have the combination of high thermal stability and exquisite photosensitivity are very uncommon. The iodonium, and even to a greater extent, the sulfonium salts have this combination of properties. It is really this combination of thermal stability, photosensitivity, solubility in common organic solvents, synthetic accessibility and low toxicity that have made the sulfonium salts so valuable. It was the recognition of the potential of these materials and the thorough and careful study and demonstration of their application to polymerization that has made this paper a landmark publication. The paper has inspired many, including myself, to consider photoinitiated processes that would never have been considered had it not been published.

Here is a single article that is the basis for several significant industrial applications. Many high-speed industrial photopolymer coating processes derive directly from this publication. There exist processes for coating epoxy insulators on parts as small as ballast capacitors for fluorescent lights to iron cores for power generating plants, generation of decorations and writing on aluminum cans and the production of passivation layers on the inside of steel (tin) cans for example, all of which derive from this work. This paper also played an important role in a revolution that is now taking place in the micro-electronics industry.

The photolysis of onium salts in general, and sulfonium salts in particular, form the basis for a new class of high sensitivity, high resolution imaging materials, photoresists, that are used to generate the patterned layers of semiconductor, insulator and conductor materials that make up microelectronic circuits. This imaging process used to pattern these materials is currently undergoing a revolution. The materials and processes in current use have reached their resolution limits and radically new materials that provide both higher resolution and higher pro-

ductivity are being introduced. All of these new materials are based on photogeneration of acid catalysts. By far, the most preferred photoacid generators are the sulfonium salts that are the focus of this paper by Crivello and Lam.

By all measures, this paper deserves the recognition it will receive for having been chosen as one of the most important publications in the *Journal of Polymer Science* during it first 50 years of service to the polymer science community. I am pleased to have this opportunity to congratulate these authors for having made a such an important contribution to the scientific knowledge base, to my own work, and for having done so in such a thorough and well-written paper.

Photoinitiated Cationic Polymerization with Triarylsulfonium Salts

J. V. CRIVELLO and J. H. W. LAM, *General Electric Corporate Research and Development, Schenectady, New York 12301*

Synopsis

Triarylsulfonium salts $Ar_3S^+MX_n^-$ with complex metal halide anions such as BF_4^-, AsF_6^-, PF_6^-, and SbF_6^- are a new class of highly efficient photoinitiators for cationic polymerization. In this article we describe several synthetic routes to the preparation of these compounds along with their physical and spectroscopic properties. Mechanistic studies have shown that when these compounds are irradiated at wavelengths of 190–365 nm carbon–sulfur bond cleavage occurs to form radical fragments. At the same time the strong Brønsted acid HMX_n, which is the active initiator of cationic polymerization that takes place in subsequent "dark" steps, is also produced. A study of the parameters that affect the photolysis of triarylsulfonium salts is reported with a measurement of the absolute quantum yields. The cationic polymerizations of four typical monomers—styrene oxide, cyclohexene oxide, tetrahydrofuran, and 2-chloroethyl vinyl ether—with triarylsulfonium salt photoinitiators are described.

INTRODUCTION

Recently we described our work in which we used diaryliodonium salts as photoinitiators in cationic polymerizations.[1] On irradiation with ultraviolet (UV) light, diaryliodonium salts (I) which bear complex metal halide anions undergo photolysis during which the organic cation is destroyed and a powerful Brønsted acid is liberated [eq. (1)]. The strong protonic acid, in subsequent steps, efficiently initiates the polymerization of cationically polymerizable monomers:

where $MX_n^- = BF_4^-$, PF_6^-, AsF_6^-, SbF_6^-, etc.

While this work was in progress a parallel effort was being directed toward uncovering other new compounds also capable of photoinitiating cationic polymerization. This work was begun by considering those factors that contribute to the ability of diaryliodonium salts to function as photoinitiators.

The iodine atom in these salts is a positively charged electronegative heteroatom attached to two aryl groups. Resonance delocalization effects of the aromatic rings as well as back bonding through the d orbitals of the iodine serve to stabilize these compounds against thermal decomposition. These same factors serve to reduce the sensitivity of diaryliodonium salts to attack by weakly nucleophilic agents such as many polymerizable monomers. As a result diaryliodonium salts in monomer solutions are highly stable, yet on photolysis rapidly initiate polymerization.

Journal of Polymer Science: Polymer Chemistry Edition, Vol. 17, 977–999 (1979)
0360-6376/79/0017-0977$01.00

In analogy with diaryliodonium salts, we considered other aryl-substituted onium salts, particularly those of the group VI elements. Although triaryloxonium salts have been prepared and characterized, the synthetic routes are difficult and have been reported to give only low yields of products.[2] In contrast, triarylsulfonium salts (II) are easily prepared in moderate yields by Ar$_3$S$^+$X$^-$ (II) straightforward methods. Our attention was further drawn to these compounds by the work of Knapczyk and McEwen who found that sulfonium salts were photoactive.[3] Under prolonged irradiation (61 hr) in ethanol these workers observed the following fragments from triphenylsulfonium chloride:

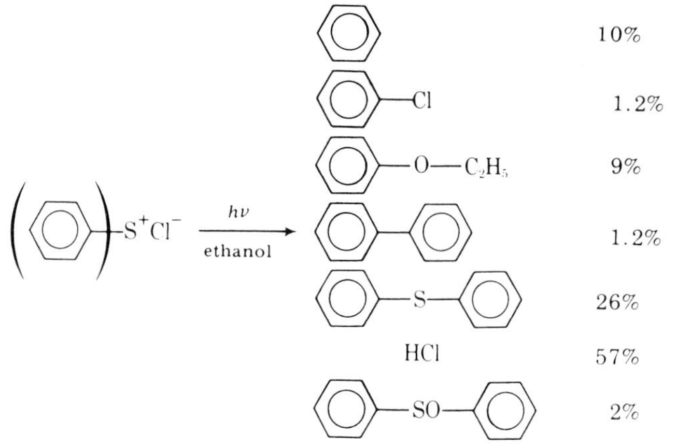

About 34% unreacted starting material was also recovered. Despite the unpromising low level of photoactivity observed in these experiments, we decided to prepare and evaluate a number of triarylsulfonium salts that have complex metal halide precursor anions as photoinitiators for cationic polymerizations.

EXPERIMENTAL

Purification of Reagents and Starting Materials

All solvents used in the photolysis and quantum yield studies were spectro grade and were used without further purification. Reagents used in the preparation of triarylsulfonium salts were reagent grade. The salts were purified according to the individual procedure for the specific salt. Cyclohexene oxide, styrene oxide, and 2-chloroethyl vinyl ether were dried over calcium hydroxide and purified by fractional distillation just before use, and tetrahydrofuran was freshly distilled from sodium naphthalene complex under dry nitrogen just before polymerization.

Preparation of Triarylsulfonium Salt Photoinitiators

Specific examples are given for each of the five methods used in the synthesis of the triarylsulfonium salt photoinitiators shown in Table I.

TABLE I
Preparation of Triarylsulfonium Salts

Cation	Anion	mp (°C)	λ_{max} (ϵ)	Elemental analysis			Method
				C	H	S	
	BF_4^-	191–193	227 (21,000)	Calcd 61.71	4.28	9.14	A
				Found 62.00	4.31	9.33	
	AsF_6^-	195–197	227 (21,000) 298 (10,000)	Calcd 47.68	3.31	7.06	A
				Found 47.78	3.41	7.06	
	PF_6^-	133–136	237 (20,400) 249 (19,700)	Calcd 56.90	4.96	6.90	A
				Found 57.05	5.03	7.09	
	AsF_6^-	…	225 (21,740) 280 (10,100)	Calcd 46.49	3.88	5.90	B
				Found 46.70	3.98	6.00	
	BF_4^-	162–165	243 (24,700) 278 (4900)	Calcd 64.28	5.35	8.16	A,E
				Found 64.32	5.41	7.91	
	AsF_6^-	111–112	275 (42,100) 287 (36,800) 307 (24,000) 263 (25,200)	Calcd 50.00 Found 50.10	3.95 4.01	6.67 7.23	D
	BF_4^-	168–169	277 (25,000) 232 (3100)	Calcd 62.99	4.14	8.84	A
				Found 62.81	4.10	8.96	
	AsF_6^-	245–251	280 (22,400) 316 (7700)	Calcd 49.39	4.62	5.48	C
				Found 49.39	4.59	5.55	

Method A—The Preparation of Triphenylsulfonium Hexafluoroarsenate[17]

Into a 500-ml, three-necked, round-bottom flask equipped with a paddle stirrer, addition funnel, thermometer, and condenser were placed 25 g (0.116 mole) of potassium iodate, 60 ml of methylene chloride, 50 ml of acetic anhydride, and 32 g (0.41 mole) of benzene. The reaction mixture was cooled to −10°C in a dry ice–acetone bath and then 25 ml of concentrated sulfuric acid was added in drops in an addition funnel. During the addition the temperature was not allowed to rise above −5°C. When the addition was complete, the temperature was maintained at −10 to −5°C for 3 hr and then allowed to rise to room temperature. After standing for 16 hr 100 ml of distilled water was slowly added to hydrolyze the remaining acetic anhydride. Then 26.5 g (0.116 mole) of potassium hexafluoroarsenate was added in 100 ml of water; the reaction mixture was stirred for 1 hr to complete the metathesis. The organic layer was separated by means of a separatory funnel and the aqueous layer was extracted twice with 25-ml portions of methylene chloride. Trituration of the combined methylene chloride solutions with diethyl ether resulted in crystallization of the diphenyliodonium hexafluoroarsenate. A yield of 36.7 g (67% theory) of the iodonium salt, mp 192–195°C, was obtained.

In a 50-ml, single-necked, round-bottom flask equipped with a magnetic stirrer, a reflux condenser, and a nitrogen bypass were placed 11.75 (0.025 mole) of diphenyliodonium hexafluoroarsenate, 4.65 g (0.025 mole) of diphenylsulfide, and 0.2 g of copper benzoate. The reaction mixture was heated and stirred under nitrogen in a silicone oil bath at 120–125°C for 3 hr. After this time the reaction mixture was poured while hot into a 150-ml beaker, whereupon crystallization occurred. The product was extracted three times with ether to remove iodobenzene and then air dried. Examination of the reaction product by nuclear magnetic resonance (NMR) and comparison with spectra of authentic material showed that the product was triphenylsulfonium hexafluoroarsenate. A yield of 10.9 g (96.5% theory) of a light tan product was obtained. Further recrystallization from 95% ethanol–water gave the analytically pure triphenylsulfonium hexafluoroarsenate (mp 194–197°C).

Method B—Preparation of Tris(4-Methoxyphenyl)sulfonium Hexafluoroarsenate

A modification of the method described by Pitt[5] was used. In a 500-ml, three-necked, round-bottom flask fitted with a mechanical stirrer, a thermometer, and a dropping funnel was placed 20 ml of $SnCl_4$ dissolved in 48.1 g (0.45 mole) of anisole. To this solution was added in drops 6.75 g (0.41 mole) of sulfur monochloride while the reaction mixture was maintained between 10 and 15°C by the use of an ice bath. When the sulfur monochloride had been added, the reaction flask was filled with a gas inlet tube and 23 g of chlorine gas was bubbled into the reaction mixture. The dark yellow solution that resulted was then hydrolyzed by the addition of 70 ml of water to yield an oil. The oil was separated from the aqueous phase and washed three times with diethyl ether to remove unreacted anisole. Next, 22.8 g (0.10 mole) of $KAsF_6$ was added in 100 ml of water, shaken with the oil, and allowed to stand overnight. The oil layer was again separated, washed with diethyl ether, and dried in a vacuum oven at 60°C. A 59.6% yield (32.3 g) of tris(4-methoxyphenyl)sulfonium hexafluoroarsenate was obtained as a light oil. The proton NMR spectrum showed a strong singlet 3.88 ppm (3 protons) and a multiplet centered at 7.53 ppm (4 protons).

Method C—Preparation of Tris(3,5-Dimethyl-4-Hydroxyphenyl)sulfonium Hexafluoroarsenate[6]

To 400 ml of carbon disulfide contained in a 2000-ml, three-necked, round-bottom flask equipped with a mechanical stirrer, a thermometer, and a condenser was added 122 g (1.0 mole) of 2,6-dimethylphenol. To the cooled solution maintained at 10°C was added 89 g (0.66 mole) of aluminum chloride in small portions. A green solution was obtained to which was slowly added 79.5 g (0.66 mole) of thionyl chloride. The temperature was maintained at ca. 10°C during this addition. A dark viscous mass was obtained which was stirred for 2 hr after completion of the addition. Next the reaction mixture was slowly poured into 1000 g of ice that contained 50 ml of concentrated HCl. The black aluminum chloride complex was rapidly hydrolyzed to a clear solution. After heating this mixture briefly in a steam bath to remove the carbon disulfide and to complete the hydrolysis the precipitated sulfonium chloride was filtered off, washed with water, and dried. The slightly brown sulfonium chloride was purified by recrystallization from methanol–water to give a nearly white crystalline product with a mp of 250–252°C (dec.). The yield of product was 74.9 g (52.4%).

To a solution of 21.5 g (0.05 mole) of tris(3,5-dimethyl-4-hydroxyphenyl) sulfonium chloride in hot ethanol was added 11.4 g (0.05 mole) of potassium hexafluoroarsenate. About 100 ml of water was added and crystallization of the sulfonium hexafluoroarsenate took place. The salt was filtered, washed thoroughly with water, then with ether, and dried. A 25.7 g yield of the product with a mp of 245–251°C was obtained. Similarly, metathesis with KPF_6, $KSbF_6$, and $NaBF_4$ yields the corresponding PF_6^-, SbF_6^-, and BF_4^- salts.

Method D—Preparation of Diphenyl-2,5-Dimethylphenyl Sulfonium Salts[9]

To 18.6 g (0.01 mole) of diphenylsulfide in 200 ml of methanol was added 28.2 g (0.11 mole) of chloramine T in 200 ml of methanol at room temperature. The mixture was permitted to stand at 50°C for 3 hr and then poured into a cold dilute sodium hydroxide solution. The precipitate was filtered, washed with water, and recrystallized from methanol–water. The yield of S,S-diphenyl-p-tolylsulfilimine (mp 111–112°C) was 94%.

A mixture of 7.0 g (0.02 mole) S,S-diphenyl-p-tolylsulfilimine and 100 ml p-xylene was stirred in a 250-ml flask under a blanket of nitrogen; 10 g (0.08 mole) of anhydrous aluminum chloride was then introduced in small portions. During the addition the temperature rose to 70–75°C with the evolution of HCl gas. After ½ hr the HCl evolution had ceased and the reaction mixture was heated for an additional 2 hr. The cooled mixture was poured into a mixture of 50 ml of 43% $HAsF_6$ and 300 g of ice. This mixture was heated briefly to complete the hydrolysis. The colorless oil obtained was washed with water and recrystallization was attempted with chloroform–ether mixtures. A colorless solid which contained 1 mole of methylene chloride of crystallization per mole of sulfonium salt (mp 86–89°C) was then obtained.

Method E—Preparation of Tris(4-Methylphenyl)sulfonium Salts[10]

In a 500-ml, three-necked, round-bottom flask, equipped with a magnetic stirrer, a reflux condenser, a gas inlet, and an addition funnel was placed 3 g of magnesium turnings in 30 ml of anhydrous ether. Slowly, 22.75 g of p-bromotoluene was added to prepare the p-tolylmagnesium bromide, and to the Grignard reagent was added 50 ml of anhydrous benzene. The reaction mixture was then heated to 70°C to remove the ether; 7.0 g (0.03 mole) of p-tolylsulfoxide in 100 ml of benzene was added in drops at 70°C. After the addition the reaction was kept at reflux for 48 hr. The reaction mixture was cooled to 5–10°C and hydrolyzed with 10 g of 48% HBF_4 in 20 ml of water. The benzene layer was separated and extracted with 100 ml of 5% HBF_4 and the two aqueous portions were combined. This aqueous solution was extracted three times with $CHCl_3$. After trituration of the $CHCl_3$ extracts with ether a white crystalline product was obtained. The yield of product (mp 162–165°C) was 19.6%.

Other tris(4-methylphenyl)sulfonium salts were prepared in the same manner by substituting HBF_4 for $HAsF_6$, HPF_6, or $HSbF_6$.

982   CRIVELLO AND LAM

Photolysis Studies

Photolyses were conducted (Fig. 1) with a Hanovia 450-W medium-pressure arc lamp with a measured output of 13,000 μW/cm^2 from 200 to 300 nm at a distance of 5 cm. Surrounding the lamp is a quartz well through which is pumped cooling water. Sample tubes were placed in a "merry-go-round" holder that is rotated continuously with a motor to provide even illumination throughout the photolysis. The entire apparatus was placed in a large thermostatted water bath which controls the temperature within 1°C.

Photodecomposition Studies

The photodecomposition of triarylsulfonium salts was carried out by irradiating 0.07M solutions of the salts in acetonitrile, 1:1 ethanol–water, and acetone in quartz sample tubes. Irradiations were terminated at approximately 20–25% conversion (7–8 min irradiation) to avoid the formation of secondary photoproducts. The photolysis products were identified by a combination of two analytical techniques. The products were initially identified by a comparison of their GLC retention times with the retention times obtained from authentic samples. Confirmation of the structural assignment was then made with a GLC-mass spectrometer.

Kinetic Studies

Kinetic studies were conducted with the photolysis apparatus described above. The sample tubes were constructed of quartz and sealed with a rubber septum through which aliquots were drawn for analysis. Samples consisted of 2–5 cm^3 of 0.07M solutions of the sulfonium salt in acetonitrile, acetone, ethanol–water,

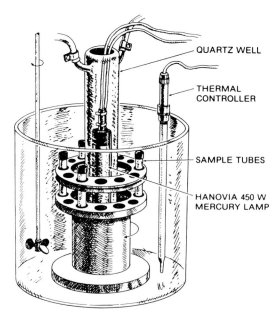

QUARTZ WELL

THERMAL CONTROLLER

SAMPLE TUBES

HANOVIA 450 W MERCURY LAMP

Fig. 1. Photolysis apparatus.

or nitromethane. During kinetic runs the tubes were withdrawn from the UV light, shielded with aluminum foil, and analyzed by GLC. In all cases the amount of diarylsulfide formed was the equivalent of a quantity of triphenylsulfonium compound that has undergone photolysis. An internal standard, usually naphthalene, biphenyl, or decalin, was included for purposes of quantitative determination of the diarylsulfide product. The addition of these compounds in no way affected the rates of photolysis of the triaryliodonium salts. The conditions under which the GLC analyses were performed are outlined below.

Solutions of photolysis products were chromatographed on a 6-ft UC, W98, $\frac{1}{8}$-in. column with a helium pressure of 40 psi. Runs were programmed from 100 to 300°C at a heating rate of 30°C/min. The column was held at its upper limit for 5 min to ensure purging of the column of all residual products.

The course of the photolysis can also be conveniently followed by NMR techniques. In this method 0.35M solutions of the triarylsulfonium salt in a deuterated solvent were periodically irradiated in a quartz NMR sample tube and the disappearance of starting material was followed by integration of the appropriate absorption bands. A Varian T-60 NMR spectrometer was used in these studies.

In the study of the effect of light intensity on the rate of photolysis of triarylsulfonium salts a commercial Rayonet apparatus was used. This device consists of 16 lamps disposed in a circle about a rotating sample holder. Regular variations in the light intensity can be achieved by removing a portion of the lamps.

Quantum Yield Determinations

The procedure used in the determination of the quantum yields of tris(4-methylphenyl)sulfonium hexafluoroarsenate was adapted from the method described by Turro and his co-workers[16] and has been described by us.[1] The method requires the use of uranyl oxalate actinometry in combination with filters to isolate the 313- and 365-nm regions of the UV spectrum.

PHOTOPOLYMERIZATIONS

The Photoinitiated Polymerization of Cyclohexene Oxide and Styrene Oxide

Styrene oxide and cyclohexene oxide were freshly distilled from calcium hydride and kept under a nitrogen blanket; 0.02M solutions of triphenylsulfonium fluoroborate, hexafluorophosphate, hexafluoroarsenate, and hexafluoroantimonate in each of the two monomers were prepared and 5-g (4.8 ml) aliquots were sealed under nitrogen into vials fitted with polyethylene-lined caps. Photopolymerization was conducted by placing the vials in the "merry-go-round" apparatus and irradiating them with a 450-W Hanovia lamp. All photopolymerizations were conducted at 25°C. The samples were quenched at various irradiation times by adding 1 ml of a methanolic solution that contained a small amount of NH_4OH into the reaction mixtures and then pouring the solutions into methanol. After filtering the coagulated polymers and washing them with methanol they were dried at 60°C *in vacuo* and weighed.

The Photoinitiated Polymerization of Tetrahydrofuran

Three-milliliter aliquots (2.66 g) of a $5 \times 10^{-3}M$ solution of triphenylsulfonium hexafluorophosphate in freshly distilled tetrahydrofuran were sealed under nitrogen into 14-mm quartz polymerization tubes fitted with rubber septum caps. The samples were irradiated in the apparatus at 25°C and terminated at the appropriate times by injecting a methanolic solution of NH_4OH into the polymerization mixture. The samples were isolated and dried as described for polycyclohexene oxide and polystyrene oxide.

The Photoinitiated Polymerization of 2-Chloroethylvinyl Ether

A solution consisting of 12.5 ml (0.123 mole) of 2-chloroethylvinyl ether freshly dried and distilled from calcium hydride and 0.3124 g (6×10^{-4} mole) of diphenyl-4-t-butylphenylsulfonium hexafluoroarsenate was diluted to 50 ml with dry methylene chloride. Four-milliliter aliquots of the solution were placed in polyethylene-capped Pyrex glass vials and irradiated at 25°C in the apparatus shown in Figure 1. The polymers were isolated as described in the preceding two experiments.

The following time/% conversion data were obtained:

Irradiation time (min)	% Conversion
5	Trace
10	92
15	87
20	88
25	92

RESULTS AND DISCUSSION

Synthesis and Characterization

Many synthetic methods have been used in the preparation of triarylsulfonium salts.[4–13] Before our work in this area, however, no single synthetic method existed that was both general in scope and gave good yields of these salts. Another experimental difficulty stems from the often intractable nature of the salts themselves, together with their general tendency to be hygroscopic or to combine molecules of a solvent into their crystalline structure. We recently reported the direct synthesis of triarylsulfonium salts by the copper-catalyzed arylation of diarylsulfides with diaryliodonium salts [eq. (3)].[17]

$$Ar_2S + Ar'_2I^+MX_n^- \xrightarrow{Cu^{II}} Ar'Ar_2S^+MX_n^- + Ar'I \qquad (3)$$

Method A

This reaction proceeds smoothly at 120–125°C and is complete within 3 hr to give good-to-excellent yields of pure triarylsulfonium salts. The method is particularly suited to the preparation of unsymmetrical compounds that are difficult to make via other techniques. Only triarylsulfonium salts with complex metal halide counterions MX_n^- such as BF_4^-, PF_6^-, AsF_6^-, and SbF_6^- can be prepared by this reaction. When diaryliodonium halides are used, no sulfonium salt product is obtained.

In addition to this new preparative method, we have also used the scheme developed by Pitt,[5] which is a method for condensation of aromatic hydrocarbons with sulfur monochloride in the presence of a Lewis acid followed by chlorination and further condensation [eq. (4)].

$$6ArH + S_2Cl_2 \xrightarrow[\text{or} \atop SnCl_4]{AlCl_3} \xrightarrow{3Cl_2} 2Ar_3S^+Cl^- + 6HCl \qquad (4)$$

Method B

We have also used the Friedel-Crafts reaction of 2,6-disubstituted phenols with thionyl chloride in the presence of aluminum chloride to prepare an interesting class of triarylsulfonium salts.[6] This method produces good yields of well-characterized crystalline salts but it seems entirely limited to phenols as

$$3 \underset{R}{\overset{R}{\bigcirc}}\!\!-OH + SOCl_2 \xrightarrow[0-20°C]{AlCl_3} (HO\!-\!\overset{R}{\underset{R}{\bigcirc}}\!-)_3 S^+Cl^- + HCl + H_2O \qquad (5)$$

Method C

substrates and cannot be applied to other more deactivated compounds.

Another synthetic method is based on sulfilimine compounds. These intermediates can be prepared by various routes; however, the method described by Oae and his co-workers and shown below is the best in terms of yield.[8] Sulfilimines can be converted to sulfonium salts by condensing them with aryl hy-

$$Ar_2S + CH_3\!-\!C_6H_4\!-\!SO_2\!-\!N\!\!\begin{array}{c}Na\\\\Cl\end{array} \xrightarrow[H^+]{CH_3OH}$$

$$NaCl + Ar_2S\!=\!N\!-\!SO_2\!-\!C_6H_4\!-\!CH_3 \qquad (6)$$

drocarbons in the presence of aluminum chloride to give the desired sulfonium salt [eq. (7)]:

$$\underset{Ar}{\overset{Ar}{\diagdown}}S\!=\!N\!-\!Ts + Ar'H \xrightarrow[\substack{\Delta \\ 2\text{ hr}}]{AlCl_3} \xrightarrow[H_2O]{HX} \underset{Ar}{\overset{Ar}{\diagdown}}S^+\!-\!Ar'X^- + Ts\!-\!NH_2 \qquad (7)$$

Method D

Method D seems to be applicable only to aromatic hydrocarbons that undergo facile electrophilic substitution; that is, toluene, xylene, and mesitylene. Arylalkyl ethers such as anisole suffered ether cleavage reactions, whereas compounds that bore deactivating functional groups failed to react.

The condensation of diarylsulfoxides with Grignard reagents [eq. (8)] provides

$$\underset{Ar}{\overset{Ar}{\diagdown}}S\!\rightarrow\!O + Ar'\!-\!MgX \xrightarrow[\Delta]{C_6H_6} \xrightarrow{HMX_n} \underset{Ar}{\overset{Ar}{\diagdown}}S^+\!-\!Ar'MX_n^- + MgXOH \qquad (8)$$

Method E

a general synthetic pathway for the synthesis of triarylsulfonium salts.[10] Although this preparation requires long reaction times (48 hr) at reflux in benzene and gives fair yields of products, it produces the desired sulfonium salts in a relatively high state of purity.

All the methods of preparation with the exception of method A give rise to sulfonium salts with halide counterions. To be effective as cationic photoinitiators these anions must somehow be converted to complex metal halide-type MX_n^- anions. The general method by which halide anions have been exchanged in sulfonium salts is in a metathetical reaction with the parent acid HMX_n or their alkali metal salts [eq. (9)]:

$$Ar_3S^+X^- + YMX_n \rightarrow Ar_3S^+MX_n^- + YX \qquad (9)$$

where X = halogen, and Y = H, K, Na, Li, etc.

Table I lists a number of triarylsulfonium salt photoinitiators that were prepared during the course of our work. Included in this table are their melting points, UV spectra, elemental analyses, and their method of preparation. Figures 2–4 show, respectively, the IR, UV, and ^{13}C-NMR of triphenylsulfonium hexafluoroarsenate which can be considered typical of this class of compounds.

In any series of triarylsulfonium salts with the same cation structure and differing only in their anions the ^{13}C-NMR and UV spectra are identical. Only slight changes are noted in their IR spectra.

Photolysis Studies

Mechanistic Studies

Only three reports of the photolysis of sulfonium salts appear in the literature,[3,14,15] and only one, described by McEwen, involves the photolysis of triarylsulfonium salts.[3] On the basis of the products formed two major primary photochemical processes can take place. The first reaction concerns the initial formation of a diarylsulfinium radical cation, and aryl radical, and an anion. Subsequent interaction of the diaryl–sulfinium radical cation with the solvent produces the diarylsulfide, a proton, and a solvent derived radical. The second photochemical reaction, which appears to be the major photochemical process

$$Ar_3S^+X^- \overset{h\nu}{\rightleftarrows} (Ar_2S^{\cdot +}, Ar\cdot)X^- \rightarrow Ar_2S^{\cdot +} + Ar\cdot + X^-$$

$$Ar_2S^{\cdot +} + R{-}H \rightarrow Ar_2S^+{-}H + R\cdot$$

$$Ar_2S^+{-}H \rightarrow Ar_2S + H^+ \qquad (10)$$

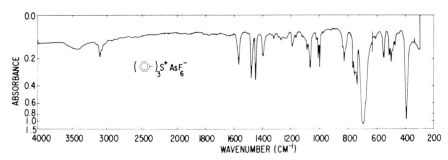

Fig. 2. Infrared spectrum (KBr disk) triphenylsulfonium hexafluoroarsenate.

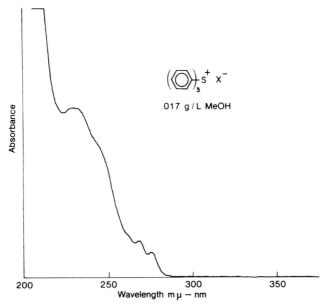

Fig. 3. Ultraviolet spectrum of triphenylsulfonium salts in methanol.

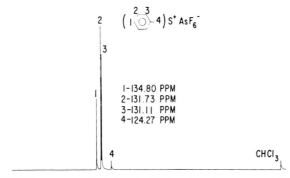

Fig. 4. ^{13}C-NMR spectrum of triphenylsulfonium hexafluoroarsenate in chloroform.

in the special case of triarylsulfonium iodides, is an electron-transfer process in which a triarylsulfur radical and a halogen atom are produced.   In the photolysis of these particular salts substantial quantities of aryl halide are produced. McEwen and his co-workers performed their photolysis experiments in methanol

$$Ar_3S^+X^- \overset{h\nu}{\rightleftarrows} Ar_3S\!-\!X \rightleftharpoons Ar_3S\cdot X\cdot \rightarrow Ar_3S\cdot + X\cdot$$

$$Ar_3S\cdot \rightleftharpoons Ar_2S + Ar\cdot$$

$$Ar\cdot + X\cdot \rightarrow Ar\!-\!X \tag{11}$$

for a total irradiation period of 61 hr.[3]   Photolysis was relatively slow under these conditions and after 61 hr considerable amounts of the starting sulfonium salt remained.   It is also conceivable that some of the many products observed during the long irradiation [eq. (2)] are the result of secondary photochemical processes.

Our own photochemical studies were conducted with triarylsulfonium salts that had anions of poor nucleophilic character derived from strong Brønsted acids such as BF_4^-, AsF_6^-, PF_6^-, or SbF_6^-. When the photolysis of a 0.07M solution of tris-4-methoxyphenylsulfonium hexafluoroarsenate in acetone was carried out with UV light in a quartz apparatus at 25°C for 15 min, the products identified by GLC-mass spectrometry were anisole, 4,4′-dimethoxydiphenylsulfide, and three isomeric dimethoxybiphenyls [eq. (12)].

The photolysis had reached approximately 25% conversion after 15 min irradiation and the photolysis was then terminated to avoid secondary photoproducts. A similar series of products was observed in the photolysis of other sulfonium salts such as triphenylsulfonium hexafluoroarsenate and triphenylsulfonium fluoroborate.

The products of photolysis can be rationalized as being derived by a radical process. The mechanism proposed by McEwen and shown in eq. (10) adequately accounts for the observed products and is the most likely pathway.

Further confirmation of the mechanism of eq. (10) was obtained by carrying out the following two experiments: anisole was shown to be derived by hydrogen abstraction from the solvent by the anisyl radical by photolyzing tris-4-methoxyphenylsulfonium hexafluoroarsenate in acetone-d_6. Analysis of the products by GLC-mass spectroscopy showed the presence of 4-deuteroanisole. Still more evidence of the proposed mechanism was obtained by performing a typical crossover experiment. Equivalent amounts of triphenylsulfonium hexafluoroarsenate and tris(4-methoxyphenyl)sulfonium hexafluoroarsenate were photolyzed together in acetone. Along with the two diarylsulfides produced were the coupling products biphenyl and 4,4′-dimethoxybiphenyl and the crossover product 4-methoxybiphenyl. This experiment demonstrates that not only are aryl radicals formed during the photolysis of triarylsulfonium salts but they have sufficient lifetimes to diffuse from the reaction site and give crossover coupling products.

A consideration of this mechanism leads to two important conclusions. Because fluorinated aromatic products are not produced from the photolysis of triarylsulfonium BF_4^-, AsF_6^-, PF_6^-, and SbF_6^- salts, it must be inferred that these anions survive the photolysis intact and that they must appear as their corresponding Brønsted acids, that is, HBF_4, $HAsF_6$, HPF_6, and $HSbF_6$. These acids must therefore be the ultimate initiators of cationic polymerization. A further implication of the mechanism detailed in eq. (10) suggests that because radical fragments are directly formed during the photolysis of triarylsulfonium salts, they should be capable of initiating radical polymerization. This, in fact, turns out to be the case; acrylic and other vinyl monomers polymerized readily when irradiated with UV light in the presence of triarylsulfonium salts. At present, we are continuing to investigate photoinitiated radical polymerizations with triarylsulfonium salt photoinitiators. The results of this work will be the topic of a forthcoming article.

Triarylsulfonium Photodecomposition Studies

A series of photolysis studies was carried out on a number of triarylsulfonium salts to determine their relative order of photosensitivity and the influences of solvent, temperature, atmosphere, and wavelength of incident light on their rates of photolysis.

Photolysis studies were carried out in the apparatus shown in Figure 1. Irradiations were performed with a Hanovia 450-W mercury arc lamp. The course of the photolysis reactions was followed by two methods: gas chromatography and NMR. Gas chromatographic analysis determinations were made by dissolving the sulfonium salt ($0.07M$) in an appropriate solvent with a nonreactive internal standard such as naphthalene, decalin, or biphenyl, performing the irradiation, and then analyzing for the diarylsulfide photolysis products. The presence of the internal standard in no way affected the course or extent of the photolysis as determined in separate blank experiments. Figure 5 shows a plot of the data obtained in a study of the photolysis of triphenylsulfonium hexafluoroarsenate in acetone. Under these conditions the photolysis is quite rapid, the reaction attains 53% completion within 20 min.

The course of the photolysis can also be conveniently followed by NMR techniques. In this method $0.35M$ solutions of the triarylsulfonium salt in a deuterated solvent were irradiated periodically in a quartz NMR sample tube and the disappearance of starting material was followed by the integration of the appropriate absorption bands. Figure 6 shows a composite NMR spectrum obtained in a study of the photolysis of tris(4-methylphenyl)sulfonium hexafluoroarsenate. As the photolysis proceeds, deep-seated changes occur in the aromatic region (7.60 ppm) and in the methyl band (2.50 ppm). Coincident with the decrease in these bands is the increase in bands due to the formation of products. Particularly useful for following the reaction rates is the rapid increase

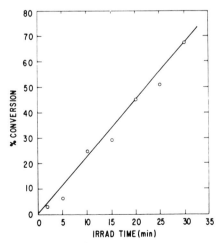

Fig. 5. Photolysis of triphenylsulfonium hexafluoroarsenate in acetone.

990   CRIVELLO AND LAM

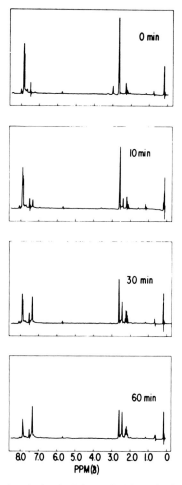

Fig. 6. NMR study of the photolysis of tris(4-methylphenyl)sulfonium hexafluoroarsenate in acetone-d_6.

in the band at 2.30 ppm which is due to the methyl protons in the major product: 4,4'-dimethyldiphenylsulfide. The NMR technique is a rapid, efficient way of evaluating the photolysis of triarylsulfonium salts that require small sample sizes.

Figure 7 compares the photolyses of triphenylsulfonium hexafluoroarsenate in two different solvents, acetone and nitromethane, determined by the GLC technique. The cutoff absorptions for these solvents are 330 and 380 nm, respectively. Photolysis is most rapid in acetone in which the transmission of UV light is greatest. Appreciable photolysis does, however, take place in nitromethane, even though the extinction coefficient of the absorption spectrum of the sulfonium salt is rather low ($\epsilon > 200$) at wavelengths greater than 330 nm. These observations suggest that a highly efficient photoprocess takes place and implies substantial quantum yields for these salts. Photolysis rates of triphenylsulfonium salts are independent of temperature. Irradiations performed at 24 and 35°C yield identical photolysis curves.

We were especially anxious to determine whether oxygen would inhibit the

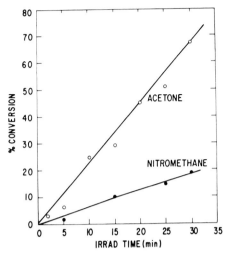

Fig. 7. Comparison of the photolysis of triphenylsulfonium hexafluoroarsenate in acetone and nitromethane.

rate of photolysis of triarylsulfonium compounds because we have already shown that the photolysis proceeds by a radical process. Oxygen is also known to be an efficient excited-state quencher in certain photochemical reactions. For purposes of comparison two photolyses of triphenylsulfonium hexafluoroarsenate were carried out. One was run in air and the other was repeatedly vacuum freeze-thaw degassed and sealed. Identical photolysis behavior was noted in both atmospheres. Analogous results were obtained in a photolysis study conducted in the presence and absence of the radical trap, 2,6-di-t-butyl-4-methyl-phenol. The lack of variation in the photolysis rates suggests that radical-chain-induced decomposition processes do not play a role in the photolysis of these compounds. By the same token, we have not found it possible to induce the decomposition of sulfonium salts by the addition of compounds that generate radicals on photolysis. In experiments in which sulfonium salts were irradiated in the presence of radical photoinitiators such as butylbenzoin ether or 2,2-diethoxyacetophenone, the photolysis rates of the sulfonium salts showed no evidence of enhancement.

Similarly, we were not able to quench the photolysis of triarylsulfonium salts with various triplet quenchers such as 1,3-pentadiene (E_T = 56.9 kcal/mole), $trans$-stilbene (E_T = 49 kcal/mole), naphthalene (E_T = 60.9 kcal/mole), and biphenyl (E_T = 70 kcal/mole).[18] Similarly, Heine, Rosenkranz, and Rudolph[19] have reported that the α-cleavage of benzoin ethers cannot be quenched with triplet quenchers. Although these data tend to indicate that fragmentation takes place in the excited singlet, it does not conclusively prove the point. More definitive information, obtainable, for example, by flash photolysis, will be necessary before the multiplicity of triarylsulfonium salt excited state can be assigned.

Figure 8 displays the results of an experiment in which the effects of variations in light intensity on the rates of photolysis of triphenylsulfonium hexafluoroarsenate were studied. The experiment was conducted in a Rayonet apparatus equipped with 16 lamps with their major emission bands at 253.7 nm. Variations

992   CRIVELLO AND LAM

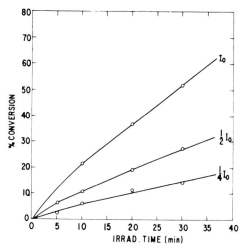

Fig. 8. Photolysis of triphenylsulfonium hexafluoroarsenate in acetone using 350-nm light at various intensities.

in the intensity were achieved by simply removing some of the lamps; that is, I = 16 lamps, $I/2$ = 8 lamps, and $I/4$ = 4 lamps. As the curves in Figure 8 indicate, a direct proportional relation exists between the intensity and the rate of photolysis.

The effect of differences in structure of the anions on the photolysis rates of tris(4-methylphenyl)sulfonium tetrafluoroborate and hexafluoroarsenate, determined by the NMR technique, is shown in Figure 9. The similarity of these two curves is apparent with the data points lying coincident on one another. Identical results were also obtained for the corresponding PF_6^- and SbF_6^- salts. These data suggest that the character of the anions examined has no effect on the photosensitivity of the sulfonium salt. Earlier observations that the NMR and UV spectra for a series of sulfonium salts with the same cationic structures but differing only in their anions are virtually identical would tend to support this conclusion.

A further implication of these results leads us to conclude that the molar amounts of different Brønsted acids which would be generated by such a series of sulfonium salts per unit time are also identical. This allows us to compare directly the efficiencies of various highly reactive acids in cationic polymerization. There is little information in the literature that relates to such initiator structures with respect to their reactivity in cationic polymerization.

The determinant factor in the photosensitivity of triarylsulfonium salts lies in the structure of the organic cation. The effects of cation structure are clearly shown in Figure 10 in which the photolysis curves of a number of sulfonium salts are displayed. Maximum photolysis rates were observed in triphenylsulfonium and tris(4-methylphenyl)sulfonium salts. The other sulfonium salts shown have substantially lower rates of photolysis. This wide variation in photolysis rates is in sharp contrast to that observed in diaryliodonium salts in which the photolysis rates for substituted compounds were similar.[1] We may postulate that the reasons for this difference in the substituent effects for iodonium and sulfonium salts lie in the greater bond strength of the C—S^+ bond versus the C—I^+

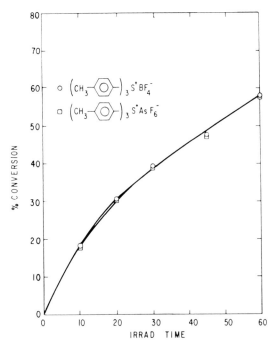

Fig. 9. NMR study of the photolysis of tris(4-methylphenyl)sulfonium salts ($0.35M$) in acetone-d_6.

bond and/or the greater degrees of freedom available in the sulfonium salts for dissipation of the photochemical energy.

Quantum Yield Measurements

The photolysis of triarylsulfonium salts is quite rapid, even at wavelengths greater than 300 nm, where the amount of light absorbed by these salts is small. This observation suggests reasonably high quantum yields for triarylsulfonium salts at these wavelengths. To substantiate this conclusion the quantum yield of a representative member of these compounds, namely, tris(4-methylphenyl)-sulfonium hexafluoroarsenate (III), was determined at 313 and 365 nm in acetonitrile.

$$\left(CH_3O-\!\!\left\langle\bigcirc\right\rangle\!\!-\right)_3\!\!S^+AsF_6^-$$

(III)

Φ 313 nm = 0.17
Φ 365 nm = 0.19

The method of Turro and his co-workers, in which various filter solutions isolate the wavelengths of interest, was used for these measurements.[16] Uranyl oxalate actinometry was used to determine the light intensity. Measurement of the light intensity by this method is subject to an estimated error of approxi-

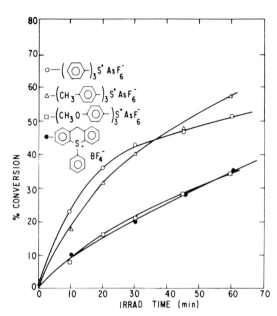

Fig. 10. NMR study of the photolysis of various triarylsulfonium salts (0.35M) in acetone-d_6.

mately 50%.[16] Hence the quantum yield values reported here should be understood to contain a certain amount of inherent error. Nevertheless, the determinations were made by averaging 3–5 separate quantum yield and light intensity measurements. In all cases the data exhibited only slight variations in their values and were highly reproducible.

The absolute quantum yields obtained, $\Phi = 0.17$ and 0.19, indicate that the photolytic cleavage of the C—S bond in triarylsulfonium salts is a highly efficient process. These results further imply that triarylsulfonium salts should be active photoinitiators of cationic polymerization.

Photoinitiated Polymerization Studies

Table I shows the structures of a number of triarylsulfonium salts that contain complex metal halide anions of the type $MX_n{}^-$ which have considerable variation in the structure of their cations. All of these compounds are photosensitive and efficient photoinitiators of cationic polymerization. Our initial investigations have shown that all known types of cationically polymerizable monomers can be polymerized with these new photoinitiators. Included among these monomers are epoxides, cyclic ethers, mono- and polyfunctional vinyl compounds, spiroesters, spirocarbonates, and cyclic siloxanes.

The mechanism by which sulfonium salt photoinitiators catalyze the polymerization of cyclic ether and olefinic monomers is given in eqs. (13)–(15).

The first step consists of the photoinduced generation of the strong protonic acid HMX_n. Cationic polymerization occurs in subsequent dark (nonphotochemical) steps. Initiation takes place by the addition of the acid to the monomer with the formation of a carbenium or an oxonium ion species. In the propagation step

Photolysis

$$Ar_3S^+MX_n^- \xrightarrow[\text{solvent-H}]{h\nu} Ar_2S + Ar{-}H + HMX_n + Ar{-}Ar \qquad (13)$$

Initiation

$$\begin{array}{l} \overset{\frown}{R\ O} \\ \text{or} \\ R{-}CH{=}CH_2 \end{array} \underset{}{\overset{HMX_n}{\rightleftharpoons}} \begin{array}{l} \overset{\frown}{R\ O}{}^{+}{-}H\ MX_n^- \\ \\ R{-}C^+H{-}CH_3\ MX_n^- \end{array} \qquad (14)$$

Propagation

$$\overset{\frown}{R\ O}{}^{+}{-}H\ MX_n^- + n\overset{\frown}{R\ O} \longrightarrow HO\ R{-}\!\left(O\ R\right)_{n-1}\!{-}\overset{+}{O}\overset{\frown}{\ R}\ MX_n^-$$

or

$$R{-}C^+H{-}CH_3MX_n^- + nR{-}CH{=}CH_2 \longrightarrow {-}\!\left(CH_2{-}\underset{R}{CH}\right)_n\!{-}CH_2{-}\underset{R}{CH^+}\ MX_n^- \qquad (15)$$

the active center propagates by the addition of monomer which results in chain growth. In consideration of this mechanism it would appear that participation of the monomer in some sort of charge transfer complex with the photoinitiator would not be necessary to produce the initiating species. These initiator–monomer species have been invoked to explain the photopolymerization of monomers such as *N*-vinylcarbazole in the presence of tetraalkylammonium tetrahaloaurate (III) salts.[20] Charge transfer complex involvement can be ruled out on the basis that triarylsulfonium salts are highly photosensitive and decompose rapidly in the presence of UV light in the absence of monomers. Furthermore, sulfonium salts can be photolyzed in inert solvent to generate catalytic amounts of strong acids which can then be added to monomers to cause their polymerization.

The possibility that initiation may occur by attack of the nucleophilic monomer on the photoexcited triarylsulfonium salt, as shown in eq. (16), can also be ruled out by the later experiment in which the photolysis and polymerization are carried out in separate steps. Additional evidence against this mechanism and

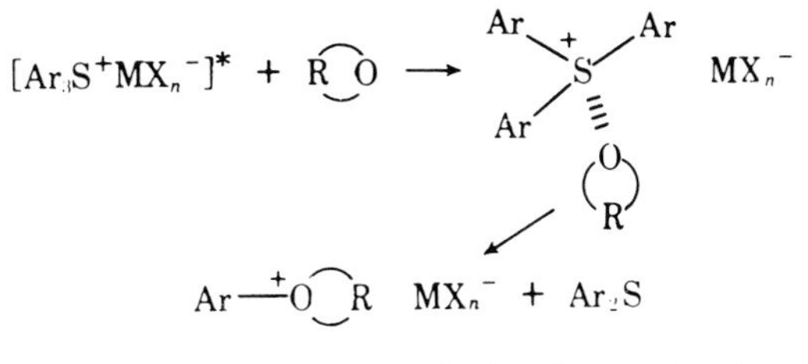

(16)

alternate mechanisms which involve initiation by arylcarbenium ions was obtained by an examination of the UV spectra of the polymers prepared by photopolymerization with triphenylsulfonium salts. Polymers prepared by this route [eq. (16)] would have phenyl ether end groups. Saegusa and his co-workers have shown that a sensitive technique for the quantitative determination of phenyl ether end groups in polyethers consists of measuring the extinction coefficient at $\lambda_{max} = 272$ nm.[21] Examination of the UV spectrum of polyethers produced by photoinitiated cationic polymerization with triphenylsulfonium salts shows the lack of an absorption band at this wavelength.

Although triarylsulfonium salts have been used to photoinitiate the cationic polymerization of a wide variety of different monomers, in this article we present data on the polymerization of only four monomers which represent the three major classes of cationically polymerizable substrates.

Oxirane compounds are highly reactive substrates in cationic polymerization and can be used as sensitive probes to investigate photoinitiator structure–reactivity relationships. All triarylsulfonium salts that have complex metal halide anions of the type MX_n^- are active in the polymerization of oxirane compounds as well as other cationically polymerizable monomers. Triarylsulfonium halides and bisulfates are not photoinitiators for cationic polymerization. The site of photoactivity therefore lies in the character of the organic portion of the molecule, whereas the corresponding catalytic reactivity in photopolymerization is dependent on the nature of the anion. Figures 11 and 12 show, respectively, the results of a study of the photoinitiated polymerizations of cyclohexene oxide and styrene oxide with triphenylsulfonium salts that have different MX_n^- anions. All the polymerizations were carried out at 25°C. Equimolar quantities of the various sulfonium salts were used in each case. Because we have shown that the photolysis rates are the same in a series of triarylsulfonium salts with the same cation structure but differing in their anions, photolysis of these salts yields identical amounts of the Brønsted acid per unit time. The differences observed in the rates of polymerization with different sulfonium salt photoinitiators, noted in Figures 11 and 12, are therefore attributable only to the efficiency of the respective Brønsted acids in cationic polymerization. Rapid, nearly explosive rates were noted in the case of triphenylsulfonium hexafluoroantimonate with both

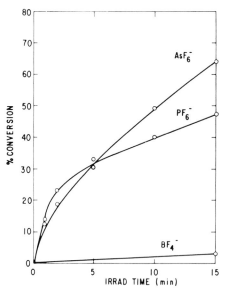

Fig. 11. Photoinitiated polymerization of cyclohexene oxide 0.02M $(C_6H_5)_3S^+MX_n^-$ as photoinitiator.

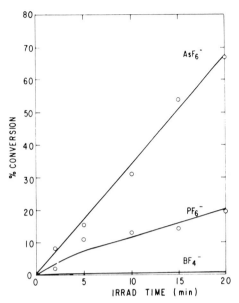

Fig. 12. Photoinitiated polymerization of styrene oxide $0.02M$ $(C_6H_5)_3S^+MX_n^-$ as photoinitiator.

monomers. Polymerization rates were so high and the exotherm so vigorous that accurate percentage of conversion versus time curves could not be obtained. The order of reactivity of triphenylsulfonium salts in the polymerization of cyclohexene oxide and styrene oxide is therefore $SbF_6^- > AsF_6^- > PF_6^- > BF_4^-$. A similar order was observed for protonic acid-catalyzed polymerization of styrene and was explained on the basis of the degree of separation in the propagating ion pair.[22,23] The larger the negatively charged ion, the more loosely it is bound to the cation and the more active the propagating cationic species in polymerization. In the series listed above the size of the anion falls in the same order as the observed order of reactivity of the sulfonium salts.

Figure 13 shows the data collected in a study of the triphenylsulfonium hexafluorophosphate-photoinitiated polymerization of tetrahydrofuran. This polymerization was carried out in bulk at $25°C$ with 5×10^{-3} mole of the sulfonium salt photoinitiator. Irradiation of the reaction mixture was conducted for 60 min in an all-quartz system; the 450-W medium-pressure mercury arc lamp was then turned off. Workup of the reaction mixture at this point gave a 13% conversion to polytetramethylene oxide. When the reaction was allowed to stand in the dark after irradiation and periodic samples had been taken, the percentage of conversion to polymer continued to rise. This experiment again demonstrates the well-known "living" nature of tetrahydrofuran cationic polymerizations. Once the Brønsted acid initiator is generated in the photolysis portion of the reaction, propagation continues in the dark until equilibrium conditions are reached. With triphenylsulfonium hexafluorophosphate as the photoinitiator, number-average molecular weights of 1,000,000 g/mole and greater, determined by gel permeation chromatography, have been obtained.

Vinyl ethers are reactive monomers in triarylsulfonium salt-photoinitiated cationic polymerization. In a typical study the polymerization of 2-chloroeth-

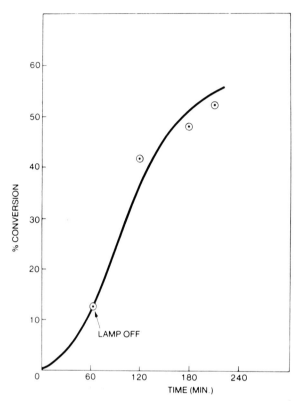

Fig. 13. Photoinitiated polymerization of tetrahydrofuran $5 \times 10^{-3} M$ $(C_6H_5)_3S^+PF_6^-$ as photoinitiator.

ylvinyl ether in methylene chloride at 25°C was conducted with diphenyl-4-*t*-butylphenylsulfonium hexafluoroarsenate. At the end of the first 5 min of irradiation only a trace of polymer could be isolated. After 10 min of irradiation, however, the yield of polymer had increased to 92%. Thus there appears to be an inhibition period in the initial 5 min during which perhaps impurities in the system are being consumed. This is followed by a rapid exothermic reaction in which the bulk of the monomer is quickly polymerized.

In conclusion, triarylsulfonium salts represent a new class of photoinitiator compounds. Because triarylsulfonium salts can be regarded as highly efficient photochemical sources of strong Brønsted acids, they are capable of photoinitiating the cationic polymerization of a wide variety of monomeric substrates among which the polymerization of styrene oxide, cyclohexene oxide, tetrahydrofuran, and 2-chloroethyl vinyl ether have been demonstrated in this article.

References

1. J. V. Crivello and J. H. W. Lam, *4th Int. Symp. Cationic Polym.*, August 1976, *J. Polym. Sci. Symp.*, **56**, 383 (1976); J. V. Crivello and J. H. W. Lam, *Macromolecules*, **10**, 1307 (1977).

2. A. N. Nesmeyanov, L. G. Makarova, and T. D. Tolstaya, *Tetrahedron*, **1**, 145 (1957).

3. J. W. Knapczyk and W. E. McEwen, *J. Org. Chem.*, **35**, 2539 (1970).

4. Houbin-Weyl, *Method. Org. Chim.*, **9**, 184 (1955).

5. H. M. Pitt, U.S. Pat. 2,807,648 (September 24, 1957).

6. W. Hahn and R. Stroh, U.S. Pat. 2,833,827 (May 6, 1958).

7. T. Oishi, M. Mori, and Y. Ban, *Chem. Pharm. Bull. Jpn.*, **19**(9), 1863 (1971).

8. K. Tsujihara, N. Furukawa, K. Oae, and S. Oae, *Bull. Chem. Soc. Jpn.*, **42**, 2631 (1969).

9. P. Manya, A. Sekera, and P. Rumpf, *Bull. Soc. Chim. Fr.*, **1**, 286 (1971).

10. D. S. Wildi, S. W. Taylor, and H. A. Potratz, *J. Am. Chem. Soc.*, **73**, 1965 (1951).

11. L. G. Makarova and A. N. Nesmeyanov, *Izv. Akad. Nauk SSSR Otd. Khim. Nauk,* 617 (1945).

12. C. Courtout and T. Y. Tung, *Compt. Rend.*, **197,** 1227 (1933).

13. G. H. Wiegand and W. E. McEwen, *J. Org. Chem.*, **33**(7), 2671 (1968).

14. T. Laird and H. Williams, *Chem. Commun.*, 561 (1969).

15. A. L. Maycock and G. A. Berchtold, *J. Org. Chem.*, **35**(8), 2532 (1970).

16. J. C. Dalton, P. A. Wriede, and N. J. Turro, *J. Am. Chem. Soc.*, **92**(5), 1318 (1970); N. J. Turro and P. A. Wriede, *J. Am. Chem. Soc.*, **92**(2), 320 (1970).

17. J. V. Crivello and J. H. W. Lam, *J. Org. Chem.*, **43**, 3055 (1978).

18. A. A. Lamola and N. J. Turro, *Energy Transfer Organic Photochemistry*, Interscience, New York, 1969, p. 106.

19. H. G. Heine, H. J. Rosenkranz, and H. Rudolph, *Angew. Chem. Int. Ed. Engl.*, **11**, 974 (1972).

20. M. Asai and S. Tazuke, *Macromolecules,* **6**(6), 818 (1973).

21. T. Saegusa and Matsumoto, *J. Polym. Sci. A-1*, **6**, 1559 (1968); T. Saegusa, S. Matsumoto, and Y. Hashimoto, *Polym. J.*, **1**, 31 (1970).

22. D. C. Pepper and P. J. Reilly, *J. Polym. Sci.*, **58**, 639 (1962).

23. N. Kanok, A. Gotoh, T. Higashimura, and S. Okamura, *Makromol. Chem.*, **63,** 115 (1963).

Received October 4, 1977

COMMENTARY

Reflections on "Living Carbocationic Polymerization. IV. Living Polymerization of Isobutylene," by R. Faust and J. P. Kennedy, *J. Polym. Sci.: Part A: Polym. Chem.,* Vol. 25, 1847 (1987)

JOSEPH P. KENNEDY and RUDOLF FAUST

Institute of Polymer Science, University of Akron, Akron, Ohio 44325-3909

They said "it cannot be done"[1] and indeed it took some 30 years[2] to catch up with anionic polymerization scientists to demonstrate that a simple olefin such as isobutylene (IB) *can* be polymerized cationically in a truly "living" manner. Re-reading our decade-old paper reminded us of the excitement we felt when we became convinced that our living polymerization of IB was real and reproducible. After having filed patent memoranda, presented our main findings at two scientific meetings, and summarized some of our data in a brief preliminary publication,[3] we proceeded to describe our work in a detailed paper, the paper we are now commenting upon (that is why this paper became the fourth in the series of "Living Carbocationic Polymerization," a series which at the latest count has swollen to 61 publications[4]).

The subject paper is significant *both* as a scientific breakthrough in polymer synthesis, and as a potential source of various commercial products with precision designed architectures. Seminal new "living" initiating systems were defined, e.g., cumyl acetate/BCl_3, in which esters in conjunction with BCl_3 were the critical and active initiating agents. This was remarkable considering that until this publication esters were regarded as "poisons" of cationic olefin polymerizations. Furthermore, the role of esters in fashioning the counteranion was recognized ("supernonnucleophilic anion") which in a larger context led to the recognition that "living" counteranions strongly influence the entire life span of growing cations (and led to "counteranion engineering"). Later work substantiated this notion and, in addition to esters, other nucleophiles were also found to be suitable initiators. Further, for the first time, the critical importance of the first order chain transfer to monomer (counteranion assisted process), in counterdistinction to the relatively unimportant second order (counteranion unassisted) process, for living polymerization was brought into sharp focus.

Also for the first time, rapid mobile equilibria between active living and dormant species in which the concentration of the living species is very low, was postulated. This concept became the cornerstone of many subsequent developments not only in living cationic but also in living radical polymerizations.

Another important discovery was that the living nature of the polymerization was unaffected by protic impurities (moisture) in the system; thus, living cationic olefin polymerizations can be carried out under conventional conditions and the use of super-dry chemicals is unnecessary.

Not too surprisingly the living polymerization of IB gave rise to quite narrow dispersity products ($M_w/M_n = 1.2$–1.5) and today, with improved techniques (use of proton traps, etc.) the preparation of close-to-Poisson distributions ($M_w/M_n = 1.03$–1.07) has become routine.

Living carbocationic polymerizations were a longstanding synthetic challenge mainly because of the promise of the many novel down-stream product potentialities. And what a rich lode this has become!

By a small extension of the syntheses described in the subject publication, most desirable di- and tri-telechelic PIBs were developed which, among other things, became intermediates for specialty PIB-based polyurethanes, polyepoxies, ionomers, etc.[5] Another large stream originating with this paper led to novel block copolymers, thermoplastic elastomers, multi-arm stars for oil additives, etc. Shortly after the first patents issued[6-10] a small venture company, Akron Cationic Polymer Development Co., was formed to produce various telechelic PIBs. Pilot plant development on polystyrene-b-PIB-b-polystyrene thermoplastic elastomers prepared by living carbocationic polymerization is still ongoing.

And last but not least, the subject publication became the fountainhead for a variety of novel designed biomaterials by living IB polymerization.[11] With very specific combinations of properties with possible biological significance, the road toward possible improved bone cements, vascular grafts, delayed drug delivery reservoirs, immunoisolation membranes, etc., has been outlined. Research is intensively pursued in these fields worldwide.

REFERENCES AND NOTES

1. D. C. Pepper, *J. Polym. Sci. Symp.,* **50,** 51 (1975).
2. See the dedication to M. Szwarc on the bottom of the first page of our paper.
3. The history together with some information relative to our U.S. and other patents has been analyzed in *Designed Polymers by Carbocationic Macromolecular Engineering: Theory and Practice,* by J. P. Kennedy and B. Ivan, Hanser, Munich, 1992, p. 35 ff.
4. J. Si and J. P. Kennedy, *Polym. Bull.,* 651 (1994).
5. An extensive analysis of these matters is described in Ref. 3.
6. J. P. Kennedy and R. Faust, U.S. Patent 4,910,321 (March 20, 1990).
7. J. P. Kennedy and M. K. Mishra, U.S. Patent 4,929,683 (May 29, 1990).
8. J. P. Kennedy, J. E. Puskas, G. Kaszas, and W. G. Hager, U.S. Patent 4,946,899 (Aug. 7, 1990).
9. J. P. Kennedy and M. K. Mishra, U.S. Patent, 5,066,730 (Nov. 19, 1991).
10. J. P. Kennedy and R. Faust, U.S. Patent 5,122,572 (June 16, 1992).
11. For a review with many references, see J. P. Kennedy, *Chemtech,* February 1994, p. 24.

PERSPECTIVE

Comments on "Living Carbocationic Polymerization, IV. Living Polymerization of Isobutylene," by R. Faust and J. P. Kennedy, *J. Polym. Sci.: Part A: Polym. Chem.*, 25, 1847 (1987)

MITSUO SAWAMOTO

Kyoto University, Kyoto 606-01, Japan

In the history of polymer chemistry, the 1980s will be remembered as an interesting decade during which a variety of living polymerizations have mushroomed, virtually simultaneously in different research groups with varying backgrounds, to cover almost all possible chain polymerization mechanisms including anionic, cationic, group transfer, metathesis, and so on.[1] First reported in 1956 by Szwarc,[2] living polymerizations are chain polymerizations that are free from chain transfer, termination, and other chain-breaking reactions and provide powerful methodologies to synthesize macromolecules of precisely controlled architectures and molecular weights, such as monodisperse, end-functionalized, block, and star-shaped polymers. Current journal literature readily attests that extensive interest in living polymerizations continues into the late 1990s, where further efforts are being directed to the development of living radical and other new polymerizations.

Among these modern living processes is included living cationic polymerization of vinyl monomers.[3] Before the 1980s, carbocationic polymerization has been regarded as one of the least controllable addition polymerizations, because its chain carriers, the growing carbocations, are inherently unstable and particularly prone to undergo chain transfer, isomerization, and other side-reactions. For a long time it has therefore been considered fundamentally difficult to achieve living polymerization therein.

The following paper by Faust and Kennedy[4] is the first full article that reported the first example of living cationic polymerization of isobutylene, initiated with a combination of 2-phenyl-2-propyl (cumyl) acetate and boron trichloride (BCl$_3$) as the initiating system. This was the beginning of the subsequent prolific publications from the Akron group for new initiating systems, living carbocationic polymerizations of other monomers, and related polymer synthesis thereby with precise structural and molecular weight control. The article also constitutes part of the earliest set of papers on well-defined living carbocationic polymerizations that almost simultaneously began to emerge in the mid-1980s for vinyl ethers[5] and other cationically polymerizable alkene monomers.[3] The worldwide rejuvenation of the field has thus been ignited in the following years.

The significance and impact of Kennedy's work include: (i) that it led to the first example of living cationic polymerization of nonpolar hydrocarbon monomers, in contrast to the polar counterparts with hetero-atom containing substituents, such as vinyl ethers and *N*-vinylcarbazole; (ii) that it involves isobutylene, the monomer of highest industrial importance in carbocationic polymerization; and (iii) that it employs a binary initiating system, with cumyl acetate (initiator) and BCl$_3$ (coinitiator), contributing to the generalization of the use of such two-component systems that is now considered essential to most of the modern living polymerizations.[1,3]

Another point of interest is that, in retrospect, the development of the cumyl acetate/BCl$_3$ and subsequently reported similar initiating systems seems to be a logical consequence, whether intended or not, from deep understanding and knowledge of isobutylene polymerization accumulated by Kennedy and his group for over four decades.[6] Thus, by a cursory search of his literature one may readily find a few preludes to the discovery, as summarized below in terms of initiating systems:

(a) Initiation Activity of Alkyl Chlorides (1968)

$$R–Cl \; + \; AlEt_2Cl \; \longrightarrow$$

(b) "Inifer" Systems (1980)

$$Cl{-}{\bigcirc}{-}Cl \; + \; BCl_3 \; \longrightarrow$$

(c) "Quasiliving" Carbocationic Polymerization (1982)

$${\bigcirc}{-}Cl \; + \; BCl_3 \; \longrightarrow$$

(d) Living Carbocationic Polymerization (1986)

$${\bigcirc}{-}OCCH_3 \; + \; BCl_3 \; \longrightarrow$$
$$\underset{O}{\overset{\parallel}{}}$$

First of all, these systems invariably consist of two components, an initiator (cationogen) and a coinitiator (Lewis acid), and their development demonstrates that proper combination of these ingredients is of prime importance in the design of controlled and living carbocationic polymerizations. For example, in 1968 Kennedy conducted a systematic study of the "activity" of alkyl chloride/Et$_2$AlCl systems in isobutylene polymerization.[7] In addition to the clear dependence of polymer yield on the structure of the alkyl groups (*tert*-butyl is the best), it is important in the current context to note that he properly employed Et$_2$AlCl, a relatively weak Lewis acid that cannot initiate polymerization with adventitious water, and thereby simplified his semiquantitative evaluation of initiator activity.

About a decade later Kennedy and Smith[8] reported, again in the *Journal of Polymer Science*, on isobutylene polymerization with the dicumyl chloride/boron trichloride (BCl$_3$) system, by which they disclosed a concept, the "inifer" (*ini*tiator-trans*fer* agent), that provided a way to selectively synthesize *tert*-chloride-capped poly(isobutylene). They proposed that the dual function of inifers results in the quantitative attachment of the terminal *tert*-chlorine; the key to this methodology is again judicious choice of inifers and the Lewis acid. This paper also marks the beginning of Kennedy's extensive use of, and perhaps inclination for, BCl$_3$ as coinitiator, which evidently forms the basis of the early phase of his living polymerization research a few years later.

The third systems, coined "quasiliving,"[9] also utilize binary initiating systems to induce polymerizations that are not living in a strict sense but indeed exhibit features (such as a progressive increase in polymer molecular weight against its yield) that are reminiscent of living polymerization. Obviously, as the term "quasiliving" implies, the 1982/83 publications also suggest that around that time, Kennedy started to devote serious efforts to achieve living carbocationic processes. Although the quasiliving polymerizations are characterized by a unique technique of continuous and slow

addition of monomer solution, most of the initiating systems employed therein involve BCl$_3$ and are in fact akin to those used for the fourth development, the living isobutylene polymerization with cumyl acetate/BCl$_3$.

This brief historical analysis is of course not intended to suggest that Kennedy's discovery[4] is simply an extension of his preceding work. In sharp contrast, it reminds one the utmost importance of persistent, painstaking, systematic, and creative fundamental research focused on a particular field, which often leads to a major breakthrough as witnessed in his particular case. Finally, after its publication just a decade ago, the first full paper on living isobutylene polymerization is still fresh and teaches me much, particularly when I consider the current view that "precision" polymerization, not of specially designed exotic monomers but rather well-known commodity monomers like isobutylene, will be critically required in the polymer science and technology of the next century.

REFERENCES AND NOTES

1. As recent general reviews for living polymerization, see: (a) O. W. Webster, *Science*, **251**, 887 (1991); (b) T. Aida, *Prog. Polym. Sci.*, **19**, 469 (1994).

2. (a) M. Szwarc, *Nature*, **178**, 1168 (1956); (b) M. Szwarc, M. Levy, and R. Milkovich, *J. Am. Chem. Soc.*, **78**, 2656 (1956).

3. As reviews for living cationic polymerization, see: (a) J. P. Kennedy and B. Iván, *Designed Polymers by Carbocationic Macromolecular Engineering: Theory and Practice*, Hanser, Munich, pp. 55 ff, 1991; (b) M. Sawamoto, *Prog. Polym. Sci.*, **16**, 111 (1991); (c) K. Matyjaszewski and M. Sawamoto, in *Cationic Polymerization: Mechanisms, Synthesis, and Applications*, K. Matyjaszewski, Ed., Marcel Dekker, New York, pp. 288–330, 1996.

4. R. Faust and J. P. Kennedy, *J. Polym. Sci.: Part A: Polym. Chem.*, **25**, 1847 (1987). A preliminary paper on the same subject: R. Faust and J. P. Kennedy, *Polym. Bull.*, **15**, 317 (1986).

5. M. Miyamoto, M. Sawamoto, and T. Higashimura, *Macromolecules*, **17**, 265 (1984).

6. (a) J. P. Kennedy, *Cationic Polymerization of Olefins: A Critical Inventory*, Wiley, New York, 1975; (b) J. P. Kennedy and E. Maréchal, *Carbocationic Polymerization*, Wiley, New York, 1982.

7. (a) J. P. Kennedy, in *Polymer Chemistry of Synthetic Elastomers*, Part 1, J. P. Kennedy and E. G. M. Törngvist, Eds., Wiley–Interscience, New York, pp. 304–308, 1968; (b) J. P. Kennedy, *J. Macromol. Sci.-Chem.*, **A3**, 885 (1969).

8. J. P. Kennedy and R. A. Smith, *J. Polym. Sci., Polym. Chem., Ed.*, **18**, 1523 (1980).

9. R. Faust, A. Fehérvári, and J. P. Kennedy, *J. Macromol. Sci.-Chem.*, **A18**, 1209 (1982–83); see also the papers compiled in the same special issue.

Living Carbocationic Polymerization. IV. Living Polymerization of Isobutylene*

R. FAUST and J. P. KENNEDY, *Institute of Polymer Science, The University of Akron, Akron, Ohio 44325*

Synopsis

Truly living polymerization of isobutylene (IB) has been achieved for the first time by the use of new initiating systems comprising organic acetate-BCl_3 complexes under conventional laboratory conditions in various solvents from -10 to $-50°C$. The overall rates of polymerization are very high, which necessitated the development of the incremental monomer addition (IMA) technique to demonstrate living systems. The living nature of the polymerizations was demonstrated by linear $\overline{M}_n$ versus grams polyisobutylene (PIB) formed plots starting at the origin and horizontal number of polymer molecules formed versus amount of polymer formed plots. $\overline{DP}_n$ obeys $[IB]/[CH_3COOR' \cdot BCl_3]$. Molecular weight distributions (MWD) are very narrow in homogeneous systems ($\overline{M}_w/\overline{M}_n = 1.2$–$1.3$) whereas somewhat broader values are obtained when the polymer precipitates out of solution ($\overline{M}_w/\overline{M}_n = 1.4$–$3.0$). The MWDs tend to narrow with increasing molecular weights, i.e., with the accumulation of precipitated polymer in the reactor. Traces of moisture do not affect the outcome of living polymerizations. In the presence of monomer both first and second order chain transfer to monomer are avoided even at $-10°C$. The diagnosis of first and second order chain transfer has been accomplished, and the first order process seems to dominate. Forced termination can be effected either by thermally decomposing the propagating complexes or by nucleophiles. In either case the end groups will be tertiary chlorides. The living polymerization of isobutylene initiated by ester $\cdot BCl_3$ complexes most likely proceeds by a two-component group transfer polymerization.

INTRODUCTION

Truly living polymerizations, i.e., polymerizations that proceed in the absence of termination and chain transfer, are a most desirable objective of the synthetic polymer chemist. Living polymerizations are of great scientific and commercial interest, and several modern industrial processes are based on living systems, e.g., cis-1,4-polybutadiene,[1] triblock polymers of styrene/butadiene/styrene,[2] and polytetrahydrofuran;[3] however, these processes are not carbocationic in nature.

Since the discovery of anionic living systems,[4,5] great efforts have been made to find conditions under which living carbocationic polymerization could be achieved, but success remains limited. In the cationic polymerization of olefins the propagating carbocations are usually unstable and therefore un-

---

*This paper is dedicated to M. Szwarc to commemorate the 30th anniversary of the discovery of anionic "living" polymerizations.

Journal of Polymer Science: Part A: Polymer Chemistry, Vol. 25, 1847–1869 (1987)
CCC 0360-3676/87/071847-23$04.00

dergo chain transfer to monomer and termination reactions. In view of these facts, in 1975 a distinguished practitioner of cationic polymerizations rather pessimistically concluded that it is "unlikely that any cationic polymerization will display living characteristics... ."[6]

The first break to this bleak prediction occurred only very recently when Higashimura, Sawamoto, et al.[7-9] discovered truly living polymerizations of vinyl ethers[7] and other cationically highly reactive monomers such as N-vinyl carbazole[8] and p-methoxy styrene[9] with the HI/I_2 initiating systems. The mechanism of these living polymerizations of vinyl ethers involves monomer insertion into an iodine activated $\sim CH-I$ bond.[7]

$$OR$$

A different recent approach is the "quasiliving" technique that has been used with conventional monomers as well, i.e., styrene and its derivatives, isobutylene (IB), vinyl ethers.[10-12] Under certain rather restrictive conditions (i.e., low temperatures, continuous slow monomer addition), quasiliving carbocationic polymerizations approach those of truly living systems; even though termination can be completely suppressed ($R_t = 0$), the rate of chain transfer to monomer does not become zero but only approaches zero.

Recently we have discovered a family of initiating systems that yield truly living polymerizations of simple olefins. This paper, the fourth in the series,[13-15] demonstrates the living polymerization of IB by tertiary ester $\cdot$ BCl_3 initiating systems and concerns information on the fundamentals of this new concept.

A Hypothesis of Living Carbocationic Polymerizations: Initiating Complexes of Organic Esters with Lewis Acids

The growing species in carbocationic polymerization is conventionally visualized as a carbenium ion accompanied by a counteranion arising from the coinitiator, e.g., $R-Cl + BCl_3 \rightleftharpoons R^{\oplus}BCl_4^{\ominus} \xrightarrow{+M} R \sim\!\!\sim M^{\oplus}BCl_4^{\ominus}$.[16] Hence, the counteranion will strongly influence (control?) every elementary event of a polymerization. The counteranion will affect both the monomolecular and bimolecular chain transfer to monomer steps and will determine the rate of termination.[16]

We hypothesized that living carbocationic polymerizations may arise in the presence of counteranions that would facilitate initiation and propagation but would not participate in chain transfer and termination events, and therefore we set out to search for counteranions that promote initiation and propagation but do not cause or assist proton elimination (i.e., chain transfer to monomer) and are sufficiently stable not to cause termination.

A few comments on ester/BCl_3 complexes are in order. Although well-characterized complexes of esters of carboxylic acids and Lewis acids, i.e., BCl_3, have been known for a long time,[17] they have not yet been used as initiating systems for cationic polymerization. A large number of ester/Lewis acid complexes have been prepared and characterized, mainly by Lappert et al.[18] The structures were determined by IR spectroscopy and other techniques, and it was found that the Lewis acids were exclusively coordinated to the acyl oxygen, e.g.

$$\begin{array}{c} X \\ | \\ B-X \\ \nearrow \quad | \\ O \quad X \\ \| \\ C \qquad CH_2 \\ \diagup \quad \diagdown \quad \diagup \quad \diagdown \\ CH_3 \qquad O \qquad CH_3 \end{array}$$

Since the $C{=}O \rightarrow MX_n$ angle is $\sim 120°$, the possibility for *cis-trans* isomerism arises. However, Lappert et al.[18] could not find a difference by NMR spectroscopy among the protons in the $(CH_3)_2CO/BF_3$ complex even at $-80°C$. The authors postulated a rapid exchange of BF_3 among the donor molecules, a process that has been demonstrated with alcohol adducts. Based mainly on infrared absorption shifts of the carbonyl frequency of ethyl acetate and its complexes with various Lewis acids, Lappert[18] proposed the following order of relative acceptor strengths of these halides: $B > Ga > Al > In$, $Sn > Ga > Si$ and $Ti > Zr$. Among the boron halides the acceptor strength order was $BBr_3 > BCl_3 > BF_3$.

The behavior of ester/BCl_3 complexes was studied by Gerrard et al.[19] These complexes decompose thermally by cleavage of the acyl-oxygen bond (primary alkyl esters) or by the alkyl-oxygen bond (secondary or tertiary alkyl esters). The decomposition temperature decreases in the order primary > secondary > tertiary. The decomposition of the *sec*-butyl acetate/BCl_3 complex yielded *sec*-butyl chloride, acetyl chloride, and a boron-containing residue. The complex of 2-*t*-butyl acetate/$3BCl_3$ started to decompose at $-25°C$ and yielded *t*-butyl chloride, acetyl chloride, and dichloro diacetyl-diborate.[19] Importantly, the thermal decomposition (98°C) of BCl_3 complexes of neopentyl acetate and isobutyl acetate yielded mainly *t*-amyl chloride and *t*-butyl chloride, respectively,[19] which we regard to be cationic processes. The decomposition of the optically active *sec*-butyl- and 1-methylheptyl acetate · BCl_3 complexes afforded the inverted *sec*-butyl- and 1-methylheptyl chlorides, most likely by an S_{Ni} process that may or may not be carbocationic.

In line with this information, we have hypothesized that the living carbocationic polymerization of IB could be initiated, for example, by the *t*-butyl acetate/BCl_3 complex, and that initiation will proceed via the $t\text{-}Bu^{\oplus}BCl_3OAc^{\ominus}$ ion pair, i.e., by the highly active *tert* cation and the expectedly highly stable $BCl_3OAc^{\ominus}$ counteranion. This paper describes the living polymerization of IB by various tertiary ester/BCl_3 complexes, a brief examination of the effect of solvents on these polymerizations, and various supporting studies. The structure of the end groups of polyisobutylenes (PIB's) prepared under various conditions in living polymerizations induced by *t*-ester/BCl_3 complexes has also been investigated. Finally an attempt has been made to explain the data in terms of a mechanism, particularly in regard to the initiation, propagation and forced termination steps.

1850　　　　　　　　FAUST AND KENNEDY

EXPERIMENTAL

Materials

Cumyl acetate (CuOAc): Acetyl chloride (18.8 g, 0.24 M) was slowly added to a solution of 27.2 g (0.2 M) 2-phenyl-2-propanol and 31.4 g (0.26 M) freshly distilled N,N-dimethyl-aniline in 110 mL anhydrous diethyl ether at such a rate as to maintain gentle reflux. After refluxing it for another 2 days, the mixture was cooled to room temperature, treated with water, and the organic layer separated. The ether solution was washed with 10% H_2SO_4 aq., water, and dried over $MgSO_4$. The vacuum-distilled product (80% 2-phenyl-2-propyl acetate, 20% unreacted 2-phenyl-2-propanol) was chromatographed on neutral Al_2O_3 (activity III) and afforded upon elution with n-pentane pure 2-phenyl-2-propyl acetate in 67% yield. *2,4,4-Trimethyl-2-pentyl acetate (TMPOAc):* 2,4,4-Trimethyl-2-pentanol was prepared from 2,4,4-trimethyl-pentene-1 according to Brown et al.[20] The vacuum-distilled pure 2,4,4-trimethyl-2-pentanol was esterified with acetyl chloride following the procedure described at the preparation of 2-phenyl-2-propanol acetate. The vacuum-distilled product was chromatographed on neutral Al_2O_3 (activity III) and afforded upon elution with n-pentane pure 2,4,4-trimethylpentyl acetate in 40% yield. *Triphenylmethyl acetate* was prepared according to Ref. 21. *t-Butyl acetate (tBuOAc) (Aldrich)* was freshly distilled from CaH_2. *Methyl chloride* was dried by passing the gas through a column packed with BaO and condensing it under nitrogen atmosphere. *Methylene chloride* was dried over molecular sieves, distilled from CaH_2, refluxed with triethylaluminum under nitrogen atmosphere overnight, and distilled on the day of experiment. *Ethyl chloride (Fluka)* was refluxed and distilled from CaH_2 under nitrogen atmosphere. *n-Hexane* was refluxed with fuming sulfuric acid, washed neutral with distilled water, dried on molecular sieves, refluxed, and distilled from CaH_2 under nitrogen atmosphere. *Isobutylene* was dried by passing the gas through a column packed with BaO and condensing it under nitrogen atmosphere. *Boron trichloride* was used as received.

Procedures

Polymerizations were carried out in a dry box under dry nitrogen. The charges were quenched by prechilled MeOH. The solvents were evaporated, and the polymer was redissolved in n-hexane. The n-hexane solution was decanted or filtered off the inorganic boron compounds, and the polymer was recovered by evaporating the n-hexane. Minute losses of the polymer were unavoidable.

In termination studies the charges and the TMPOAc · BCl_3 model compound were quenched by the addition of prechilled MeOH, Et_3N, pyridine, and $CH_3OH/NaOH$. The solvents were evaporated, the polymer dissolved in n-hexane, washed with 5% aq. sulfuric acid, followed by distilled water until neutral, dried over $CaCl_2$, and filtered, and the polymer (or the model compound) was recovered by evaporating the n-hexane.

The infrared spectra were recorded on a Perkin Elmer Model 521 spectrophotometer. A pair of Wilmad matched 1.0 mm liquid cells with NaCl windows were used. Only the C=O stretching frequency was followed owing to warm up and decomposition caused by the IR beam.

Molecular weights were determined using a Waters high-pressure GPC instrument (Model 6000A Pump, a series of five μ-Styragel columns (10^5, 10^4, 10^3, 500 and 100 Å), Differential Refractometer 2401, UV Absorbance Detector Model 440, and a calibration curve made by well-fractionated polyisobutylene standards.

^{1}H-NMR spectra were taken by a Varian T-60 Spectrometer using appropriate solutions and TMS standards.

RESULTS AND DISCUSSION

Preliminary Experiments

Various tertiary esters, i.e., t-butyl acetate (tBuOAc), 2,4,4-trimethylpentyl-2-acetate (TMPOAc), and cumyl acetate (CuOAc) in conjunction with BCl_3 in polar diluents (i.e., CH_3Cl and CH_2Cl_2) readily initiate the polymerization of IB and yield high conversions in the range from -10 to $-50°C$. Initiation occurs only in tricomponent systems, i.e., organic acetate/BCl_3/IB, regardless of the introduction sequence. "Control" experiments, i.e., experiments carried out by mixing IB with BCl_3 in CH_3Cl, yield insignificant conversions ($< 7\%$) and relatively higher molecular weights under essentially identical conditions. Control experiments carried out under the same conditions with CH_2Cl_2 yield much higher conversions; however, these results are in line with previous experience[22] that initiation of IB polymerization with BCl_3 in CH_2Cl_2 is due to traces of protic impurities (most likely moisture) that, short of rigorous high vacuum drying,[23] are extremely hard to remove. Nonetheless even in these controls we obtained much higher molecular weights and the conversions were below 100%; in contrast, in the presence of the various acetates, conversions were invariably complete.

Table I shows the results of representative scouting experiments. Polymerizations ensued instantaneously upon BCl_3 addition (appearance of haziness). In several experiments conversions reached close to 100% 1 min after BCl_3 addition (footnote c in Table I). Nonetheless, we have terminated most reactions arbitrarily 30 min after BCl_3 introduction. Thus experiment time and polymerization time may be quite different.

In all experiments carried out in the presence of acetates, conversions were invariably 100% (except in one case in which a very low t-BuOAc concentration was used, see line 5 in Table I). The I_{eff}'s [$= N$ (number of polymer molecules) $\times 100/I_0$] obtained with t-BuOAc are much lower than 100%, most likely because ion generation is slow with this ester. I_{eff}'s were invariably $\sim 100\%$ with TMPOAc and CuOAc.

We were much encouraged that the molecular weights were independent of the temperature and solvent at a given ester concentration, [CH_3COOR^t], and that in a few cases $\overline{M}_w/\overline{M}_n$ were below 2.0. Significantly, ^{1}H-NMR spectroscopic analysis of the low molecular weight products obtained even at $-10°C$ showed the absence of terminal unsaturation in PIB, i.e., indicating the absence of proton elimination.

TABLE I
Demonstration of Initiating Activity of Various Tert-acetate · BCl_3 Complexes Under Various Conditions[a]

	Conditions				Results			
CH_3COOR^t								
R^t	Conc.	Diluent	Temp. (°C)	Conv. (%)	$\bar{M}_n$ (g/mol)	$\bar{M}_w$ (g/mol)	$\bar{M}_w/\bar{M}_n$	I_{eff} (%)
—	—	CH_3Cl	−50	6.6	116400	230000	2.0	—
t-Bu	2.8×10^{-2}	CH_3Cl	−50	~100[b]	7370	15000	2.0	28
t-Bu	5.6×10^{-2}	CH_3Cl	−50	~100	4240	7210	1.7	24
—	—	CH_3Cl	−30	6.0	36000	72000	2.0	—
t-Bu	5.6×10^{-3}	CH_3Cl	−30	13.5	2800	4200	1.5	44
t-Bu	2.8×10^{-2}	CH_3Cl	−30	~100	4650	8930	1.9	35
t-Bu	5.6×10^{-2}	CH_3Cl	−30	~100	2950	5590	1.9	34
—	—	CH_2Cl_2	−30	75	49800	103300	2.1	—
t-Bu	5.6×10^{-3}	CH_2Cl_2	−30	~100	27500	66800	2.4	36
t-Bu	2.8×10^{-2}	CH_2Cl_2	−30	~100	4230	10420	2.4	48
t-Bu	5.6×10^{-2}	CH_2Cl_2	−30	~100	2470	5950	2.4	42
—	—	CH_2Cl_2	−10	68	28600	80000	2.8	—
t-Bu	2.8×10^{-2}	CH_2Cl_2	−10	~100	4370	9200	2.1	47
t-Bu	5.6×10^{-2}	CH_2Cl_2	−10	~100	2040	5090	2.5	51
—	—	CH_3Cl	−50	6.6	116400	230000	2.0	—
TMP	5.6×10^{-3}	CH_3Cl	−50	~100	10500	21500	2.0	97
TMP	2.8×10^{-2}	CH_3Cl	−50	~100	1800	6300	3.6	121
TMP	5.6×10^{-2}	CH_3Cl	−50	~100	1400	3400	2.4	80
—	—	CH_3Cl	−30	6.0	36000	72000	2.0	—
TMP	5.6×10^{-3}	CH_3Cl	−30	~100	10770	15150	1.5	94
TMP	2.8×10^{-2}	CH_3Cl	−30	~100	2090	5320	2.5	103
TMP	5.6×10^{-2}	CH_3Cl	−30	~100	1290	2800	2.2	88
—	—	CH_3Cl	−30	6.0	36000	72000	2.0	—
Cu	5.6×10^{-3}	CH_3Cl	−30	~100[c]	8500	13600	1.6	115
Cu	2.8×10^{-2}	CH_3Cl	−30	~100	2000	3400	1.7	104
Cu	5.6×10^{-2}	CH_3Cl	−30	~100	1160	2820	2.4	96
—	—	C_2H_5Cl	−30	2.0	—	—	—	—
Cu	5.6×10^{-3}	C_2H_5Cl	−30	~100[c]	9500	14100	1.5	107
Cu	2.8×10^{-2}	C_2H_5Cl	−30	~100	2390	3980	1.7	90
Cu	5.6×10^{-2}	C_2H_5Cl	−30	~100	1260	2050	1.6	91

[a] $[BCl_3] = 2.6 \times 10^{-1}$ mol/L, $[M] = 1.0$ mol/L, time = 30 min.
[b] Polymerization time was 1 h.
[c] Conversion was 100% after 1 min.

LIVING CARBOCATIONIC POLYMERIZATION. IV   1853

A closer analysis of the data also showed that the molecular weights ($\overline{M}_n$) were inversely proportional to [CH_3COOR^t]; indeed with TMPOAc and CuOAc $\overline{DP}_n$ was equal with [M]/[CH_3COOR^t]. The [BCl_3] was deemed to be of lesser importance: an excess was used to ensure complete CH_3COOR^t utilization. In view of the controls, "free" BCl_3 was viewed to be of diminished significance.

A closer examination of the molecular weight data obtained in experiments using CH_2Cl_2 and CH_3Cl indicate that the effect of moisture impurities is most likely unimportant. The molecular weights obtained in CH_2Cl_2 and CH_3Cl are similar, although the level of moisture in these diluents is quite different (note the high conversions obtained in CH_2Cl_2 in contrast to the insignificant conversions found in CH_3Cl; cf. Table I). The concentration of CH_3COOR^t and moisture in CH_2Cl_2 are probably of the same order of magnitude ($\sim 10^{-3}$ M). This implies that the effect of the $H_2O \cdot BCl_3$ is negligible. We theorize that the $CH_3COOR^t \cdot BCl_3$ complexes give rise to the $R^{t\oplus} \cdot BCl_3CH_3COO^{\ominus}$ initiating species in the presence of which the equilibrium

$$H_2O \cdot BCl_3 \rightleftharpoons H^{\oplus}BCl_3OH^{\ominus} \tag{1}$$

is displaced toward the unionized species so that initiation by H_2O does not occur. This proposition is supported by the observation that polymerization of IB is absent in CH_2Cl_2 in the presence of $(C_6H_5)_3C^{\oplus}BCl_3OAc^{\ominus}$.[24] The trityl cation is too stable to initiate IB polymerization,[16] and ion generation by eq. (1) from $H_2O \cdot BCl_3$ is probably suppressed by $(C_6H_5)_3C^{\oplus} \cdot BCl_3OAc^{\ominus}$.

In line with these observations, the living polymerization of IB induced by organic ester $\cdot$ BCl_3 complexes can be carried out reproducibly under conventional laboratory conditions, and the use of super-dry chemicals is unnecessary.

Experimental Proof for the Living Polymerization of Isobutylene in the Presence of $CH_3COOR^t \cdot BCl_3$ Complexes

A diagnostic proof of living polymerizations is linear $\overline{M}_n$ versus conversion plots starting from the origin. If initiation is instantaneous and faster than propagation ($R_i > R_p$), the molecular weight distributions will approach the Poisson distribution (i.e., $\overline{M}_w/\overline{M}_n \sim 1.0$). To ensure rapid ion generation initially we chose to work with the TMPOAc $\cdot$ BCl_3 and CuOAc $\cdot$ BCl_3 systems since the expected cations from these initiators $(CH_3)_3CCH_2C^{\oplus}(CH_3)_2(TMP^{\oplus})$ and $C_6H_5C^{\oplus}(CH_3)_2(Cu^{\oplus})$, respectively, have shown to be efficient initiating entities for IB polymerizations (see also Table I).[25,26] The use of $TMP^{\oplus}$ as the initiating entity is particularly advantageous since the structure of this cation and that of the propagating polyisobutylene carbenium ion $PIB^{\oplus}$ are for all practical purposes identical. Indeed the addition of BCl_3 to TMPOAc/IB or CuOAc/IB charges dissolved in CH_2Cl_2 caused immediate polymerizations at $-30°C$, and the reactions proceeded to complete conversions within a few seconds. Essentially similar observations have been made by the use of CH_3Cl diluent.

Since the polymerizations were too fast for kinetic investigations, i.e., to withdraw a series of samples from a single reactor for conversion/$\overline{M}_n$ de-

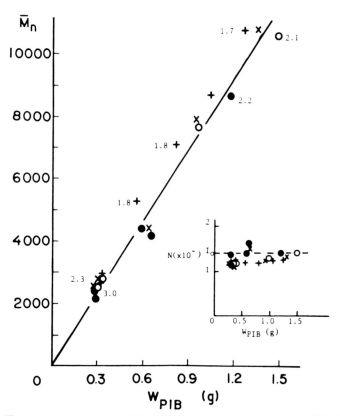

Fig. 1. $\overline{M}_n$ and N, the number of PIB chains (insert), versus the weight of PIB formed W_{PIB} in the TMPOAc/BCl_3/IB polymerization system using the incremental monomer addition (IMA) technique at $-30°C$: CH_2Cl_2 ($\bigcirc$, $\bullet$) and CH_3Cl ($+$, X), $[I_0] = 5.6 \times 10^{-3}$ M, $[BCl_3] = 2.8 \times 10^{-1}$ M. Numbers indicate $\overline{M}_w/\overline{M}_n$ values.

terminations, we were compelled to carry out experiments by the use of a series of reactors. Thus from five to six large test tubes or three-neck flasks were charged with CuOAc/IB or TMPOAc/IB in CH_2Cl_2 or CH_3Cl diluents, and at zero time an appropriate amount of BCl_3 coinitiator was added to all the charges at $-30°C$. The systems were strongly mixed (turbomix) and replaced in a constant temperature bath at $-30°C$. After an arbitrary 30 min the polymerization was quenched in the first reactor (addition of a few milliliters prechilled methanol), whereas an additional quantity of IB was added to all the remaining reactors. After 30 min the reaction in the second reactor was quenched, and a further quantity of IB was added to the rest of the reactors. This procedure, which we term the "incremental monomer addition" (IMA) technique, was continued until all the charges in the series had been quenched. Subsequently, the amounts of polymers formed (conversions) were determined by gravimetry, and $\overline{M}_n$ and $\overline{M}_w/\overline{M}_n$ were determined by GPC.

Figure 1 shows a representative $\overline{M}_n$ versus grams polymer formed plot obtained in a series of experiments carried out with TMPOAc in CH_3Cl and CH_2Cl_2 diluents at $-30°C$. The different symbols indicate replicate experi-

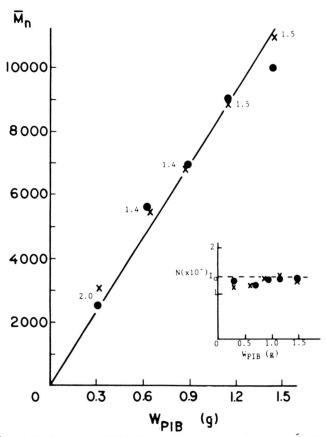

Fig. 2. $\overline{M}_n$ and N, the number of PIB chains (insert), versus the weight of PIB formed W_{PIB} in the CuOAc/BCl$_3$/IB/CH$_2$Cl$_2$ polymerization system using the IMA technique at $-30°C$: $[I_0] = 5.6 \times 10^{-3}$ M, $[BCl_3] = 2.8 \times 10^{-1}$ M. Numbers indicate $\overline{M}_w/\overline{M}_n$ values.

ments. Similarly, Figures 2 and 3 show the results obtained with the CuOAc/BCl$_3$ system using CH$_2$Cl$_2$ and CH$_3$Cl diluents at $-30°C$, respectively. The inserts in Figures 1–3 show the number of PIB chains in these systems (N) as a function of polymer formed. (The N was obtained from g/$\overline{M}_n$). The linear $\overline{M}_n$ and horizontal N versus grams PIB plots as shown in Figures 1–3 strongly suggest living polymerizations. The fact that the $\overline{M}_w/\overline{M}_n$ values decrease with incremental monomer addition (i.e., by increasing molecular weights) suggests that growth of the living species persists throughout the experiment.

The molecular weight dispersities ($\overline{M}_w/\overline{M}_n$ values) of PIB's obtained with the TMPOAc/BCl$_3$ system are broader than those produced by the CuOAc/BCl$_3$ system. Evidently the rate of initiation with TMPOAc/BCl$_3$ is lower than that with the CuOAc/BCl$_3$ system. Indeed the $\overline{M}_w/\overline{M}_n$ of the initially formed PIB was appreciably lower when the individual components were premixed (aged) and the monomer was added to the preformed TMPOAc · BCl$_3$ complex (e.g., 2.3 versus 3.7 in CH$_3$Cl). Importantly the molecular weight distributions decreased with each incremental monomer addition in this series of experiments.

1856   FAUST AND KENNEDY

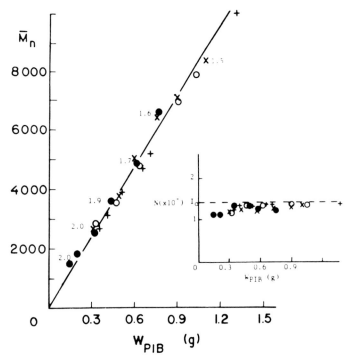

Fig. 3. $\overline{M}_n$ and N, the number of PIB chains (insert), versus the weight of PIB formed W_{PIB} in the $CuOAc/BCl_3/IB/CH_3Cl$ polymerization system using the IMA technique at $-30°C$: $[I_0] = 5.6 \times 10^{-3}$ M, $[BCl_3] = 2.8 \times 10^{-1}$ M. Numbers indicate $\overline{M}_w/\overline{M}_n$ values.

The fact that the $\overline{M}_n$ of the first formed PIB's in which BCl_3 was added to $TMPOAc/BCl_3$ charges and in which the monomer was added to the preformed $TMPOAc \cdot BCl_3$ complexes, are equal within experimental error indicates that the rate of formation of the initiating $TMPOAc \cdot BCl_3$ complexes is extremely high and that initiation follows very rapidly. Moreover, experiments have been carried out in which $TMPOAc \cdot BCl_3$ complexes were prepared (mixing $TMPOAc$ plus BCl_3 in CH_3Cl or CH_2Cl_2 at $-30°C$), and after 30 min of aging various amounts of IB were added; immediate polymerization to close to 100% conversion ensued, and the $\overline{DP}_n$ of the product obeyed $[IB]/[TMPOAc \cdot BCl_3]$. These experiments suggest that the $TMPOAc \cdot BCl_3$ complex is an active "living PIB dimer" and that the structure of this dimer and that of the propagating end of the living PIB polymer are essentially identical.

According to the data shown in Figures 1–3, the molecular weight distributions gradually decrease with each incremental monomer addition despite the heterogeneous nature of the polymerizations. Evidently dormant species are absent, or the rate of exchange between dormant and growing species is much faster than that of propagation. Experiments carried out in homogeneous media (mixed n-hexane/CH_3Cl systems) specifically to test the influence of rapid precipitation on MWD yielded close to Poisson distributions. Additional experiments with styrene derivatives (to be reported later) that yield CH_2Cl_2 and CH_3Cl-soluble polymers readily yielded $\overline{M}_w/\overline{M}_n = 1.05–1.2$.

LIVING CARBOCATIONIC POLYMERIZATION. IV   1857

The Effect of Diluent

The data in the previous sections indicate that the living polymerization of IB can be induced by TMPOAc · BCl$_3$ and CuOAc · BCl$_3$ complexes in CH$_2$Cl$_2$ and CH$_3$Cl diluents (cf. Figs. 1–3). Since PIB with $\overline{M}_n$ higher than about 3000 g/mol is insoluble in CH$_2$Cl$_2$ and CH$_3$Cl, these polymerizations tend to be heterogeneous. This circumstance and the very rapid propagation (relative to initiation) in these diluents results in somewhat broader than Poisson molecular weight distributions. Experiments carried out in homogeneous systems (i.e., in n-hexane-CH$_3$Cl mixtures) readily yielded $\overline{M}_w/\overline{M}_n$ = 1.2–1.3.

To overcome problems caused by system heterogeneity and to study the effect of solvents in general, experiments have been carried out by the use of various CH$_2$Cl$_2$/n-hexane mixtures, C$_2$H$_5$Cl and HCCl$_3$. Since the polymerization rates were extremely high even by the use of 60/40 v/v CH$_3$Cl/n-C$_6$H$_{14}$ mixtures (i.e., polymerizations were complete in 60 s at $-30°$C), we used the incremental monomer addition technique (cf. the previous section) to study the $\overline{M}_n$ versus grams PIB formed profile.

Figure 4 shows some of the results. All the experiments were homogeneous, and the initial $\overline{M}_w/\overline{M}_n$ values were indeed smaller than those obtained in the

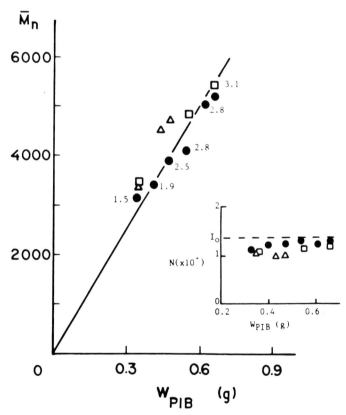

Fig. 4. $\overline{M}_n$ and N, the number of PIB chains (insert), versus the weight of PIB formed W_{PIB} in the CuOAc/BCl$_3$/IB polymerization system using the IMA technique at $-30°$C with different solvents: C$_2$H$_5$Cl (●), CH$_2$Cl$_2$/n-hexane 80v/20v (□), CH$_2$Cl$_2$/n-hexane 60v/40v (△) [I_0] = 5.6 × 10^{-3} M, [BCl$_3$] = 2.8 × 10^{-1} M. Numbers indicate $\overline{M}_w/\overline{M}_n$ values.

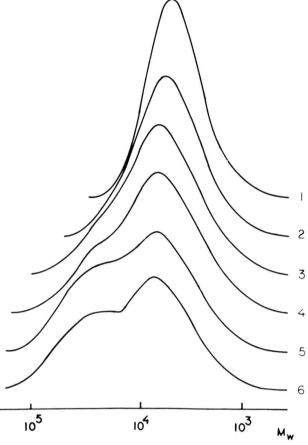

Fig. 5.   MWD of PIBs obtained by the use of the $CuOAc/BCl_3/IB/C_2H_5Cl$ system using the IMA technique at $-30°C$; $[I_0] = 5.6 \times 10^{-3}$ M, $[BCl_3] = 2.8 \times 10^{-1}$ M.

more polar CH_2Cl_2 or CH_3Cl systems. However, in the later course of the experiment the MWD's broadened and finally became bimodal. The findings obtained with C_2H_5Cl are shown in Figures 4 ($\overline{M}_w/\overline{M}_n$ values) and 5. This broadening of the MWD was even more pronounced with 80/20 and 60/40 v/v CH_2Cl_2/n-C_6H_{14}. The broadening of MWD with decreasing solvent polarity is conceivably due to the presence and accumulation of a second (most likely less dissociated or less active) propagating species. Broad distributions that ultimately become bimodal suggest an increasingly slower equilibrium between a dormant and active growing species. Importantly, however, this equilibrium can be controlled, i.e., the equilibrium is reached slowly as indicated by the narrow distributions obtained with C_2H_5Cl in which all the monomer was present at initiation (cf. the last three entries in Table I). Evidently the polymerization is complete before the extent of dormant species would increase to a significant level (determined by the equilibrium). In these instances theoretical molecular weights ($\overline{DP}_n = [M]/[CH_3COOR^t]$) and narrow MWD's were achieved at ~100% conversions. Evidently, then, theoretical $\overline{M}_n$, conversion, and $\overline{M}_w/\overline{M}_n$ values can be obtained not only in polar diluents

LIVING CARBOCATIONIC POLYMERIZATION. IV   1859

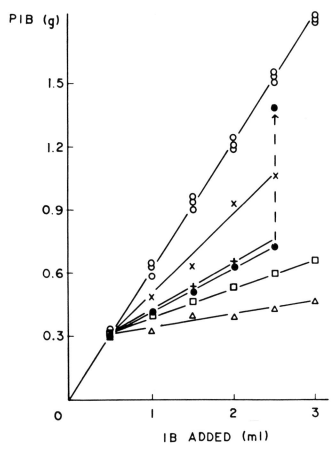

Fig. 6. The weight of PIB versus cumulative amount of IB added by the use of various solvent systems at $-30°C$. CH_2Cl_2 (O), CH_3Cl (●), CH_2Cl_2/n-hexane $80v/20v$ (+), C_2H_5Cl (□), CH_2Cl_2/n-hexane $60v/40v$ (△) (time between monomer additions 30 min), CH_3Cl (X) (time between monomer addition 60 min). $[I_0] = 5.6 \times 10^{-3}$ M, $[BCl_3] = 2.8 \times 10^{-1}$ M.

but under select conditions even in less polar media. A series of experiments had also been carried out by the use of $HCCl_3$; however, the conversions were so low that experimentation with this solvent was discontinued (high or ~100% conversions were obtained only in the presence of relatively high CuOAc concentrations). The reason(s) for these observations may be the same as that discussed above in conjunction with the nonpolar solvent mixtures.

It was of interest to determine the amount of PIB formed as a function of IB added in various solvents and solvent mixtures by the use of the incremental monomer addition technique. Figure 6 shows the results obtained with the CuOAc/BCl_3 system, but similar results have also been obtained with TMPOAc/BCl_3. Since the time elapsed between individual monomer additions was always the same, an arbitrary 30 mins, the grams PIB formed reflects the relative polymerization rates. The results obtained in quadruplicate experiments by the use of CH_2Cl_2 diluent are theoretical, within what is considered to be experimental error (i.e., unavoidable minute losses of PIB during workup; see the Experimental Section). Similarly, the amount of PIB formed in the first reactor, i.e., in the charge that contained all the monomer at the moment the BCl_3 was added, is theoretical (~100% conversions) irrespective of the particular solvent used. From this point on the amount of PIB that arises at each incremental monomer addition is mainly determined by the nature of the solvent. It appears that the rate is directly related to the overall polarity. By this measure CH_2Cl_2 is more polar than CH_3Cl, which is in turn more polar than C_2H_5Cl.

The absolute amount of grams PIB formed is a function of conditions, e.g., time, temperature, concentrations, and heterogeneity. Specifically, the less than theoretical quantities (< 100% conversions) of PIB formed in CH_3Cl are most likely due to polymer precipitation, which must retard propagation (decreased diffusion). Indeed in an experiment in which the precipitated system was not quenched after 30 min but was allowed to "live" for an additional 180 min (see dotted line in Fig. 6), the yields gradually increased and finally reached a close-to-theoretical value. Otherwise, in C_2H_5Cl and CH_2Cl_2/n-C_6H_{14} mixtures the limited quantities formed (lower rates) are most likely due to the formation of less-dissociated (dormant?) species. Interestingly, close-to-theoretical values have been obtained by the use of CH_2Cl_2, although the system appears to be heterogeneous in this diluent.

The Effect of Temperature: First- and Second-order Chain Transfer to Monomer

Figure 7 shows the $\overline{M}_n$ versus grams PIB plots obtained with the $CuOAc/BCl_3/IB/CH_2Cl_2$ system using the IMA technique at $-30°C$ and $-10°C$. The linear plot characteristic of the $-30°C$ experiment together with the insert, showing that the number of growing chains remains constant throughout the series, indicates a living system. In contrast the experiment at $-10°C$ yielded a kinked $\overline{M}_n$ versus grams PIB plot, and, according to the insert, the number of PIB molecules increases monotonically, which indicates the appearance of chain transfer to monomer at this higher temperature.

These findings led to some valuable insights into the nature of chain transfer to monomer processes in general. Conventionally, chain transfer to monomer is visualized to occur by β-proton elimination either by a second-order bimolecular or first-order monomolecular process:[16]

where $k_{tr,^2M}$ and $k_{tr,^1M}$ are the rate constants of the second- and first-order chain transfer to monomer, respectively. First we assumed that by employing

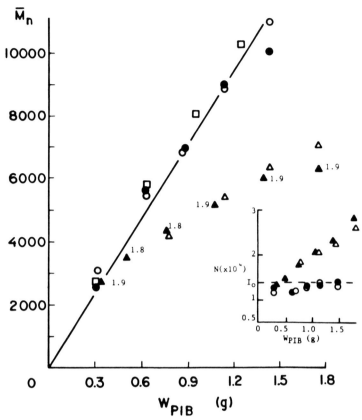

Fig. 7. $\overline{M}_n$ and N, the number of PIB chains (insert), versus the weight of PIB formed W_{PIB} in the $CuOAc/BCl_3/IB/CH_2Cl_2$ polymerization system using the IMA technique at $-30°C$ ($\circ, \bullet$), $-10°C$ ($\triangle, \blacktriangle$), and with all the monomer present at the addition of BCl_3 ($\square$). [I_0] = 5.6 × 10^{-3} M, [BCl_3] = 2.8 × 10^{-1} M.

the IMA technique at $-10°C$ the more important pathway is by the second-order process and set out to elucidate the rate constant ratio $k_{tr,^2M}/k_p$ by the following treatment.

If chain transfer occurs by the second-order route, and since termination is absent,

$$\overline{DP}_n = \frac{[M_0] - [M_t]}{[I_0] - [I_t] + \int_0^t k_{tr,^2M}[M_n^\oplus][M]\,dt} \qquad (2)$$

where [M_0] and [M_t] are the initial and final IB concentrations, [I_0] and [I_t] are the initial and final acetate concentrations, respectively, [$M_n^\oplus$] is the concentration of propagating chains, and [M] is the momentary concentration of monomer. Assuming instantaneous initiation and 100% conversion,

$$\overline{DP}_n = \frac{[M_0]}{[I_0] + \int_0^t k_{tr,^2M}[M_n^\oplus][M]\,dt} \qquad (3)$$

TABLE II
Polymerization of Isobutylene at Various Monomer Concentrations[a]

M (mol/L)	Conv. (%)	$\overline{M}_n$	$\overline{M}_w$	$\overline{M}_w/\overline{M}_n$	I_{eff} (%)
0.225	100	2640	5000	1.9	92
0.45	100	3900	5800	1.7	81
0.676	100	8200	13900	1.7	85
0.9	100	10400	16320	1.6	88

[a] Cumyl acetate initiator $[I] = 5.6 \cdot 10^{-3}$ M, $[BCl_3] = 2.6 \cdot 10^{-1}$ M, CH_2Cl_2, $-10°C$.

Since

$$-dt = \frac{d[M]}{k_p[M_n^{\oplus}][M]} \tag{4}$$

we get

$$\overline{DP}_n = \frac{[M_0]}{[I_0] + \int_{M_0}^{0} \frac{k_{tr,^2M}}{k_p} \cdot d[M]} \tag{5}$$

thus

$$\overline{DP}_n = \frac{[M_0]}{[I_0] + \frac{k_{tr,^2M}}{k_p} \cdot [M_0]} \tag{6}$$

and

$$\overline{DP}_n^{-1} = \frac{[I_0]}{[M_0]} + \frac{k_{tr.^2M}}{k_p} \tag{7}$$

A series of experiments have been designed to test this expression. Thus IB was polymerized at $-10°C$ by the CuOAc/BCl$_3$ system with varying $[M_0]$ at $[I_0]$ = constant. Table II shows the results and Figure 8 the corresponding $1/\overline{DP}_n$ versus $1/[M_0]$ plot. Obviously the straight line can be smoothly back-extrapolated to the origin, and there is no evidence for an intercept. The slope of the plot yields $[I_0]$ within experimental error. According to this evidence $k_{tr,^2M}/k_p = 0$, i.e., bimolecular chain transfer to monomer is absent, and by inference, chain transfer at $-10°C$ is most likely due to the mono-molecular process. This is the first time that one is able to distinguish experimentally with some measure of confidence between monomolecular and bimolecular chain transfer processes.

The "All Monomer in" Technique and Its Significance for the High Temperature Synthesis of Living Polymers

Significantly, the first data point (duplicates) obtained at $-10°C$ by the IMA technique also fall on the theoretical line in Figure 7, indicating the

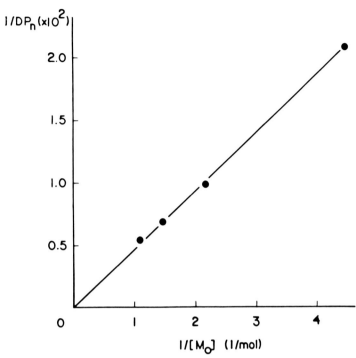

Fig. 8. $1/\overline{DP}_n$ versus $1/M_0$ plot for the CuOAc/BCl$_3$/IB/CH$_2$Cl$_2$ system at $-10°$C. $[I_0] = 5.6 \times 10^{-3}$ M, [BCl$_3$] = 2.8×10^{-1} M.

absence of chain transfer to monomer even at this unusually high temperature. The reader is reminded that these first data were obtained in the experiment in which the BCl$_3$ was added to monomer/CuOAc charges, whereas the subsequent data were collected in experiments in which increasing amounts of monomer were added to living CuOAc/BCl$_3$/IB charges. Thus the question arises as to why chain transfer is absent when BCl$_3$ is added to CH$_3$COOR^t/IB charges but is present when additional IB is added to active CH$_3$COOR^t/BCl$_3$/IB charges, i.e., to charges 2, 3, etc., in the IMA series. The answer to this puzzle becomes apparent by considering the long delay (i.e., 30 min) that elapses between incremental monomer additions. Evidently the polymerization is living in nature even at $-10°$C *as long as sufficient monomer is present* (i.e., when BCl$_3$ is added last), because the rate of the only operational chain transfer to monomer step, i.e., the monomolecular process, is negligible compared with that of propagation. Owing to the very high polymerization rate, the monomer disappears in a few seconds by living polymerization; however, given a sufficient length of time (i.e., 30 min) in the absence of monomer the living chain ends will slowly undergo proton elimination to the counteranion, and thus the stage is set for monomolecular chain transfer to monomer (see scheme above). Thus it is not too surprising that living polymerizations more readily occur by the IMA technique at $-30°$C because monomolecular chain transfer is presumably "frozen out" at this lower temperature level; however, living systems may still be obtained at $-10°$C or even at higher temperatures provided sufficient monomer is initially available in the system, and the polymerization is over before the much slower monomolecular chain transfer to monomer process becomes noticeable.

A direct experiment indicates that this is indeed the case, and living polymerization can be obtained even at $-10°C$, provided the charge contains all the monomer when initiation occurs, i.e., by the "all monomer in" (AMI) technique. In this experiment we added BCl_3 to a series of IB/CuOAc charges in CH_2Cl_2 at $-10°C$ by the AMI technique. The conversions were complete, and as shown by the plot in Figure 7, the $\bar{M}_n$ versus grams monomer added plot is close to theoretical. This important lead has been followed up, and recently we are routinely carrying out various living cationic polymerizations by batch and continuous techniques at previously unattainable high temperatures.

Following this train of thought, the incremental monomer addition technique is a more sensitive method to demonstrate the existence of living systems than the conventional intermittent monomer withdrawal technique because only by the use of the former technique can the monomolecular chain transfer process be diagnosed.

A Comment in Regard to the Generality of the Mayo Equation

These conclusions have a bearing on the interpretation of data obtained by the popular Mayo analysis.[27] In line with the above considerations, the Mayo equation $1/\overline{DP}_n = k_{tr,M}/k_p + (k_t/k_p)1/[M]$ should be expanded as follows:

$$\frac{1}{\overline{DP}_n} = \frac{k_{tr,^1M}[M_n^\oplus] + k_{tr,^2M}[M_n^\oplus][M]}{k_p[M_n^\oplus][M]} + \frac{k_t[M_n^\oplus]}{k_p[M_n^\oplus][M]}$$

and thus

$$\frac{1}{\overline{DP}_n} = \frac{k_{tr,^2M}}{k_p} + \frac{k_{tr,^1M} + k_t}{k_p} \cdot \frac{1}{[M]}$$

In other words, the absence of an intercept of such an expanded Mayo plot would indicate the absence of only the second-order chain transfer to monomer process; however, the slope of the plot may still contain the first-order component, or the presence of an intercept would suggest not only the operational presence of the second-order process but also that the first-order process is buried in the intercept. Since the first-order process is operational even when the second-order process is absent (see above), it appears that, in general, the first-order process is probably more important than the second-order one in IB polymerizations. Similar conclusions have been derived from work done with α-methylstyrene polymerization in the presence of proton traps.[28]

CONCLUSIONS: MECHANISTIC SPECULATIONS

Initiation and Propagation

Evidence has been presented that demonstrates for the first time the truly living polymerization of IB induced by $TMPOAc \cdot BCl_3$ and $CuOAc \cdot BCl_3$ complexes in CH_2Cl_2 and CH_3Cl diluents. The PIBs so obtained do not contain terminal unsaturation, which is evidence for the absence of chain transfer to monomer. PIB molecular weights are controlled by the relative ratio of monomer over initiator (i.e., CH_3COOR^t) concentrations, i.e., $\overline{DP}_n = [M_0]/[I_0]$. The $\overline{M}_w/\overline{M}_n = 1.3$–$2.0$, most likely due to rapid propagation relative to initiation and physical effects, i.e., system heterogeneity and insufficient stirring (heat effect). Homogeneous systems designed to give narrower MWDs have indeed yielded $\overline{M}_w/\overline{M}_n = 1.2$–$1.3$. Other monomers (styrene derivatives) that give soluble polymers gave $\overline{M}_w/\overline{M}_n = 1.05$–$1.2$.

Protic impurities (moisture) have very little effect on these living polymerizations, and the experiments can be carried out under conventional laboratory conditions by the use of conventionally dried solvents under a blanket of dry nitrogen, for example. An attempt is made to rationalize these results by the use of information gleaned from the literature in regard to the nature of ester/BCl_3 complexes.[18,19] Thus the following mechanism may explain some of the features of living IB polymerization by, for example, the TMPOAc/BCl_3 system ($R = -CH_2-C(CH_3)_3$):

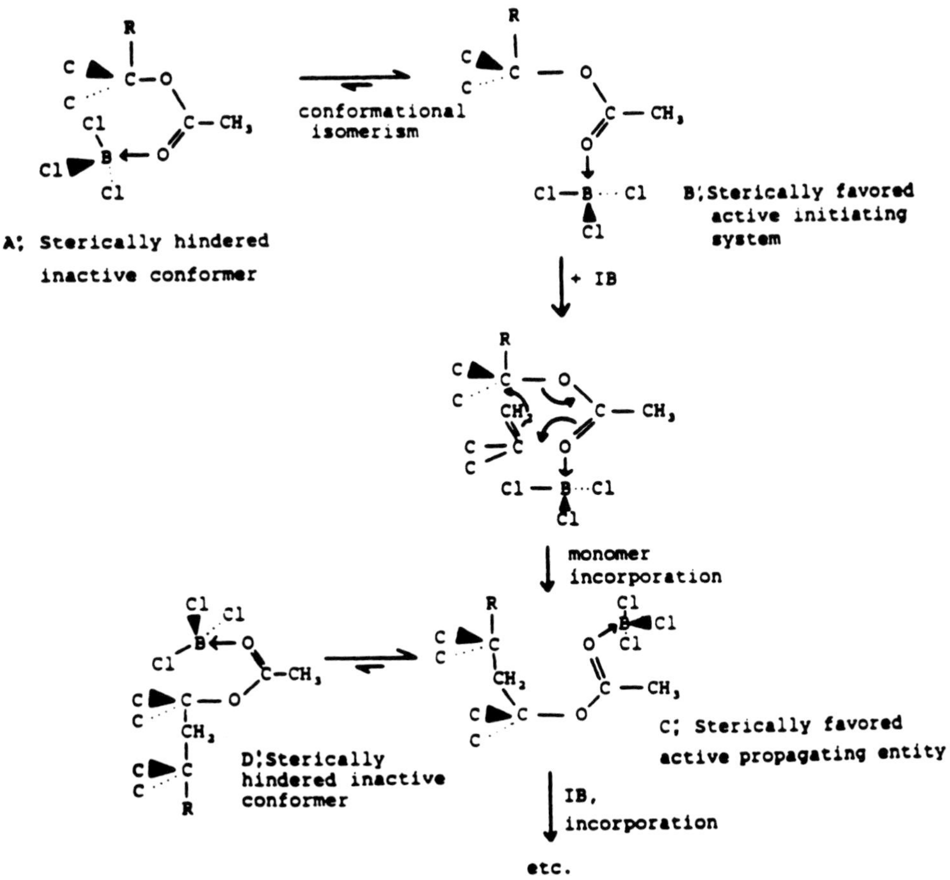

The rate of complex formation between organic acetates and BCl_3 is very rapid. The site of complexation is the acyl oxygen,[18] and equilibration between the two possible conformational isomers A and B is most likely also very rapid. Inspection of molecular models would indicate that conformer B is sterically more favored to incorporate incoming monomer. Monomer incorporation would yield a tertiary acetate · BCl_3 complex structurally identical to the initiating complex. In this sense this system is similar to a group transfer polymerization.[29]

The nature (i.e., ionicity) of the propagating site is obscure. The facts that chain transfer to monomer and termination are absent indicate that the active species are not free ions or highly dissociated ion pairs encountered in conventional systems. The large rate-enhancing effect of polar diluents (CH_2Cl_2, CH_3Cl, C_2H_5Cl) may be due to the presence of a dormant (nondissociated) and an active (more or less ionic) species connected by a mobile equilibrium. The polar diluent accelerates equilibration and/or accentuates the ionic character of the active species:

To elucidate the nature of the propagating species, the common ion effect was studied using the TMPOAc/BCl_3 initiating system in conjunction with the $(C_6H_5)_3COAc$/BCl_3 combination as the common ion precursor. Table III shows the results. The rate of polymerization was considerably reduced in the presence of $(C_6H_5)_3COAc$/BCl_3. The effect is quite dramatic at the lowest acetate concentrations (Experiment 4 in Table III). With higher concentrations (Experiment 5) and/or lower relative $(C_6H_5)_3CCOAc$/TMPOAc amounts (Experiment 6), the effect is masked by the high conversions. The $\overline{M}_w/\overline{M}_n$ values are very low (virtually Poisson distributions) in the presence of TMPOAc/$(C_6H_5)_3COAc = 1.0$ (Experiments 4 and 5).

TABLE III
The Common Ion Effect: Isobutylene Polymerization in the Absence and Presence
of $(C_6H_5)_3 COAc/BCl_3$

Exp.	TMPOAc (mol/L)	$(C_6H_5)_3 COAc$ (mol/L)	Conv. (%)	$\overline{M}_n$	$\overline{M}_w/\overline{M}_n$	I_{eff} (%)
1	5.6×10^{-3}	—	100	10770	1.5	94
2	2.8×10^{-2}	—	100	2090	2.5	103
3	5.6×10^{-2}	—	100	1290	2.2	88
4	5.6×10^{-3}	5.6×10^{-3}	9	840	1.17	100
5	2.8×10^{-2}	2.8×10^{-2}	90	2090	1.1	93
6	5.6×10^{-2}	1.1×10^{-2}	100	1200	1.4	94

[a] $[BCl_3] = 2.6 \times 10^{-1}$ mol/L, [IB] = 1.0 mol/L, CH_3Cl diluent, $-30°C$ 30 min.

Evidently the common ion strongly affects the rate (it lowers it) and the molecular weight distributions (it narrows them). The reduction of the rate indicates a shift of ionic propagating species toward less-dissociated entities. Similarly the narrowing of the MWDs is interpreted to reflect a shift away from several kinds of ionic (but not free) propagating species toward a more uniform less-dissociated entity. We speculate that propagation (insertion) may involve several kinds of stretched $\overset{\delta}{C}{}^{\oplus} \cdots \overset{\delta}{O}{}^{\ominus}$ bonds (see above) and that the addition of the common ion drives the equilibrium connecting these toward a more uniform less-dissociated species.

Termination

While the living species may remain active for a considerable length of time, if termination is desired (e.g., for terminal functionalization) it could be brought about either by increasing the temperature above the decomposition temperature of the propagating ester $\cdot BCl_3$ complex or by the addition of strong nucleophiles, e.g., alcohols and amines. We have examined the structure of the end groups of PIBs obtained in living systems after the active ester $\cdot BCl_3$ complex was decomposed by heating to room temperature or by the addition of various amines, CH_3OH, etc. We have also carried out model experiments with the TMPOAc $\cdot BCl_3$ complex to corroborate the results obtained in polymerization experiments (see Experimental section for details).

Surprisingly all the polymerization experiments irrespective of the particular killing method employed yielded t-chloro end groups, i.e.,

$$R \sim PIB \sim CH_2 - \underset{\underset{CH_3}{|}}{\overset{\overset{CH_3}{|}}{C}} - CH_2 - \underset{\underset{CH_3}{|}}{\overset{\overset{CH_3}{|}}{C}} - Cl$$

where $R = (CH_3)_3C-$, $(CH_3)_3CCH_2C(CH_3)_2-$, or $C_6H_5(CH_3)_2C-$ depending on whether the initator was t-BuOAc, TMPOAc, or CuOAc, respectively. The analysis of the end groups has been carried out by ^{1}H-NMR spectroscopy by the use of well-established methods.[30]

The results of these polymerization experiments have been fully corroborated by model experiments in which the TMPOAc $\cdot BCl_3$ complex was decomposed with nucleophiles in the same manner that the polymerization experiments were "killed." According to ^{1}H-NMR analysis the product was invariably TMP-Cl [^{1}H-NMR δ (ppm) = 1.05 (methyl, 9H), 1.65 (methyl, 6H), 1.85 (methylene, 2H)]:

$$CH_3 - \underset{\underset{CH_3}{|}}{\overset{\overset{CH_3}{|}}{C}} - CH_2 - \underset{\underset{CH_3}{|}}{\overset{\overset{CH_3}{|}}{C}} - Cl$$

Additional model thermal decomposition experiments have been carried out with t-BuOAc $\cdot BCl_3$ and TMPOAc $\cdot BCl_3$ complexes mimicking the active

PIB chain end. Thus we have followed the disappearance of carbonyl stretching frequency at $\nu_{C=O} = 1570$ cm^{-1} characteristic of acetate $\cdot$ BCl$_3$ complexes and the appearance of the free C=O bond in acetyl chloride at $\nu_{C=O} = 1800$ cm^{-1} (acetyl chloride is one of the decomposition products of the above complexes) in CH$_2$Cl$_2$ at different temperatures. The complexes did not decompose at $-10°$C after 1 h; however, decomposition occurred at room temperature, and after 1 h about 50% of the complexes decomposed. The decomposition products were acetyl chloride and the corresponding alkyl chloride, i.e., t-BuCl and TMP-Cl, respectively. These findings are in line with the results of Gerrard et al.[18]

In line with these data and observations, inspection of molecular models provided some insight into the mechanism of forced termination. We speculate that termination could occur thermally or by strong nucleophiles by the following intramolecular S$_{N_i}$ process:

One of the Cl's of the BCl$_3$ is in the proximity to the propagating tertiary carbon, and either heating or nucleophilic attack on the complexed B center may induce the above rearrangement. Previous work on the decomposition of acetate $\cdot$ BCl$_3$ complexes[19] and our polymer end group and model studies indicate the formation of the expected tertiary alkyl chlorides together with the boron-containing species.

The synthesis of terminal tertiary chlorines by this simple route is a valuable new technique for the preparation of end-functional polymers. Future publications in this series will be concerned with the synthesis of various telechelic PIBs incorporating a variety of head, in-chain, and end structures by this technique.

Support by the Polymer Program of the NSF (Grant DMR-841617) is gratefully acknowledged.

References

1. W. M. Saltman, *Encyclopedia of Polymer Science and Technology*, John Wiley and Sons, New York, 1965, Vol. 2.

2. R. Milkovich, British Patent 1,000.090 (1965).

3. S. Penczek, P. Kubisa, and K. Matyjaszewski, Cationic Ring Opening Polymerization. Advances in Polymer Science, **68/69**, 1, 1985.

4. M. Szwarc, *Nature* **178**, 1168 (1956).

5. M. Szwarc, M. Levy, and R. Milkovich, *J. Am. Chem. Soc.* **78**, 2656 (1956).

6. D. C. Pepper, *J. Polym. Sci., Symp. No. 50*, 51 (1975).

7. T. Higashimura, M. Miyamoto, and M. Sawamoto, *Macromolecules* **18**, 611 (1985).

8. T. Higashimura, H. Teranishi, and M. Sawamoto, *Polymer J.* **12**, 383 (1980).

9. T. Higashimura and O. Kishiro, *Polymer J.* **9**, 87 (1977).

LIVING CARBOCATIONIC POLYMERIZATION. IV   1869

10. R. Faust, A. Fehérvári, and J. P. Kennedy, *J. Macromol. Sci. Chem.* **A18** 1209 (1982).

11. J. Puskás, G. Kaszás, J. P. Kennedy, T. Kelen, and F. Tüdös, *J. Macromol Sci. Chem.* 1229 (1982).

12. M. Sawamoto and J. P. Kennedy, *J. Macromol. Sci. Chem.* **A18**, 1275 (1982).

13. R. Faust and J. P. Kennedy, International Symposium on Recent Advances in Polyolefins, Abstract, Macr. 0029, ACS Meeting 1985, Chicago.

14. R. Faust and J. P. Kennedy, Symposium on Unconventional Mechanisms and Methods of Polymerization, Abstract, Poly. 13, ACS Meeting 1986, New York.

15. R. Faust and J. P. Kennedy, *Polym. Bull.* **15**, 317 (1986).

16. J. P. Kennedy and E. Maréchal, *Carbocationic Polymerization*, John Wiley-Interscience, New York, 1982.

17. N. Demarcay, *Bull. Soc. Chim. France*, **20**, 127 (1873).

18. M. F. Lappert, *J. Chem. Soc.* 817 (1961); *ibid.* 542 (1961).

19. W. Gerrard and M. A. Wheelans, *J. Chem. Soc.* 4196 (1956).

20. H. C. Brown and P. J. Geoghegan, *J. Org. Chem.* **35**, 1844 (1970).

21. C. G. Swain, T. E. C. Knee, and A. MacLachlan, *J. Am. Chem. Soc.* **82**, 6101 (1960).

22. J. P. Kennedy, S. Y. Huang, and S. C. Feinberg, *J. Polym. Sci.* **18**, 2801 (1977).

23. J. P. Kennedy and F. Y. Chen, *Polym. Bull.* **15**, 201 (1986).

24. J. P. Kennedy and A. Fehérvári, unpublished data (1985).

25. J. P. Kennedy and M. Hiza, *Polym. Bull.* **8**, 557 (1982).

26. G. Kaszás, M. Györ, and J. P. Kennedy, *J. Macromol. Sci. Chem.* **A18**, 1367 (1982–1983).

27. F. A. Mayo, R. A. Gregg, and M. S. Matheson, *J. Am. Chem. Soc.* **73**, 1691 (1951).

28. J. P. Kennedy and R. T. Chou, *J. Macromol. Sci. Chem.*, **A18**, 17 (1982).

29. O. W. Webster, W. R. Hertler, D. Y. Sogah, W. B. Farnham, and T. V. Rajan-Babu, *J. Am. Chem. Soc.* **105**, 5706 (1983).

30. A. Fehérvári and J. P. Kennedy, *J. Macromol. Sci. Chem.* **A15**, 215 (1981).

Received April 7, 1986.
Accepted September 4, 1986.

A Half-Century of the

JOURNAL OF

Polymer Science

Part B
POLYMER PHYSICS

Edited by
ERIC J. AMIS

COMMENTARY

Reflections on "Simple Presentation of Network Theory of Rubber, with a Discussion of Other Theories," Hubert M. James and Eugene Guth, *J. Polym. Sci.,* IV, 153 (1949)

A. S. LODGE

The Bannatek Co. Inc., Madison, Wisconsin 53705

Because of the difficulty of treating many-body interactions mathematically, analytical molecular theories of macroscopic properties of condensed phases are, in general, simply too hard to handle. One exception is the perfect ionic crystal, where the lattice regularity furnishes sufficient simplicity to enable a realistic theory to be developed. The only other exception of which I am aware is the lightly crosslinked amorphous elastomer above the glass transition temperature. In this case, the dominant contribution to the extra stress comes from network strands' thermal motion, which is much easier to handle than intermolecular forces. In the first instance, the main properties to be described are the low magnitude of the shear modulus, its increase with temperature, elastic recoil, and the forms of stress–strain relations for different types of finite strain. The novel features to be treated are the statistical mechanics of a single covalent-bonded network molecule, which extends throughout the macroscopic sample, and its thermal motion.

Kuhn,[1] Wall,[2] and Treloar[3] dealt with the thermal motion of a single strand whose ends were assumed to undergo the same affine motion as the points on the boundary of a macroscopic elastomer sample. They used the Gaussian function of components of the strand end-to-end vector to describe the configuration density. According to these early theories, the shear modulus has the correct order of magnitude and dependence on temperature. Flory and Rehner[4] took a small step away from a single-strand theories by considering a four-strand model.

None of these theories, however, can be regarded as giving an acceptable molecular explanation of elastic recoil. The remarkable phenomenon of elastic recoil involves an affine motion of the whole macroscopic sample, at least in the case of prime interest, namely, that in which the prior stress is homogeneous. One needs, therefore, a valid molecular model which will enable one to deduce that there is a global affine motion when the stress is made zero; no global affine motion assumption can be admitted. Moreover, it is essential to have a model which describes the interconnections of strands because unconnected strands would retract about unmoving mass centers and could not, therefore, generate any recoil in the macroscopic sample.

James and Guth were the first to consider the full molecular network; they restricted their affine displacement assumption to the sample boundary points. They proved that the average positions of network junctions shared the affine motion of the boundary points. They also showed that the network junctions fluctuated about their average positions with amplitudes comparable to those of the rest of the strand points, but that, nevertheless, for some but not all calculations, it was safe to ignore the junction fluctuations and treat the junctions as if they were located at their ensemble-averaged positions. They also abandoned the early theories' incompressibility assumption; to represent the combined effects of all the secondary force interactions between molecules, they simply added to the Helmholtz free energy a function of density and temperature alone. Although the form of this function was not calculated, its presence is essential: it determines the bulk modulus, and its omission leads to the un-

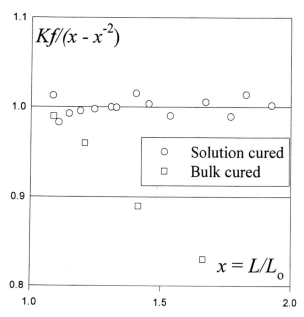

Figure 1. Elongation data are consistent with the Gaussian network theory of James and Guth for polyisoprene crosslinked in solution,[10] but not for polyisoprene crosslinked in the absence of solvent.[11] L, L_o denote filament lengths when the tension has values f, 0, respectively.

acceptable conclusion that the stress-free state is unstable.[5]

From the James and Guth assumptions, it follows in particular that, for a network of given connectivity at a given temperature, there is a unique stress-free state.[6] Thus, this main property characteristic of crosslinked elastomers is described by the network theory of James and Guth.

Furthermore, the James and Guth theory, with its extensions for describing swollen elastomers and stress–birefringence behavior, gives a fair-to-good description of a variety of properties; even the well-known disagreement with the form of the tension-length relation in simple elongation is absent when appropriate crosslinking methods are employed (Fig. 1).

It is also interesting that the early single-strand/network dichotomy finds an echo in present-day theories for polymeric liquids: the tube models consider a single polymer molecule which does not affect its surroundings and cannot therefore generate elastic recoil.[7] Recoil was not mentioned in the first tube theory monograph.[8] The temporary-junction network models, on the other hand, can describe recovery.[7,9]

REFERENCES AND NOTES

1. W. Kuhn, *Kolloid Z.*, **76**, 258 (1936).
2. F. T. Wall, *J. Chem. Phys.*, **10**, 132 (1942).
3. L. R. G. Treloar, *Trans. Faraday Soc.*, **39** (1943).
4. P. J. Flory and J. Rehner, Jr., *J. Chem. Phys.*, **11**, 512 (1943).
5. K. Freed, *J. Chem. Phys.*, **55**, 5588 (1971).
6. A. S. Lodge, *Proc. VIIth Int. Congr. Rheol.*, Klason and Kubat (eds.), Chalmers University of Technology, Gothenburg, 1976, p. 79.
7. A. S. Lodge, *Rheol. Acta*, **28**, 351 (1989).
8. M. Doi and S. F. Edwards, *Theory of Polymer Dynamics*, Oxford University Press, New York, 1986.
9. A. S. Lodge, *Kolloid Z.*, **171**, 46 (1960).
10. C. Price, G. Allen, F. de Candia, M. C. Kirkham, and A. Subramaniam, *Polymer*, 11, 486 (1970).
11. R. S. Rivlin and D. W. Saunders, *Philos. Trans. Roy. Soc. London, Sec. A*, **243**, 251 (1951).

PERSPECTIVE

Comments on "Simple Presentation of Network Theory of Rubber, with a Discussion of Other Theories," by Hubert M. James and Eugene Guth, *J. Polym. Sci.,* IV, 153 (1949)

FERENC HORKAY and GREGORY B. McKENNA

Polymers Division, NIST, Gaithersburg, Maryland 20899

It seems now to be generally agreed that soft rubberlike materials consist of long flexible molecules more or less completely linked into a coherent network by chemical bonds formed during cure.

This is how James and Guth began their 1949 paper[1] that summarized succinctly the important developments of what has now come to be referred to as the Phantom network model of rubber elasticity. Today we might add that "it seems now to be generally agreed that the crosslink junctions of rubber networks 'have a Brownian motion comparable to that of other elements of the network'. " Such was the importance of the James and Guth contributions[1-6] to our understanding of the behavior of crosslinked rubber networks.

The theory of rubber elasticity is of foremost importance to interpretation of mechanical and thermodynamic properties of polymers from a molecular point of view. Its origins mark the beginning of molecular theories of physical properties of macromolecules. To understand the deformation of rubbers, the modeling of the macromolecular chain was probably the most important idea. The rubber elasticity theory is based directly on the concept of randomly coiling chains that was theoretically described by Guth and Mark[7] and by Kuhn.[8] In the earlier theories of rubber the connectivity of the network chains was not taken into account explicitly. The theory of James and Guth[1-6] was a pioneering effort in polymer physics to deal with many-body problems (i.e., interdependent chain configurations) inherent

in the statistical description of permanently bonded network structures. They introduced the concept of fluctuations of the junctions about their mean (equilibrium) positions. The existence of such equilibrium positions describes the essential difference between crosslinked networks and uncrosslinked chains. James and Guth developed their model, which is now called the phantom model, in a rigorous, self-consistent way that has proven to be the basis for several subsequent attempts to treat similar problems.

During the same years a competing theory, called the affine model, was also developed by Wall and Flory.[9,10] The fundamental difference between the affine and the phantom models is that in the latter crosslinks experience thermal fluctuations (Brownian motion) about their mean positions, while in the former the crosslink positions are considered to be fixed. The affine model takes no account of thermal fluctuations while the phantom model allows unrestricted motion to the network junctions. In the latter case junction fluctuations are subject only to the constraint of fixed connectivity patterns around the crosslinks; interchain obstructions, entanglements, etc., are completely ignored. Flory's model appears to agree with experimental data for small deformations, whereas the James–Guth prediction is approached at large extensions. Neither model describes the non-constancy of the reduced stress.

The disagreement was not resolved until 1976, when Flory[11] presented a new version of the theory of rubber elasticity. The explicit postulate of Flory's new theory was that fluctuations are dependent upon the extent of network deformation: specifically fluctuations in the direction of molecular stretching increase with increasing deformation. Thus, the affine model gives way to the phantom model in going from zero to infinite deformation. Today the idea of fluctuations is widely accepted and the concept has been applied not just to the junctions but also to the entire chain.[12,13]

Although all network theories are built with specific assumptions based upon the motion of the junction points, until very recently there were no experiments that probed directly the crosslink dynamics. This situation changed with the advent of neutron spin echo spectroscopy. Neutron spin echo measurements[14,15] were used to directly determine the spatial extent and relaxation dynamics of crosslink fluctuations. It was found that, qualitatively, theoretical prediction and experimental results agree. The experimentally observed fluctuation ranges of the crosslinks are, however, smaller in the real network than in the phantom network, a result anticipated in the work of James and Guth.[1] This result unambiguously supports the original idea of junction fluctuations underlying the phantom network model of James and Guth.

While the resolution of still existing controversial issues in rubber elasticity (e.g., the entanglement problem) will have to wait for the accumulation of more experimental data on well defined model systems and probably new ideas, it is clear that the work of James and Guth has made, and continues to make, an important contribution to the advancement of polymer science in general and to the better understanding of the unique properties of elastomeric networks in particular. The *Journal of Polymer Science* has been a part of this continuing story.

REFERENCES AND NOTES

1. H. M. James and E. J. Guth, *J. Polym. Sci.,* **IV,** 153 (1949).
2. H. M. James and E. Guth, *J. Ind. Eng. Chem.,* **33,** 624 (1941).
3. H. M. James and E. Guth, *J. Chem. Phys.,* **11,** 455 (1943).
4. H. M. James and E. Guth, *J. Appl. Phys.,* **15,** 294 (1944).
5. H. M. James and E. Guth, *J. Chem. Phys.,* **15,** 669 (1947).
6. H. M. James and E. Guth, *J. Chem. Phys.,* **21,** 1039 (1953).
7. E. Guth and H. Mark, *Monatch.,* **65,** 93 (1934).
8. W. Kuhn, *Kolloid Z.,* **68,** 2 (1934).
9. P. J. Flory and J. Rehner, *J. Chem. Phys.,* **11,** 512 (1943).
10. F. T. Wall, *J. Chem. Phys.,* **10,** 485 (1942).
11. P. J. Flory, *Proc. Roy. Soc.,* **A351,** 51 (1976).
12. P. J. Flory and B. Erman, *Macromolecules,* **15,** 800 (1982).
13. B. Erman and L. Monnerie, *Macromolecules,* **22,** 3342 (1989).
14. R. Oeser, B. Ewen, D. Richter, and B. Farago, *Phys. Rev. Lett.,* **60,** 1041 (1988).
15. B. Ewen and D. Richter, in *Elastomeric Polymer Networks,* J. E. Mark and B. Erman (eds.), Prentice Hall, Englewood Cliffs, NJ, 1992, p. 220.

Journal of Polymer Science   Vol. IV, pp. 153-182, 1949

Simple Presentation of Network Theory of Rubber, with a Discussion of Other Theories

HUBERT M. JAMES, *Purdue University, Lafayette, Indiana, and*
EUGENE GUTH, *University of Notre Dame, Indiana*

It seems now to be generally agreed that soft rubberlike materials consist of long flexible molecules more or less completely linked into a coherent network by chemical bonds formed during cure. These bonds suppress, for the most part, the plasticity that the liquidlike mass of molecules would otherwise have, but leave the molecules free to take on a great many configurations of essentially the same energy under the influence of thermal agitation. The tendency of stretched rubber to retract is then understood as a kinetic phenomenon, like the tendency of a gas to expand: it is the tendency of a system to assume the form of maximum entropy when the internal energy is essentially independent to form.

There is no similar agreement as to what constitutes an adequate theoretical treatment of such materials. In particular, there exists a wide variety of formalisms for the derivation of the stress-strain relation of an ideal soft rubber. These are usually referred to as network theories, though the only treatment that actually deals with a general network of flexible molecular chains is that of the authors (1). Other treatments have been based on the consideration of individual elements or small groups of elements from networks, concerning the behavior of which special assumptions were made, or they have proceeded on the basis of general ideas that involve no reference whatever to the network structure of the material. The relation of these theories to the general network theory of rubber is the subject of the present paper.

As background for the discussion of other theories we shall first develop the theory of rubber, considered as a random network of long flexible molecules, in a particularly simple way.* The ease

---

*The treatment given here starts from the same basis as that of reference 1, but proceeds in a simpler way to a more restricted result-the derivation of the stress-strain relation. This analysis concentrates attention on the entropy, as in the custom of other writers on the subject; our earlier one shifted attention at the earliest possible moment from the entropy to the average forces exerted by the network segments, since this seemed to give a better picture of the detailed relations between network elements. The significance of our earlier analysis of the general rubber network has been widely overlooked, partly because its detailed character led to its being put in an Appendix, and partly because of its unconventional approach to the

153

154   HUBERT M. JAMES AND EUGENE GUTH

with which a network of general form can be treated will make it evident that there is little need to base network theories of rubber on the use of more special models. Next, we shall examine an idea that appears in many discussions of rubber which employ simplified models— the idea that the junctions of the rubber network can be treated as if they were fixed in space. We shall show that this picture of the situation is quite unrealistic: the junctions have a Brownian motion comparable to that of any portion of the intervening molecular segments. The common assumption to the contrary does not affect the results of some types of calculations, but it is inadmissible in the treatment of other problems. Finally, we shall show that the theory of Wall (2), which employs no special model, is based on postulates that are inconsistent with the network structure of rubber.

I. ELEMENTARY PRESENTATION OF NETWORK THEORY OF RUBBER

We start the formulation of our picture of rubber by the consideration of milled rubber. All rubbers—natural or synthetic—are composed of very long flexible molecules. When raw rubber is milled, such structure as formerly existed in it is broken down, and the milled rubber becomes essentially a liquidlike mass of flexible rubber molecules. This mass is highly viscous, because the tendency of the long flexible molecules to become tangled with each other impedes their relative motion, but, as with any liquid, there is no definite form to which it will return when external forces are removed.

The process of cure introduces into this mass the elements of structure which hold the molecules in fixed relations to each other and suppress the plasticity. It consists in the introduction of strong and definite bonds between the long molecules which eventually link them into a coherent network extending throughout the whole mass; it is the definite structure of this network that determines the definite form of the material under zero external force. This network will be highly irregular, being formed by the introduction of bonds essentially at random, but on the average it will be homogeneous and isotropic when the material is unstretched. Not all of the rubber molecules will be actively involved in it. There will be many side chains and loose ends, as well as material not bound into the network, part of it extractable by solvents, part consisting of snarls inextricably entangled with the network.

In lightly cured materials there will be relatively few of these bonds introduced. On the average, each molecule will be involved in only a few bonds; a relatively small portion of each molecule will be involved in the formation of intermolecular bonds. Only near these bonds will the molecules be brought into definite relations to their

---

subject. In particular, we must call attention again to the fact that the simple three-dimensional network employed for purposes of visualization in the body of reference 1 was introduced only after it was proved to be equivalent for the purposes at hand to a network of arbitrary form. This equivalence also follows easily from the treatment in Section I of the present paper.

neighbors. Over most of their great length the molecules will exert the same forces on, and be in the same relation to, adjacent molecules as in the liquidlike milled rubber. In particular, the Brownian motion of the molecules will not be suppressed. They will move past each other and take on a great variety of configurations, just as they would if immersed in a foreign liquid—the only restraints on their Brownian motion being those impressed by the relatively few fixed bonds between the molecules.

We shall consider a soft rubberlike substance, then, as a liquidlike mass of flexible molecules subject to the permanent constraints implied by their union into a coherent network by bonds formed at relatively few points. It is the presence of these bonds—which tend to control the form of the material—that differentiates rubberlike materials from liquids. It is the small number of these bonds—and their weak control of the form of the material through the entropy rather than through the internal energy—that differentiates rubberlike materials from ordinary solids.

We shall discuss the behavior of rubber by consideration of two thermodynamic functions: the internal energy U and the entropy S. From these, one can determine the force Z tending to change a dimension L of the system, as:

$$Z(L,T) = (\partial U/\partial L)_T - T(\partial S/\partial L)_T \qquad (1)$$

where the derivatives are to be taken with the perpendicular dimensions constant (cf. equations 18 and 19).

Aside from a constant contribution due to the formation of intermolecular bonds, the internal energy of a soft, uncrystallized, unstretched rubber is the same as that of a liquid of rubber molecules. We shall assume that the internal energy of rubber, like that of any liquid, will not depend at all on the form of the material, but only on its volume and temperature. In doing so, we shall neglect the change in potential energy of the molecules when they are bent or extended, up to the point where they are completely straightened out and further extension will involve stretching of valence bonds in the molecule; in other words, we treat the molecules as being perfectly flexible. We also neglect changes in the internal energy due to the increasing alignment of the molecules as the material is stretched. This is permissible only as a first approximation, and for not too great extensions of the material.* We write, then, as for a liquid:

$$U = U(V,T) \qquad (2)$$

U has a sharp minimum with respect to variation of V at the normal volume of the material; the volume of the material under given external forces is largely determined by U. The exact form of the function U (V,T) will depend on the forces with which the molecules interact, and will thus depend, for instance, on the number and character of the side chains and polar structures which may be present in

---

*It is to be emphasized that the internal energy depends strongly on the form of the material when crystallization takes place, as when the material is subject to large extension. We shall not consider such cases here.

the molecules. Such features of the structure, which vary from one rubberlike material to the next, will in effect determine the compressibility and the coefficient of thermal expansion for the material. For this treatment of rubberlike elasticity it is not necessary to consider these intermolecular forces in detail; they enter only through the compressibility and coefficient of thermal expansion, which appear in the theory as observable constants.

As with gases and liquids, the entropy of such a system of flexible chains has, in statistical terms, a relatively simple form. It is characteristic of these systems that there are accessible to the components an enormous number of configurations, all with essentially the same low potential energy. Configurations of higher potential energy are also possible, as when molecular chains are stretched or the separations of atoms are otherwise changed, but these occur with relatively small probability. Under these circumstances the entropy can be expressed, with sufficiently good approximation, as the sum of two parts. The first, S_1, associated with the kinetic energy of the network or the thermal capacity of the material, is primarily a function of the temperature, but may depend slightly on the volume. The second is related to the number, C, of low energy configurations accessible to the system:

$$S_2 = k \ln C \tag{3}$$

The number of configurations accessible to the network will depend upon the external form of the material; C, and thus S_2, will be a function of the form of the material, but will be independent of its temperature. For a rectangular parallelepiped of rubber, with edges of lengths L_x, L_y, L_z:

$$S = S_1(V,T) + k \ln C(L_x, L_y, L_z) \tag{4}$$

The first of these terms will need no detailed discussion here.

In enumerating the configurations possible for a network, and evaluating the second term in the entropy, some simplification of the physical picture is unavoidable. Any simplified model must, however, give an adequate representation of the following features of the real system: (1) the flexibility of the molecules, which makes it possible for each constituent segment of the network to take on a great number of configurations; (2) the constraints implied by the union of the chains into a network, which make it impossible for the chains to take on all configurations independently; and (3) the volume-filling property of the molecular chains, which prevents any two of them from occupying the same region in space simultaneously.

In the mathematical treatment of the system, the flexibility of the molecules enters through the functions that give the number of configurations possible for each individual section of chain when its ends are fixed at various separations. These functions we shall call the configuration functions for the chain segments. Now, whatever the structure of a flexible chain, its configuration function may be taken as of Gaussian form:

$$c(r) = C_1 e^{-C_2 r^2} \tag{5}$$

r being the separation of the ends, under just two conditions: (1) the chain must be long as compared to any part of it with appreciable stiffness, and (2) only separations of the ends which are sufficiently small compared to the total length of the chain shall be considered (3). In the present discussion we shall restrict our attention to lightly cured materials in which all chain segments between junctions in the network are long, and shall consider only moderate extensions of the material, such that the segments are, on the average, not highly extended. Accordingly, without regard for any special characteristics of the molecular chains, we can assume with good approximation that every segment of the network has a configuration function of Gaussian form.

The constant C_2 in the configuration function will vary with character and length of the chain. It is often convenient to characterize each chain not by the constant C_2, but by a parameter N defined by:

$$C_2 = \frac{3}{2Nl^2} \qquad (6)$$

writing:

$$c(r) = C_1 \exp \left\{ - \frac{3}{2Nl^2} r^2 \right\} \qquad (7)$$

This expression is of the form of the configuration function, for small extensions, of a chain of N independent links of fixed length l (3a). Without in any way restricting the generality of our discussion, we can apply the same formalism in dealing with other types of flexible chain, such as occur in actual rubbers. The length l is a parameter characteristic of a material and N may vary from segment to segment.

In calculating the number of configurations possible for the molecular chains in a piece of rubber, the essential constraints are those implied by the linking of these chains into a network. Of these we shall take full account.

In addition, there are the constraints, which we shall term steric hindrances, that arise from the fact that two molecules cannot occupy the same volume in space or pass through each other. This volume-filling property of the molecules makes itself felt in its control of the volume of the material, largely through the internal energy. It also affects the entropy by eliminating from consideration all configurations of the network in which two portions of a chain occupy the same volume. These steric hindrances would have a complicated effect in a detailed enumeration of the configurations possible for the network, and it is necessary to neglect them. Justification of this approximation is an important step in the development of the theory of rubberlike materials. The argument required is given in Appendix A to avoid undue interruption of the main argument at this point. Fortunately, the result of the discussion is simple: in lightly cured materials, not overly stretched, the steric hindrances change the number of configurations possible for the network by a constant factor. This factor, independent of the form of the material, does not affect the forces exerted by the material, and can be neglected here.

To summarize, in computing the entropy of rubber we shall consider it as an arbitrary network of chains with Gaussian configuration functions, subject to all the constraints implied by their union into a coherent network, but free from other constraints. Such a network we shall call a Gaussian network.

We now define certain terms needed for the precise statement

158   HUBERT M. JAMES AND EUGENE GUTH

of our methods and results. These are illustrated in Figure 1.

By the "fixed points" of the network we mean the points at which it is immobilized by external constraints. It is at these points that the network exerts forces on the external system, and, conversely, that the external system exerts forces on the network. The positions of the fixed points will determine the external form of the material.

By the active portion of the network we mean that portion of the network that stretches between the fixed points and contributes to the forces exerted by the network on these points. A portion of chain is part of the active network if and only if it is part of some non-retracing path along the chains from one fixed point to another. The rest of the material may be described as unattached material and loose ends, the latter category including perhaps quite complicated chain structures attached to the network at a single point.

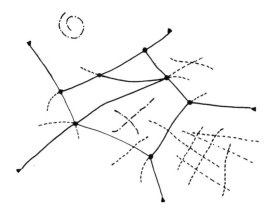

Fig. 1. Terminology for description of network. The segments are to be considered as bonded together wherever they cross. Segments of active network (solid line); loose ends (broken line); unattached material (−·−·); fixed points (▲); junctions of active network (●).

The term "junction" typically includes all points of bonding between molecules; a "segment" of the network is a portion of a molecular chain that extends between two adjacent junctions of the network. In considering the active network alone, we shall consider as junctions only the points of union of segments of the active network; we exclude from consideration, for instance, the points at which loose ends are bonded to a segment of the active network.

Now the essential characteristic of the active part of the network, as we have defined it, is that it contributes to the entropy a term that depends upon the positions of the fixed points, that is, upon the external form of the material. In the case of a system of independent flexible chains, the number of configurations possible for the system as a whole is the product of the number of configurations possible for each of the independent parts. The entropy, essentially the logarithm of this number, will be, correspondingly, a sum of contributions from the independent molecules, and will be independent of the form of the material unless each molecule is somehow subjected to a constraint depending on the form of the material. In the case of a network struc-

NETWORK THEORY OF RUBBER   **159**

ture, the number of configurations possible for each of the loose ends and for each portion of unattached material will be independent of the positions of the fixed points. The number of configurations possible for the system as a whole will involve as factors each of these numbers, together with the number of configurations possible for the active network, which does depend on the position of the fixed points. The entropy, correspondingly, is the sum of contributions from the loose ends and the unattached material, which will be independent of the form of the material, and a contribution from the active network, which will depend on the form. As is evident from equation 1, it is only the latter part of the entropy that will contribute to the forces exerted by the system against external constraints. In computing such forces it is thus possible to consider the network as stripped down to its active part.

The number of configurations possible is easily computed for an arbitrary active network. First of all, one can immediately write the number of configurations consistent with any given set of positions of the fixed points and junctions of the network. Let the positions of these points be indicated by Cartesian coordinates $(x_\tau, y_\tau, z_\tau,)$ or the vector r_τ. When it is necessary to distinguish junctions from fixed points we shall use numerical or letter subscripts for the former, and subscripts $\alpha, \beta, \ldots$ for the latter. Let the chain segment joining the points (x_τ, y_τ, z_τ) and (x_ν, y_ν, z_ν) be one of $N_{\tau\nu}$ links, with configuration function:

$$c(|r_\tau - r_\nu|) = C_{\tau\nu} \exp\{-\frac{3}{2N_{\tau\nu}1^2}[(x_\tau - x_\nu)^2 + (y_\tau - y_\nu)^2 + (z_\tau - z_\nu)^2]\} \qquad (8)$$

Given the positions of the junctions and fixed points, the chain segments are otherwise independent, in the present approximation of neglected steric hindrances. The number of configurations possible for the network is then simply the product of the number of configurations possible for the individual segments:

$$C(r_\alpha, r_\beta, \ldots r_1, r_2, \ldots) = C_0 \exp\{-\frac{3}{21^2}\sum_\tau \sum_\nu \frac{1}{N_{\tau\nu}}[(x_\tau - x_\nu)^2 + (y_\tau - y_\nu)^2 + (z_\tau - z_\nu)^2]\} \quad (9)$$

the sum being over all pairs of points (fixed points or junctions) that are connected by chain segments.

To determine the total number of configurations of the network consistent with given positions of the fixed points only, we now have to sum $C(r_\alpha, r_\beta \ldots r_1, r_2 \ldots)$ over all possible sets of positions of the junctions. In the approximation in which all configuration functions are of Gaussian form, each junction can take on any position in space, whatever the positions of other junctions.* The total number of configurations as a function of the positions of the fixed points is then:

$$\mathfrak{C}(r_\alpha, r_\beta, \cdots) = \int dx_1 \int dy_1 \int dz_1 \int dx_2 \int dy_2 \int dz_2 \cdots C(r_\alpha, r_\beta, \cdots r_1, r_2, \cdots) \quad (10)$$

the integrations being over the coordinates of all network junctions.

Complete evaluation of this integral is possible (4) but not necessary for the purposes of this argument. We note that the integral

---

*In the actual network the separation of adjacent junctions is limited by the length of the chain segment connecting them. With the equivalent Gaussian network, greater separations are possible, but the "configurations" possible for these separations are so few that it makes no difference whether we count them or not.

can be expressed as the product of three factors involving, respectively, the x, y, and z coordinates. The x factor is:

$$\int dx_1 \int dx_2 \int dx_3 \cdots \exp\{-\frac{3}{2l^2}\sum_{\tau>\nu}\sum \frac{1}{N_{\tau\nu}}(x_\tau-x_\nu)^2\}$$

The effect of the integrations on the form of the integrand is easily ascertained. If we single out the coordinate x_1 for special attention, we can rewrite the integrand, completing the square of the terms in x_1 in the exponent in the familiar way, to obtain:

$$\exp\{-\frac{3}{2l^2}(\sum_\tau \frac{1}{N_1\tau})[x_1-L(x_a,\cdots x_2,\cdots)]^2-\frac{3}{2l^2}Q(x_a,\cdots x_2,\cdots)\}$$

where L is a linear form and Q a quadratic form in all x's but x_1. Integration over x_1 affects only the first factor, replacing it by a constant multiplier; the multiple integral becomes:

$$k_1\int dx_2 \int dx_3 \cdots \exp \{-\frac{3}{2l^2}Q(x_a,\cdots x_2, x_3,\cdots)\}$$

Similarly, integration over x_2, x_3,... eliminates the variable in question from the integrand, but leaves this as an exponential of a quadratic form in the remaining variables. After integration over the coordinates of all junctions there then remains an exponential of a quadratic form in the coordinates of the fixed points only, which can be written as:

$$C_0\exp \{-\frac{3}{2l^2}\sum_{a>\beta}\sum \frac{1}{N'_{a\beta}}(x_a-x_\beta)^2\}$$

the sum being over all pairs of fixed points. The $N'_{\alpha\beta}$ are constants, having the character of the number of links in a chain, which in general may depend on all the constants $N_{\tau\nu}$ (8). The y and z integrals are of exactly the same form, and the results of the integrations are the same, x being replaced by y or z throughout. Thus it is evident that:

$$\mathfrak{C}(r_a, r_\beta,\ldots) = C_1\exp \{-\frac{3}{2l^2}\sum_{a>\beta}\sum \frac{1}{N'_{a\beta}}[(x_a-x_\beta)^2 + (y_a-y_\beta)^2 + (z_a-z_\beta)^2]\} \tag{11}$$

The form of this result is independent of the character of the network, which makes itself felt only through the constants C_1 and $N'_{\alpha\beta}$. It will be noted that this expression is the same as that which would be found if the network consisted simply of independent chains of $N'_{\alpha\beta}$ links running between the fixed points α and β. As concerns entropy and entropy forces, then, an arbitrary Gaussian network can be replaced by a corresponding set of independent Gaussian chains running between each pair of fixed points. This simplified equivalent network is conveniently employed in certain types of computation.

The special case to which we shall now devote our attention is that of a piece of rubber which at a standard temperature and in the absence of external forces is a unit cube, but which is subjected to uniform stretching parallel to the edges until these have lengths L_x, L_y, L_z. If all external forces are applied normal to the surfaces, the fixed points will lie on these surfaces in positions which change in proportion to the corresponding dimensions of the material. If three of the faces of the parallelepiped are taken as the coordinate planes:

NETWORK THEORY OF RUBBER 161

$$x_a = x_a^{(0)} L_x \qquad y_a = y_a^{(0)} L_y \qquad z_a = z_a^{(0)} L_z \tag{12}$$

$x_\alpha^{(0)}$, $y_\alpha^{(0)}$, $z_\alpha^{(0)}$ being the coordinates of the αth fixed point when the material is unstretched and at the standard temperature.* The number of configurations possible for the network, now a function of L_x, L_y, L_z, then becomes simply:

$$\mathfrak{C}(L_x, L_y, L_z) = K_1 \exp \{-(K_x L_x^2 + K_y L_y^2 + K_z L_z^2)\} \tag{13}$$

where:
$$K_x = \frac{3}{2l^2} \sum_{a > \beta} \sum \frac{1}{N_{a\beta}'} (x_a^{(0)} - x_\beta^{(0)})^2 \tag{14}$$

and K_y and K_z have similar forms. If the rubber is isotropic on the average, there will be nothing to distinguish the three coordinates directions, and one will have $K_x = K_y = K_z = K/2$. We then write:

$$\mathfrak{C}(L_x, L_y, L_z) = K_1 \exp \{-\frac{K}{2}(L_x^2 + L_y^2 + L_z^2)\} \tag{15}$$

This equation, and more generally equation 11, states a very simple and interesting result: when an arbitrary network of Gaussian chains is subjected to uniform stretching in three coordinate directions, by corresponding displacement of its fixed points, the number of configurations possible for the system as a whole is a Gaussian function of its external dimensions. This result is not at all affected by the presence of loose ends or unattached material in the system, nor will these change K or the entropy forces exerted by the network.

Equations 4 and 15 yield the following expression for the entropy of a rectangular parallelepiped of rubber:

$$S = S_1(V, T) - \frac{kK}{2}(L_x^2 + L_y^2 + L_z^2) \tag{16}$$

where the constant $k \ln K_1$ has been absorbed into S_1.

Having arrived at expressions for U and S, we can derive the stress–strain relation for an ideal rubberlike material. We consider as before a rectangular parallelepiped of volume:

$$V = L_x L_y L_z \tag{17}$$

originally a unit cube, subject to total outward forces X, Y, and Z on the faces perpendicular to the three coordinate directions. The work done by the system when its dimensions undergo small changes is:

$$dW = -X \, dL_x - Y \, dL_y - Z \, dL_z \tag{18a}$$

the change in internal energy is:

$$dU = T \, dS + X \, dL_x + Y \, dL_y + Z \, dL_z \tag{18b}$$

and the change in the free energy ($F = U - TS$) is:

$$dF = -S \, dT + X \, dL_x + Y \, dL_y + Z \, dL_z \tag{18c}$$

---

*These relations can be derived as consequences of the assumption that all forces are normal to the surfaces. In order to keep the argument here as simple as possible one may consider these equations as describing one possible type of surface constraint.

162   H U B E R T   M.   J A M E S   A N D   E U G E N E   G U T H

Thus:

$$X = \left(\frac{\partial F}{\partial L_x}\right)_{L_y, L_z, T} = \left(\frac{\partial U}{\partial L_x}\right)_{L_y, L_z, T} - T\left(\frac{\partial S}{\partial L_x}\right)_{L_y, L_z, T} \qquad (19)$$

Similar equations hold for **Y** and **Z**. From equations 2 and 17 it follows that:

$$\left(\frac{\partial U}{\partial L_x}\right)_{L_y, L_z, T} = \left(\frac{\partial V}{\partial L_x}\right)_{L_y, L_z, T}\left(\frac{\partial U}{\partial V}\right)_T = L_y L_z \left(\frac{\partial U}{\partial V}\right)_T$$

Similarly:

$$\left(\frac{\partial S_1}{\partial L_x}\right)_{L_y, L_z, T} = L_y L_z \left(\frac{\partial S_1}{\partial V}\right)_T$$

By use of equation 16, equation 19 then becomes:

$$X = -\left\{-\left(\frac{\partial U}{\partial V}\right)_T + T\left(\frac{\partial S_1}{\partial V}\right)_T\right\}L_y L_z - kTKL_x \qquad (20)$$

Now the quantity in brackets involves only terms which occur also in the internal energy and entropy of the liquidlike uncured material; it is in fact just the pressure in such a liquid:*

$$P = -\left(\frac{\partial U}{\partial V}\right)_T + T\left(\frac{\partial S_1}{\partial V}\right)_T \qquad (21)$$

We then write:

$$X = -PL_y L_z + KkTL_x \qquad (22a)$$

$$Y = -PL_z L_x + KkTL_y \qquad (22b)$$

$$Z = -PL_x L_t + KkTL_x \qquad (22c)$$

These equations express the forces exerted by the material, -**X**, -**Y**, -**Z**, as sums of two parts—the outward forces exerted by a hydrostatic pressure P acting on the respective face areas, and inward pulls (entropy forces exerted by the network) proportional to the extension of the material in the given direction.**

In the case of unilateral stretch in the z direction of material free from external pressure, one has $L_x = L_y$, X = Y = 0. Then, from equations 22a and 17, one has:

$$P = P_n = KkT/L_z \qquad (23)$$

Thus, there is a hydrostatic pressure inside the material even when it is subject to no external pressure. This is due to the tendency of the network to take on configurations of higher entropy and smaller volume; it is an internal pressure arising from the tendency of the network to contract. Equation 22c, by use of equations 17 and 23, then becomes:

$$Z = KkT\left[L_z - \frac{V}{L_z^2}\right] \qquad (24)$$

When an external pressure P_e is applied to the system, we

---

*The difference in sign in equations 19 and 21 is due to the fact that P gives a force tending to decrease **V**, whereas **X** is a force tending to increase L_x.

**The origin of these forces has been discussed from the point of view of kinetic theory in Reference 1.

may divide the force on the x-faces of the cuboid into two parts, one of hydrostatic origin, $-P_e L_y L_z$, and one of other origin, X':

$$X = X' - P_e L_y L_z$$

Equation 22a then becomes:

$$X' = - (P - P_e) L_y L_z + KkTL_x \qquad (22a')$$

Similar equations hold for Y' and Z'. If the material is subject to unilateral stretch under external pressure ($X' = Y' = 0$) one has as before:

$$P - P_e = P_n = KkT/L_z \qquad (23')$$

and:

$$Z' = KkT [L_z - V/L_z^2] \qquad (24')$$

Thus, equation 23 gives the internal pressure due to the network, and equation 24 gives the force required to bring the cuboid to length L_z, whether or not the material is subject to a hydrostatic pressure in addition. This result is valid whatever the change in volume of the material on stretching, or on change in temperature or hydrostatic pressure. As a first approximation one may write:

$$V = V_0 [1 + \alpha (T - T_0) - \beta P] \qquad (25)$$

or:

$$V = V_0 [1 + \alpha (T - T_0) - \beta (P_e + P_n)]$$

The compressibility of rubber is of the order of that of typical liquids, and the change of V with P_n during stretch is unimportant. On the other hand, the temperature dependence of the volume is of importance for the understanding of the thermoelastic properties of rubber.

Equation 24 gives a form of the stress–strain relation that is characteristic of molecular networks, so long as they can be treated as Gaussian. In particular, it is independent of the structure of the network. The magnitude of the forces, however, does depend on the structure of the network; it is changed, for instance, if the structure is modified by further cure of the material. To determine the constant K requires specification of the network structure, either in detail or in some statistical sense. The most realistic approach to the determination of the constant K or the rigidity of the material, as it depends on the factors which describe the state of cure of the material, seems to lie in a consideration of the process by which physically occurring networks are built up as cure proceeds. We have developed the theory of this process elsewhere (5), and have there discussed fully the relation of this treatment of the problem to that employed by other authors. Comments on this point will also be found in following sections.

II. MODIFICATION OF THEORY BY ASSUMPTION THAT NETWORK JUNCTIONS ARE FIXED IN SPACE

In the theoretical treatment of so complex a system as a piece of rubber it is always necessary to replace the physical system by a simplified model. In the preceding section we have discussed a model of rubber which is idealized in three principal ways: it is assumed (a) that the internal energy is independent of the form of the material, (b) that each segment of the network has a Gaussian configuration func-

164 HUBERT M. JAMES AND EUGENE GUTH

tion, and (c) that the steric hindrances between the network segments change the entropy only by an additive constant (see Appendix A). On the other hand, full account is taken of the fact that the molecular seg—ments are linked into a network, without its being necessary to assume that this network has any special form.

Other theories of rubber that employ definite models of the system have avoided consideration of the rubber network as a whole by introducing additional assumptions. The resulting simplifications of the model all depend on the more or less extensive use of the idea (explicit or implied) that the junctions of the network can be considered to occupy fixed positions in space–positions which change in some definite way as the material is deformed.

The idea that the network junctions occupy fixed positions in space has been most concretely expressed by Kuhn. His earlier papers (6) deal with a model of rubber which, in effect, consists of flexible molecular chains with fixed ends, mid-points, and quarter-points. In his more recent papers (7,8) he pictures bulk rubber as a network of molecular chains, but states (7) that the points at which the chains are linked together to form the network thereby "become fixed and are practically excluded from taking part in the Brownian motion... Only the chain elements lying between these knots can move freely to some extent." On this assumption, his earlier formalism is applicable to the new model without change, the fixed points of the molecular chains in his earlier theory being identified with the network junctions, or with other points where equivalent constraints are effective (7).

Flory and Rehner (9) have employed the same idea as a simplifying assumption, without insisting on its close correspondence to physical reality. They consider a portion of a network, or "cell," consisting of four molecular segments joined to each other at one end and with the other ends fixed at the four vertices of a regular tetrahedron. Thus, they consider a central junction that is not fixed in position, and four junctions that are fixed at the vertices of the tetrahedron. The idea of fixed network junctions is also implicit in the work of Treloar (10), who has modified Kuhn's formalism by treating only the end points of the chains as fixed.

In Section III we shall show that it is very far from true that the junctions are fixed in space; in fact, they have a Brownian motion that is almost as extensive as that of other elements of the network. On the other hand, we shall there give arguments which indicate that one can nevertheless compute the entropy correctly on the basis of this assumption. In the present section we shall therefore carry through such a derivation, one which is simpler than those usually given, and which seems to us to make clearer the essential ideas of the calculation.

If the junctions of the network are essentially fixed, each segment of the network will have a fixed extension determined by the external form of the material alone. If, as in Section I, we consider a unit cube of rubber stretched into a cuboid with edges $L_x, L_y,$ and $L_z,$ then the extensions of the $\tau\nu$ segment will be:

$$r_{\tau\nu}(L_x, L_y, L_z) = \{(x_\tau - x_\nu)^2 + (y_\tau - y_\nu)^2 + (z_\tau - z_\nu)^2\}^{1/2} \qquad (26)$$

If steric hindrances between the segment are neglected, these segments become effectively independent systems: the number of configurations possible for the whole network can be computed simply as the product of the numbers of configurations which can be assumed by the component segments:

$$\mathfrak{C}(L_x, L_y, L_z) = C_0 \prod_{\tau > \nu} \prod \exp \left\{ -\frac{3}{2 l^2 N_{\tau \nu}} r_{\tau \nu}^2 \right\} \qquad (27)$$

The entropy of the system is then:

$$S(L_x, L_y, L_z) = S_1(V, T) + k \ln \mathfrak{C}(L_x, L_y, L_z)$$

$$= S_1(V, T) + k \ln C_0 - \frac{3k}{2 l^2} \sum_{\tau > \nu} \sum \frac{1}{N_{\tau \nu}} r_{\tau \nu}^2 (L_x, L_y, L_z) \qquad (28)$$

where the last and most important term is the sum of contributions from the individual segments, as though they were independent. The next essential assumption in the calculation, common to all theories in which the junctions are treated as fixed in space, is that the coordinates of all junctions change according to the law:

$$x_\tau = x_\tau^{(o)} L_x \qquad\qquad y_\tau = y_\tau^{(o)} L_y \qquad\qquad z_\tau = z_\tau^{(o)} L_z \qquad (29)$$

when the material is deformed. In Section I, similar equations described the motion of the points of the network that were subject to external constraint; here, the assumption applies to all junctions, however remote from external constraints. By equations 26 and 29, equation 28 becomes:

$$S(L_x, L_y, L_z) = S_1(V, T) - k \{ K_x L_x^2 + K_y L_y^2 + K_z L_z^2 \} \qquad (30)$$

where:

$$K_x = \frac{3}{2 l^2} \sum_{\tau > \nu} \sum \frac{1}{N_{\tau \nu}} (x_\tau^{(o)} - x_\nu^{(o)})^2 \qquad (31)$$

K_y and K_z have similar forms. The constant $k \ln C_0$ has been absorbed into S_1. If the undeformed material is isotropic on the average, then $K_x = K_y = K_z = K/2$, and one has again equation 16.

$$S(L_x, L_y, L_z) = S_1(V, T) - \frac{kK}{2} (L_x^2 + L_y^2 + L_z^2) \qquad (16)$$

To establish a definite value of K it is necessary to introduce further assumptions into the theory. It follows from equation 31 and the assumption of isotropy that:

$$K = \frac{2}{3}(K_x + K_y + K_z)$$

$$= \frac{1}{l^2} \sum_{\tau > \nu} \sum \frac{1}{N_{\tau \nu}} \{ (x_\tau^{(o)} - x_\nu^{(o)})^2 + (y_\tau^{(o)} - y_\nu^{(o)})^2 + (z_\tau^{(o)} - z_\nu^{(o)})^2 \} \qquad (32)$$

$$= \frac{1}{l^2} \sum_{\tau > \nu} \sum [r_{\tau \nu}^{(o)}]^2 / N_{\tau \nu}$$

where $r_{\tau \nu}^{(o)}$ is the extension of the $\tau \nu$ segment when the material is unstretched. Here K appears as a sum of contributions from the component segments of the network. If we define:

$$\lambda_{\tau \nu}^{(o)} = r_{\tau \nu}^{(o)} / l N_{\tau \nu} \qquad (33)$$

the fractional extension of the $\tau \nu$ segment in the unstretched material, we can also write:

$$K = \sum_{\tau > \nu} \sum N_{\tau \nu} [\lambda_{\tau \nu}^{(o)}]^2 \qquad (34)$$

It is thus evident that K does not depend on the detailed structure of the network, but only on the number of segments with a given number of links, and on their fractional extensions when the material is unstretched. If F(N)dN is the number of segments per unit volume for which the number of links lies between N and N + dN, and $G(N,\lambda)\,d\lambda$ is the fraction of these segments which have fractional extensions between λ and $\lambda + d\lambda$, then:

$$K = \int_0^\infty dN \int_0^1 d\lambda \quad F(N)\; G(N,\lambda)\; N\lambda^2 \tag{35}$$

Theories which assume fixed network junctions usually include some additional hypotheses which serve to fix the distribution functions $F(N)$ and $G(N,\lambda)$. The most common assumption, originally due to Wall (2), is that the extensions of the molecules have the same distribution as would be found in a system of unconstrained molecules subject to thermal agitation. From this assumption it follows that:

$$G(N,\lambda) = \frac{4}{\pi^{1/2}}\left(\frac{3N}{2}\right)^{3/2} \lambda^2 \; \exp\{-\frac{3}{2}N\lambda^2\} \tag{36}$$

On carrying out the integration in equation 35 one finds K = G, the total number of segments per unit volume in the network, whatever the form of F(N). Other assumptions lead to different forms of $G(N,\lambda)$, and to other values of K. The significance of this approach to the determination of K, and of the rigidity of a material, is discussed at length in Section II of Reference 5.

III. ARE NETWORK JUNCTIONS REALLY FIXED IN SPACE?

It now becomes necessary to examine the justification of Kuhn's assumption that the network junctions occupy effectively fixed positions in space. Kuhn gives no detailed argument in support of this idea; his acceptance of it seems to be based on a qualitative picture of the situation. We shall now show that this assumption is very unrealistic—that the junctions, in fact, have a Brownian motion comparable to that of other elements of the network.

The coupling together of two elements of a network to form a new junction of the network does decrease the extent of the Brownian motion of these elements. This can be discussed quantitatively as follows.

It has been shown by James (4) that the number of configurations of an arbitrary Gaussian network consistent with element i having coordinates x_i, y_i, and z_i is:

$$\mathfrak{C}_i(x_i, y_i, z_i) = c_1 \exp \{-\frac{3}{2l^2\eta_i}\;[(x_i-x_{oi})^2 + (y_i-y_{oi})^2 + (z_i-z_{oi})^2]\} \tag{37}$$

where x_{oi}, y_{oi}, z_{oi}, and η_i are constants. The probability that the Brownian motion of the network will bring element i to the point (x_i, y_i, z_i) is proportional to $\mathfrak{C}_i$; if we write:

$$P_i(x_i, y_i, z_i) = \{3/2\pi l^2 \eta_i\}^{3/2} \exp \{-(3/2l^2\eta_i)[(x_i-x_{oi})^2 + (y_i-y_{oi})^2 + (z_i-z_{oi})^2]\} \tag{38}$$

then the probability that element i will lie in a volume dV about the point x_i, y_i, z_i is:

$$dP = P_i (x_i, y_i, z_i) \; dV \tag{39}$$

Obviously, then, x_{0i}, y_{0i}, z_{0i} are the most probable values of these co-ordinates. The rms distance of the element from its most probable position is $\eta_i^{1/2}l$; thus $\eta_i^{1/2}$ will serve as a measure of the extent of the Brownian motion of the element.

Let j be a second element of the network, with a mean deviation $\eta_j^{1/2}l$ from a most probable position (x_{0j}, y_{0j}, z_{0j}); its probability distribution will have the form of equation 38, except for a change in subscripts. Now let elements i and j be linked together permanently by a bond formed when they happen to lie near each other. Together, these elements will then constitute a junction in a modified network, with a probability distribution function:

$$P(x,y,z) = \{3/2\pi l^2 \eta_{i+j}\}^{3/2} \exp \{-(3/2l^2 \eta_{i+j})[(x-x_0)^2 + (y-y_0)^2 + (z-z_0)^2]\} \quad (40)$$

The most probable position of the new junction, (x_0, y_0, z_0), can be shown to lie on the line between the former most probable positions of the separate elements; the fluctuations in position of the junction are proportional to $\eta_{i+j}^{1/2}$.

It is easy to show that, if the elements i and j had independent Brownian motion before they were bonded together, then:

$$\frac{1}{\eta_{i+j}} = \frac{1}{\eta_i} + \frac{1}{\eta_j} \quad (41)$$

In general, however, these two elements will have been more or less closely coupled together in the network structure, and equation 41 will not apply. In such cases the value of η_{i+j} depends on η_i and η_j, but also on the fluctuation in the relative coordinates of elements i and j in the original network. It is shown in reference (4) that this fluctuation is described by a Gaussian distribution function:

$$R_{ij}(\xi, \eta, \zeta) = \left\{\frac{3}{2\pi l^2 \eta_{ij}}\right\}^{3/2} \exp \{-(3/2l^2 \eta_{ij})[(\xi-\xi_0)^2 + (\eta-\eta_0)^2 + (\zeta-\zeta_0)^2]\} \quad (42)$$

where ξ, η, and ζ are the relative coordinates of the elements and ξ_0, η_0, and ζ_0 are the most probable values of these relative coordinates [$\eta_0 = x_{0i} - x_{0j}$, etc.] The fluctuations in the relative positions of the elements are thus proportional to $\eta_{ij}^{1/2}$. It can be shown that:

$$\eta_i + \eta_j \geq \eta_{ij} \geq |\eta_i - \eta_j| \quad (43)$$

When the elements are independent,

$$\eta_{ij} = \eta_i + \eta_j \quad (44)$$

The calculation of η_{i+j} in the general case can be made as follows. Before the bond is formed between elements i and j, the number of configurations of the network for which x_i and x_j have specified values in proportional to:

$$\exp \left\{ -\frac{6}{l^2} \frac{\eta_j(x_i - x_{0i})^2 + [\eta_{ij} - \eta_i - \eta_j](x_i - x_{0i})(x_j - x_{0j}) + \eta_i(x_j - x_{0j})^2}{2\eta_i\eta_j + 2\eta_i\eta_{ij} + 2\eta_j\eta_{ij} - \eta_i^2 - \eta_j^2 - \eta_{ij}^2} \right\} \quad (45)$$

This follows easily from equation 6.2 of reference 4, by use of equations 5.4, 6.6, and 7.6. After the bond is formed, only those configurations are possible for which $x_i = x_j = x$. On making this substitu-

168   HUBERT M. JAMES AND EUGENE GUTH

tion into equation 45 and rearranging terms, one finds that their number is proportional to:

$$\exp\left\{-\frac{6}{l^2}\frac{\bar{\eta}_{ij}\left\langle x-(1/2\bar{\eta}_{ij})([\bar{\eta}_{ij}+\bar{\eta}_j-\bar{\eta}_i]x_{0i}+[\bar{\eta}_{ij}+\bar{\eta}_i-\bar{\eta}_j]x_{0j})\right\rangle^2}{2\bar{\eta}_i\bar{\eta}_j+2\bar{\eta}_i\bar{\eta}_{ij}+2\bar{\eta}_j\bar{\eta}_{ij}-\bar{\eta}_i^2-\bar{\eta}_j^2-\bar{\eta}_{ij}^2}\right\} \tag{46}$$

The probability of finding the new junction with coordinates x is proportional to this number; the probability of finding it with coordinates x, y, z is the product of three such factors, differing only in the substitution of y or z for x. Thus, the normalized probability distribution can be written as in equation 40, with:

$$x_0=([\bar{\eta}_{ij}+\bar{\eta}_j-\bar{\eta}_i]x_{0i}+[\bar{\eta}_{ij}+\bar{\eta}_i-\bar{\eta}_j]x_{0j})/2\bar{\eta}_{ij} \tag{47}$$

and:

$$\bar{\eta}_{i+j}=(2\bar{\eta}_i\bar{\eta}_j+2\bar{\eta}_i\bar{\eta}_{ij}+2\bar{\eta}_j\bar{\eta}_{ij}-\bar{\eta}_i^2-\bar{\eta}_j^2-\bar{\eta}_{ij}^2)/4\bar{\eta}_{ij}$$

$$=\bar{\eta}_i-(\bar{\eta}_{ij}+\bar{\eta}_i-\bar{\eta}_j)^2/4\bar{\eta}_{ij} \tag{48}$$

$$=\bar{\eta}_j-(\bar{\eta}_{ij}+\bar{\eta}_j-\bar{\eta}_i)^2/4\bar{\eta}_{ij}$$

It is thus proved that $\bar{\eta}_{ij}$ is always less than $\bar{\eta}_i$ or $\bar{\eta}_j$ and that the Brownian motion of the new junction is less than that of the unbonded elements.

Formation of a bond between two elements of a network will not merely reduce the Brownian motion of the bonded elements: it will also reduce the Brownian motion of all nearby elements of the network. Kuhn assumes that the Brownian motion of the junctions of the network will become very much less than that of elements of the intervening chain segments, but this does not follow from the foregoing proof. This point also is susceptible to a general quantitative discussion. Let points i and j be junctions of the network connected by a segment with M links, and let k be an element of this intervening segment, distant from junction i by N_i links and from junction j by $N_j = M - N_i$ links. The distribution function for this intermediate element is of the form of equation 38, with the subscript k replacing the subscript i. The intermediate element will thus have a Brownian motion proportional to $\bar{\eta}_k^{1/2}$ about a most probable position, which is easily seen by the methods in Section VIII of reference 4 to be given by:

$$x_{0k}=(N_j x_{0i}+N_i x_{0j})/(N_i+N_j),\text{ etc.} \tag{49}$$

It is shown in Appendix B of the present paper that:

$$\bar{\eta}_k=\frac{N_iN_j}{N_i+N_j}\left\{1+\frac{\bar{\eta}_i}{N_i}+\frac{\bar{\eta}_j}{N_j}-\frac{\bar{\eta}_{ij}}{N_i+N_j}\right\} \tag{50}$$

$\bar{\eta}_k$ approaches $\bar{\eta}_i$ as $N_i \to 0$, and approaches $\bar{\eta}_j$ as $N_j \to 0$; between these limits it passes through a maximum.

A convenient illustration of the relative magnitudes of these quantities is provided by the regular cubic network discussed in Section IX of reference 4. This consists of segments, each containing M links, joined together with the connectivity of a cubic lattice; each junction, then, is the point of union of six segments. It is shown in Reference 4 that for junctions of this network not too near the fixed points one has:

$$\bar{\eta}_i\cong\bar{\eta}_j\cong M/4 \tag{51}$$

If i and j are adjacent junctions:

$$\eta_{ij} \cong (M/2)(1-\pi^{-1}) \qquad (52)$$

Since $\eta_i = \eta_j$ for this network, η_k takes on its maximum value for the element midway between the junctions: $N_i = N_j = M/2$. For this element one has, by equation 50:

$$\eta_k \cong \frac{3 + 1/\pi}{8} M = 1.66\eta_i \qquad (53)$$

It follows easily that in this network the mean fluctuation of the junction coordinates is only 23% less than that of the midpoints of the segments. In addition, it should be emphasized that the difference would be expected to be greater in this model, where each junction is under constraint by six segments, than in a network in which only three or four segments end at each junction. In general, then, one cannot expect that the junctions of molecular networks will have a Brownian motion which is very much less than that of other elements of the network.

It is, nevertheless, a striking fact that one obtains the same stress-strain relation whether or not one assumes that the junctions are essentially fixed in space. The form of the relation is always the same; the identity of the results extends even to the magnitude of the stress if the assumed fixed positions of the junctions are identical with their actual most probable positions.*

The identity of these results was first indicated by us, in the Appendix of an earlier paper (1). It was there shown that the junctions of an arbitrary Gaussian network have most probable positions which vary according to equation 29, when the fixed points of the network are subjected to displacements of the same homogeneous type. It was also shown that the forces exerted by the network will be unchanged if any movable junction (or other point) of the network is fixed at what would otherwise be only its most probable position. It follows that the computed stress will be unchanged if one treats any or all of the actually movable junctions as fixed at its most probable position. This detailed analysis of the system is useful for many purposes, but it makes the identity of the results of the two types of calculation appear as a consequence of special properties of the Gaussian function.

An understanding of the identity of these results can be based on more general grounds by making use of the idea of distributions of configurations (11). One can divide all space into domains, not too many in number, such that there will be an equal probability that junction i will lie in each of these domains; we may denote these domains by $V_{i\tau}$, $\tau = 1,2,3\ldots$ A different division of space into domains $V_{j\tau}$ appropriate to junction j can be made. A "distribution of configurations" of the network will include all configurations in which junction i lies in a specified domain $V_{i\tau}$, junction j in a specified domain $V_{j\nu}$, and

---

*For instance, equation 17 of reference 5, derived on the basis of the general network theory, is identical in form with equation 34 of this paper, derived on the assumption that the junctions are fixed; in the first equation $\lambda_{\tau\nu}^{(0)}$ is defined in terms of the most probable positions of movable junctions τ and ν, whereas in the second it is correspondingly defined in terms of the positions of fixed junctions.

170   HUBERT M. JAMES AND EUGENE GUTH

so on. If the number of configurations possible for the network segments is sufficiently large compared with the number of junctions in the network, the total number of distributions will be small compared to the number of possible configurations of the network. Now the most probable configuration of the junctions is that in which each junction lies at its most probable position (4). The most probable distribution will, correspondingly, be that in which each junction lies in the domain which includes its most probable position. The number of configurations in this distribution will differ from the total number of configurations by a factor which is very small as compared with that total number. In computing the entropy, proportional to the logarithm of the number of configurations, one will then make only a small error if one counts only the configurations in the most probable distribution-configurations in which each junction lies near its most probable position. In the derivation of results that are valid for networks with small numbers of junctions (such as those with which we are here concerned) it is possible to carry this process to the limit of treating each junction as fixed at its most probable position: with such networks one can reasonably define distributions in which the position of each junction is very accurately specified. In other words, it is possible to treat the junctions as fixed in deriving results that depend on the statistical behavior of the links as components of the segments, but not on the statistical behavior of segments as components of the network. One may compare such a treatment to a calculation of the pressure of a gas based on the idea that one will find exactly half of the molecules in a given half of the volume at any given time. The result of the calculation will be correct, though the assumed restriction has no analog in nature, and one is in fact considering only a small fraction of the configurations which the system can assume.

The possibility of treating the network junctions as fixed in calculations of this type does not imply that a similar procedure will be possible when other properties of the system are to be discussed. This assumption should be avoided—as indeed it can be without much trouble—in discussing details in the behavior of the network. It would, for instance, be quite out of place in a discussion of the change of network structure during cure.

IV. WALL'S THEORY OF RUBBER

Wall (12-14) has derived a stress-strain relation for rubber:

$$Z = GkT \left[L_Z - 1/L_Z^2 \right] \tag{54}$$

without making use of any definite model of the system. His theory is often referred to as a network theory of rubber, and Wall himself (13,14) indicates that his calculations apply to rubber considered as a molecular network.

Wall's theory is not a typical application of statistical mechanics. In a statistical theory one usually starts from a definite model of the physical system, with specified statistical weights, constraints, potential fields, and so on. On this basis one then derives distribution functions which describe the system statistically; from these one can finally compute average or observable properties of the system. Wall, on the other hand, does not describe his model, and postulates rather than describes his distribution functions. His postulates do not have a general validity which would bestow a like general validity on his results; instead, they correspond to a special model of unspecified type, the nature of which one can only infer from the character of the postulates. We shall now show that Wall's mathematical postulates are inconsistent with the network structure of rubber, and that his theory can lead to results different from those given by the network theory.

Wall's procedure is indicated in most detail in the one-dimensional treatment of his first paper, but even there it is not completely explicit. In essence it involves the following steps. The notation is that of Wall's paper (12).

(1) It is assumed that the molecules are of uniform length, with Gaussian configuration functions.

(2) It is assumed that when the bulk material is unstretched the distribution function for extensions of the chains (displacements between the end points) is of the same form as the configuration function, but normalized to 1. In the one-dimensional case the distribution of extensions x is given by:

$$p(x) = \frac{\beta}{\pi^{1/2}} \exp \{-\beta^2 x^2\} \tag{55}$$

(3) It is postulated that when the material is stretched the distribution function is modified in a particular way (13): "The components of the lengths of the individual molecules will change in the same ratio as does the corresponding dimension of the piece of rubber." In the one dimensional case the distribution function becomes:

$$p'(x) = \frac{\beta}{\alpha\pi^{1/2}} \exp \{-\beta^2 x^2/\alpha^2\} \tag{56}$$

α being the ratio of the extended length to the unstretched length of the material.

(4) The relative probability of the states described by these distribution functions (i.e., the relative number of configurations of the system consistent with the specified distributions of chain extension) is then computed as follows: The range of values of the extensions is divided into small regions, the ith of which, about $x = x_i$, is of length Δx_i. Then:

$$p_i = p(x_i)\Delta x_i \tag{57}$$

is the relative number of configurations for a chain of extension x_i to within Δx_i. When the material is unstretched, the average number of chains out of a total number N, which have extensions in the range Δx_i, is:

$$n_i = Np_i = Np(x_i)\Delta x_i \tag{58}$$

When the material is stretched, the average number of chains with extensions in the range Δx_i is:

$$s_i = N\,p'(x_i)\Delta x_i \tag{59}$$

The relative probability of the distribution (equation 56) for the stretched material is then computed as:

$$P = N! \prod_i \frac{p_i^{s_i}}{s_i!} \tag{60}$$

172   HUBERT M. JAMES AND EUGENE GUTH

which reduces, as stretch of the material is reduced to zero, to:

$$P_0 = N! \prod_i \frac{p_i^{n_i}}{n_i!} \tag{61}$$

(5) The change of entropy of the material on stretching is then taken to be:

$$S - S_0 = k \ln (P/P_0) \tag{62}$$

We have now to examine some of the implications of this formalism.

The distribution functions of Wall's theory imply that there exist certain unspecified constraints on the extension of the molecules. Since $p(x)$ is of the same form as the configuration function, it would describe the distribution of extensions for a system of independent, unconstrained molecules; this is not, however, a possible model for the system, for in the absence of constraints on or between the molecules the distribution function would be independent of the form of the material, and the theory would predict no restoring forces. The function $p(x)$ will also describe a model formed by freezing the ends of the molecules into fixed positions, either simultaneously or at random times. This same model will also yield Wall's postulated distribution functions for the stretched material ($p'(x)$ in the one-dimensional case) if the fixed positions of the ends of the chains are supposed to change according to equation 29. Indeed, the quotation from Wall's second paper (13), as given under (3) above, seems to indicate that he had such a model in mind.

Wall's distribution functions would also describe a network with fixed junctions formed by the union of his "molecules" at their ends—a model in which each "molecule" becomes a segment of the network. Wall's postulates do not apply to a general model of this type, for they establish also the distribution of the fixed fractional extensions of the segments, which, in the three-dimensional case, is essentially that of equation 36. They thus serve to determine the value of the factor K in the stress–strain relation, as discussed in Section II.

The further development of Wall's calculation is, however, inconsistent with any of the models mentioned above. If each molecule has a definite extension, changing in a definite way when the material is stretched, the relative number of configurations accessible to the system is (cf. eq. 27):

$$P = \prod_i p_i^{s_i} \tag{63}$$

In equation 60 there appears an additional factor $(N!/\prod s_i!)$. This is the number of different ways in which the specified set of chain extensions can be assigned to the molecules, provided that each molecule can take on any extension independently of the extensions of the other molecules. The presence of this factor is inconsistent with the assumption of fixed extensions for the segments. It is also inconsistent with the assumption that the molecules are joined together to form a network, for their connection into a network would establish certain relations between the segment extensions.

In his third paper (14) Wall has attempted to show that the distributions of equations 55 and 56 apply to a system of molecules linked to form a network with movable junctions. However, his discussion and conclusions appear to be incorrect. The distribution function for the components ξ, η, and ζ of the extension of any segment of any Gaussian network is of the form of equation 42. It is a Gaussian distribution about mean values ξ_0, η_0, and ζ_0 of the components; these mean values change as the network is stretched, but the breadth of the distribution, determined by η_{ij}, does not. Wall's distribution is of a quite different character, being a Gaussian distribution about fixed mean values (0 for each component), which changes in breadth when the material is stretched.

The argument in Wall's third paper (14) will be indicated briefly. Wall restricts his consideration to the simplest possible network, a one-dimensional chain of N molecules with identical Gaussian configuration functions, like that of equation 55. The distribution function for molecular extensions describes the most probable distribution of molecular extensions consistent with the fixed total extension of the chain, which will here be called L. The relative number of molecular configurations consistent with the assignment of N_0 molecules to the range Δx_0 of extensions, N_i molecules to the range Δx_i etc., is:

$$W = \frac{N!}{N_0! \; N_1! \; N_2! \ldots} \; p_0^{N_0} \; p_1^{N_1} p_2^{N_2} \ldots \tag{64}$$

To determine the distribution function $N_i(x_i)$ one has to choose the N_i so as to maximize this expression, subject to two conditions of constraint: the total number of molecules is N, and the total extension of the chain is L. In symbols,

$$N = \sum_i N_i \tag{65}$$

and:

$$L = \sum_i N_i x_i \tag{66}$$

Solution of this problem leads to the correct distribution function:

$$p''(x) = \frac{\beta}{\pi^{1/2}} \exp\left\{- \beta^2 (x - L/N)^2\right\} \tag{67}$$

Instead, Wall replaces equation 66, obviously required by his statement of the problem, by a different condition of constraint:

$$L^2 = C \sum_i N_i x_i^2 \tag{68}$$

and finds the erroneous distribution function:

$$p'(x) = \left(\frac{CN}{2\pi L^2}\right)^{1/2} \exp\left(- \frac{CN}{2L^2}x^2\right) \tag{69}$$

which has a quite different character. Wall arrives at equation 68 in the following way. In a preliminary discussion of the chain of molecules he concludes that:

$$\left\langle x_i^2 \right\rangle_{av} = L^2/N \tag{70}$$

where $\left\langle x_i^2 \right\rangle_{av}$ is the mean square extension of any molecule in the chain. This he rewrites, with the addition of an arbitrary constant factor C, to obtain equation 69. However, in deriving equation 70, he neglects terms which are individually small (of the order of L^2/N^2) but so numerous that in the aggregate they far outweigh the terms

174   HUBERT M. JAMES AND EUGENE GUTH

which he retains. An exact discussion of the problem yields:

$$\left\langle x_i^2 \right\rangle_{av} = \frac{L^2}{N^2} + \frac{1}{2\beta^2} \tag{71}$$

which differs from equation 70 by a large and variable factor. It will be noted that equations 68 and 70 imply that when the chain of molecules has total extension zero every molecule in the chain must have total extension zero—an obviously false result. Correspondingly, the distribution function of equation 69 gives for every molecule $\left\langle x_i \right\rangle_{av} = 0$, whereas the correct value must be L/N; this value is of course given by equation 67.

The additional factor in equation 60, as compared to equation 63, leads to the presence in Wall's theory of a term in the entropy which has no analog in the theory of a Gaussian network. In the one-dimensional case the network theory, with Wall's distribution functions for a system of G molecules, gives:

$$S - S_0 = -\frac{G}{2}\alpha^2 \tag{72}$$

whereas Wall's computation yields:

$$S - S_0 = G \ln \alpha - \frac{G}{2}\alpha^2 \tag{73}$$

Correspondingly, the one-dimensional network theory leads to the stress-strain relation:

$$Z = GkT\alpha \tag{74}$$

but Wall's computation yields:

$$Z = GkT(\alpha - \frac{1}{\alpha}) \tag{75}$$

Wall's theory thus leads to a non-zero length for the system under zero external force. On the other hand, the network theory gives zero length for zero external force; with a one-dimensional network, as with a single Gaussian chain, it requires a non-zero average force to keep the ends of the system at a fixed finite separation, however small.* Neither one-dimensional model is suitable for discussion of the characteristics of the real three-dimensional networks; when the theory is extended to three dimensions it is the network model that corresponds more closely to reality.

_____

*Thus, when equation 63 is used, $\alpha = 1$ does not represent the case of zero external force. Some confusion has arisen in this connection. Because a molecule under zero external force has non-zero mean square extension, it has been assumed that a calculation of the stress-strain relation must yield a non-zero "natural length" for the molecule under zero external force. Whether or not this is the case will depend on how terms are defined. If the stress-strain relation is defined in terms of the rms extension of a system under constant external force one will find a non-zero "length" for zero force. If, however, one defines the stress-strain relation in forms of the average force required to maintain a constant extension, then the force to zero only as the length goes to zero. It is the latter point of view that is implicit in the calculation of the entropy as a function of the fixed extension of the molecular chain, and the derivation of the force Z from this entropy.

In three dimensions, Wall's method of calculation would again yield, in general, an additional term in the entropy:

$$\Delta S = G \ln (\alpha_x \alpha_y \alpha_z) \tag{76}$$

where α_x, α_y, α_z are the relative extensions in the x,y,z directions, respectively. This term does not appear in Wall's second paper, being suppressed by his immediate assumption that the material is incompressible: $\alpha_x \alpha_y \alpha_z = 1$. Thus, Wall's treatment yields the same result as the network theory in this special ease. The difference between the theories would become evident if compressibility were taken into account, and would be particularly important in a discussion of the swelling of rubber.

It is clear that Wall's formalism cannot be interpreted in terms of any system consisting of flexible molecular chains alone. We have not been able to devise any model to which it can properly be applied. We believe, therefore, that Wall's method of calculation, and particularly the use of equations 60 and 61, should be avoided as lacking physical significance.

This criticism applies also to the work of Kuhn and Kuhn (8), who have pictured rubber as a system of molecular segments with movable ends, and have calculated its entropy by a method similar to that of Wall. They divide the entropy into two parts, one associated with the orientation of the segments, the other with their extensions, and calculate these quantities as though each segment could take on any given orientation and extension with the same probability as any other segment. It should hardly need further emphasis that this is not the case if the segments are connected into a network. The calculation of Kuhn and Kuhn, like that of Wall, yields the same result as the network theory for the special case of a distortion at constant volume.

We have elsewhere (5) discussed our objections to Wall's assumed distribution functions (such as that of equation 56) as descriptions of the mean fractional extensions of segments in physically occurring networks.

APPENDIX A

Justification of Neglect of Steric Hindrances

In using a Gaussian network as a model for the actual molecular network of rubber, one is not neglecting steric hindrances entirely. The fixed points of the model, representing the points of the real network to which external forces are applied, have been constrained to lie on the external surfaces of the material—in positions which are in fact largely determined by the volume-filling property of the molecules through the form of U(V,T). When computing the entropy of the system by enumerating configurations of the network, one is consequently limited to the counting of network configurations which extend through a volume in space determined by the steric hindrances. Hence one does not consider, in a way which would make senseless the outcome of the calculation, the very numerous configurations of very small volume which might be assumed by an idealized network without volume-filling properties and without fixed points. In other words, calculation with this model includes enumeration of only those configurations with the correct average density of chain elements in space.

The steric hindrances of real molecular chains do not merely exclude configurations of the system in which the average density is abnormal; they prevent also the realization of all configurations of the system in which any two portions of molecular chain occupy the same volume in space. In our treatment of the idealized network the details of the configurations are not specified completely enough to permit exclusion of these configurations, which are consequently included in the counting. We have now to indicate the consequences of thus taking into account the volume-filling properties of the molecules on the average, but not in detail.

For this discussion we distinguish between transient and permanent constraints. By a "transient constraint" we mean one that does not permanently exclude any configuration of the system that would otherwise be possible, nor give to it a weight different from that of any other possible configuration with the same energy. Typical of transient constraints are those imposed upon a flexible molecular chain immersed in an (idealized) liquid of foreign, or even similar, molecules. The molecules of this liquid subject the given molecule to constraints which vary from instant to instant, excluding by their volume-filling properties now this configuration of the molecule, now that; nevertheless, in the course of a sufficiently long period, they would permit this given molecule to take on all the configurations which it could assume in their absence, and with the same relative probability. A "permanent constraint," on the other hand, is one which permanently changes the relative probability of occurrence of certain configurations or makes them impossible of realization. Union of molecules into a coherent network, or the establishment of fixed points of a molecular network, are examples of permanent constraints.

A single flexible molecular chain with ends fixed at separation L exerts against these constraints an average force Z given by the familiar equation:

$$Z = -kT \, \frac{d}{dL} \ln c(L)$$

where $c(L)$ is the number of configurations consistent with the given L — or, more generally, the sum of the statistical weights of the possible configurations. The fixing of L is of course a permanent constraint on the chain. If other constraints are present they will modify $c(L)$ by excluding some configurations from consideration, or changing their statistical weights. If these constraints have an effect which varies in time, $c(L)$, like Z, is to be taken as an average value over a long period.

An important special class of constraints has the property of changing $c(L)$ by a constant factor, as by excluding from consideration a constant fraction of the configurations that would otherwise be possible, or by changing their statistical weights by a constant factor; such constraints leave $\langle Z \rangle_{av}$ completely unchanged. Transient constraints, as defined above, are of this type. Thus, Z will be the same for a chain whether it is in thermal equilibrium with a tenuous gas, or with a surrounding liquid, so long as the period of averaging the force is sufficiently long for the constraints imposed by the liquid to have a tran-

NETWORK THEORY OF RUBBER   177

sient character. The result is the same when one averages, not over a considerable time, but over a large number of similar chains and a correspondingly shorter time—as one actually does in observing bulk rubber.

In a liquid of rubber molecules the volume-filling property of the molecules causes them to exert constraints on each other, but only transient ones. This situation is largely unaltered when the material is vulcanized. To the extent to which each chain moves as in a fluid composed of the other chains, our treatment is complete: equation 7 continues to give the relative number of configurations of each segment as a function of the separation of its ends, and the computation proceeds as before. To this approximation the volume-filling property of the chains reduces the number of configurations of the network by a factor which is independent of the form of the material; it modifies only the value of the constant K_1 in equation 15, and not the forces exerted by the system.

A more detailed, and in some respects more satisfying, analysis of the effects of volume-filling can be carried out by the methods of the Appendix in a previous paper (1), which fixes attention on the forces acting within the network, as well as on those exerted by the network against external constraints. It is there shown that:

(a) The mean position of each junction of a Gaussian network is the position at which it would be subject to zero average force (due to the chains which enter it) if it were fixed there, and

(b) The average force which the network exerts on any fixed point can be computed as the sum of the average forces exerted on it by all the segments which enter it, when all junctions of the network are treated as fixed at their most probable positions.

To the approximation in which any segment is subject only to transient constraints by the liquidlike mass of other segments in which it is embedded, the average force exerted by that segment is unchanged by the presence of the other segments. If this assumption is extended to all segments, it follows from (a) that the mean positions of the network junctions are unchanged by these transient constraints, and then from (b) that the average force on each fixed point of the network—and hence the net force which we have computed in section I—is unchanged.

However, not all the constraints arising from steric hindrances are of transient character. For instance, when two molecules are linked at a junction there will arise steric hindrances of the permanent type; the effect of these will be, essentially, to reduce the flexibility of the molecular segments in the immediate neighborhood of the junction. So long as one restricts attention to lightly vulcanized materials, in which only a relatively small portion of the total chain length is near to any junction, one is justified in neglecting this type of steric hindrance. On the other hand, it will become an extremely important factor in the behavior of highly vulcanized hard rubbers.

A more interesting type of steric hindrance comes into play when the network is formed as illustrated in Figure 2. Sketches (a) and (b) illustrate possible structures in which chains of the same length connect similar junctions in essentially different ways. In our discussion such different structures have received identical treatment, even though, considered in detail, the configurations possible for one structure are impossible for the other. The effect of neglecting such

178   HUBERT M. JAMES AND EUGENE GUTH

steric hindrances will vary markedly with the structure of the network and its condition of stretch. If the segments are long as compared with the distances between their ends, as illustrated in Figure
2(c), the constraints exerted by one segment on the other will be almost completely of the transient type, causing but little change in the
forces exerted by the network. This is the situation typical of lightly
vulcanized materials at small extensions. If one segment is tightly
drawn over the other, as in Figure 2(d), the effect of the entanglement
will be almost the same as if the molecules were bonded together in
the region of their crossing. Such a condition is very unlikely to occur
in an unstretched material. The relation of the molecules is one of
very low probability, one unlikely to occur in the material undergoing
vulcanization and correspondingly unlikely to be built into the network
as it is formed. On the other hand, a situation of type (c) can be converted into one of type (d) if the material is highly extended. Our neglect of these steric hindrances can thus be justified for lightly vulcanized materials not too highly extended; at high extensions these steric
hindrances may contribute markedly to the rigidity of the material.
This is, however, a range in which the configuration functions of the
segments can no longer be treated as Gaussian, and the theory outlined in this paper cannot be expected to apply.

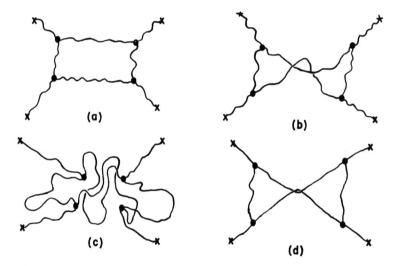

Fig. 2. Diagram illustrating action of steric hindrances. In
(b),(c),(d), the lower chain is to be understood as passing
around the upper one. Fixed points (×); junctions (●).

APPENDIX B

Brownian Motion of Intermediate Points of a Chain

The Brownian motion of the intermediate points of a chain segment can be related to the Brownian motion of its ends by an analysis
of the sort employed earlier (4), and in section III of this paper. A
more compact proof will be given here. The notation is that of a previous paper (4).

We consider a segment with $N_i + N_j$ links. Let the terminal elements i and j of the segment be assigned the highest number, n - 1 and n, in the numbering of the junctions of the network. Without loss of generality it can be assumed that the segment with which we are concerned is the only one which passes from i to j without passing through some other junction; if any other connections between i and j exist, one can pick some element of each segment for treatment as a junction, as is explained elsewhere (4). The lower right corner of the Γ determinant then has the following appearance:

$$\Gamma = \begin{vmatrix} \cdot & \cdot & \cdot & \cdot & \cdot & \cdot & \cdot & \cdot & \cdot \\ \cdot & x & & x & & x & \\ \cdot & & & & & & \\ \cdot & x & \xi + \dfrac{1}{N_i + N_j} & & -\dfrac{1}{N_i + N_j} \\ \cdot & & & & & \\ \cdot & x & -\dfrac{1}{N_i + N_j} & & \eta + \dfrac{1}{N_i + N_j} \\ \cdot \end{vmatrix} \qquad (B.1)$$

The terms $1/(N_i + N_j)$ in the matrix elements arise from the segment connecting elements i and j; the values of ξ and η depend on what other segments end on elements i and j, respectively.

Now let us treat the intermediate element k as a junction, the $(n + 1)^{st}$ in the network. When this is done, the determinant Γ for the network will have a different form. We call this new determinant G. Its lower right corner has the following appearance:

$$G = \begin{vmatrix} \cdot & \cdot & \cdot & \cdot & \cdot & \cdot & \cdot & \cdot & \cdot \\ \cdot & x & x & x & 0 \\ \cdot & x & \xi + \dfrac{1}{N_i} & 0 & -\dfrac{1}{N_i} \\ \cdot & x & 0 & \eta + \dfrac{1}{N_j} & -\dfrac{1}{N_j} \\ \cdot & 0 & -\dfrac{1}{N_i} & -\dfrac{1}{N_j} & \dfrac{1}{N_i} + \dfrac{1}{N_j} \end{vmatrix} \qquad (B.2)$$

Junction n - 1 (element i) is no longer connected directly to junction n, but is connected to junction n + 1 by a chain of N_i links-hence the zero in the next to last column, and the entry $-1/N_i$ in the last column of row n - 1. Junction n + 1 (element k) is connected only to junctions n - 1 and n; all elements in the last row and column not shown above are zero.

We now wish to compute the value of η for element k—that is, η_{n+1} when the junctions are numbered as above. Denoting by $G_{ij\cdots}^{kl\cdots}$ the determinant obtained from G by striking out rows i,j,... and columns k, l, ..., and multiplying by $(-1)^{i+j+k+l+\cdots}$, we have, by equation 5.4 of Reference 4,

$$\eta_{n+1} = G_{n+1}^{n+1}/G \qquad (B.3)$$

By adding $N_j/(N_i + N_j)$ times the last column to G to the n - 1st column, and $N_i/(N_i + N_j)$ times the last column to the nth column, one can reduce to 0 the off-diagonal elements in the last row. This same

180   H U B E R T   M .   J A M E S   A N D   E U G E N E   G U T H

manipulation makes G identical with Γ, except for the presence of the last row and column. It follows at once that:

$$G = \left(\frac{1}{N_i} + \frac{1}{N_j}\right) \Gamma \tag{B.4}$$

We now write G_{n+1}^{n+1} in a form which emphasizes its relation to Γ:

$$G_{n+1}^{n+1} = \begin{vmatrix} \cdot & \cdot & \cdot & \cdot & \cdot & \cdot & \cdot & \cdot & \cdot \\ \cdot & x & & x & & & x & \\ \cdot & x & \xi + \dfrac{1}{N_i+N_j} + \alpha & & -\dfrac{1}{N_i+N_j} + \beta & \\ \cdot & x & -\dfrac{1}{N_i+N_j} + \beta & & \eta + \dfrac{1}{N_i+N_j} + \gamma & \end{vmatrix} \tag{B.5}$$

where:

$$\alpha = \frac{N_j}{N_i(N_i + N_j)}$$

$$\beta = \frac{1}{N_i + N_j} \tag{B.6}$$

$$\gamma = \frac{N_i}{N_j(N_i + N_j)}$$

All elements of G_{n+1}^{n+1} not shown above are identical with the corresponding elements in Γ. It is evident that G_{n+1}^{n+1} depends on α, β, γ in the following way:

$$G_{n+1}^{n+1} = K + A\alpha + B\gamma + C\alpha\gamma + D\beta + E\beta^2 \tag{B.7}$$

where $K, A, \ldots$ are independent of α, β, γ. Inspection of the determinant shows that:

$$K = \Gamma \quad A = \Gamma_i^i \quad B = \Gamma_j^j \quad C = \Gamma_{ij}^{ij} \quad D = 2\Gamma_i^j \quad E = -\Gamma_{ij}^{ij} \tag{B.8}$$

The terms in $\alpha\gamma$ and β^2 cancel, and we have:

$$G_{n+1}^{n+1} = \Gamma + \frac{N_j\Gamma_i^i}{N_i(N_i + N_j)} + \frac{N_i\Gamma_j^j}{N_j(N_i + N_j)} + \frac{2\Gamma_i^j}{N_i + N_j} \tag{B.9}$$

Substituting equations B.4 and B.9 into B.3, and, remembering that:

$$\eta_i = \Gamma_i^i/\Gamma \tag{B.10}$$

$$\eta_i + \eta_j - \eta_{ij} = 2\Gamma_i^j/\Gamma \tag{B.11}$$

by equations 5.4 and 7.7 of an earlier paper (4), one obtains:

$$\eta_{n+1} = \frac{N_i N_j}{N_i + N_j}\left\{ 1 + \frac{\eta_{n-1}}{N_i} + \frac{\eta_n}{N_j} - \frac{\eta_{n,n-1}}{N_i + N_j} \right\} \tag{B.12}$$

This is the same, except for notation, as equation 50.

Acknowledgment

We would like to express our appreciation for support in part of this work by ONR Navy Contract N6-ori-83.

REFERENCES

1. H. M. James and E. Guth, J. Chem. Phys., 11, 455 (1943).
2. F. T. Wall, J. Chem. Phys., 10, 485 (1942).
3. H. M. James and E. Guth, J. Chem. Phys., 11, 472-3 (1943).
3a. E. Guth and H. Mark, Monatsh., 65, 93 (1934).
4. H. M. James, J. Chem. Phys., 15, 651 (1947).
5. H. M. James and E. Guth, J. Chem. Phys., 15, 669 (1947).
6. See, for example, W. Kuhn, Kolloid Z., 76, 258 (1936).
7. W. Kuhn, J. Polymer Sci., 1, 380 (1946).
8. W. Kuhn and H. Kuhn, Helv. Chim. Acta, 29, 1615 (1946).
9. P. J. Flory and J. Rehner, J. Chem. Phys., 11, 512 (1943).
10. R. L.G. Treloar, Trans. Faraday Soc., 39, 36 (1943).
11. See, for instance, Mayer and Mayer, "Statistical Mechanics." Wiley, New York, 1940, Chapt. 3.
12. F. T. Wall, J. Chem. Phys., 10, 132 (1942).
13. F. T. Wall, ibid., 10, 485 (1942).
14. F. T. Wall, ibid., 11, 527 (1943).

Synopsis

The approximations implicit in the use of the Gaussian network model for soft rubber are discussed. It is shown that the form of the stress–strain curve can be derived for this model simply, and without special assumptions about the form or behavior of the network. The common assumption that the network junctions are fixed, or can be treated as fixed, is discussed. It is shown that this picture of the situation is unrealistic: the junctions have a Brownian motion comparable to that of any portion of the intervening molecular segments. The introduction of this assumption is not generally admissible, but it will not affect the outcome of certain types of calculation; in particular, one can foresee that it need not affect the calculated form of the stress-strain curve. A particularly simple and straightforward calculation of the network entropy on this basis is given. Wall's theory of rubber is analysed. It is shown that Wall's postulates are not consistent with the network structure of rubber, and in general lead to different results.

Résumé

Les approximations, inclues dans l'utilisation d'un modèle en réseau de Gauss pour le caoutchouc, sont soumises à discussion. L'allure des courbes tension-déformation peut être dérivée de ce modèle simplement, sans préjuger de la forme ou du comportement de ce réseau. En discutant la supposition généralement admise, que les points de jonction du réseau sont fixes ou peuvent être considérés tels, on constate que cette interprétation ne correspond pas à la réalité:les jonctions possèdent un mouvement Brownien comparable à celui de chaque partie des segments moléculaires présents. L'introduction de cette hypothèse n'est pas généralement admissible; certains résultats de calculs toutefois restent inchangés; en particulier, on peut prévoir que la forme calculée de la courbe tension-déformation n'en sera pas affectée. Un calcul particulièrement simple et nouveau

182   HUBERT M. JAMES AND EUGENE GUTH

de l'entropie du réseau dans ces conditions est donné. La théorie du caoutchouc de Wall est analysée; ses postulats sont incompatibles avec la structure en réseau du caoutchouc, et amènent en général à des résultats différents.

Zusammenfassung

Es werden die Vernachlässigungen, die sich für die Anwendung der Gausschen Netzwerktheorie auf Weichgummi ergeben, diskutiert. Es wird gezeigt, dass die Gestalt der Zug-Spannungskurve für dieses Modell einfach, und ohne spezielle Voraussetzungen für die Form und das Verhalten der Netzwerke abgeleitet werden kann. Es wird die allgemein übliche Annahme, dass die Knotenpunkte festgelegt sind- oder als festgelegt betrachtet werden können, diskutiert. Es wird gezeigt, dass diese Vorstellung unrealistisch ist. Die Knotenpunkte haben eine Brownsche Bewegung, die der jeden Teiles der sie verbindenden molekularen Segmente vergleichbar ist. Die Einführung dieser Annahme ist nicht allgemein zulässig, aber wenn man est tut, wird das resultat gewisser Berechnungen nicht beeinflusst werden; im besonderen kann man voraussehen, dass die berechnete Form der Zug-Spannungskurve unbeeinflusst bleibt. Auf dieser Grundlage wird eine besonders einfache und elegante Berechnung der Netzwerkentropie gegeben. Walls Gummitheorie wird analysiert. Es wird gezeigt, dass die Postulate dieser Theorie nicht in Einklang mit der Netzwerkstruktur von Kautschuk stehen, und im allgemeinen zu verschiedenen Resultaten führen.

Received February 19, 1948.

COMMENTARY AND PERSPECTIVE

Reflections and Comments on "Molecular Configuration and Thermodynamic Parameters" from Intrinsic Viscosities by Paul J. Flory and Thomas G Fox, *J. of Polym. Sci.*, 5, 745 (1950)

EDWARD F. CASASSA AND GUY C. BERRY

Department of Chemistry, Carnegie Mellon University, Pittsburgh, Pennsylvania 15213

This note by Flory and Fox[1] gives a succinct presentation of their treatment of the intrinsic viscosity of polymer solutions. It draws together and enlarges upon their ideas from contemporaneous papers[2-5] and puts them in a framework that, albeit with modifications, still provides practical approximations for polymer characterization and has also stimulated much theoretical study.

By 1950, the intrinsic viscosity $[\eta]$, obviously commended by ease of measurement, had long been the most common tool for characterization of soluble polymers, even though theoretical understanding remained inadequate. Early controversies[6] about interpretation of experiments had centered on the relation between $[\eta]$ and molecular weight M, and attempts to justify sometimes deficient experimental data in terms of molecular size and shape. Four decades earlier, Einstein had shown that $[\eta]$ for a suspension of solid spheres is independent of their size; yet it was apparent that high polymers gave far larger values that increased markedly with molecular weight. Attempts to clarify the situation in terms of hydrodynamic behavior were seriously hampered by the ideas that typical polymers would have a rodlike conformation in solution and that they would be free-draining, even if they were not rodlike (i.e., there would be no entrainment of solvent by a polymer chain moving in a shear field). Debye showed that $[\eta]$ is proportional to molecular weight for a free-draining coil.[7] More realistic attempts at a theory for coiling polymers with intramolecular hydrodynamic interactions were given by Debye and

Bueche, who modeled the polymer chain as a sphere containing a uniform concentration of chain segments, and Kirkwood and Riseman, who refined the model by adopting a more accurate Gaussian density distribution of segments.[7] These treatments afford qualitatively similar predictions, with a crossover from nondraining behavior for large M to a partially draining molecule at low M.

With their characteristic discernment of the essentials of a physical situation from secondary complications, Flory and Fox made two simplifying assumptions. (a) Coiling polymers in the molecular weight range of practical interest exhibit the asymptotic behavior of the Kirkwood-Riseman theory: i.e., the chains are essentially nondraining, the chain domain with included solvent constituting effectively an Einstein sphere with a radius proportional to the root-mean-square radius of gyration R_G of the chain.[3] (b) Flory's uniform expansion approximation[2,7,8] holds: the excluded volume effect in a good solvent causes every unperturbed random-flight averaged chain dimension to be expanded by the same linear factor α, so that $R_G^2 = \alpha^2 R_{G0}^2 = R_L^2/6 = \alpha^2 R_{L0}^2/6$, for example, where R_L is the rms end-to-end distance of the chain. With assumptions (a) and (b), the hydrodynamic volume $[\eta]M$ of the molecule is given by

$$[\eta]M = \Phi R_L^3 = \Phi' R_G^3 \qquad (1)$$

The factor Φ' (or Φ) is a universal hydrodynamic constant for high molecular weight flexible linear polymers—the limiting (maximum) value at high molecular weight of a parameter appearing in the

Kirkwood-Riseman theory. Flory and Fox took $\Phi'(R_{G0}^2/M)^{3/2}$ to be independent of molecular weight and solvent, but possibly weakly temperature dependent. According to eq. (1), molecular dimensions can be obtained directly from intrinsic viscosity measurements if Φ' is known. Conversely, if Φ' is truly "universal," an empirical value obtained from a system for which M and R_G are known, say from diffraction measurements, can be applied to other systems. Theoretical calculations of Φ' have yielded a range of values.[7,9,10]

Flory and Fox use eq. (1) to eliminate α from Flory's thermodynamic equation[2,8]:

$$\alpha^5 - \alpha^3 = Cz \tag{2}$$

where C is a universal constant and z is a dimensionless, temperature-dependent interaction parameter proportional to $(M/R_{G0}^2)^{3/2}$ and to $M^{1/2}$. The resulting form shows that the intercept of a plot of $([\eta]^2/M)^{1/3}$ versus $M/[\eta]$ is $(\Phi')^{2/3}R_{G0}^2/M$. Thus viscosity measurements on a series of homologous polymers in a common solvent at a fixed temperature can be used to obtain unperturbed chain dimensions in systems for which the unperturbed (theta) condition, $\alpha = 1$, may not be empirically accessible.

Together, eqs. (1) and (2) form the basis of the Flory-Fox analysis of intrinsic viscosity to obtain the size of coiling polymers in solution in good and in poor solvents. They have served as the starting point for a large number of investigations to estimate the unperturbed dimensions and thermodynamic parameters from data on $[\eta]$ over forty-plus years. These include studies of

- Excluded volume interactions, including the role of temperature,
- Non-Gaussian chain statistics,
- Branched chain configurations, and
- Electrostatic interactions.

In discussing these developments, it is convenient to represent the unperturbed dimensions in terms of the persistence length $\hat{a}$ and chain contour length $L = M/M_L$ (with mass per unit length M_L) as $6R_{G0}^2 = R_{L0}^2 = 2\hat{a}L$ for high molecular weight linear flexible chain polymers, and $12R_{G0}^2 = R_{L0}^2 = L^2$ for rodlike chains, with intermediate values dependent on $L/\hat{a}$ for semiflexible chains.[7,10,11] Thus, for linear chains, $R_G^2 = (\hat{a}L/3)\alpha^2$ for $L/\hat{a} \gg 1$, but $R_G^2 = L^2/12$ for rodlike chains, with both α and the unperturbed dimensions dependent on $L/\hat{a}$ for semiflexible chains.[7] Excluded volume effects are expressed as functions of the thermodynamic interaction param-

eter z proportional to $(d_T/\hat{a})(L/\hat{a})^{1/2}$, with a thermodynamic diameter d_T dependent on temperature and the polymer/solvent pair. For temperatures T near the (upper) Flory theta temperature Θ, $d_T(T) = (d_T)_0(1 - \Theta/T)$ where $(d_T)_0$ approximates the geometric diameter of the chain repeat unit.[10] Studies of the temperature dependence of $[\eta]$ from the upper to the lower critical solution temperatures require a more complex expression for $d_T(T)$ than that given above, including for example, a form with a maximum in d_T at intermediate temperatures.[10] Finally, Φ' is considered in the form

$$\Phi' = \pi N_A K R_H/R_G \tag{3}$$

with a hydrodynamic function K and a hydrodynamic radius R_H;[10] the latter may be determined independently from the translational diffusion. In general, models for threadlike molecules give K in the range from unity (rods and low molecular weight flexible linear chains) to 10/3 (high molecular weight flexible linear chains). By contrast, R_H incorporates much of the hydrodynamics that can cause Φ' to depend markedly on $L/\hat{a}$, for example. Thus, for linear flexible chains, $R_H/R_G \approx 2/3$ for $L/\hat{a} \gg 1$; but $R_H \propto L$ for $L/\hat{a} \ll 1$, owing to a crossover from "nondraining" to "free-draining" hydrodynamics.[7,9,10]

A wide variety of formulations have been used for $\alpha(z)$ in studies based on the method of Flory and Fox to deduce $\hat{a}$ from data on $[\eta]$ as a function of M, for M high enough that Φ' is essentially constant (e.g., see references 10–12 and citations therein). The following relation[11] may be used to represent most of the relations for $\alpha(z)$ to within a few percent with suitable k and μ:

$$\alpha \approx \{1 + \hat{z} + k\,\hat{z}^2\}^\mu \tag{4}$$

where $\hat{z} = \{A(L/\hat{a}) \cdot C/2\mu\}z$; $A(L/\hat{a})$ tends to zero for $L/\hat{a} \ll 1$, and to unity for $L/\hat{a} \gg 1$. A number of values have been proposed for C, including values based on the perturbation theory for α.[7,10,11] A treatment based on renormalization results gave $\mu \approx 0.0886$ and $k \approx 0.2$,[11] and a close fit to eq. (2) obtains with $\mu = 0.1$ and $k = 0.04$. A relation with $\mu = 1/3$ and $k = 0$ has been widely used since its introduction in 1963,[13] but has been challenged by a number of investigators.[7,9–11] In alternative treatments, the factor α^3 in eq. 1 is replaced by $\alpha^{3-\epsilon}$, with ϵ a parameter to account for the dependence of Φ' on z,[7] mostly reflecting the difference in the dependence of K on z.

The Flory-Fox relation is not directly useful for semiflexible chains (intermediate $L/\hat{a}$) since the dependence of Φ' on $L/\hat{a}$ introduces a variation of $[\eta]/L^{1/2}$ with $L/\hat{a}$ that is not included in expressions for α (the dependence of K on $L/\hat{a}$ is much weaker).[7,9,10,12] For example, with decreasing $L/\hat{a}$, α tends to unity, but R_H/R_G and, to a small extent, K decrease with decreasing $L/\hat{a}$.[10] This is compensated by a decreased excluded volume effect as, for example, $A(L/\hat{a})$ tends to zero with decreasing $L/\hat{a}$, in addition to the direct effect on z. Care must be taken in this regime to account for both excluded volume effects and non-Gaussian statistics.

In the absence of excluded volume effects, extension of the Flory-Fox approximation to branched flexible chains is simply accomplished using $R_G^2 = \{\hat{a}L/3\}g$, where g depends on the chain branching configuration, and $R_H = \{2(\hat{a}L/3)^{1/2}/3\}h$, with $h \approx g^{1/2}$. The preceding gives $[\eta] \propto \Phi'g^{3/2}l^{1/2}$ or $[\eta] \propto KghL^{1/2} \approx Kg^{3/2}L^{1/2}$. Theoretical calculations for branched Gaussian chains[7,12,14] give $\Phi'/\Phi'_{\text{LIN}} \approx K/K_{\text{LIN}} \approx g^m$ with m approximately in the range 1 (star-shaped chains) to 0.5 (comb-shaped chains), demonstrating the limitations of the Flory-Fox relation and the complex and differing effects of hydrodynamic interactions on $[\eta]$ and R_H for these cases.

Application to polyelectrolytes requires a formulation for the dependence of $\hat{a}$ and d_H on electrostatic parameters such as the Bjerrum length, the linear charge density, and the Debye screening length.[12,15] Electrostatic interactions are well screened at high concentration of supporting electrolyte, resulting in behavior much like that of uncharged chains for flexible chain polyelectrolytes. Enhanced intramolecular interactions develop at low ionic strength, generally requiring recognition of the role of semiflexible chain statistics on $\hat{a}$ and d_T, as well as effects on both R_H and R_G leading to a variable Φ', and limited usefulness of a direct application of the Flory-Fox relation, even if an appropriate relation for α is used.

REFERENCES AND NOTES

1. P. J. Flory and T. G Fox, *J. Polym. Sci.*, **5**, 745 (1950).
2. P. J. Flory, *J. Chem. Phys.*, **17**, 303 (1949).
3. T. G Fox and P. J. Flory, *J. Phys. Colloid Chem.*, **53**, 197 (1949).
4. P. J. Flory and T. G Fox, *J. Am. Chem. Soc.*, **73**, 1904 (1951).
5. P. J. Flory and T. G Fox, *J. Am. Chem. Soc.*, **73**, 1909 (1951).
6. H. Morawetz, *Polymers: The Origins and Growth of a Science,* Wiley, New York, 1985.
7. H. Yamakawa, *Modern Theory of Polymer Solutions,* Harper and Row, New York, 1971.
8. P. J. Flory, *Principles of Polymer Chemistry,* Cornell University Press, Ithaca, NY, 1953.
9. K. F. Freed, *Renormalization Group Theory of Macromolecules,* Wiley, New York, 1987.
10. E. F. Casassa and G. C. Berry, in *Comprehensive Polymer Science;* Vol. 2, G. Allen (ed.), Pergamon Press, New York, 1988, p. 71.
11. M. Muthukumar, in *Comprehensive Polymer Science;* Vol. 2, G. Allen (ed.), Pergamon Press, New York, 1988, p. 1.
12. M. Bohdanecký and J. Kovář, *Viscosity of Polymer Solutions;* 2, Elsevier, Amsterdam, 1982.
13. M. Kurata and W. H. Stockmayer, *Adv. Polym. Sci.,* **3**, 196 (1963).
14. J. D. Ferry, *Viscoelastic Properties of Polymers,* 3rd ed., Wiley, New York, 1980.
15. C. Wei-Berk and G. C. Berry, *J. Polym. Sci.: Part B: Polym. Phys.,* **28**, 1873 (1990).

Molecular Configuration and Thermodynamic Parameters from Intrinsic Viscosities*

In previous publications[1,2] we pointed out that the intrinsic viscosity of a polymer in solution should be proportional to the ratio of the volume occupied by the molecule to the molecular weight M throughout the molecular weight range ordinarily of interest. As a convenient measure of the volume of a polymer molecule one may take the cube of the root-mean-square distance $(\overline{r^2})^{1/2}$ from beginning to end of the chain. In the absence of interactions of the segments with their environment, this distance, then designated as $(\overline{r_0^2})^{1/2}$, is proportional to the square root of the chain length, from which it follows that the intrinsic viscosity $[\eta]$ under such conditions should be proportional to $M^{1/2}$. In general, as the combined result of long range interactions between segments (the "volume filling" effect) and of segment–solvent interaction, the molecular dimensions will be increased appreciably.[1,3] Consequently, it is appropriate to write:

$$[\eta] = KM^{1/2}\alpha^3 \tag{1}$$

where $\alpha = (\overline{r^2}/\overline{r_0^2})^{1/2}$ represents the linear expansion factor arising from the above-mentioned interactions,[1,3] and:

$$K = \Phi(\overline{r_0^2}/M)^{3/2} \tag{2}$$

where Φ is a constant which should be *the same for all polymers irrespective of the solvent*.[3] According to the theory of Kirkwood and Riseman,[4] $\Phi \cong 3.6 \times 10^{21}$.[3]

The interaction of the chain segments with their environment may be treated as a mixing problem which leads to:[3]

$$\alpha^5 - \alpha^3 = 2\psi_1 C_M(1 - \Theta/T)M^{1/2} \tag{3}$$

where ψ_1 is an entropy of dilution factor[3,5] (equal to $1/2$ according to the simple lattice treatment[6,7]), Θ represents the temperature at which the second virial coefficient vanishes for the given solvent–polymer pair,[2,3,5,7] and:

$$C_M = (27/2^{5/2}\pi^{3/2}\mathrm{N})(\bar{v}^2/\mathrm{v_1})(M/\overline{r_0^2})^{3/2}$$
$$= 1.4 \times 10^{-24}(\bar{v}^2/\mathrm{v_1})(\Phi/K) \tag{4}[3]$$

where N is Avogadro's number and $\bar{v}$ and $\mathrm{v_1}$ are the partial specific volume of the polymer and the molar volume of the solvent, respectively. The heat of dilution parameter $\kappa_1 = BV_1/RT$ does not appear in (3) but is given by $\kappa_1 = \Theta\psi_1/T$.

Intrinsic viscosities of polyisobutylenes covering a wide molecular weight range[2] and measured with unusual precision as functions of temperature in a wide variety of solvents[8] are in quantitative agreement with equations

* This investigation was carried out at Cornell University in connection with the Government Research Program on Synthetic Rubber under contract with the Office of Rubber Reserve, Reconstruction Finance Corporation.

(1) and (3). The value of K was found to be independent of the solvent and of molecular weight as required by theory. Preliminary measurements on polystyrene,[9] supplemented with other data published recently,[10-12] also confirm the above relationships. In fact, no deviations from the theory which exceed experimental error have yet been observed in any instance, except at quite low molecular weights. We wish to call attention to the apparent general validity of the above relations and to indicate preferred procedures for their use.

Ordinarily it is advantageous to begin with the determination of the intrinsic viscosity in a poor solvent at the absolute temperature $T = \Theta$. Θ is conveniently established by plotting the critical miscibility temperatures for several fractions against $1/M^{1/2}$ and finding the intercept, which equals Θ.[3,7] At $T = \Theta$, $\alpha = 1$ according to equation (3), and K is readily calculated from $[\eta]$ if M is known. If similar determinations are carried out in other poor solvents having suitably different Θ values, K may be obtained as a function of T. If this procedure is adopted the molecular weight of only one fraction is required; those of all other samples may, in principle at least, be calculated from their intrinsic viscosities at $T = \Theta$ and the value of K. If intrinsic viscosity measurements in a poor solvent at $T = \Theta$ are impractical, as may be true if the melting point of the polymer is high, K may be ascertained from the intercept of a plot of $[\eta]^{2/3}/M^{1/3}$ against $M/[\eta]$ using measurements made in a given solvent at a fixed temperature.[3] High accuracy over a wide molecular weight range is required for the successful application of this method. Having established K, the value of α may be calculated from the intrinsic viscosity in any solvent. The slopes and intercepts of linear plots of $(K_T/K_0)(\alpha^5 - \alpha^3)/M^{1/2}$ against $1/T$ lead to $\psi_1 C_M$ and Θ for the polymer in each solvent. The ratio (K_T/K_0) of K at temperature T to its value at a reference temperature T_0 is introduced in view of the slight temperature dependence of C_M, which, according to equation (4), is inversely related to the temperature coefficient of K.

Eliminating K from equations (1) and (2) and recalling that $\overline{r^2} = \alpha^2 \overline{r_0^2}$:

$$\Phi = M[\eta]/(\overline{r^2})^{3/2} \qquad (5)$$

which suggests at once a means for establishing the value of the universal intrinsic viscosity constant Φ. Thus, the values of $\overline{r^2}$ deduced from measurements of the angular dissymmetry of light scattering by Zimm and coworkers[10] for polystyrenes of various molecular weights in several solvents, in conjunction with intrinsic viscosities in the same solvents, yield $\Phi \cong 2.1 \times 10^{21}$.[9] A similar value is indicated[8] by results of Kunst[11] on polyisobutylene solutions. Accurate assignment of the proper value for Φ is of the utmost importance, for it then becomes possible to calculate $\overline{r^2}$ from $[\eta]$ and M. Knowledge of Φ permits also the calculation of the important ratio $\overline{r_0^2}/M$ from K; the influence of hindrance to free rotation on chain dimensions is readily computed from $\overline{r_0^2}/M$.[3] In Table I are given values of K for several polymers obtained from recent investigations. For both polyisobutylene[8] and polystyrene,[9] K, and therefore $\overline{r_0^2}/M$ as well, appears to decrease gradually with increase in temperature (see Table I) indicating that barriers to free rotation are preferentially overcome in the direction of less extended configurations as the temperature is raised.

TABLE I

VALUES OF K AND CALCULATED DIMENSIONS OF THE RANDOMLY COILED POLYMER CHAINS UNPERTURBED BY THERMODYNAMIC INTERACTIONS

Polymer	T, °C.	$K \times 10^4$	$(\overline{r_0^2}/M) \times 10^{17}$	$\sqrt{\overline{r_0^2}/\overline{r_0^2}}$ (free rot.)[a]
Polyisobutylene[8]	25	10.6	6.32	1.93
	100	9.3	5.80	1.85
Polystyrene[9]	34	8.0	5.25	2.40
	70	7.3	4.24	2.32
Polymethyl methacrylate[b]	31.5	6.5	4.6	2.20

[a] $\overline{r_0^2}$ (free rotation) $= \overline{r_0^2}$ calculated assuming free rotation from the bond length (1.54 A.) and the fixed bond angle (109.5°).

[b] This value of K was obtained by a measurement of $[\eta]$ for a polymethyl methacrylate fraction of $M = 8.7 \times 10^5$ in di-n-propyl ketone at $T = \theta = 31.5$°C.

Assuming tentatively that the theoretical expression (4) is exact, C_M may be computed from $\overline{r_0^2}/M$ obtained in the above manner. ψ_1 and κ_1 may then be computed from $\psi_1 C_M$ and θ. Values of ψ_1 thus obtained are less than $1/2$ in all cases investigated thus far and are variable from one solvent to another over a tenfold range. These parameters should be independently determined from thermodynamic measurements in the future.

We believe that the procedures outlined above afford a simple and effective means for the investigation of both polymer chain configuration and the thermodynamics of polymer–solvent interaction. With respect to the former, the intrinsic viscosity procedure offers the advantage of greater sensitivity than other methods since it depends on the volume occupied by the polymer chain rather than on its linear dimension.

References

1. P. J. Flory, *J. Chem. Phys.*, **17**, 303 (1949).
2. T. G. Fox, Jr., and P. J. Flory, *J. Phys. & Colloid Chem.*, **53**, 197 (1949).
3. P. J. Flory and T. G. Fox, Jr., *to be published*.
4. J. G. Kirkwood and J. Riseman, *J. Chem. Phys.*, **16**, 560 (1948).
5. P. J. Flory and W. R. Krigbaum, *ibid.*, **18**, 1086 (1950).
6. M. L. Huggins, *J. Phys. Chem.*, **46**, 151 (1942); *Ann. N. Y. Acad. Sci.*, **43**, 1 (1942).
7. P. J. Flory, *J. Chem. Phys.*, **10**, 51 (1942).
8. T. G. Fox, Jr., and P. J. Flory, *to be published*.
9. T. G. Fox, Jr., and P. J. Flory, *to be published*.
10. P. Outer, C. I. Carr, and B. H. Zimm, *J. Chem. Phys.*, **18**, 830 (1950).
11. E. D. Kunst, *Rec. trav. chim.*, **69**, 125 (1950).
12. L. H. Cragg and J. E. Simkins, *Can. J. Research*, **B27**, 961 (1949).

PAUL J. FLORY
THOMAS G FOX, JR.†

Department of Chemistry
Cornell University
Ithaca, New York

Received September 25, 1950

† Present address: Rohm and Haas Company, Inc., Philadelphia, Pa.

COMMENTARY

Reflections on "The Counterion Distribution in Solutions of Rod-Shaped Polyelectrolytes," by Turner Alfrey, Jr., Paul W. Berg, and Herbert Morawetz, *J. Polym. Sci.,* VII, 543 (1951)

HERBERT MORAWETZ

Polytechnic University, 6 Metrotech Center, Brooklyn, New York 11201

The short paper by Alfrey, Berg, and myself summarized a portion of the theoretical part of my Ph.D. thesis on which I worked under the direction of Turner Alfrey. He suggested to me a theoretical study of the counterion distribution around rod-shaped polyions, which was supposed to facilitate the escape of the counterions from the polyion attraction. Since Alfrey had a very large number of graduate students who would have to line up in front of his office for a chance of a brief interview, I could not expect a great deal of help from him. Luckily, Paul Berg, a colleague of my wife who was studying for a Ph.D. in mathematics, took an interest in my problem and gave me crucial help in solving the Poisson–Boltzmann equation for cylindrical symmetry. It was a great surprise that an analytical solution could be obtained for the case of cylindrical symmetry without the use of the assumption that $\varepsilon\psi/kT \ll 1$, which Debye had to use for spherical symmetry. The unexpected feature of the results was the fact that for a polyion charge corresponding to that of poly(acrylic acid) at full ionization a sizable fraction of the counterions remained close to the polyion even at extreme dilution of the system. This contradicted the hitherto accepted view that dilution could separate the polyion from counterions. (It took me a month to compute the results using tables of trigonometric functions and logarithms—a task that could be done in a few hours with a hand calculator).

I took a draft of the thesis to the 1950 Gordon Research Conference to have it approved by Turner. Since the conference was chaired by Raymond Fuoss, who was deeply involved with polyelectrolytes, Turner suggested that I make a brief presentation. This was the first time that Herman Mark heard me speak and I believe that it induced him to offer me a faculty position at Poly's Polymer Research Institute when Alfrey left during the following year.

When I came as a postdoc to the Harvard Medical School protein laboratory I shared a bench with Ephraim Katchalski, who told me that his brother Aharon's student, Shneior Lifson, was working on the same problem. We obtained, of course, the same result, but it transpired later that our solution had long been known, since it was identical (except for boundary conditions) to the distribution of electrons around a heated wire.

I tried to derive an expression for the activity coefficients of the counterions, but J. J. Hermans pointed out a mistake in my approach. This discouraged me so much that I gave up writing up my results for publication. Eventually, I received a letter from Fuoss, who was spending a sabbatical at the Weizmann Institute. He wrote that he was preparing a paper (which happened largely to be based on Lifson's thesis) which he "seemed to remember to be similar to my work" and he advised me to write up my work "so as to protect my priority." This then was how the paper in *J. Polym. Sci.,* **VII,** 543 (1951) came into existence.

PERSPECTIVE

Comments on "The Counterion Distribution in Solutions of Rod-Shaped Polyelectrolytes," by Turner Alfrey, Jr., Paul W. Berg, and Herbert Morawetz, *J. Polym. Sci.*, VII, 543 (1951)

GERALD S. MANNING

Department of Chemistry, Rutgers University, New Brunswick, New Jersey 08903

The cylindrical cell model for salt-free polyelectrolyte solutions on the Poisson–Boltzmann level generates a rare example of an important nonlinear differential equation for which the exact solution is known. The solution was given independently by Alfrey, Berg, and Morawetz[1] and by Fuoss, Katchalsky, and Lifson.[2] The free energy and associated derivatives obtained from the exact Poisson–Boltzmann potential were immediately successful in providing a realistic description of electrochemical properties of salt-free charged polymer systems, and use of this theoretical breakthrough has continued with no sign of abatement for almost half a century.

A most interesting property of the formula for the exact Poisson–Boltzmann potential was reported by Zimm and Le Bret[3,4] about 10 years ago. For cylindrical charge densities greater than a certain critical value the formula implies emergence of a fraction of counterions which remains within finite distances from the cylinder even in the face of indefinite dilution. The critical charge density and the "bound" fraction of counterions coincide with the corresponding quantities contained in an alternate formulation for the polyelectrolyte free energy, counterion condensation theory.[5] Thus, Poisson–Boltzmann and counterion condensation theories can be reconciled.

The strong contribution of Professor Morawetz to Poisson–Boltzmann theory has indeed proved its viability and practical importance. It stands as a landmark in polyelectrolyte science, whose increasing numbers of practitioners guarantee it an interested and appreciative audience for years to come.

REFERENCES AND NOTES

1. T. Alfrey, P. W. Berg, and H. Morawetz, *J. Polym. Sci.*, **7**, 543 (1951).
2. R. M. Fuoss, A. Katchalsky, and S. Lifson, *Proc. Nat. Acad. Sci.*, **37**, 579 (1951).
3. B. H. Zimm and M. Le Bret, *J. Biomol. Struct. Dynam.*, **1**, 461 (1983).
4. B. H. Zimm and M. Le Bret, *Biopolymers*, **23**, 287 (1984).
5. G. S. Manning, *J. Chem. Phys.*, **51**, 924 (1969).

JOURNAL OF POLYMER SCIENCE   VOL. VII, NO. 5, PAGES 543–547

The Counterion Distribution in Solutions of Rod-Shaped Polyelectrolytes*

TURNER ALFREY, JR.,† PAUL W. BERG, and HERBERT MORA-
WETZ, *Department of Chemistry, Polytechnic Institute of Brooklyn and
Institute for Mathematics and Mechanics, New York University*

In any theory of the titration behavior of polyelectrolytes, a knowledge of the counterion distribution function is required for the calculation of the electrical free energy of ionization.   The problem has been examined by Hermans and Overbeek[1] for flexible polyelectrolyte molecules at low extension of the polymer coil.   Künzle suggested a treatment[2] which can be applied over a wider range of molecular configurations.

In both of these treatments the procedure of the Debye-Hückel theory is followed in that the ratio $\epsilon\psi/kT$ (where ϵ is the electronic charge, ψ the electrostatic potential, k Boltzmann's constant, and T the temperature) is assumed to be small compared with unity.   This assumption is hardly applicable to highly charged polyions, and it will be shown that for a special case an explicit solution can be obtained without the Debye-Hückel approximation.

Physical Model

The present contribution deals with the counterion distribution of partially neutralized polymeric acids in solutions containing no electrolytes except for the polyanions and their monovalent counterions.   The polyanions are represented by thin rods which are so long that end effects can be neglected, and the charges are considered to be spread uniformly over their surfaces.   The rod is the axis of a cylindrical region containing all the counterions belonging to it.   Neighboring chains are considered to be essentially parallel, and the time-average charge distribution is assumed to have cylindrical symmetry.

General Expression for Counterion Distribution

We shall use the following notation:

    a = radius of the rod representing the polyanion
    b = distance between ionizable groups on the polyanion
    r = distance from the axis of the polyanion

* H. Morawetz, Doctoral Dissertation, Polytechnic Institute of Brooklyn, 1950.
† Present address: Dow Chemical Co., Midland, Michigan.

α = degree of ionization
c = base molarity (normality) of polymeric acid
R = radius of cylinder containing the counterions of one polyanion
D = dielectric constant of the solution
ϵ = electronic charge
k = Boltzmann's constant
N = Avogadro's number
T = temperature
ρ = local charge density
ψ = electrostatic potential

Since the space occupied by a polyion and its counterions must be electrically neutral as a whole, we have the relation:

$$R = \sqrt{1/\pi cb} \tag{1}$$

To obtain the distribution of counterions around a charged rod, we proceed in a manner analogous to the treatment of the ion cloud in simple electrolyte solutions. The charge distribution is cylindrically symmetrical in our case, so that the Poisson equation assumes the form:

$$\frac{1}{r} \frac{d}{dr} \left(r \frac{d\psi}{dr} \right) = \frac{-4\pi}{D} \rho\,(r) \tag{2}$$

Between $r = a$ and $r = R$ the charge density $\rho(r)$ is due to counterions only. Their local density is governed by the Boltzmann distribution law. Thus, if we set the electrostatic potential at zero at points where the local charge density equals the average charge density ρ_0 of the bulk of the solution, we obtain:

$$\rho = \rho_0\, e^{\psi\epsilon/kT} \tag{3}$$

where:

$$\rho_0 = -Nc\alpha\epsilon/1000 \tag{4}$$

Combining (2), (3), and (4):

$$\frac{1}{r} \frac{d}{dr} \left(r \frac{d\psi}{dr} \right) = \frac{4\pi}{D} \cdot \frac{Nc\alpha\epsilon}{1000} \cdot e^{\psi\epsilon/Tk} \tag{5}$$

This differential equation has a solution of the form:

$$\psi(r) = \frac{1}{B} \left(\ln \frac{2\delta^2}{AB} - 2 \ln r - 2 \ln \cos\,(\delta[\ln r + \beta]) \right) \tag{6}$$

where δ and β are constants to be determined from the boundary conditions, and:

$$A = \frac{4\pi}{D} \cdot \frac{Nc\alpha\epsilon}{1000} \tag{7}$$

$$B = \epsilon/kT \tag{8}$$

The following boundary conditions are used: (a) The total charge on

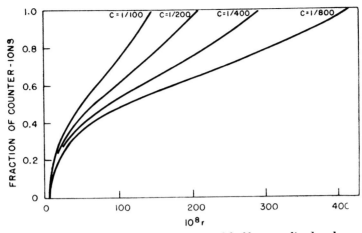

Fig. 2. Integral counterion distribution of half-neutralized polyacrylic acid at various solution concentrations ($a = 6 \times 10^{-8}$).

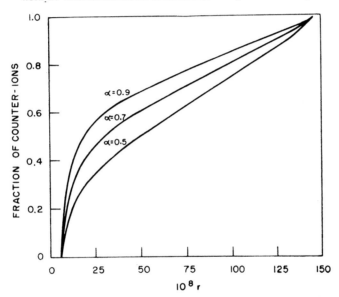

Fig. 3. Integral counterion distribution in 0.01 N polyacrylic acid at various degrees of neutralization ($a = 6 \times 10^{-8}$).

of the present treatment. The distance between carboxyl groups is $b = 2.5 \times 10^{-8}$ cm. For the radius of the rod a value of $a = 6 \times 10^{-8}$ cm. was chosen, but it was found that the counterion distribution is rather insensitive to this quantity. Calculations were performed for $T = 290$ and $D = 80$.

Figure 1 shows the electrostatic potential as a function of the distance from the polyanion. Figures 2 and 3 give the counterion distribution as a function of solution concentration and degree of neutralization of the polymeric acid. It can be seen that a large fraction of the counterions are conthe counterions is equal in magnitude and opposite in sign to the charge on the polyion, or:

$$-\alpha\epsilon/b = \int_{r=a}^{r=R} 2\pi r \rho(r)\, dr \tag{9}$$

(b) The electrical field vanishes at the boundary of the cylinder enclosing the counterions of a given polyion, due to the time-average cylindrical symmetry of neighboring polyions and their counterions. Thus:

$$(d\psi/dr)_{r=R} = 0 \tag{10}$$

If we set:

$$\delta(\ln R + \beta) = \xi \tag{11}$$

then the boundary conditions lead to:

$$\delta = \cot \xi \tag{12}$$

$$\xi = \tan^{-1}\left[\frac{1 - (\alpha\epsilon^2/bDkT)}{\delta}\right] + \delta \ln \frac{R}{a} \tag{13}$$

Numerical Evaluation

Kuhn, Künzle, and Katchalsky showed[3] that in dilute solutions polyacrylic acid molecules are essentially completely extended between the degrees of neutralization $\alpha = 0.5$ and $\alpha = 0.9$. Although this conclusion has lately been questioned,[4] the charge density of such extended polyacrylic acid molecules was taken as the basis for a numerical evaluation of results

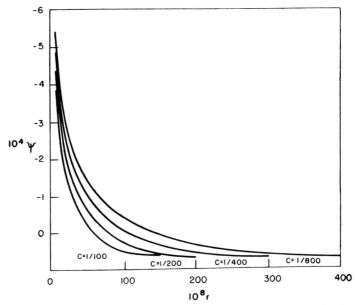

Fig. 1. Electrostatic potential in solutions of half-neutralized polyacrylic acid ($a = 6 \times 10^{-8}$).

centrated at short distances from the polyanion even at comparatively high dilution.

Discussion

The high counterion densities calculated for the neighborhood of polymeric acids, even at the dilution used by Kuhn, Künzle, and Katchalsky in their experiments, throw doubt on their assumption that counterions did not affect appreciably the configuration of the flexible polyelectrolyte. As a result, it is questionable whether the polymer was really sufficiently extended to be approximated by the rigid rod used as a model in our treatment.

However, tobacco mosaic virus is a polyelectrolyte with a rigid rodlike shape and the relations derived may be applicable to the behavior of its solutions. It is interesting to note that tobacco mosaic virus molecules are actually observed to assume positions parallel to each other in hexagonal close packing,[5] which would probably result in only minor modifications of the counterion distribution calculated on the basis of cylindrical symmetry.

References

1. J. J. Hermans and J. T. G. Overbeek, *Bull. soc. chim. Belg.*, No. 57, 154 (1948).
2. O. Künzle, *Proc. Intern. Colloquium on Macromolecules*, p. 296. Centen, Amsterdam, 1950.
3. W. Kuhn, O. Künzle, and A. Katchalsky, *Helv. Chim. Acta*, 31, 1984 (1948).
4. A. Oth and P. Doty, *J. Phys. & Colloid Chem.*, in press.
5. J. D. Bernal and I. Fankuchen, *J. Gen. Physiol.*, 45, 111 (1941).

Synopsis

For solutions containing rod-shaped polyelectrolytes and their counterions only, the Poisson-Boltzmann equation can be solved explicitly without the usual assumption of the Deybe-Hückel theory. At charge densities corresponding to half-neutralized polyacrylic acid a large proportion of monovalent counterions is located close to the polyion even in solutions as dilute as 1/800 base molar.

Résumé

Pour des solutions contenant des polyélectrolytes sous formes de bâtonnets et leurs ions opposés, l'équation de Poisson-Boltzmann peut être résolue explicitement sans faire appel aux hypothèses habituelles de la théorie de Debye-Hückel. Aux densités de charge correspondant à un acide polyacrylique demi-neutralisé, une large proportion d'ions opposés monovalents est localisée à proximité du polyion, même en solutions aussi diluées que 1/800 molaire.

Zusammenfassung

Für Lösungen, die nur stäbchenförmige Polyelektrolyte und deren Gegenionen enthalten, kann die Poisson-Boltzmann Gleichung vollkommen ohne die übliche Voraussetzung der Debye-Hückel Theorie gelöst werden. Bei Ladungsdichten, die halb-neutralisierter Polyacrylsäure entsprechen, ist ein grosser Anteil der monovalenten Gegenionen nahe dem Polyion lokalisiert, selbst in sehr verdünnten Lösungen, so wie 1/800 basenmolar.

Received July 9, 1951

COMMENTARY

Reflections on "The General Theory of Irreversible Processes in Solutions of Macromolecules," by John G. Kirkwood, *J. Polym. Sci.,* XII, 1 (1954)

MARSHAL FIXMAN

Department of Chemistry, Colorado State University, Fort Collins, CO 80523

The elegance and clarity of this paper make it worth reading today. Still, Kirkwood might not be entirely pleased to find that his paper has more than historical value, some four decades after its publication. He would be puzzled to learn what little progress has been made along the path he mapped out. But despite the beauty of Riemannian geometry and its seeming relevance to the problems of polymer dynamics, the approach has been taken up only sporadically and for purely theoretical objectives. Actual computations with generalized coordinates, i.e., with dihedral angle coordinates, have been restricted to very small systems. Many bad basis functions, the Gaussian or Rouse modes, have been easier to work with than fewer good ones, just as in quantum chemistry. Perhaps Kirkwood would concede the difficulty of solving partial differential equations in spaces with nonorthogonal basis vectors, and with metric tensors that are sufficiently opaque to have caused him one error in this paper (later corrected). But he would also be slightly impatient; good students are supposed to overcome difficulties and not just complain about them.

This paper exemplifies the style of Kirkwood's approach to a field of science, a formal style which had a seductive influence on a generation of theoretical chemistry students in the 40s and 50s, myself included. I would like to sketch, from my limited perspective, the scientific atmosphere of the time when the paper was written.

In late 1953 I had completed my doctoral work with Stockmayer at MIT, and moved to Yale to begin postdoctoral research with Kirkwood. He had recently arrived from CalTech to become Chairman of the Yale Chemistry Department, and to add to the distinction that Onsager had already brought to Yale theoretical chemistry.

Kirkwood had visited MIT during the previous year and I recall quite vividly, even now, a brief exchange I had with him in one of the long corridors of Building 6. I asked him what he thought of Flory's recent theory of the excluded volume effect and the use of that theory for the interpretation of intrinsic viscosities. "Theory?" he said with noticeable impatience, "Do you mean Flory's hypothesis?" At the time I took this to be a fearful omen of standards I would not meet. But Kirkwood at Yale was patient at all meetings but the last, when I confessed that my wit was insufficient to evade the persistent calls of the military draft.

In 1954 Kirkwood doubted the existence of a significant excluded volume effect. I recall his praise of an author who had just written a paper demonstrating its absence by perturbation theory. When I presented my doctoral research at a Yale seminar, covering similar ground but supporting Flory, Kirkwood was sceptical and Onsager was actively (but temporarily) opposed. Their scepticism was not based solely on a distaste for the crude segment cloud model that Flory had used. Of more personal impact, Kirkwood was not satisfied with the approximate theories of intrinsic viscosity then available, the Kirkwood–Riseman and Debye–Bucche theories. On the basis of these theories Flory had concluded that the observed dependence of hydro-dynamic radius on molecular weight was to be explained by excluded volume effects rather than by hydrodynamic screening. At that time experimental results from Rayleigh scattering were still sparse, the Zimm theory of chain dynamics with hydrodynamic interaction was not yet published, quasi-elastic light scattering, renormalization theory, and computer simulations of large systems lay far in the future.

Much is now known that was the subject of hot debate in 1954, but this sword, wrought by a great scientist and teacher, still lies in the stone.

PERSPECTIVE

Comments on "The General Theory of Irreversible Processes in Solutions of Macromolecules," by John G. Kirkwood, *J. Polym. Sci.,* XII, 1 (1954)

JACK DOUGLAS

Polymers Division, National Institutes of Standards and Technology, Gaithersburg, MD 20899

The problem of describing the transport properties of polymer chains which are dynamically evolving through a wide range of configurations through Brownian motion of the chain segments and the chain as a whole is cast in terms of a diffusion equation acting within molecular configuration space. Kirkwood's elegant formulation of polymer chain dynamics, utilizing methods of Riemann geometry, generalized Kramers' previous formulation of polymer chain dynamics in the absence of intrachain hydrodynamic interactions.[1] The hydrodynamic interaction modeling of polymers, based on the idealized Oseen tensor, also had a precedent in the work of Burgers[2] so that Kirkwood's contribution represents a natural logical synthesis of earlier work. Kirkwood's review article spans a wide range of applications—viscoelastic response of polymer solutions to applied stresses, flow birefringence and the Kerr effect, and dielectric polarization and dispersion.

Although the general conceptual framework developed by Kirkwood and its earlier precedents provide an appealing framework for treating a wide range of hydrodynamic problems involving suspensions of flexible particles, this theory is difficult to apply in its complete generality to concrete polymer models[3] and a variety of approximations were necessarily introduced by Kirkwood for reasons of mathematical expediency. Kirkwood's calculations, for example, involved a number of "preaveraging" approximations which are still active subjects of investigation and frustration because adequate analytic methods to avoid certain of these approxima-

tions remain elusive. The preaveraging approximations included an "angular averaging" approximation of the Oseen tensor to reduce the tensorially defined hydrodynamic interaction to a scalar interaction. Recent calculations[4] suggest that this approximation is actually rather minor in comparison with the "configurational preaveraging" approximation[5,6] which can be on the order of a 10% error for common transport properties such as the intrinsic viscosity and the diffusion coefficient. It is also now appreciated that Kirkwood's often cited expression for the diffusion coefficient of a flexible polymer [eq. (9) of Kirkwood's paper] is incorrect due to a faulty transformation between the configuration space and coordinate space representations of the hydrodynamically interacting polymer.[7] Akcasu showed[8] that Kirkwood's expression for the diffusion coefficient is actually correct for the "short time" diffusion coefficient (assuming configurational and angular preaveraging approximations) and found that the error involved in Kirkwood's "approximation" is less than 2% for flexible chains. Wang et al.[6] later showed that Kirkwood's approximate expression for the diffusion coefficient results from an additional "contour preaveraging" approximation within the configurationally preaveraged Kirkwood-Riseman (rigid body) theory. The various preaveraging approximations introduced by Kirkwood remain an active topic of investigation and the configurational preaveraging approximation remains a quantitative limitation on the accuracy of analytic calculations.

The subtle technical difficulties in Kirkwood's calculations reflect the complexity of his general Riemann geometry formulation of polymer dynamics and it was no doubt a great relief to many polymer science theorists when Zimm recognized[9] that Kirkwood's general theory could be recast into a more tractable form using chain normal coordinates similar to those introduced shortly before by Rouse[10] for an idealized "bead-spring" model of polymer chains without intramolecular hydodynamics interactions. The Rouse theory in turn had its antecedent in terms of Kuhn's coarse grained modeling of polymer chains in terms of statistical segments which naturally led to a dynamical analog involving chains of harmonic oscillators to represent the dynamical polymer chain segments. Zimm's specialized form of the general Kirkwood theory in terms of a normal coordinate representation has become a cornerstone of polymer solution theory and the influence of Kirkwood's *Journal of Polymer Science* article is evident from Zimm's prominent citation of this work. Bixon[3] and Zwanzig[11] later established the foundations of Zimm's important conceptual advance of the Kirkwood theory and these developments have recently led to more molecularly faithful modeling of polymers through the incorporation of statistical information related to the detailed chain structure through the Rouse–Zimm force constant matrix. This type of "optimized Rouse–Zimm" model has recently been refined in applications to protein dynamics involving inhomogeneous polymer structure, complicated bond rotation potentials, excluded volume interactions, and other physically important features such as mode-coupling effects.[12] The Kirkwood formalism, as modified by Zimm, continues to evolve into an increasingly realistic model of polymer dynamics which should ultimately have applications to protein folding, the dynamics of polymer adsorption, polymer collapse in poor solvents, and other important dynamical processes of polymers in solution. It should also be mentioned that the Kirkwood theory also developed into a tractable theory of rigid particle hydrodynamics[13] and justified earlier, more heuristic, treatments of rigid body hydrodynamics by Kirkwood and Riseman.[14] Yamakawa and Yamaki[13] have discussed the conceptual foundations of this rigid body approach to polymer hydrodynamics, which has also led to many applications. Interestingly, the configurational preaveraging errors seem to be different in the Rouse–Zimm theory of ideally flexible chains and the Kirkwood–Riseman theory of perfectly rigid chains, suggesting that the degree of configurational preaveraging error depends on the degree of dynamic rigidity of the polymer chain [see ref. 6b].

Although the full Riemann geometry formulation of polymer hydodynamics developed by Kirkwood is rather uncommonly utilized in the polymer literature at present, this formalism has recently been revived by Brenner and co-workers[15] in an effort to avoid angular and configurational preaveraging approximations and also the point source ("bead") model of intrachain hydrodynamic interactions. These calculations, which are restricted by their technical difficulty to simple models such as a pair of tethered spheres, indicate that preaveraging calculations miss important Taylor-dispersion-type phenomena associated with the dynamical fluctuations of flexible bodies. This work shows that the original Kirkwood formalism continues to have importance in developing a greater understanding of the dynamics of polymer solutions.

REFERENCES AND NOTES

1. a) H. A. Kramers, *J. Chem. Phys.*, **14**, 415 (1946). b) J. G. Kirkwood, *Rec. Trav. Chim.*, **68**, 649 (1949).
2. J. M. Burgers in *Second Report on Viscosity and Plasticity,* North Holland, Amsterdam, 1938, p. 113–184.
3. M. Bixon, *J. Chem. Phys.*, **58**, 1459 (1973).
4. J. F. Douglas, H.-X. Zhou, and J. B. Hubbard, *Phys. Rev. E*, **49**, 5319 (1994).
5. a) B. H. Zimm, *Macromolecules,* **13**, 592 (1980). b) J. G. de la Torre, A. Jiminez, and J. Friere, *Macromolecules,* **15**, 148 (1982).
6. a) S.-Q. Wang, J. F. Douglas, and K. F. Freed, *J. Chem. Phys.*, **85**, 3674 (1986), **87**, 1346 (1987). b) J. F. Douglas and K. F. Freed, *Macromolecules,* **27**, 6088 (1994).
7. a) Y. Ikeda, *Kobayashi Rigaku Kenkyusho Hokoku,* **6**, 44 (1956). b) R. Zwanzig, *J. Chem. Phys.*, **45**, 1858 (1966).
8. A. Z. Akcasu, *Macromolecules,* **15**, 1321 (1982).
9. B. H. Zimm, *J. Chem. Phys.*, **24**, 269 (1956).
10. P. E. Rouse, Jr., *J. Chem. Phys.*, **21**, 1272 (1953).
11. a) R. Zwanzig, *J. Chem. Phys.*, **60**, 2717 (1974). b) A. Perico and M. Guenza, *J. Chem. Phys.*, **83**, 3103 (1985); **84**, 510 (1986).
12. a) X. Y. Chang and K. F. Freed, *J. Chem. Phys.*, **99**, 8016 (1993). b) Y. Hu, J. M. Macinnis, B. J. Cherayil, G. R. Fleming, K. F. Freed, A. Perico, *J. Chem. Phys.*, **93**, 822 (1990).
13. H. Yamakawa and J. Yamaki, *J. Chem. Phys.*, **58**, 2049 (1973).
14. a) J. G. Kirkwood and J. Riseman, *J. Chem. Phys.*, **16**, 565 (1948). b) J. G. Kirkwood, *Macromolecules*, P. L. Auer (ed.), (Gordon and Breach, New York, 1967.
15. a) A. Nadim and H. Brenner, *Phys. Chem. Hyd.*, **11**, 315 (1989). b) H. Brenner, A. Nadim, S. Haber, *J. Fluid. Mech.*, **183**, 511 (1987). c) S. Haber, H. Brenner, M. Shapiro, *J. Chem. Phys.*, **92**, 5569 (1990).

JOURNAL OF POLYMER SCIENCE   VOL. XII, PAGES 1–14 (1954)

SYMPOSIUM ON MACROMOLECULES

Stockholm-Uppsala, 1953

PART I. Properties of Macromolecules

The General Theory of Irreversible Processes in Solutions of Macromolecules

JOHN G. KIRKWOOD, *Sterling Chemistry Laboratory, Yale University, New Haven, Connecticut*

INTRODUCTION

The course of irreversible processes in solutions of macromolecules is determined by the hydrodynamic forces which the large molecules exert on the solvent and by the Brownian drift in configuration which the molecules experience under the combined action of the frictional hydrodynamic forces and of external fields of force, gravitational, centrifugal, electric, or magnetic. Perturbations in the flow pattern of the solvent by the macromolecules, which determine the visco-elastic behavior of their solutions, are most conveniently treated by the Oseen method[1] which is based upon solutions of the Navier-Stokes equation possessing singularities appropriate to the frictional forces exerted on the solvent by the segments of a macromolecule. Macromolecular Brownian motion may be described by a distribution function satisfying a generalized equation of forced diffusion in molecular configuration space.

In an earlier investigation,[2] we have endeavored to formulate a unified statistical mechanical theory of irreversible processes in solutions of macromolecules, based upon these concepts. Following Kramers,[3] the methods of Riemannian geometry were employed to formulate a generalized theory of Brownian motion in molecular configuration space leading to a generalized diffusion equation. The Oseen method was used in the analysis of the perturbations produced in the flow pattern of the solvent and for the determination of the components of the molecular diffusion tensor. The molecular relaxation time spectrum was constructed from the eigenvalues of the diffusion operator, and the perturbations in the distribution function in molecular configuration space were expanded in the corresponding eigenfunctions. It is our purpose here to review the general theory and to outline its application to the analysis of a set of structurally significant

1

phenomena, visco-elastic response to applied stresses, flow birefringence and the Kerr effect, and dielectric polarization and dispersion.

Since we shall not attempt to take molecular interaction into account, the applicability of our results is limited to highly dilute solutions. Furthermore, the schematic character of the molecular model which we employ limits the structural details which may be deduced by means of the theory from experimental data relating to irreversible processes. Further advances in the theory should be directed toward the development of methods for treating molecular interaction and toward the exploration of the properties of more detailed molecular models.

GENERALIZED MACROMOLECULAR BROWNIAN MOTION

We shall adopt the general feature of the pearl necklace model to describe the hydrodynamic behavior of a macromolecule in solution in a solvent of low molecular weight. The molecule is regarded as an array of n identical structural units, attached to a rigid or flexible framework, and immersed in a structureless fluid continuum, which is supposed to possess the viscosity coefficient, refractive index, and dielectric constant of the solvent in bulk. Each structural unit of the molecule is assumed to exert a hydrodynamic force on the solvent which is proportional to its velocity relative to the local particle velocity of the fluid, with a friction constant ζ, in accordance with the theory of Brownian motion. Due to structural restraints, fixed bond angles, bond lengths, etc., the molecule will in general possess a number of degrees of freedom, ν, which is less than $3n$. Three of its degrees of freedom will be translational, associated with the coordinates specifying the position of its center of mass, three will be rotational, associated with coordinates specifying its orientation relative to an external frame of reference, and ν-6 will be internal degrees of freedom associated with coordinates specifying the configuration of the n structural units relative to each other. In the case of rigid molecules, such as the globular proteins, the significant degrees of freedom are the six degrees of translational and rotational freedom. In the case of flexible molecules, such as the high polymers, there are additional degrees of freedom describing internal configuration, specified, for example, by the angles between the planes determined by successive pairs of bonds of the skeletal chain of the polymer molecule. In general, because of the small amplitude of the motion associated with them, we can ignore the influence of vibrational degrees of freedom on the hydrodynamic behavior of solutions of macromolecules. We shall refer to the ν-dimensional molecular configuration space as m-space, a point in which will be specified by a conveniently selected set of generalized coordinates, $q^1 \ldots q^\nu$. The m-space is a sub-space of the complete $3n$-dimensional configuration space of the n structural elements, which will be referred to as e-space. If $\mathbf{R}$ is the $3n$-dimensional vector specifying the position of the n structural elements in e-space, we may span m-space by the following set of covariant vectors, $\mathbf{a}_\alpha$:

$$\mathbf{a}_\alpha = \sum_{l=1}^{n} (\partial \mathbf{R}^l / \partial q^\alpha)$$

$$\mathbf{R} = \sum_{l=1}^{n} \mathbf{R}^l \tag{1}$$

where $\mathbf{R}^l$ is the projection of $\mathbf{R}$ on the 3-dimensional space of element l, and the derivatives are to be taken at constant values of all other q^β and subject to the structural restraints characteristic of the molecule. The metric tensor of the m-space is given by:

$$g_{\alpha\beta} = \sum_{l=1}^{n} \frac{\partial \mathbf{R}^l}{\partial q^\alpha} \cdot \frac{\partial \mathbf{R}^l}{\partial q^\beta}$$

$$\mathbf{a}^\alpha = \sum_\beta g^{\alpha\beta} \mathbf{a}_\beta$$

$$g^{\alpha\beta} = \frac{|g|_{\alpha\beta}}{g}; \qquad g = |g_{\alpha\beta}| \tag{2}$$

where $\mathbf{a}^\alpha$ is the contravariant vector reciprocal to $\mathbf{a}_\alpha$ and $|g|_{\alpha\beta}$ is the appropriate minor of the determinant. The explicit determination of the elements $g_{\alpha\beta}$ of the metric tensor depends upon a knowledge of the structural details of the molecule.

The probability density $f(q,t)$ of the ensemble describing the statistical mechanical behavior of the system, macromolecule-solvent, determines the observed value of a function $\varphi(q)$ of the coordinates q as the average value:

$$\bar{\varphi}(t) = \int \ldots \int \sqrt{g} \; \varphi(g) f(q,t) \prod_\alpha dq^\alpha \tag{3}$$

where the integration extends over all of m-space. In the canonical ensemble appropriate for thermodynamic equilibrium, the probability density $f(q,t)$ reduces to:

$$f^0(q) = e^{\beta[A_0 - W_0(q)]} \tag{4}$$

$$\beta = 1/kT$$

where A_0 is the configurational free energy of the molecule and $W_0(q)$ the potential of average force associated with its internal degrees of freedom. For systems departing from equilibrium, the probability density, $f(q,t)$, is determined by the generalized diffusion equation:

$$\sum_{\substack{\alpha,\beta \\ =1}}^{\nu} \frac{1}{\sqrt{g}} \frac{\partial \sqrt{g}}{\partial q^\beta} \left(D^{\alpha\beta} \frac{\partial f}{\partial q^\alpha} + \frac{D^{\alpha\beta}}{kT} \frac{\partial W_0}{\partial q^\alpha} f \right) - \frac{\partial f}{\partial t} =$$

$$\sum_{\substack{\alpha,\beta \\ =1}}^{\nu} \frac{1}{\sqrt{g}} \frac{\partial \sqrt{g}}{g \; \partial q^\beta} \left(\frac{D^{\alpha\beta}}{kT} X_\alpha f + g^{\alpha\beta} v_\alpha^0 f \right); \int \ldots \int \sqrt{g} f(q,t) \prod_\alpha dq^\alpha = 1 \tag{5}$$

where the X_α are the covariant components of external force, for example, external electric, magnetic, or gravitational fields, and the v_α^0 are the covariant components of the hydrodynamic particle velocity of the solvent, unperturbed by the presence of the macromolecule. The quantities $D^{\alpha\beta}$ are the contravariant components of the diffusion tensor in m-space, given by:

$$D^{\alpha\beta} = kT[(g^{\alpha\beta}/\zeta) + T^{\alpha\beta}]; \qquad T^{\alpha\beta} = \sum_{\substack{l \pm s \\ =1}}^{n} T_{ls}^{\alpha\beta}$$

$$T_{ls}^{\alpha\beta} = \frac{1}{8\pi\eta_0 R_{ls}} \sum_{\substack{\sigma,\tau \\ =1}}^{\nu} g^{\alpha\sigma} g^{\beta\tau} \left[\frac{\partial \mathbf{R}_l}{\partial q_\sigma} \cdot \frac{\partial \mathbf{R}_s}{\partial q_\tau} + \frac{1}{R_{ls}^2} \left(\mathbf{R}_{ls} \cdot \frac{\partial \mathbf{R}_l}{\partial q^\sigma} \right) \left(\mathbf{R}_{ls} \cdot \frac{\partial \mathbf{R}_s}{\partial q^\tau} \right) \right] \quad (6)$$

where the $T^{\alpha\beta}$ are the contravariant components of the Oseen hydrodynamic interaction tensor, η_0 is the viscosity coefficient of the solvent, and ζ is the friction constant associated with the motion of an isolated structural unit of the molecule in the solvent. The vectors $\mathbf{R}_l$ and $\mathbf{R}_s$ specify the positions of the structural units l and s in a common 3-dimensional space and R_{ls} is the distance between them in the specified configuration.

The determination of the general elements $D^{\alpha\beta}$ of the diffusion tensor requires detailed knowledge of the molecular structure. However, the translational components, associated with coordinates q^1, q^2, q^3, specifying the position of the center of mass in an external rectangular coordinate system, may be expressed in general form, by virtue of the relations:

$$g_{\alpha\beta} = n\delta_{\alpha\beta}$$

$$g^{\alpha\beta} = \frac{\delta_{\alpha\beta}}{n}; \qquad \alpha = 1, 2, 3; \beta = 1, \ldots, \nu \quad (7)$$

Equations (6) and (7) lead to the simple result:

$$D^{\alpha\beta} = kT \left[\frac{\delta_{\alpha\beta}}{n\zeta} + \frac{1}{8\pi n^2 \eta_0} \sum_{\substack{l \pm s \\ =1}}^{n} \left(\frac{\delta_{\alpha\beta}}{R_{ls}} + \frac{X_{ls}^\alpha X_{ls}^\beta}{R_{ls}^3} \right) \right] \qquad \alpha, \beta = 1, 2, 3 \quad (8)$$

where X_{ls}^α is the component of the distance between units l and s along the rectangular axis α of the coordinate system to which the position of the center of mass is referred. The mean translational diffusion constant, $\bar{D}$, is equal to one-third of the trace of $D^{\alpha\beta}$, averaged over the internal coordinates:

$$\bar{D} = kT \left[\frac{1}{n\zeta} + \frac{1}{6\pi n^2 \eta_0} \sum_{\substack{l,s \\ =1}}^{n} \overline{\left(\frac{1}{R_{ls}} \right)} \right] \quad (9)$$

The sedimentation constant, s, is then given by:

$$s = \frac{M}{nN} \left[\frac{1}{\zeta} + \frac{1}{6\pi n \eta_0} \sum_{\substack{l,s \\ =1}}^{n} \overline{\left(\frac{1}{R_{ls}} \right)} \right] \quad (10)$$

where M is the molecular weight and N is Avogadro's number. For the randomly coiled polymer molecule, consisting of n statistical units of length b, we have:

$$\bar{D} = \frac{kT}{n\zeta}\left(1 + \frac{8}{3}\lambda_0 n^{1/2}\right)$$

$$s = \frac{M_0}{N\zeta}\left(1 + \frac{8}{3}\lambda_0 n^{1/2}\right)$$

$$\lambda_0 = \zeta/\sqrt{6\pi^3\eta_0 b} \tag{11}$$

where M_0 is the molecular weight of the statistical unit. This result agrees with that obtained by Kirkwood and Riseman[4] by a less rigorous method. Except in sedimentation processes, the translational coordinates are redundant, since $f(q,t)$ is independent of them, and equations (3) and (5) are valid in the space of the internal coordinates of the molecule with ν equal to the total number of degrees of freedom less three, with $f(q,t)$ normalized to unity in the internal space alone.

We shall now investigate the solution of equation (5) by the methods of perturbation theory. We first transform equation (5) in the following manner:

$$f(q,t) = e^{\beta[A_0 - W_0]/2}\, \rho(q,t)$$

$$L\rho - (\partial\rho/\partial t) = -Q\rho$$

$$L = \sum_{\alpha,\beta}\frac{1}{\sqrt{g}}\frac{\partial\sqrt{g}}{\partial q^\alpha}\left(D^{\alpha\beta}\frac{\partial}{\partial q^\beta}\right) + U \tag{12}$$

$$U = \frac{1}{2kT}\sum_{\alpha,\beta}\frac{1}{\sqrt{g}}\frac{\partial\sqrt{g}}{\partial q^\alpha}\left(D^{\alpha\beta}\frac{\partial W_0}{\partial q^\beta}\right) - \frac{1}{(2kT)^2}\sum_{\alpha,\beta}D^{\alpha\beta}\frac{\partial W_0}{\partial q^\alpha}\frac{\partial W_0}{\partial q^\beta}$$

$$Q(q,t) = -e^{-\beta[A_0 - W_0]/2}\sum_{\alpha,\beta}\frac{1}{\sqrt{g}}\frac{\partial\sqrt{g}}{\partial q^\beta}\left\{\left[\frac{D^{\alpha\beta}}{kT}X_\alpha + g^{\alpha\beta}v^0_\alpha\right]e^{\beta[A_0 - W_0]/2}(\quad)\right\}$$

The differential operator L is self-adjoint, and therefore possesses a complete orthonormal set of eigenfunctions ψ_λ:

$$L\psi_\lambda + \lambda\psi_\lambda = 0 \tag{13}$$

with negative eigenvalues, $-\lambda$, subject to the boundary conditions of single-valuedness and integrability in m-space. We now suppose that the operator Q may be expanded in powers of a parameter γ, for example, the rate shear of the velocity field $\mathbf{v}^0$ in the solvent, or the strength of an externally applied electric field:

$$Q = \gamma Q^{(1)} + \gamma^2 Q^{(2)} + \ldots \tag{14}$$

We further suppose that the function ρ may be similarly expanded:

$$\rho(q,t) = f^{0\,1/2} + \sum_{s=1}^{\infty}\rho^{(s)}\gamma^s$$

$$f^{0\,1/2} = e^{\beta[A_0 - W_0]/2} \tag{15}$$

where the first term corresponds to thermodynamic equilibrium and the remaining terms describe the departure of the distribution function from its equilibrium value as a result of the perturbation Q, given by equation (14). Equation (12), (14), and (15) then yield the following system of inhomogeneous differential equations for the functions $\rho^{(s)}$

$$L\rho^{(1)} - (\partial\rho^{(1)}/\partial t) = -Q^{(1)}f^{0\,1/2}$$

$$L\rho^{(2)} - (\partial\rho^{(2)}/\partial t) = -Q^{(1)}\rho^{(1)} - Q^{(2)}f^{0\,1/2}$$

$$L\rho^{(s)} - (\partial\rho^{(s)}/\partial t) = -Q^{(1)}\rho^{(s-1)} - Q^{(2)}\rho^{s-2} + \ldots \qquad (15)$$

The function $\rho^{(1)}(q,t)$ may be expanded in the eigenfunctions $\psi_\lambda(q)$ of the operator L in the form:

$$\rho^{(1)}(q,t) = \int_{-\infty}^{+\infty} G^{(1)}(q,\omega)e^{i\omega t}d\omega$$

$$G^{(1)}(q,\omega) = \sum_\lambda \frac{B_\lambda^{(1)}(\omega)}{\lambda + i\omega}\psi_\lambda(q)$$

$$B_\lambda^{(1)} = (1/2\pi)\int_{-0}^{+\infty}(Q^{(1)}f^{0\,1/2})_\lambda e^{-i\omega t}dt$$

$$(Q^{(1)}f^{0\,1/2})_\lambda = \int\ldots\int\sqrt{g}\psi_\lambda^*(q)Q^{(1)}f^{0\,1/2}\prod_\alpha dq^\alpha \qquad (16)$$

Carrying the perturbation calculation to the second order, we obtain for the function $\rho^{(2)}(q,t)$:

$$\rho^{(2)}(q,t) = \int_{-\infty}^{+\infty} G^{(2)}(q,\omega)e^{i\omega t}\,d\omega$$

$$G^{(2)}(q,\omega) = \sum_\lambda \frac{G_\lambda^{(2)}(\omega)}{\lambda + i\omega}\psi_\lambda(q)$$

$$G_\lambda^{(2)}(\omega) = \sum_{\lambda'}\int_{-\infty}^{+\infty}\frac{K_{\lambda\lambda'}^{(1)}(\omega - \omega')B_\lambda^{(1)}(\omega')}{\lambda' + i\omega'}\,d\omega' + B_\lambda^{(2)}(\omega)$$

$$K_{\lambda\lambda'}^{(1)}(\omega) = (1/2\pi)\int_{-\infty}^{+\infty}Q_{\lambda\lambda'}^{(1)}e^{-i\omega t}\,dt$$

$$Q_{\lambda\lambda'}^{(1)} = \int\ldots\int\sqrt{g}\psi_\lambda^*Q^{(1)}\psi_{\lambda'}\prod_\alpha dq^\alpha$$

$$B_\lambda^{(2)} = (1/2\pi)\int_{-\infty}^{+\infty}(Q^{(2)}f^{0\,1/2})_\lambda e^{-i\omega t}\,dt \qquad (17)$$

Perturbations of higher order may be obtained by similar methods. Mean values of functions $\varphi(q)$, equation (3), may now be expanded in the form:

$$\bar\varphi(t) = \bar\varphi^0 + \gamma\bar\varphi^{(1)} + \gamma^2\bar\varphi^{(2)} + \ldots$$

$$\bar\varphi^0 = \int\ldots\int\sqrt{g}\varphi e^{\beta[A_0 - W_0]}\prod_\alpha dq^\alpha$$

$$\bar\varphi^{(1)} = \int\ldots\int\sqrt{g}\varphi e^{\beta[A_0 - W_0]/2}\rho^{(1)}\prod_\alpha dq^\alpha$$

$$\bar\varphi^{(2)} = \int\ldots\int\sqrt{g}\varphi e^{\beta[A_0 - W_0]/2}\rho^{(2)}\prod_\alpha dq^\alpha \qquad (18)$$

where $\bar{\varphi}^0$ corresponds to thermodynamic equilibrium. The techniques useful in determining the eigenfunctions ψ_λ and eigenvalues $-\lambda$ of the operator L are identical with those employed in the solution of the Schrödinger equation in quantum mechanics. The reciprocals of the λ-spectrum constitute the relaxation time spectrum, $\tau = 1/\lambda$, of the system, a set of real positive number, determined by the diffusion tensor $D^{\alpha\beta}$.

VISCO-ELASTIC PARAMETERS

The visco-elastic properties of solutions of macromolecules are determined by the hydrodynamic forces, $-\mathbf{F}_l$, which the structural units of the molecule exert on the solvent. These forces produce perturbations in the velocity field of the solvent, which may manifest themselves not only in an increment in the viscosity coefficient but also by imparting a rigidity to the solution for time dependent rates of strain. The intrinsic viscosity $[\eta]$ of the solution is related to the hydrodynamic forces $\mathbf{F}_l$, exerted by the fluid on the structural elements according to the following relation:[4]

$$[\eta] = N\Phi/100M\eta_0$$

$$\Phi = -\frac{1}{\dot{\epsilon}^2} \sum_{l=1}^{n} \overline{\mathbf{F}_l \cdot \mathbf{V}_l^0} \tag{19}$$

where $\dot{\epsilon}$ is the magnitude of the rate of shear, and $\mathbf{V}_l^0$ is the unperturbed velocity of the solvent at the point of location of structural element l relative to the velocity of the center of gravity of the molecule. For a simple alternating shear of frequency $\omega/2\pi$, in direction $\mathbf{e}_x$ propagated in the direction $\mathbf{e}_y$, we may write:

$$\mathbf{V}_l^0 = \dot{\epsilon}(\mathbf{R}_{ol} \cdot \mathbf{e}_y)\mathbf{e}_x e^{i\omega t} \tag{20}$$

where $\mathbf{R}_{ol}$ is the position of element l relative to the molecular center of mass. According to the Oseen method, as employed by Burgers[1] and by Kirkwood and Riseman,[4] the forces $\mathbf{F}_l$, regarded as vectors in the common 3-space of all structural elements, have been shown to satisfy the following set of linear equations:

$$\mathbf{F}_l + \zeta \sum_{s=1}^{n} \mathbf{T}_{ls} \cdot \mathbf{F}_s = \zeta(\mathbf{V}_l^0 - \mathbf{U}_l)$$

$$\mathbf{T}_{ls} = \frac{1}{8\pi\eta_0 \mathbf{R}_{ls}} \left(1 + \frac{\mathbf{R}_{ls}\mathbf{R}_{ls}}{R_{ls}^2}\right) \tag{21}$$

where $\mathbf{R}_{ls}$ is the vector distance between elements l and s, $\mathbf{1}$ is unit tensor, $\mathbf{U}_l$ is the velocity of element l, and ζ is the friction constant of a single element. The velocities $\mathbf{U}_l$ are determined by balancing the components of the hydrodynamic force in m-space by the corresponding components of the diffusion forces, according to the theory of Brownian motion, to obtain:

$$\mathbf{U}_l = \sum_\alpha v^{0\alpha} \frac{\partial \mathbf{R}_{ol}}{\partial q^\alpha} - \sum_{\alpha,\beta} D^{\alpha\beta} \frac{\partial \log f}{\partial q^\beta} \frac{\partial \mathbf{R}_{ol}}{\partial q^\alpha} \tag{22}$$

where f is the probability density in m-space, which satisfies equation (3). For alternating shearing strain f exhibits, according to equations (12), (15), and (16), a lag in phase relative to the rate of shear in the solvent. As a consequence, the function Φ possesses both real and imaginary parts. and thus $[\eta]$ also possesses both real and imaginary parts:

$$[\eta] = [\eta'] - i[\eta'']$$
$$[\mu] = \lim_{c \to 0} \mu/c = \omega\eta_0[\eta''] \tag{23}$$

where the real part $[\eta']$ is the observed intrinsic viscosity, and the imaginary part determines the intrinsic rigidity $[\mu]$, experimentally observed by Mason and Baker[8] and their collaborators. A detailed application of the theory which has been outlined here to the rectilinear pearl necklace molecule has been carried out by Kirkwood and Auer.[5] For a molecule composed of a set of hydrodynamically resisting elements spaced at equal intervals b on a straight line segment of length L, they obtain:

$$[\eta'] = \frac{\pi N b L^2}{9000 M_0 \log (L/b)} \left(1 + \frac{3}{1 + \omega^2\tau^2}\right)$$
$$[\mu] = \frac{6NkT}{1000M} \frac{\omega^2\tau^2}{1 + \omega^2\tau^2}$$
$$\tau = \pi\eta_0 L^3/18kT \log (L/b) \tag{24}$$

For stationary viscous flow, $\omega = 0$, the rigidity of the solution vanishes and $[\eta']$ becomes equal to the asymptotic form of Simha's formula[6] for the prolate ellipsoid of eccentricity approaching unity.

BIREFRINGENCE

When macromolecules dissolved in optically isotropic solvents are subjected to external torques, hydrodynamic, electric, or magnetic, their solutions become birefringent. This phenomenon is known as flow birefringence when the torque is hydrodynamic and as the Kerr effect when the torque is produced by an external electric field. In this section, we shall briefly describe the phenomenon of macromolecular birefringence in terms of the general theory.

We suppose each structural unit of the molecule to possess a polarizability tensor increment α_l, dependent on the coordinates q_α, in excess of the polarizability of the solvent which it displaces. The average polarizability tensor $\bar{\alpha}$ of the entire molecule is then given by equation (3) in the form:

$$\alpha = \sum_{l=1}^{n} \alpha_l$$
$$\bar{\alpha} = \int \ldots \int \sqrt{g}\,\alpha f(q,t) \prod_\alpha dq^\alpha \tag{25}$$

In flow birefringence produced by a simple shear or in the Kerr effect produced by a homogeneous electric field, $\bar{\alpha}$ is uniaxial. If we denote by the unit vector $\mathbf{e}_x$ the direction of the streamlines of the solvent velocity field in the first instance or the direction of the electric field in the second, the birefringence, $n_1 - n_2$, equal to the difference of the principal refractive indices in the x,y plane, is given by:

$$n_1 - n_2 = \frac{4\pi}{3}\left(\frac{n_0^2 + 2}{3}\right)^2 c\,[(\bar\alpha_{xx} - \bar\alpha_{yy})^2 + 4\bar\alpha_{xy}^2]^{1/2} \tag{26}$$

where n_0 is the refractive index of the solvent, c is the number of macromolecules in unit volume, and $\bar\alpha_{xx}$, $\bar\alpha_{yy}$, $\bar\alpha_{xy}$ are the appropriate components of the mean polarizability tensor. The tangent of the extinction angle χ between $\mathbf{e}_x$ and the principal axis $\mathbf{e}_2$ of the dielectric constant tensor in the x-y plane is related to $\bar\alpha_{xx}$, $\bar\alpha_{yy}$, and $\bar\alpha_{xy}$ in the following manner:

$$\tan\chi = \frac{[(\bar\alpha_{xx} - \bar\alpha_{yy})^2 + 4\bar\alpha_{xy}^2]^{1/2} + \bar\alpha_{xx} - \bar\alpha_{yy}}{2\bar\alpha_{xy}}$$

$$= 1 + \frac{\bar\alpha_{xx} - \bar\alpha_{yy}}{2\bar\alpha_{xy}} + \cdots$$

$$\chi = \frac{\pi}{4} + \frac{\bar\alpha_{xx} - \bar\alpha_{yy}}{2\bar\alpha_{xy}} + \cdots \tag{27}$$

We consider first the case of flow birefringence produced by a constant rate of shear in the solvent, corresponding to the velocity field of equation (20) with ω equal to zero. We set the perturbation parameter γ of equation (14) equal to the rate of shear $\dot\epsilon$, and obtain from the last of equation (12):

$$Q = \dot\epsilon Q^{(1)}$$

$$Q^{(1)} = -e^{-\beta[A_0 - W_0]/2}\sum_{\alpha,\beta}\frac{1}{\sqrt{g}}\frac{\partial\sqrt{g}}{\partial q^\beta}\left[g^{\alpha\beta}(\mathbf{R}_{ol}\cdot\mathbf{e}_y)\left(\mathbf{e}_x\cdot\frac{\partial\mathbf{R}_{ol}}{\partial q^\alpha}\right)f^{0\,1/2}(\ )\right] \tag{28}$$

For the stationary case, the distribution functions $\rho^{(1)}$ and $\rho^{(2)}$ of equations (16) and (17) reduce to:

$$\rho^{(1)}(q) = \sum_\lambda\frac{(Q^{(1)}f^{0\,1/2})_\lambda}{\lambda}\psi_\lambda(q)$$

$$\rho^{(2)}(q) = \sum_{\lambda,\lambda'}\frac{Q^{(1)}_{\lambda\lambda'}(Q^{(1)}f^{0\,1/2})_{\lambda'}}{\lambda\lambda'}\psi_\lambda(q) \tag{29}$$

With these distribution functions, we obtain from equations (18) and (25):

$$\bar\alpha_{xy} = \dot\epsilon\sum_\lambda\frac{(\alpha_{xy}f^{0\,1/2})_\lambda(Q^{(1)}f^{0\,1/2})_\lambda}{\lambda} + \cdots$$

$$\bar\alpha_{xx} - \bar\alpha_{yy} = \dot\epsilon^2\sum_{\lambda,\lambda'}\frac{[(\alpha_{xx} - \alpha_{yy})f^{0\,1/2}]_\lambda Q^{(1)}_{\lambda\lambda'}(Q^{(1)}f^{0\,1/2})_{\lambda'}}{\lambda\lambda'} \tag{30}$$

The mean values of equation (30), when substituted in equations (26) and (27), yield the desired expansions of the birefringence and extinction angle in powers of the rate of shear, as sums over the relaxation time spectrum of the molecule.

Birefringence due to the Kerr effect produced by an alternating electric field of frequency ω, with direction $\mathbf{e}_x$ may be treated in the following manner. The internal electric field $\mathbf{E'}$, acting on the macromolecule in a nonpolar solvent is:

$$\mathbf{E'} = \frac{\epsilon_0 + 2}{3}\,\mathbf{E}$$

$$\mathbf{E} = E_0\mathbf{e}_x \cos \omega t \tag{31}$$

where E_0 is the amplitude of the external field and ϵ_0 is the dielectric constant of the solvent. The applied field produces generalized torques X_α in m-space which may be derived from a potential, V:

$$V = -\frac{\epsilon_0 + 2}{3}\,\mathbf{\mu}\cdot\mathbf{e}_x E_0 \cos \omega t - \frac{1}{2}\left(\frac{\epsilon_0 + 2}{3}\right)^2 (\mathbf{e}_x\cdot\mathbf{\alpha}\cdot\mathbf{e}_x)E^2 \cos^2 \omega t$$

$$X_\alpha = -\frac{\partial V}{\partial q^\alpha} \tag{32}$$

where $\mathbf{\mu}$ is the electric dipole moment of the molecule, equal to the sum of the dipole moments $\mathbf{\mu}_i$ of its structural units, and $\mathbf{\alpha}$ is again the polarizability tensor. Setting the perturbation parameter of equation (14) equal to the amplitude, E_0, of the electric field, we obtain from equations (12) and (32) the following perturbation operator:

$$Q = Q^{(1)}E_0 + Q^{(2)}E_0^2$$

$$Q^{(1)} = -\frac{e^{-\beta[A_0 - W_0]/2}}{kT}\frac{\epsilon_0 + 2}{3}\sum_{\alpha,\beta}\frac{1}{\sqrt{g}}\frac{\partial\sqrt{g}}{\partial q^\beta}\left[D^{\alpha\beta}\frac{\partial\mathbf{\mu}\cdot\mathbf{e}_x}{\partial q^\alpha}f^{0\,1/2}(\ )\right]$$

$$Q^{(2)} = -\frac{e^{-\beta[A_0 - W_0]/2}}{2kT}\left(\frac{\epsilon_0 + 2}{3}\right)^2\sum_{\alpha,\beta}\frac{1}{\sqrt{g}}\frac{\partial\sqrt{g}}{\partial q^\beta}\left[D^{\alpha\beta}\frac{\partial\mathbf{e}_x\cdot\mathbf{\alpha}\cdot\mathbf{e}_x}{\partial q^\alpha}f^{0\,1/2}(\ )\right]$$

$$\tag{33}$$

Equations (17), (18), and (33) lead to the appropriate mean values of $\bar{\alpha}_{xx}$, $\bar{\alpha}_{yy}$, and $\bar{\alpha}_{xy}$, and these in turn yield the birefringence and extinction angle when substituted in equations (26) and (27). The rather cumbersome resulting expressions will not be displayed here. The birefringence is proportional to the square of the amplitude of the applied field and the extinction angle is found to be $\pi/2$, since $\bar{\alpha}_{xy}$ vanishes by symmetry for all field strengths. Thus the two principal axes of the dielectric constant tensor remain parallel and perpendicular to the direction of the applied field. The relaxation of the Kerr effects arising from $Q^{(1)}$ and $Q^{(2)}$, the contributions of the permanent and induced electric moments, exhibit entirely different frequency dependence, with the induced component the only surviving one at very high frequencies.

DIELECTRIC POLARIZATION AND DISPERSION

We shall briefly review the application of the general theory to the analysis of dielectric polarization and dispersion of solutions of polar macromolecules. The principal results of the previous treatment of the problem, with special reference to polar polymers, by Kirkwood and Fuoss[7] will be shown to be a consequence of the general theory of irreversible processes.

We shall suppose the solution to be polarized by a homogeneous alternating electric field of frequency $\omega/2\pi$ acting in the $\mathbf{e}_z$ direction, equal to the real part of the complex field $\mathbf{E}$:

$$\mathbf{E} = \mathbf{e}_z E_0 e^{i\omega t} \tag{34}$$

The internal field $\mathbf{E}'$ acting on a macromolecule will be approximated by the Lorentz field in the case of nonpolar solvents and by the Onsager field in the case of polar solvents.

$$\mathbf{E}' = E_0' \mathbf{e}_z e^{i\omega}$$

$$E_0'/E_0 = \frac{\epsilon_0 + 2}{3}; \quad \text{nonpolar solvent}$$

$$E_0'/E_0 = \frac{3\epsilon_0}{2\epsilon_0 + 2}; \quad \text{polar solvent} \tag{35}$$

where ϵ_0 is the real dielectric constant of the solvent, which will be assumed to have negligible dispersion at frequency ω.

The complex dielectric constant increment, Δ, per molecule is defined by the relation:

$$\Delta = \lim_{c \to 0} (d\epsilon/dc)$$

$$\epsilon = \epsilon' - i\epsilon'' \tag{36}$$

$$\Delta = \Delta' - i\Delta''$$

where ϵ is the complex dielectric constant of the solution and c is the number of macromolecules in unit volume. If we ignore the optical contribution to its polarization and also assume that the polarization of the solvent displaced by a macromolecule is negligible, the dielectric increment Δ is related to the average component of the macromolecular dipole moment $\overline{\mathbf{\mu} \cdot \mathbf{e}_z}$ in the direction of the applied field in the following manner:

$$\Delta = \sigma\mu_x^{(1)}$$

$$\overline{\mathbf{\mu}\cdot\mathbf{e}_x} = \mu_x^{(1)}E_0'e^{i\omega t} + O(E_0'^2)$$

$$\sigma = 4\pi\left(\frac{\epsilon_0 + 2}{3}\right)^2;\quad \text{nonpolar solvent}$$

$$\sigma = \frac{12\pi\epsilon_0^2}{2\epsilon_0^2 + 1};\quad \text{polar solvent} \tag{37}$$

The appropriate perturbation function, adequate for the calculation of $\mu_x^{(1)}$ is obtained from equations (12) and (14) by setting the parameter γ equal to the amplitude E_0' of the internal field.

$$Q = Q^{(1)}E_0'$$

$$X_\alpha = \frac{\partial(\mathbf{\mu}\cdot\mathbf{e}_x)}{\partial q^\alpha}E_0'\,e^{i\omega t}$$

$$Q^{(1)} = -\frac{1}{f^{0\,1/2}kT}\sum_{\alpha,\beta}\frac{1}{\sqrt{g}}\frac{\partial\sqrt{g}}{\partial q^\beta}\left[D^{\alpha\beta}\frac{\partial(\mathbf{\mu}\cdot\mathbf{e}_x)}{\partial q^\alpha}f^{0\,1/2}(\ )\right]$$

$$Q^{(1)}f^{0\,1/2} = -\frac{1}{kT}L[f^{0\,1/2}(\mathbf{\mu}\cdot\mathbf{e}_x)]e^{i\omega t} \tag{38}$$

We then obtain from equations (16), (18), and (27):

$$\mu_x^{(1)} = \frac{1}{kT}\sum_\lambda\frac{\lambda(\widetilde{\mathbf{\mu}}\cdot\mathbf{e}_x)_\lambda^*(\widetilde{\mathbf{\mu}}\cdot\mathbf{e}_x)_\lambda}{\lambda + iw}$$

$$= \frac{1}{3kT}\sum_\lambda\frac{\widetilde{\mathbf{\mu}}_\lambda^*\cdot\widetilde{\mathbf{\mu}}\lambda}{1 + i\omega\tau_\lambda}$$

$$\widetilde{\mathbf{\mu}} = f^{0\,1/2}\mathbf{\mu} \qquad \tau_\lambda = 1/\lambda$$

$$\sum\widetilde{\mathbf{\mu}}_\lambda^*\cdot\widetilde{\mathbf{\mu}}_\lambda = \int\cdots\int\sqrt{g}\,f^0(q)\mu^2\prod_\alpha dq^\alpha = \overline{\mu^{20}} \tag{39}$$

where $\overline{\mu^{20}}$ is the mean square electric moment of the macromolecule in the absence of the external field.   The real and complex parts of the dielectric constant increment, Δ, are then given by:

$$\Delta' = \frac{\sigma}{3kT}\sum_\lambda\frac{\widetilde{\mathbf{\mu}}_\lambda^*\cdot\widetilde{\mathbf{\mu}}_\lambda}{1 + \omega^2\tau_\lambda^2}$$

$$\Delta'' = \frac{\sigma}{3kT}\sum_\lambda\frac{\omega\tau\widetilde{\mathbf{\mu}}_\lambda^*\cdot\widetilde{\mathbf{\mu}}_\lambda}{1 + \omega^2\tau_\lambda^2} \tag{40}$$

Equation (40) generalizes the results of the earlier treatment of Kirkwood and Fuoss and presents them as a consequence of the general theory.

IRREVERSIBLE PROCESSES IN MACROMOLECULE SOLUTIONS   13

References

1. J. M. Burgers, *Second Report on Viscosity and Plasticity*, Amsterdam Academy of Sciences, Nordemann, 1938, Chap. III.
2. J. G. Kirkwood, *Rec. trav chim.*, **68**, 649 (1949).
3. H. A. Kramers, *J. Chem. Phys.*, **16**, 565 (1948).
4. J. G. Kirkwood and J. Riseman, *J. Chem. Phys.*, **16**, 565 (1948).
5. J. G. Kirkwood and P. Auer, *J. Chem. Phys.*, **19**, 281 (1951).
6. R. Simha, *J. Phys. Chem.*, **44**, 25 (1940).
7. J. G. Kirkwood and R. M. Fuoss, *J. Chem. Phys.*, **9**, 329 (1941).
8. Mason, Baker, McSkenin, and Heiss, *Phys. Rev.*, **73**, 1074 (1948).

Synopsis

The general theory of irreversible processes in solutions of macromolecules, previously formulated by the author, is reviewed.   The theory is based upon the Oseen method for determining the perturbation in the hydrodynamic flow pattern produced by the frictional forces exerted by the macromolecule on the solvent, and on a generalized theory of Brownian motion in molecular configuration space.   Applications of theory to viscoelastic behavior, flow birefringence, and the Kerr effect, and to dielectric dispersion are presented in outline.

Résumé

La théorie de procès irréversibles dans les solutions de macromolécules, plus tôt présentée par l'auteur, est résumée.   Elle est fondée sur la méthode d'Oseen pour la détermination des variations de l'écoulement hydrodynamique causées par des forces de friction, qui sont exercées sur le dissolvant par la macromolécule, ainsi que sur une théorie généralisée du mouvement brownien dans l'espace de la configuration moléculaire.

Les applications de la théorie à la visco-elasticité, à la biréfrigence d'écoulement et l'effet Kerr ainsi qu'à la dispersion diélectrique ont été ébauchées.

Zusammenfassung

Die vom Verfasser früher aufgestellte Theorie irreversibler Processe in Lösungen von Makromolekeln wird revidiert.   Die Theorie gründet sich auf Oseens Methode zur Bestimmung der Störungen im Verlaufe der hydrodynamischen Strömung, welche durch von der Makromolekel auf das Lösungsmittel ausgeübte Reibungskräfte verursacht sind, sowie auf eine von allgemeinste Theorie der Brownschen Bewegung im Raume molekularer Anordnung.

Anwendungen der Theorie auf visco-elastisches Verhalten, Strömungsdoppelbrechung und Kerreffect sowie dielektrische Dispersion werden in Umrissen vorgelegt.

Received January 16, 1953

Discussion

Professor W. Kuhn (*Basel*):   For some years we have been particularly interested in the irreversible processes connected with the orientation of geometrically anisotropic particles and with the orientation and deformation of chain molecules in streaming solutions.   (W. Kuhn and H. Kuhn, *Helv. Chim. Acta*, 37, 97 (1944); 38, 1533 (1945); 39, 71 (1946).)

In the case of very elongated particles in a streaming solution a *partial orientation* of the long axis is taking place, an orientation which gives rise to angular diffusion of the particle axis which is analogous to the ordinary diffusion of particles in a liquid in which a concentration gradient exists.   Taking this irreversible (angular) diffusion process into account increases the heat developed per unit time and volume by a factor of 2.

14   J. G. KIRKWOOD

That is, the intrinsic viscosity is twice as high (compared with the case in which only
the hydrodynamic effects with neglect of the Brownian movement are taken into account) if the orientation of the particle axis by the field of flow and the irreversible diffusion processes competing with the orienting influences are taken into account.   This
factor 2 is of interest.   It disappears in the case of very high velocity gradients where
Brownian movement becomes negligible, a circumstance which explains the decrease of
viscosity with increasing rate of shear in very dilute solutions of geometrically anisotropic particles.

For less elongated particles, the factor is smaller than 2 and becomes equal to 1 in the
case of spheres.   The factor 2 is therefore characteristic for small rate of shear and for
very elongated particles.   It would be interesting to know whether 2 is a general maximum value of the factor by which the heat development per unit of time is increased,
taking the Brownian movement into account, above the value which would occur if the
effects of the Brownian movement were neglected.   The general theory developed by
Dr. Kirkwood might give an answer to this question.

Professor J. G. Kirkwood (*Yale*): Dr. Kuhn was the first to point out that the
hydrodynamic torque on a macromolecule is not zero but equal to the rotatory diffusion
torque, and to show that the intrinsic viscosity is significantly affected by the gradient
of the orientational distribution function.   This was also recognized by Simha (*J. Phys.
Chem.*, **44**, 25 (1940)).   However, both Simha and Kirkwood and Auer (*J. Chem. Phys.*,
3, 281 (1951)) find that the intrinsic viscosity of extremely elongated molecules is four
times the value obtained with the neglect of rotatory Brownian motion, whereas Dr.
Kuhn finds a factor of two.   I do not know the reason for this discrepancy.

COMMENTARY

Reflections on "The Melting Points of Chain Polymers" by C. W. Bunn, *J. Polym. Sci.,* XVI, 323 (1955)

A. KELLER

H. H. Wills Physics Laboratory, University of Bristol, Tyndall Avenue, Bristol BS8 1TL, UK

Precisely 40 years ago polymers were still upcoming materials with a seemingly unlimited horizon. Most of their applications at the time were structural, i.e., in one form or another they were supposed to support load. This they could only do, at least above the glass transition, if they were crystalline. They lost this ability on melting, hence, understandably, the melting point was of central interest: specifically to foresee the melting point in the synthesis of new compounds, and specially to be able to aim for high melting points, which was desirable for most purposes. An understanding of the relation between chemical constitution and melting point was therefore imperative. The present paper by C. W. Bunn was a principal milestone in this direction.

The paper sets about its monumental task by approaching the subject from two angles. From the one side it attempts *a priori* predictions from basic principles taken from the physical science of small molecules by first considering the monomer (or rather the chemical repeat unit along the chain) and then applying the conclusions to the corresponding polymer. From the other side, more heuristically, the author collates all existing knowledge on melting points of crystalline polymers available at the time and scrutinizes for relationships between melting and chemical structures. The two lines of approach were then combined and tested to see how far predicting schemes and observational experiences matched. In most cases they did, but in some they failed to do so. The latter cases were then subjected to further scrutiny for yet other rational connections

not initially recognized. Indeed, in many cases such connections arose as a consequence of the repeat units linking up into chains.

Here I can only sample the numerous accomplishments that arose from the above approach, from the present viewpoint often self-evident, but at the time pioneering. Along the heuristic line such is the systematic display of melting behavior in homologous series and the correlation between series of different basic chemical types. Along the predictive line there is the identification of two determining factors: molecular interaction, expressed as cohesive energy density (CED), and chain flexibility.

The former increases and the latter decreases the melting point, as demonstrated by quantitative or semi-quantitative estimates. The success in explaining these observations on the above basis is impressive. Foremost is the realization that the competition between the two factors, CED and flexibility, can explain features which appeared incomprehensible at the time, e.g., the minima in T_m as a function of the frequency of O atoms along the chain in linear polyethers and polyesters, the origin of low melting points of unsaturated elastomers, and more subtly, the large differences in between the trans and cis conformations. Further, there is the recognition of the odd–even variation in polyamides and the transference of the well established conceptions of the end group packing effect in n-alkanes to the anchoring effect of dipolar chain portions within the long chain polyamides. Other features recognized include the stiffening influence of phenyl groups, shape effects, and the effect of symmetry in the disposition of non-identical groups along the chains.

While we may marvel at the range and scope of the achievement, as it was then, it is also salutary to note the limitations arising from lack of information and availability of materials now taken for granted. Thus, isotactic polyolefins (just about to be discovered when the paper was submitted) clearly do not fall into any of the classifications and schemes contained in the paper with their melting points completely inexplicable at the time.

(Or, just as a thought, do we understand them now?) Neither were liquid crystal polymers in sight, which obviated, to a certain extent at least, the consideration of the free energy of the molten state, thus allowing (to a good approximation— even if no such qualification is made in the paper) the attribution of melting behavior essentially to the intrinsic features of the crystalline state alone.

JOURNAL OF POLYMER SCIENCE   VOL. XVI, PAGES 323–343 (1955)

The Melting Points of Chain Polymers

C. W. BUNN, *Imperial Chemical Industries, Ltd., Plastics Division, Welwyn Garden City, Herts, England*

INTRODUCTION

When a crystalline polymer is heated, the first-order change from a crystalline solid to an amorphous phase (which may be a viscous liquid or a rubber-like solid, depending on the length of the molecules) is undoubtedly the same type of molecular process as the melting of a monomeric substance. It is true that there are two points of difference in the behavior of polymers and monomers: no long-chain polymer is completely crystalline, and melting occurs over an appreciable range of temperature, not sharply as in monomers; nevertheless, the change in the x-ray diffraction pattern from a sharp pattern characteristic of three-dimensional order to a very diffuse pattern similar to that of a monomeric liquid, together with the absorption of a considerable latent heat of melting, indicates essentially similar molecular processes, and in attempting to understand the melting points of polymers in relation to their molecular structures, it is justifiable and indeed highly desirable to include monomeric substances in the discussion and to link up the two classes of substances by tracing the influence of molecular length of the melting point.

The advantages of this approach are empirical rather than theoretical, for it must be admitted that even for monomeric crystals of the simplest substances the process of melting is not at all well understood from any fundamental theoretical standpoint. At the present time there is little hope of predicting theoretically the relation between chemical constitution and the melting point, even for simple monomeric compounds. Nevertheless, it is now possible, by empirical comparative methods, to discern in a qualitative manner the main molecular characteristics which determine the melting points of monomeric compounds, and this knowledge is of value in approaching the melting points of the long-chain polymers.

THEORIES OF MELTING

Before proceeding in this way, it is worth while glancing at the two main theoretical approaches to melting which have been made in recent years. Lindemann[1] suggested that a crystal melts when the amplitude of vibration of the molecules (which increases with rise of temperature) reaches a critical magnitude related to the distance between neighboring molecules; at this

323

point the crystal, so to speak, shakes itself to pieces. Lennard-Jones and Devonshire,[2] on the other hand, focus attention on the increase of lattice defects in crystals as the temperature increases, and show that the lattice eventually becomes unstable with respect to the liquid phase as the proportion of defects grows. The type of lattice defect envisaged—transfer of molecules to interstitial sites—seems very unlikely to occur; but the particular type of defect is perhaps not so important as the general concept of defects; certainly in most crystals, defects such as vacant lattice sites, or molecules in correct sites but wrong orientations, are likely to be much more common. Of course, the distortions due to thermal waves through the crystal are also defects, and so perhaps the defect theory, considered quite generally, does embrace both approaches; but one should distinguish between distortion waves going through the crystal and isolated displacements and misorientations of molecules. It is worth remarking that, in high polymer crystals, one cannot displace or rotate a chain unit without dragging the rest of the chain (which extends through the entire crystal) with it to some extent; a rigid chain would have to move as a whole; in a more flexible molecule, distortion at one point would produce waves along the chain. We are therefore inevitably led to focus attention on thermal wave motions in the crystals. Furthermore, if the melting point depends on the amplitude of the thermal vibrations, then the degree of flexibility of any molecule, monomeric or polymeric, must influence the melting point strongly; in a very flexible molecule small portions of the molecule vibrate semiindependently, so that the melting point would be like that of the small vibrating units—that is, it would be much lower than that of a comparable rigid molecule. Therefore, in considering the melting points of monomeric crystals, it is necessary to distinguish between rigid and flexible molecules, and, for simplicity, to deal with rigid molecules first.

The melting point, T_m, is of course equal to $\Delta H/\Delta S$, the ratio of latent heat of melting to entropy of melting; but at the present time the approach to melting points by way of this expression does not appear profitable. To understand or predict melting points by this approach, it would be necessary to predict, from the chemical constitution, first of all the latent heat of melting, and, *quite independently*, the entropy of melting. The relations between these quantities are at present even more obscure than the direct relation of melting point to chemical constitution. For this reason, this aspect of the subject is by-passed, and direct correlations between chemical constitution and melting temperatures are sought.

FACTORS CONTROLLING MELTING POINTS OF MONOMERS

In monomeric substances, the melting point for a rigid molecule (one in which no large changes of configuration such as are produced by rotations around single bonds are possible) would be expected to depend on the total forces holding the molecule in place in the lattice, *i.e.*, the cohesion energy,

E, which can be measured experimentally as the molar latent heat of evaporation, L (E is taken as $L - RT$, the second term being a small correction for the work done on expansion). If this were all, we should expect a simple relation between melting and boiling points, since the boiling point is directly proportional to the molar latent heat of evaporation of the liquid at the boiling point (Trouton's rule), which in turn is approximately proportional to the latent heat of evaporation of the crystal. In actual fact, the melting points of rigid molecules vary very widely, some being nearly equal to the boiling points while others are less than half the absolute boiling points. The melting points of all rigid molecules for which I have been able to find the necessary data (several hundreds) lie in the broad band ABCD in Figure 1, in which the horizontal axis gives cohesion energy

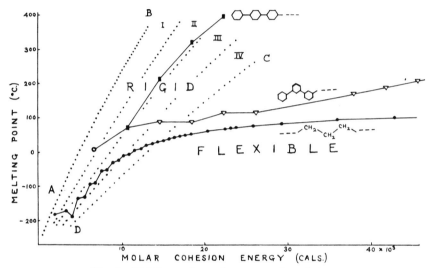

Fig. 1. Melting points of monomeric substances.

at the boiling point. (The cohesion energy figures were obtained from the latent heats of boiling, where these were available; or from the boiling point and Trouton's rule; or, failing this, from an additive scheme (details given later) in which each functional chemical group is associated with a definite cohesion energy increment—a scheme which works fairly well for all substances for which it can be tested.) A plot of melting points against boiling points would look very similar to this.

An attempt to discern empirically what molecular characteristics determine the position of any substance in this broad band showed that there is a convincing correlation between the melting point and the *symmetry of general over-all shape;* not molecular symmetry in the strict sense—it is general over-all shape that is important. All molecules of approximately spherical shape (ranging from the rare gases, through SF[6] and the like, to cage molecules like camphor derivatives) are at the top of the band, with melting points nearly equal to their boiling points; most cylindrical molecules are in the first subdivision, flat symmetrical molecules in the second, flat unsymmetrical molecules in the third, and asymmetric molecules in the fourth. The tendency for spherical molecules to have high melting points is already recognized[3]; but it does not appear to be generally realized that the correlation is remarkably good—for there are no exceptions in this group—all the substances fall nearly on a straight line; nor is it realized that the influence of shape extends through the entire shape-spectrum. Another general tendency is that within any one type of shape-symmetry, a skeletal shape with pronounced hollows in the outline depresses the melting point (details will be published elsewhere).

The mechanism whereby low shape-symmetry depresses the melting point is a matter for speculation. I shall offer one simple suggestion. Melting is a catastrophic breakdown of a highly organized vibrating system. Vibrating systems are notoriously susceptible to resonance and feedback effects: any feedback is liable to lead to a great build-up of amplitude until the whole system gets out of control and disintegrates. In crystals composed of nonspherical molecules, rotation of one molecule affects its neighbors (or some of them) by cog-wheel effects. One of the consequences of this is that impulses may work round in circles and thus produce feedback effects; and the lower the shape-symmetry of the molecules, the greater the number of different types of rotation which can lead to such feedback effects; an asymmetric molecule presents an unsymmetric outline from all points of view and thus may have cog-wheel effects by rotation round any direction. Consequently, the lower the shape-symmetry, the greater the chance of feedback effects and therefore the greater the amplitude of the vibrations at any temperature and the lower the melting point (for a given cohesion energy). This would account not only for the general effect of shape-symmetry but also for the depressing effect of skeletal shape.

One other possible influence should be mentioned. Since molecular shape influences the melting point strongly, it might be supposed that the effect has something to do with the efficiency of packing of the molecules in the crystal. Kitaigorodski[4] has shown that the proportion of the space occupied by the molecules does vary considerably in different organic crystals; and a low molecular packing efficiency in a crystal might lower the melting point. Some calculations of this quantity show, however, that this factor does not account for the large effects observed. The less symmetrical molecules do not necessarily pack more openly than highly symmetrical ones; packing efficiency does not depend on symmetry but on highly specific shape characteristics. Possibly the smaller differences between the melting points of substances of the same symmetry type and similar cohesion energy may be influenced by molecular packing efficiency; but the large effects of shape-symmetry do not appear to depend on it.

Flexible chain molecules melt lower than comparable rigid molecules, as we should expect. This is well illustrated by the series of normal paraffin hydrocarbons, also shown in Figure 1: with increasing molecular length, the melting point at first rises (steadily, except for the well-known alternation of odd and even members in the shorter molecules), but eventually becomes independent of molecular length, approaching the limiting value of 136.5°C.[5] for high molecular polymethylene (which would be far off to the right of the diagram). But the most striking illustration of the effect of molecular flexibility is given by the polyphenyls (also shown in Fig. 1): when the rings are linked *para* fashion, so that they are bound to remain in a straight line even though individual rings rotate, the melting point rises rapidly with the number of rings (almost as rapidly as in the fused-ring series, naphthalene-anthracene, etc.), but when the rings are linked *meta* fashion so that rotation round bonds causes large changes of configuration just as in the paraffin hydrocarbons, the melting point rises only very slowly with the number of rings.[6] *para*-Pentaphenyl melts at 395°C., while the *meta* compound melts at 112°C. The *para* seven-ring compound melts at 545°C., while in the *meta* series even the one with sixteen rings only melts at 321°C. The longer flexible molecules lie, in the diagram, far below even the least symmetrical rigid molecules which have similar cohesion energies.

FACTORS CONTROLLING THE MELTING POINTS OF HIGH POLYMERS

The melting point of each chemically different long-chain polymer may be regarded as the limiting melting point of the appropriate series of shorter chained substances. Consequently, if the relation between melting point and molecular length is of the same general nature in the different series, we should expect that high or low melting points in the polymers would be associated with corresponding high or low melting points in the relevant short-chain monomers. This is borne out by the facts, in the few cases

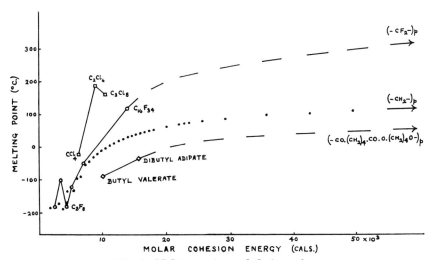

Fig. 2. Melting points of chain series.

for which data are available. Polytetrafluorethylene, $(-CF_2-)_p$, melts at 330°C., very much higher than polymethylene, $(-CH_2-)_p$, and correspondingly the shorter fluorocarbons melt higher than the corresponding paraffins (see Fig. 2). The aliphatic polyesters like polytetramethylene adipate melt lower than polymethylene; and corresponding short-chain esters like dibutyl adipate and butyl valerate (Fig. 2 again) also melt lower than normal hydrocarbons of the same cohesion energy (or even of the same chain length).

The conclusions drawn for rigid monomeric molecules lead us to expect that the melting points of high polymer crystals will depend on some function of the cohesion energy of the molecules, on the degree of flexibility of the chains, and on certain shape effects—probably the degree of departure from cylindrical shape. There may of course be other factors, but these three are likely to be the most important. To make further progress along these lines, it is necessary to consider what molecular characteristics may appropriately be used as indications of the magnitudes of the cohesion energy and flexibility factors.

It is obvious that the cohesion energy of the whole molecule, which was used in discussing monomers, is not appropriate for polymers, since for long molecules of any one series the melting point is independent of molecular length. The reason for this independence is not merely that a polymer crystal contains only segments of molecules (any one molecule threads its way through several crystals); it is also that even in one crystal different parts of the same molecule vibrate semiindependently—or, rather, they influence each other by wave motions just as do the separate molecules in monomer crystals. The cohesion energy and flexibility factors are both intimately concerned here, and it would appear that for a simple type of chain like polymethylene, $(-CH_2-)_p$, it would be appropriate to refer to the cohesion energy per chain unit (in this case CH_2) and to the energy required to rotate round the chain bonds, since this is the chief source of flexibility. (The energies needed for bending or stretching bonds are much higher and need not be considered.) For more complex molecules in which the chain units are not all of the same type, there are inevitable difficulties; where there are two or more different chain units (as in polyesters, with CH_2, $C=O$, and O chain units, or polyvinyl compounds generally, with CH_2 and CHR) the cohesion energies for the different chain units are likely to be very dissimilar. It may be appropriate to refer to the average cohesion energy per chain unit. Similarly, in all such molecules the energy required for rotation round the chain bonds is not the same at each bond; the average energy for rotation may not be a significant quantity; in some circumstances the most easily rotating bonds may dominate the situation (see later discussion of chain esters).

Estimation of Intermolecular Forces

The cohesion energies of chain units in polymer molecules may be estimated by an additive system, based on the properties of substances composed of small molecules, in which a definite increment is associated with each functional group of atoms. More than one such scheme has been proposed in the past[7,8] and figures relevant to some of the well-known polymers

TABLE I

CoHESION ENERGY AND VOLUME INCREMENTS FOR COMMON GROUPS

Group	Cohesion energy, cal. per mole	Volume, cc. per mole
—CH$_2$—	680	21.8
—CH$_3$	1700	27.8
—C$_6$H$_4$—	3900	83.9
—C$_6$H$_5$	5400	89.9
—CH=CH—	1700	32.0
—C(CH$_3$)=CH—	2400	53.8
—CH=CH$_2$	2700	37.8
—C≡CH	2750	—
—CH(CH$_3$)—	1360	42.8
—C(CH$_3$)$_2$—	1900	65.4
—CH(C$_6$H$_5$)	4300	105.7
—CF$_2$—	760	34.0
—CF$_3$	1800	46.1
—CCl$_2$—	3100	53.4
—CHCl—	2360	37.6
—Cl	2800	21.8
—Br	3100	30.5
—I	4200	40.5
—CO—	2660	21.6
—O—	1000	7.3
—CO—O— (ester)	2900	28.9
—CO—O—CO— (anhydride)	3900	50.5
—OH (alcohol)	5800	14.9
—COOH	5600	36.5
—CHOH—	5100	30.7
—CH(CO·OCH$_3$)—	3500	72.5
—CH(O·CO·CH$_3$)—	3500	72.5
—S—	2200	—
—SH	3380	—
—NH$_2$ (amine)	3100	—
—NH— (amine)	1500	14.6
—CO·NH—	8500	36.2
—CO·NH$_2$	8500	42.2
—O·CO·NH—	8740	43.5

have been given by Mark.[9] My own system is based on the fact that the cohesion energy ($L - RT$) of a small-molecule liquid at the boiling point can be represented to a fair approximation as the sum of increments due to the constituent functional groups. (It is best not to attempt to subdivide as far as atoms.) Table I gives figures for some of the common groups. These figures are appropriate for the molecular volumes at the boiling point, which, be it noted, are also (for small molecules) additive functions of the volume increments of the atomic groups, to a similar degree of approximation. To obtain the cohesion energy for any molecule, it is only necessary to add the increments for the constituent groups. There is no reason why these figures should not be used for high polymer molecules, if it is remembered that they are appropriate for a molecular volume equal to the sum of the corresponding volume increments given in Table I. The molecular volumes in actual polymer specimens are smaller than these sums, and the cohesion energies therefore greater. A correction for the smaller molecular volumes can be made: for hydrocarbons, it is found empirically that if E_1 and E_2 are the molar cohesion energies at molar volumes V_1 and V_2, respectively, then:

$$\frac{E_1 - E_2}{E_2} = 1.73\frac{V_2 - V_1}{V_1}$$

and this is probably also true for other substances in which intermolecular forces are of van der Waals' type. (The expression holds whether there is a change from liquid to solid or not.) For the discussion of melting points, cohesion energies of different substances in some standard state should be compared; it would not be appropriate to compare them at room temperature—in fact, the only satisfactory basis for comparison is at the absolute zero of temperature; but since there are not sufficient data on heats of vaporization at low temperatures and the extrapolation from values at room temperature and above would be a long one, this is scarcely practicable. Probably, as a general indication of the magnitude of intermolecular forces, comparison of the cohesion energies at the "boiling point" volumes is sufficient. Figures for a number of polymers on this basis are given in Table II. (Since the zero point volumes of monomers are, according to the limited available information, an approximately constant fraction (0.7) of the boiling point volumes, the zero point cohesion energies would be higher than the "boiling point" cohesion energies by an approximately constant factor (about 1.7).)

Molecular Flexibility. In small molecules like ethane, H_3C—CH_3, and n-butane:

$$\begin{array}{ccc} CH_2 & & CH_3 \\ \diagup & \diagdown & \diagup \\ CH_3 & & CH_2 \end{array}$$

rotation round the central bond is accompanied by energy changes which pass through three minima (at the staggered positions 120° apart) and three maxima. In ethane the potential barriers to rotation (heights of the maxima) are all equal, the heights being about 3000 cal. per mole; in n-butane the energy for the plane zigzag configuration is lower than the other two minima by 800 cal., but the barriers are again about 3000 cal. in height. In a long unbranched CH_2 chain the situation is presumably similar to that in n-butane.

The vibrations of such a chain molecule in a crystal are largely made up of rotations round the chain bonds. It is likely that rotations of 120° to

TABLE II

Cohesion Energies of Polymers

Chemical unit	E	No. of chain units	Average E per chain unit	M.p., °C. (Ref.)
—CH_2—	680	1	680	136.5 (5)
—CF_2—	760	1	760	330
—$CHCl \cdot CH_2$—	3040	2	1520	—
—$CCl_2 \cdot CH_2$—	3780	2	1890	200 (10)
—$C(CH_3)_2 \cdot CH_2$—	2580	2	1290	0
—$(CH_2)_{16}CO \cdot NH$—	19380	18	1076	149 (11)
—$(CH_2)_4CO \cdot NH(CH_2)_6NH \cdot CO$—	23800	14	1700	265 (12)
—$(CH_2)_8CO \cdot NH(CH_2)_6NH \cdot CO$—	26520	18	1474	215 (12)
—$(CH_2)_nCO \cdot NH$—	$5800 + 680n$	$n + 2$	—	—
—$(CH_2)_4CO \cdot O(CH_2)_2O \cdot CO$—	9880	10	988	52 (13)
—$(CH_2)_nCO \cdot O$—	$2900 + 680n$	$n + 2$	—	—
—$C_6H_4 \cdot CO \cdot O(CH_2)_2O \cdot CO$—	11060	5	2212	264 (12)
—$C_6H_4 \cdot CO \cdot O(CH_2)_nO \cdot CO$—	$9700 + 680n$	$n + 3$	—	—
—$C_6H_4 \cdot C_6H_4 \cdot CO \cdot O(CH_2)_2O \cdot CO$—	14900	5	2980	346 (12)
—$C_6H_4 \cdot C_6H_4 \cdot CO \cdot O(CH_2)_6O \cdot CO$—	17620	9	1958	214 (12)
—$CH_2 \cdot C_6H_4 \cdot CH_2 \cdot CO \cdot O \cdot CH_2 \cdot$- $C_6H_4 \cdot CH_2 \cdot O \cdot CO$—	16320	10	1632	146 (12)
—$C_6H_4 \cdot CO \cdot O(CH_2)_2O \cdot CO \cdot C_6H_4 \cdot$- $O(CH_2)_2O$—	18320	10	1832	240 (14)
—$C_6H_4 \cdot CO \cdot O(CH_2)_nO \cdot CO \cdot C_6H_4 \cdot$- $O(CH_2)_2O$—	$16960 + 680n$	$n + 8$	—	—
—$(CH_2)_nO \cdot CO \cdot NH$—	$8740 + 680n$	$n + 3$	—	(12,15)
—$(CH_2)_nO$—	$1000 + 680n$	$n + 1$	—	(12)
—$(CH_2)_nS$—	$2200 + 680n$	$n + 1$	—	(16)
—$(CH_2)_nS_2$—	$4400 + 680n$	$n + 2$	—	—
—$C(CH_3){=}CH(CH_2)_2$—	3760	3	1250	$\begin{cases} cis & 20 \\ trans & 60 \end{cases}$

the next energy minimum seldom occur, because owing to the zigzag configuration such large rotations would produce gross distortions, too large to be tolerated in the structure; the rotations which occur most frequently are therefore represented by oscillations in the lowest trough of the energy diagram. Since the energy necessary to climb up the sides of the trough is some function of the height of the next maximum (a cosine curve is usually assumed), it is reasonable to take the magnitude of this energy barrier as an indication of the degree of flexibility of the molecule. Figures for energy barriers in several other types of small molecules have been collected by McCoubrey and Ubbelohde[17]; it is significant that they are of the same order of magnitude (1000–5000 cal. per mole) as the cohesion energy figures for chain units; this is the reason why these two factors— intermolecular forces and molecular flexibility—are of equal importance in determining the melting points (and many other properties) of high polymers.

The amount of information on potential barriers relevant to the principal types of chain polymers is unfortunately scanty, and the figures are admittedly very rough[23]: in some cases, two estimates for the same molecule differ by a factor of 2 or more, so that these are of little value for the discussion of individual polymers. For this reason it is best at present to proceed by empirical comparative methods, while keeping these general concepts in mind. We shall now discuss some of the well-known polymers against this background. It will be shown that reference to the melting points of monomeric substances often throws light on those of high polymers.

MELTING POINTS OF POLYMER SERIES

Polymethylene and Polytetrafluoroethylene. The cohesion energy of a CF_2 group is very nearly the same as that of a CH_2 group (see Table I), hence the very much higher melting point of the fluorocarbon polymer cannot be accounted for by any difference of intermolecular forces. We look, therefore, for a difference of molecular flexibility. The potential barrier for rotation in hexafluoroethane, $F_3C \cdot CF_3$, is reported[18] to be 4350 cal. per mole—considerably higher than that of ethane, 3000 cal. per mole; the difference is thus in the right direction to explain the high melting point of the fluorocarbon polymer, assuming that similar figures apply to chain molecules. The ratio of the two figures (1.45) is about the same as the ratio of the absolute melting points of the two polymers (603/408 = 1.48); but the agreement is probably fortuitous. The absolute melting point would not be expected to be directly proportional to the potential barrier to rotation, for a chain polymer with zero barrier to rotation would not have a melting point near absolute zero.

Shape effects should be considered. It has been pointed out that owing to the fact that fluorine atoms are larger than hydrogen atoms, the surface of the fluorocarbon chain is more filled in, and the chain is in fact more nearly cylindrical, than that of the hydrocarbon chain; and it has been suggested[19] that this molecular feature is responsible for certain unusual properties of polytetrafluoroethylene (the disorder transition at 20–30°C., and the very low coefficient of friction). Is the cylindrical shape also responsible for the very high melting point? It is true that among rigid monomers substances with cylindrical molecules melt higher than those with flat lengthy molecules of compact shape and high symmetry; but this degree of difference of shape is associated with only moderate differences of melting point (50° or less); and since the difference between the shapes of hydrocarbon and fluorocarbon chains is less still, it probably accounts for very little of the difference between the melting points of the hydrocarbon and fluorocarbon polymers. The difference must therefore be due chiefly (as suggested above) to the great rigidity of the fluorocarbon chain. To what extent this greater rigidity is due to the greater size of the fluorine atoms (leading to hindered rotation), or to more subtle bond-orientation effects, is not known.

Polyamides. The melting points of many different polyamides are now known.   They are all higher than that of polymethylene, and there is a general rise of melting point with the proportion of amide groups in the chain; this presumably means that there is no great difference of chain flexibility, so that the much higher cohesion energies due to the hydrogen-bonding amide groups have a dominant influence.   There is, however, in a plot of melting points against proportion of amide groups (which is equivalent to a plot against cohesion energy per chain unit), a considerable spread in the array of points (for details, see refs. 20, 21), and it is a striking fact that polyamides in which there are even numbers of CH_2 groups between the amide groups melt high, those with odd numbers of CH_2 groups melt low, and those with mixed odd and even sequences of CH_2 groups (such as even diamine-odd diacid polymers and *vice versa*) melt at intermediate temperatures.   It is also interesting that the "even" type polymers fall approximately on a straight line (reproduced, on the present basis of cohesion energy per chain unit, in Fig. 3):   evidently, other things being equal,

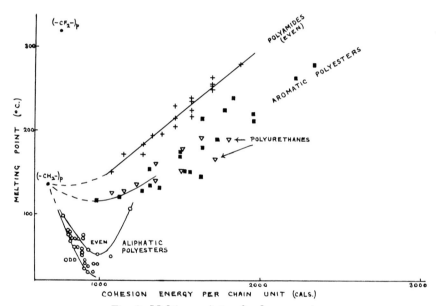

Fig. 3. Melting points of polymer series.

there is a linear relation between melting point and cohesion energy per chain unit, at any rate for those polymers in which the amide groups are not too sparsely placed.   This straight line, however, does not pass through the point for polymethylene; nor does it pass through the absolute zero of temperature—it cuts the temperature axis at about $-50\,°C$.   The fact that the line passes below the point for polymethylene may mean that certain bonds in polyamide molecules are a little more flexible than those in a CH_2 chain.   There is no independent evidence on the barriers to rotation in such molecules; the CO—NH bonds appear to have some double-bond character, which is usually associated with additional rigidity, and so any additional ease of rotation must be in the adjacent CO—CH$_2$ and NH—CH$_2$ bonds; although estimates of barriers in simple ketones and amines have been made, these may not be relevant to bonds adjacent to an amide group.

It has been definitely shown[21] that in the crystal structure of one of the "odd" polyamides—polycaproamide—all the possible hydrogen bonds are made normally just as in the "even" polymers; it seems likely, therefore, that the low melting points of "odd" members in general are not due to any failure of hydrogen bond formation. This contrast between odd and even polymers recalls the alternation of melting points of odd and even chain monomers (with even members always having the higher melting points) which runs through the whole of organic chemistry; and it is particularly interesting to find that while in the normal paraffin series the melting point curves for odd and even members are only a few degrees apart, the difference is very much increased for all chains having heavy or strongly bound groups on both ends; it is about 50°C. for the lower dicarboxylic acids and diphenyl derivatives, decreasing with increasing number of CH$_2$ groups. The situation is similar in the polyamides—there are CH$_2$ sequences between the amide groups which are strongly bound to neighboring molecules; and here again the difference between odd and even members decreases with increasing length of the CH$_2$ sequences. The quite general occurrence of the phenomenon in such diverse groups of monomers as acids, amides, chlorides, bromides, iodides, and phenyl derivatives rules out any specific influences of crystal structure, and suggests a fundamental cause such as the effect of an odd or even number of units on the pattern of vibrations in the crystals; it appears that the anchoring effect of the heavy groups on the ends of the chain is more effective when the chain is even than when it is odd. It may be remarked that the end bonds connecting the CH$_2$ chain to the anchoring groups are parallel to each other in an even chain, but are at an angle of 110° in an odd chain; if melting is due to the uprooting of the anchoring groups by the more freely vibrating CH$_2$ chain, we must ask why this is more easily done when the end bonds are inclined to each other than when they are parallel. The answer to this question is not obvious, but it may well be connected with other effects referred to below which appear to depend on whether rotatable bonds are parallel or inclined.

Polyesters. Aliphatic polyesters melt lower than polymethylene, and there is a general decrease of melting point as the proportion of ester groups in the chain increases, in spite of the increase of intermolecular forces by the polar ester groups. (A detailed diagram has been given elsewhere,[20] but the points are reproduced in Fig. 3.) This suggests immediately that in the ester group there is a bond round which rotation is considerably easier than in a CH$_2$ chain. Figures for potential barriers to rotation in simple substances ((CH$_3$)$_2$CO, 1400 cal. per mole; (CH$_3$)$_2$O, 2500–3100 cal. per mole) suggest that the CH$_2$—CO bond is the easily rotating one, but there are several reasons for rejecting this suggestion and attributing especially easy rotation to the O—CH$_2$ bond: (*1*) Monomeric chain ketones melt a little higher than the corresponding normal hydrocarbons; on a molar cohesion energy basis they fall nearly on the same curve as the hydrocarbons, suggesting that CH_2—CO and CH_2—CH_2 bonds rotate equally easily. (2) All chain ethers, monomeric as well as polymeric, melt lower than the corresponding hydrocarbons; this is the clearest indication of easy rotation about the O—CH_2 bond. (3) In the crystal structure of polyethylene terephthalate the ester group

$$\overset{CO}{\diagdown}\underset{O}{\diagdown}\overset{CH_2}{\diagup}$$

is nearly planar, whereas in polyethylene adipate there is considerable departure from planarity by rotation of 80° about the O—CH_2 bond; such a difference, brought about presumably by packing requirements in the two crystals, suggests easy rotation about this bond.[22] This conclusion implies either that the estimates of potential barriers in acetone and ether are seriously in error, or that these estimates which apply to the rotation of methyl groups do not apply to chain molecules with groupings

$$\overset{CH_2}{\diagup}\underset{CO}{\diagdown}\diagup \quad or \quad \overset{CH_2}{\diagup}\underset{O}{\diagdown}\diagup$$

It should also be mentioned that since chain anhydrides R·CO·O·CO·R melt lower than hydrocarbons, the CO—O bond may also be easily rotatable.

There is one other generalization which throws an interesting light on the question of flexibility of chain esters. The melting points of monomeric chain esters with one ester group in an unbranched hydrocarbon chain show (Fig. 4) that for a given chain length the melting point decreases as the ester group moves from the end toward the center of the chain. (There are not enough data to demonstrate this for any one series of isomeric

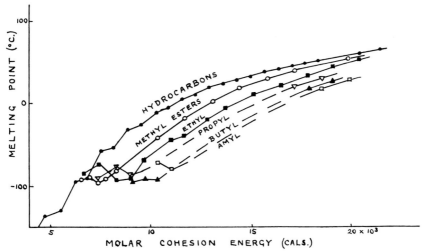

Fig. 4. Melting points of monomeric chain esters.

esters, $CH_3(CH_2)_m \cdot CO \cdot O \cdot (CH_2)_n \cdot CH_3$, where $m + n = $ constant, but the assembled data for a variety of esters in Figure 4 show it clearly enough.) This is exactly what one would expect if there is easy flexibility of the molecule at the ester group: a hinge near the end of the molecule would make little difference to its vibrations, but a hinge at the center divides the molecule into two half-size units which would vibrate with greater amplitude than a full-length unhinged molecule. The same is true for chains containing two ester groups: the lowest melting point is attained when the ester groups divide the chain into three equal portions. There are indications of similar effects in polymeric esters: the melting point (70–76° C.) of poly(ethylene sebacate), $(-O(CH_2)_2O \cdot CO(CH_2)_8CO-)_p$, in which there are alternate long and short segments, is higher than that (56–59° C.) of its isomer poly(hexamethylene adipate), $(-O(CH_2)_6O \cdot CO(CH_2)_4CO-)_p$, in which the segments are more nearly equal in length.

Polyesters in which there are odd CH_2 sequences melt lower than isomers with even sequences.[20] The differences are, however, smaller and less consistent than in the polyamides; this may be connected with the fact that the cohesion energy of an ester group is much smaller than that of an amide group, for it will be recalled that in monomeric chain compounds the differences between odd and even melting points show up most clearly for molecules with the most strongly bound end groups. The aliphatic polyesters with the highest proportions of ester groups show extreme divergences from the rest[20]; these might be regarded as strong odd-even effects coming in rather suddenly, or possibly they are special effects due to the very close proximity of the dipolar groups, which by their interaction may stiffen the even molecules (where the dipoles are oppositely directed in a zigzag molecule and therefore give this configuration additional stability) and increase the flexibility of the odd ones in which the dipoles would be parallel and therefore would repel each other, decreasing the stability of the zigzag configuration.

Aromatic polyesters melt very much higher than aliphatic ones; here the flexibility of the ester group is more than offset by the presence of the rigid benzene ring. In Figure 3 the melting points of several series are plotted on the basis of cohesion energy per chain unit. CH_2, O, CO, and C_6H_4 are each taken as single chain units, except when a CO group is directly joined to a benzene ring, when the whole group

in the terephthalates, is taken as a single chain unit which owes its rigidity to resonance effects. On this basis the melting points[12,14] of the series $-CO \cdot C_6H_4COO \cdot (CH_2)_nO-$ (terephthalates), $-CO \cdot C_6H_4 \cdot O(CH_2)_mO \cdot C_6H_4 \cdot CO \cdot O(CH_2)_nO-$, $-CO \cdot C_6H_4 \cdot C_6H_4 \cdot CO \cdot O(CH_2)_nO-$ (in which the diphenyl unit is taken as a single chain unit), and the single substance $-CO \cdot CH_2 \cdot C_6H_4 \cdot CO \cdot O \cdot CH_2 \cdot C_6H_4 \cdot CH_2 \cdot O-$ fit fairly well into a coherent band, and this is taken as some justification of the method of representation and the ideas on which it is based.   Where there are long sequences of CH_2 groups, the melting points tend toward that of polymethylene, although it is important to note that there is a minimum in the curve—that is, some of these polyesters melt a little lower than polymethylene.   There are again marked odd-even effects[12]—in any series the polymers with even CH_2 sequences melt high, those with odd sequences melt low.

Polyurethans.   These polymers, containing the grouping —O·CO·-NH—, form hydrogen bonds like the polyamides, giving high cohesion energy, but at the same time they contain O—CH_2 bonds like the polyesters, so that the molecular flexibility would be expected to be greater than in the polyamides.   The melting points are, as might be expected, intermediate between those of the polyamides and aliphatic polyesters; and again those with even CH_2 sequences melt higher than those with odd sequences.[20]   In Figure 3 they are plotted on the same basis as the other groups of polymers; they fit into the same band as the aromatic polyesters.

Polyethers and Polythioethers.   When CH_2 groups in a chain are replaced by oxygen or sulfur atoms, the melting point is at first reduced, in spite of the increase of cohesion energy (see Table III); this applies to monomeric and polymeric substances alike, and suggests that the S—CH_2 bond, like the O—CH_2 bond, is easily rotatable.   The barrier to rotation in $(CH_3)_2S$, 2000 cal. per mole,[17] is consistent with this conclusion (if the estimate can be trusted), for it is lower than that in $(CH_3)_2CH_2$.   For higher proportions of sulfur there is an increase of melting point.   Disulfide polymers containing two linked sulfur atoms in the chain reach a lower minimum than the thioethers, indicating that the S—S link is still more flexible; this appears to be consistent with the fact that "plastic sulfur," a chain polymer which crystallizes at room temperature only on stretching, evidently has a melting point (if indeed it crystallizes at all on cooling without stretching) below room temperature, in spite of its high cohesion energy (2000 cal. per S atom).   There is, however, a problem here: the melting points in the series $(CH_2)_nS_2$, after going through a minimum, rise again to 113° C. in $(CH_2)_2S_2$; if plastic sulfur is regarded as the end member of this series, there is again a fall.   No explanation can be offered for this.

TABLE III

MELTING POINTS OF SULFIDE POLYMERS

Group (Ref.)	$n = 6$	5	4	3	2
—$(CH_2)_nS$— (24)	68°	65°	67°	61°	145°
—$(CH_2)_nS_2$— (24)	—	44°	39°	67°	113°
	$m,n = 10,6$	6,6	4,6	2,6	2,2
—$(CH_2)_mS(CH_2)_nS$— (25)	76–78°	71–76°	65–67°	82–86°	190°

Unsaturated Polymers.   The natural isoprene polymers rubber and gutta-percha, $(—CH_2—C(CH_3)=CH·CH_2—)_p$, as well as polychloroprene, $(—CH_2—CCl=CH—CH_2—)_p$, and 1,4-polybutadiene itself $(—CH_2—$

338   C. W. BUNN

CH=CH—CH$_2$—), all melt much lower than polymethylene; this is consistent with the fact that monomeric chain compounds containing (nonconjugated) double bonds melt lower than the corresponding saturated
compounds, and suggests strongly that the lowering of the melting point
is due to a property of the bond structure rather than to specific stereochemical or shape effects.   The double-bonded group itself is, of course,
very rigid, but in several small molecules such as propene, CH$_3$·CH=CH$_2$,
and isobutene, (CH$_3$)$_2$C=CH$_2$, the estimated potential barriers to rotation
of the methyl groups are lower than in saturated hydrocarbons.[17]   Although all such estimates are rough, the fact that low values (450–2100 cal.
per mole) have been consistently obtained for several unsaturated hydrocarbons suggests that in this case, even if individual figures are very
approximate, we may accept as established the generalization that in such
molecules rotation round a (nominally) single carbon-carbon bond is easier
when there is an adjacent double bond than in a saturated compound.
If this applies to chain molecules, it provides an explanation of the low
melting points in question.   (I suggested this in 1942,[26] but the evidence
is now stronger.)   There are, of course, conflicting factors—the rigidity
of the double-bonded group itself, opposed by the easy rotation about
adjacent single bonds.   In some situations it seems likely that the latter
effect would be dominant; for instance, in a monomeric paraffin molecule
with one double bond at the center of the chain, the presence of one rigid
link would make little difference to the vibrations of the whole semistiff
molecule, but the adjacent easily rotating bonds would have the effect of
hinging the molecule at the center and thus allowing much greater vibrations
of the two halves (as in the case of chain esters).   In the polymers in which
every fourth bond is a double bond, the outcome is less certain, although it
may be remarked that for every chain bond which is stiffened, there are

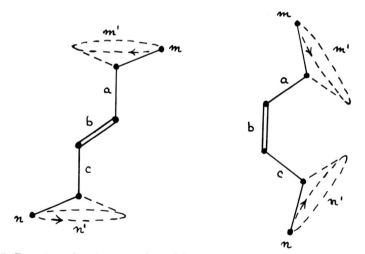

Fig. 5. Rotation of a given number of degrees round bonds *a* and *c* distorts a
cis molecule more than a *trans* molecule.

two adjacent bonds round which rotation is easier; and in any case local distortions of the molecule, which are perhaps dominant for melting, can be greater when easily rotating bonds are present, so that in a sense the flexibility of a molecule is that of its most easily rotating bonds.

cis-Polyisoprene (rubber) melts lower than the *trans* isomer, gutta-percha. Although it has been suggested[26] that this may be due to a specific stereochemical difference which may make rotations round certain bonds easier in the *cis* isomer, the considerations brought forward in the present paper suggest other possibilities. One is a shape effect: the *cis* isomer is flatter (further removed from cylindrical in cross section) and less smooth in outline than the *trans* isomer. The other is that in the *cis* isomer the rotatable chain bonds adjacent to the double bond are inclined to each other at an angle of 70°, whereas in the *trans* isomer they are parallel; even if the barriers to rotation were identical in the two isomers, the net flexibility of the *cis* molecule would be greater than that of the *trans* molecule, for a geometrical reason illustrated in Figure 5. Rotation round bonds *a* and *c* of a given number of degrees distorts the *cis* molecule (right)—*i.e.*, shortens the distance *mn* to *m'n'*—more than the *trans* molecule (left), simply because the rotatable bonds are inclined in the *cis* and parallel in the *trans*. The fact that *cis* chain compounds in general melt lower than the *trans* isomers (in a great variety of monomers as well as in polymers) suggests that this geometrically enhanced flexibility due to inclined rotating bonds, rather than any specific stereochemical or shape effects, is the explanation.

The differences between *cis* and *trans* compounds are related to the striking differences noted earlier between the *meta* and *para* polyphenols: the *para*-linked molecules remain linear even when rotation occurs round all the links, while similar rotations in the *meta*-linked molecules cause gross distortions. This is an extreme case because in *para*-linked molecules the rotating links are in a straight line. The *trans* double-bonded molecules are intermediate because the rotating links, although parallel, are not in a straight line.

CONCLUDING REMARKS

The conclusion just reached leads on to the suspicion that the odd-even effects in the polyamides and other series may be due to similar causes; for at the ends of an odd CH_2 sequence, the bonds to the amide or ester groups are inclined at an angle of 112°, whereas at the ends of an even sequence they are parallel, the odd and even sequences being analogous to *cis* and *trans* double-bonded groups. If melting depends on uprooting the strongly bound groups, we have to ask why an odd CH_2 sequence does this more effectively than an even sequence. The wave motions along a chain molecule are made up of rotations round the successive chain bonds; suppose that at a given temperature a certain maximum rotation round each bond is possible; when an impulse travels along a chain, its amplitude is due to the resultant of individual amplitudes of rotation. A heavy or strongly bound group will of course damp down the motions at the end of a CH₂ sequence, but the forces tending to displace the heavy group must depend on the motions in the CH_2 chain.   The relative displacement of the groups held by the end bonds will be greater for an odd than for an even sequence for the same geometrical reason as in the case of *cis* and *trans* compounds (Fig. 5).   The fact that the odd-even effects die out as the CH_2 sequence lengthens is also to be expected on this geometrical theory. This seems a reasonable interpretation of the odd-even effects not only in polymers but in monomeric chain compounds generally.

One other general effect deserves additional comment.   In various series of polymers which may be regarded as CH_2 chains in which CH_2 groups are replaced at varying regular intervals by other groups, the graph of melting point against cohesion energy per chain unit shows a minimum: the polymers with sparsely placed strongly bound groups actually melt lower than polymethylene.   This has been discussed by Izard,[14] who connects this depression with Flory's treatment[27] of the depression of the melting point by copolymerization.   This does not seem justified; Flory's treatment is based on the assumption that in a copolymer, only the homogeneous stretches of the main component crystallize; the proportions of homogeneous stretches which can form crystals are determined statistically, assuming a random succession of units in the copolymer.   The effect is somewhat similar to the depression of the melting point of a monomer by an impurity which does not enter into the crystals.   But in the homopolymers considered here, the different chain units are all included in the crystals, and it is not justifiable to regard them as copolymers and apply Flory's treatment.   Indeed, one series which shows no minimum (copolymers of hexamethylene sebacamide and sebacate) is explained quite reasonably by Izard as due to mixed crystal formation; it is therefore not consistent to expect homopolymer series to show minima.   The explanation of the minima which some homopolymer series do show is not to be found in a copolymer theory.   The explanation suggested by the present treatment is that the depression of the melting point is due to the introduction of easily rotating bonds.   The insertion of ester groups sparsely at regular intervals in a polymethylene chain does increase the average cohesion energy per chain unit, but only a little; and it is entirely reasonable that the additional flexibility effect due to the introduction of easily rotating bonds should a have much greater depressing effect on the melting point than the elevating effect of a slightly increased cohesion energy. However, for a fuller understanding of this phenomenon, and indeed of the whole subject of polymer melting points, a theory expressing the quantitative interaction of cohesion energy and molecular flexibility factors is required.   Any attempt to develop such a theory might well be focused on the fact that the minimum for aliphatic polyesters is much deeper than for aromatic polyesters and polyurethans; indeed the fact that aliphatic and aromatic polyesters do not fit on the same curve, in the treatment adopted here, certainly calls for comment.   The clue is presumably that if we compare an aliphatic with an aromatic polyester of the same cohesion energy per chain unit, the aliphatic chain contains a much higher proportion of easily rotating O—C bonds, so that the over-all flexibility of the molecule is greater.

Shape effects do not appear to be very important in the series of polymers considered here, probably because most of them consist of slender molecules which are roughly cylindrical in shape. Shape effects like those which are prominent in monomers must be expected for polymer molecules having large side groups; a shape having large projections and depressions would be expected to depress the melting point, not necessarily owing to bad packing but more probably to vibration interaction of interlocking molecules.

References

1. F. A. Lindemann, *Physik. Z.*, **11**, 609 (1910).

2. J. E. Lennard-Jones and A. F. Devonshire, *Proc. Roy. Soc.*, **A169**, 317 (1939); **A170**, 464 (1939).

3. W. O. Baker and C. P. Smyth, *Ann. N. Y. Acad. Sci.*, **40**, 447 (1940).

4. A. Kitaigorodski, *Acta Physicochim. U. R. S. S.*, **22**, 309 (1947).

5. L. Mandelkern, M. Hellmann, D. W. Brown, D. E. Roberts, and F. A. Quinn, *J. Am. Chem. Soc.*, **75**, 4093 (1953).

6. G. Egloff, *Properties of Hydrocarbons*.

7. M. Dunkel, *Z. physik. Chem.*, **A138**, 42 (1928).

8. P. A. Small, *J. Applied Chem.*, **3**, 71 (1953).

9. H. Mark, *Physical Chemistry of High Polymeric Systems*, Interscience.

10. R. C. Reinhardt, *Ind. Eng. Chem.*, **35**, 422 (1943).

11. D. D. Coffmann, N. L. Cox, E. L. Martin, W. E. Mochel, and F. J. van Natta, *J. Polymer Sci.*, **3**, 85 (1948).

12. R. Hill and E. E. Walker, *J. Polymer Sci.*, **3**, 609 (1948).

13. C. S. Fuller and C. L. Erickson, *J. Am. Chem. Soc.*, **59**, 344 (1937).

14. E. F. Izard, *J. Polymer Sci.*, **8**, 503 (1952).

15. R. Thiebault, *Compt. rend. journ. intern. plastiques*, **1949**, 75.

16. C. S. Marvel and R. R. Chambers, *J. Am. Chem. Soc.*, **70**, 993 (1948).

17. J. C. McCoubrey and A. R. Ubbelohde, *Quart. Revs. Chem. Soc.*, **5**, 364 (1951).

18. E. L. Pace and J. G. Aston, *J. Am. Chem. Soc.*, **70**, 566 (1948).

19. C. W. Bunn and E. R. Howells, *Nature*, **174**, 549 (1954).

20. C. W. Bunn, Chapter 12 of *Fibres from Synthetic Polymers*, ed. R. Hill, Elsevier, 1953.

21. D. R. Holmes, C. W. Bunn, and D. J. Smith, *J. Polymer Sci.*, in press.

22. R. de P. Daubeny, C. W. Bunn, and C. J. Brown, *Proc. Roy. Soc.*, **A226**, 531 (1954).

23. E. Blade and G. E. Kimball, *J. Chem. Phys.*, **18**, 630 (1950).

24. L. S. Rayner, private communication.

25. C. S. Marvel and R. R. Chambers, *J. Am. Chem. Soc.*, **70**, 993 (1948).

26. C. W. Bunn, *Proc. Roy. Soc.*, **A180**, 67, 82.

27. P. J. Flory, *J. Chem. Phys.*, **17**, 223 (1949).

Synopsis

The molecular characteristics which determine the melting points of high polymer crystals are considered, and it is shown that the properties of monomeric crystals often throw light on those of the polymers. The principal factors controlling melting points appear to be molar cohesion energy (of the whole molecule for monomers, or per chain unit for polymers), molecular flexibility (due to rotation round bonds), and molecular shape effects.   Figures for the cohesion energy increments of a number of chain units and substituent groups are given, and melting points of polymer series are correlated with cohesion energy per chain unit.   The flexibility factor is less easy to assess; barriers to rotation in appropriate monomer molecules are relevant, but available data are very rough.   The approach therefore is mainly by empirical and comparative methods. When plotted against cohesion energy per chain unit, the melting points of various series of aromatic polyesters and polyurethans fall within the same band, while those of the polyamides lie on the whole higher and those of the aliphatic polyesters, polyethers, polythioethers and polydisulfides much lower.   The differences are attributed to difference of molecular flexibility arising from the presence of easily rotating O—C, S—C and S—S bonds.   The low melting points of rubber and other unsaturated polymers are attributed to the fact (which can now be regarded as definitely established by independent evidence) that rotation round single bonds which are adjacent to double C=C bonds is easier than in saturated chains.   Easily rotating bonds which are inclined to each other, as in *cis* isomers, confer greater chain flexibility than the parallel bonds in *trans* isomers, and thus lead to lower melting points.   The marked odd-even effects in saturated molecules which run through the whole of organic chemistry (the even members always melting higher than the odd) are attributed to similar effects arising from the fact that the end bonds of an odd CH_2 sequence are inclined to each other while those at the ends of an even sequence are parallel.

Résumé

Les caractéristiques moléculaires qui déterminent les points de fusion de hauts polymères cristallisés ont été étudiées; on montre que les propriétés du cristal monomérique offre fréquemment un éclaircissement concernant les propriétés correspondantes du polymère.   Les facteurs principaux controlant les points de fusion sont l'énergie de cohésion moléculaire (de la molécule entière dans le cas des monomères, de l'unité périodique dans le cas des polymères), la flexibilité moléculaire (due à la rotation autour des liaisons existantes), et les effets dus à la forme moléculaire.   Des valeurs sont indiquées pour les incréments d'énergie de cohésion pour un certain nombre d'unités périodiques et pour certains substituants; les points de fusion de plusieurs polymères ont été reliés à cette énergie de cohésion par unité périodique.   Le facteur de flexibilité est plus délicat; les empêchements à la rotation dans des molécules de monomères appropriés sont évidents; les résultats obtenus jusqu'ici sont très approximatifs, et proviennent de méthodes purement empiriques et comparatives.   Si on porte en diagramme les points de fusion en regard de l'énergie de cohésion de l'unité périodique, on constate que de nombreuses séries de polyesters et polyuréthanes aromatiques tombent dans un même domaine, tandis que les points figuratifs des polyamides se situent tous à une région plus élevée; ceux des polyesters, polyéthers, polythioéthers et polydisulfures aliphatiques se situent beaucoup plus bas.   Les différences sont attribuées à des différences de flexibilité moléculaire résultant de la présence de liaisons mobiles O—C, S—C et S—S.   Les points de fusion bas des caoutchoucs et d'autres polymères non-saturés sont à attribuer au fait (qui peut être considéré maintenant comme définitivement établi pour des motifs totalement indépendants) que la rotation autour des liaisons simples qui sont adjacentes aux doubles liaisons est plus facile que dans une chaîne saturée ordinaire.   Des liens aisément mobiles et inclinés les uns par rapport aux autres, ainsi qu'il en est le cas pour les isomères-*cis*, confèrent une plus grande flexibilité des chaînes que les liens parallèles entre eux, qui sont présens dans les isomères-*trans*, dont les points de fusion sont plus élevés.   Les effets d'alternance pair-impair bien connus dans les molécules saturées et qui se retrouvent dans toute la chimie organique (les membres à nombre pair fondent toujours plus haut que ceux à nombre impair) doivent être attribués à des effets semblables; les liaisons terminales dans une séquence impaire de groupes CH_2 sont inclinées les unes par rapport aux autres, tandis qu'elles sont parallèles dans le cas d'un nombre pair de mêmes groupes.

Zusammenfassung

Die Molekulareigenschaften, die den Schmelzpunkt von Hochpolymerkristallen bestimmen, werden betrachtet, und es wird gezeigt, dass die Eigenschaften monomerer Kristalle oft Erklärungen über die Eigenschaften der Polymere geben. Die Hauptfaktoren, die den Schmelzpunkt kontrollieren, sind molare Kohäsionsenergie (des ganzen Moleküls für Monomere, oder pro Ketteneinheit für Polymere), molekulare Biegsamkeit (durch Rotation um Bindungen herum bedingt) und molekulare Form-Effekte. Es werden Zahlen für die Inkremente der Kohäsionsenergie einer Anzahl von Ketteneinheiten und substituierenden Gruppen gegeben, und Schmelzpunkte von Polymerreihen werden mit der Kohäsionsenergie pro Ketteneinheit in Beziehung gebracht. Der Biegsamkeitsfaktor ist weniger leicht zu bestimmen; Schranken der Rotation sind in geeigneten Monomermolekülen bezeichnend, aber die erhaltbaren Daten sind sehr ungenau. Die Annäherung wird deshalb hauptsächlich durch empirische und vergleichende Methoden vorgenommen. Bei graphischer Darstellung gegen die Kohäsionsenergie pro Ketteneinheit fallen die Schmelzpunkte der verschiedenen Reihen aromatischer Polyester und Polyurethane innerhalb des gleichen Bandes, während die der Polyamide im allgemeinen höher und die der aliphatischen Polyester, Polyäther, Polythioäther und Polydisulfide viel tiefer liegen. Die Unterschiede werden den Unterschieden der molekularen Biegsamkeit zugeschrieben, die aus der Gegenwart von leicht rotierenden O—S, S—C und S—S Bindungen entstehen. Die niedrigen Schmelzpunkte von Kautschuk und anderen ungesättigten Polymeren werden der Tatsache zugeschrieben (welche jetzt als endgültig und durch unabhängige Beweise festgestellt betrachtet werden kann), dass Rotation um einfache Bindungen herum, welche benachbart zu doppelten C=C Bindungen stehen, einfacher als in gesättigten Ketten ist. Leicht rotierende Bindungen, welche gegeneinander geneigt sind, wie in *cis*-Isomeren, verleihen grössere Kettenbiegsamkeit als die parallelen Bindungen in *trans*-Isomeren, und führen so zu niedrigeren Schmelzpunkten. Die starken ungerade-gerade Effekte in gesättigten Molekülen, die in der gesamten organischen Chemie bestehen (die geraden Glieder schmelzen immer höher als die ungeraden) werden ähnlichen Effekten zugeschrieben, die von der Tatsache herrühren, dass die Endbindungen einer ungeraden CH_2-Sequenz gegeneinander geneigt sind, während die am Ende einer geraden Sequenz parallel sind.

Received December 1, 1954

COMMENTARY

Reflections on "Copolymers in Dilute Solution. I. Preliminary Results for Styrene-Methyl Methacrylate," by W. H. Stockmayer, L. D. Moore, Jr., M. Fixman, and B. N. Epstein, *J. Polym. Sci.*, XVI, 517 (1955)

W. H. STOCKMAYER

Department of Chemistry, Dartmouth College, Hanover, NH 03755

It is a great honor to join in the 50th birthday celebration of the *Journal of Polymer Science*, with a paper that originally helped to celebrate Herman Mark's 60th birthday. However, as General Sherman once said, "No three witnesses of a simple brawl can agree on all the details," and one can only hope that the "mystic chords of memory" are not too cacophonous.

Our studies at MIT of light scattering and viscosity of copolymer solutions were begun in 1950 by Louis Moore, the first of the three graduate students who appear as co-authors of this paper. Light scattering was then a relatively new technique. Our instrument was built by Harry E. Stanley (father of theoretician H. Eugene Stanley) in 1948–49, with a design modifying that of Zimm.[1] We benefitted greatly from the advice of three friends who were already experienced practitioners of the art: Bruno Zimm, Paul Doty, and Fred Billmeyer. In 1952 Marshall Fixman, whose theoretical powers were already evident, was nevertheless bludgeoned into fractionating a copolymer and then designing and building a differential refractometer. This instrument lived until 1965 when it was mangled by a bumbling Dartmouth undergraduate, while Stanley's photometer underwent several overhaulings and was still viable when my former student and later colleague Bob Cleland retired it in the late eighties.

The choice of copolymers for study must have reflected the subconscious influence of my mentor and senior colleague George Scatchard, one of the giants in the field of liquid solutions. The reasons for our choice of styrene/methyl(methacrylate) copolymers are spelled out in the paper.

I believe that there are two main contributions in the paper. The first is embodied in eq. (3). We were aware of the observations by Tremblay et al.[2] that scattering by styrene/butadiene copolymer solutions exceeded that predicted by the standard Debye formula. The correct qualitative interpretation given by these authors was unaccompanied by a quantitative discussion, which we therefore supplied. As stated, the aim in our own experimental work was to minimize the extra scattering [last two terms in eq. (3)]. The systematic exploitation of these extra terms for assaying non-uniformity of copolymer composition was initiated a few years later in the laboratory of Henri Benoit.[3]

Our second main contribution was the application of eq. (6) to our data to obtain values of the interaction parameter χ between the unlike segment pairs in the copolymer. We considered the derivation of eq. (6) to be transparent, and devoted some space to discussing its limitations, but we did not observe that in fact this equation had already appeared in an appendix to a much earlier paper by Simha and Branson[4] on copolymerization kinetics. My students probably had never read this paper, but I was well acquainted with the main text and must surely have read the appendix. I cannot account for my lapse of memory, but take this occasion to regret that it occurred and also that Robert Simha was not a referee.

Of course eq. (6) is only a first approximation, since no account is taken of concentration fluctuations[5] or of second-neighbor effects dependent on the sequence distribution.[6] We also remark that eq. (5) for the unperturbed dimensions is a naive approximation that turned out to be adequate for our system but at times can prove dangerous.[7]

Our paper bears the numeral I, but no further contributions have appeared in the literature, though in fact two later papers were presented at ACS meetings. In these unpublished works, by Ben Epstein and by Allan Shultz (then a postdoc in the group), light-scattering and viscosity measurements were extended to styrene/methacrylate copolymers of 25 and 75% compositions, and including theta solvents. Further, extensive cloud-point data were obtained and interpreted in the manner developed by Shultz and Flory[8] for polystyrene. I take most of the blame for the failure to submit these results for publication. In general, they support the roughly quantitative achievements displayed in Paper I.

With the development of anionic living polymerization by Michael Szwarc and others, it became possible to conduct far more sophisticated investigations than those described here.

REFERENCES AND NOTES

1. B. H. Zimm, *J. Chem. Phys.*, **16**, 1099 (1948).
2. R. Tremblay, M. Rinfret, and R. Rivest, *J. Chem. Phys.*, **20**, 523 (1952).
3. W. Bushuk and H. Benoit, *Can. J. Chem.*, **36**, 1616 (1958).
4. R. Simha and H. Branson, *J. Chem. Phys.*, **12**, 253 (1944).
5. M. Olvera de la Cruz, S. F. Edwards, and I. C. Sanchez, *J. Chem. Phys.*, **89**, 1704 (1988).
6. A. C. Balazs, I. C. Sanchez, I. R. Epstein, F. E. Karasz, and W. J. MacKnight, *Macromolecules*, **18**, 2188 (1985).
7. W. Silberszyc, *J. Polym. Sci., Part B, Polymer Letters*, **1**, 577 (1963).
8. A. R. Shultz and P. J. Flory, *J. Amer. Chem. Soc.*, **75**, 3888 (1953).

PERSPECTIVE

Comments on "Copolymers in Dilute Solution. I. Preliminary Results for Styrene-Methyl Methacrylate," by W. H. Stockmayer, L. D. Moore, Jr., M. Fixman, and B. N. Epstein, *J. Polym. Sci.*, XVI, 517 (1955)

TIM LODGE

Department of Chemistry and Department of Chemical Engineering & Materials Science, University of Minnesota, Minneapolis, MN 55455-0431

Stockmayer, Moore, Fixman, and Epstein provided one of the earliest quantitative studies of copolymers in dilute solutions.[1] The central themes of the work are deceptively simple: Given two monomeric components A and B in one polymer, how may the resulting properties be described in terms of the properties of the respective A and B homopolymers? What new features result from the interactions between A and B units? Forty years later these broad issues are still vigorously pursued. The paper describes light scattering and intrinsic viscosity measurements on a series of styrene-methylmethacrylate (ST-MMA) statistical copolymers in a variety of solvents. It represents a substantial amount of painstaking experimental labor, including the development and execution of a suitable fractionation scheme to prepare a range of molecular weights with roughly constant mean composition. The writing style is a model of clarity and concision; the authors rigorously suppress the natural desire to display all the data.

The presence of two monomers necessitates at least one new parameter, here taken to be the Flory-Huggins χ_{AB}. Usually χ_{AB} is positive, and thus favors coil expansion relative to the constituent homopolymers. This is embodied in eq. (6), whereby the statistical copolymer is viewed as a homopolymer with an effective χ that is lower than the average of χ_{AS} and χ_{BS} (the subscript S denotes solvent). The data convincingly demonstrate the enhanced expansion, as $[\eta]$ for a given molecular weight is generally about 25% larger than for either homopolymer in a given solvent. Although one might anticipate this enhancement to be greater in solvents that are poor for both components, the data in dioxane, a mutual good solvent, still indicate significant expansion. Similar results on the same system were reported in 1970 in a detailed study by Kotaka and coworkers.[2] More recently, Kent et al. examined the behavior of ST-MMA copolymers by dynamic light scattering (DLS), and found relative coil expansion in butanone, but not in the mutual good solvent THF.[3]

The theme of enhanced coil expansion, and modified conformational properties, has been explored extensively for block copolymers. Here, the possibility of intramolecular segregation of the two blocks has been raised.[4] The experimental situation has been rather controversial over the years, but the results appear to have converged on the conclusion that there is, in fact, little effect on the dimensions of the two blocks, or on the separation of their centers of mass, at least for systems such as ST-MMA with rather small χ_{AB}.[3] This is not true in the melt, however, where interesting block copolymer stretching effects have been discussed.[5,6]

On the basis of eq. (6), Stockmayer et al. note the possibility of a solvent dissolving the copolymer while being a non-solvent for both components.[1]

This effect should require delicate adjustment of experimental conditions to be realized. For example, if the solvent were a theta solvent for A ($\chi_{AS} = 0.5$) and a non-solvent for B (say $\chi_{BS} = 0.6$), and the composition were 50 : 50, then χ_{AB} would need to exceed 0.20; most studies on bulk copolymers indicate χ_{AB} to lie in the range 0.001 to 0.1. However, this phenomenon was clearly demonstrated by Kotaka and coworkers.[2] They determined cyclohexanol to be a theta solvent for PS at 83°C, for PMMA at 79°C, and for (50 : 50) statistical copolymers at 62°C.

Stockmayer et al. also proposed and demonstrated a method to determine χ_{AB} from measurements of the second virial coefficient and $[\eta]$.[1] Despite the many well-known quantitative limitations of the Flory-Huggins theory that underlies this analysis, they obtain values in remarkably good agreement with values obtained from PS-PMMA block copolymer melts,[7] and with their own previous measurements on ternary (PS/PMMA/solvent) systems.[8] This strategy has recently been adapted to DLS, which can also be utilized to obtain χ_{AB} from ternary or copolymer solutions.[9-11]

Equation 3 of the paper quantifies the by then already established extra scattering from copolymers, due to compositional heterogeneity;[1] this was subsequently exploited by Benoit and coworkers to characterize composition distributions.[12] Very recently, this phenomenon has also resurfaced in DLS, where a novel diffusive slow mode in block copolymer liquids has been attributed to small chain-to-chain variations in composition.[13,14] Interestingly, this mode provides ready access to the translational diffusion coefficient in non-dilute solutions, without recourse to other, more elaborate, forms of labeling.

Detailed studies of statistical copolymers in dilute solution have not been particularly numerous in the intervening 40 years, presumably for two reasons. First, the advent of anionic polymerization made the production of narrow molecular weight distribution block copolymers feasible, whereas statistical copolymers must always be fractionated extensively when polydispersity is an issue. Second, block copolymers have heretofore proven more intriguing, due to the ready formation of interesting mesophases, e.g., microphase-separated structures in the bulk, and micelles in solution. Nevertheless, statistical copolymers retain their practical importance as relatively inexpensive materials with tunable properties, and as potential blend compatibilizers. And, increasingly, attention is being focused on multiblock copolymers, which can display a rich variety of thermodynamic properties,[15] and which form an appealing intermediate class between statistical and block copolymers.

REFERENCES AND NOTES

1. W. H. Stockmayer, L. D. Moore, Jr., M. Fixman, and B. N. Epstein, *J. Polym. Sci.*, **16**, 517 (1955).
2. T. Kotaka, T. Tanaka, H. Ohnuma, Y. Murakami, and H. Inagaki, *Polymer J. (Japan)*, **1**, 245 (1970).
3. M. S. Kent, M. Tirrell, and T. P. Lodge, *J. Polym. Sci., Polym. Phys. Ed.*, **32**, 1927 (1994).
4. S. F. Edwards, *J. Phys. A: Math., Nucl. Gen.*, **7**, 332 (1974).
5. K. Almdal, J. H. Rosedale, F. S. Bates, G. D. Wignall, and G. H. Fredrickson, *Phys. Rev. Lett.*, **65**, 1112 (1990).
6. J.-L. Barrat and G. H. Fredrickson, *J. Chem. Phys.*, **95**, 1281 (1991).
7. T. P. Russell, *Macromolecules*, **26**, 5819 (1993).
8. W. H. Stockmayer and H. E. Stanley, *J. Chem. Phys.*, **18**, 153 (1950).
9. J. Desbrières, R. Borsali, M. Rinaudo, and M. Milas, *Macromolecules*, **26**, 2592 (1993).
10. M. Benmouna, H. Benoit, M. Duval, and A. Z. Akcasu, *Macromolecules*, **20**, 1107 (1987).
11. M. Benmouna, H. Benoit, R. Borsali, and M. Duval, *Macromolecules*, **20**, 2620 (1987).
12. W. Bushuk and H. Benoit, *Can. J. Chem.*, **36**, 1616 (1958).
13. C. Pan, W. Maurer, Z. Liu, T. P. Lodge, P. Stepanek, E. D. von Meerwall, and H. Watanabe, *Macromolecules*, **28**, 1643 (1995).
14. T. Jian, S. H. Anastasiadis, A. N. Semenov, G. Fytas, K. Adachi, and T. Kotaka, *Macromolecules*, **27**, 4762 (1994).
15. G. H. Fredrickson and S. T. Milner, *Phys. Rev. Lett.*, **67**, 835 (1991).

JOURNAL OF POLYMER SCIENCE   VOL. XVI, PAGES 517–530 (1955)

Copolymers in Dilute Solution.   I.   Preliminary Results for Styrene-Methyl Methacrylate*

W. H. STOCKMAYER, L. D. MOORE, JR., M. FIXMAN, and B. N. EPSTEIN, *Department of Chemistry, Massachusetts Institute of Technology, Cambridge, Massachusetts*

INTRODUCTION

The mechanism of copolymerization and the bulk properties of copolymers have both been the subjects of extensive investigation, but little attention seems to have been paid thus far to the properties of copolymers in dilute solution.   Several years ago a study of the dilute-solution behavior of copolymers was begun in this laboratory, and in this first communication we outline a simple theoretical approach and report some experimental results for copolymers of styrene and methyl methacrylate.

The proper objective of a physicochemical investigation of copolymers in solution is identical with that of the study of ordinary liquid solutions. In each case it is sought to explain the properties of the composite system in terms of those of the components, with the introduction of a minimum number of new parameters.   In particularly fortunate circumstances no additional parameters may be required, as for example in the well-known van Laar-Scatchard-Hildebrand theory of nonpolar liquid solutions,[1] but more generally at least one extra quantity will be needed to express the effects of the interactions between the unlike molecules (or monomer units) present in the composite systems.   In this paper it is indicated that such a treatment of copolymers may be achieved in a useful form, although many more experimental results will be needed to test it thoroughly.

Styrene-methyl methacrylate copolymers were selected for initial study for several reasons.   The properties of the two parent homopolymers have been extensively investigated in many solvents.   The interaction between these two polymers has also been studied,[2,3] and intrinsic viscosities of one sample of the copolymer in a number of solvents have already been reported.[4]   Finally, the monomer reactivity ratios[5] of the system are well known and such that nearly equimolal "azeotropic" copolymers with negligible heterogeneity of composition are easily prepared.   We have made and fractionated such copolymers and examined the light scattering and viscosity of their solutions in several solvents.

* Work supported in part by Office of Ordnance Research.

THEORETICAL

We first investigate the applicability of the usual Einstein-Debye equation for Rayleigh scattering to copolymer solutions. This is necessary because there is extra scattering when the sample is not uniform in composition.[6] At vanishing concentrations, the turbidity due to a multicomponent solute may be written:

$$\tau = H' \Sigma_i \zeta_i^2 M_i c_i \tag{1}$$

where $H' \equiv 32\pi^3 n^2/3N\lambda_0^4$ in the usual notation and c_i, M_i, ζ_i are weight concentration, molecular weight, and specific refractive index increment, respectively, of component i. Now for a binary copolymer it may be assumed (except possibly when the two kinds of monomer unit differ greatly in refractivity and the nonuniformity of composition is unusually large) that ζ_i depends linearly on the composition x_i and of course not on the molecular weight, so that:

$$\zeta_i = \zeta_0 + b\Delta x_i; \qquad \Delta x_i \equiv x_i - x_0 \tag{2}$$

where $\zeta_0 = \Sigma c_i \zeta_i/\Sigma c_i$ is the observed refractive index increment of the whole copolymer, $x_0 = \Sigma x_i c_i/\Sigma c_i$ is the over-all composition, and b is a constant which can be estimated from the values of ζ for the related homopolymers. Equation (1) then becomes:

$$\tau/H'c = \zeta_0^2 \langle M \rangle + 2\zeta_0 b \langle M\Delta x \rangle + b^2 \langle M(\Delta x)^2 \rangle \tag{3}$$

where the quantities in angular brackets are weight averages. It is seen that some turbidity remains even at vanishing ζ_0, and that for small ζ_0 the molecular weight calculated from the first term alone will be too high.[6]

It has been pointed out[6] that the above effect offers a light-scattering method of assaying the nonuniformity of composition in copolymers. We have not explored this possibility, but rather have worked with copolymers for which the extra terms of equation (3) are minimized, namely, azeotropes (or, more generally, low-conversion copolymers of any composition). For such cases, we find with the aid of distribution formulas[7] previously given:

$$\langle M\Delta x \rangle = 0; \qquad \langle M(\Delta x)^2 \rangle = \kappa M_0 x(1-x) \tag{4}$$

where M_0 is the average molecular weight of a monomer unit and κ is a parameter of the order of magnitude of unity. Clearly the extra scattering is completely negligible in such copolymers for all reasonable values of ζ_0, b, and $\langle M \rangle$.

The angular dependence of the light scattered by a copolymer solution should also differ from that of a simple polymer, but it is easily seen physically that the difference must be negligible when the average lengths of sequences of like monomer units (and hence the distances over which correlations in composition are appreciable) are much less than the wave length of the light, as is surely true for all synthetic copolymers (whether "random" or "block") thus far prepared. This statement can be confirmed by a tedious calculation of the scattering function which need not be given here. Thus, the light-scattering behavior of any reasonably homogeneous copolymer cannot in any way be distinguished (except when ζ_0 is very small) from that of a homopolymer, and the experimental results may be treated in the same way.

We turn now to the question of how the dilute-solution properties of a copolymer, such as intrinsic viscosity $[\eta]$, osmotic second virial coefficient A_2 and mean square end-to-end distance $\overline{r^2}$ can be related to those of the parent hompolymers. It is apparent that a simple linear average will not suffice; for in a binary copolymer made up of A and B units there are A-B interactions entirely absent in either a polymer of pure A or one of pure B. In most cases, the net effect of these interactions is repulsive, and the two homopolymers are incompatible.[2] It is qualitatively obvious that these extra repulsions will expand the polymer coils further and cause $[\eta]$, A_2, and $\overline{r^2}$ all to be larger. In other words, a given solvent will usually be "better" for the copolymer than would be expected from its averaged behavior toward the homopolymers; indeed, if the extra repulsions are sufficiently large it should be possible to find solvents for copolymers which are nonsolvents for both of the parent homopolymers.

Quantitatively, the problem has two aspects. The first concerns the "unperturbed" average dimensions of the copolymer chain such as would be observed at the Flory[8] temperature Θ, where random-flight statistics prevail. If the chain conformed to an idealized random-flight model and consisted of two different kinds of independent statistical chain elements, the unperturbed mean square end-to-end distance would obey the simple relation:

$$\overline{r_0^2}/M = w_A(\overline{r_0^2}/M)_A + w_B(\overline{r_0^2}/M)_B \tag{5}$$

where w_A, w_B are weight fractions and the quantities in parentheses are each characteristic of a single type of chain element. Actually, the values of $\overline{r_0^2}/M$ for polymer chains may be influenced by interactions extending over a few skeletal chain atoms, so that equation (5) may be inadequate for real copolymer chains in which the average sequence lengths are short, although it should surely suffice for block copolymers. We note, however, that the values of $\overline{r_0^2}/M$ thus far observed[9] for various synthetic polymers are quite similar, which suggests that equation (5) will not be seriously in error.

The second part of the problem has to do with the "long range" intra- and intermolecular interactions which control the expansion factor α, the intrinsic viscosity, and the second virial coefficient. Here several routes may be followed, but it is perhaps easiest to start with the familiar lattice model.[8] If we assume that (a) the chains are much longer than the average sequence lengths, (b) only nearest-neighbor interactions need be considered, and (c) solvent molecules and both kinds of monomer units have the same volume, we then find the interaction parameter χ_1 of a binary copolymer with a pure solvent to be:

$$\chi_1 = \phi_A \chi_{1A} + \phi_B \chi_{1B} - \phi_A \phi_B \chi_{AB} \tag{6}$$

where ϕ_A and ϕ_B are the mole fractions of A and B in the copolymer, χ_{1A} and χ_{1B} are the interaction parameters for the homopolymers of A and B, respectively, in the same solvent, and χ_{AB} is a new parameter characterizing the A-B interactions.* It is seen that χ_1 is smaller and the solvent therefore "better" when χ_{AB} is positive, as for repulsive A-B interactions.

When assumption (c) fails, as it generally must, we may hope to salvage equation (6) by defining a "segment" in the usual way[8] as equal in volume to a solvent molecule, and by taking ϕ_A and ϕ_B as volume fractions. The limitations of this procedure are apparent. If the average sequence lengths in a copolymer are short, one cannot distinguish just two distinct kinds of copolymer "segments," and in general χ_1 will depend in a more complicated way on the composition. Moreover, even if the quadratic form of equation (6) were empirically adequate, the interpretation of χ_{AB} might remain complicated. Reduction of the segment volume to avoid this trouble will only lead to serious failure of assumption (b). It appears that equation (6) will be most nearly correct for block copolymers which still conform to assumption (a).

If one takes the optimistic view that the above difficulties are not great, the parameter χ_{AB} can be identified with a similar quantity which characterizes the A-B interactions in the ternary system solvent-homopolymer A-homopolymer B, and which can be evaluated from measurements of osmotic pressure or light scattering[3,10] or of phase equilibria[2,11,12] in such systems. Moreover, on literal interpretation of the model, χ_{AB} should depend on the solvent only through its molal volume. Thus, in principle, the thermodynamic and frictional properties of copolymers could be completely predicted, with the aid of equations (5) and (6) and the attendant theories, from measurements only on the related homopolymers. The test of this sweeping but tempting statement awaits accumulation of a body of experimental results.

Other approaches to the interaction problem lead to similar predictions. For example, some theories of intra- and intermolecular interactions in polymer solutions[13,14] invite direct consideration of "cluster integrals" β between pairs of chain elements or segments. In copolymers with two kinds of segments, the average β will be a quadratic function of composition. Since β is given in terms of Flory's symbols[8] by:

$$\beta = 2JV_1^2/\bar{v}^2 = V_1(1 - 2\chi_1) = 2V_1\psi_1(1 - \Theta/T) \tag{7}$$

where $\bar{v}$ is partial specific volume of polymer, V_1 molecular volume of solvent, and ψ_1 and Θ the well-known Flory parameters, it is seen that the equivalent of equation (6) is obtained, with the same limitations. In general, both ψ_1 and $\psi_1\Theta$ should be quadratic functions of copolymer composition, but it seems likely[15] that the quadratic term in ψ_1 will usually be

---

* To make χ_{AB} independent of chain length, we omit a factor x_i from Flory's definition (reference 8, Eq. XII-21), writing simply $\chi_{ij} = z\Delta w_{ij}/kT$. There is no danger of ambiguity except in mixed solvents. See also reference 8, page 549.

small. If this were so, χ_{AB} for *nonpolar* copolymers could be expressed in terms of "solubility parameters"[1] as:

$$\chi_{AB} = V_1(\delta_A - \delta_B)^2/kT \tag{8}$$

which is the equation used by Scott[15] in his discussion of the self-compatibility of copolymers.

EXPERIMENTAL

Polymerization. Three different samples of copolymer were prepared from the azeotropic monomer composition, 46.0 weight per cent methyl methacrylate as calculated from the reactivity ratios[5] at 60°. Commercial monomers were freshly vacuum-distilled and outgassed by the conventional freezing and thawing technique under nitrogen atmosphere before polymerization in sealed vessels at 60°. For the first two copolymers, coded C and MS, the initiator was 5.4×10^{-4} molar benzoyl peroxide, which gave 10% conversion after 24 hours to polymers with $M_w \sim 10^6$. The third sample, 11 L, was of much lower average molecular weight ($M_w \simeq 1.2 \times 10^5$); it was prepared and fractionated by Dr. A. R. Shultz and will be described elsewhere.

Fractionation. Although the spread of chemical compositions in azeotropic or low-conversion copolymers is already very small, the influence of composition on solubility so much exceeds that of molecular weight that some care in choice of a fractionation procedure is advisable. Numerous solvent-precipitant pairs were therefore empirically examined by observing the volume fraction γ of precipitant required for incipient phase separation at room temperature in 1% solutions of polystyrene and polymethyl methacrylate fractions of comparable molecular weight. Examples of unsuitable systems are butanone-methanol ($\gamma = 0.13$ for polystyrene and 0.57 for polymethyl methacrylate, respectively) and toluene-hexane ($\gamma = 0.42$ and 0.13). The best system found was butanone-diisopropyl ether ($\gamma = 0.51$ and 0.48), which was used for all the fractionations.

Copolymer C was fractionated essentially by the method of Flory,[16] the initial precipitations being from approximately 1% solutions and the second precipitations from about 0.1% solutions. Thirteen fractions were thus isolated, each being finally dissolved once more in butanone and precipitated with methanol before drying to constant weight under vacuum over P_2O_5 at 55°. The third, sixth, and thirteenth (last) fractions were analyzed, with results shown in Table I. It appears that negligible separation according to composition took place (the slightly high results probably indicating some residual solvent), and a wide range of molecular weights was in fact obtained.

Copolymers MS and 11 L were fractionated by the alternative procedure of first isolating five or six large fractions and then dividing each of these into three or four sub-fractions. Since the initial separations were here made from more dilute solutions (about 0.2%), these fractionations were doubtless more efficient than those of copolymer C.

TABLE I

COPOLYMER FRACTIONATION

Sample	Weight per cent methyl methacrylate[a]
Monomer feed	46.0
Unfractionated copolymer C	45.8
Fractions: CF3	46.4
CF6	46.4
CF13	47.7

[a] From combustion analyses, M. I. T. Microchemical Laboratory.

Fractions of polystyrene were kindly furnished by Dr. E. H. Merz (Monsanto Chemical Co.) and Professor P. Doty (Harvard University) and were of well-known origin.[17] Polymethyl methacrylate fractions, probably rather broad, were obtained from Dr. E. W. Patterson (M. I. T. Plastics Laboratory), who had prepared them from special polymers furnished by Rohm and Haas Co.

Solvents and Refractive Index Increments. Solvents used for light scattering and viscosity measurements were specially purified by

TABLE II

LIGHT SCATTERING AND VISCOSITY RESULTS[a]

Fraction	$10^{-6} M_w$	Solvent[b]	$10^4 A_2$	$10^{10}(\overline{r^2})_z$	$10^8(\overline{r^2})_w^{1/2}$	$[\eta]$	$10^{-21} \Phi$
			Copolymers				
CF3	1.90	D (0.133*)	3.6	3.38	1.68	4.11	1.7
		C (0.098*)	2.4	2.29	1.38	3.57	2.6
		B (0.172)	1.5	2.16	1.34	2.52	2.0
		N (0.148)	1.2	1.82	1.23	2.13	2.2
CF5	1.33	D	3.8	1.97	1.28	3.11	2.0
		C	2.3	1.64	1.17	2.76	2.4
		B	1.8	1.25	1.02	2.08	2.6
		N	1.5	1.10	0.96	1.69	2.5
			Polystyrene				
M5	0.92	D (0.171)	2.8	1.13	—	1.96	
		C (0.144*)	2.6	1.15	—	1.97	
H	0.90	D	2.9	1.13	—	1.90	
		B (0.220)	1.3	0.79	—	1.05	
			Polymethyl Methacrylate				
P 10-3	2.0	D (0.075*)	3.0	3.35	—	3.42	
		N (0.100*)	2.5	2.74	—	2.88	
P 10-4	1.22	D	3.1	1.90	—	2.31	
		N	2.6	1.82	—	1.93	
		B (0.111)	1.6	1.22	—	1.60	

[a] Units: A_2, ml. mole/g.^{2}; $\overline{r^2}$, cm.^{2}; $[\eta]$, dl./g.
[b] D = dioxane, C = carbon tetrachloride, B = butanone, N = nitroethane. The figures in parentheses are the refractive index increments, dn/dc, in ml./g., for green light, those marked with an asterisk being calculated[18] values and the others measured values.

Mr. E. J. Curtis of the M. I. T. Chemistry Department, the final products being middle cuts from 60-plate distillations. Refractive index increments for green mercury light were measured with a Brice-Phoenix refractometer (courtesy Professor P. Doty) or with a refractometer constructed in this laboratory. Since the measured values were generally in good agreement with the Gladstone-Dale rule as applied by Zimm,[18] increments for several of the systems were calculated in this way. The measured and calculated values we have used are indicated in Table II.

Light Scattering. The procedure followed in this laboratory for cleaning solutions and measuring turbidities has been described by Shultz.[19] All the measurements here reported were made at $25 \pm 1°$C. with green Hg light, for which theRayleigh ratio of benzene, the primary standard, was taken[20] as 16.3×10^{-6}. Reciprocal reduced intensity plots for the same polymer in several solvents were drawn to a common intercept at zero angle and concentration, i.e., to the same value of M_w. This procedure did no great violence to any of the results, thus indicating that the refractive index increments used were reasonably self-consistent; however, values of the osmotic second virial coefficients A_2 are thereby rendered less certain, so that no elaborate procedures in these cases (results of Table II only) were used to assess the effect of the third virial coefficient in producing curvature. For the more extensive series of measurements on copolymers in butanone (Table III), the curvature was estimated with the aid of an approximate theory.[21]

TABLE III

Copolymers in Butanone

Fraction	$10^{-6} M_w$	$10^4 A_2$	$10^{10}(\overline{r^2})_z$	$10^5(\overline{r^2})_w^{1/2}$	$[\eta]$	$10^{-21} \Phi$
MS 1-1	2.27	1.4$_5$	2.62	1.48	3.03	2.1
CF3	1.90	1.5	2.16	1.34	2.52	2.0
MS 1-2	1.69	1.5$_8$	1.72	1.26	2.46	2.1
CF5	1.33	1.8	1.25	1.02	2.08	2.6
MS 2-2	1.23	1.7$_0$	1.13	1.02	2.03	2.4
MS 2-3	0.95	1.7$_5$	0.73	0.82	1.70	2.9
MS 3-2	0.71	1.9$_3$	—	—	1.32	
MS 4-3	0.31	2.2$_1$	—	—	0.71	
11 L-3	0.205	2.3$_8$	—	—	0.59	
11 L-7	0.094	2.8$_7$	—	—	0.35	
11 L-9	0.049	3.2	—	—	0.23	
UMS[a]	0.97	1.5$_5$	1.15	0.87*	1.70	2.5*

[a] Unfractionated copolymer: $(\overline{r^2})_w$ calculated assuming $z = 1$ in equation (9).

Average dimensions of the coils are primarily reported as z-average mean square end-to-end distances $(\overline{r^2})_z$, in conformity with convention for linear polymers. Weight-average root mean square lengths $(\overline{r^2})_w^{1/2}$ were obtained from these by the formula:

$$(\overline{r^2})_w = (\overline{r^2})_z (z + 1)/(z + 2) \tag{9}$$

where z is the parameter in the distribution formula used by Zimm.[18] The value $z = 10$ was used for all the MS copolymer fractions except MS 1-1, which is a top cut; this choice has been shown to be adequate for polyvinyl acetate subjected to a similar fractionation.[19]   The figure $z = 4$, corresponding to less efficient fractionation, was used for copolymer fractions MS 1-1, CF3, and CF5.   Although these assignments of z are rather arbitrary, they have less influence on $(\overline{r^2})_w$ than is suggested by equation (9), since a partly compensating effect arises in the Zimm method[18] of obtaining the limiting tangent to the reciprocal intensity curve.

Viscosity Measurements.   An Ostwald-Fenske viscometer calibrated for kinetic energy corrections was used at 25°, temperature being controlled to 0.05° during measurements.   A modified viscometer with a tenfold smaller maximum rate of shear was also employed for the higher moelcular weights, and indicated shear corrections to $[\eta]$ to be negligible. Four concentrations of each polymer were measured and the results extrapolated in the usual manner.

RESULTS AND DISCUSSION

Table II gives the results of light scattering and viscosity measurements in four different solvents on two copolymer fractions and on several fractions of polystyrene and polymethyl methacrylate.   The solvents were chosen to differ in their behavior toward the homopolymers.   Thus, dioxane is a good solvent for both; carbon tetrachloride is a good solvent for polystyrene but rather poor for polymethyl methacrylate; nitroethane is a good solvent for the methacrylate but a nonsolvent for polystyrene; and butanone is fair for methacrylate but rather poor for polystyrene.

First examining the behavior of the copolymer fractions CF3 and CF5 alone, we observe that A_1, $\overline{r^2}$, and $[\eta]$ all change in the same direction on going from one solvent to another.   As in the case of polystyrene,[22] plots of $(\overline{r_w^2}/M_w)^{1/2}$ against A_2 are roughly linear, as shown in Figure 1.   The

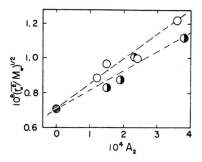

Fig. 1. **Root mean square dimensions and virial coefficients of copolymer fractions CF3 (open circles) and CF5 (half filled circles) in several solvents (Table II).** The cross-hatched circle at $A_2 = 0$ is calculated from equation (5).

last column of Table II gives values of the Flory-Fox viscosity constant Φ, as calculated from the formula:

$$\Phi = [\eta]M_w/(\overline{r_w^2})^{3/2} \tag{10}$$

which neglects a small correction[19] for the effect of heterogeneity of the fractions. The average value for these copolymers and those of Table III is $\Phi = 2.3 \times 10^{21}$, in good agreement with the figures reported for other polymers. Thus the copolymers behave in every way like homopolymers, and their copolymeric nature is not revealed by any internal evidence.

When the copolymers are compared to the homopolymers, however, the effect of the repulsions between the unlike units is easily seen. We may

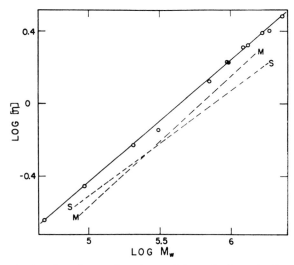

Fig. 2. **Intrinsic viscosity-molecular weight relations in butanone:** Open circles copolymer fractions; solid circle, unfractionated copolymer. Solid line, equation (11). Dashed lines, polystyrene (S, ref. 22) and polymethyl methacrylate (M, ref. 23).

compare, for example, the second virial coefficients of fractions CF5, M5, H, and P 10-4, all of which are close enough in molecular weight so that the relatively minor dependence of A_2 on M can be ignored. It is seen that in dioxane, butanone, and nitroethane (in which A_2 for polystyrene must be negative) the value of A_2 for the copolymer is higher than the average of the values for the homopolymers. Similar behavior can be detected in $\overline{r^2}/M$.

A wider range of copolymer molecular weights has been studied in the single solvent butanone, the results being given in Table III. The second virial coefficients are seen to increase regularly with decreasing molecular weight, as for homopolymers. The intrinsic viscosities are plotted logarithmically against molecular weights in Figure 2, where the solid line through the points corresponds to the equation:

$$[\eta] = 1.54 \times 10^{-4} M_w^{0.675} \tag{11}$$

The dashed lines in this figure are drawn from the equations for intrinsic viscosities in butanone given for polystyrene by Outer, Carr, and Zimm[22] and for polymethyl methacrylate by Bischoff and Desreux.[23]   Once more it is clear that extra intramolecular repulsions act in the copolymer.

Turning to more quantitative treatment of the results, we first investigate the applicability of equation (5).   From the values listed by Flory[9] for the unperturbed dimensions of the homopolymers, we find by this equation a figure of $(\overline{r_0^2}/M)^{1/2} = 7.1 \times 10^{-9}$ cm. (mole/gram)$^{1/2}$ for the copolymer containing 46 weight per cent methyl methacrylate.   The cross-hatched circle at $A_2 = 0$ in Figure 1 corresponds to this value, and is seen at least not to be inconsistent with the measurements in four different solvents.   A better test is afforded by the data of Table III.   Following Flory and Fox,[24] we plot $[\eta]^{2/3}/M^{1/3}$ against $M/[\eta]$ and obtain a satisfactory straight line, as shown in Figure 3.   The intercept of this line at vanishing $M/[\eta]$ should equal $K^{2/3}$, where K is defined by:

$$K = \Phi \, (\overline{r_0^2}/M)^{3/2} = [\eta]/M^{1/2}\alpha^3 \tag{12}$$

The solid circle in Figure 3 corresponds to $K = 7.5 \times 10^{-4}$, which is the value calculated from the above equation with Flory's preferred[8] figure of $\Phi = 2.1 \times 10^{21}$ and the value of $\overline{r_0^2}/M$ obtained from equation (5).   Although it would be still better to have measurements on the copolymer at a Θ-point, it appears clear that equation (5) is quite satisfactory for styrene-methyl methacrylate copolymers.

We may now employ two different methods to obtain interaction parameters χ_1 from the results of Tables II and III.   Both are far from satisfactory because of deficiencies in existing theory.   The first method depends on intrinsic viscosities alone, and uses Flory's expression[8] for the expansion factor:

$$\alpha^5 - \alpha^3 = 2C_M M^{1/2}\psi_1(1 - \Theta/T) = 2C_M M^{1/2}(1/2 - \chi_1)$$
$$2C_M = (9M/2\pi\overline{r_0^2})^{3/2}\overline{v}^2/NV_1 \tag{13}$$

which, together with equation (12) amounts to setting the slope of the line in Figure 3 equal to $2C_M K^{5/3}(1/2 - \chi_1)$.   A comparison of equation (13) with more rigorous theories[14] of the expansion factor indicates[25] that the values of $(1/2 - \chi_1)$ thus obtained are too low by a factor of about 2.   The second method depends primarily on the virial coefficients, given according to Flory and Krigbaum[8] by:

$$A_2 = (\overline{v}^2/NV_1)\psi_1(1 - \Theta/T)F(X) \tag{14a}$$

with:

$$X = 2(\alpha^2 - 1) \tag{14b}$$

In this procedure, X is computed from the intrinsic viscosity by equations (12) and (14b), and the interaction parameter is then obtained from A_2 by (14a).   It is known from both experiment[26] and theory[27] that the Flory-

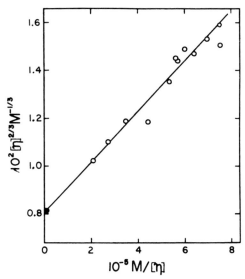

Fig. 3. Flory-Fox plot of intrinsic viscosities of copolymers in butanone. The solid circle is the intercept calculated by equations (5) and (12).

Krigbaum function $F(X)$ is too large and varies too little with molecular weight. This is also easily seen from our results, since the values of $(1/2 - \chi_1)$ obtained exhibit a steady decrease as M increases. The results given by the second method are therefore also too low, though by an unknown factor. We nevertheless restrict ourselves at this time to the use of published theories, and obtain by the above procedures the figures given in Table IV. The values for dioxane are all derived from Table II, and for the copolymer in butanone from Table III. For polystyrene in butanone we have used the data of Outer, Carr, and Zimm[22] and for polymethyl methacrylate in butanone those of Bischoff and Desreux[23] and of Casassa.[28] It is seen that the agreement between the two methods is at least as good for the copolymer as for the homopolymers, and may be considered satisfactory at present in view of the fundamental difficulties involved.*

TABLE IV

INTERACTION PARAMETERS

Solvent	Polymer	$(1/2 - \chi_1)$ From $[\eta]$	From A_2
Butanone......	Polystyrene	0.006	0.023
	Polymethyl methacrylate	0.022	0.035
	Copolymer	0.025	0.040
Dioxane.......	Polystyrene	0.025	0.037
	Polymethyl methacrylate	0.047	0.057
	Copolymer	0.057	0.060

* See Flory's remarks on pp. 625–626 of reference 8.

Finally, we extract χ_{AB} from the figures of Table IV by means of equation (6). The precision is very low, since an error of ± 0.003 in each of the values of χ_1 produces a probable error of ± 0.02 in χ_{AB}. The results are shown in Table V together with those earlier reported[3] for ternary systems, which are likewise of low precision and limited theoretical significance. Since the solvents involved have nearly equal molal volumes, a single value of χ_{AB} should be found if the most optimistic form of the theory were correct. The figures for the copolymer in butanone and dioxane are the same within experimental error, but those for the ternary systems seem definitely lower, probably for the reasons given earlier. It is amusing to note that equation (8), with solubility parameters of 9.1 for polystyrene[29] and 9.5 for polymethyl methacrylate,[4] gives $\chi_{AB} \simeq 0.025$, which is at least of the right order of magnitude.

TABLE V

Method	Solvent	χ_{AB}
Copolymer		
$[\eta]$.....................	Butanone	0.05
A_2.....................	"	0.05
$[\eta]$.....................	Dioxane	0.08
A_2.....................	"	0.06
Ternary system, A_2.............	Butanone	0.02
Ternary system, phase separation.................	Benzene	< 0.015

It is clear that many more measurements on both copolymers and ternary systems are needed to give definite answers to these questions, and we hope to provide these in future work.

We are indebted to H. E. Stanley, E. F. Casassa, and A. R. Shultz for material assistance and valuable discussions. We thank Swift and Company for a grant in aid of the initial part of this work, and the Office of Ordnance Research for later support.

References

1. See, for example, J. H. Hildebrand and R. L. Scott, *The Solubility of Non-Electrolytes*, Reinhold, New York, 1950, Chap. VII.

2. A. Dobry and F. Boyer-Kawenoki, *J. Polymer Sci.* **2**, 90 (1947).

3. W. H. Stockmayer and H. E. Stanley, *J. Chem. Phys.* **18**, 153 (1950).

4. T. Alfrey, A. I. Goldberg, and J. A. Price, *J. Colloid Sci.* **5**, 251 (1950).

5. F. R. Mayo and C. Walling, *Chem. Revs.* **46**, 191 (1950).

6. R. Tremblay, M. Rinfret, and R. Rivest, *J. Chem. Phys.* **20**, 523 (1952).

7. W. H. Stockmayer, *J. Chem. Phys.* **13**, 199 (1945).

8. P. J. Flory, *Principles of Polymer Chemistry*, Cornell Univ. Press, Ithaca, N. Y., 1953, Chaps. XII–XIV.

9. Reference 8, Table XXXIX, p. 618.

10. W. R. Krigbaum and P. J. Flory, *J. Chem. Phys.* **20**, 873 (1952).

11. R. L. Scott, *J. Chem. Phys.* **17**, 279 (1949).

12. H. Tompa, *Trans. Faraday Soc.* **45**, 1142 (1949).

13. B. H. Zimm, *J. Chem. Phys.* **14**, 164 (1946).

14. B. H. Zimm, W. H. Stockmayer, and M. Fixman, *J. Chem. Phys.* **21**, 1716 (1953).

15. R. L. Scott, *J. Polymer Sci.* **9**, 423 (1952).
16. P. J. Flory, *J. Am. Chem. Soc.* **65**, 372 (1943).
17. E. H. Merz and R. W. Raetz, *J. Polymer Sci.* **5**, 587 (1950).
18. B. H. Zimm, *J. Chem. Phys.* **16**, 1099 (1948).
19. A. R. Shultz, *J. Am. Chem. Soc.* **76**, 3422 (1954).
20. C. I. Carr, Jr., and B. H. Zimm, *J. Chem. Phys.* **18**, 1616 (1950).
21. W. H. Stockmayer and E. F. Casassa, *J. Chem. Phys.* **20**, 1560 (1954).
22. P. Outer, C. I. Carr, Jr., and B. H. Zimm, *J. Chem. Phys.* **18**, 830 (1950).
23. J. Bischoff and V. Desreux, *J. Polymer Sci.* **10**, 437 (1953).
24. P. J. Flory and T. G. Fox, Jr., *J. Polymer Sci.* **5**, 745 (1950).
25. W. H. Stockmayer, to be published.
26. W. R. Krigbaum and P. J. Flory, *J. Am. Chem. Soc.* **75**, 1775 (1953).
27. M. Fixman, Ph.D. Thesis, M. I. T. (1953).
28. E. F. Casassa, Ph.D. Thesis, M. I. T. (1952).
29. Reference 1, Chap. XX, Table 5, p. 387.

Synopsis

The dilute solution properties of copolymers are briefly discussed in relation to those of the parent homopolymers. It is shown that copolymer molecules are usually more expanded in solution than would be expected from the averaged behavior of the pure polymers, because of repulsive interactions between the unlike monomer units. A thermodynamic parameter χ_{AB} characterizing these interactions can be derived from measurements of the dilute solution properties of copolymers. In favorable cases this parameter can be independently evaluated from studies of ternary systems composed of the two parent homopolymers and a solvent, thus allowing prediction of the behavior of the copolymer. Light scattering and viscosity measurements on fractions of approximately equimolal copolymers of styrene and methyl methacrylate are presented and analyzed. The values of χ_{AB} deduced from the results in two solvents agree satisfactorily with each other, but are somewhat larger than those earlier obtained from measurements on ternary systems.

Résumé

Les propriétés de solutions diluées de copolymères sont brièvement discutées en comparaison avec celles des homopolymères correspondants. On montre ainsi que les molécules de copolyméres sont en moyenne plus allongées en solution qu'on ne s'y attendrait au départ du comportement moyen des polymères purs, par suite des interactions répulsives qui s'exercent entre unités monomériques différentes. Un paramètre thermodynamique χ_{AB} caractérisant ces interactions peut être défini aux dépens de mesures des propriétés des solutions diluées des copolymères. Dans les cas favorables ce paramètre peut être évalué indépendamment au départ d'études de systèmes ternaires, composés des deux homopolymères et d'un solvant; ceci permet ainsi de prévoir le comportement du copolymère. Des mesures de diffusion de lumière et de viscosité ont été effectuées à cet effet sur des fractions de copolymères contenant des quantités approximativement équimolaires de styrène et de méthacrylate de méthyle. Les valeurs de χ_{AB} déduites de l'analyse de ces résultats dans le cas de deux solvants sont en accord réciproque; ils sont toutefois quelque peu supérieures à celles trouvées précédemment au départ de mesures effectuées sur des systémes ternaires.

Zusammenfassung

Die Eigenschaften in verdünnter Lösung von Copolymeren werden kurz in Bezug zu den Eigenschaften der Eltern-Homopolymere diskutiert. Es wird gezeigt, dass Copolymermoleküle im allgemeinen in Lösung ausgebreiteter sind, als aus dem üblichen Verhalten der reinen Polymere erwartet werden würde, da zwischen den ungleichen Mono-

STOCKMAYER, MOORE, FIXMAN, AND EPSTEIN

mereinheiten abstossende Interaktionen bestehen. Ein thermodynamischer Parameter χ_{AB}, der diese Interaktionen charakterisiert, kann aus Messungen von Eigenschaften in verdünnten Lösungen von Copolymeren abgeleitet werden. In günstigen Fällen kann dieser Parameter unabhängig aus Untersuchungen von ternären Systemen bestimmt werden, die aus zwei Eltern-Homopolymeren und einem Lösungsmittel zusammengesetzt sind; dies ermöglicht Voraussage des Verhaltens des Copolymers. Es werden Lichtstreuungs- und Viskositätsmessungen an Fraktionen von ungefähr äquimolalen Copolymeren von Styrol und Methylmethacrylat gegeben und analysiert. Die Werte von χ_{AB}, die aus den Resultaten in zwei Lösungsmitteln abgeleitet wurden, stimmen zufriedenstellend miteinander überein, aber sie sind etwas grösser als die früher aus Messungen an ternären Systemen erhaltenen.

Received December 7, 1954

COMMENTARY

Reflections on "The Growth of Single Crystals of Linear Polyethylene," by P. H. Till, J. Polym. Sci., XXIV, 301 (1957)

Chain Folding at Dupont in the Mid-'50s

P. H. GEIL

Department of Materials Science and Engineering, University of Illinois, 1304 W. Green St., Urbana, IL 61801

It was in the early spring of 1956 that I visited the Polychemicals Department at the Dupont Experimental Station in Wilmington, Delaware, on an interview trip as I neared completion of my Ph.D. in the area of biophysics. Although I was sure we were going to go west, the visit to several East Coast sites permitted a scenic tour with my wife that was too good to turn down. It was during this visit, before the near simultaneous publication of any of the three seminal papers[1-3] in the area, that my hosts showed me micrographs of the "higher density" (e.g., as we had to label it later[4]) polyethylene single crystals, their electron diffraction patterns, and its interpretation in terms of chain folding. There may even have been some suggestion of the applicability of a lamellar model to melt crystallized material based on the never published micrographs by W. Peck and V. Kaye (at Tennessee Eastman in Tennessee) of lamellar structures in melt crystallized branched

polyethylene[5] (see Figs. IV-15 and VII-24, ref. 6). With their "promise" that if I accepted their offer my first assignment* would be an examination of spherulite growth and structure, I not only agreed but used our small angle x-ray equipment at the University of Wisconsin to take some scans of samples with different thermal histories before finishing my degree.

Those micrographs were taken by Paul Till, probably, as far as I have been able to discover, at his own instigation. Paul was the Polychemical Department's "service microscopist" in the analytical section, "down the hill" along the Brandywine where he had an RCA microscope. I don't recall that I met him during my visit; in fact, he probably had left Dupont and science before I arrived in January 1957. I know of no other publication of his from his Dupont days.

The paper had all the evidence of folding, and it was being discussed during my visit by at least some of my hosts, but there was no mention of folding in the paper—why not? The paper was not even submitted until November 1956, at least eight months after my visit. As an example of the thinking of many at Dupont at the time, my first assignment was in the area of melt crystallization, but not to extend Paul Till's work. Rather it was to support and prepare for publication a model of my supervisor for the growth of ringed spherulites of polyethylene.[7]

---

* This lasted six months following which the goals were changed to an area of more "practical" concern, environmental stress cracking. Clearly, however, one couldn't understand ESC without understanding deformation mechanisms, which required a knowledge of the original morphology, which brought me, at least, back to the original goal. Such arguements permitted survival of both my research and that of what was, at the time, probably the best polymer physics research group in the world to continue for five years before its collapse and disbursement.

Based on the auto-orientation model of Schuur,[8] string models for fringed micelle crystals were sealed between lantern slide glasses for use in an ACS talk, the abstract being submitted in the fall of 1957. The information from all I have talked to, then and recently, is that Till's supervisor thought the idea of folding too outrageous to be postulated in a paper from Dupont, both delaying the publication so that it came out about the same time as the totally independent papers of Keller[2] and Fischer[3] and having all mention of folding deleted before being cleared. Similar concerns may have resulted in lack of publication of the Peck and Kaye paper, the explanation for the observed excellent lamellar morphology in the talk being given in terms of Keller's earlier helically twisted fibrils model.[9]

Such thinking was not restricted to the two supervisors involved, however. Simultaneously with the preparation of the material for the ACS meeting, and following submission of the Abstract, I was gradually learning EM techniques from Ken Symons and Bob Scott (the then service microscopists in Polychemicals and Textile Fibers Departments) and "taking over" the Phillips 100 microscope that Ed Clark, my office mate, had obtained before I arrived. It was at this time the examination of polyoxymethylene morphology began, a project that was at least reasonably justifiable in terms of both Dupont interest and relevance to the assigned project. Single crystals,[9] hedrites,[10] and lamellar spherulites[11] were all observed before presentation of the ACS paper. This work, I think, soon convinced most Dupont personnel of the validity and significance of lamellae, but convincing the outside world took considerably longer. The ACS paper was presented, 15 minutes on the fringed micelle model and five minutes on

the POM observations, with the conclusion that if the last five minutes was correct the first 15 were nonsense. In the discussion period I was informed by Prof. Flory, from the audience, that what ever the validity of the first 15 minutes, the last five were nonsense, a disagreement that never was fully resolved. The last five, at least, convinced my supervisor since he presented the hedrite work at the Cooperstown, New York, International Conference on Crystal Growth meeting in the summer of 1957.[12] The rest is history.

REFERENCES AND NOTES

1. P. H. Till, *J. Polym. Sci.*, **24**, 301 (1957).
2. A. Keller, *Phil. Mag.*, **2**, 1171 (1957).
3. E. W. Fischer and Z. Naturforsch., **12a**, 753 (1957).
4. P. H. Geil, N. K. J. Symons, and R. G. Scott, *J. Appl. Phys.*, **30**, 1516 (1959). This term was used up to at least 1961, because of patent litigation.
5. V. Peck and W. Kaye, paper presented at EMSA meeting, abst. in *J. Appl. Phys.*, **25**, 1485 (1954). A copy of the text of the talk and the figures was available at Dupont.
6. P. H. Geil, *Polymer Single Crystals*, Wiley-Interscience, New York 1963.
7. C. F. Hammer and P. H. Geil paper presented at ACS meeting, San Francisco, April 1958.
8. G. Schuur, *J. Polym. Sci.*, **11**, 385 (1953) and Rubber Stichting commun. # 276, Delft 1955.
9. P. H. Geil, N. K. J. Symons, and R. G. Scott, *J. Appl. Phys.*, **30**, 1516 (1959).
10. P. H. Geil, in *Growth and Perfection of Crystals*, R. H. Doremus, B. W. Roberts, and D. Turnbull (eds.), Wiley, New York, 1958.
11. P. H. Geil, *J. Polym. Sci.*, **47**, 65 (1960).
12. A. Keller and J. R. S. Waring, *J. Polym. Sci.*, **17**, 447 (1955) and references therein.

PERSPECTIVE

Comments on "The Growth of Single Crystals of Linear Polyethylene," by P. H. Till, Jr., J. Polym. Sci., XXIV, 301 (1957)

F. KHOURY

Polymers Division, National Institute of Standards and Technology, Gaithersburg, MD 20899

Till's paper was the first of three independent reports published in 1957 describing the growth habits of dilute solution grown crystals of polyethylene as revealed by electron microscopy. The two other papers, by Keller[1] and Fischer,[2] acknowledged Till's publication and results, and variously confirmed, complemented, or supplement them. There is no question that these three papers spurred a revolution in the field of the crystallization and morphology of polymers. Before dwelling on the specifics of Till's contribution it is pertinent to recall, albeit very briefly, the general state of knowledge concerning the morphology of crystallizable polymers circa the early-to-mid-1950s.[3] It had become evident by then that flexible-chain polymers crystallize from the melt, over a large range of under-coolings, in the form of spherulites which are consisted of a spherically symmetrical radiating array of long and narrow structural units (often described at the time as fibrillar) in which the chain molecules are oriented preferentially parallel to the tangential direction. Such crystallization and morphological organization was beyond accounting in terms of the fringe-micellar model of the fine structure of semi-crystalline polymers that had prevailed up to about the mid-to-late 1940s. A particularly vexing problem was providing a satisfactory explanation of the origins of the tangential orientation of the chains in the constituent radiating structural units in spherulites.

In retrospect, given the ease with which it has since been found to grow lamellar polymer crystals from solution, the dearth of reported investigations on the crystallization of polymers from dilute solutions prior to about 1957 reflects a then prevailing apparent "mind-set" concerning limitations in achievable levels of crystalline order in polymers.

As indicated by Till, Schlesinger and Leeper[4] described in 1951 solution grown single crystal-like structures of gutta (mol. wt. 16,000–18,000 trans-polyisoprene from gutta percha). These structures were large enough to be seen in the optical microscope. Their examination was limited to a characterization of their optical anisotropy. Details the fine structure and molecular orientation characteristics in these objects were not pursued further at the time.

Till's report on polyethylene crystals was presaged by a short note published by Jaccodine[5] in 1955. The focus of Jaccodine's interest was the role of dislocations in crystal growth. Referring to a similarity with previous work on paraffin crystals,[6] Jaccodine's limited report of observations in the electron microscope of low molecular weight linear polyethylene (average about 10,000) crystallized from xylene and benzene indicated that screw dislocations in lozenge shaped lamellar growths resulted in the spiral growths of additional lamellae. Till's electron microscopic study on unfractionated linear polyethylene (Mw = 150,000), and fractions, demonstrated convincingly the common tendency of these materials to crystallize from dilute solutions in the form of single crystals consisting of thin lamellae which are about 10 nm thick and whose lateral growth faces are [110]. Till indicated that branched polyethylene aggregates grown from solution also exhibited layer-like growth. He also referred in his paper to two studies[7,8] reporting that the constituent radiating structural units in polyethylene spherulites are lamellar or layer-like rather than fibrillar. The following concluding statement in Till's paper certainly evidences his appreciation of the more general relevance and importance of elucidating the origins of the observed lamellar crystallization habits: "By an investigation of the disposition of the individual polymer chains in the layers forming such structures · · · and the organization of what appear to be similar sheets in spherulites we hope that a model of spherulitic structure can be developed to explain the various optical microscopic and x-ray diffraction observations on polyethylene spherulites."

The matter of lamellar crystallization was resolved soon thereafter by Keller[1] who concluded in his seminal paper that it is a habit consequent to the manifestation of chain folding. It is interesting to note that the since much studied phenomenon of chain folding as a fundamental characteristic of the crystallization of flexible chain polymers, had been invoked by Storks[9] back in 1938 in a suggested explanation of results of an electron diffraction study of thin films of gutta percha. In that paper which had been long neglected until mentioned by Keller,[1] Storks had proposed the possibility of chain folding in an attempt to reconcile the fact that the chains were oriented perpendicular to the gutta percha films which were considerably thinner that the average chain length.

Even after more than thirty years following the events of 1957 experimental and theoretical studies on various aspects of the growth habits of polyethylene crystals have continued in earnest. An area of interest, for example, has been the diversity of the lateral growth habits exhibited by solution and melt grown crystals of this polymer (e.g., refs. 10–13, which are far from inclusive).

REFERENCES AND NOTES

1. A. Keller, *Phil. Mag.,* **2,** 1171 (1957).
2. E. W. Fischer, *Z. Naturforch.,* **12a,** 753 (1957).
3. The following two references, published after 1957, provide a historical perspective setting pre- and immediate post-1957 developments in the crystallization and morphology of polymers in relation to one another: (a) A. Keller, in "Growth and Perfection of Crystals" (R. H. Doremus, B. W. Roberts, D. Turnbull, eds.), pp. 499–528, Wiley, New York (1958). (b) P. H. Geil, "Polymer Single Crystals," Wiley, New York (1963).
4. W. Schlesinger and H. M. Leeper, *J. Polym. Sci.,* **9,** 203 (1953).
5. R. Jaccodine, *Nature,* **176,** 305 (1955).
6. I. M. Dawson and V. Vand, *Nature,* **167,** 476 (1951).
7. V. Peck and W. Kaye, *J. Appl. Phys.,* **25,** 1464 (1954).
8. G. C. Claver, Jr., R. Buchdahl, and R. L. Miller, *J. Polym. Sci.,* **20,** 202 (1956).
9. K. H. Storks, *J. Am. Chem. Soc.,* **60,** 1753 (1938).
10. M. L. Mansfield, *Polymer,* **29,** 1755 (1988).
11. R. L. Miller and J. D. Hoffman, *Polymer,* **32,** 963 (1991).
12. J. J. Point and D. Villers, *J. Crystal Growth,* **114,** 228 (1991).
13. A. Toda, *Colloid Polym. Sci.,* **270,** 667 (1992).

The Growth of Single Crystals of Linear Polyethylene

Most organic compounds when cooled from the melt or precipitated from solutions exhibit a tendency to assume an ordered crystalline arrangement. Many polymers when properly annealed also develop a crystallinity as revealed by x-ray diffraction. However, owing to their polydisperse nature, the crystallinity is somewhat different in degree and nature from that found in simpler compounds. We would expect the following unique features of high polymers to be responsible for these differences: (1) chain entanglement due to the length of high polymer chains; (2) heterogeneous chain length; (3) side chain branching; (4) randomness in head-head, tail-tail, head-tail polymerization; (5) randomness in disposition of bulky substituents around carbon atoms.

By using linear polyethylene, the structural obstacles (3), (4), and (5) are eliminated, and crystallization from very dilute solution greatly reduces the obstacle of chain entanglement. Solution crystallization of polymeric materials has been considered by some investigators, but on the whole this powerful technique of examining the ideal crystallization tendency of polymers has received little attention. Schlesinger and Leeper have reported the growth of single crystals of the natural polymer alpha-gutta ($\bar{M}_w = 16,000$).[1] They verified the single crystals by examination under crossed Nicols in an optical microscope. More recently Jaccodine reported the interesting observation of spiral growth steps in low molec-

302 JOURNAL OF POLYMER SCIENCE   VOL. XXIV, ISSUE NO. 106 (1957)

ular weight ($\overline{M}_w$ = 10,000) Phillips linear polyethylene (Marlex 20) grown from xylene solution.[2]

Thus, by crystallization of linear polyethylene from dilute solutions, we might expect to more nearly realize the growth of single polymer crystals, suffering only from the effect of heterogeneity of chain length. Figure 1 is an electron micrograph of linear polyethylene ($\overline{M}_w$ = 150,000) crystallized from 0.06% xylene solution. The crystallization was performed by cooling the xylene solution under room temperature gradient from its

Fig. 1. Linear polyethylene ($\overline{M}_w$ = 150,000) crystallized from 0.06% xylene solution.   (Chromium shadowed, 30°) (7200X)

Fig. 2. Selected area electron diffraction pattern of single crystal of linear polyethylene crystallized from 0.06% xylene solution.   (80 kv.)

boiling point to room temperature. The structure is crystalline in appearance, consisting of stacked layers of thin plates about 100 A. in thickness in a staggered array which form a pyramid with a lozenge shaped base. All of the precipitated polymer examined was found to be in this form. The crystalline nature of these structures was settled by selected area electron diffraction. A diffraction pattern of a single structure is shown in Figure 2 and demonstrates that these are single polymer crystals. It appears that the distribution of chain lengths (molecular weight) in the polymer does not present any real obstacle in the crystallization process. Further evidence of this fact is furnished by the success we have had in growing single crystals of dimensions greater than 50 microns by employing a very slow crystallization process.

The effect of chain entanglement has been studied by crystallizing linear polyethylene from more concentrated solutions under otherwise identical experimental conditions as in the dilute solution crystallization. Figure 3 is an electron micrograph of linear polyethylene crystallized from 0.6% xylene solution. The structure would appear to consist of intertwining fibrils which are several hundred angstrom units in diameter; however, it is possible that the fibrils are edges of one layer on top of another as observed by Claver in solvent-cast films of polyethylene.[3]

The influence of side chain branching was investigated. Figure 4 is an electron micrograph of polyethylene containing 2.1 methyls per 100 carbon atoms crystallized from 0.06% xylene solution. A general tendency toward a layer-type growth is evidenced, but the presence of side chains on the parent chain has prevented the attainment of the single crystals found with linear polyethylene.

Fig. 3. Linear polyethylene ($\overline{M}_w$ = 150,000) crystallized from 0.60% xylene solution. (3500X)

Fig. 4. Polyethylene (2.1 methyls per 100 carbon atoms) crystallized from 0.06% xylene solution. (3500X)

304 JOURNAL OF POLYMER SCIENCE   VOL. XXIV, ISSUE NO. 106 (1957)

While heterogeneity of chain length does not seriously impair the growth of single crystals, the range and mean of the molecular chain length (molecular weight) has a profound effect on the nature of crystallization. A series of various molecular weight fractions of linear polyethylene, obtained by selective precipitation from xylene with dimethyl phthalate, were crystallized from 0.05% xylene solution and studied in the electron microscope. Figure 5 is an electron micrograph of a crystal of the fraction $\bar{M}_w = 850$. These crystals grew by a dislocation mechanism as evidenced by the screw dislocation visible in the micrograph. All the solution grown

Fig. 5. Low molecular weight fraction of linear polyethylene ($\bar{M}_w = 850$) crystallized from 0.05% xylene solution.   (6000X)

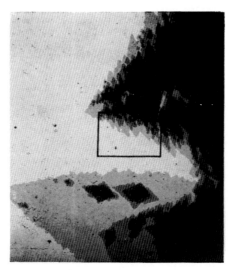

Fig. 6. Intermediate molecular weight fraction of linear polyethylene ($\bar{M}_w = 29{,}000$) crystallized from 0.05% xylene solution.   (4200X)

crystals of this fraction were either very thin single sheets or dislocation grown multilayer crystals and are almost identical in appearance to the solution grown crystals of pure long chain normal paraffins reported by Dawson.[4-6] The fundamental difference is that while Dawson employed pure compounds, these are formed from a heterogeneous mixture. Dislocation growth is also found in crystals of the fraction $\overline{M}_w = 18,000$. A transition in mechanism from a dislocation to a layer mechanism is found in going from the fraction $\overline{M}_w = 18,000$ to the next cut from our fractionation, $\overline{M}_w = 29,000$. Figure 6 is an electron micrograph of a solution grown crystal of the fraction $\overline{M}_w = 29,000$, showing a layer growth similar to that found for unfractionated polymer (Fig. 1). The distinctive "spruce tree" perforation observable in Figure 6 is characteristic of a dendritic crystal. The complexity of the dendrite was found to increase for higher molecular weight fractions.

In both types of growth, the exposed faces of the resulting crystals are the 110 faces. The acute angle of lozenge-shaped crystals, such as that shown in Figure 5, averages about 69°, although individual values ranging from 66 to 74° have been found. The value of 69° is in good agreement with the value of 67.5° calculated from the intersection of the 110 plates of the orthorhombic unit cell as determined by Bunn.[7] Dawson found a comparable value for crystallized long chain normal paraffins.[4] A similar observation is noted in Figure 7 which is an enlargement of the bounded area in Figure 6. The angles of the saw tooth edge of a single layer of the

Fig. 7. Enlargement of bounded area of Figure 6 showing angle formed by intersection of 110 planes. (16,000X)

306 JOURNAL OF POLYMER SCIENCE   VOL. XXIV, ISSUE NO. 106 (1957)

lamellar crystal have a value of 67°, in excellent agreement with the calculated 67.5° value.

It appears from our studies that under proper experimental conditions a high polymer can demonstrate the same tendency to crystallize in the form of single crystals as shown by simpler organic molecules.   The growth of such crystals proceeds by different methods, dependent on the size and heterogeneity of length of the molecules.   We feel that such studies of the ideal crystallization behavior of polymers will lead to a better understanding of the bulk crystallization from the melt.   That such an extension may be practical is indicated by the studies of Peck of slowly annealed thin films of polyethylene in which he observed lamellar structures consisting of sheets about 100 A. thick within well developed spherulites.[8] These structures closely resemble the ones we observed.   By an investigation of the disposition of the individual polymer chains in the layers forming such structures as shown in Figure 1, and the organization of what appear to be similar sheets in spherulites, we hope that a model of spherulitic structure can be developed to explain the various optical microscopic and x-ray diffraction observations on polyethylene spherulites.

I am indebted to Mr. R. G. Scott of the Pioneering Research Laboratory, E. I. du Pont de Nemours & Company for the election diffraction pattern.

References

1. W. Schlesinger and H. M. Leeper, *J. Polymer Sci.*, **11**, 203 (1953).
2. R. Jaccodine, *Nature*, **176**, 305 (1955).
3. G. C. Claver, Jr., R. Buchdahl, and R. L. Miller, *J. Polymer Sci.*, **20**, 202 (1956).
4. I. M. Dawson and V. Vand, *Proc. Royal Soc. (London)*, **A206,** 555 (1951).
5. I. M. Dawson, *ibid.*, **A214,** 72 (1952).
6. I. M. Dawson and N. G. Anderson, *ibid.*, **A218,** 255 (1953).
7. C. W. Bunn, *Trans. Faraday Soc.*, **35,** 482 (1939).
8. V. Peck and W. Kaye, *J. Appl. Phys.*, **25,** 1465 (1954).

P. H. TILL, JR.

Research Division
Polychemicals Department
E. I. du Pont de Nemours & Company
Wilmington, Delaware

Received November 28, 1956

p-306.lay

COMMENTARY

Reflections on "Dynamics of Branched Polymer Molecules in Dilute Solution," by Bruno H. Zimm and Ralph W. Kilb, *J. Polym. Sci.*, XXXVII, 19 (1959)

BRUNO H. ZIMM

Department of Chemistry and Biochemistry, University of California (San Diego), La Jolla, CA 92093-0314

The work was done in what was then called the General Electric Research Laboratory, GE's center for long-range research in Schenectady, New York. GE was developing and using polymers for industrial and engineering applications. The applications of many of these polymers, for example, silicone and polyester resins, depended on gelation by crosslinking in solution as the initial stage of the curing process. It was natural that several of us in the Laboratory became involved in a study of the related physical chemistry.

Five physical chemists, Joseph Bianchi, Julian H. Gibbs, Ralph W. Kilb, Fraser P. Price, and myself occupied one laboratory room. We were experimenting on the effect of solvent dilution on the gelation of polymerizing mixtures containing multifunctional reactants. (This work led to a set of four papers published in the *Journal of Physical Chemistry* in 1958.) Such systems were expected to contain many branched molecules. Comparison of the light-scattering molecular weight with the intrinsic viscosity was the obvious method for detecting branching, but the interpretation was uncertain. Paul Flory's viscosity theory gave $[\eta]$ scaling as R^3, where R was the rms radius of the molecule. (How the language has changed! At that time it would never have occurred to us to use the word "scaling" in the above sense; we might have said "varying.") The Schaefgen–Flory experiments on polyester star molecules[1]

and the Thurmond–Zimm experiments on styrene-divinyl benzene copolymers[2] did not fit the R^3 theory well.

I had extended Rouse's and F. Bueche's normal-mode theories for polymer dynamics in 1954, and I had applied normal-modes theory to the viscosity of branched molecules in 1955, but had not got around to writing it up. Ralph Kilb and I decided to collaborate to finish the work and write a paper. Ralph had a chemical–physics background from working on molecular spectroscopy at Harvard with E. Bright Wilson and enjoyed dealing with matrix algebra, which was the mathematical tool used in the calculations.

When we examined a series of models of molecules of the same M but differing amounts of branching, we found that $[\eta]$ scaled with R, which agreed with the experiments much better than the R^3 scaling. (The first power was exact only in the limit of infinite molecular weight for "perfect stars," that is, stars with all arms of equal length, but seemed to be a good approximation for other shapes.)

The question immediately arose: Why the first power rather than the obvious third power? The naive scaling argument says that since the intrinsic viscosity has dimensions of reciprocal concentration (volume over mass), while the obvious length parameter is R, hence $[\eta]$ should scale as R^3, and this is indeed what happens when molecular weight is the independent variable and the number of branches is held constant. (Actually, there is another length parameter involved; this is the Stokes, or hydrodynamic, radius of the individual segments, but this is independent of R, so consideration of it has no effect on the R^3 scaling.)

The explanation apparently comes from the effect of branching on the distribution of segments in space. In the nondraining case the viscosity depends on the sum of the hydrodynamic interactions between all pairs of segments, the interaction of one pair varying as the reciprocal of the distance between the members of the pair. The distribution of pair distances is not the same in a branched molecule as in a linear molecule of the same molecular weight; for example, in the star molecule with f branches there are $f(f-1)/2$ distances between the chain ends whereas in the linear molecule there is only one such distance. Thus the situation is more complicated than can be described by simple dimensional scaling. Even so, the extent to which the naive scaling argument is wrong is surprising.

REFERENCES AND NOTES

1. Schaefgen and Flory, *J. Amer. Chem. Soc.,* **70,** 2709 (1948).
2. Thurmond and Zimm, *J. Polym. Sci.,* **8,** 477 (1952).

Dynamics of Branched Polymer Molecules in Dilute Solution

BRUNO H. ZIMM and RALPH W. KILB, *General Electric Research Laboratory, Schenectady, New York*

I. INTRODUCTION

Out of a great deal of speculation about the properties of branched, high polymer chain molecules, at least one fact has been definitely established experimentally: the intrinsic and bulk viscosities are less than those of linear molecules of the same molecular weight.[1-3] The theory of the bulk viscosity of even linear polymers is still rather uncertain, but the theory of the intrinsic viscosity of linear polymers has developed to the point of quite fair agreement with experiment.[4-7] In this paper we attempt to extend the hydrodynamic theory of the intrinsic viscosity to branched chains.

After the initial success of the viscosity theory for linear chains, which showed that the intrinsic viscosity, $[\eta]$, was proportional to the cube of the root-mean-square radius of the molecule, R, divided by the molecular weight, M, Thurmond and one of the present authors[2] attempted to use the same relation for branched molecules. In fact, if we assume that the Flory-Fox relation,

$$[\eta] = \Phi R^3/M$$

is valid, where Φ is a universal constant, then it immediately follows that the ratio of the intrinsic viscosities of a branched and a linear molecule of the same molecular weight is $g^{3/2}$, where g is the ratio of the mean-square radii of the branched and unbranched molecules. It is not difficult to calculate g theoretically for many diverse cases.[8]

However, this attempt had only modest success in a quantitative way. Later, Stockmayer and Fixman[9] carried out a calculation of the translational friction constant for branched chains, and showed that the true friction constant was much closer to that of the linear molecule than a theory based on a formula corresponding to the Fox-Flory relation would suggest. They interpreted this to mean that the factor $g^{3/2}$ also overestimates the effect of branching on viscosity, as the experimental results had already indicated.

Stockmayer and Fixman attributed the failure of the Flory-Fox relation to the fact that the root-mean-square radius R is not simply related to the hydrodynamic behavior of the molecule when the shape of the molecule

19

is changed, since the shape of the distribution of the parts of the molecule about the center of mass is also changed.   The pattern of flow of the solvent through the molecule is too complex to be taken account of properly by a simple average of the square of the radius.

In this paper we carry out some direct calculations of the intrinsic viscosity of certain model branched chains.   We carry out essentially the same calculation as that of Kirkwood and Riseman[4] for linear chains, but the actual method is the adaptation of the normal coordinate method of Rouse[10] developed by one of the present authors.[7]   This method has both the advantage of greater simplicity and of yielding the relaxation behavior without additional labor.

Ham[11] has recently published calculations for the viscosity in which the Rouse method is used but hydrodynamic interaction is neglected.   In this paper we extend the method to include the interaction.

II. FORMULATION OF THE PROBLEM

It will be assumed that the polymer consists of branches radiating from a central segment.   The model chosen to represent this molecule is similar to that of our previous paper[7] (hereafter referred to as I).   This consists of $N + 1$ beads connected by identical Hookean springs (which represent a Gaussian distribution of the beads with respect to one another) suspended in a viscous liquid.   The hydrodynamic interaction of the beads will be described by the Kirkwood-Riseman approximation.[4]

The branches are labeled with index numbers represented by the letter b and which range from 1 to f.   The segments in each branch are indexed by numbers represented by i_b or j_b, running consecutively from 1, at the attached end, to m_b at the free end (m_b may differ on the various branches).

As shown in I, the x component of the average forces exerted on the liquid by one of the beads is:

$$F_{xm,b} = -kT \frac{\partial \ln \psi}{\partial x_{m,b}} - \frac{3kT}{b^2} (x_{m,b} - x_{m-1,b}) \qquad (j_b = m_b) \qquad (1)$$

$$F_{xj,b} = -kT \frac{\partial \ln \psi}{\partial x_{j,b}} - \frac{3kT}{b^2} (-x_{j-1,b} + 2x_{j,b} - x_{j+1,b}) \quad (0 < j_b < m_b)$$
$$(2)$$

whereas, for the central bead, we now obtain

$$F_{x0} = -kT \frac{\partial \ln \psi}{\partial x_0} - \frac{3kT}{b^2} (fx_0 - x_{1,1} - x_{1,2} - \ldots - x_{1,f}) \qquad (3)$$

where $\psi(x_0, x_{1,1}, \ldots, z_{m,f})$ is the distribution function for the beads, $3kT/b^2$ is the Hookean spring constant, and b is the root-mean-square length of the spring constant (not to be confused with the subscript b).

The hydrodynamic interaction of two beads depends on the distance between them, which will be approximated by their average distance.

The interaction coefficients for two beads on the same branch are given by:

$$T_{ji} = 1/(6\pi^3)^{1/2}\eta b(|j_b - i_b|)^{1/2} \tag{4}$$

or, if the beads are on different branches, by

$$T_{ji} = 1/(6\pi^3)^{1/2}\eta b(|j_b + i_b|)^{1/2} \tag{5}$$

where η is the viscosity of the liquid.

Define $D = kT/\rho$, $\sigma = 3kT/b^2\rho$; here, ρ is the friction constant of one bead, and $\mathbf{v}_x$ is the unperturbed liquid velocity vector.

$$H_{jj} = 1 \tag{6a}$$

$$H_{jk} = \rho T_{jk} \qquad j \neq k \tag{6b}$$

$$\mathbf{x}^T = (x_0, x_{1,1}, x_{2,1}, \ldots, x_{1,2}, \ldots, x_{m-1,f}, x_{m,f}) \tag{7}$$

$$\mathbf{A} = \begin{bmatrix} f & -1..0......0 & -1..0..... & -1..0...... & -1......0 \\ -1 & & & & \\ 0 & \mathbf{A_1} & 0 & 0 & 0 \\ \vdots & & & & \\ 0 & & & & \\ \hline -1 & & & & \\ 0 & 0 & \mathbf{A_2} & 0 & 0 \\ \hline -1 & & & & \\ 0 & 0 & 0 & \ddots & 0 \\ \vdots & & & & \\ \hline -1 & & & & \\ 0 & 0 & 0 & 0 & \mathbf{A_f} \\ \vdots & & & & \end{bmatrix} \tag{8a}$$

$$\mathbf{A}_b = \begin{bmatrix} 2 & -1 & 0............0 \\ -1 & 2 & -1............0 \\ 0 & -1 & 2 & & \\ \vdots & & & \ddots & \\ & & & 2 & -1 \\ 0 & & & -1 & 1 \end{bmatrix} \tag{8b}$$

where $\mathbf{A}$ is an $(N + 1) \times (N + 1)$ matrix, and the $\mathbf{A}_b$ are $m_b \times m_b$ submatrices.

Applying the equation of continuity, one obtains the linear second-order differential equation for ψ:

$$\frac{\partial\psi}{\partial t} = \sum_{u=x,y,z} \left\{ -\left(\frac{\partial\psi}{\partial\mathbf{u}}\right)^T \cdot \mathbf{v}_u - \psi\left(\frac{\partial}{\partial\mathbf{u}}\right)^T \cdot \mathbf{v}_u + D\left(\frac{\partial}{\partial\mathbf{u}}\right)^T \cdot \mathbf{H} \cdot \frac{\partial\psi}{\partial\mathbf{u}} \right.$$
$$\left. + \sigma\left(\frac{\partial\psi}{\partial\mathbf{u}}\right)^T \cdot \mathbf{H} \cdot \mathbf{A} \cdot \mathbf{u} + \sigma\psi\left(\frac{\partial}{\partial\mathbf{u}}\right)^T \cdot \mathbf{H} \cdot \mathbf{A} \cdot \mathbf{u} \right\} \tag{9}$$

III. TRANSFORMATION TO NORMAL COORDINATES

We wish to transform to normal coordinates which diagonalize the quadratic forms occurring in eq. (9). Such normal coordinates exist if it is possible simultaneously to diagonalize $\mathbf{H} \cdot \mathbf{A}$, $\mathbf{H}$, and $\mathbf{A}$ by the transformations

$$\mathbf{Q}^{-1} \cdot \mathbf{H} \cdot \mathbf{A} \cdot \mathbf{Q} = \Lambda \tag{10a}$$

$$\mathbf{Q}^T \cdot \mathbf{A} \cdot \mathbf{Q} = \mathbf{M} \tag{10b}$$

$$\mathbf{Q}^{-1} \cdot \mathbf{H} \cdot \mathbf{Q}^{-1T} = \mathbf{N} = \mathbf{M}^{-1}\Lambda \tag{10c}$$

where Λ, $\mathbf{M}$, and $\mathbf{N}$ are diagonal matrices with respective diagonal elements λ_k, μ_k, and $\nu_k \cdot \mathbf{Q}$ is the matrix whose $N + 1$ columns are the eigenvectors α_k of $\mathbf{H} \cdot \mathbf{A}$. It is, in fact, sufficient to show that $\mathbf{H} \cdot \mathbf{A}$ and $\mathbf{A}$ may be diagonalized, since:

$$
\begin{aligned}
\Lambda &= \mathbf{Q}^{-1} \, \mathbf{H} \cdot \mathbf{A} \cdot \mathbf{Q} \\
&= \mathbf{Q}^{-1} \cdot \mathbf{H} \cdot \mathbf{Q}^{-1T} \cdot \mathbf{Q}^T \cdot \mathbf{A} \cdot \mathbf{Q} \\
&= \mathbf{Q}^{-1} \cdot \mathbf{H} \cdot \mathbf{Q}^{-1T} \cdot \mathbf{M}
\end{aligned}
\tag{10d}
$$

which is equivalent to eq. (10c).

As shown in I, one is always able to do these diagonalizations if the eigenvalues λ_k of $\mathbf{H} \cdot \mathbf{A}$ are distinct. For a branched polymer with unequal branches, or at most two equal branches, it turns out that the λ_k are distinct. Therefore, in such cases, eq. (9) may be diagonalized by the methods already known.

If n of the f branches are equal ($n \geqslant 3$), we will find that certain eigenvalues are $(n - 1)$-fold degenerate. In this case, our former proof[7] is not complete. However, the theorem still holds, as shown in Appendix A.

The normal coordinates ξ, η, ζ corresponding to $\mathbf{x}, \mathbf{y}, \mathbf{z}$ are given by:

$$\mathbf{x} = \mathbf{Q} \cdot \xi \tag{11a}$$

$$\xi = \mathbf{Q}^{-1} \cdot \mathbf{x} \tag{11b}$$

and similarly for $\mathbf{y}$ and $\mathbf{z}$.

IV. THE EIGENVALUE PROBLEM

The central problem is the solution of the eigenvalue equation:

$$\mathbf{H} \cdot \mathbf{A} \cdot \alpha = \lambda_k \alpha \tag{12}$$

We are interested in the case of large N and small values of λ_k. In this case the components $\alpha_{j,b}$ of α vary slowly with the index j on branch b so that they may be represented by a continuous function $\alpha(s_b)$, which is defined by

$$\alpha_{j,b} = \sqrt{\frac{p}{N}} \, \alpha_b(s) \tag{13}$$

DYNAMICS OF BRANCHED POLYMER MOLECULES   **23**

$$s = pj/N \qquad 0 \leqslant s \leqslant pm_b/N \qquad (14a)$$

$$r = pi/N \qquad 0 \leqslant r \leqslant pm_b/N \qquad (14b)$$

and p is a small integer, introduced for later convenience.

The problem now confronting us is the representation of the operators **A** and **H**.

Consider the effect of **A** on $\alpha_b(s)$. For a bead in the interior of a branch, one finds:

$$\mathbf{A}\alpha_b(s) = -\alpha_b\left(s - \frac{p}{N}\right) + 2\alpha_b(s) - \alpha_b\left(s + \frac{p}{N}\right) \qquad (15a)$$

For the terminal bead:

$$\mathbf{A}\alpha_b(s) = -\alpha_b\left(pm_b/N - \frac{p}{N}\right) + \alpha_b(pm_b/N) \qquad (15b)$$

And for the central bead:

$$\mathbf{A}\alpha_0(0) = f\alpha_0(0) - \alpha_1\,(p/N) - \alpha_2(p/N) - \ldots - \alpha_f(p/N) \quad (15c)$$

where $\alpha_0(0)$ is the value of the eigenfunction at the branch point.

Since $\alpha_b(s)$ is a continuous function and p/N is a small fraction for N large, we expand the right hand sides of the above equations in a Taylor's series about their central values and obtain (neglecting higher order terms):

$$\mathbf{A}\alpha_b(s) = -\frac{p^2}{N^2}\,\alpha_b''(s) \qquad (16a)$$

$$\mathbf{A}\alpha_b(pm_b/N) = \frac{p}{N}\,\alpha_b'(pm_b/N) - \frac{p^2}{2N^2}\,\alpha_b''(pm_b/N) \qquad (16b)$$

$$\mathbf{A}\alpha_0(0) = [f\alpha_0(0) - \alpha_1(0) - \ldots - \alpha_f(0)]$$

$$- \frac{p}{N}\sum_{b=1}^{f}\alpha_b'(0) - \frac{p^2}{2N^2}\sum_{b=1}^{f}\alpha_b''(0) \quad (16c)$$

Since the eigenfunction is assumed to be slowly varying, it follows that:

$$\alpha_0(0) = \alpha_b(0) \qquad b = 1, \ldots, f \qquad (17a)$$

and therefore the bracketed expression in eq. (16c) is zero.

If we now apply the boundary conditions

$$\alpha_b'(pm_b/N) = -\frac{p}{2N}\,\alpha_b''(pm_b/N) \qquad (17b)$$

$$\sum_{b=1}^{f}\alpha_b'(0) = -\frac{p}{2N}\sum_{b=1}^{f}\alpha_b''(0) \qquad (17c)$$

then the operator **A** may be represented by $(p^2/N^2)d^2/ds^2$ over the whole range.

But since $\alpha_b(s)$ is a continuous function, $\alpha_b''(s)$ will be of finite magnitude everywhere. For large $\mathbf{N}$, it will then be sufficient to replace eqs. (17b) and (17c) by:

$$\alpha_b'(pm_b/N) = 0 \tag{17b'}$$

$$\sum_{b=1}^{f} \alpha_b'(0) = 0 \tag{17c'}$$

The continuous representation of $\mathbf{A}$ then consists of these boundary conditions and the following:

$$\mathbf{A}\alpha_b(s) = -\frac{p^2}{N^2} \int \delta(r - s) \frac{d^2\alpha_b(s)}{ds^2} \, ds$$

$$= -\frac{p^2}{N^2} \alpha_b''(r) \tag{18}$$

Similarly, one obtains for $\mathbf{H} \cdot \mathbf{A}$

$$\mathbf{H} \cdot \mathbf{A}\alpha_b(s) = -\frac{p^2}{N^2} \alpha_b''(r) - \frac{h2^{1/2}p^{3/2}}{N^2} \int_c^{pm_b/N} \frac{\alpha_b''(s)ds}{(|r - s|)^{1/2}}$$

$$- \frac{h2^{1/2}p^{3/2}}{N^2} \sum_{b' \neq b} \int_0^{pm_{b'}/N} \frac{\alpha_{b'}''(s)ds}{(|r + s|)^{1/2}} \tag{19}$$

$$h = \frac{N^{1/2}\rho}{(12\pi^3)^{1/2}b\eta} \tag{20}$$

with the same boundary conditions as for $\mathbf{A}$.

Two limiting cases can be distinguished; these are the well-known "free-draining" and "nondraining" cases. For the former, $\mathbf{H} \equiv 1$ (i.e., $h = 0$), and so one obtains from eq. (19) and eq. (12)

$$\alpha_b''(r) = -\frac{N^2}{p^2} \lambda_K \alpha_b(r) \tag{21}$$

Solutions of this equation are:

$$\alpha_{Kb}(r) = \sin\left(\frac{\pi K r}{p}\right) \tag{22a}$$

or

$$\alpha_{Kb}(r) = \cos\left(\frac{\pi K r}{p}\right) \tag{22b}$$

with

$$\lambda_K = \frac{\pi^2 K^2}{N^2} \tag{22c}$$

The values of K will be determined by the boundary conditions for a particular polymer molecule.

We note that for the free-draining case, the values of μ_K are equal to λ_K. Since this case has been treated adequately elsewhere by this[11] as well as a different[8] method, it is used in this paper only for checking the formulas.

For the nondraining case, i.e., $h \gg 1$, one obtains from eq. (19)

$$\int_0^{p m_b/N} \frac{\alpha_b''(s)ds}{(|r-s|)^{1/2}} + \sum_{b' \neq b} \int_0^{p m_{b'}/N} \frac{\alpha_{b'}''(s)ds}{(|r+s|)^{1/2}} = -\lambda' \alpha_b(r) \quad (23)$$

where

$$\lambda' = \frac{N^2}{4h}\left(\frac{2}{p}\right)^{3/2}\lambda \quad (24)$$

This integral equation may be solved by the technique presented in a previous paper.[12]

A fair approximation of the above integral equation is given by:

$$\int_{-\infty}^{\infty} \frac{\alpha_b''(s)ds}{(|r-s|)^{1/2}} = -\lambda' \alpha_b(r) \quad (25)$$

with the same boundary conditions as above. As may be verified by trial in eq. (25), the solutions are

$$\alpha_{Kb}(r) = \sin\frac{\pi K r}{p} \quad (26a)$$

$$\alpha_{Kb}(r) = \cos\frac{\pi K r}{p} \quad (26b)$$

$$\lambda_K' = \frac{N^2}{4h}\left(\frac{2}{p}\right)^{3/2}\lambda = \frac{\pi^2}{2}\left(\frac{2K}{p}\right)^{3/2} \quad (26c)$$

Equation (26c) gives a useful estimate of eigenvalues where high accuracy is not required.

V. VISCOSITY AND BIREFRINGENCE

Sections III and IV of our previous papers[7] carry through identically for a branched polymer. Thus we obtain

$$[\eta] = (N_a b^2 \rho / 6M\eta)\sum_{k=1}^{N} 1/\lambda_k(1 + i\omega\tau_k) \quad (27)$$

where N_a is Avogadro's number, M the molecular weight, $\tau_k = 1/2\sigma\lambda_k$, and ω the frequency of the applied sinusoidal shear κ:

$$\kappa = \kappa_0 e^{i\omega t} \quad (28)$$

For extinction angle χ in steady flow ($\omega = 0$), one has

$$\tan 2\chi = \sum_{k=1}^{N}\tau_k \bigg/ \sum_{k=1}^{N}\tau_k^2 \quad (29)$$

The magnitude of the birefringence is proportional to the $[\eta]$.[7]   The energy stored under conditions of steady state flow is[11,13]

$$\mathfrak{J} = \sum \tau_k{}^2 / (\sum \tau_k)^2 \qquad (30)$$

In the above equations, λ_k always appears in the denominator of the sums. Therefore the smallest eigenvalues are the most important, and, in fact, only the first few values of λ_k contribute significantly. This is fortunate, since the higher eigenvalues have rapidly varying eigenvectors which are poorly represented by the above continuous eigenfunctions and operators. Since the higher eigenvalues contribute negligibly to the above sums, for convenience we may sum $k = 1$ to $k = \infty$, rather than $k = 1$ to $k = N$.

For the case of steady flow, $\omega = 0$, the sum

$$\sum_{k=1}^{N} \frac{1}{\lambda_k} \approx \sum_{k=1}^{\infty} \frac{1}{\lambda_k} \qquad (31)$$

appears in the intrinsic viscosity. For the free-draining case, the intrinsic viscosity is proportional[14] to the mean square radius R^2, which may be calculated by another method[8] also. A direct proof that $\sum 1/\lambda_k$ is proportional to R^2 is given in Appendix B. Therefore, the above sum provides a check on the procedures presented here.

For convenience, define g to be the ratio of R^2 of the branched molecule to that for a linear molecule of equal number of units. Thus, from eq. (75) of the previous paper,[7]

$$g = \left(\sum_{k=1}^{\infty} 1/\lambda_k \right)_{\text{branched}} \left(\sum_{k=1}^{\infty} 1/\lambda_k \right)_{\text{linear}}^{-1}$$

$$= \frac{6}{N^2} \left(\sum_{k=1}^{\infty} 1/\lambda_k \right)_{\text{branched}} \qquad (32)$$

VI. POLYMER WITH BRANCHES OF EQUAL LENGTH

Free-Draining Polymer

For the case of f branches of equal length, we have $m_b = N/f$. Set p equal to f. Thus $0 \leq r \leq 1$ and the wavefunctions for the free-draining polymer must be constructed from $\sin(\pi K r/f)$ and $\cos(\pi K r/f)$ with the boundary conditions:

$$\alpha_{K0}(0) = \alpha_{Kb}(0) \qquad b = 1, 2, \ldots, f \qquad (33a)$$

$$\sum_{b=1}^{f} \alpha_{Kb}'(0) = 0 \qquad (33b)$$

$$\alpha_{Kb}'(1) = 0 \qquad b = 1, 2, \ldots, f \qquad (33c)$$

We proceed to find the wave functions by inspection, taking care to assure the independence of the solutions. Completeness will be checked by comparing the derived value of g in eq. (32) with that found by another method by Zimm and Stockmayer.[8]

(a) Symmetric Functions

These functions are nonzero at the branch point, and therefore, by eq. (33a), all branches must be excited. Thus

$$\alpha_{Kb}(r) = \cos \frac{\pi K r}{f} \qquad b = 1, 2, \ldots, f \qquad (34)$$

The boundary condition of eq. (33b) is automatically satisfied by choice of the cosine function, while eq. (33c) determines the parameter K:

$$K = \frac{kf}{2} \qquad k = 2, 4, 6, \ldots \qquad (35)$$

The symmetric functions are therefore:

$$\alpha_{k1}(r) = \alpha_{k2}(r) = \ldots = \alpha_{kf}(r) = \cos\left(\frac{\pi k r}{2}\right) \qquad k = 2, 4, 6, \ldots \qquad (36a)$$

$$\lambda_k = \frac{\pi^2 f^2 k^2}{4N^2} \qquad k = 2, 4, 6, \ldots \qquad (36b)$$

(b) Antisymmetric Functions

These functions have nodes at the branch point, thus satisfying equation (33a) automatically. If we excite two branches only, in an antisymmetric manner, equation (33b) is satisfied. For a polymer of f equal branches, $(f - 1)$ independent eigenfunctions (with the degenerate eigenvalue λ_k) can be formed:

$$\alpha_{k1}(r) = -\alpha_{kb'}(r) = \sin \frac{\pi k r}{2} \qquad (37a)$$

$\alpha_{kb''}(r) = 0$ for $b'' \neq 1$ or b', where $b' = 2, 3, \ldots, f$. The degenerate eigenvalue is

$$\lambda_k = \frac{\pi^2 f^2 k^2}{4N^2} \qquad k = 1, 3, 5, \ldots \qquad (37b)$$

Since these degenerate eigenfunctions have only two branches excited at a time with a node at the branch point, these functions and eigenvalues are the same as the odd functions for a linear molecule of $2N/f$ units. This observation is useful in obtaining the eigenfunctions in the nondraining case.

Of course, any set of $(f - 1)$ independent linear combinations of the above $(f - 1)$ functions may be substituted for those given.

To check for completeness, we evaluate g:

$$\sum 1/\lambda_k = \frac{4N^2}{\pi^2 f^2}\left[(f-1)\sum_{\text{odd }k} 1/k^2 + \sum_{\text{even }k} 1/k^2\right]$$

$$= \frac{N^2}{6f^2}[3(f-1)+1] \tag{38}$$

where the sums have been evaluated by using properties of the Riemann Zeta Function.[15]

Calculating g, we find:

$$g = (3f-2)/f^2 \tag{39}$$

which is identical with the value of Stockmayer and Zimm,[8] and thus the above set of eigenfunctions appears to be complete.

One also finds:

$$\sum 1/\lambda_k^2 = \frac{N^4}{90f^4}[15(f-1)+1] \tag{40}$$

whereas for a linear molecule[7]:

$$\left(\sum 1/\lambda_k^2\right)_{\text{linear}} = \frac{N^4}{90} \tag{41}$$

The intrinsic viscosity, extinction angle, stored energy function, and lowest relaxation time at zero rate of shear for a linear molecule are denoted by $[\eta]_{02}$, χ_{02}, $\mathfrak{S}_{02}$, and $\tau_{1,2}$, respectively; these are denoted, for a branched polymer, by $[\eta]_{0f}$, χ_{0f}, and $\mathfrak{S}_{1,f}$. Evaluating the ratio of these quantities for the free-draining case, we find:

$$g = [\eta]_{0f}/[\eta]_{02}$$

$$= (3f-2)/f^2 \tag{39}$$

$$\tan 2\chi_{0f}/\tan 2\chi_{02} = f^2(3f-2)/(15f-14) \tag{42}$$

$$\mathfrak{S}_{0f}/\mathfrak{S}_{02} = (15f-14)/(3f-2)^2 \tag{43}$$

$$\tau_{1f}/\tau_{1,2} = 4/f^2 \tag{44}$$

Nondraining Polymer

For the nondraining case, we must satisfy the integral equation from eq. (23),

$$\int_0^1 \frac{\alpha_b''(s)ds}{(|r-s|)^{1/2}} + \sum_{b'\neq b}^f \int_0^1 \frac{\alpha_{b'}''(s)ds}{(|r+s|)^{1/2}} = -\lambda'\alpha_b(r) \tag{45a}$$

$$\lambda' = \frac{\lambda N^2}{4h}\left(\frac{2}{f}\right)^{3/2} \tag{45b}$$

The boundary conditions are the same as in the free-draining case.

DYNAMICS OF BRANCHED POLYMER MOLECULES   **29**

(a) Antisymmetric Functions

Just as in the free-draining case, we observe that solutions may be obtained by the excitation of two branches only, provided the functions have a node at the branch point. These functions again are $(f - 1)$-fold degenerate.

We have, then,

$$\alpha_1(r) = -\alpha_{b'}(r) \tag{46}$$

$\alpha_{b''}(r) = 0$ for $b'' \neq 1$ or b'. Since $\alpha_1(r)$ is zero at the branch point, it must be expansible in a sine series. Thus

$$\alpha_1(-r) = -\alpha_1(r) \tag{47}$$

Substituting eqs. (46) and (47) in eq. (45a) yields

$$\int_0^1 \frac{\alpha_1''(s)ds}{(|r - s|)^{1/2}} - \int_0^1 \frac{\alpha_1''(s)ds}{(|r + s|)^{1/2}} = \int_{-1}^1 \frac{\alpha''(s)ds}{(|r - s|)^{1/2}} = -\lambda'\alpha_1(r) \tag{48a}$$

with boundary conditions:

$$\alpha'(\pm 1) = 0 \tag{48b}$$
$$\alpha(0) = 0$$

But this eigenvalue problem is identical with that solved for the linear molecule,[12] if we restrict ourselves to those functions which have nodes at the center of the linear molecule, i.e., the sine series; therefore:

$$\lambda_{f,k}' = \lambda_{2,k}' \tag{48c}$$

where only the odd integers are to be used in $\lambda_{2,k}'$. Here $\lambda_{f,k}'$ is $(f - 1)$-fold degenerate.

(b) Symmetric Functions

(i) Exact Solution. If all branches are excited, we may expand the eigenfunctions in the Fourier series:

$$a_{k,b}(r) = {}^k a_0 + \sum_{m=2}^\infty {}^k a_m \cos\left(\frac{\pi m r}{2}\right)$$
$$k \text{ and } m \text{ even} \tag{49}$$

TABLE I
The Eigenvalues[a] $\lambda_{f,k}'$

k	$\lambda_{2,k}'$	$\lambda_{4,k}'$	$\lambda_{8,k}'$	$\pi^2 k^{3/2}/2$
1	4.04	4.04	4.04	—
2	12.79	14.74	16.43	13.95
3	24.2	24.2	24.2	—
4	37.9	40.67	44.82	39.47
5	53.5	53.5	53.5	—
6	70.7	74.39	81.98	72.52
7	89.4	89.4	89.4	—
8	—	114.96	129.4	111.66

[a] For odd k, $\lambda_{f,k}'$ is $(f - 1)$-fold degenerate; for even k, $\lambda_{f,k}'$ is nondegenerate.

Proceeding as in the linear case,[12] one may solve for the nondegenerate eigenvalues $\lambda_{f,k}'$, $k = 2, 4, \ldots$

This procedure has been carried out for polymers of four and of eight equal branches. The eigenvalues are listed in Table I.

The sum $\sum 1/\lambda_k$ converges rather slowly. Consequently one needs to calculate many values of $\lambda_{f,k}'$. However, this is a lengthy task if it is to be done by the above exact method for all $\lambda_{f,k}'$. Fortunately, however, the above exact values quickly converge to the following approximate values for all the lowest $\lambda_{f,k}'$.

From eq. (26b), an approximate eigenfunction is found from

$$\alpha_{Kb}(r) = \cos \frac{\pi K r}{f} \quad b = 1, 2, \ldots, f \quad \text{(50a)}$$

with boundary condition $\alpha_{Kb}'(1) = 0$. Thus

$$K = \frac{kf}{2} \qquad k = 2, 4, 6, \ldots \quad \text{(50b)}$$

$$\lambda_{fk}' = \frac{\pi^2}{2}\left(\frac{2K}{f}\right)^{3/2} = \frac{\pi^2}{2} k^{3/2} \qquad k = 2, 4, 6, \ldots \quad \text{(50c)}$$

The various sums of λ_{fk}' were obtained by summing the approximate λ_{fk}' by means of the Riemann Zeta Function and correcting for the differences between these values and the exact values of λ_{fk}' at low k.

We observe that the lowest eigenvalue is $(f - 1)$-fold degenerate, and therefore the nondegenerate eigenvalues contribute relatively little to the various sums at higher f values.

(ii) Approximate Solution. One notes from Table I that, even in the nondegenerate case, λ_{2k}' is roughly equal to λ_{fk}'. Thus, for most purposes, it suffices to use λ_{2k}' throughout, especially at higher f values where the degenerate eigenvalues are quite predominant in any case.

For example:

$$\sum 1/\lambda_k = \frac{N^2}{4h}\left(\frac{2}{f}\right)^{3/2}\left[(f-1)\sum_{\text{odd } k}\frac{1}{\lambda_{2k}'} + \sum_{\text{even } k}\frac{1}{\lambda_{fk}'}\right]$$

$$\approx \frac{N^2}{4h}\left(\frac{2}{f}\right)^{3/2}\left[(f-1)\sum_{\text{odd } k}\frac{1}{\lambda_{2k}'} + \sum_{\text{even } k}\frac{1}{\lambda_{2k}'}\right]$$

$$\approx \frac{N^2}{4h}\left(\frac{2}{f}\right)^{3/2}[0.390(f-1) + 0.196] \quad \text{(51)}$$

Similarly,

$$\sum\frac{1}{\lambda_k^2} \approx \frac{N^4}{2h^2 f^3}[0.06361(f-1) + 0.00721] \quad \text{(52)}$$

Denote the nondraining case by primes. Then:

$$g' = [\eta]_{0f}'/[\eta]_{02}'$$

$$g' \approx \left(\frac{2}{f}\right)^{3/2} [0.390(f-1) + 0.196]/0.586 \tag{53}$$

$$\tan 2\chi_{0f}' \tan 2\chi_{02}' \approx \left(\frac{f}{2}\right)^{3/2} [0.390(f-1)$$
$$+ 0.196]/8.27[0.0636(f-1) + 0.0072] \tag{54}$$

$$\mathfrak{I}_{0f}'/\mathfrak{I}_{02}' \approx \tan 2\chi_{02}'/g' \tan 2\chi_{0f}' \tag{55}$$

$$\tau_{1f}'/\tau_{1,2}' = \left(\frac{2}{f}\right)^{3/2} \text{(exact value)} \tag{56}$$

In Table II, a comparison is given between the approximate value for g and the exact value for $f = 4$ and $f = 8$. The discrepancy is quite negligible. One also finds from Table II that:

$$g' \approx g^{1/2} \tag{57}$$

$$\therefore [\eta]_{0f}' \approx g^{1/2}[\eta]_{02}' \tag{58}$$

TABLE II
Computed Results for Molecules with f Equal Arms.
Approximate Values from Equations (53) and (54).

	Exact values				Approximate values	
f	g	$g^{1/2}$	g'	$\tau_{1,f}'/\tau_{1,2}'$	g'	$\tan 2\chi_{0f}'/\tan 2\chi_{02}'$
2	1	1	1	1	1	1
3	0.778	0.882	—	0.544	0.907	1.61
4	0.625	0.790	0.814	0.353	0.823	2.36
8	0.344	0.586	0.625	0.125	0.632	6.24

This is a very useful observation, since g may be calculated relatively easily,[8] whereas g' requires very extensive computation for complex molecules. Note that $g^{1/2}$ appears, rather than $g^{3/2}$ which has previously been suggested.[2]

Equation (57) has some theoretical basis. For large f one finds from eq. (53) that $g' \approx 1.88f^{-1/2}$. Similarly, eq. (39) gives $g^{1/2} \approx 1.73f^{-1/2}$. Thus $g' \approx g^{1/2}$.

VII. POLYMER WITH TWO LONG AND EIGHT SHORT BRANCHES

Consider a polymer with eight equal branches and two longer branches. Let the long branches be four times the length of the short branches. Let $p = 16$. Label the long branches by $b = 1$ and $b = 2$, and the others by $b = 3, \ldots, 10$. Thus $m_1 = m_2 = 4N/16$ and $m_3 = m_4 = \ldots = m_{10} = N/16$. The boundary conditions and variable ranges are:

$$\alpha_1'(4) = \alpha_2'(4) = 0 \qquad 0 \leqslant r \leqslant 4 \tag{59a}$$

$$\alpha_3'(1) = \alpha_1'(1) = \ldots = \alpha_{10}'(1) = 0$$

$$0 \leqslant r \leqslant 1 \qquad (59b)$$

$$\sum_{b=1}^{10} \alpha_b'(0) = 0 \qquad (59c)$$

$$\alpha_b(0) = \alpha_{b'}(0) \qquad \text{for all } b \text{ and } b' \qquad (59d)$$

We now sketch the various modes of excitation for this molecule. Let us first study the free-draining case.

Free-Draining Polymer

Antisymmetric A Modes of the Long Branches

$$\alpha_1(r) = -\alpha_2(r) = \sin \frac{\pi k r}{8} \quad k = 1, 3, 5, \ldots \qquad (60a)$$

$$\alpha_i(r) = 0 \qquad i = 3, \ldots, 10 \qquad (60b)$$

$$\lambda_k = \frac{4\pi^2 k^2}{N^2} \qquad k = 1, 3, 5, \ldots \qquad (60c)$$

where λ_k is nondegenerate.

Antisymmetric B Modes of the Short Branches

$$\alpha_3(r) = -\alpha_b(r) = \sin \frac{\pi k r}{2} \quad b = 4, \ldots, 10 \qquad (61a)$$

$$\alpha_{b'}(r) = 0 \qquad \text{for } b' \neq 3 \text{ or } b \qquad (61b)$$

$$\lambda_k = \frac{64\pi^2 k^2}{N^2} \qquad k = 1, 3, 5, \ldots \qquad (61c)$$

where λ_k is 7-fold degenerate.

For symmetric modes, all branches are excited. Let:

$$\alpha_1(r) = \alpha_2(r) = A_l \cos \frac{\pi K r}{16} + B_l \sin \frac{\pi K r}{16} \qquad (62a)$$

$$\alpha_3(r) = \alpha_4(r) = \ldots = \alpha_{10}(r) = A_s \cos \frac{\pi K r}{16} + B_s \sin \frac{\pi K r}{16} \qquad (62b)$$

The boundary conditions lead to:

$$A_l = A_s$$

$$B_l = -4B_s$$

$$\cot\left(\frac{\pi K}{4}\right) = A_l/B_l$$

$$\cot\left(\frac{\pi K}{16}\right) = A_s/B_s$$

Combining, we obtain the eigenvalue condition:

$$4 \cot \frac{\pi K}{4} = -\cot \frac{\pi K}{16} \qquad (63)$$

D *Modes* (*Symmetric*)

One set of solutions of eq. (63) is:

$$K = 16k \qquad k = 1, 2, 3, \ldots \qquad (64)$$

Therefore

$$\lambda_k = \frac{256 \pi^2 k^2}{N^2} \qquad k = 1, 2, 3, \ldots \qquad (65)$$

where λ_k is nondegenerate.

$C_1, C_2, C_3, and C_4$ *Modes*

Further solutions of the transcendental equation (63) are:

C_1:
$$K = 16(k + 0.1590)$$
$$k = 1, 2, 3, \ldots \qquad (66a)$$

C_2:
$$K = 16(k - 0.1590)$$
$$k = 1, 2, 3, \ldots \qquad (66b)$$

C_3:
$$K = 16(k + 0.3827)$$
$$k = 1, 2, 3, \ldots \qquad (66c)$$

C_4:
$$K = 16(k - 0.3827)$$
$$k = 1, 2, 3, \ldots \qquad (66d)$$

The corresponding eigenvalues, which are nondegenerate, may be obtained from:

$$\lambda_K = \frac{\pi^2 K^2}{N^2} \qquad (67)$$

Calculating g, we find $g = 0.4026$. This is in excellent agreement with the value of 0.4023 found by Zimm and Stockmayer,[8] and we may therefore presume we have the complete set of eigenfunctions.

The properties of the eigenfunctions for the nondraining case will be similar.

Nondraining Polymer

Antisymmetric A *Modes of the Long Branches*

$$\alpha_1(r) = -\alpha_2(r) \qquad 0 \leqslant r \leqslant 4 \qquad (68a)$$

$$\alpha_b(r) = 0 \qquad b = 3, \ldots, 10 \qquad (68b)$$

Substituting in eq. (23) and changing variables to $s = 4x$ and $r = 4y$, we obtain:

$$\int_0^1 \frac{\alpha_1''(x)dx}{(|y - x|)^{1/2}} + \int_0^1 \frac{\alpha_2''(x)dx}{(|y + x|)^{1/2}} = -8\lambda'\alpha_1(x) \tag{68c}$$

This is exactly analogous to the linear molecule, for those functions which have a node in the center (k odd), thus

$$\lambda' = {}^1\!/_8\, \lambda_{2,k}' \qquad k = 1, 3, 5, \ldots \tag{68d}$$

$$\lambda = \frac{4h}{N^2}\left(\frac{16}{2}\right)^{1/2}\lambda' = \frac{8 \cdot 2^{1/2}h}{N^2}\,\lambda_{2,k}'$$

$$k = 1, 3, 5, \ldots \tag{68e}$$

Antisymmetric B Modes of the Short Branches

$$\alpha_3(r) = -\alpha_b(r)$$

$$0 \leqslant r \leqslant 1,\, b = 4, \ldots, 10 \tag{69a}$$

$$\alpha_{b'}(r) = 0 \qquad \text{for } b' \neq 3 \text{ or } b \tag{69b}$$

Substituting in eq. (23), we find:

$$\int_0^1 \frac{\alpha_3''(s)ds}{(|r - s|)^{1/2}} + \int_0^1 \frac{\alpha_b''(s)ds}{(|r + s|)^{1/2}} = -\lambda'\alpha_3(r) \tag{69c}$$

This is again analogous to the linear molecule, and so $\lambda' = \lambda_{2,k}'$ with odd k.

$$\lambda = \frac{4h}{N^2}\left(\frac{16}{2}\right)^{3/2}\lambda' = \frac{64 \cdot 2^{1/2}h}{N^2}\,\lambda_{2,k}'$$

$$k = 1, 3, 5, \ldots \tag{69d}$$

Here λ is 7-fold degenerate.

For the symmetric modes, all branches are excited. Although the corresponding integral equation could be solved in a manner analogous to that in Section VI, this would be a very lengthy task. Noting that the antisymmetric modes make the major contribution to $\sum 1/\lambda$, etc. and that the approximate solutions in the equal branch case are quite close to the exact values, it will suffice to use the approximate solutions of eqs. (26) for the symmetric modes.

D *Modes*

Using the values $K = 16k$, k integer, in eq. (26c), we find:

$$\lambda_{K}' = \frac{N^2}{4h}\left(\frac{2}{16}\right)^{3/2}\lambda = \frac{\pi^2}{2}\,(2k)^{3/2}$$

$$k = 1, 2, 3, \ldots \tag{70}$$

C_1, C_2, C_3, and C_4 Modes

The values K from eqs. (66a–d) are substituted in

$$\lambda_k' = \frac{N^2}{4h}\left(\frac{2}{16}\right)^{3/2}\lambda = \frac{\pi^2}{2}\left(\frac{K}{8}\right)^{3/2} \tag{71}$$

We may now calculate g'. The contribution of the various modes to g' is listed in Table III. Just as in the equal branch case, g' is nearly equal to $g^{1/2}$ ($g' = 0.616$ and $g^{1/2} = 0.634$).

TABLE III
Contribution of the Various Modes to the Sums g and g' for the
Molecule of Section VII

	Free-draining	Nondraining
A modes	0.187	0.236
B modes	0.082	0.206
C modes	0.129	0.160
D modes	0.004	0.014
	$g = 0.402$	$g' = 0.616$

VIII. APPLICATIONS

We have just seen that g', the ratio of the intrinsic viscosity of a branched molecule to that of a linear molecule of the same molecular weight, appears to be approximately equal to $g^{1/2}$, where g is the corresponding ratio of the mean square radii. This conclusion is based on calculations of g' and $g^{1/2}$ for "perfect stars," i.e., molecules with one branch point and various numbers of arms of equal length, as well as one calculation for an "imperfect star," a molecule with arms of two different lengths. Therefore we propose as an approximate empirical formula that $g' = g^{1/2}$ for all star-shaped molecules, with the additional hypothesis that *all* branched molecules, of whatever shape, will obey the same formula to a degree of approximation sufficiently good for practical purposes.

There is some experimental evidence in favor of this formula. First there is the careful work of Schaefgen and Flory[1] on the viscosities and molecular weights of a series of linear and multichain polyamides. The multichain molecules were of the four- or eight-armed, imperfect star type, but with a distribution of sizes and arm lengths in each sample; hence our calculations do not apply directly, but the empirical formula $g' = g^{1/2}$ may be tried. The comparison is shown in Figure 1. The topmost solid line was drawn through the points of the linear samples, and the other two lines were displaced downward by the factors $g^{1/2}$ appropriate for tetra- and octachain molecules with random chain lengths; these values of g, which had been computed previously,[8] were 0.800 and 0.533, respectively. The agreement is certainly within the experimental error. On the other hand, the $g^{1/2}$ formula, which is represented by the dashed lines, is clearly very poor.

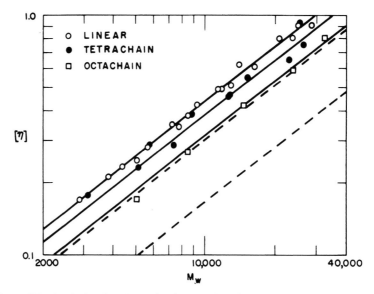

Fig. 1. The intrinsic viscosity-molecular weight relation for linear and branched polyamides; data from Schaefgen and Flory (ref. 1). Solid lines are the theory of this paper (viscosity varying as $g^{1/2}$); dashed lines assume that the viscosity varies as $g^{3/2}$.

A second comparison is possible with the viscosity-molecular weight relation obtained by Thurmond[2] with a series of fractions from a copolymer of styrene and divinylbenzene that had been polymerized nearly to gelation. From the theory of the polymerization of multifunctional reactants,[16] it is possible to predict the distribution of branch units (in this case divinylbenzene units) among the fractions of different molecular weights; the viscosity then follows from our empirical $g^{1/2}$ relation together with the values of g previously computed theoretically.[8] However, since the quantitative interpretation in the paper by Thurmond and Zimm[2] was based on the incorrect $g^{3/2}$ formula, some reworking of their results is necessary.

TABLE IV

Experimental Viscosity Ratios, g', and Computed Average Number of Branch Units per Molecule, $\bar{n}_x$, for the Polystyrene Fractions of Thurmond and Zimm (ref. 2).

Fraction	g'	$\bar{n}_x$
S-1a	0.61	12.5
S-2a	0.76	3.9
S-3a	0.80	2.8
S-4a	0.93	0.75
S-5a	0.93	0.75
S-6a	1.00	0.0
weight-average $\bar{n}_x = 3.36$		

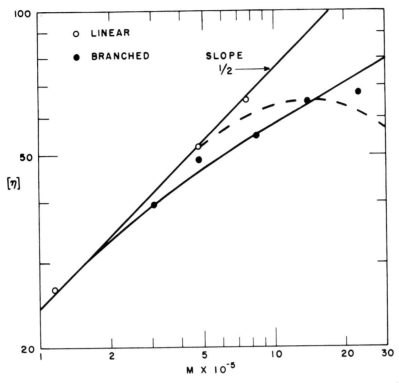

Fig. 2. The intrinsic viscosity-molecular weight relation for linear and branched polystyrene fractions; data from Thurmond and Zimm (ref. 2). Solid straight line has the theoretical slope for the linear fractions and is drawn through the experimental points. The solid curved line is derived from the theory of this paper; the dashed line assumes the viscosity varies as $g^{3/2}$.

We follow exactly the procedure and notation of the previous paper,[2] except for the use of $g^{1/2}$ where $g^{3/2}$ occurred before. Table IV corresponds to a part of Table VII of the previous paper, giving the experimental values of the viscosity ratio, g', and the values of the average number of branch units per molecule, $\bar{n}_x$ in each fraction of the polymer computed from the observed g', assuming $g' = g^{1/2}$, and using the theory of Zimm and Stockmayer[8] for the relation between g and $\bar{n}_x$. The weight-average number of branch units per molecule in the whole polymer is then found by averaging the $\bar{n}_x$ and is found to be 3.36 (compared to the former value of 0.68). From eq. (7) of reference (2), the crosslinking parameter γ is now found to be 0.77, which is much closer to the expected value of a polymer at gelation, unity, than the former value of 0.406. This is an indication of improvement in the viscosity theory.

Further, we can now reverse the process to compute $[\eta]$ as a function of M. First we compute $\bar{n}_x$ theoretically using our new value of γ in exactly the same way as used to compute Table I of reference (2); then, with the theoretical relation between $\bar{n}_x$ and g and our empirical relation that g'

equals $g^{1/2}$, we have g'; finally, since the intrinsic viscosity is a known function of molecular weight for linear polystyrenes, we have $[\eta]$ as a function of M for the branched polymers. The result is shown as the solid curve in Figure 2. The dashed curve is the old relation based on $g^{1/2}$. The agreement with the experimental points is now quite satisfactory, in contrast to the former results.

Thus we are encouraged to feel some confidence in the square-root relation. Some typical values of g', as calculated in this way from the values of g given in reference (8), are shown in Table V. These are for molecules

TABLE V

Viscosity Ratio, g', as a Function of the Average Frequency of Branch Units with Random Branching

Average number, m, of branch units per molecule	g'
A. Trifunctional units	
0.5	0.976
1	0.955
2	0.916
3	0.885
4	0.85
5	0.83
10	0.76
15	0.71
50	0.57
B. Tetrafunctional units	
0.25	0.976
0.5	0.954
1	0.914
2	0.84
3	0.79
6	0.70
15	0.59

with an average of m tri- or tetrafunctional branch units per molecule distributed at random. Other values may be computed from the approximate interpolation formulas for trifunctional units,

$$g' = [(1 + m/7)^{1/2} + 4m/9\pi]^{-1/4} \tag{72}$$

and, for tetrafunctional units,

$$g' = [(1 + m/6)^{1/2} + 4m/3\pi]^{-1/4} \tag{73}$$

These relations lead to an interesting dependence of intrinsic viscosity on the fourth root of the molecular weight for polymers containing very many branch units. In such polymers one expects that the number of branch units per molecule is proportional to the molecular weight. From eqs.

(72) and (73) we see that g' then varies as $M^{-1/4}$.   If, then, we have the common relation for linear polymers,

$$[\eta] = KM^a \tag{74}$$

where a is usually between 0.5 and 0.7, we should get

$$[\eta] = K'M^{a-1/4} \tag{75}$$

for highly branched polymers.   In the particular case of an "ideal" or "theta" solvent,[17] where a is $1/2$, the molecular weight exponent becomes $1/4$.   Exponents of 0.21 and 0.28 have been observed experimentally with silicone systems containing large numbers of trifunctional units in ideal solvents,[18] and similar exponents occur in polyester systems[20] and with the natural polysaccharide dextran.[19]

APPENDIX A

Let $\mathbf{S}$ be a matrix which diagonalizes $\mathbf{H} \cdot \mathbf{A}$:

$$\mathbf{S}^{-1}\mathbf{H} \cdot \mathbf{A} \cdot \mathbf{S} = \Lambda \tag{76}$$

However, since some eigenvalues of $\mathbf{H} \cdot \mathbf{A}$ are degenerate, $\mathbf{S}$ is no longer unique.   In particular, we note that the $(n-1)$-fold degenerate eigenvectors $\mathbf{s}_k{}^I$, $\mathbf{s}_k{}^{II}$, ..., related to the $(n-1)$-fold degenerate eigenvalue λ_k span an $(n-1)$-dimensional subspace of our $(N+1)$-dimensional space. Since any vector in the subspace is an eigenvector with eigenvalue λ_k, it follows that all eigenvectors $\mathbf{s}_m$ associated with $\lambda_m \neq \lambda_k$ must lie outside this subspace, and consequently any unitary transformation within the subspace will leave the eigenvectors $\mathbf{s}_m$ and eigenvalues λ_m unchanged.   Furthermore, the unitary transformation will also leave λ_k unchanged.

Let us apply $\mathbf{S}$ to $\mathbf{A}$:

$$\mathbf{S}^T \cdot \mathbf{A} \cdot \mathbf{S} \tag{77}$$

From our former proof,[7] it still follows that

$$\mathbf{s}_m{}^T \cdot \mathbf{A} \cdot \mathbf{s}_k = 0 \tag{78}$$

but, for two degenerate eigenvectors $\mathbf{s}_k{}^I$ and $\mathbf{s}_k{}^{II}$ the value of $\mathbf{s}_k{}^{IT} \cdot \mathbf{A} \cdot \mathbf{s}_k{}^{II}$ is not necessarily zero.   Therefore $\mathbf{A}$ is diagonalized by $\mathbf{S}$, except for submatrices of size $(n-1)$ by $(n-1)$.   We note, however, that $\mathbf{S}^T \cdot \mathbf{A} \cdot \mathbf{S}$ is symmetric, and therefore may be diagonalized by a unitary (in fact, orthogonal) transformation $\mathbf{U}$, where

$$\mathbf{U}^T = \mathbf{U}^{-1} \tag{79}$$

Because $\mathbf{S}^T \cdot \mathbf{A} \cdot \mathbf{S}$ is already diagonal except in the various $(n-1)$ subspaces, $\mathbf{U}$ acts only in these subspaces, and therefore commutes with Λ. Thus, we have

$$\mathbf{U}^{-1} \cdot \Lambda \cdot \mathbf{U} = \mathbf{U}^{-1} \cdot \mathbf{U} \cdot \Lambda = \Lambda \tag{80}$$

$$\mathbf{U}^{-1} \cdot \mathbf{S}^T \cdot \mathbf{A} \cdot \mathbf{S} \cdot \mathbf{U} = \mathbf{U}^T \cdot \mathbf{S}^T \cdot \mathbf{A} \cdot \mathbf{S} \cdot \mathbf{U} = \mathbf{M} \tag{81}$$

where Λ and $\mathbf{M}$ are diagonal.   Let

$$\mathbf{Q} = \mathbf{S} \cdot \mathbf{U} \tag{82}$$

then, eqs. (10) obtain as before.

APPENDIX B

Let us derive the mean square radius:

$$(N + 1)R^2 = \sum_{i=0}^{N} [(x_i - x_{cm})^2 + (y_i - y_{cm})^2 + (z_i - z_{cm})^2] \tag{83}$$

where x_{cm}, y_{cm}, and z_{cm} are the coordinates of the center of mass.   From our previous paper,[7]

$$x_{cm} = N^{1/2} \xi_0 = Q_{i0} \xi_0 \tag{84}$$

and, from eq. (11a) for x_i, we find

$$\sum_{i=0}^{N} (x_i - x_{cm})^2 = \sum_{k=0}^{N} (\sum_{k=1}^{N} Q_{ik}\xi_k)^2$$

$$= \sum_{i=0}^{N} \sum_{k=1}^{N} \sum_{l=1}^{N} Q_{ik}Q_{il}\xi_k\xi_l \tag{85}$$

$$= \sum_{k=1}^{N} \sum_{l=1}^{N} (\mathbf{Q}^T\mathbf{Q})_{kl}\xi_k\xi_l$$

For the free-draining case, however, $\mathbf{H} \equiv 1$, and so $\mathbf{Q}$ is a unitary matrix. Therefore $\mathbf{Q}^T\mathbf{Q}$ is the unit matrix, and we obtain:

$$\sum_{i=0}^{N} (x_i - x_{cm})^2 = \sum_{k=1}^{N} \xi_k{}^2 \tag{86a}$$

Similarly:

$$\sum_{i=0}^{N} (y_i - y_{cm})^2 = \sum_{k=1}^{N} \eta_k{}^2 \tag{86b}$$

$$\sum_{i=0}^{N} (z_i - z_{cm})^2 = \sum_{k=1}^{N} \zeta_k{}^2 \tag{86c}$$

Averaging these quantities with ψ and using eqs. (47), (48), and (49) from the previous paper,[7] one finds, for the case of zero shear rate,

$$(N + 1)R^2 = \frac{3D}{\sigma} \sum_{k=1}^{N} \frac{1}{\mu_k}$$

$$= \frac{3D}{\sigma} \sum_{k=1}^{N} \frac{1}{\lambda_k} \tag{87}$$

where λ_k has been substituted for μ_k, since the two are equal for the free-draining case.

References

1. J. R. Schaefgen and P. J. Flory, *J. Am. Chem. Soc.*, **70**, 2709 (1948).
2. C. D. Thurmond and B. H. Zimm, *J. Polymer Sci.*, **8**, 477 (1952).
3. A. Charlesby, *J. Polymer Sci.*, **17**, 379 (1955).
4. J. G. Kirkwood and J. Riseman, *J. Chem. Phys.*, **16**, 565 (1948).
5. P. J. Flory and T. G. Fox, Jr., *J. Polymer Sci.*, **5**, 745 (1950).
6. P. L. Auer and C. S. Gardner, *J. Chem. Phys.*, **23**, 1545 (1955).
7. B. H. Zimm, *J. Chem. Phys.*, **24**, 269 (1956).
8. B. H. Zimm and W. H. Stockmayer, *J. Chem. Phys.*, **17**, 1301 (1949).
9. W. H. Stockmayer and M. Fixman, *Ann. N. Y. Acad. Sci.*, **57**, 334 (1953).
10. P. E. Rouse, Jr., *J. Chem. Phys.*, **21**, 1272 (1953).
11. J. S. Ham, *J. Chem. Phys.*, **26**, 625 (1957).
12. B. H. Zimm, G. M. Roe, and L. F. Epstein, *J. Chem. Phys.*, **24**, 279 (1956).
13. J. D. Ferry, M. L. Williams, and D. M. Stern, *J. Phys. Chem.*, **58**, 987 (1954).
14. P. Debye, *J. Chem. Phys.*, **14**, 636 (1946).
15. E. Jahnke and F. Emde, *Tables of Functions*, Dover, New York, 1945, p. 269.
16. W. H. Stockmayer, *J. Chem. Phys.*, **11**, 45 (1943); **12**, 125 (1944).
17. P. J. Flory, *Principles of Polymer Chemistry*, Cornell Univ. Press, Ithaca, N. Y., 1953.
18. F. P. Price, S. G. Martin, and J. P. Bianchi, *J. Polymer Sci.*, **22**, 41 (1956).
19. M. Wales, P. A. Marshall, and S. G. Weissberg, *J. Polymer Sci.*, **10**, 229 (1953).
20. F. P. Price, J. H. Gibbs, and B. H. Zimm, *J. Phys. Chem.*, **62**, 972 (1958).

Synopsis

Theoretical formulas for the intrinsic viscosity and viscoelastic properties of some model branched molecules in dilute solution are calculated by means of the normal coordinate method of Rouse modified to include hydrodynamic interactions. The calculations are exact except for the usual approximation of the hydrodynamic interactions by the Kirkwood-Riseman formula. The ratio of the intrinsic viscosity of a branched molecule to that of a linear molecule of the same weight is found to vary almost as the square root of the ratio of the mean square radii, instead of as the latter ratio to three-halves power, as has been postulated before. It is proposed that this square root relation is applicable in general to branched molecules of all types. Several sets of experimental data in the literature are shown to agree well with this hypothesis.

Résumé

Les formules théoriques de viscosité intrinsèque et de viscoélasticité de certaines molécules ramifiées en solution diluée sont calculées au moyen de la méthode de coordonnées normales de Rouse modifiée afin d'y introduire les interactions hydrodynamiques. Les calculs sont exacts excepté pour les approximations usuelles d'interaction hydrodynamique par le formule de Kirkwood-Riseman. Le rapport de viscosité intrinseque d'un polymère ramifié à celui d'une molécule linéaire de même poids moléculaire varie proportionnellement à la racine carrée du rapport des carrés moyens de leurs rayons, et non selon la puissance trois-demi de ce rapport-ci comme il a été postulé. Cette relation de la racine carrée semble être applicable aux molécules ramifiées de différents types. Plusieurs résultats expérimentaux de la littérature semblent vérifier cette hypothese.

Zusammenfassung

Theoretische Formeln werden für die Viskositätszahl und die viskoelastischen Eigenschaften einiger Modelle für verzweigte Molekel in verdünnter Lösung auf Grund der Normalkoordinaten-Methode von Rouse berechnet, die zur Berücksichtigung hydro-

42   B. H. ZIMM AND R. W. KILB

dynamischer Wechselwirkungen modifiziert wird. Mit Ausnahme der üblichen Näherung für die hydrodynamische Wechselwirkung durch die Formel von Kirkwood und Riseman werden die Rechnungen streng durchgeführt. Es wird gefunden, dass das Verhältnis der Viskositätszahl einer verzweigten zu der einer linearen Molekel vom gleichen Gewicht sich nahezu mit der Wurzel aus dem Verhältnis der mittleren Radienquadrate, anstatt, wie früher angenommen worden war, mit der drei-halbten Potenz desselben ändert. Es wird die Vermutung ausgesprochen, dass die Quadratwurzel-Bezeichnung allgemein auf verzweigte Molekel aller Art angewendet werden kann. Mehrere Reihen von experimentellen Daten aus der Literatur stimmen, wie gezeigt wird, mit dieser Hypothese gut überein.

Received March 18, 1958

COMMENTARY

Reflections on "The Dynamic Birefringence of High Polymers," by Richard S. Stein, Shigeharu Onogi, and Daniel A. Keedy, *J. Polym. Sci.,* 57, 801 (1962)

RICHARD S. STEIN

University of Massachusetts, Amherst, MA 01003

The ideas leading to this paper arose during the period of the conception and birth of the *Journal of Polymer Science* and the "glory days" of Brooklyn Poly when I published a review paper on the theory of light scattering in its predecessor, the *Polymer Bulletin* (with Bruno Zimm and Paul Doty), which appeared for one year in 1945. Graduate studies followed with the late Professor Arthur Tobolsky at Princeton who made the challenging proposal, "Can't we find a way to paint a molecule in a solid polymer red so that we can observe what it is doing?" Much of my scientific life has been devoted to answering that question.

These studies in the late 1940s came soon after Werner Kuhn (with F. Grün) and L. R. G. Treloar published molecular theories of the birefringence of rubbers with the interesting prediction that the stress-optical coefficient (SOC), the ratio of birefringence to stress, should be a constant for a crosslinked rubber, independent of the number of crosslinks. Apparatus for such studies was not common at the time, so Arthur arranged for me to use an old photoelasticity apparatus in the basement of the Princeton Engineering Building. We verified these predictions using several rubbers being studied by the Princeton group and extended these to looking at the chemorheology of natural rubber and showed that as stress decayed due to the oxidative scission of chains, birefringence did likewise, so the SOC remained constant. At this time, Tobolsky (mostly with the late Rod Andrews) was conducting stress relaxation studies of the well-characterized polyisobutylene samples provided by the Esso (now Exxon) laboratory. We made the surprising observation that even if there were no crosslinks (a situation not covered by the theories), the SOC remained constant. In an effort to account for this, the concept of "physical crosslinks" (entanglements?) was proposed, and a paper on this was published with Ken Scott. The real understanding of this did not come until relatively recently with Masao Doi's extension of the Doi–Edwards and DeGennes reptation model for polymer dynamics.

An obvious extension was to look at crystalline polymers. P. H. Hermans had published his pioneering studies on cellulose and proposed the use of the "orientation function" to describe crystalline orientation. Polyethylene was a rapidly developing new polymer, having relatively simple molecular structure, so the extension of these concepts to polyethylene was attempted with a fellow graduate student, Sam Krimm, and led to a description of the total birefringence in terms of the contribution from the crystalline and amorphous phases. It was apparent that the mechanical response of crystalline polymers was more complex than that of amorphous ones in that the response mechanisms of the crystalline and amorphous regions would be different. Birefringence, along with the other rheo-optical methods subsequently developed, presented the opportunity to resolve these contributions.

Tobolsky's group was actively studying polymer stress relaxation, paralleling the viscoelasticity studies at Wisconsin by the group of John Ferry, both of whom were developing the early ideas of Herbert Leaderman for linear viscoelasticity theory. The application of birefringence to complement these studies was desirable. However, trying to measure stress and birefringence simultaneously before the days of lasers, solid state electronics, and computers to log data proved a challenge. I remember feeling like Charlie Chaplin in *Modern Times,* trying to measure stress with a mechanical balance and birefringence with a Babinet compensator as both were changing with time. There was not time to write down data, so numbers were shouted into a tape recorder to be sorted out later!

After coming to the University of Massachusetts in 1950, I began a program of optical studies of polymers in which the techniques of birefringence, x-ray diffraction, and solid state light scattering were combined to study the morphology and deformation of polymers. Shortly thereafter, Professor Hiromichi Kawai of Kyoto University, Japan, at the suggestion of Leaderman, arrived in Amherst as a visiting scientist and we began cooperative work on the use of static birefringence. Kawai was followed by Professor Shigaharu "Dennis" Onogi, also from Kyoto University. The advantages of steady-state vibrational (dynamic) measurements, pioneered by Ferry's group, became apparent, so efforts were made to extend these to birefringence measurements. Dennis's work, with the able help of Dan Keedy, led to the present paper. We needed a way to subject a sample to a vibrational strain, and being an advocate of first doing "quick and dirty" experiments, I purchased a hacksaw attachment to my home utility drill which did the job. We published some early experiments on polyethylene made using this crude apparatus, which encouraged us to design and have built the much more elegant version used in this paper. With Dennis's Japanese connections, we arranged for this to be built in Japan and to be shipped to Amherst. However, it did not arrive as scheduled. We traced the shipment as far as JFK Airport in New York, but somehow, it disappeared between there and Hartford. We never solved the mystery of where it went, but we think, being around Christmas time, some little boy may have received an unusual present!

This loss led to a few months delay, getting a second one built in Japan, which ultimately did arrive in Hartford, but wrapped in a kimono! Perhaps this accounts for the first one going astray. Its arrival led to more high quality results. Experiments were complemented with theoretical efforts to parallel linear viscoelasticity theory with its optical analog. Onogi was followed by a number of other Japanese physicists, including Sadao Hoshino, Kiichiro Sasaguri, and Ryo Yamada, who continued with many interesting experiments. Sometime later, Bryan Read, from the UK National Physical Laboratories, spent a year with us and extended studies to the glass transition region and initiating work later continued at the NPL.

T. Oda, a Kawai student, visited and conducted some exploratory experiments on time-dependent x-ray diffraction during the relaxation of polyethylene. This inspired a second visit by Kawai to carry out vibrational experiments on polyethylene using x-ray diffraction, again using a Japanese-built apparatus. Upon bringing the technique back to Kyoto, he had built a much better apparatus, leading me to send one of my graduate students, Bob Cembrola, to his laboratory to carry out part of his Ph.D. thesis work.

Later on, Takeji Hashimoto (now Professor at Kyoto University) joined us for Ph.D. studies and with graduate student Bob Prud'homme (now Professor at Laval University) and extended studies to vibrational light scattering. Then, Garth Wilkes (now Professor at VPI) made a heroic effort (too early for the technology available then) to do the same for infrared dichroism. More recently success was achieved here by Shunji Nomura at Kyoto Institute of Technology and Isao Noda and his group at Proctor and Gamble.

The development of many of these rheo-optical techniques occurred before modern computers, lasers, and area detectors were available and presented a great experimental challenge. It is good to see them progress using these newer "state-of-the-art" tools in the able hands of good scientists like Gerry Fuller and Julie Kornfield. Fortunately, the UMass tradition is continuing with Roger Porter having effectively used birefringence to characterize orientation in ultra-strong polymers, Shaw Ling Hsu's able use of infrared dichroism and Raman spectroscopy polarization, Jan von Egmond, Fuller's former student, having joined our Chemical Engineering Department, and Tom Russell, my former student, joining Polymer Science and Engineering. I look forward to having Natailia Pogodina, daughter of the Russian pioneer of birefringence, Professor V. Tsvetvov, join my group this summer.

PERSPECTIVE

Comments on "The Dynamic Birefringence of High Polymers," by Richard S. Stein, Shigeharu Onogi, and Daniel A. Keedy, *J. Polym. Sci.,* 57, 801 (1962)

GERALD G. FULLER

Department of Chemical Engineering, Stanford University, Stanford, CA 94305-5025

This contribution by Stein and coworkers[1] represents an important, early development in rheo-optics and the use of optical methods to provide microstructural information that can compliment mechanical rheometry. The objective of the work was to measure the strain-optical coefficient of low density polyethylene as a function of frequency and strain during oscillatory deformation of the sample. To accomplish this goal, however, required a significant advance in experimental development that allowed the simultaneous acquisition of mechanical and optical data. This seminal paper provided the foundation for subsequent rheo-optical studies and a recent resurgence of optical measurements applied to oscillatory flows.

The principal finding of this paper was that the stress-optic coefficient was a decreasing function of frequency, and this observation was compared with a phenomenological "spring-and-dashpot" model. Using this model as a guide, the authors argued that this dispersion in the stress-optic coefficient was most likely due to relaxation phenomena associated with the separate dynamics of the amorphous and crystalline regions of the polymer. In this respect, this paper is an excellent display of the use of rheo-optics to elucidate microstructural relaxation mechanisms.

This paper has served as a benchmark for a rich history of applications of dynamic rheo-optic measurements that have ensued since its appearance more than 30 years ago. Certainly much of the finest examples of these applications have continued to emanate from the laboratory of Professor Stein himself, but many others have adopted this methodology and have extended it to include other optical interactions as well as to use it to study orientation dynamics in all classes of polymer liquids.

The instrument consisted of a crossed polarizer design with the polarizers oriented at $\pm 45°$ relative to the stretching direction. To optimize the signal in the presence of a large offset birefringence from the sample at rest, Stein and coworkers devised an ingenious procedure involving the insertion of a second, identical static sample at orthogonal orientation combined with a quarter waveplate. However, the basic working equation to extract the dynamic birefringence, eq. (50) of ref. 1, is essentially the same used in more recent work, such as that due to Schrag and coworkers[2] who have used oscillatory flow birefringence to study the orientational dynamics of dilute solutions of polystyrene. The apparatus developed by these later researchers used a sophisticated signal averaging scheme to greatly enhance the signal-to-noise ratio. In that work both linear chains and star chain polymers were investigated and the results were compared against calculations using the Zimm bead-and-spring model.

These studies also revealed a frequency dependence to the stress-optical coefficient. The interpretation offered for this effect suggested that a coupling exists between the polymer and the solvent molecules that influences the local polarizability of the chains. These results and those of the original work of Stein et al. demonstrate that when the stress-optical coefficient becomes frequency-dependent, the dynamic birefringence measurement is of most value in elucidating microstructural relaxation processes. Other recent applications of dynamic birefringence are its use by Osaki and coworkers[3] to study the orientational dynamics of polymer chains below the glass transition temperature and the study of compatible blends by Zawada et al.[4]

Dynamic birefringence measurements have been extended to include other optical interactions such as infrared dichroism,[5] Raman scattering,[6] and scattering dichroism.[7] The former two phenomena are vibrational spectroscopies that enable the extraction of the orientation of specific chemical moieties. Scattering dichroism, on the other hand, provides information on the anisotropy of much larger length scale structures, such as concentration fluctuations.

REFERENCES AND NOTES

1. R. S. Stein, S. Onogi, and D. A. Keedy, *J. Polym. Sci.,* **57,** 801 (1962).
2. J. Miller and J. Schrag, *Macromolecules,* **8,** 361 (1975); T. P. Lodge, J. Miller, and J. Schrag, *J. Polym. Sci., Polym. Phys. Ed.,* **20,** 1409 (1982).
3. T. Inoue, H. Hayashihara, H. Okamoto, and K. Osaki, *Macromolecules,* **24,** 5670 (1991); T. Inoue, H. Okamoto, and K. Osaki, *J. Polym. Sci., Polym. Phys. Ed.,* **30,** 409 (1992).
4. J. A. Zawada, G. G. Fuller, R. H. Colby, L. J. Fetters, and J. Roovers, *Macromolecules,* **27,** 6851 (1994); J. A. Zawada, G. G. Fuller, R. H. Colby, L. J. Fetters, and J. Roovers, *Macromolecules,* **27,** 6861 (1994).
5. I. Noda, *J. Am. Chem. Soc.,* **111,** 8116 (1989); I. Noda, *Applied Spectroscopy,* **44,** 550 (1990).
6. K. Huang, L. A. Archer, and G. G. Fuller, *Macromolecules,* **29,** 966 (1996).
7. J. Lai and G. G. Fuller, *J. Rheol.,* **40,** 153 (1996).

JOURNAL OF POLYMER SCIENCE   VOL. 57, PAGES 801–821 (1962)

The Dynamic Birefringence of High Polymers

RICHARD S. STEIN, SHIGEHARU ONOGI,* and DANIEL A. KEEDY, *Department of Chemistry, University of Massachusetts, Amherst, Massachusetts*

INTRODUCTION

The stress and strain-optical coefficients have been used as a means for studying elastic deformation mechanisms of polymers in equilibrium and during relaxation experiments.[1,2] Measurements in a dynamic experiment have the capability of yielding more information, since these make accessible the contributions of molecular mechanisms having relaxation times too short to contribute to static experiments.

A preliminary apparatus for the determination of the dynamic birefringence of polymer films has been described, and preliminary results for polyethylene have been published.[3] These data indicate that the strain-optical coefficient is dependent upon frequency. In this paper, an extension of the conventional linear theory of distribution of relaxation times[4] to the description of the strain-optical effect is made.

PART I. THEORY

The Birefringence in Static Relaxation

The Maxwell element consisting of a spring of modulus E and dashpot of tensile viscosity η connected in series (Fig. 1) is the basis for this calculation. The viscosity parameter is replaced by the relaxation time $\tau = \eta/E$ and E is expressed as a function of τ, $E(\tau)$. The differential equation of the Maxwell element is[5]

$$\frac{dS(\tau)}{dt} = \frac{1}{E(\tau)} \frac{df(\tau)}{dt} + \frac{1}{\tau E(\tau)} f(\tau) \tag{1}$$

where $S(\tau)$ is the total displacement of the element and $f(\tau)$ is the force upon it. In a relaxation experiment where S is instantaneously increased to a value S_0 and then held constant, the solution is

$$f(\tau) = S_0 E(\tau) e^{-t/\tau} \tag{2}$$

While the total displacement remains constant during the relaxation ex-

* On leave from the Department of Polymer Chemistry, Kyoto University, Kyoto, Japan.

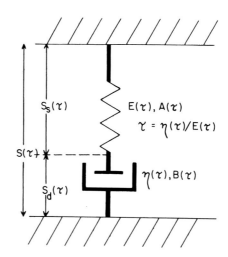

Fig. 1. The Maxwell element with modulus $E(\tau)$, viscosity $\eta(\tau)$, relaxation time τ, and strain-optical coefficients, $A(\tau)$ and $B(\tau)$.

periment, the displacements of the spring, $S_s(\tau)$, and the dashpot $S_d(\tau)$ individually change subject to the restriction

$$S(\tau) = S_s(\tau) + S_d(\tau) \tag{3}$$

The displacement of the dashpot may be obtained by integrating the equation for its motion

$$f(\tau) = \eta dS_d(\tau)/dt \tag{4}$$

Use of eq. (2) for $f(\tau)$ gives

$$dS_d(\tau)/dt = [S_0E(\tau)/\eta]\, e^{-t/\tau} \tag{5}$$
$$= (S_0/\tau)\, e^{-t/\tau}$$

and

$$S_d(\tau) = (S_0/\tau) \int_0^t e^{-t/\tau}\, dt \tag{6}$$
$$= S_0(1 - e^{-t/\tau})$$

$S_s(\tau)$ may then be obtained from eq. (3).

$$S_s(\tau) = S_0 - S_d(\tau) \tag{7}$$
$$= S_0 - S_0(1 - e^{-t/\tau})$$
$$= S_0\, e^{-t/\tau}$$

Two strain-optical coefficients are introduced. One of these, $A(\tau)$, characterizes the spring,

$$\Delta_s(\tau) = A(\tau)S_s(\tau) \tag{8}$$

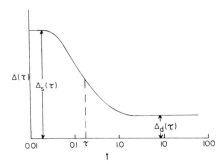

Fig. 2. The variation of $\Delta(\tau)$ with ln τ for a single Maxwell element for which $A(\tau) > B(\tau)$.

while the other, $B(\tau)$, characterizes the dashpot,

$$\Delta_d(\tau) = B(\tau)S_d(\tau) \tag{9}$$

$\Delta_s(\tau)$ and $\Delta_d(\tau)$ are the contributions to the birefringence of this Maxwell element by the spring and dashpot, respectively. The total birefringence contributed by this element is assumed to be the sum of these contributions, giving

$$\Delta(\tau) = \Delta_s(\tau) + \Delta_d(\tau)$$
$$= S_0[A(\tau)e^{-t/\tau} + B(\tau)(1 - e^{-t/\tau})]$$
$$= S_0\{B(\tau) + [A(\tau) - B(\tau)]\, e^{-t/\tau}\} \tag{10}$$

For a polymer consisting of a distribution of such elements, the total birefringence is assumed to be the sum of that of all of the elements. Replacing the summation by integration, one obtains

$$\Delta = \int_{\tau=0}^{\infty} \Delta'(\tau)d\tau$$
$$= S_0\{\int_0^{\infty} B'(\tau)d\tau + \int_0^{\infty} [A'(\tau) - B(\tau)]\, e^{-t/\tau}\, d\tau\} \tag{11}$$

Thus, the birefringence relaxation of such a polymer may be interpreted in terms of two distribution functions, $A'(\tau)$ and $B'(\tau)$.

The variation of $\Delta(\tau)$ with ln t for a single Maxwell element is shown in Figure 2 for the case of $A(\tau) > B(\tau)$. $\Delta(\tau)$ falls off from $A(\tau)$ at short times to $B(\tau)$ at long times. The inflection occurs at $t = \tau$, and changes in τ result in curves which shift horizontally as with stress relaxation curves. If $B(\tau) > A(\tau)$, the birefringence increases rather than decreases with time.

With a distribution of Maxwell elements, the time range during which the birefringence changes may be increased as in Figure 3, where $A(\tau) > B(\tau)$ for all of the elements. If the distribution contains some elements for which $A(\tau) > B(\tau)$ and others for which $B(\tau) > A(\tau)$, the birefringence may first decrease and then increase with time (or vice versa) (Fig. 4). Changes of this type have been observed with poly(ethyl acrylate)[6] and polystyrene[7] near their transition points.

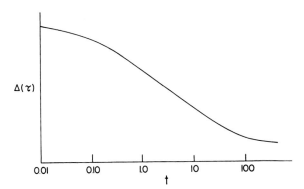

Fig. 3. The variation of Δ with $\ln \tau$ for a distribution of Maxwell elements, all of which have $A(\tau) > B(\tau)$.

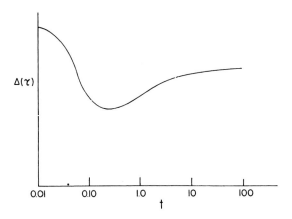

Fig. 4. The variation of Δ with $\ln \tau$ for a distribution of Maxwell elements, where $A(\tau) > B(\tau)$ for small values of τ and $B(\tau) > A(\tau)$ for large values of τ.

By using methods similar to those used to obtain relaxation spectra from stress relaxation data,[8] eq. (11) may be inverted.   If eq. (11) is expressed in terms of $\ln \tau$ rather than τ, where

$$A'_0(\ln \tau) \, d \ln \tau = A'(\tau) d\tau$$

$$B'_0(\ln \tau) \, d \ln \tau = B'(\tau) d\tau \qquad (12)$$

and is differentiated with respect to $\ln t$, one obtains

$$-\partial(\Delta/S_0)/\partial \ln t = \int_{-\infty}^{\infty} [A'_0(\ln \tau) - B'_0(\ln \tau)] \, (t/\tau) \, e^{-t/\tau} \, d \ln \tau \quad (13)$$

$(t/\tau)e^{-t/\tau}$ is a quantity which is zero at both $t/\tau = 0$ and $t/\tau = \infty$ and goes through a maximum at $t/\tau = 1$.   If $[A'_0 - B'_0]$ is not rapidly varying in this region where $(t/\tau)e^{-t/\tau}$ is large, then as an approximation, it may be taken outside of the integral sign, giving

$$-\partial(\Delta/S_0)/\partial \ln t \cong [A'_0(\ln \tau) - B'_0(\ln \tau)] \int_{-\infty}^{\infty} (t/\tau) e^{-t/\tau} d \ln \tau$$

$$= [A'_0(\ln \tau) - B'_0(\ln \tau)] \tag{14}$$

Thus, as a first approximation

$$A'_0(\ln \tau) - B'_0(\ln \tau) \cong -\left(\frac{\partial \Delta/S_0}{\partial \ln t}\right)_{t=\tau} \tag{15}$$

Higher order approximations involving higher derivatives of (Δ/S_0) may be obtained as with stress-relaxation inversion.[7]

Birefringence in Dynamic Relaxation

With a Maxwell element subjected to a forced vibration of angular frequency ω and amplitude S_0, one may substitute in eq. (1)

$$S(\tau) = S_0 e^{-i\omega t} \tag{16}$$

and

$$f(\tau) = A e^{-i\omega t} \tag{17}$$

where A is a complex amplitude of the strain. On substituting and expressing the real part of $f(\tau)$ in trigonometric form, this gives

$$f(\tau) = S_0 E(\tau) \left[\frac{\omega^2 \tau^2}{1 + \omega^2 \tau^2} \cos \omega t - \frac{\omega \tau}{1 + \omega^2 \tau^2} \sin \omega t\right] \tag{18}$$

On substituting into eq. (4) one obtains

$$\frac{dS_d(\tau)}{dt} = \frac{S_0}{\tau} \left[\frac{\omega^2 \tau^2}{1 + \omega^2 \tau^2} \cos \omega t - \frac{\omega \tau}{1 + \omega^2 \tau^2} \sin \omega t\right] \tag{19}$$

On integrating from an initial time such that $S_d(\tau) = 0$ at $t = 0$, one obtains

$$S_d(\tau) = S_0 \left[\frac{\omega \tau}{1 + \omega^2 \tau^2} \sin \omega t + \frac{1}{1 + \omega^2 \tau^2} \cos \omega t\right] \tag{20}$$

On substituting eqs. (16) and (20) in eq. (3), eq. (21) is obtained:

$$S_s(\tau) = S_0 \left[\frac{\omega^2 \tau^2}{1 + \omega^2 \tau^2} \cos \omega t - \frac{\omega \tau}{1 + \omega^2 \tau^2} \sin \omega t\right] \tag{21}$$

Then, using eqs. (8) and (9), we have

$$\Delta(\tau) = \Delta_s(\tau) + \Delta_d(\tau)$$

$$= S_0 \left(\left[\frac{A(\tau)\omega^2 \tau^2 + B(\tau)}{1 + \omega^2 \tau^2}\right] \cos \omega t\right.$$

$$\left. + \left\{\frac{[B(\tau) - A(\tau)]\omega \tau}{1 + \omega^2 \tau^2}\right\} \sin \omega t\right) \tag{22}$$

At $\omega \tau = 0$, this gives

$$\Delta(\tau) = S_0 B(\tau) \cos \omega t \tag{23}$$

while as $\omega\tau$ approaches infinity

$$\Delta(\tau) = S_0 A(\tau) \cos \omega t \qquad (24)$$

Thus at very low frequencies only the strain-optical coefficient of the dash-pot contributes, while at high frequencies only that of the spring does. At both very high and very low frequencies, the birefringence is in phase with the strain. At intermediate frequencies, both strain-optical coefficients contribute and the birefringence differs in phase from the strain. The phase shift is more apparent if the equation is written in the form

$$\Delta(\tau) = \Delta_0(\tau) \cos (\omega t - \alpha) \qquad (25)$$

where $\Delta_0(\tau)$ is the amplitude of the birefringence and α is the phase difference. On expanding eq. (25)

$$\Delta(\tau) = [\Delta_0(\tau) \cos \alpha] \cos \omega t + [\Delta_0(\tau) \sin \alpha] \sin \omega t \qquad (26)$$

It follows from comparison with eq. (22) that

$$\Delta_0(\tau) \cos \alpha = \frac{A(\tau)\omega^2\tau^2 + B(\tau)}{1 + \omega^2\tau^2} \qquad (27)$$

$$\Delta_0(\tau) \sin \alpha = \frac{[B(\tau) - A(\tau)] \omega\tau}{1 + \omega^2\tau^2} \qquad (28)$$

Therefore

$$\Delta_0(\tau) = [\Delta_0(\tau)^2 \cos^2\alpha + D_0(\tau)^2 \sin^2\alpha]^{1/2}$$
$$= S_0 \left(\left[\frac{A(\tau)\omega^2\tau^2 + B(\tau)}{1 + \omega^2\tau^2} \right]^2 + \left\{ \frac{[B(\tau) - A(\tau)]\omega\tau}{1 + \omega^2\tau^2} \right\}^2 \right)^{1/2} \qquad (29)$$

and

$$\tan \alpha = \frac{\Delta_0(\tau) \sin \alpha}{\Delta_0(\tau) \cos \alpha}$$
$$= \frac{[B(\tau) - A(\tau)] \omega\tau}{A(\tau)\omega^2\tau^2 + B(\tau)} \qquad (30)$$

On rearranging eq. (29), one obtains for the dynamic strain-optical coefficient, $K(\omega)$

$$K(\omega, \tau) = \Delta_0(\tau)/S_0$$
$$= \left\{ [A(\tau)^2\omega^2\tau^2 + B(\tau)^2]/(1 + \omega^2\tau^2) \right\}^{1/2} \qquad (31)$$

A plot of $K(\omega, \tau)$ against $\ln \omega$ is given in Figure 5 for the case where $B(\tau) > A(\tau)$. A corresponding plot of α against $\ln \omega$ is given in Figure 6. α goes through a maximum at a frequency given by $\omega\tau = [B(\tau)/A(\tau)]^{1/2}$ which is in a region of rapidly changing $K(\omega, \tau)$.

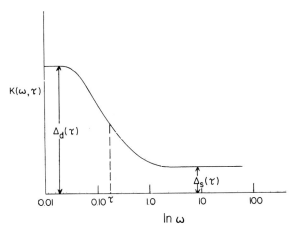

Fig. 5. The variation of $K(\omega,\tau)$ with $\ln \omega$ for a Maxwell element subjected to periodic strain for the case where $B(\tau) > A(\tau)$.

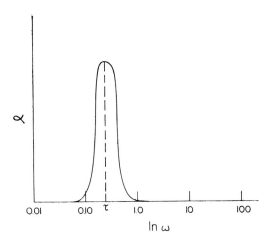

Fig. 6. The variation of the phase angle α with $\ln \omega$ for a Maxwell element subjected to periodic strain.

For the case of a distribution of Maxwell elements, eq. (22) becomes

$$\Delta = \int_0^\infty \Delta'(\tau)\, d\tau$$

$$= S_0 \left(\int_0^\infty \left[\frac{A'(\tau)\omega^2\tau^2 + B'(\tau)}{1 + \omega^2\tau^2} \right] d\tau \cos \omega t \right.$$

$$\left. + \int_0^\infty \left\{ \frac{[B'(\tau) - A'(\tau)]\,\omega\tau}{1 + \omega^2\tau^2} \right\} d\tau \sin \omega t \right) \qquad (22a)$$

By a sequence of steps like those of eqs. (25) through (29), one gets

$$K(\omega) = \left(\left[\int_0^\infty \frac{A'(\tau)\omega^2\tau^2 + B'(\tau)}{1 + \omega^2\tau^2} \, d\tau \right]^2 \right.$$
$$\left. + \left\{ \int_0^\infty \frac{[B'(\tau) - A'(\tau)]\,\omega\tau}{1 + \omega^2\tau^2} \, d\tau \right\}^2 \right)^{1/2} \quad (32)$$

and

$$\tan \alpha = \left\{ \int_0^\infty \frac{[B'(\tau) - A'(\tau)]\,\omega\tau}{1 + \omega^2\tau^2} \, d\tau \right\} \Big/ \left[\int_0^\infty \frac{A'(\tau)\omega^2\tau^2 + B'(\tau)}{1 + \omega^2\tau^2} \, d\tau \right] \quad (33)$$

The effect of a distribution of strain-optical coefficients is usually to broaden the frequency region over which $K(\omega)$ changes or to produce a number of regions of variation corresponding to a number of different relaxation mechanisms.   As with stress relaxation, it is possible for $K(\omega)$ to increase or decrease with frequency in a particular frequency range depending upon whether $B(\tau)$ or $A(\tau)$ is greater for the relaxation mechanisms prevailing in that region.

These equations for dynamic strain-optical coefficient may be inverted to obtain the strain-optical coefficient spectra following the method used for dynamic modulus.[8,9]   Equations (32) and (33) may be combined to give

$$C(\omega) = \frac{K(\omega)}{[1 + \tan^2\alpha(\omega)]^{1/2}}$$
$$= \int_{-\infty}^\infty \frac{A'_0(\ln \tau)\omega^2\tau^2 + B'_0(\ln \tau)}{1 + \omega^2\tau^2} \, d\ln \tau \quad (34)$$

where again, A'_0 and B'_0 are the strain-optical distribution functions expressed in terms of $\ln \tau$.   Then

$$\frac{\partial C(\omega)}{\partial \ln(1/\omega)} = -\int_{-\infty}^\infty x \frac{\partial}{\partial x} \left[\frac{A'_0 x^2 + B'_0}{1 + x^2} \right] d\ln \tau \quad (35)$$

where $x = \omega\tau$.   This gives, upon differentiating,

$$-\frac{\partial C(\omega)}{\partial \ln(1/\omega)} = 2\int_{-\infty}^\infty (A'_0 - B'_0) \frac{x^2}{(1 + x^2)^2} \, d\ln \tau \quad (36)$$

The function $x^2/(1 + x^2)^2$ has a maximum at $x = 1$ and is zero for large and small x. If $(A_0 - B_0)$ does not vary too rapidly in the region where $x^2/(1 + x^2)^2$ is large, it may be taken outside the integral to give

$$-\frac{\partial C(\omega)}{\partial \ln(1/\omega)} \cong 2(A'_0 - B'_0) \int_0^\infty \frac{x}{(1 + x^2)^2} \, dx \quad (37)$$
$$= (A'_0 - B'_0)$$

and

$$[A'_0(\ln \tau) - B'_0(\ln \tau)]_{\tau = 1/\omega} \cong \frac{\partial\{K(\omega)/[1 + \tan^2\alpha(\omega)]^{1/2}\}}{\partial[\ln \omega]} \quad (38)$$

as a first approximation. Better approximations involving additional terms with higher derivatives may be obtained. (The second approximation involving the second derivative does not contribute because of the symmetrical form of the function $x^2/(1 + x^2)^2$. The first additional term is the one involving the third derivative.)

As with the stress-relaxation inversion, only the combination of functions $[A'_0(\ln \tau) - B'_0 (\ln \tau)]$ is obtained. A method for separating this into the individual functions has not been developed.

Molecular Interpretations

The spring and dashpot of the Maxwell element are analogs of certain molecular processes which may be described by viscoelastic equations that are equivalent to these describing the Maxwell elements. The springs may represent any molecular process exhibiting instantaneous linear elastic response. This might include the instantaneous orientation of amorphous chains, the deformation of bonds in amorphous regions, and the deformation and orientation of crystals. Each of these processes would have a characteristic restoring force and strain-optical coefficient. In general, there is no relationship between the restoring force (spring constant) and strain-optical coefficient. The slipping of the dashpot might correspond to delayed amorphous region or crystallite orientation and viscous deformation of crystals.

In some simple cases, only a single element is necessary. For example, the birefringence relaxation accompanying the stress relaxation of cross-linked natural rubber at constant length has been described by an exponential decay toward zero stress and birefringence.[10] In this case, the spring represents the entropy elasticity of the rubber chains, while the dashpot represents the oxidative scission of those chains. The stress-optical coefficient, $B(\tau)$, of this dashpot is zero, while for the spring, $A(\tau)$ is proportional to the stress according to the Kuhn-Treloar theory of rubber birefringence[11,12] and has only one value for the single relaxation time involved.

Thus, from eq. (10)

$$\Delta = AS_0 e^{-t/\tau} \tag{39}$$

where

$$AS_0 = [(\bar{n}^2 + 2)^2/\bar{n}] (2\pi/45) N_c(b_1 - b_2) [\alpha^2 - (1/\alpha)] \tag{40}$$

$\bar{n}$ is the average refractive index, N_c the number of active network chains per cubic centimeter, b_1 and b_2 the principal polarizabilities of the statistical segment, and α the degree of elongation $[S_0/S(\text{unstretched})]$.

In the case of a linear rubber above its transition point, the stress and birefringence are found to decay toward zero over many cycles of log time. The ratio of stress to birefringence remains constant during the relaxation.[10] The ratio appears to be the same as for the same rubber in the uncrosslinked state.

The stress σ of such a rubber may be expressed in the form[13]

$$\sigma = S_0 \int_{-\infty}^{\infty} E_0 (\ln \tau) \, e^{-t/\tau} \, d \ln \tau \qquad (41)$$

where $E_0(\ln \tau)$ is the stress relaxation spectrum function, which has been determined for many polymers. These polymers relax by a viscous flow process toward a nonbirefringent state. Thus, $B(\tau)$ must be zero for all of the elements. There will be a distribution of $A(\tau)$ corresponding to the distribution of normal modes of chain motion.[14,15] Therefore

$$\Delta = \int_{-\infty}^{\infty} A'_0(\ln \tau) \, e^{-t/\tau} \, d \ln \tau \qquad (42)$$

The result that Δ/σ is independent of time indicates that

$$A_0(\ln \tau) = KE_0(\ln \tau) \qquad (43)$$

where K, the stress-optical coefficient, is the same for all τ. Substituting in eq. (42) gives

$$\Delta = K \int_{-\infty}^{\infty} E_0(\ln \tau) \, e^{-t/\tau} \, d \ln \tau = K\sigma \qquad (44)$$

This indicates that the stress-optical coefficient given by the Kuhn[10]-Treloar[11] theory

$$K = (2\pi/45kT) \, [(\bar{n}^2 + 2)^2/\bar{n}] \, (b_1 - b_2) \qquad (45)$$

is the same for all of the observed normal modes. This result is reasonable since the chain length does not appear in the Kuhn-Treloar expression as long as the length is great enough so that the chain obeys Gaussian statistics. It is probable that for short τ or high frequency where the length of moving chain unit becomes comparable with the statistical segment length, K will deviate from this constant value. It is also likely that deviations will occur when the frequency excites resonance vibrations of side groups.

When a polymer is cooled to a temperature close to T_g, new relaxation mechanisms become apparent. At low temperature, the chains are not able to orient, and a high modulus results from rotation about primary bonds, bond bending, and stretching. The birefringence resulting from this may be different in magnitude and sign from that arising from orientation.[6,7] At temperatures close to T_g, this distortional contribution relaxes out as chain orientation increases and contributes its characteristic birefringence. It is apparent that these differing molecular mechanisms would be represented by distributions of elements with varying strain-optical coefficients.

With crystalline polymers, differing relaxation mechanisms have differing strain-optical coefficients. For example, the viscous flow of chains in disordered regions contributes differently to birefringence than does relaxation accompanying plastic flow of crystalline regions.

PART II. EXPERIMENTAL

Apparatus

The apparatus is designed to measure the dynamic birefringence and strain of a sample subjected to forced tensile vibration. The sample is vibrated mechanically at low frequencies (5.4×10^{-3}–10 cycles/sec.) by a motor-transmission-cam assembly and at high frequencies (10–10^4 cycles/sec.) by an electromagnetic transducer and electronic oscillator. Birefringence is simultaneously measured by means of determining the transmission of light by the sample when placed between crossed polaroids.

The mechanical system for vibrating the sample at low frequencies is shown in Figures 7 and 8. A 1-h.p. motor (A) drives a variable speed hydraulic transmission (B) (Groban Supply No. X115A). The power is then transmitted to the mechanical vibrator (Iwamoto Mfg. Co., Kyoto, Japan), through a drive shaft (C), a system of step pulleys, and a worm gear reducer (E). The frequency control is achieved by changing the pulley belts to different ratios and adjusting the speed control of the hydraulic transmission. One can either power the mechanical vibrator directly from pulley (1) to (3) or by going through the worm gear reducer and having pulley (2) drive (3), depending upon the desired frequency.

The mechanical vibrator is shown in Figure 8, and consists of two arms (A) which are mounted on cam-driven slides (C). The cams (B) are rotated by means of a gear system contained within the vibrator. The two

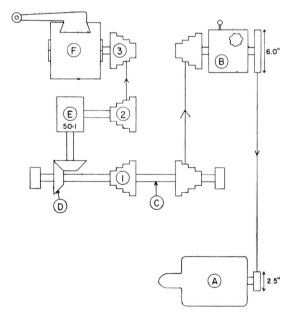

Fig. 7. The mechanical vibrator transmission: (A) 1 h.p. motor; (B) variable speed hydraulic transmission; (C) drive shaft; (D) helical bevel gears; (E) worm gear reducer; (F) mechanical vibrator; (1), (2), and (3) step pulleys.

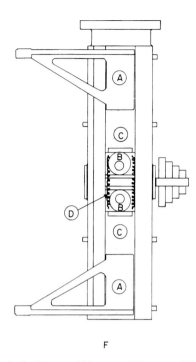

F

Fig. 8. The mechanical vibrator: (A) arms; (B) cams; (C) slides; (D) springs.

slides are connected together by springs (D) which hold the slides tightly against the cams, producing uniform sinusoidal vibration of the arms. The amplitude of vibration is controlled by the eccentricity of the cams. There are five interchangeable pairs of cams for producing vibrations of amplitude of 0.25, 0.5, 1.0, 2.0, and 4.0 mm. The sample is mounted between the two arms of the vibrator with clamps. Vibration at high frequency is electromagnetically developed by using a vibration generator (Goodman Model 390A) driven by a 120-w. power oscillator (Goodman Type D120). The apparatus is converted to the high frequency range by removing the lower arm of the mechanical vibrator and sliding the vibration generator into place. The sample is then mounted between the upper arm of the mechanical vibrator and the vibration generator.

The optics and measuring systems are given in Figure 9. An A100H4 mercury vapor lamp serves as the light source. This lamp, operated on d.c. power from a motor-generator at 110 v. which is filtered using a condenser choke system (60 mfd) (General Electric Co. Ltd. of England, Choke No. Z1878), is started by applying 160 v. and using a Teslar coil to strike the arc. A 200-ohm variable resistance is placed in series with the lamp.

The light beam is rendered parallel by condensing lens (B), monochromatic by a 5460 A. interference filter (C). The beam is polarized at 45° to stretching direction by a polaroid (J filter) (D), passes through the sample (G) and analyzer (polaroid) (I) oriented perpendicular to the polarizer.

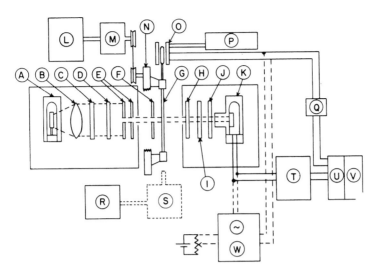

Fig. 9. The optical system: (A) mercury vapor lamp GE AH-4; (B) condensing lens; (C) interference filter, 5460λ; (D) polaroid (polarizer); (E) slits; (F) slit; (G) sample; (H) ¹/₄ wave plate; (I) polaroid (analyzer); (J) neutral density filter (O.D. 2.0); (K) photomultiplier tube (1P21); (L) transmission system (Fig. 7); (M) worm gear reducer; (N) arm of mechanical vibrator (Fig. 8); (O) linear variable differential transformer; (P) oscillator (10,000 cycles/sec.); (Q) rectifier; (R) Goodman power oscillator D120; (S) Goodman vibration generator 390A; (T) D-C amplifier (General Radio); (U) Sanborn preamplifier; (V) Sanborn two-channel recorder; (W) Tektronik dual-beam oscilloscope.

A quarterwave plate (H) is mounted between the sample and the analyzer to reduce the initial retardation of the sample. The beam then passes through a Wratten (Kodak) neutral density filter (J) of optical density 2.0 and into a battery-powered 1P21 photomultiplier tube (K).

For the low frequency range measurement the core of a Schaevitz linear variable differential transformer (O) is fastened to the upper arm of the mechanical vibrator. The LVDT is powered by a General Radio oscillator type No. 1210–C (P) at 10,000 cycles/sec. The LVDT's output is rectified (Q) and the signal fed into an amplifier (U), Sanborn Model 15–1500 low level preamplifier. This signal then is recorded on one channel of a Sanborn Model No. 152–100B two-channel recorder (V). The LVDT serves to monitor the strain of the sample. The output of the photomultiplier (K) is amplified with a General Radio d.c. amplifier (T) Type No. 1230-A. The signal then enters the Sanborn system, where the initial static signal is suppressed to zero, amplified, and recorded on the other channel of the Sanborn recorder. The recorder reading is a function of the birefringence and is calibrated against it using a Babinet compensator.

In the case of the high frequency range, the core of the LVDT (O) is fastened directly to the moving rod of the vibration generator, and its output signal is measured directly on a Tektronix Dual-Beam oscilloscope (W) type 502. The signal from the 1P21 is also traced and measured on the other beam of this oscilloscope. The initial static signal is suppressed to zero by using the difference amplifier stage of the oscilloscope and an opposing voltage developed from a potentiometer circuit.

In each case, the phase of the strain signals can be compared over the entire frequency range to that of the light intensity signal.

The upper frequency of operation is limited by the ability to detect the birefringence at the small amplitudes of vibration that may be excited. This is determined by the strain optical coefficient and thickness of the sample and appears to be about 400 cycles/sec. for 0.1 mm.-thick polyethylene where the maximum amplitude of vibration is 0.25 mm.

Calibration

The LVDT is calibrated in terms of strain by measuring its voltage output as a function of the position of the core, using a dial gage. The transformer is used in the linear region of its calibration. At the 10,000 cycles/sec., the calibration constant was 0.760 v./mm. For a typical sample length of 80 mm., strains down to 0.325% may be measured at high frequency. At maximum power of the Goodman vibrator, this limiting strain is reached at a frequency of 300 cycles/sec.

The calibration of the birefringence apparatus is somewhat more complex. A birefringent polymer sample, when placed between crossed polaroids, will absorb and reflect light as well as transmit it. For perfect polarizers, transmission will occur as a result of (a) the presence of unoriented but locally birefringent crystals or spherulites, (b) birefringence of the sample resulting from initial orientation, internal strains and static applied strain, and (c) birefringence arising from the applied dynamic strain. The intensity of the transmitted light is given by

$$I_t = \frac{1}{2} I_0 \kappa \{ [1 - \cos (2\pi d/\lambda) \, \Delta] + T \} \tag{46}$$

I_0 is the intensity of incident light, κ is the transmittance given by

$$\kappa = (1 - a) (1 - r) e^{-\tau d} \tag{47}$$

where a is the true absorbance, r the reflectance of the sample, and τ is the turbidity of the sample for scattering. d is the thickness of the sample and λ the wavelength of light in vacuum (expressed in the same units as d). Δ is the birefringence defined by

$$\Delta = n_{||} - n_\perp \tag{48}$$

$n_{||}$ and $n_\perp$ are the refractive indices parallel to and perpendicular to the direction of vibration, respectively. T is the transmission of the sample resulting from unoriented inclusions having local birefringence.[16,17]

The birefringence Δ consists of three components

$$\Delta = \Delta_0 + \Delta_i + \Delta_d \tag{49}$$

Δ_0 is the birefringence resulting from initial orientation and strain, Δ_i is the birefringence of any optical elements (such as quarterwave plates) inserted in the optical train, and Δ_d is the dynamic birefringence resulting from the applied dynamic strain.

If eq. (49) is substituted into eq. (46) and the sum of cosines is expanded, one obtains

$$I_t = \frac{1}{2} I_0 \kappa \left\{ 1 - \cos\left[\frac{2\pi d}{\lambda}(\Delta_0 + \Delta_i)\right]\cos\left[\frac{2\pi d}{\lambda}\Delta_d\right] \right.$$
$$\left. + \sin\left[\frac{2\pi d}{\lambda}(\Delta_0 + \Delta_i)\right]\sin\left[\frac{2\pi d}{\lambda}\Delta_d\right] + T \right\} \quad (50)$$

If this equation is differentiated with respect to the applied dynamic strain, S_d, and κ, d, Δ_0, and T are assumed independent of S_d, one obtains

$$\frac{\partial I_t}{\partial S_d} = \frac{1}{2} I_0 \kappa \left\{ \sin\left[\frac{2\pi d}{\lambda}(\Delta_0 + \Delta_t + \Delta_d)\right]\frac{2\pi d}{\lambda}\right\}\frac{\partial \Delta_d}{\partial S_d} \quad (51)$$

It is seen that the proportionality factor relating change in birefringence to change in transmitted intensity is not constant but depends upon the total birefringence of the system. To obtain maximum sensitivity, it is desirable to make this factor as large as possible. This occurs when

$$(d/\lambda)(\Delta_0 + \Delta_t + \Delta_d) = \frac{1}{4}, \frac{3}{4}, \frac{5}{4}, \text{ etc.} \quad (52)$$

This may be accomplished by adjusting the birefringence of the inserted retarding element, Δ_t. A procedure for accomplishing this is to insert a second sample identical with the first but rotated through 90° about its normal. This compensates for Δ_0. Then, by inserting a quarterwave plate into the beam, $(d/\lambda)\Delta_t = \frac{1}{4}$, so $(d/\lambda)(\Delta_0 + \Delta_t)$ then equals $\frac{1}{4}$, and maximum sensitivity will be achieved for small Δ_d. The sensitivity is zero when eq. (52) is 0, $\frac{1}{2}$, 1, $\frac{3}{2}$, etc.

For small strains, Δ_d is negligible as compared with $(\Delta_0 + \Delta_i)$ and the proportionality factor between intensity change and birefringence change is a constant given by $(\pi d/\lambda) I_0 \kappa \sin[(\pi d/\lambda)(\Delta_0 + \Delta_i)]$. If the photomultiplier, amplifier, and indicator (recorder or oscilloscope) are operating in a linear range with a proportionality constant Q between its reading ρ and the transmitted intensity I_t, then the amplitude of the dynamic birefringence response $(\Delta_d)_{max}$ is related to the maximum deflection of the detector ρ_{max} by the equation

$$\rho_{max} = [\frac{1}{2} Q I_0 \kappa (2\pi d/\lambda)\sin(2\pi d/\lambda)(\Delta_0 + \Delta_i)](\Delta_d)_{max} \quad (53)$$
$$= \sigma(\Delta_d)_{max}$$

To summarize the conclusions from this equation and the preceding discussion, to assure maximum sensitivity of detection of birefringence change, σ, it is desirable to (a) have Q, I_0, κ, and d as large as possible and (b) have $(d/\lambda)(\Delta_0 + \Delta_t)$ close to $\frac{1}{4}$, $\frac{3}{4}$, etc.. This must be done subject to linearity considerations. For example, one cannot increase the incident intensity, I_0, indiscriminately because there will always be a static component of transmitted intensity [at the position of maximum sensitivity where $(d/\lambda)\,(\Delta_0 + \Delta_i) = {}^1/_4$] equal to

$$I_{t_0} = {}^1/_2\,I_0\,\kappa\,(1 + T) \tag{54}$$

according to eq. (46) which results from the static transmittance T and the transmittance resulting from Δ_0 and Δ_i.   Thus

$$\rho = {}^1/_2\,QI_0\,\kappa\left\{(1 + T) + \left[\frac{2\pi d}{\lambda}\sin\frac{2\pi d}{\lambda}(\Delta_0 + \Delta_i)\right]\Delta_d\right\} \tag{55}$$

If I_0 is made too large, ρ will exceed the linear range of the photomultiplier-amplifier-detector combination and Q will decrease.   It is best for T to be as small as possible so as to minimize this effect.

A theory for T for a spherulitic polymer has been proposed by Price[16] and gives

$$T = \frac{1}{4}\,(1 - e^{-\eta y^2}) - \frac{\eta e^{-\eta y^2}}{4}\,[2(1-\cos y) - 2y\sin y]$$

$$+ \frac{\eta^2 e^{-\eta y^2}}{8}\,[(y^2 - 1)\cos 2y - 2y\sin 2y + 2\cos y + 2y\sin 2y - 1] \tag{56}$$

where

$$\eta = N/y^2 = Na\lambda/4\pi(1 - [n_1/n_2]) \tag{57}$$

and

$$y = (2\pi/\lambda)\,[1 - (n_1/n_2)]r \tag{58}$$

N is the number of spherulites of radius r per square centimeter of cross-sectional area, and n_1 and n_2 are the principal refractive indices. λ is the wavelength of light (in the material), and $a = 4\pi r^2$.   Equation (56) is fitted well by the approximation

$$T = A\,(1 - e^{-Na}) \tag{59}$$

It is apparent that large T's will result from samples having large numbers of large spherulites.   Such samples will be difficult to study for this reason and also because appreciable scattering will occur which will reduce κ as well as depolarize the transmitted elliptically polarized light.

In the derivation of eq. (53), the variation of thickness with elongation is neglected.   Even if the birefringence of a sample does not vary with strain, its transmission may change with dynamic strain because of the variation in retardation accompanying the thickness change.   A correction for this for small strains is

$$\rho_{\max} = \sigma\left[(\Delta_d)_{\max} + \Delta_0\left(\frac{1}{S_{\max}{}^{1/_2}} - 1\right)\right] \tag{60}$$

so that

$$(\Delta_d)_{\max} = \frac{\max}{\sigma} - \Delta_0\left(\frac{1}{S_{\max}{}^{1/_2}} - 1\right) \tag{61}$$

This correction becomes greater with increasing Δ_0 and may be large for samples having appreciable static birefringence.

At large strains, σ will not be constant because the $(\Delta_0 + \Delta_t)$ term of eq. (53) should then be replaced by $(\Delta_0 + \Delta_t + \Delta_d)$. If Δ_t has been chosen for maximum sensitivity, the contribution of Δ_d will cause a decrease in σ with increasing strain and will consequently lead to nonlinearity. If Δ_d is sufficiently large (so that $(d/\lambda)\Delta_d$ becomes equal to or greater than $1/4$), σ will pass through zero and become negative. Such nonlinearity will result in severe distortion of the intensity–time curve from its normal sinusoidal form.

In practice, the calibration of intensity in terms of birefringence is carried out experimentally by measuring the intensity of light transmitted through samples for which the birefringence has been measured using a Babinet compensator. From eq. (53) it is apparent that the calibration constant, σ, depends upon the incident light intensity, the high voltage applied to the photomultiplier, the amplifier gain, and the thickness of the sample. Thus, these variables must be controlled when comparing the calibration with the measurement. For a typical DuPont Alathon 10 polyethylene sample, σ is 4.44 v. per refractive index unit of birefringence when operating at maximum sensitivity. The upper frequency limit of 300–400 cycles results from the dynamic birefringence response on the oscilloscope becoming comparable with the noise level of the photomultiplier.

Procedures

In performing measurements, it is necessary for the lamp, electronics, and mechanical transmission to operate for a "warm-up" period of 15–30 min. to assure stability. Sample sizes of 8 to 10 cm. length, 1 to 1.5 cm. width, and 4 mils thick were used in these experiments. In order that the samples be taut between clamps, they were subject to an initial static strain of 2%. (The results were not dependent upon this static strain for small strains.) The amplitude of dynamic strain was 2% for the mechanically excited vibrations and varied from 0.5 to 0.16% over the frequency range of 10 to 300 cycles of electromagnetic vibration.

A fatigue effect was observed where the strain-optical coefficient was dependent upon the duration of vibration. This might be associated with static relaxation or with structural changes accompanying vibration. Measurements were made at sufficient time intervals after starting vibration so that a steady state was reached.

Results

The variation of the dynamic strain-optical coefficient with frequency for an unstretched sample of low density polyethylene (DuPont Alathon 10) is given in Figure 10. One observes a very appreciable decrease in the coefficient with increasing frequency. The phase angle between strain and birefringence was observed, but any deviation from 0° was too small to detect (less than 5°).

818   R. S. STEIN, S. ONOGI, AND D. A. KEEDY

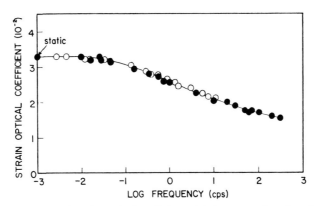

Fig. 10. The variation of strain-optical coefficient with frequency for a low density polyethylene sample.

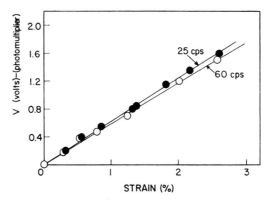

Fig. 11. The variation of the amplitude of the photomultiplier response with strain amplitudes at two frequencies.

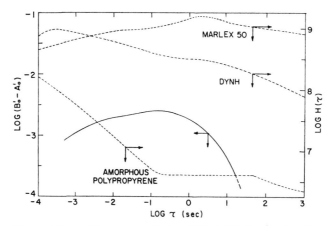

Fig. 12. The variation of the strain-optical coefficient distribution function with relaxation time for a low density polyethylene sample.   It is compared with the stress relaxation spectra $H(\tau)$ for high and low density polyethylene and amorphous polypropylene.

The possibility that the stress-optical coefficient may vary with frequency because the strain amplitude also varied with frequency was checked by determining the variation of birefringence with strain at two frequencies (Fig. 11). The proportionality between the amplitude of the photomultiplier response and the strain amplitude indicates that the strain-optical coefficient was independent of frequency over the range of strains in our experiments.

The strain-optical coefficient distribution function $[A'_0 (\ln \tau) - B_0' (\ln \tau)]$ was calculated from the slope of a plot of the strain-optical coefficient, $K(\omega)$, against ω using eq. (38) (assuming $\alpha(\omega) = 0$) and is plotted in Figure 12. This function shows a much greater variation than does the relaxation spectrum $H(\tau)$ which is shown by dotted lines on the same plot for a similar low-density polyethylene sample.[18] The optical function exhibits a maximum at a relaxation time of about 0.15 sec. where the relaxation spectrum also shows a less pronounced maximum. It appears likely that this maximum is associated with a relaxation time for crystal orientation. Measurements on samples of differing degree of crystallinity (to be reported elsewhere) substantiate this association.

It appears that simultaneous measurement of optical and mechanical spectra are capable of yielding more information about structural changes than are obtainable from mechanical spectra alone, and will further elucidate the molecular nature of these processes.

Supported in part by a grant from the Directorate of Chemical Sciences of the Air Force Office of Scientific Research.

References

1. Treloar, L. R. G., *The Physics of Rubber Elasticity*, Oxford Univ. Press, 1949.

2. Stein, R. S., in H. A. Stuart, Ed., *Die Physik der Hoch Polymeren*, Vol. IV, Springer, Berlin, Chapt. 5.

3. Onogi, S., D. A. Keedy, and R. S. Stein, *J. Polymer Sci.*, **50**, 153 (1961).

4. See, for example, J. D. Ferry, *Viscoelastic Properties of Polymers*, Wiley, New York, 1961.

5. Maxwell, J. C., *Phil. Trans.*, **157**, 49 (1867).

6. Stein, R. S., S. Krimm, and A. V. Tobolsky, *Textile Research J.*, **19**, 8 (1949).

7. Kolsky, H., and A. C. Shearman, *Proc. Phys. Soc.*, **55**, 383 (1943).

8. See, for example, R. D. Andrews, *Ind. Eng. Chem.*, **44**, 707 (1952).

9. Kuhn, W., O. Künzle, and A. Preissmann, *Helv. Chim. Acta*, **30**, 307, 464, 839 (1947).

10. Stein, R. S., and A. V. Tobolsky, *Textile Research J.*, **18**, 302 (1948).

11. Kuhn, W., and F. Grün, *Kolloid-Z.*, **101**, 248 (1942).

12. Treloar, L. R. G., *Trans. Faraday Soc.*, **43**, 277 (1947).

13. See, for example, J. D. Ferry, *Viscoelastic Properties of Polymers*, Wiley, New York, 1961, p. 45.

14. P. Rouse, *J. Chem. Phys.*, **21**, 1272 (1953).

15. Bueche, F., *J. Chem. Phys.*, **20**, 1959 (1952).

16. Price, F., General Electric Research Laboratory Report No. RL-774, Nov. 1952.

17. Magill, J. H., *Polymer*, **2**, 221 (1961).

18. Faucher, J. A., *Trans. Soc. Rheology*, **3**, 81 (1959).

Synopsis

The simultaneous measurement of the birefringence and strain of a sample subjected to sinusoidal strain is proposed as a means for elucidating the molecular mechanism of the response of high polymers to mechanical deformation. A linear phenomenological theory is proposed in which strain-optical coefficient distribution functions are assigned to the elastic and viscous members of a distribution of Maxwell elements. An apparatus has been constructed to determine the dynamic strain-optical coefficient at frequencies between 0 and 600 cycles/sec. by means of mechanical excitation of vibration at low frequencies and electromagnetic at high frequencies. The dynamic birefringences of polyethylene has been studied at room temperature over this frequency range.

Résumé

La mesure simultanée de la biréfringence et de la tension d'un échantillon soumis à une tension sinusoïdale est proposée comme moyen pour élucider le mécanisme moléculaire de réponse des hauts polymères à la déformation mécanique. On propose une théorie phénoménologique linéaire dans laquelle des fonctions de distribution tension-coefficient optique sont attribuées aux terms d'élasticité et de viscosité, d'une distribution d'éléments de Maxwell. On a construit un appareil pour déterminer cette relation dynamique tension-coefficient optique à des fréquences comprises entre 0 et 600 cycles/sec, en utilisant une excitation mécanique de vibration aux fréquences basses et électromagnétique pour les hautes fréquences.

Zusammenfassung

Die gleichzeitige Messung der Doppelbrechung und Verformung einer Probe unter sinusförmig variierender Beanspruchung wird zur Aufklärung des molekularen Mechanismus des Verhaltens von Hochpolymeren bei mechanischer Deformation angewendet. Es wird eine lineare, phänomenologische Theorie vorgeschlagen, in welcher die Verteilungsfunktion des spannungsoptischen Koeffizienten mit den elastischen und viskosen Komponenten einer Verteilung von Maxwellelementen in Beziehung gebracht werden. Es wurde ein Apparat zur Messung der dynamischen spannungsopitschen Koeffizienten bei Frequenzen von 0 bis 600 Hertz konstruiert; die niedrigen Frequenzen wurden mechanisch, die hohen elektromagnetisch angeregt. Die dynamische Doppelbrechung von Polyäthylen wurde in diesem Frequenzbereich bei Raumtemperatur untersucht.

Discussion

C. T. O'Konski (*Department of Chemistry, University of California, Berkley, Calif.*): What do you think of the possibility of measuring the degree of crystallinity by this technique?.

S. Onogi: It is a fairly complex way of quantitatively determining crystallinity. The dynamic birefringence, will also depend on many other variables such as elongation, temperature, and orientation. However, it should be possible to obtain qualitative indications of crystallization in this way, particularly if the variation of dynamic birefringence with static strain is studied.

E. E. Gruber (*General Tire and Rubber Company, Akron, Ohio*): Would not this technique be applicable to a study of the development of crystallinity with extension of crosslinked rubbers, particularly as a function of frequency or rate of extension?

R. Stein: Yes, it would be.

D. Kutscha: *1.* How high in frequency can you go with this technique? *2.* How thick were your films?

R. Stein: *1.* The upper frequency is limited by the sensitivity of the birefringence measurement since the amplitude of vibration decreases with increasing frequency. At present, this frequency limit occurs at about 300 cycles/sec.; we believe this may be extended to a few thousand. *2.* 0.005 to 0.01 in. Sensitivity ordinarily increases with increasing thickness because of the greater retardation produced by a given birefringence. For polyethylene, however, sample thickness imposes an upper limit on thickness.

D. Atack (*Montreal, Canada*): *1.* Is there any significant heating of the specimen during the straining, particularly at high temperatures? *2.* Have you tested for linear viscoelastic behavior of the specimens to the full extent of the strain employed?

R. Stein: *1.* The power input (about 100 w.) is kept constant as frequency is changed, so that a steady state of heating results. While sample temperatures were not measured, the samples did not feel warm to the hand after testing, so heating was probably limited to a few degrees. A drift of strain-optical coefficient with time was observed during the first 5 min. of test, which may be associated with the initial approach to steady-state temperature. After this, reproducible results were obtained in experiments with both ascending and descending frequency. *2.* We had found that the dynamic strain-optical coefficient is independent of strain up to the maximum strain of our present dynamic experiments (3%). I suspect that nonlinear behavior will result at appreciably larger strains.

COMMENTARY

Reflections on "A Universal Calibration for Gel Permeation Chromatography," by Z. Grubisic, P. Rempp, and H. Benoit, *J. Polym. Sci.*, 5, 753 (1967)

HENRI C. BENOIT

Institut Charles Sadron (CRM-EAHP), 6, rue Boussingault, F-67083 Strasbourg Cedex, France

In 1964 the *Journal of Polymer Science* published a paper, written by J. C. Moore[1] which was entitled Gel Permeation Chromatography; it was a kind of revolution in the experimental methods to study polymer solutions, since it did allow the measurement of polydispersity much faster than by the classical batch or column fractionation, which were the only possibilities at the time. Since our lab was specialized in the field of polymer solutions, this technique was highly desirable and I decided to acquire the machine manufactured by the Waters company. This was not easy and with the help of industry, we were able, at the end of 1965, to install this equipment in the Centre de Recherche sur les Macromolecules and to hire a student, Zlatka Grubisic, to run the machine and prepare her doctoral thesis on the applications of this technique.

The first step, when you use a set of G.P.C. columns, is to establish a calibration curve relating the elution volume to the molecular weight (MW) of various samples of narrow polydispersity (usually polystyrene). This procedure is perfect when you study polystyrene but there is a problem when you want to study another polymer, since there is no reason for the curve established for polystyrene to be valid for other polymers. Since it is practically impossible to make a new calibration curve each time a new polymer is studied, one has to find a way to use the polystyrene calibration curve for all polymers. The first method is to assume that, regardless of the nature of the polymer, the calibration curve

obtained for polystyrene is valid. This is still used in some cases but the authors are cautious and call this "the equivalent polystyrene MW," which means that this MW is really the MW of a polystyrene having the same elution time.

This is not satisfactory since the relation with the equivalent polystyrene molecular weight and the real MW is unknown and, due to the importance of this problem for users of G.P.C., the scientific community was looking for a better solution. One has to bear in mind that, at that time, nobody had clear ideas about the mechanism of the separation. Some believed that it was a hydrodynamic effect, some others that it was a size exclusion effect.

The first approach that was suggested was to take the contour length of the polymer for amorphous polymers (or the length of the crystal in the direction of the chain) and to say that two polymers having the same length had the same elution volume. Evidently there was no reason for this to be the case and, due to the difference between amorphous and crystalline polymers, it could just have been qualitative. This was not satisfying and other different methods were suggested. Our feeling was that it was the dimensions in solution which were important and my first idea was to use the radius of gyration as determined by light scattering. Unfortunately, this quantity can only be measured for rather large molecular weights and one could never have enough points for a precise determination. I therefore decided, after discussions with Paul Rempp, to use (for commodity reasons) the intrinsic viscosity as a measure of size, or, more precisely, the product $[\eta]M$ since, following Einstein, it is what one usually calls the hydrodynamic volume of the coil, a measure of the dimensions of the coil in solution. We had at our disposal a large number of samples prepared via anionic polymerization by the group of P. Rempp and doctoral student Z. Grubisic. They checked whether, when plotting $[\eta]M$ as a function of the elution volume for different polymers, one obtained a unique curve. The result was astonishingly good and published first in a French journal.[2] Since it did not attract any attention I felt obliged to publish a short note as a *Polymer* letter, which is the only paper referred to in the literature.

In fact, the first paper was more complete and showed that for the few samples for which we had determined the radii of gyration by light scattering, viscosity was a much better parameter. More surprising was that, for branched polymers, the $[\eta]M$ plot was far better. This led me to conclude that separation was due to the flow process and to the perturbation of the flow by the macromolecules. We had to wait until E. Casassa published a very important paper[3] showing that one can interpret the universal calibration by considering just equilibrium between free molecules and molecules confined in the pores. Since the Casassa paper, nothing new has been added to our understanding of this phenomenon and, if advances have been made on this technique, they are essentially of technical nature, making available better columns and better detectors.

It is interesting to understand why this method of calibration is called Universal Calibration, which at first might seem presumptuous. I am not the father of this expression; it had been used before by a scientist who was proposing an inaccurate method. My feeling was that the universal calibration was just temporary and that, with the use of multidetectors like measuring viscosimetry and light scattering, no calibration would be needed since the result of a chromatography would give a complete characterization of the fractions. In fact this is not the case: the sensitivity of the actual detectors is such that there is only a small range of sizes where the results of both techniques are precise enough to be used simultaneously and Universal Calibration is still in use in many laboratories. This shows that a simple idea with only little theoretical support can survive for a long time, a fact that I would never have suspected.

REFERENCES AND NOTES

1. J. C. Moore, *J. Polym. Sci.*, **A-2,** 835 (1964).
2. H. Benoit, Z. Grubisic, P. Rempp, D. Decker and J. G. Zilliox, *J. Chim. Phys.*, **63,** 1507 (1966).
3. E. F. Casassa, *J. Polym. Sci.*, **B-5,** 773, (1967).

PERSPECTIVE

Comments on "A Universal Calibration for Gel Permeation Chromatography," by Z. Grubisic, P. Rempp, and H. Benoit, *J. Polym. Sci.,* 5, 753 (1967)

HOWARD G. BARTH

DuPont Company, Central Research & Development, Experimental Station, P.O. Box 80228, Wilmington, Delaware 19880-0228

Size exclusion chromatography (SEC), often referred to as gel permeation chromatography (GPC) for synthetic polymers and gel filtration chromatography (GFC) for biopolymers, is unquestionably the most popular analytical procedure for rapidly determining molecular weight distributions (MWD) of macromolecules. Forty years ago, Lathe and Ruthven[1] used grains of starch as the column packing to size separate carbohydrates, polysaccharides, amino acids, and proteins. Several years later, Porath and Flodin[2] introduced crosslinked dextran (Sephadex) for the size separation of biopolymers, which gave birth to GFC as a viable technique. In an attempt to produce stable, hydrophobic packings for the characterization of synthetic polymers, Moore,[3] whose work was first reported in this Journal, synthesized a series of crosslinked polystyrene gels of different pore sizes. Using these packings with a flow-through differential refractometer built by Waters Associates, he was able to demonstrate that polystyrene and poly(ethylene glycol) samples could be separated on the basis of their molecular weight, and proposed the term "gel permeation chromatography." The use of flowthrough detectors and stable packings, combined with the speed and simplicity of an SEC experiment, enabled biochemists and polymer chemists to separate macromolecules on the basis of size within hours rather than weeks to months of laborious fractionation. (It was not until the mid-70s with the introduction of high-performance packings that this time was reduced to minutes.)

The second major development that had occurred during this time period was the work by Casassa,[4-8] who had formulated a theoretical model of SEC using statistical mechanical calculations. (It is of interest to note that the first paper in this series was also published in this Journal[4]). In essence, his random-coil solute model, which assumed cylindrically shaped pores, predicted that SEC is an equilibrium, entropy-controlled, size exclusion process that is independent of temperature and is governed by the ratio of the radius of gyration of a polymer to the average pore size of the packing. Casassa's model was subsequently verified experimentally by Yau and Malone.[9]

Although researchers recognized the fact that the separation was based on a molecular size parameter, there was a great deal of uncertainty as to which was the correct size parameter to use. Furthermore, although SEC quickly had become a well established technique, it was nevertheless a relative method that required column calibration, that is, $\log M$ vs. V_r, where M is the molecular weight and V_r is the elution volume of a polymer standard. The only way of obtaining accurate molecular weight distributions of a given sample was to calibrate an SEC column with polymer standards of identical chemical composition to that of the sample. Furthermore, if the sample were branched or chemically heterogeneous (as in the case of copolymers and blends), molecular weight data could be erroneous.

In 1967, the concept of universal calibration was proposed by Henri Benoit and co-workers[10] to address these concerns. In this simple, but elegant, paper this group demonstrated that if the logarithm of the viscometric hydrodynamic volume $[\eta]M$, in which $[\eta]$ is the intrinsic viscosity and M is the molecular weight of the polymer standard, is used instead of $\log M$ for SEC calibration, the elution volume of all polymers, *independent of chemical com-*

position or architecture, will fall on the same line (Fig. 1). It is of interest to note that the universal calibration curve in the 1967 paper is, without doubt, the most reproduced figure in all of chromatography!) Thus, one could calibrate a given SEC system with, for example, a series of polystyrene standards of known M and $[\eta]$. Provided that the Mark-Houwink coefficients a_s and K_s of the sample are known in the given mobile phase at the appropriate temperature, the molecular weight of the sample M_s at each elution volume increment could be calculated:

$$\log M_s = [1/(1 + a_s)]\log(K/K_s)$$
$$+ [(1 + a)/(1 + a_s)]\log M$$

where a and K are the Mark–Houwink coefficients of the polystyrene standards. Not only did Benoit's universal calibration concept give researchers a highly useful calibration approach, but it firmly established the size separation mechanism governing SEC.

During these intervening years, with some exceptions,[11] universal calibration has proven to be a valid and useful concept, provided that enthalpic interactions between polymer and packing are absent, and that injected concentrations are sufficiently low to avoid macromolecular crowding effects. Nevertheless, one of the practical limitations of universal calibration used to be the need for accurate Mark–Houwink coefficients of the sample. Furthermore, if the sample were chemically heterogeneous, there would not be a single set of Mark–Houwink coefficients, but variable values.

To resolve these difficulties, online viscometric detectors were developed,[12] whereby the intrinsic viscosity at each elution volume can now be determined directly. Through the "magic" of universal calibration, the corresponding molecular weights are readily obtained. In modern SEC systems, online viscometric detection, combined with Benoit's universal calibration procedure, has provided a quantum leap in the development of SEC. If universal calibration is valid for a given system, one can measure not only absolute MWD, but also Mark–Houwink coefficients, and distributive properties of long-chain branching and hydrodynamic volume of a polymer in a single SEC analysis.[12] Furthermore, as proposed by Yau,[13] SEC viscometry with universal calibration can be used to calculate the radius of gyration distribution of linear polymers, by the use of the Flory-Fox equation.[14,15]

The universal calibration concept of Benoit and co-workers[10] has now made it possible, for the first time, to complete polymer molecular characterizations in less than 1 h.

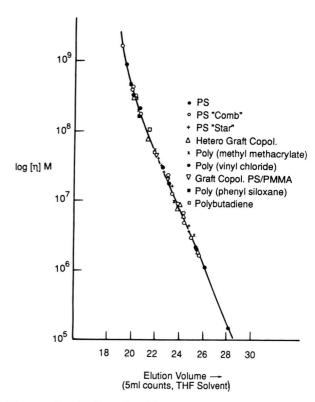

Figure 1. Universal calibration demonstrating that molecular hydrodynamic volume $[\eta]M$ governs SEC separation.[10]

REFERENCES AND NOTES

1. G. H. Lathe and C. R. J. Ruthven, *Biochem. J.,* **62,** 665 (1956).
2. J. Porath and P. Flodin, *Nature,* **183,** 1657 (1959).
3. J. C. Moore, *J. Polym. Sci., Part A,* **2,** 835 (1964).
4. E. F. Casassa, *J. Polym. Sci., Part B,* **5,** 773 (1967).
5. E. F. Casassa and Y. Tagami, *Macromolecules,* **2,** 14 (1969).
6. E. F. Casassa, *J. Phys. Chem.,* **75,** 3929 (1971).
7. E. F. Casassa, *Sep. Sci.,* **6,** 305 (1971).
8. E. F. Casassa, *Macromolecules,* **9,** 182 (1976).
9. W. W. Yau and C. P. Malone, *Polym. Prepr., Am. Chem. Soc., Div. Polym. Chem.,* **12**(2), 797 (1971).
10. Z. Grubisic, P. Rempp, and H. Benoit, *J. Polym. Sci., Part B,* **5,** 753 (1967).
11. P. L. Dubin and J. M. Principi, *Macromolecules,* **22,** 1891 (1989).
12. C. Jackson and H. G. Barth, in *Handbook of Size Exclusion Chromatography,* C.-S. Wu, Ed., Dekker, NY, 1995, Chapter 4.
13. W. W. Yau, *Chemtracts-Macromol. Chem.,* **1,** 1 (1990).
14. P. J. Flory and T. G. Fox, Jr. *J. Am. Chem. Soc.,* **73,** 1904 (1951).
15. O. B. Ptitsyn and Y. E. Eizner, *Sov. Phys. Tech. Phys.,* **4,** 1020 (1960).

POLYMER LETTERS VOL. 5, PP. 753–759 (1967)

A UNIVERSAL CALIBRATION FOR GEL PERMEATION CHROMATOGRAPHY

Gel permeation chromatography is one of the most powerful techniques for characterizing the polydispersity of polymeric materials (1). A versatile commercial apparatus (2) has been used successfully in numerous laboratories on various problems of molecular weight distributions. But one of the difficulties still unsolved is the problem of calibration, i.e., the relation between elution volume and molecular weight.

Some authors (2) have assumed that retention time depends on the contour length of the molecular chain. Others think that it is more reasonable to use the radius of gyration or some average volume of the polymer molecule as the calibration parameter.

In a recent paper (3) we have reported GPC retention times of a series of polystyrenes exhibiting different molecular structures: linear, star-shaped, and comb-like. All were of known molecular weight and of low polydispersity. Obviously the conventional calibration method, where the logarithm of molecular weight is plotted against elution volume, does not yield a universal curve for all the samples. Retention times for branched samples were always larger than those for the linear homologs of the same molecular weight. We therefore tried using the hydrodynamic volume as the calibration parameter.

According to the Einstein viscosity law one can write

$$[\eta] = K \, (V/M) \tag{1}$$

where $[\eta]$ is the limiting viscosity index, V the hydrodynamic volume of the particles, M their molecular weight, and K a constant. This equation shows that the product $[\eta]M$ is a direct measure of the hydrodynamic volume of the particles and suggests the use of log $[\eta]M$, instead of log M, in the calibration of the chromatograms. With this type of plot, all our experimental points fall on the same curve. This assertion is supported by the following considerations. It is well known that for branched polymers one can write

$$[\eta]M = \phi R_L^3 \, g^x \tag{2}$$

or

$$[\eta]M = \phi' R^3 \, g^{x - 3/2} \tag{3}$$

where R is the actual radius of gyration, g is the parameter introduced

753

by Zimm and Stockmayer (4), ϕ is a universal constant, and x is an exponent ranging from 1/2 to 3/2 depending upon the theory used.

If this is true, it means that it is the hydrodynamic volume—obtained from viscometric data—which determines retention in the chromatographic columns. But if the hydrodynamic volume is the parameter responsible for GPC retention, the above calibration should be valid for any polymer, regardless of its chemical nature as well as of its morphological structure: it should be universal, and thus be characteristic for any given set of columns and elution solvent at a given temperature. We have measured GPC elution volumes of a number of polymer samples chosen for their small polydispersity. The polymers, obtained in most cases through anionic reaction mechanisms, were: polystyrene (PS), poly(methyl methacrylate) (PMM), polybutadiene, and poly(vinyl chloride) samples which can be considered as having linear molecular structure; block and graft copolymers of styrene and methyl methacrylate; star-shaped polystyrenes and poly(methyl methacrylates); "heterograft" copolymers composed of a three-block sequence PMM–PS–PMM, with PS grafts on the PMM blocks.

Molecular weights M were determined by light scattering (Sofica apparatus) in suitable solvents; control of polydispersity rendered it necessary, sometimes, to get number-average values through osmometry (Mechrolab equipment).

Intrinsic viscosities were measured, for all samples, on tetrahydrofuran solutions in a capillary viscometer at $25°C$.

The gel permeation chromatography experiments were carried out on a Waters machine equipped with 4 columns (10^6, 10^5, 10^4, 9×10^2 A.), at room temperature. The solvent was tetrahydrofuran. Injection time was 2 min. and pumping rate was always 1 ml./min. All the chromatograms were quite sharp and almost symmetric. The maximum of the peak was taken as the elution volume for each of the tested polymers and the log of the product $[\eta]M$ was plotted as a function of this elution volume, which was measured in 5-ml. increments (counts).

The results are collected in Table I, and a plot is shown in Figure 1.

It can be seen that all of the experimental points lie on a single curve. This confirms our hypothesis according to which the viscometric hydrodynamic volume, characterized by $[\eta]M$, determines retention in the chromatographic column. It is interesting to note that this universal calibration curve takes into account interactions of all types—those between polymer and solvent, and in the case of copolymers the hetero-contact interactions—which are included in $[\eta]$. This explains why molecular weight calibration curves, i.e., log M vs. elution volume, established for each homologous series exhibit different slopes (Fig. 2): in a given solvent the viscosity laws for different polymers have different exponents.

TABLE I

Sample	Shape	$\bar{M}_w$	$[\eta]$	$[\eta] \times \bar{M}_w$	Elution vol. counts, 5 ml.
Polystyrene PS	Linear	13,000	11.5	1.49×10^5	28.2
		68,000	31.3	2.13×10^6	25.4
		213,000	81.5	1.74×10^7	23.1
		290,000	96.7	2.80×10^7	22.6
		934,000	220.5	2.06×10^8	20.7
		2,230,000	391	8.7×10^8	19.6
Poly(methyl methacrylate) PMMA	Linear	170,000	54.4	9.25×10^6	23.5
		570,000	96.0	5.47×10^7	22.0
		947,000	190	1.80×10^8	20.6
		1,277,000	216	2.76×10^8	20.4
Poly(vinyl chloride) PVC	Linear	27,000	40	1.08×10^6	26.1
		40,000	42.3	1.69×10^6	25.6
		46,500	42	1.95×10^6	25.5
		85,000	78	6.6×10^6	24.3
		100,000	83	8.3×10^6	24
Polybutadiene	Linear	390,000	255	9.94×10^7	21.5
Polystyrene PS	Comb	147,000	19	2.91×10^6	25.0
		294,000	24.4	7.17×10^6	23.9
		550,000	41.4	2.28×10^7	23.0
		2,040,000	86.5	1.76×10^8	20.75
		3,600,000	116.5	4.19×10^8	21.0
		11,200,000	160	1.68×10^9	19.2

Polystyrene PS	Star	146,000	23.5	3.43×10^6	24.9
		176,000	23	4.05×10^6	24.75
		409,000	59.8	2.45×10^7	22.7
		590,000	67.5	3.98×10^7	22.3
PS/PMMA graft copolymer	Comb	460,000	41.8	1.92×10^7	23.0
		830,000	57.5	4.77×10^7	22.05
PS/PMMA	Heterograft[a]	122,000	15	1.83×10^6	25.6
		235,000	24.8	5.83×10^6	24.3
		308,000	23.7	7.29×10^6	23.8
		311,000	27.3	8.50×10^6	24.2
		1,400,000	56.3	7.88×10^7	21.5
		3,630,000	81.6	2.96×10^8	20.2
Poly(phenyl siloxane) TP-62-4	Ladder	5,200,000	318	1.65×10^9	20.7
TP-61		840,000	380	3.19×10^8	20.2
			325	2.73×10^8	20.2
TP-62		1,720,000	280	4.81×10^8	20
			256	4.40×10^8	20
PS/PMM statistical copolymer	Linear	234,000	89.6	2.09×10^7	23.0

[a] See text.

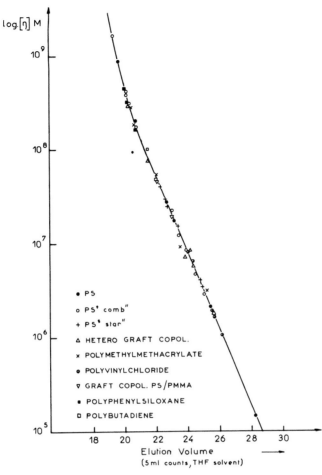

log.[η] M

10^9

10^8

10^7

10^6

10^5

• PS
○ PS' comb"
+ PS' star"
△ HETERO GRAFT COPOL.
× POLYMETHYLMETHACRYLATE
● POLYVINYLCHLORIDE
▽ GRAFT COPOL. PS/PMMA
▪ POLYPHENYLSILOXANE
□ POLYBUTADIENE

18   20   22   24   26   28   30

Elution Volume
(5 ml counts, THF solvent)

Figure 1.

Recently it has been shown (4) that the log [η]M plot also furnishes
a good fit for polystyrene and for poly(L-benzyl glutamate) in dimethylformamide. It is well known that the shape of these molecules is quite
different: the first is a coil, the other a rigid rodlike molecule. Therefore, this result suggests that our calibration is independent of the
shape of the molecules and is valid for elongated particles as well.

Recently, Meyerhoff (5) published his own results in this field, and
according to him the plots of log $[\eta]^{1/3} M^{1/2}$ vs. elution volume should
yield parallels for different polymeric series. Although the theoretical
basis of this plot is not very convincing, we plotted our own experimental results according to Meyerhoff and obtained for each type of
polymer a different straight line, but all the lines were parallel. Thus

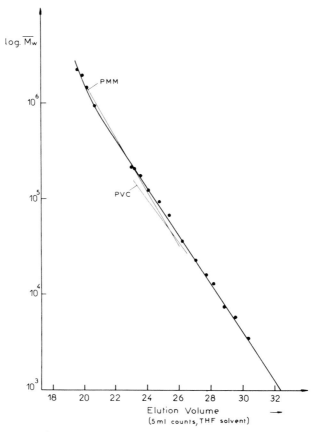

Fig. 2. GPC calibration: polystyrene/THF.

it is demonstrated that the calibration proposed by Meyerhoff cannot be considered universal, since different polymeric series yield different curves for the same set of columns, the same solvent, and the same temperature.

The question now arises if it would not be possible to obtain the same kind of results using the radius of gyration. From our data on graft polymers this seems not to be the case. The value of x in eq. (3) is often assumed to be 1/2 (Zimm and Kilb) (7). Since in our samples g values range from 1 to 0.2, there is for the most highly branched polymers a factor of five between hydrodynamic and geometrical volumes, which makes it impossible to have a good fit in both representations.

It is somewhat surprising to obtain such a good fit with the volume obtained viscometrically, and one could ask why another hydrodynamic volume, such as that obtained from translational brownian motion, could not be used. If one looks at the molecules moving in front of the pores in a velocity gradient, just as in a viscosity experiment, one could assume that it is the volume perturbing the flow that also governs the entry of molecules into the pores. But this is merely a hypothesis.

Even if this explanation is not valid, this new method of calibration seems promising. GPC results can be considered as a combination of molecular weight and viscosity. Owing to the ease of both GPC and viscosity experiments, this method can be used for molecular weight determinations on unknown polymers, and it should be especially useful and efficient for polymers soluble only at elevated temperatures.

We thank Mrs. Decker, Messrs. Curchod, Guyot, Gallot, and Zilliox for some of the samples we have used.

References

(1) J. C. Moore, J. Polymer Sci. A, 2, 835 (1954).

(2) L. E. Maley, in Analysis and Fractionation of Polymers (J. Polymer Sci. C, 8), J. Mitchell, Jr. and F. W. Billmeyer, Jr., Eds., Interscience, New York, 1965, p. 253.

(3) H. Benoit, Z. Grubisic, P. Rempp, D. Decker, and J. G. Zilliox, J. Chim. Phys., 63, 1507 (1966).

(4) B. H. Zimm and W. H. Stockmayer, J. Chem. Phys., 17, 1301 (1949).

(5) Z. Grubisic, L. Reibel, and G. Spach, Compt. Rend., in press.

(6) G. Meyerhoff, Makromol. Chem., 89, 282 (1965);
G. Meyerhoff, Ber. Bunsenges. Physik., 69, 866 (1965).

(7) B. Zimm and R. W. Kilb, J. Polymer Sci., 37, 19 (1959).

Z. Grubisic
P. Rempp
H. Benoit

Centre de Recherches sur les Macromolécules
Strasbourg, France

Received March 27, 1967
Revised May 22, 1967

COMMENTARY

Reflections on "Theory of Block Copolymers. I. Domain Formation in A–B Block Copolymers," by D. J. Meier, J. Polym. Sci., C26, 81 (1969)

DALE J. MEIER

Michigan Molecular Institute, Midland, MI 48640

I am naturally very pleased and honored that one of my papers has been selected as part of the 50th birthday celebration of the *Journal of Polymer Science.*

The work described in the paper was done while I was a member of the Chemical Physics Department at the Shell Development Company in Emeryville, California. The incentive for the work grew out of the important discovery made by Milkovich and Holden[1] at Shell's Torrance Laboratory in the early 1960s of the tri-block thermoplastic elastomers. Although I was not directly involved at that time in Shell's development project, I was in frequent contact with the group at Torrance and thus aware very early on of the discovery. I was also aware of the literature on block copolymers, particularly that of the Strasbourg group (Sadron, Skoulios, Gallot, etc.) who had shown[2] by the early 60s that A-B block copolymers form mesomorphic structures (which I called domains).

Given this knowledge of domain structures, it was not too great a leap of imagination for me to conclude that they were responsible for the virtual crosslinking of the S–B–S tri-block copolymers. However, the factors that governed their formation and their size and shape were not as obvious at first. The critical basis for the theory was established later in realizing that the size of the domains must be related to chain dimensions through a space-filling (uniform density) requirement. This realization actually occurred while I was mindlessly chiseling some excess wood while sculpting.

In earlier interactions with a lubricants group at Emeryville, I became involved with their work on polymeric dispersants, and eventually developed a theory for such material.[3] The theory was based upon the diffusion equation to generate the chain statistics of chains constrained by barriers (Di Marzio[4]). I later recognized that the physics involved was the same for the block copolymer problem where the constraints forcing chains to be in restricted regions of space came from polymer/polymer incompatibility. In all of this work, I benefited greatly from interaction with my colleagues, particularly Tom Schatzki and Sol Davison. Tom was an excellent and informed critic, and in one of the talks I gave in a department seminar, his comment was "Dale, you don't know what you are talking about!" (eventually we straightened one another out). Discussions of the theory and comparisons of theory and experiment were also greatly helped by my interaction which began in the late 60s with the Kyoto group (H. Kawai, T. Hashimoto, and T. Inoue).

Subsequent years have seen an amazing development of the block copolymer field in both theoretical and commercial aspects. Particular note should be made of the early theories which followed this paper, i.e., those of Helfand and coworkers,[5] and of Noolandi and coworkers,[6] both of whom removed some of the rough edges of the theory by developing a self-consistent field approach. Also noteworthy is the theory of Leibler[7] who first treated in a realistic way the order-disorder transition. However, even these improved theories have been merely the forerunner of ever-newer theories. Hardly a month passes at the present time without a new one appearing.

REFERENCES AND NOTES

1. G. Holden and R. Milkovich, U.S. Patent 3,365,765. Aug. 1964.
2. C. Sadron, *Angew. Chem.*, International Edition **2**, 249 (1963).
3. D. J. Meier, *J. Phys. Chem.*, **71**, 1861 (1967).
4. E. A. Di Marzio, *J. Chem. Phys.*, **423**, 2101 (1965).
5. E. Helfand and Z. R. Wasserman, *Macromolecules*, **9**, 879 (1976).
6. J. Noolandi and K. M. Hong, *Ferroelectrics*, **30**, 117 (1980).
7. L. Leibler, *Macromolecules*, **13**, 1602 (1980).

PERSPECTIVE

Comments on "Theory of Block Copolymers. I. Domain Formation in A–B Block Copolymers," by D. J. Meier, *J. Polym. Sci.,* C26, 81 (1969)

TAKEJI HASHIMOTO

Department of Polymer Chemistry, Graduate School of Engineering, Kyoto University, Kyoto 606-01, Japan

At the time this seminal paper by Meier was published, many papers on the colloidal properties of block copolymer solutions had already appeared. The studies involved monomolecular and polymolecular micelles in solution, the possibility of intramolecular phase separation, polymeric oil-in-oil emulsions in systems composed of an A-B block copolymer, a homopolymer and/or B homopolymer in a solvent, and so on.[1] However, little had been done on the morphology of bulk block copolymers. The state of the art at that stage may be seen in the volumes based on some symposia on block copolymers, e.g., those at the California Institute of Technology in 1967,[2] in which the Meier's paper was presented, the ACS National Meetings in New York City (1969)[3] and Chicago (1970).[4] There were quite a number of papers on the thermal and mechanical behavior of block copolymers as thermoplastic elastomers.[2–4] These papers suggested the existence of "microdomain" structures as a consequence of the "microphase separation" between constituent block chains of A and B. However, unequivocal identifications of the microdomain morphologies in bulk block copolymers could not be achieved until the osmium tetraoxide staining method was first introduced to this field by Kato[5] in 1967. This technique coupled with transmission electron microscopy on ultrathin sections clearly revealed the phase-separated microdomain structures that were typically a few hundred angstrom in size.[6–9] Moreover, it showed qualitatively that size and shape of the structures depended on molecular weights of the constituent block chains. These findings naturally led people to

recognize that block copolymers can provide "tailor-made" supermolecular structures and properties. However, how the domain morphology and its thermodynamic stability depend on molecular and thermodynamic variables was still totally unexplored, though understanding this point is crucial for the rational design of the structures and hence properties of block copolymers.

Meier developed the pioneering theory of microdomain morphology in block copolymers and gave a rigorous free energy expression for the microdomain formation.[10] The theory describes criteria for the formation of spherical microdomains and their size in terms of molecular and thermodynamic variables. Further developments of this theory for other microdomain morphologies were reviewed by him.[11] The appearance of his theory was quite timely and stimulated basic experimental studies of (i) the domain size as a function of molecular weight, temperature, and block copolymer concentration when the neutral solvents are used; (ii) the thickness of the interface between two coexisting microdomains; (iii) a long-range order of the microdomains; (iv) morphology as a function of molecular weight ratios of the constituent blocks, and so on.[12–14]

Meier's theory also stimulated further developments of fundamental statistical mechanical theories of microdomain morphology, such as those by Helfand,[15] Helfand and Wasserman,[16] Leibler,[17] Noolandi and Hong,[18] Semenov,[19] Ohta and Kawasaki,[20] Kawasaki and Kawakatsu,[21] Muthukumar and his coworkers,[22] and Matsen et al.[23] His theory predicted that the critical block molecular weights required for domain formation are many-fold greater than required for phase separation of a simple mixture of the constituent blocks. This aspect of his study eventually led to another seminal theory of block copolymers, namely, Leibler's Landau-type theory of the phase transition in block copolymer melts—so-called microphase separation transition (MST) or order-disorder transition (ODT).[17] Leibler's mean-field theory was further generalized by Fredrickson-Helfand[24] and Hohenberg-Swift.[25] Both groups elucidated that effects of random thermal noise alter the nature of ODT from a second-order to a first-order phase transition, the theoretical prediction of which have been experimentally verified.[26-28] Thus Meier's pioneering theory was the progenitor of modern theories of the behavior of block copolymers, not only in the field of polymer science but also in other fields such as mathematical physics, soft-condensed matter physics, physical chemistry, and so on.

REFERENCES AND NOTES

1. See for example, a review article by G. E. Molau, in ref. 3 (below).
2. J. Moacasin, G. Holden, and N. W. Tschoegl, Eds., *Block Copolymers, J. Polym. Sci.,* **C26** (1969).
3. S. L. Aggarwal, Ed., *Block Polymer,* Plenum, New York (1970).
4. G. E. Molau, Ed., *Colloidal and Morphological Behavior of Block and Graft Copolymers,* Plenum, New York (1971).
5. K. Kato, *Polym. Eng. Sci.,* **7,** 38 (1967).
6. H. Hendus, K. H. Illers, and E. Popte, *Kolloid-Z., u. z. Polym.,* **216–217,** 110 (1967).
7. J. F. Beecher, L. Marker, R. D. Bradford, and S. L. Aggarwal, *Polym. Prepr.,* **8,** 1532 (1967).
8. M. Matsuo, T. Ueno, H. Horino, S. Chujyo, and H. Asai, *Polymer,* **9,** 425 (1968).
9. T. Inoue, T. Soen, H. Kawai, M. Fukatsu, and M. Kurata, *J. Polym. Sci. Part B: Polym. Phys. Ed.,* **6,** 75 (1968).
10. D. J. Meier, *J. Polym. Sci.,* **C26,** 81 (1969); Much more primitive and qualitative treatments, based on somewhat similar ideas to those given by Meier, were

presented by Inoue et al. in the same year for spherical, cylindrical, and lamellar microdomains: T. Inoue, T. Soen, T. Hashimoto and H. Kawai, *J. Polym. Sci.,* **7,** 1283 (1969).
11. D. J. Meier, in *Thermoplastic Elastomers,* N. R. Legge, G. Holden and H. E. Schroeder, Eds., Hanser, Munich (1987), Chapter 11. Second ed. in press.
12. See for example, T. Hashimoto, M. Shibayama, M. Fujimura, and H. Kawai, in *Block Copolymers-Science and Technology,* D. J. Meier, Ed., MMI Press Symp. Series, Vol. 3, Harwood, 1983, and references cited therein: T. Hashimoto, in the Book of ref. 11, Chapter 12, and references cited therein.
13. B. M. Gallot, *Advances in Polym. Sci.,* **29,** 85 (1978).
14. See for example, H. Hasegawa, H. Tanaka, K. Yamasaki, and T. Hashimoto, *Macromolecules,* **20,** 1651 (1987).
15. E. Helfand, *Macromolecules,* **8,** 552 (1975).
16. E. Helfand and Z. R. Wasserman, *Macromolecules,* **9,** 879 (1976); **11,** 960 (1978); **13,** 994 (1980).
17. L. Leibler, *Macromolecules,* **13,** 1602 (1980).
18. J. Noolandi and K. M. Hong, *Ferroelectrics,* **30,** 117 (1980); K. M. Hong and J. Noolandi, *Macromolecules,* **14,** 727 (1981).
19. A. N. Semenov, *Sov. Phys.–JETP (Engl. Transl.),* **61,** 733 (1985).
20. T. Ohta and K. Kawasaki, *Macromolecules,* **19,** 2621 (1986).
21. K. Kawasaki and T. Kawakatsu, *Macromolecules,* **23,** 4006 (1990).
22. J. Melenkevitz and M. Muthukumar, *Macromolecules,* **24,** 4199 (1991); R. L. Lescanec and M. Muthukumar, *Macromolecules,* **26,** 3908 (1993).
23. M. W. Matsen and F. S. Bates, *Macromolecules,* **29,** 1091 (1996), and references cited therein.
24. G. H. Fredrickson and E. Helfand, *J. Chem. Phys.,* **87,** 697 (1987).
25. P. C. Hohenberg and J. B. Swift, *Phys. Rev. E.,* **52,** 1828 (1995).
26. F. S. Bates, J. H. Rosedale, and G. H. Fredrickson, *J. Chem. Phys.,* **92,** 6255 (1990).
27. F. S. Bates and G. H. Fredrickson, *Annu. Rev. Phys. Chem.,* **41,** 525 (1990), and references cited therein.
28. N. Sakamoto and T. Hashimoto, *Macromolecules,* **28,** 6825 (1995).

J. POLYMER SCI.: PART C NO. 26, PP. 81–98 (1969)

Theory of Block Copolymers. I. Domain Formation in A–B Block Copolymers

D. J. MEIER, *Shell Development Company, Emeryville, California 94608*

Synopsis

Microphase separation occurs in many block copolymers to give domain structures. In this first paper in a series dealing with domain formation and the consequences thereof, a theory is presented for the formation of spherical domains in A–B block copolymers. The theory establishes criteria for the formation of domains and their size in terms of molecular and thermodynamic variables. It is shown that the considerable loss in configurational entropy due to the constraints on the spacial placement of chains in a domain structure requires that the critical block molecular weights required for domain formation are many-fold greater than required for phase separation of a simple mixture of the component blocks. The relation between domain radius R and molecular dimensions is obtained from the requirement that space in the domain must be filled with a constant density of segments. Segment densities are evaluated from solutions of the diffusion equation, treating the constraints on chain placement as boundary value problems. This gives the relationship $R = 4/3 \langle L^2 \rangle^{1/2}$, where $\langle L^2 \rangle^{1/2}$ is the root-mean-square end-to-end chain length. Because of chain perturbations in a domain system, $\langle L^2 \rangle^{1/2}$ is larger than the unperturbed value $\langle L^2 \rangle_0^{1/2}$ normally expected for bulk polymers. A means to evaluate the perturbations is shown. The agreement between the predictions of the present theory and the limited published experimental information appears quite satisfactory.

INTRODUCTION

A remarkable increase in interest in block copolymers has occurred in the past few years, probably inspired by the realization that new types of technologically important materials are possible by use of block copolymers, e.g., thermoplastic elastomers (1). The thermally reversible, physical "crosslinking" that occurs in the thermoplastic elastomers is generally recognized to result from a unique type of microscopic phase separation in which complete aggregation of the separate phases does not occur, in contrast to ordinary phase separation. The formation, size, shape, etc., of the microscopic phase regions, which will be called

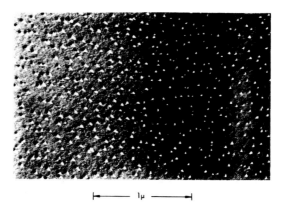

Fig. 1. Electron micrograph of S-I film.

domains, are functions of such molecular properties of a block copolymer as the nature of the block components, molecular weights, distribution of the blocks, etc. This paper is the first in a series concerning
block copolymers in which theories of domain formation and the consequences thereof will be developed.

In this paper, we confine our attention to the simplest type of block
copolymer, namely an A–B type in which the molecular weight of one
component, A, is much less than that of the other. This stipulation ensures that the A-component will be the dispersed, domain-forming component and also fixes the domain shape. Our theoretical work on domain
shapes (to be presented in a subsequent paper) and experimental evidence both show that the equilibrium domain shape is spherical when
the component block molecular weights are greatly different and the
components are amorphous. Other shapes, e.g., planar, cylindrical, etc.,
can, however, be the stable forms under other conditions (2,3). Figure
1 shows an example of the spherical domains formed from an amorphous
block copolymer in which the block molecular weights are greatly different. This figure is an electron micrograph of a very thin film of a sty-
rene-isoprene (S-I) block copolymer of 15,000–75,000 block molecular
weights. The film was prepared by evaporation of a very dilute benzene
solution of the block copolymer, followed by shadowing, and shows
spherical domains of polystyrene projecting above the surface of the
film. The domain diameters are approximately 260 Å.

MODEL AND APPROACH

The model domain structure to be treated in this paper is shown in
Figure 2 (in which only a few of the chains that make up a domain are
shown). The domain is assumed to be spherical, consisting predomi-
nately of the A component and is imbedded in a matrix of the B compo-

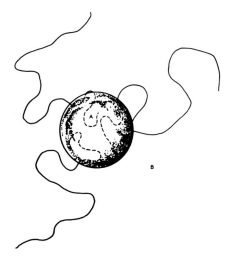

Fig. 2. Model domain structure.

nent. It can be shown that the "phases" that form in a domain system will be essentially pure, just as with phase separation of two mixed homopolymers in which the phases that form are almost pure homopolymers. Only at the domain surface will there be a thin region in which A and B segments remain mixed. This region will contain the junction between the A and B blocks. The surface of a domain is relatively well defined, as confirmed by electron microscopy.

The following additional assumptions are made for simplicity; none are thought to be unduly restrictive nor unrealistic: (a) random-flight statistics are applicable, with perturbations allowed for by the use of the familiar isotropic chain expansion parameter a (4), (b) A and B segments are equal in size, (c) the polymers are amorphous, and (d) the block molecular weights of each component are uniform.

The basic difference in a treatment of block copolymers are opposed to simple homopolymers or random copolymers is the additional complication that arises from the constraints that restrict the components to separate regions of space, i.e., in the domain structure the A component is constrained to stay within the domain and the B component to stay out. These constraints, of course, restrict the number of configurations available to the chain with a concomitant reduction in the entropy of the system. Rather than evaluate the number of configurations (and hence the entropy) with these constraints by the usual lattice-type model, we avoid this almost hopelessly complex approach by treating the constraints as a boundary value problem, using the diffusion equation to generate the applicable chain statistics (5,6).

DOMAIN SIZE

It has been mentioned that the size of a domain is fixed by molecular chain dimensions. The reason is obvious as Figure 2 will show. If the A—B junction is fixed near the domain surface, the domain cannot grow without limit since then space near the center of the domain could not be filled with A segments. A vacuole or region of low density is a region of very high energy and will not occur. In fact, we take as the criterion to evaluate domain sizes in terms of molecular variables that the density of segments in the domain be constant. In the use of this criterion, we are assuming that of the terms relating the free energy of the system to domain size the term associated with density is of overriding importance.

We require an expression for the (number) density of segments within the domain as a function of the ratio of domain size to molecular chain dimensions. That ratio which gives the most constant density of segments will then be taken as the predicted relationship between domain sizes and molecular dimensions.

We obtain the segment number density in the domain by first solving the diffusion equation for the probability $W(n; \bar{r}, \bar{r}', R)$ of finding the free end of a subchain of n elements at $\bar{r}$ when the first end is fixed at $\bar{r}'$ and all segments are constrained to stay within the spherical region of radius R. The origin of the coordinate system will correspond to the center of a domain. The number density $\rho(\sigma_A; \bar{r}, \bar{r}', R)$ of an A chain having a total of σ_A segments is then obtained by summing $W(n; \bar{r}, \bar{r}', R)$ over n, i.e.,

$$\rho(\sigma_A; \bar{r}, \bar{r}', R) = \sum_{n=1}^{\sigma_A} W(n; \bar{r}, \bar{r}', R) \tag{1}$$

The diffusion equation in the form applicable to the present problem is

$$\frac{\partial W(n; \bar{r}, \bar{r}', R)}{\partial n} = \frac{l^2}{6} \nabla^2 W(n; \bar{r}, \bar{r}', R) \tag{2}$$

where l is the length of a statistical segment. The boundary condition for this problem is $W(n; R, \bar{r}', R) = 0$, which removes from the ensemble of configurations those in which any segment of the chain has reached R, i.e., the surface is an absorbing barrier. The remaining configurations are given proper statistical weight by renormalization (5).

Equation (1) gives the number density of a single chain in the domain space. The total segment density $\Omega(\sigma_A; \bar{r}, R)$ is obtained by summing at $\bar{r}$ the number densities $\rho(\sigma_A; \bar{r}, \bar{r}', R)$ of the many chains that make up a domain. This summation requires specification of the placement $\bar{r}'$ of each chain origin. We have carried out the summation for 24 chains

whose origins were equidistant from one another on the surface of a
sphere, with results showing that the angular variation in total segment
density becomes very small for this packing density of chains because
of overlap of adjacent molecules. Since the packing density in a domain will, in general, be even greater than that used here, we may ignore the angular variables and consider only the radial variation in total
segment density. With this simplification, the total number segment density Ω at r becomes from eqs. (1) and (2)

$$
\Omega(\sigma_A; r, R) = \frac{\eta_A}{4R^3} \sum_{n=1}^{\sigma_A} \frac{\sum_{m=1}^{\infty} \frac{R}{r} \sin\left(\frac{m\pi r}{R}\right) \sin\left(\frac{m\pi r'}{R}\right) \exp\{-\sigma_A l^2 m^2 \pi^2/6R^2\}}{\sum_{p=1}^{\infty} (-1)^p \frac{1}{p} \sin\left(\frac{p\pi r'}{R}\right) \exp\{-\sigma_A l^2 p^2 \pi^2/6R^2\}}
$$

(3)

where η_A is the number of A chains in the domain.

Equation (3) has been evaluated with an IBM 7040 computer for vari-
ous values of r/R, r'/R, and $\sigma_A l^2/R^2$ and for $\sigma_A = 20$ and $\sigma_A = 100$ statistical elements. Normalized chain densities $\Omega' = (4R^3/\sigma_A \eta_A)\Omega$ were
found to be independent of these values of σ_A. Since our domain model
places the origins r' of the A chains near the domain surface, we have
restricted r'/R to the range 0.8–1.0. For r' R > 0.9, segment densities
were found to be negligibly dependent on the value chosen for r'/R.
Since we shall later take r'/R to be greater than 0.9, we shall show results for only one value of r'/R, namely 1.0, to avoid clutter in the figure.

Figure 3 shows the relative segment densities $\Omega'(\sigma_A; r, R) = (4R^3/\sigma_A \eta_A)\Omega(\sigma_A; r, R)$ as a function of the radius r/R and for several values
of $(\sigma_A l^2)^{1/2}/R$, i.e., the ratio of the rms end-to-end chain distance of a
free chain to the domain radius R. Also shown in the figure is a curve
giving the desired relative density of segments that would be obtained
if the density of segments in a sphere were constant to r/R = 0.9 and
then linearly decreased to zero at r/R = 1.0. The region 0.9 < r/R < 1.0
is taken to represent the interfacial region in which A and B segments
are intermixed, with the A segment density decreasing from its uniform
value characteristic of the "pure" A interior of the domain to zero at
the surface. The B segment density, of course, increases in this model
from zero at r/R = 0.9 to its "pure" value at r/R = 1.0. The particular
value r/R = 0.95, which gives the thickness $\Delta R/R = 0.1$ to the interfacial region, is used here to be consistent with a later choice, but the
value has little influence on the present evaluation of the relationship
between domain size and molecular dimensions.

In Figure 3, we see that the present model does not give the desired
constant density of segments throughout the domain for any value of
$(\sigma_A l^2)^{1/2}/R$. Obviously in a real domain there must be chain perturba-

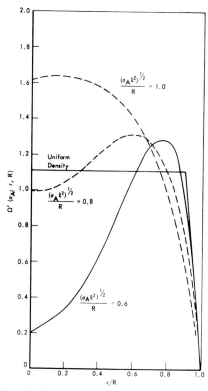

Fig. 3. Relative density of segments.

tions which smooth out these variations in segment densities that occur for purely random-flight statistics. The problem is now to choose that value of $(\sigma_A l^2)^{1/2}/R$ which minimizes the chain perturbations or movement of segments required to obtain constant segment density. A rigorous solution of this problem would be quite difficult, although possible by using the diffusion equation modified by a biasing potential. Rather than introduce this complexity, we take a simplified approach and minimize the absolute deviations from the desired density Ω'_0 over the domain, i.e., the expression

$$\int_{r/R=0}^{1} | \Omega'(\sigma_A; r, R) - \Omega'_0 | \left(\frac{r^2}{R^2}\right) d(r/R)$$

is minimized as a function of $(\sigma_A l^2)^{1/2}/R$. The absolute value of the density difference is minimized rather than, say, the mean-square difference, since the number of segments that must be moved achieve constant density is directly proportional to the density difference. The above function has a minimum for $(\sigma_A l^2)^{1/2}/R \simeq 0.75$, which is now adopted as giving the desired relationship between domain radii and chain dimensions, i.e., $R = 4/3 \, (\sigma_A l^2)^{1/2}$. However, this expression is still not adequate to enable prediction of domain radii in terms of molecular weights, even though the constant K in the relationship $\langle \sigma l^2 \rangle_0^{1/2} = KM^{1/2}$ between unperturbed rms end-to-end chain dimensions for a bulk polymer and molecular weight is known. The domain system has additional chain perturbations which do not exist in a bulk polymer and which must be evaluated before predictions of domain size can be made. If the ratio of perturbed to unperturbed chain dimensions is represented by α (4), then we may write

$$R = \frac{4}{3} \, (\sigma_A l^2)^{1/2} = \frac{4}{3} \, \alpha \, (\sigma_A l^2)_0^{1/2} = \frac{4}{3} \, \alpha \, KM_A^{1/2} \qquad (4)$$

A method to evaluate α is presented in the following section which then allows prediction of R as a function of M_A (if K is known).

THERMODYNAMICS OF A–B BLOCK COPOLYMERS

The free energy difference ΔG between a random mixture of block copolymer molecules and the domain system can be separated into several entropic and enthalpic contributions. First, the restriction on the placement of the A–B junctions to the interfacial regions of domains decreases the entropy relative to random placement. This entropy decrease will be termed the "placement entropy" difference ΔS_p and will be evaluated by a lattice model. Second, the constraints on the placement of the A and B segments in the domain system (to the inside and outside regions of the domains, respectively) also reduces the entropy of the domain system relative to a random mixture. This entropy difference will be termed the "restricted volume" entropy difference ΔS_v and will be evaluated by generation of the applicable chain statistics with the diffusion equation. Third, the perturbation of chain dimensions in the domain system from their random-flight values also decreases the entropy. This will be called the "elasticity entropy" difference ΔS_{el} and will be taken from standard elasticity theory.

The enthalpy difference ΔH between the domain and random mixture systems will be taken as the heat of mixing of a simple mixture of A and B molecules, i.e., the fact that the component blocks are joined together in the block copolymer is ignored. In a random mixture of A–B molecules, the effect of the A–B junction on the relative number of like- and unlike-segment interactions can hardly extend more than a few segments away from the junction. The errors introduced by ignoring this effect are trivial compared to those inherent in the pair-interaction model (4) used to evaluate ΔH.

The "residual" interaction of the A and B segments at the domain surface will be treated as a surface free energy G_s and characterized by an interfacial tension γ. This interaction energy could, in principle, also be obtained from the pair-interaction model, but would require knowledge of the distribution of A and B segments in the interfacial region. It appears simpler to treat the interaction as a surface energy.

Placement Entropy Difference ΔS_p

We evaluate here the entropy difference between the random placement of one segment per molecule (the junction segment) on a lattice and the placement of the segment on a lattice of domains in which the segment is restricted to a domain surface. It is emphasized that at this point we are concerned with only one segment per molecule and are not concerned with the configurational entropy of the remaining segments.

A random mixture of N_{AB} copolymer molecules having σ_A and σ_B A and B segments, respectively, has $N_{AB}(\sigma_A + \sigma_B)$ total lattice sites available. The number of possible sites available for the first segment of the ith molecule after $i - 1$ molecules have been placed on the lattice is $(N_{AB} - i + 1)(\sigma_A + \sigma_B)$. Thus the total number of distinctive ways Ω_1 of placing one segment each of N_{AB} identical molecules on the lattice is

$$\Omega_1 = \frac{1}{N_{AB}!} \prod_{i=1}^{N_{AB}} (N_{AB} - i + 1)(\sigma_A + \sigma_B) = (\sigma_A + \sigma_B)^{N_{AB}}$$

and the entropy S_1 associated with Ω_1 is $S_1 = N_{AB}k \ln (\sigma_A + \sigma_B)$.

In the domain system, we assume that the interfacial region of a domain is divided into η cells, where η is the number of molecules in a domain, and only one AB junction will occupy the lattice sites within a cell, i.e., multiple occupancy by the origins of the A and B chains (their junction) is prohibited. This assumption appears reasonable since the density of segments from a given molecule is greatest near the chain origin and thus to maintain constant density the origins will tend to be as far away from one another as possible. If the junction segment is restricted to an interfacial region of thickness ΔR, then each molecule added to the system will fill $3\sigma_A \Delta R/R$ sites. This follows from the number of lattice sites on a domain surface $4\pi R^2 \Delta R/v$, where v is the volume required for a segment, and the domain volume $4/3 \pi R^3 = \eta \sigma_A v$. In this last equation, we have neglected the small contribution of B segments to the domain volume. After $i - 1$ molecules have been placed in domains, the number of sites available on the domain surfaces to place the ith molecule is merely $(N_{AB} - i + 1)(3\sigma_A \Delta R/R)$. If the probability of placing a junction segment on a lattice site within a cell is equivalent for all sites, then the number of ways Ω_2 of placing the junction segments of N_{AB} identical molecules on the domain surfaces becomes

$$\Omega_2 = \frac{1}{N_{AB}!} \prod_{i=1}^{N_{AB}} (N_{AB} - i + 1)(3\sigma_A \, \Delta R/R) = (3\sigma_A \, \Delta R/R)^{N_{AB}}$$

and the entropy becomes

$$S_2 = kN_{AB} \ln (3\sigma_A \, \Delta R/R) \tag{5}$$

This probably represents an upper limit to the placement entropy on the domain surface since it appears likely that our assumption of equal probability of all sites within a cell overestimates Ω_2 because of the tendency of the chain origins to avoid one another. The lower limit of Ω_2 is of the order of one representing the extreme case where the junction segments are fixed at specific sites on the surface. It will also be noted that we have neglected the minor contribution to the entropy in the domain system arising from the possible arrangements of domains in space. It is easy to show that the entropy gained from this source is of the order of k/η per molecule, and is negligible since η is a relatively large number for systems of interest.

The placement entropy difference ΔS_p equals $S_2 - S_1$ or

$$\Delta S_p = kN_{AB} \ln \frac{3\sigma_A \, \Delta R}{(\sigma_A + \sigma_B)R} \tag{6a}$$

when the lattice sites in a cell are equally accessible and

$$\Delta S_p = - kN_{AB} \ln (\sigma_A + \sigma_B) \tag{6b}$$

as the lower bound when only one site per cell can be occupied by the junction.

Restricted Volume Entropy Difference ΔS_v

In the preceeding section, the entropy change associated with the nonrandom placement (i.e., on the domain surface) of the junction segment was evaluated without regard to the configurational statistics of the remaining segments of the chain. In this section we consider the remaining segments and determine the change in entropy resulting when constraints are applied to keep the A segments within the domain and the B segments outside. As has been mentioned previously, the diffusion equation offers a means to evaluate this entropy change; the constraints become boundary values in the solution. In the present case, since we wish to remove those configurations which have chain elements across the domain boundary, the boundary is taken as a completely absorbing barrier. For the A chains, the diffusion equation gives (7) the probability $Q(\sigma_A; \bar{r}, \bar{r}', r < R)$ (per unit **volume) that** the second (free) end of the chain will be found at $\bar{r}$ when the **fixed end is at** $\bar{r}'$ and all segments are at $r < R$ as

$$Q(\sigma_A; \bar{r}, \bar{r}', r < R) =$$

$$\sum_{m=0}^{\infty} \sum_{\beta} \exp\{-\sigma_A l^2 \beta^2/6\} \frac{j_m(\beta r) \, j_m(\beta r') \, (2m+1) \, P_m(\mu)}{[j_m'(\beta R)]^2}$$

where $j_m(z)$ is the spherical Bessel function of order m, $P_m(\mu)$ is the Legendre polynomial of order m, $\mu = \cos\theta$, $j_m'(z) = dj_m(z)/dz$, and the β's are the positive roots of $j_m(\beta R) = 0$.

Correspondingly, for the B chains the probability that the free end of the chain is at $\bar{r}$ when the fixed end is at $\bar{r}'$ and all segments are at $r > R$ is

$$Q(\sigma_B; \bar{r}, \bar{r}', r > R) = \frac{1}{4\pi r^{1/2} r'^{1/2}} \sum_{m=0}^{\infty} (2m+1) \, Pm(\mu) \cdot$$

$$\int_0^{\infty} \frac{C_{m+1/2}(ur) \, C_{m+1/2}(ur') \, \exp\{-\sigma_B l^2 u^2/6\}}{J^2_{m+1/2}(uR) + Y^2_{m+1/2}(uR)} \, du$$

where $J_m(z)$ and $Y_m(z)$ are Bessel functions of the first and second kinds, respectively, of order m, and

$$C_{m+1/2}(z) = J_{m+1/2}(z) \, Y_{m+1/2}(uR) - Y_{m+1/2}(z) \, J_{m+1/2}(uR)$$

In the above equations the free ends of the chains are at the particular locations $\bar{r}$. Since in the domain system the free A chain end may be anywhere within the domain and the free B end may be anywhere outside of the domain, the above equations are integrated over the accessible volume to remove the constraints on the free ends. Thus, the desired probability $P(\sigma_A; r', r < R)$ that all σ_A chain elements are inside the domain when the chain origin is at r' becomes

$$P(\sigma_A; r', r < R) = \int Q(\sigma_A; \bar{r}, \bar{r}', r < R) \, d^3\bar{r}$$

$$= 2 \sum_{i=1}^{\infty} (-1)^{i+1} j_0(i\pi r'/R) \exp\{-i^2 \pi^2 \sigma_A l^2/6R^2\} \qquad (7a)$$

The probability $P(\sigma_B; r', r > R)$ that all σ_B chain elements are outside the domain becomes

$$P(\sigma_B; r', r > R) = \int Q(\sigma_B; \bar{r}, \bar{r}', r > R) d^3\bar{r}$$

$$= 1 - \frac{R}{r'} \text{Erfc} \left[\left(\frac{3}{2\sigma_B l^2} \right)^{1/2} (r' - R) \right] \quad (7b)$$

where Erfc $(z) = 1 - \text{Erf}(z)$ and Erf (z) is the error function.

The loss in entropy due to restricted volume ΔS_v is from eqs. (7a) and (7b)

$$\Delta S_v = N_{AB}k \left[\ln P(\sigma_A; r', r < R) + \ln P(\sigma_B; r', r > R) \right] \quad (8)$$

and is easily evaluated given molecular sizes, domain sizes, and the placement r′ of the origins (junction segment) of the A and B chains.* Results are shown in Table I for various values of r′/R and for various ratios of the molecular sizes of the A and B blocks. For the domain radius R we have used $R = (4/3)(\sigma_A l^2)^{1/2}$ as established in a previous section.

Elasticity Entropy Difference ΔS_{el}

The positive interfacial free energy in the domain system will tend to cause an increase in domain dimensions. However, the increase in dimensions can occur only if the average dimensions of chains in the domain are increased. This increase in dimensions over the unperturbed random flight values can be characterized by α (4), the ratio of perturbed to unperturbed end-to-end chain distances, and gives an entropy decrease (4) (for chains that have one free end) of

$$\Delta S_{el} = -3/2 \, N_{AB}k \, (\alpha^2 - 1 - 2 \ln \alpha) \quad (9)$$

In an earlier section dealing with domain sizes, it was noted that other chain perturbations must occur if a constant density of segments in the domain is to be attained. We neglect the entropy decrease due to these perturbations for two reasons: (a) the fraction of the total number of

---

*It will be noted that the same R is used in the steps leading to eqs. (6) and (7) for both the A and B chains. Actually, in our model, the values of R used for the two types of chains should differ by ΔR, where ΔR is the thickness of the interfacial region ($\Delta R \simeq 2 |R - r'|$). However, electron micrographs of domains in block polymers show that the interface is relatively sharp, indicating that $\Delta R/R$ is small. The same R may then be used for both chains and also identified as the domain radius. In obtaining eq. (7b), we have neglected the volume excluded to a B chain by domains surrounding the one being considered. This can be shown to introduce a negligible error for the present model.

TABLE I

Restricted Volume Entropy Decrease

r'/R	$P(\sigma_A; r', r < R)$	$\dfrac{\sigma_B l^2}{\sigma_A l^2} \sim \dfrac{M_B}{M_A}$	$P(\sigma_B; r', r > R)$	$\Delta S_v/N_{AB}k$
0.90	0.0914	1	0.284	−3.7
		6	0.175	−4.2
		10	0.158	−4.3
		∞	0.100	−4.7
0.95	0.0440	1	0.142	−4.4
		6	0.0875	−4.8
		10	0.079	−4.9
		∞	0.050	−5.4

segments that must move to achieve constant density and the movement required (perturbation) are both small and (b) a satisfactory treatment of this (small) entropy decrease is not apparent.

Enthalpy Change ΔH

It has been mentioned that the enthalpy change for domain formation from a random mixture of block copolymer molecules will be taken as the negative of the heat of mixing of a simple mixture of the component blocks of the block copolymer. The pair-interaction model then gives ΔH as

$$\Delta H = N_{AB}kT\chi\phi_A \tag{10}$$

where χ is the Flory interaction parameter (4) and ϕ_A is the volume fraction of A segments, $\phi_A = \sigma_A/(\sigma_A + \sigma_B)$.

Surface Free Energy

If the residual interaction of the A and B segments at the domain interface is characterized by an interfacial energy γ, the total surface free energy of the domain system becomes $G_s = \eta_d 4\pi R^2\gamma$, where η_d is the total number of domains. Using our previously established relationship between R and $(\sigma_A l^2)^{1/2}$, i.e., $R = (4/3)(\sigma_A l^2)^{1/2} = (4/3)\,\alpha K M_A^{1/2}$, we obtain

THEORY OF BLOCK COPOLYMERS. I   93

$$G_s = \frac{9}{4} \frac{N_{AB} M_A^{1/2} \gamma}{\overline{A} \rho \alpha K} \tag{11}$$

where $\overline{A}$ is Avogadro's Number and ρ is the density.

Free Energy of Domain Formation ΔG_d

The free energy change associated with the formation of domains from a random mixture of A–B molecules is from eqs. (6a), (8), (9), (10), and (11).

$$\Delta G_d/N_{AB} kT = \ln \left(\frac{\sigma_A + \sigma_B}{3\sigma_A} \right) \left(\frac{R}{\Delta R} \right) - \ln P(\sigma_A; r', r < R)$$

$$- \ln P(\sigma_B; r', r > R) + \frac{3}{2}(a^2 - 1 - 2 \ln a) + \frac{9 M_A^{1/2} \gamma}{4 a K \overline{A} \rho kT}$$

$$- \chi \sigma_A /(\sigma_A + \sigma_B) \tag{12}$$

where we have used the expression for the minimum placement entropy difference, i.e., eq. (6a).

The chain expansion parameter a appears only in the terms relating to the interfacial energy (tending to increase a) and the elastic free energy (tending to decrease a). The equilibrium value a_m is obtained by differentiation and is

$$a^3{}_m - a_m = \frac{3}{4} M_A^{1/2} \gamma / K k \rho \overline{A} T \tag{13}$$

which then leads to the minimum value of ΔG_d as

$$(\Delta G_d/N_{AB} kT)_{min} = \ln \left(\frac{\sigma_A + \sigma_B}{3\sigma_A} \right) \left(\frac{R}{\Delta R} \right) - \ln P(\sigma_A; r', r < R) -$$

$$\ln P(\sigma_B; r', r > R) + \frac{9}{2}(a^2{}_m - 1) - 3 \ln a_m - \chi \sigma_A /(\sigma_A + \sigma_B) \tag{14}$$

In order to evaluate a_m, we require values of the interfacial tension γ. As far as the author is aware, only one report of the interfacial tensions between pairs of polymers has appeared in the literature (8). The polymer pairs and the results reported were nylon–polystyrene (5–6 dynes/cm), nylon–polyethylene (6.4 dynes/cm), and polyethylenephthalate–polyethylene (15 dynes/cm). In view of the lack of data on other systems of interest, we shall merely show a_m for several representative val-

TABLE II

Equilibrium Expansion Factors α_m

M_A	α_m		
	$\gamma = 1$ dyne/cm	$\gamma = 5$ dynes/cm	$\gamma = 15$ dynes/cm
10^3	1.04	1.18	1.40
10^4	1.12	1.43	1.86
3×10^4	1.20	1.62	2.16
10^5	1.31	1.87	2.56

ues of γ. Table II shows such α_m data for $\gamma = 1$, 5, and 15 dynes/cm for various values of M_A. In the use of eq. (13), we have taken T = 400°K, $\rho = 1$ g/cm^3 and K = 7.5×10^{-9} (an average value from a number of investigators (9–13) for polystyrene).

PREDICTED RADII OF POLYSTYRENE DOMAINS

With the data in Table II and with eq. (4), we may predict the size of domains as a function of molecular weight. Figure 4 shows such results for polystyrene domains (the results shown are calculated for 400°K but they are not very sensitive to temperature). The curves are nearly linear on a log–log plot, and give the following values of the exponent β in an equation of the type R = kM^β as a function of γ.

γ, dynes/cm	β
1	0.55
5	0.60
15	0.65

There are few data concerning domain sizes in the literature and no data concerning γ for those polymers for which domain sizes have been reported. Hence, the direct comparison of theory and experiment is impossible at the present time. However, it is still of interest to see if the theoretical results agree with experiment for reasonable values of γ. The radii of the domains shown in Figure 1 are approximately 130 Å. This size would be predicted if γ were slightly less than 1 dyne/cm, a not unreasonable value for this system since the domains shown were formed in the presence of a third component which was a good solvent for both component blocks and hence might be expected to lead a low value for the interfacial free energy. However, this "agreement" may

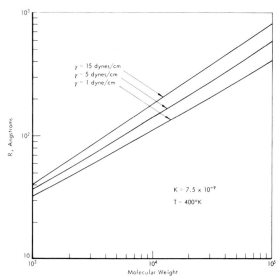

Fig. 4. Predicted polystyrene domain radii as a
function of molecular weight.

be somewhat fortuitous since this theory does not treat the possible effects on domain size of the third (solvent) component that is later removed.

CRITERION FOR DOMAIN FORMATION

It is, of course, necessary that ΔG_d be negative if domains are to form. Thus, from equation (14) the criterion for domain formation is

$$\chi \sigma_A/(\sigma_A + \sigma_B) > \ln\left(\frac{\sigma_A + \sigma_B}{3\sigma_A}\right)\left(\frac{R}{\Delta R}\right) - \ln P(\sigma_A; r', r < R)$$

$$- \ln P(\sigma_B; r', r > R) + \frac{9}{2}(a^2_m - 1) - 3\ln a_m \tag{15}$$

The left-hand side of this inequality is almost directly proportional to the molecular weight of the A block (when $\sigma_B \gg \sigma_A$, as assumed), while the right-hand side is only a slowly varying function of molecular weight. Thus, there will be a critical molecular weight of the A block above which the inequality is satisfied and domains will form. However, if predictions are to be made of critical molecular weights, values of χ (as well as γ, K, etc., must be known). Unfortunately, there are few literature data of χ for pairs of polymers and the few results that have been given differ greatly among themselves. For example, two results have been reported for the polybutadiene–polystyrene system (14,15). The reported values of $\chi/\overline{V}$, where $\overline{V}$ is the molar volume, differ by a factor of more than 200, and are obviously of little value in making predictions. In order to circumvent this problem, we shall compare the ratios of predicted critical molecular weights for domain formation with predicted critical molecular weights for phase separation of a simple mixture of A and B homopolymers. In this way, we eliminate the necessity that χ be known.

For a simple mixture of homopolymers, Flory-Huggins theory (4,16) gives the free energy of mixing as

$$\Delta G_m = kT(n_A \ln \phi_A + n_B \ln \phi_B + \chi n_B \phi_A) \tag{16}$$

where n_A and n_B are the numbers of A and B molecules, respectively. Since we wish to compare phase separation under conditions which are similar to those that occur in domain formation (equal numbers of A and B molecules) we take $n_A = n_B$ and find that $\Delta G_m = 0$ when

$$\chi \sigma_A/(\sigma_A + \sigma_B) = \ln \left\{ \frac{(\sigma_A + \sigma_B)^2}{\sigma_A \, \sigma_B} \right\} \tag{17}$$

If we now divide eq. (15) by eq. (17), we find the critical ratio of $\sigma_A{}^d$ for domain formation to $\sigma_A{}^m$ for phase separation of the simple mixture to be

$$\frac{\sigma_A{}^d}{\sigma_A{}^m} = \frac{\ln \left(\frac{\sigma_B}{3\sigma_A{}^d} \right) \left(\frac{R}{\Delta R} \right) - \ln P(\sigma_A; r', r < R) - \ln P(\sigma_B; r', r > R)}{\ln (\sigma_B/\sigma_A{}^m)}$$

$$+ \frac{\frac{9}{2}(\alpha_m{}^2 - 1) - 3 \ln \alpha_m}{\ln (\sigma_B/\sigma_A{}^m)} \tag{18}$$

where we have taken $\sigma_B \gg \sigma_A$. In Table II, we see that for molecular weights M_A below about 10^4 and for γ less than 5 dynes/cm, α_m will be less than about 1.5. Thus, under these conditions, the term $(9/2)(\alpha_m{}^2 - 1) - 3 \ln \alpha_m$ will vary between 0 and 4.41. In Table I we see that $\ln P(\sigma_A; r', r < R) + \ln P(\sigma_B; r', r > R)$ ($\equiv \Delta S_v/N_{AB}k$) is not a sensitive function of the ratio σ_B/σ_A and for $\Delta R/R = 0.10$ ($r'/R = 0.95$) their sum may be taken as -5.0. With these values eq. (18) becomes after collecting numerical values

$$\frac{\sigma_A{}^d}{\sigma_A{}^m} = \frac{\ln (\sigma_B/\sigma_A{}^d) + (6.3 - 10.7)}{\ln (\sigma_B/\sigma_A{}^m)} \tag{19}$$

where the term $(6.3 - 10.7)$ represents a range of values arising when γ is between 0 and 5 dynes/cm. For σ_B in the range $10^2 - 10^4$ ($M_B \approx 10^4 - 10^6$), eq. (19) indicates that the ratio of critical molecular weights will be between about $2.5 - 5$ (and would be even larger for larger values of γ). The much larger molecular weight of the A block required for domain formation than for simple phase separation is, of course, the consequence of the additional configurational constraints and interfacial energies of a domain system.

In his studies on the miscibility of polystyrene and polybutadiene, Paxton (15) reported that equal weight mixtures of polystyrene of $M_n = 2720$ and polybutadiene of $M_n = 1100$ (both polymers unfractionated) were not miscible. Assuming the applicability of the Flory-Huggins equation [eq. (16)] here, the minimum value of χ/M_B required for Paxton's results is 2×10^{-3}. With this minimum value of χ, we then find the results shown in Table III for the critical molecular weights M_c of polystyrene for phase separation from polybutadiene when <u>equal numbers of molecules</u> are mixed and the molecular weight of the polybutadiene is as shown.

TABLE III

Critical Molecular Weights for Phase Separation of Polystyrene and Polybutadiene $\chi/M_B = 2 \times 10^{-3}$

M_B polybutadiene	M_c polystyrene
10^4	1100
10^5	1970
10^6	2900

Since we have predicted that the critical molecular weights for domain formation would be at least $2.5 - 5$ times larger than for simple phase separation, we would then predict that the critical molecular weights for domain formation in a styrene–butadiene block copolymer would be between about 5000 and 10,000 when the polybutadiene block molecular weight is of the order of 50,000. Data presented by Holden, Bishop and Legge (1) give some confirmation to these predictions. Their data show that the tensile strengths of A–B–A block copolymers of styrene and butadiene change from 150 psi for a $6000 - 81,000 - 6000$ molecular weight polymer to 3350 psi for a $10,000 - 53,000 - 10,000$ molecular weight polymer, i.e., a 20-fold increase in tensile strength when the polystyrene molecular weight is changed only from 6000 to 10,000. We interpret these data as evidence for the onset of domain formation in that molecular weight range, as predicted by present theory.

Although their data were obtained with A—B—A block copolymers and present theory deals with A—B polymers, it will be shown in a following paper that the thermodynamics of the A—B—A system is not greatly different from the A—B system and hence the comparison of critical molecular weights from a theory of A—B polymers with data from A—B—A polymers is valid.

The author expresses his appreciation to Dr. S. Davison of these laboratories for many stimulating discussions about block copolymers and to Mr. N. 'A. Ross for the electron micrograph, Figure 1.

References

(1) G. Holden, E. T. Bishop, and N. R. Legge, paper presented at the International Rubber Conference sponsored by the Institution of the Rubber Industry, Brighton, England, 1967.

(2) C. Sadron, Pure Appl. Chem., 4, 347 (1962).

(3) H. Hendus, K. H. Illers, and E. Ropte, Kolloid-Z. Z. Polymere, 216-217, 110 (1967).

(4) P. J. Flory, Principles of Polymer Chemistry, Cornell Univ. Press, Ithaca, N. Y., 1953.

(5) E. A. Di Marzio, J. Chem. Phys., 42, 2101 (1965).

(6) D. J. Meier, J. Phys. Chem., 71, 1861 (1967).

(7) H. S. Carslaw and J. C. Jaeger, Conduction of Heat in Solids, Clarendon, Oxford, 1959.

(8) D. C. Chappelear, paper presented at Amer. Chem. Soc. Meeting, Chicago, September, 1964.

(9) J. Oth and V. Desreux, Bull. Soc. Chim. Belges., 66, 303 (1957).

(10) H. G. Elias and O. Etter, Makromol. Chem., 66, 56 (1963).

(11) A. R. Shultz and P. J. Flory, J. Polymer Sci., 15, 231 (1955).

(12) T. A. Orofino and J. W. Mickey, J. Chem. Phys., 38, 2512 (1963).

(13) G. V. Schulz and H. Baumann, Makromol. Chem., 60, 120 (1963).

(14) G. Allen, G. Gee, and J. P. Nicholson, Polymer, 1, 56 (1960).

(15) T. R. Paxton, J. Appl. Polymer Sci., 7, 1499 (1963).

(16) M. L. Huggins, J. Phys. Chem., 46, 151 (1942).

COMMENTARY

Reflections on "Theory of the Interface between Immiscible Polymers," by Eugene Helfand and Yukiko Tagami, *J. Polym. Sci., Polym. Lett.,* 9, 741 (1971)

EUGENE HELFAND

44 Riceman Road, Berkeley Heights, NJ 07922

Because of the wide interest in polymers at Bell Laboratories I had decided in 1969 to switch into that field. I must add that I was also attracted by the fact that the pace of advances in polymer theory in the 1960s seemed to have slowed in comparison to earlier decades, so I expected there might be great opportunity. That fundamental interest in polymers had waned is attested to by a story that De Gennes subsequently told me. He had actually begun to work on polymers in the mid 60s. When he presented his work and tried to interest experimentalists in doing related experiments he found that his research evoked little response. This was in contrast to what occurred in response to work he had begun on liquid crystals. For this reason, the latter topic commanded his attention first. He returned to polymers only late in the decade.

My choice of subject matter for research in polymer science was strongly influenced by the desire to make my work relevant to industry, and AT&T in particular. To me this meant that I should try to steer away from solution properties and toward the bulk state. It was evident that blends were becoming an important way of modifying polymers, e.g., impact modified glasses. Also, there was burgeoning interest in block copolymers. My feeling was that the interface had to play an important role in determining the properties of both these types of materials.

During my investigation of prior research on polymer interfaces I came across a curious report[1] asserting that polymer/polymer interfaces were ob-served to be thousands of Ångstroms. It is said that everyone believes an experiment except the author, while no one believes a theory, except the author. So I believed the result and, to see how it could be, I tried some dimensional analysis. My feeling was that the interfacial scale had to be proportional to the effective monomer length, b. To find such a large interface one needed a very large dimensionless factor. One immediately thinks of the degree of polymerization, N. It was hard to see how N could play a major role, since the interface would have a structure even with infinite degree of polymerization. On the other hand, the interaction parameter, χ, had to be important. For the result to be independent of the way the monomer is defined,[2] the combination of b and χ had to be $b/\chi^{1/2}$. However, this combination still could not yield a thousand Ångstrom interface.

I had studied Edward's approach to the excluded volume problem using functional integration,[3] a technique I had already used on two problems in critical phenomena. (Mel Lax once opined that when you are young you learn a few techniques, and then you use them throughout your career, no matter what the problem.) I proceeded with a functional integral formulation for random walks near the interface, but included the repulsion between unlike polymers. I was stunned, however, to discover that the resulting equation had no solution. The interface seemed to be falling apart. The problem was resolved when I realized that this amounted to the creation of a low overall density region between the phases, strongly disfavored by the incompressibility and cohesion of bulk polymer. In dilute solution this was not a stringent requirement, since the regions of low polymer concentration could be compensated by higher solvent concentration. What was necessary for bulk polymer was to put in a term representing the free energy increase of the slightly lower density in the interface. For zero compressibility this amounts to a constraint, namely, the sum of the reduced density of the two polymers must remain constant. Now this is so self-evident I can not imagine how it ever puzzled me.

Once we had the result we sought a more physical explanation, namely, the self-consistent-field argument used to derive the fundamental equation in the accompanying reprint. At a later time, trying to explain the results on an even simpler level for an article in *Accounts of Chemical Research,*[4] I came up with a more physical explanation yet. This constant effort to make one's work more and more comprehensible is a valuable exercise not only for one's colleagues, but even more so for the author.

My guess is that the observance of ultra-wide interfaces[1] was probably due to a poor preparation of the interface, with some mechanical mixing having occurred. (This brings one back to the question of how to rank the following in order of increasing scientific trustworthiness: a) experimentalists; b) theoreticians; c) oneself.)

Let me reflect on how this work on polymer/polymer interfaces and my subsequent research on block copolymers fit into the Zeitgeist. In the early 70s there was a sudden understanding of the way that polymer physics related to the field theoretic approaches already entrenched as the way to handle other condensed matter systems. Fortunately, many of the advances could be explained in physical terms, so the theoreticians and experimentalists remained strongly coupled. The era also saw a much greater theoretical emphasis placed on bulk polymer problems, especially interfaces and other inhomogeneous circumstances. The difficulties of basic polymer science in obtaining support, recognition, personnel, etc., continued into the 80s. A number of factors served to improve the climate for such research in the 80s and up to the present. They include the turn of science toward societal relevance; the growth of material science and "complex fluids" as disciplines; Flory's and De Gennes' Nobel prizes; the entry of prominent physicists into the field; and efforts on the part of the polymer community to increase awareness of the challenging and important problems.

REFERENCES AND NOTES

1. S. S. Voyutskii, A. N. Damenskii, and N. M. Fodiman, *Mekhan. Polimerov, Akad. Nauk Latv., SSR,* **3,** 446 (1966).
2. E. Helfand and A. M. Sapse, *J. Chem. Phys.,* **62,** 1327 (1975).
3. S. F. Edwards, *Proc. Phys. Soc.,* **85,** 613 (1965).
4. E. Helfand, *Acc. Chem. Res.,* **8,** 295 (1975).

PERSPECTIVE

Comments on "Theory of the Interface between Immiscible Polymers," by Eugene Helfand and Yukiko Tagami, *J. Polym. Sci., Polym. Lett.*, 9, 741 (1971)

GLENN H. FREDRICKSON

Departments of Chemical Engineering and Materials, University of California, Santa Barbara, CA 93106

The Letter by Helfand and Tagami[1] announcing their landmark theoretical analysis of the interfacial tension and structure of polymer–polymer melt interfaces has not only had lasting impact in guiding experimental studies, but provided the underpinnings for a whole host of theoretical advances in inhomogeneous polymeric systems. Prior theoretical tork on polymer interfaces had dealt with adsorbed layers from solution,[2] but largely overlooked the important case of polymer melt interfaces. The particular case treated by Helfand and Tagami, the *internal* polymer-polymer melt interface, has proved to be of increasing importance in the intervening twenty-five years as polymer technologists have turned to blends in order to meet their materials performance needs.

Helfand and Tagomi adopted the self-consistent field formalism of Edwards to the case of a symmetrical, flat interface between two molten, flexible homopolymers. Other than energetic contacts between type A and B monomers, described by a Flory χ parameter, they recognized the importance of cohesive forces that maintain nearly uniform density in polymer melts. To capture this, they included a second term in the mean field potential experienced by a monomer that provides a harmonic energy penalty for local (total) density variations. The harmonic "spring constant" was inversely related to the melt compressibility κ. Having prescribed the form of the interactions, Helfand and Tagami then followed the Edwards' procedure of generating statistical weights for the two types of chains via the solution of modified diffusion equations. The potentials "modifying" the two equations represented the total chemical potentials experienced by a monomer of each species (i.e.,

the sum of the dissimilar-interaction and cohesive potentials). To close the equations and render the formalism self-consistent, Helfand and Tagami then wrote bilinear expressions relating the partition functions obtained from the solution of the diffusion equations to the monomer species densities, which entered linearly in the species potentials.

Having mathematically framed the problem, a very important ingredient in the Helfand-Tagami work was the recognition that a full solution to the nonlinear diffusion equations was not required; rather, only the steady state (ground state approximation) solution was needed to capture the interfacial thermodynamics in the limit of infinite molecular weight. Helfand and Tagami found an *analytical* solution to the steady state problem in the incompressible limit, deduced the familiar "tanh" form of the interfacial composition profile, and extracted an interfacial width $\xi = 2b/(6\chi)^{1/2}$. The interfacial tension was obtained by a thermodynamic integration, yielding the famous formula:

$$\gamma = k_B T b \rho_0 (\chi/6)^{1/2} \tag{1}$$

where b is the statistical segment length and ρ_0 is the uniform number density of monomers. Helfand and Tagami then went on to use this formula to predict interfacial tensions of three polymer pairs for which the necessary data was available at the time and tensions had been experimentally measured. The agreement between the predicted and experimental values of the tension was remarkable.

There have been a number of important extensions of the Helfand–Tagami theory in the twenty five years since their initial letter appeared. A more detailed version of the theory was published the following year,[3] which discussed finite compressibility corrections and dilution effects associated with the addition of a third solvent component. A second follow-up paper by Helfand and Sapse[4] extended the approach to asymmetric melt interfaces in which the two pure polymer components differ in statistical segment lengths, b_K, and/or segment volumes, v_K (K = A or B). Of particular interest is that the pure-component invariant parameters $\beta_K^2 \equiv b_K^2/(6v_K)$, first highlighted by Helfand and Sapse, have very recently been linked with bulk and surface thermodynamic properties of polyolefins[5] and shown to be strongly correlated with the entanglement molecular weight in a large class of polymers.[6]

Another active area of research in recent years has been to extend the Helfand-Tagami theory to examine finite molecular weight corrections to eq. (1). The earliest work in this area was led by Helfand[7] and yielded $1/(\chi N)$ corrections to eq. (1) (N is the polymerization index); subsequent theoretical studies have refined the numerical prefactor in this correction.[8] A second contemporary development has been a reformulation of the Helfand–Tagami theory around the Kratky–Porod wormlike chain model.[9] This extension provides a quantitative means of assessing interfacial thermodynamics in situations where the intrisic backbone rigidity of the chains is comparable to the interfacial thickness. Such considerations have become increasingly important with the advent of high performance polymer blends containing conjugated or liquid crystalline components.

While the above discussion has emphasized the significance of Helfand and Tagami's work for the interpretation of experimental measurements of interfacial thermodynamics and on subsequent developments in the theory of polymer melt interfaces, perhaps its most far-reaching implications have been for inhomogeneous macromolecular systems such as *block copolymers* and *polymer brushes*. Indeed, in their original letter,[1] Helfand and Tagami hinted that their formalism could be used to examine block copolymer microphases. This was followed by an important series of papers by Helfand and Wasserman making this connection explicit.[10] Helfand later went on to develop a very elegant and general mean field theory of inhomogeneous polymeric systems[11] that serves as the framework for most theoretical calculations on block copolymers today. In particular, the extensive numerical studies of single and multicomponent block copolymer phase diagrams carried out by Hong, Noolandi, and Whitmore[12] were based on the Helfand formalism. Most recently, Matsen and Schick[13] developed a powerful new nu-

merical implementation of the Helfand self-consistent field theory that lends itself to the study of complex block copolymer microphases such as the gyroid phase and the ordered bicontinuous double-diamond phase. Finally, it should be noted that the active field of polymer brush physics[14] is indebted to the Helfand formalism for several of its theoretical tools. In particular, the Scheutjens–Fleer numerical scheme,[15] commonly employed to address problems of tethered chains in colloid science, was likely inspired by the theoretical developments of Helfand and his coworkers.

Overall, it is apparent that Helfand and Tagami's 1971 *Journal of Polymer Science* Letter will continue to broadly impact polymer science and engineering practice for some time to come. Their beautiful contribution should serve to encourage future generations of polymer theorists.

REFERENCES AND NOTES

1. E. Helfand and Y. Tagami, *J. Polym. Sci., Polym. Letts.*, **9**, 741 (1971).
2. J. Pouchly, *Collection Czech. Chem. Commun.*, **28**, 1804 (1963); E. A. DiMarzio, *J. Chem. Phys.*, **42**, 2101 (1965); A. Vrij, *J. Polym. Sci., A-2*, **6**, 1919 (1968); S. F. Edwards, *Proc. Phys. Soc.*, **85**, 613 (1965); P. G. de Gennes, *Rep. Prog. Phys.*, **32**, 187 (1969).
3. E. Helfand and Y. Tagami, *J. Chem. Phys.*, **56**, 3592 (1972).
4. E. Helfand and A. M. Sapse, *J. Chem. Phys.*, **62**, 1327 (1975).
5. F. S. Bates and G. H. Fredrickson, *Macromolecules*, **27**, 1065 (1994).
6. L. J. Fetters, D. J. Lohse, D. Richter, T. A. Witten, and A. Zirkel, *Macromolecules*, **27**, 4639 (1994).
7. E. Helfand, S. M. Bhattacharjee, and G. H. Fredrickson, *J. Chem. Phys.*, **91**, 7200 (1989); D. Broseta, G. H. Fredrickson, E. Helfand, and L. Leibler, *Macromolecules*, **23**, 132 (1990).
8. H. Tang and K. F. Freed, *J. Chem. Phys.*, **94**, 6307 (1991); A. V. Ermoshkin and A. N. Semenov, *Macromolecules,* in press.
9. D. C. Morse and G. H. Fredrickson, *Phys. Rev. Lett.*, **73**, 3235 (1994).
10. E. Helfand and Z. R. Wasserman, in *Developments in Block Copolymers-1*, I. Goodman, Ed., Applied Science, NY, 1982, pg. 99.
11. E. Helfand, *J. Chem. Phys.*, **62**, 999 (1975).
12. K. M. Hong and J. Noolandi, *Macromolecules*, **14**, 727 (1981); J. D. Vavasour and M. D. Whitmore, *Macromolecules*, **25**, 5477 (1992).
13. M. W. Matsen and M. Schick, *Phys. Rev. Lett.*, **72**, 2660 (1994).
14. S. T. Milner, *Science*, **251**, 905 (1991).
15. J. M. H. M. Scheutjens and G. J. Fleer, *J. Phys. Chem.*, **83**, 1619 (1979).

POLYMER LETTERS   VOL. 9, PP. 741–746 (1971)

THEORY OF THE INTERFACE BETWEEN IMMISCIBLE POLYMERS

This letter describes a theory of the interface between immiscible polymers A and B. There are forces at work in the junction which tend to drive the A and B molecules apart, but this separation must be done in such a way as to prevent a gap from opening between the phases. Furthermore, we must balance these energetic forces on, let us say, an A molecule against, what may be termed, an entropic force. The latter is the tendency of A to penetrate into the B phase because of the numerous configurations of the A molecule which do so.

To solve the energy-weighted configurational problem we employ a mean field theory. From the walk statistics it is possible to determine density profiles of A and B. However the mean field used is a function of these densities, so the theory involves a self-consistent-field calculation.

Fundamental aspects of the statistical treatment of polymers near an interface, or in other inhomogeneous circumstances, have been developed by Pouchlý (1), DiMarzio (2), Vrij (3), and Edwards (4). Our theory builds on these principles.

For the present purposes we shall assume that the polymers A and B are similar in that they both have the following characteristics:

1) Degree of polymerization Z, which we eventually allow to approach infinity

2) Effective length b per monomer unit (typically 6 or 7 Å), chosen so that the mean-square end-to-end distance is Zb^2

3) Density ρ_0 when pure, typically 10^{-2} times Avogadro's number monomer units/cm^3

4) Compressibility κ about 5×10^{-11} cm^2/dyne.

For the calculations below we use the geometric mean when properties are not actually identical.

The measure of incompatibility between A and B is the parameter χ, familiar in the mixture theory of van Laar, Hildebrand, Flory, and Huggins. For polymer pairs, χ has generally not been too accurately determined. From various sources we estimate χ as 0.03 for polystyrene (PS)/polybutadiene (PB) (5), and $\chi \approx 0.01$ for polystyrene/polymethyl methacrylate (PMMA) (6). These are typical values for moderately incompatible polymers.

The effective field $w_A(\mathbf{r})$ on a segment of polymer A is the work of adding that segment at the point $\mathbf{r}$ where the densities are $\rho_A(\mathbf{r})$, $\rho_B(\mathbf{r})$, less the work of adding the segment to bulk A. The fact that contacts between A and B are less favorable than AA contacts leads to a work contribution

$$kT \chi \rho_B(\mathbf{r})/\rho_0$$

741

742   POLYMER LETTERS

This tends to drive the two phases apart, and would open a hole between the phases if it were not for the force of cohesion. By this we mean the resistance of the polymeric material to deviations of the total density, $\rho_A(\mathbf{r}) + \rho_B(\mathbf{r})$, from ρ_0. Thus, the work also contains a term

$$(1/\rho_0^2 \kappa)[\rho_A(\mathbf{r}) + \rho_B(\mathbf{r}) - \rho_0]$$

The inverse compressibility is the proper measure of the tendency to attract polymer into regions with $\rho_A + \rho_B < \rho_0$, and repel polymer when the total density is greater than ρ_0.

So far, $w_A(\mathbf{r})$ has been regarded as dependent only on the local densities $\rho_K(\mathbf{r})$, $K = A, B$. As a result of the finite range of the interactions there are also nonlocal terms (proportional to $\nabla^2 \rho_K(\mathbf{r})$ to a first approximation). However, for the systems under consideration, the thickness of the surface will turn out to be many times the typical range of interatomic forces. These nonlocal terms are then minor compared with the nonlocal effect associated with the continuity of the chain, and will henceforth be neglected.

Up to this point we have discussed the nonconfigurational work of adding a segment of A. If we wish to add Zt consecutive segments ($0 < t \leq 1$) we must solve a configurational problem. Let $q_A(\mathbf{r},t)$ be the ratio of the partition function of an A polymer of Zt segments which begins at $\mathbf{r}$ (and is in the effective field) to the partition function in the bulk A phase. This function satisfies a modified diffusion equation (1,2,4):

$$\frac{1}{Z}\frac{\partial}{\partial t} q_A(x,t) = \frac{b^2}{6}\frac{\partial^2}{\partial x^2} q_A(x,t) - \frac{w_A(x)}{kT} q_A(x,t) \tag{1}$$

$$\frac{w_A(x)}{kT} = \chi \frac{\rho_B(x)}{\rho_0} + \zeta\left[\frac{\rho_A(x)}{\rho_0} + \frac{\rho_B(x)}{\rho_0} - 1\right] \tag{2}$$

$$\zeta \equiv 1/\rho_0 kT\kappa \tag{3}$$

with the initial condition

$$q_A(x,0) = 1. \tag{4}$$

We adopt a geometry where the surface centers at $x = 0$, the region to the right is rich in A, and that to the left is rich in B. The boundary conditions then are

$$q_A(\infty,t) = 1 \tag{5}$$

$$q_A(-\infty,t) = 0 \tag{6}$$

The quantity $q_A(\mathbf{r},t)$ is also the ratio of the density at $\mathbf{r}$ of the initial segments of A chains of length (in units of segments) Zt to this density in bulk A. The ratio of the density at $\mathbf{r}$ of the chain segment Zt, down a chain Z segments long, to this density in bulk is

$$q_A(\mathbf{r},t)\,q_A(\mathbf{r},1-t).$$

This follows from the fact that the segment at Zt may be regarded as the origin of two independent walks, one of length Zt, the other of length $Z(1-t)$. The overall segment density at $\mathbf{r}$ is

$$\rho_A(\mathbf{r}) = \rho_0 \int_0^1 dt\; q_A(\mathbf{r},t)\,q_A(\mathbf{r},1-t) \tag{7}$$

Equations (1–7), with analogous relations for B, are the self-consistent set to be solved for the density profiles.

These equations, generally too difficult to handle analytically, can be simplified by virtue of two physical aspects of the problem. The first is that in bulk, correlations between remote parts of the same chain are unimportant, so that we can study the asymptotic properties for degree of polymerization approaching infinity. Then q_K become independent of t, and the equations reduce to

$$\rho_K(x) = \rho_0 q_K^2(x) \tag{8}$$

$$0 = \frac{b^2}{6}\frac{d^2}{dx^2} q_K - \chi q_{K'}^2 q_K - \zeta(q_K^2 + q_{K'}^2 - 1)q_K,$$

$$K = A,B \text{ and } K' = B,A. \tag{9}$$

The second simplification arises from the fact that χ/ζ is quite small, 10^{-3} or less. A solution to eqs. (8) and (9), valid for $\chi/\zeta \to 0$, is

$$\rho_A(x) = \rho_0 \{1 + \exp[-2(6\chi)^{1/2} x/b]\}^{-1} \tag{10}$$

$$\rho_B(x) = \rho_0 \{1 + \exp[2(6\chi)^{1/2} x/b]\}^{-1} \tag{11}$$

As a measure of the surface thickness we may take

$$\rho_0 [d\rho_A/dx|_{x=0}]^{-1} = 2b/(6\chi)^{1/2}.$$

For PS/PB this is about 30 Å, which we would say is large enough to be comfortably within the limit of applicability of mean field theory. For PS/PMMA we find a surface thickness of about 50 Å.

To determine the surface free energy per unit area (interfacial tension) we must devise some coupling procedure which continuously brings the system from a state with known free energy to the state of interest. As a coupling

TABLE I

Comparison of Calculated and Measured Interfacial Tensions

Polymer Pair	Interfacial Tension, dyne/cm		χ	b, Å (geom. mean)	Spec. vol., cm³/mole (geom. mean)
	calc.	exp. (8)			
PS/PMMA	1.0	1.5	0.01	6.5	96
PMMA/PnBMA	2.0	1.8	0.07	6.1	114
PnBMA/PVA	1.9	1.9	0.05	6.3	107

scheme we employ a scaling of the parameter χ from 0+ to the full value. The surface free energy is determined by integrating the surface energy along this path:

$$F_{surf} = \frac{kT\chi}{\rho_0} \int_0^1 d\lambda \int_{-\infty}^{\infty} dx \, \rho_A(x;\lambda\chi)\rho_B(x;\lambda\chi) \tag{12}$$

For the density profiles of eqs. (10) and (11) we find

$$F_{surf} = (\chi/6)^{1/2} b\rho_0 kT. \tag{13}$$

There have been measurements recently of the interfacial tension between polymer melts by Roe (7) and Wu (8). A detailed comparison with the theory is difficult because of the uncertainty of values for the parameter χ, and because of the present limitation of the theory to polymers with identical properties. Doing our best to circumvent these limitations we have attempted theoretical estimates for the polymer pairs PS/PMMA, PMMA/PnBMA, and PnBMA/PVA (PnBMA = poly-n-butyl methacrylate and PVA - polyvinyl acetate). The results are summarized in Table I. The values of χ are based on measurement of the Hildebrand δ parameter by swelling (6). The temperature is taken as 150°C (although b and δ are determined from lower temperature data).

In a forthcoming paper we shall present detailed derivations of these results, extension to finite χ/ζ, inclusion of nonlocal terms, and a discussion of the effect of adding a third monomeric component. Topics under investigation are molecular weight effects, extension to nonsymmetric A/B pairs, and a similar study of block copolymer microphases.

The authors are grateful to Dr. R. J. Roe for enlightening discussions and suggestions.

References

(1) J. Pouchlý, Collection Czech. Chem. Commun., 28, 1804 (1963).

(2) E. A. DiMarzio, J. Chem. Phys., 42, 2101 (1965).

(3) A. Vrij, J. Polymer Sci., A-2, 6, 1919 (1968).

(4) S. F. Edwards, Proc. Phys. Soc., 85, 613 (1965).

(5) D. McIntyre, "Thermodynamic and Morphological Parameters in Triblock Polymers," XXIII IUPAC Congress, Boston, July 29, 1971.

(6) D. Mangaraj, S. K. Bhatnagar, and S. B. Rath, Makromol. Chem., 67, 75 (1963); D. Mangaraj, S. Patra, and S. Rashid, Makromol. Chem., 65, 39 (1963); D. Mangaraj, S. Patra, P. C. Roy, and S. K. Bhatnagar, Makromol. Chem., 84, 225 (1965).

(7) R. J. Roe, J. Colloid Interfac. Sci., 31, 228 (1969).

746   POLYMER LETTERS

(8) S. Wu, J. Phys. Chem., $\underline{74}$, 632 (1970).

Eugene Helfand
Yukiko Tagami

Bell Telephone Laboratories, Inc.
Murray Hill, New Jersey 07974

Received August 6, 1971

COMMENTARY

Reflections on "Quasielastic Scattering by Dilute Polymer Solutions," by A. Z. Akcasu and H. Gurol, *J. Polym. Sci., Polym. Phys. Ed.,* 14, 1 (1976)

A. Z. AKCASU

Dept. of Nuclear Engineering, University of Michigan, Ann Arbor, MI 48109

It is a great honor for me to take part in the 50th birthday celebration of the *Journal of Polymer Science,* with a paper that was my first contribution to the field of polymers. It was originally presented at the IUPAP meeting at Budapest in the Summer of 1975, and then published in *JPS* in 1976.

I became interested in polymers in 1973 when I attended a lecture on polymer solution dynamics by professor G. C. Summerfield, who had just came back from his sabbatical leave in France at Saclay, where exciting progress was being made on polymers. Previously, I had been working on the calculation of time-correlation functions in many-body systems, such as simple liquids and plasmas, using Zwanzig–Moro projection operator formalism. In this formalism, an exact equation of motion for a set of dynamical variables describing a macroscopic state of a many-body system is derived from the Lionville equation with projection operator technique. This equation is referred to as the generalized Langevin (GLE) equation due its resemblance to the original Langevin equation for Brownian diffusion. The GLE is used, among its other applications, to obtain exact equations for the time-correlation functions of the dynamical variables, in particular, for the density–density time correlation function, which is the intermediate scattering function $S(q, t)$, in the Fouries space.

After hearing Professor Summerfield's talk, I began thinking about the possibility of applying the projection operator formalism to polymer solutions, by choosing the monomer density $\rho(q, t)$ as the dynamical variable. Since the effect of the solvent on the polymer molecules is treated macroscopically in the conventional Kirkwood–Riseman description, as I had soon found out, the time evolution of the polymer configurations is known only in a probabilistic sense. One had first to obtain a Liouville-like equation which would assign a time dependence to $\rho(q, t)$ in an average sense, consistently with the approximate Kirkwood–Riseman description, in order to apply the projection operator formalism. It took me quite some time to obtain such an equation (eq. 14 in the paper) although I later found out that similar equations for other variables had already existed in the literature. Once an equation for the time evolution of $\rho(q, t)$ was obtained, it was easy to write down the GLE for $S(q, t)$ in terms of what is now called the first cumulant $\Omega(q)$, and the memory function. The expressions of $\Omega(q)$, given in equations 22 and 23 were the main contributions of this paper. They made it possible during the late 70s and early 80s to calculate the first cumulant, with and without pre-averaging the Oseen tensor, under various experimental conditions and for various chain architecture's as function of the wave number q, concentration and temperature. At this stage, I started collaborating with Dr. H. Gürol, who was then a graduate student in our department. We investigated together the variation of $\Omega(q)$ as function of q in the entire q-range, and found out that $\Omega(q)/q^2$ displays an S-shaped curve. In particular, we discovered that $\Omega(q)$ behaves as q^3 in the intermediate q-range, as was previously shown by de Gennes with scaling argument. However, we were able to obtain the numerical front factor also in our approach. We were excited about and encouraged by this observation. We estimated the front factor approximately as $\frac{1}{18}$ (see eq. 36) because we were not able to calculate the double integrals involved in the expression of $\Omega(q)$ in the intermediate q-region. About two years after the publication of this paper, Professor W. H. Stockmayer, whom I had not yet met, pointed out to me in a letter, among other things, that this front factor was exactly $\frac{1}{16}$. Since he had not sent me his derivations, I spent a whole night to verify his result, and wrote to him immediately that I also was able to reproduce it. Professor Stockmayer's and, later, Dr. C. C. Han's attention to this work contributed significantly to its early recognition.

In conclusion, while writing this paper, I experienced all the frustration, excitement, and joy of entering a new field without a proper background. I have never regretted taking this step.

PERSPECTIVE

Comments on "Quasielastic Scattering by Dilute Polymer Solutions," by A. Z. Akcasu and H. Gurol, *J. Polym. Sci., Polym. Phys. Ed.,* 14, 1 (1976)

CHARLES C. HAN

Polymers Division, National Institute of Standards & Technology, Gaithersburg, MD 20899

I was introduced to this topic when Professor Akcasu presented his work in the 1977 International Conference on Small Angle Scattering in Gatlinburg, Tennessee.

At that time, I was working on the dynamic light scattering of polymer solutions. Specifically, I was interested in the internal modes of single chain dynamics. From Professor Akcasu's presentation, it was clear that the q dependence of the characteristic frequency (or the first cumulant), $\Omega(q)$, and the shape of the intermediate scattering function, $S(q, t)$, are the two well-defined and measurable quantities which one should concentrate on. The idea of analyzing the relaxation spectrum in terms of multiple exponentials from the time correlation function was a formidable task at that time with a clipped single photon correlator or a mini computer based correlator/spectrum analyzer. The interpretation of scattering experiments in terms of $\Omega(q)$ seemed to be an attractive alternative.

I came back from the conference and read the Akcasu–Gurol paper, which describes a procedure of using the projection operator technique to obtain the density–density correlation function and its first cumulant of a Zimm chain, with and without preaveraging the Oseen tensor. Although its title referred to $S(q, \omega)$, the Fourier transform of the $S(q, t)$, the paper contained an exact expression of the $S(q, t)$ from which $S(q, \omega)$ was calculated. The $S(q, t)$ is directly measurable by dynamic light scattering and neutron spin-echo experiments. Hence, the first cumulant could be extracted with sufficient accuracy from the measured $S(q, t)$ in terms of its initial slope,

and the prediction in the Akcasu–Gurol theory as to the q-dependence of the $\Omega(q)$ could be tested.

I have to admit that the detailed calculation presented in the paper is abstract and some mathematical manipulations are beyond what I can follow, but the results are simple and revealing. The most important aspect of this paper is the use of the projection operator technique to project out the unobserved variables, and obtain directly the density-density correlation function which is measurable, instead of trying to obtain a more detailed description in terms of non-measurable quantities such as the space-time distribution of all monomers. This approach has provided a direct link between theories and experiments.

I looked into some data we had in hand at that time and called Ziya and told him that the first cumulant, $\Omega(q)$, when plotted as $\Omega(q)/q^2$ indeed showed an upturn toward q^3 dependence as the dR_g values became larger than one, as predicted in their paper. At that time, Ziya was on sabbatical leave at University of Maryland. He came to NIST (NBS then) the following day to discuss my findings. This was the beginning of our long-lasting collaboration. In the following year, Professor Julia Higgins from the Imperial College joined us, and we spent a lot of time discussing her neutron spin echo results which showed a transition from q^3 to q^2 behavior in the high-q region where $qL > 1$, also as predicted in the Akcasu–Gurol paper. I think those were exciting times in the development of polymer dynamics studied by scattering techniques. I remember very clearly a remark I made to Ziya at the time, "Ziya, your theory will be used in the interpretation of polymer dynamics for the next ten years." I didn't know I could be so wrong at that time. Twenty years later, the Akcasu–Gurol theory is still the theory for the interpretation of polymer dynamics.

This 1976 paper of Akcasu and Gurol is abstract and formal as are many other theoretical papers which aim at developing general formalisms. However, the importance of a paper is reflected in its subsequent impact and influence.

Direct Development

The physics of looking at the center of mass diffusion in the small q region ($qR_g < 1$) is reflected in its q^2 dependence of the relaxation rate. In the intermediate region ($qR_g > 1$ and $qL < 1$), the first cumulant calculation of Akcasu clearly shows the transition from the center of mass motion to the internal chain dynamics. This feature has also been demonstrated by deGennes[1] and Dubois–Voilette and deGennes[2] and also earlier by Pecora[3] when the appropriate limits are taken. However, the Akcasu–Gurol calculation demonstrated in a very simple and elegant way that not only should one observe this q^2 to q^3 (or q^4 for Rouse case) transition, but at even larger q region ($qL > 1$), the first cumulant should give a transition from q^3 (or q^4) back to q^2.[4] This is certainly consistent with the neutron spin–echo measurement.[5]

Another advantage of the A–G theory is that the first cumulant can be calculated for single chain dynamics with or without the excluded volume effect and with or without pre-averaging the Oseen tensor for hydrodynamic interaction.[4–8] This offers the opportunity for a direct test of these two crucial effects with experimental measurements which have remained important issues through the years (see also the March issue, **34**, p. 593 of the 50 years *JPS* article). From this comparison,[9–11] the deficiency of Oseen representation of the hydrodynamic interaction is clear. It was commented by H. Fujita[12] that this is one of the most significant findings in recent studies of dilute polymer solutions.

For the stiff chain case, the Akcasu–Gurol formalism has been extended by Schmidt and Stockmayer in the calculation of the first cumulant of a semi-flexible rod.[13–15]

Indirect Development

The Akcasu–Gurol approach was later extended to multi-component systems,[16] and polyelectrolyte solutions.[17,18] In 1986, in conjunction with the random phase approximation, the Akcasu–Gurol approach was extended to homopolymer blends and block copolymer melts. Some key theoretical and experi-

mental work[19–26] has established the importance of this theoretical foundation, especially in the study of block copolymer mixtures and block copolymer and homopolymer mixtures.

The most significant influence of this paper is probably the way we think when we deal with the subject of polymer dynamics in terms of coherent scattering techniques. Immediately, we think about the eigenmodes, the first cumulant and the intermediate structure factor. Most importantly, we think about the measurable quantities before we try to decompose the normal modes or even the space-time dependence of the monomers. Although, this step has bypassed the details of the exact space-time description of the monomers by only projecting out a simpler, however, measurable quantity, this has allowed a direct comparison of models and experiments.

With new developments of various constrained Laplace transformations together with new correlator technology for obtaining more precise long time correlation data, decomposition of relaxation modes in simple cases (two or three groups of modes) has been successful. Nevertheless, the projection operator technique is still the method of calculating the eigenmodes of the coupled dynamics of a multicomponent system which requires a matrix representation. The impact of using A–G theory in more complicated cases such as block copolymers, mixtures, and the bulk state is yet to come.

REFERENCES AND NOTES

1. P. G. deGennes, *Physics*, **3**, 37 (1967).
2. E. Dubois–Voilette and P. G. deGennes, *Physics*, **3**, 181 (1967).
3. R. Pecora, *J. Chem. Phys.*, **43**, 1562 (1965); **49**, 1032 (1968).
4. A. Z. Akcasu, M. Benmouna, and C. C. Han, *Polymer*, **13**, 409 (1980).
5. J. S. Higgins, K. T. Ma, K. Nicholson, J. B. Hayter, K. Dodgson, and J. A. Semlyen, **24**, 793 (1983).
6. M. Benmouna and A. Z. Akcasu, *Macromolecules*, **13**, 409 (1980).
7. M. Benmouna and A. Z. Akcasu, *Macromolecules*, **11**, 1187 (1978).
8. A. Z. Akcasu and M. Benmouna, *Macromolecules*, **11**, 1193 (1978).
9. C. C. Han and A. Z. Akcasu, *Macromolecules*, **14**, 1080 (1981).
10. T. P. Lodge, C. C. Han, and A. Z. Akcasu, *Macromolecules*, **16**, 1180 (1983).
11. Y. Tsunashima, N. Nemoto, and M. Kurata, *Macromolecules*, **17**, 425 (1984).
12. H. Fujita, "Polymer Solutions," Elsevier, (1990).

13. M. Schmidt and W. H. Stockmayer, *Macromolecules,* **17,** 509 (1984).

14. A. Z. Akcasu, B. Hammouda, W. H. Stockmayer, and G. Tanaka, *J. Chem. Phys.,* **85,** 4734 (1986).

15. M. Benmouna, A. Z. Akcasu, and M. Daoud, *Macromolecules,* **13,** 1703 (1980).

16. A. Z. Akcasu, B. Hammouda, T. P. Lodge, and C. C. Han, *Macromolecules,* **17,** 759 (1984).

17. A. Z. Akcasu, M. Benmouna, and B. Hammouda, *J. Chem. Phys.,* **80,** 2762 (1984).

18. M. Benmouna, A. Z. Akcasu, and B. Hammouda, *J. Chem. Phys.,* **85,** 1912 (1986).

19. L. Giebel, R. Borsali, E. H. Fischer, and G. Meier, *Macromolecules,* **23,** 4054 (1990).

20. M. Benmouna, H. Benoit, M. Duval, and A. Z. Akcasu, *Macromolecules,* **20,** 1987 (1987).

21. A. Z. Akcasu, M. Benmouna, and H. Benoit, *Polymer,* **27,** 1935 (1986).

22. A. Z. Akcasu, R. Klein, and B. Hammouda, *Macromolecules,* **26,** 4136 (1993).

23. B. Hammouda, *Macromolecules,* **26,** 4800 (1993).

24. C. Pan, M. Maurer, Z. Liu, T. P. Lodge, P. Stepanek, E. D. Von Meerwall, and H. Watanabe, *Macromolecules,* **28,** 1643 (1995).

25. Z. Liu, C. Pan, T. P. Lodge, and P. Stepanek, *Macromolecules,* **28,** 3221 (1995).

26. P. Stepanek and T. P. Lodge, *Macromolecules,* **29,** 1244 (1996).

JOURNAL OF POLYMER SCIENCE: Polymer Physics Edition   VOL. 14, 1–10 (1976)

Quasielastic Scattering by Dilute Polymer Solutions*

Z. AKCASU and H. GUROL, *Department of Nuclear Engineering,
The University of Michigan, Ann Arbor, Michigan 48105*

Synopsis

The scattering law $S(\mathbf{k},w)$ for dilute polymer solutions is obtained from Kirkwood's diffusion equation via the projection operator technique. The width $\Omega(k)$ of $S(\mathbf{k},w)$ is obtained for all k without replacing the Oseen tensor by its average (as is done in the Rouse–Zimm model) using the "spring-bead" model ignoring memory effects. For small $(ka\sqrt{N} \ll 1)$ and large $(ka \gg 1)$ values of k we find $\Omega = 0.195\, k^2/\beta\, a\eta_0\sqrt{N}$ and $\Omega = k^2/\beta\xi$, respectively, indicating that the width is governed mainly by the viscosity η_0 for small k values and by the friction coefficient ξ for large k values. For intermediate k values which are of importance in neutron scattering we find that in the Rouse limit $\Omega = k^4 a^2/12\beta\xi$. When the hydrodynamic effects are included, $\Omega(k)$ becomes $0.055\, k^3/\beta\eta_0$. Using the Rouse–Zimm model, it is seen that the effect of pre-averaging the Oseen tensor is to underestimate the half-width $\Omega(k)$. The implications of the theoretical predictions for scattering experiments are discussed.

INTRODUCTION

In this paper we investigate the quasielastic scattering of neutrons by dilute polymer solutions using Mori's projection operator technique.[1] This problem was first attacked by de Gennes[2] in the free-draining limit, and later by Dubois–Violette and de Gennes[3] including hydrodynamic interactions. They investigated, in particular, the k-dependence of the half-width of the quasielastic peak. Jannink and Summerfield[4] extended their work to include the effect of the average of the cosine of the bond angle on the width in the case of incoherent neutron scattering. The projection operator method was first applied to polymer dynamics by Bixon.[5] He studied the theoretical foundations of the Rouse–Zimm model using a ring polymer, and calculated the dynamic viscosity of dilute polymer solutions. Zwanzig[6] extended Bixon's work by an application of linear response theory, and presented a procedure to construct the "best possible" Rouse–Zimm model for an arbitrary polymer from Kirkwood's generalized diffusion equation. More recently Yamakawa, Tanaka, and Stockmayer[7] presented a molecular–theoretical basis to the Kirkwood equation by applying the projection operator method to the whole system including the polymer and solvent molecules, and calculated the intrinsic viscosity through correlation function formalism.

In this paper we essentially apply Zwanzig's formulation—with an alternative derivation of it—to the calculation of the density–density correlation function in dilute polymer solutions. In particular, we investigate the k-de-

* A summary of this work was presented at the IUPAP International Conference on Statistical Physics, August 25–29, 1975.

pendence of the half-width of the quasielastic peak in detail for coherent scattering using the spring-bead model for the polymer. However, we do not replace the Oseen tensor by its average when we include the hydrodynamic interactions as is done in the Rouse–Zimm model. To see the effects of preaveraging the Oseen tensor, we have also carried out the calculations using the Rouse–Zimm model, and compared Ω in these two cases. We compare our results with those obtained previously by others.[2,3]

THEORY

We consider an isolated polymer molecule which is idealized by a chain of $(N + 1)$ beads (subunits), interacting through a molecular potential $U(\mathbf{R}_0, \ldots, \mathbf{R}_N)$ where $\mathbf{R}_j$ is the location of the jth bead. The polymer is immersed in a solvent with a viscosity coefficient η_0. The solvent exerts a frictional force on each bead with a friction coefficient ξ, and a random force through molecular bombardment. The resulting random evolution of the polymer can be described, in a probabilistic sense, by a time-dependent distribution function $\psi(\mathbf{R}_0, \ldots, \mathbf{R}_N, t) = \psi(t)$. The equilibrium state of the polymer is characterized by the Boltzmann distribution

$$\psi_0 = z^{-1} \exp\left(-U/k_B T\right)$$

We assume that $\psi(t)$ satisfies Kirkwood's generalized diffusion equation (Fokker–Planck equation), which, as shown by Bixon,[5] can be written as

$$\frac{\partial f}{\partial t} = -\mathcal{L}f \tag{2}$$

where $f(t)$ is the normalized distribution function defined by

$$\psi(t) = \psi_0 f(t) \tag{3}$$

The operator $\mathcal{L}$ in eq. (2) is

$$\mathcal{L} = \sum_{j,l=0}^{N} [(1/k_B T)\nabla_j U j, \cdot \mathbf{D}_{jl} \cdot \nabla_l - \mathbf{D}_{jl}:\nabla_j\nabla_l] \tag{4}$$

where $\mathbf{D}_{jl}$ is the diffusion tensor

$$\mathbf{D}_{jl} = k_B T[\mathbf{T}_{jl} + (1/\xi)\mathbf{I}\delta_{jl}] \tag{5}$$

$\mathbf{T}_{jl}$ is the Oseen tensor

$$\mathbf{T}_{jl} = (1/8\pi\eta_0 R_{jl}^3)[\mathbf{I}R_{jl}^2 + \mathbf{R}_{jl}\mathbf{R}_{jl}] \tag{6}$$

which accounts for the hydrodynamic interaction among the beads. Bixon[5] also showed that $\mathcal{L}$ is hermitian with respect to the scalar product

$$(A,B) = \int d\mathbf{R}^N \psi_0 AB^*$$

The density–density correlation function of the polymer, which is the main theme of this paper, can be defined as

$$\langle \rho(t)\rho^*(0)\rangle = d\mathbf{r}^N \int d\mathbf{R}^N \rho^*(\mathbf{r}^N)\rho(\mathbf{R}^N)\psi_2(\mathbf{r}^N,0;\mathbf{R}^N,t) \tag{7}$$

where $\psi_2 \, d\mathbf{r}^N \, d\mathbf{R}^N$ is the joint probability of finding the polymer in $d\mathbf{r}^N$ about $\mathbf{r}^N$ at $t = 0$ and in $d\mathbf{R}^N$ about $\mathbf{R}^N$ at time t in the polymer configuration space. The polymer density $\rho(\mathbf{R}^N)$ is defined in Fourier space by

$$\rho(\mathbf{R}^N) = \sum_{j=0}^{N} \exp(i\mathbf{k} \cdot \mathbf{R}_j) \tag{8}$$

The time dependence of $\rho(t)$ in $\langle \rho(t)\rho^*(0) \rangle$ is governed by the Liouville equation for the whole system consisting of the polymer and solvent molecules. The average $\langle \ldots \rangle$ is taken over the equilibrium distribution, again, for the entire system. The projection operator method is conventionally applied to the true equation of motion (Liouville equation) of a dynamical variable, e.g., $\rho(t)$. Since the effect of the solvent on the polymer is treated macroscopically in the Kirkwood equation, the time evolution of the polymer is known only in a probabilistic sense, as implied by eq. (2). We must therefore obtain a Liouville-like equation which will assign a fictitious time dependence to $\rho(t)$, consistently with the approximate dynamical description by the Kirkwood equation, in order to apply the projection operator technique. For this purpose we express the joint probability density as a product of the conditional distribution ψ_c and $\psi(\mathbf{r}^N,0)$

$$\psi_2(\mathbf{r}^N,0;\mathbf{R}^N,t) = \psi_c(\mathbf{R}^N,t|\mathbf{r}^N,0)\psi(\mathbf{r}^N,0) \tag{9}$$

We can replace the distribution $\psi(\mathbf{r}^N,0)$ at $t = 0$ by the equilibrium distribution because we are interested only in correlations in equilibrium. The conditional distribution ψ_c, as a function of $\mathbf{R}^N$ and t, satisfies the same diffusion equation as $\psi(\mathbf{R}^N,t)$, with the initial condition

$$\psi_c(\mathbf{R}^N,0|\mathbf{r}^N,0) = \delta(\mathbf{R}^N - \mathbf{r}^N) \tag{10}$$

Hence the time dependence of f_c defined by $\psi_c = \psi_0(\mathbf{R}^N) f_c(\mathbf{R}^N,t)$ is governed by eq. (2) with the initial condition $\delta(\mathbf{R}^N - \mathbf{r}^N)/\psi_0(\mathbf{R}^N)$, i.e., $f_c(t) = \exp(-t\mathcal{L})f_c(0)$. Substitution of this result into eqs. (10) and (9) yields the time evolution of ψ_2 as

$$\psi_2(\mathbf{r}^N,0;\mathbf{R}^N,t) = \psi_0(\mathbf{R}^N) \exp(-t\mathcal{L})\delta(\mathbf{R}^N - \mathbf{r}^N) \tag{11}$$

where the operator $\mathcal{L}$ operates on $\mathbf{R}^N$. The density–density correlation function becomes

$$\langle \rho(t)\rho^*(0) \rangle = \int d\mathbf{r}^N \rho^*(\mathbf{r}^N) \int d\mathbf{R}^N \psi_0(\mathbf{R}^N)\rho(\mathbf{R}^N) \exp(-t\mathcal{L})\delta(\mathbf{R}^N - \mathbf{r}^N)$$

Using the hermiticity of $\mathcal{L}$ in the $\mathbf{R}^N$-integration we obtain

$$\langle \rho(t)\rho^*(0) \rangle = \int d\mathbf{r}^N \psi_0(\mathbf{r}^N)\rho^*(\mathbf{r}^N) \exp(-t\mathcal{L})\rho(\mathbf{r}^N) \tag{12}$$

The righthand side of eq. (12) can be expressed as the scalar product of $\rho(0)$ and $\rho(t)$ as

$$\langle \rho(t)\rho^*(0) \rangle = (\rho(t),\rho(0)) \tag{13}$$

provided we assign a time dependence to $\rho(\mathbf{r}^N)$ according to

$$\dot{\rho}(t) = -\mathcal{L}\rho(t) \qquad t > 0 \tag{14}$$

Let us emphasize here the distinction between the true time dependence of $\rho(t)$ given by the Liouville equation for the whole system and the "averaged" (over the random conformational changes caused by the random solvent interactions) time dependence represented by eq. (14). The latter is the appropriate time dependence for calculating the time correlations of the polymer as a scalar product of polymer variables, without any reference to solvent dynamics, as also pointed out by Zwanzig.[6]

We can now apply Mori's projection operator method to eq. (14). Since we are interested in the density correlations we chose the projection operator as

$$Pg(t) = (g(t),\rho)(\rho,\rho)^{-1}\rho$$

where ρ denotes $\rho(0)$. Pursuing the usual procedure of projection operator techniques we obtain the familiar generalized Langevin equation

$$\dot{\Gamma}(t) + \Omega\Gamma(t) - \int_0^t du\,\phi(u)\Gamma(t-u) = 0 \qquad t > 0, \tag{15}$$

for the normalized density–density correlation function $\Gamma(t)$

$$\Gamma(t) = (\rho(t),\rho)(\rho,\rho)^{-1} \tag{16}$$

In eq. (15), Ω and $\phi(t)$ are given by

$$\Omega = (\rho,\mathcal{L}\rho)(\rho,\rho)^{-1} \tag{17}$$

and

$$\phi(t) = (\rho,\mathcal{L}\exp(-t(1-\text{P})\mathcal{L})(1-P)\mathcal{L}\rho)/(\rho,\rho) \tag{18}$$

We note that Ω corresponds to the "frequency" matrix in the Mori's generalized Langevin equation. In time-reversible systems, this term vanishes when only a single variable A is considered, because $\langle A^*,iLA\rangle = \langle A^*,\dot{A}\rangle = 0$ as a result of the opposite parities of A and $\dot{A}$. This is not the case in eq. (15) because the parity rule is not applicable to $(\rho,\mathcal{L}\rho)$, since the starting equation $\dot{\rho} = -\mathcal{L}\rho$ is not time-reversible. Another consequence of this is that the "frequency" term is no longer imaginary in contrast with time-reversible systems. Consequently, it does not seem to be appropriate to refer to Ω and $\phi(t)$ as "frequency" and "damping" in the present context. However, $\phi(t)$ still represents memory effects.

The normalized scattering law $S(\mathbf{k},w)$ follows from eq. (15) as

$$S(\mathbf{k},w) = 2Re\,\{\text{Lim}_{s\to iw}[s + \Omega - \bar{\phi}(s)]^{-1}\} \tag{19}$$

where $\bar{\phi}(s)$ is the Laplace transform of the memory function $\phi(t)$. The rest of the paper is devoted to a study of $S(\mathbf{k},w)$.

THE WIDTH OF THE QUASIELASTIC PEAK

We shall first ignore the memory effects by setting $\bar{\phi}(s) = 0$ in eq. (19), and investigate the predictions of the resulting approximate description, in which $S(\mathbf{k},w)$ is a Lorentzian, centered at $w = 0$ with a k-dependent half-width $\Omega(k)$. In order to determine $\Omega(k)$ explicitly, we must specify the molecular potential $U(\mathbf{R}^N)$. We adopt the spring-bead model in which

$$U(\mathbf{R}^N) = (3k_BT/2a^2)\sum_{j=1}^{N}|\mathbf{R}_j - \mathbf{R}_{j-1}|^2 \tag{20}$$

where a^2 is the mean-square distance between two consecutive beads. The equilibrium distribution $\psi_0(\mathbf{R}^N)$ is of course a Gaussian in this model. In particular, the distribution of $\mathbf{R}_{jl} = \mathbf{R}_l - \mathbf{R}_j$ is

$$P(\mathbf{R}_{jl}) = (3/2\pi a^2|j - l|)^{3/2} \exp\left[-(3/2a^2|j - l|)|\mathbf{R}_{jl}|^2\right] \qquad (21)$$

The matrix element $(\rho, \mathcal{L}\rho)$ in the expression of $\Omega(k)$ in eq. (17) reduces after some manipulations to

$$(\rho, \mathcal{L}\rho) = \sum_{j,l=0}^{N'} \langle \mathbf{D}_{jl} \exp(i\mathbf{k} \cdot \mathbf{R}_{jl})\rangle : \mathbf{kk} + \sum_{l=0}^{N} \langle \mathbf{D}_{ll}\rangle : \mathbf{kk} \qquad (22)$$

where the averages on the righthand side are over the Gaussian distribution in eq. (21). Note that eq. (22) involves the average of the Oseen Tensor $\mathbf{T}_{jl}$ as well as $\mathbf{T}_{jl} \exp(i\mathbf{k} \cdot \mathbf{R}_{jl})$. The calculation of the latter would be simplified greatly if we replaced the Oseen tensor by its preaveraged value, as is done in the Rouse–Zimm model. However, we were able to carry out the calculations in eq. (22) without this simplification. The result is found to be

$$(\rho, \mathcal{L}\rho) = \frac{Nk^2}{\beta\xi}\left\{1 + \left(\frac{\xi}{8\pi\eta_0}\right)\frac{k}{N\sqrt{\pi}}\sum_{j,m=0}^{N'} X_{jm}^{-2}\left[-X_{jm}^{-1}\right.\right.$$
$$\left.\left. + (2 + X_{jm}^{-2})\exp(-X_{jm}^{-2})\int_0^{X_{jm}} \exp(u^2)du\right]\right\} \qquad (23)$$

where we have replaced $(N + 1)$ by N, and defined

$$X_{jm} \equiv (k/2)(2a^2|j - m|/3)^{1/2}$$

The Rouse–Zimm model simply gives for eq. (23),

$$(\rho, \mathcal{L}\rho) = \left(\frac{\xi}{\eta_0 a}\right)k^2\left(\frac{k_BT}{\xi}\right)\left(\frac{2}{6^{1/2}\pi^{3/2}}\right)\sum_{m=1}^{N}(N + 1 - m)\frac{1}{m^{1/2}}\exp\left\{\frac{-(ka)^2m}{6}\right\}$$
$$+ (N + 1)k^2\frac{k_BT}{\xi} \qquad (24)$$

The double summation appearing in eq. (23) can be reduced to a single summation, using for any arbitrary function,

$$\sum_{j,m=0}^{N'} A(|j - m|) = 2\sum_{m=1}^{N}(N + 1 - m)A(m) \qquad (25)$$

The final expression for $\Omega(k)$ is found to be

$$\Omega(k) = k^2(k_BT/\xi)\left\{1 + (\xi/\eta_0 a)(ka/4\pi^{3/2})\sum_{q=1}^{N-1}\left(1 - \frac{q}{N}\right)\right.$$
$$\left. \times X_q^{-2}\left[-X_q^{-1} + (2 + X_q^{-2})\exp(-X_q^2)\int_0^{X_q} du \exp(u^2)\right]\right\}$$
$$\times \{1 + 2(e^{\alpha} - 1)^{-1}[1 - (1 - e^{-\alpha N})/N(1 - e^{-\alpha})]\}^{-1} \qquad (26a)$$

where

$$X_q^2 \equiv (q/6)(ka)^2 \qquad \text{and} \qquad \alpha \equiv (ka)^2/6 \qquad (26b)$$

The denominator of eq. (26a) is the static correlation function (ρ, ρ).

In the context of the spring-bead model, eq. (26a) is valid for all values of ka and N. We observe that

$$\Omega(k)/(k_BTk^2/\xi) \equiv F(ka, \xi/\eta_0 a, N)$$

6   AKCASU AND GUROL

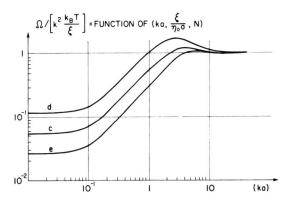

Fig. 1. Plot of $\Omega(k)/k^2(k_B T/\xi)$ vs. ka using eq. (26a). $N = 10^3$; curve c, $\xi/\eta_0 a = 3\pi$; curve d, $\xi/\eta_0 a = 6\pi$; curve e, $\xi/\eta_0 a = 3\pi/2$.

depends only on ka, N and the ratio $(\xi/\eta_0 a)$. The latter is a measure of the relative strength of the hydrodynamic interactions. Figure 1 shows a plot of $F(ka,\xi/\eta_0 a,N)$ versus ka for a fixed N and for three values of $(\xi/\eta_0 a)$. It is computed from eq. (26a). The value 3π for $(\xi/\eta_0 a)$ corresponds to Stoke's expression for the friction coefficient, $\xi = 3\pi\eta_0 a$, for a sphere of diameter a. To see the effect of preaveraging the Oseen tensor on $\Omega(k)$, we also plotted F in Figure 2 using the Rouse–Zimm model. One observes that the latter underestimate $\Omega(k)$, in particular in the region $ka \approx 1$. The difference is more pronounced when the hydrodynamic interactions are large. In Figure 2 we also present F in the Rouse limit for comparison. One observes in Figure 1, that there are three distinguishable regions of ka where F has a simple ka-dependence.

Small-K Region ($ka\sqrt{N} \ll 1$)

In this region Dawson's integral $\exp(-x^2)\int_0^x \exp(u^2)du$ appearing in eq. (26) can be approximated for each value of q by its small-x expansion, $x_q(1 -$

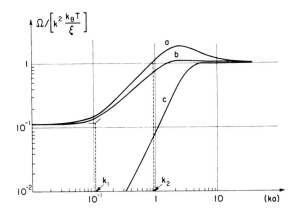

Fig. 2. Plot of $\Omega(k)/k^2(k_B T/\xi)$ vs. ka using eq. (26a). $N = 10^3$; curve a, $\xi/\eta_0 a = 6\pi$, nonaveraged Oseen tensor; curve b, $\xi/\eta_0 a = 6\pi$, averaged Oseen tensor; curve c, Rouse limit: $\xi/\eta_0 a = 0$; behavior governed by $\langle\rho\rho^*\rangle^{-1}$.

$2x_q^2/3$). With this approximation, and replacing the q-summation by an integration, we obtain

$$\Omega(k) = k^2(k_B T/\xi N)[1 + (\xi/\eta_0 a)(4/9\pi)(6N/\pi)^{1/2}] \tag{27}$$

The Rouse limit, where the hydrodynamic interactions are neglected, follows from eq. (27) when

$$(\xi/\eta_0 a) \ll 5.11/\sqrt{N} \tag{28}$$

as

$$\Omega(k) = k^2 k_B T/\xi N \tag{29}$$

When the hydrodynamic effects are dominant (large N), eq. (27) gives

$$\Omega(k) = 0.195 k^2 (k_B T/\eta_0 a\sqrt{N}) \tag{30}$$

From eq. (30) we infer the diffusion coefficient at infinite dilution to be

$$D = 0.195 (k_B T/\eta_0) \langle R^2 \rangle^{-1/2} \tag{31a}$$

where $\langle R^2 \rangle = Na^2$, and denotes the mean-square end-to-end distance of the Gaussian chain. This expression is identical to that given by Yamakawa[8] (p. 272). The fact that $\Omega(k)$ depends on $\langle R^2 \rangle$, rather than on N and a separately, is expected because the probing radiation sees the entire molecule in the small-k limit. In interpreting scattering experiments, $\langle R^2 \rangle$ is to be chosen as the mean-square end-to-end distance of the ideal chain under consideration. Using $\langle R^2 \rangle = MA^2$, where M is the molecular weight of the polymer, and A is a parameter depending on the mean-square end-to-end distance in the freely rotating state and the conformation factor, we can express D as

$$D = (0.195 \, k_B T/A\eta_0) \, M^{-0.5} \tag{31b}$$

The values of A for various polymer solutions at the theta temperature are tabulated by Yamakawa[8] (Table VII.5). For polystyrene in cyclohexane at 34.8°C, we find $A = 2.25 \sqrt{9.01} \times 10^{-9} = 0.675$ Å. With this value and $\eta_0 = 0.75 \times 10^{-2}$ poise, $T = 34.8$°C in eq. (31b), we obtain $D = 1.63 \times 10^{-4} \, M^{-0.5}$ cm^2/sec. King et al. found experimentally $D = 1.3 \times 10^{-4} \, M^{-0.497}$ cm^2/sec for polystyrene in cyclohexane at 35°C, which agrees with the theoretical prediction quite well.

Large-K Limit ($ka \gg 1$)

In this limit Dawson's integral can be approximated for all q in eq. (26) by $(1/2X_q) + (1/4X_q^3)$. This approximation leads to

$$\Omega(k) = (k^2 k_B T/\xi)[1 + (6/\pi)^{3/2}(\xi/\eta_0 a)(ka)^{-4}] \tag{32}$$

Again the Rouse limit is recaptured when in addition to $ka \gg 1$

$$(\xi/\eta_0 a) \ll 2.64 (ka)^4 \tag{33}$$

However, $\Omega(k)$ approaches $k^2 k_B T/\xi$, for any value of $(\xi/\eta_0 a)$, for sufficiently large values of ka. This trend is clearly seen in Figure 1. It implies physically that the probing radiation sees the individual subunits for large ka. The large-k limit can only be expected to give a trend of what to expect in an experiment when k is increased because the spring-bead model stresses only the statistical aspects of the chain as a whole. Significantly different results might be expected in this k-region if the details of a subunit were accounted for.

Intermediate-K Region ($ka \ll 1$ and $ka\sqrt{N} \gg 1$)

This is the region of current interest, especially in view of on-going quasi-elastic neutron scattering experiments. To obtain an analytic expression for the half-width in this region, the summation in eq. (26a) was broken up into two parts as $q \le N_c - 1$ and $q \ge N_c$, with the cut-off N_c determined by $1 = ka(N_c/6)^{1/2}$. Using the small-x and large-x expansions in these partial sums, we obtained

$$\Omega(k) = k^2(k_BT/\xi)[(ka)^2/12 + (\xi/\eta_0 a)(ka)(11/36\pi^{3/2})] \tag{34}$$

The Rouse limit is obtained when the first term dominates the second, i.e., $ka \gg 0.66\,(\xi/\eta_0 a)$, as

$$\Omega(k) = k^4(a^2 k_BT/12\xi) \tag{35}$$

A physically more interesting case is that in which the hydrodynamic interactions are dominant, i.e., $ka \ll 0.66\,(\xi/\eta_0 a)$. Then eq. (34) reduces to

$$\Omega(k) = 0.055(k_BT/\eta_0)k^3 \tag{36}$$

Equations (35) and (36) are in agreement with expressions of de Gennes,[2] and Dubois–Violette and de Gennes.[3] The k^3- and k^4-dependences can be understood qualitatively in terms of the number of beads within a wavelength, i.e., $N \approx (1/ka)^2$. If we assume that this subchain diffuses independently of the rest of the molecule, then we may use eq. (30) to calculate $\Omega(k)$ with N replaced by $(1/ka)^2$. This leads to the k^3-dependence. In the Rouse limit we use eq. (29) to obtain the k^4-dependence.

We may also obtain the Rouse limit in all k-regions from eq. (26a) with $(\xi/\eta_0 a) \to 0$:

$$\Omega(k) = k^2(k_BT/\xi)\{1 + 2(e^\alpha - 1)^{-1}[1 - (1 - e^{-\alpha N})/N(1 - e^{-\alpha})]\}^{-1} \tag{37}$$

where $\alpha = (ka)^2/6$. Evidently $\Omega(k)$ is proportional to $k^2/(\rho,\rho)$ in this limit.

Finally we should point out that eq. (26) for the half-width of the quasi-elastic peak not only predicts the asymptotic regions obtained previously by others, but also provides information on the transition k-regions, as presented in Figure 1.

EXPERIMENTAL IMPLICATIONS OF THE THEORY

The above theory predicts that the shape of $\Omega(k)$ versus k depends on the parameters T, η_0, a, N, and ξ. The temperature T and viscosity η_0 are solvent parameters which are known independently of the polymer molecule. The parameters a, N, and ξ, on the other hand, characterize a Gaussian chain, which models the actual molecule. As such they are not known *a priori*, except for $a\sqrt{N}$ which must be equal to the end-to-end distance of the molecule. Since ξ is defined as the friction coefficient for a bead in the gaussian chain, which represents a subunit of the molecule with end-to-end distance a, it must be determined consistently with a.

In an actual experiment one would obtain $\Omega(k)$ as a function of k for a given solution. We discuss how such an experiment might be interpreted in terms of the above theoretical results, and the information one might obtain about the polymer molecule. Let us assume that in the experimental $\Omega(k)/k^2$ curve one can distinguish three regions, as predicted by the theory.

By using the appropriate expressions for $\Omega(k)$ in the small and intermediate-k regions (cf. eqs. (31) and (36)), one finds that the asymptotes of these two regions intersect at $k_1 = 3.54/a\sqrt{N}$ (see Figure 2). Then, experimentally determining k_1 we immediately obtain the end-to-end distance $a\sqrt{N}$ for the entire molecule. Of course one could also obtain $a\sqrt{N}$ from the measured $\Omega(k)$ in the small-k region using $\Omega(k) = 0.195k^2 k_B T/\eta_0 a\sqrt{N}$, provided T/η_0 is known. It should also be noted that one can obtain $k_B T/\eta_0$ from the slope of $\Omega(k)$ in the intermediate region where $\Omega(k)/k^2 = 0.055 (k_B T/\eta_0)k$ holds. At any rate, one concludes that in principle, the measurement of $\Omega(k)$ in the small and intermediate-k regions yields $a\sqrt{N}$ and $(k_B T/\eta_0)$.

The interpretation of the experiment in the transition region between the intermediate and large-k limits is less clear because of the fact that the Gaussian chain model should eventually break down when k is increased. The theory predicts that the intersection of the asymptotes in these two regions intersect at $k_2 = 18.18 (\eta_0/\xi)$. Therefore if k_2 can be determined experimentally, one can infer a value for ξ. A more detailed comparison of the theoretical $\Omega(k)/k^2$ with the experimental one would yield not only ξ, but also an a. For example, one can show that $F(ka,\xi/\eta_0 a,N)$ in the intermediate and large-k regions behaves as $F(ka,\xi/\eta_0 a,\infty)$, and $dF/d(ka)$ depends only on ka. Then, the location of the maximum of $\Omega(k)/k^2$ determines a independently. In other words, if the polymer molecule were indeed a gaussian chain, then a and ξ would be determined consistently from the experiment. However, since the actual molecule will start deviating from the gaussian behavior for wave numbers $ka > 1$, the a and ξ obtained in an experiment from the transition region should be interpreted as the minimum subunit size and the associated friction coefficient that exhibits Gaussian behavior. The existence, if any, of a plateau in an experiment would then indicate that sufficiently small sections of the molecule behave more rigidly than given by the spring-bead model, at least for $ka \approx 1$.

CONCLUSIONS

We have calculated the half-width of the scattering law by setting the memory kernel $\phi(t)$ equal to zero. The resulting expression for $S(\mathbf{k},w)$ then becomes Lorentzian with a half-width $\Omega(k)$, which we studied in some detail. The question remains of whether the effect of $\phi(t)$ is important. To answer this question, we calculated (Gurol[10]) $\phi(t)$ in the limit of weak bead-to-bead interactions using the Rouse–Zimm model (i.e., spring-bead model plus pre-averaging the Oseen tensor), for large k values. We found that in this limit the memory function is exponentially decaying, and has a negligible effect on the width. However we have not been able to assess its effect in the other more realistic cases. The agreement between our results and those obtained previously by others using different techniques may suggest that the memory effects are probably small in all cases. This remains to be eventually checked by experiments.

One may extend the projection operator technique presented here to a multicomponent description of the polymer by introducing other variables in addition to the density ρ. For example a 2×2 matrix description is obtained by chosing ρ and $\mathcal{L}\rho$ as two variables. When the memory effects are neglected in such a description, one usually gets a more accurate result for the same correlation function than in a single-component description also without memory. Such a procedure is currently being studied, and is expected to yield a scattering law as a superposition of Lorentzians.

This work was supported by the National Science Foundation. The authors express their thanks to Professors G. C. Summerfield and R. Ullman, and to Mr. R. Adler for many stimulating discussions.

References

1. H. Mori, *Prog. Theor. Phys.*, **33**, 423 (1965).
2. P.-G. de Gennes, *Physics*, **3**, 37 (1967).
3. E. Dubois–Violette and P.-G. de Gennes, *Physics*, **3**, 181 (1967).
4. G. Jannink and G. C. Summerfield, IAEA-SM-155/C-2 (1972).
5. M. Bixon, *J. Chem. Phys.*, **58**, 1459 (1973).
6. R. Zwanzig, *J. Chem. Phys.*, **60**, 2717 (1974).
7. H. Yamakawa, G. Tanaka, and W. H. Stockmayer, *J. Chem. Phys.*, **61**, 4535 (1974).
8. H. Yamakawa, *Modern Theory of Polymer Solutions*, Harper and Row, New York, 1971.
9. T. A. King, A. Knox, N. I. Lee, and J. D. G. McAdam, *Polymer*, **14**, 151 (1973).
10. H. Gurol, Doctoral Dissertation, Dept. of Nuclear Engineering, Univ. of Michigan (1975).

Received May 28, 1975
Revised September 9, 1975

COMMENTARY

Reflections on "Rheology of Concentrated Solutions of poly(γ-benzyl-glutamate)," by Gabor Kiss and Roger S. Porter, *J. Polym. Sci., Polym. Symp.,* 65, 193 (1978)

GABOR KISS*

Bellcore, 445 South St., Morristown, NJ 07960

Roger Porter and I first observed the Negative Normal Force effect more than 20 years ago, in 1975. At the time, we were investigating the effect of concentration and shear rate on viscosity of lyotropic polymers, subsequent to early reports[1,2] that the onset of anisotropy, due to volume filling and packing considerations,[3,4] caused a reduction in viscosity as concentration increased. Although a regime of decreasing viscosity with increasing concentration was unprecedented for polymer solutions, this behavior was in fact not unexpected, since the analogous decrease in viscosity with decreased temperature at the isotropic-to-nematic transition had already been observed with low molecular weight liquid crystals.[5,6]

We hoped to find interesting science, but we were not expecting to uncover a phenomenon which was so unforeseen, so counter-intuitive that we ourselves initially assumed it was an artifact. We soon learned that, indeed, the inertial contribution to the normal force can produce a positive-to-negative transition in isotropic polymer solutions, as shown in Figure 1.[7]

No doubt, many experienced rheologists, upon hearing of our negative N_1 observations, assumed that we had fallen into this trap. Needless to say, all of the data we published in the late 70s was already corrected for the inertial contribution. In any case, the second sign change in N_1, from negative to positive again, could not have been so caused.

We felt privileged, therefore, to participate in the discovery of previously unknown territory in the field of rheology, and to draw the first crude maps. Our contribution was to be in the right place at the right time, to keep an open mind, and to make careful observations. The intellectual content of our contribution was not titanic, but we were willing to take risks, such as pushing shear rate beyond the point at which sample ejection occurs (resolved by extrapolating back to the time of initiation of shear), and removing large amplitude, high frequency oscillations with a highly damped recorder (later by using electronic filtering). We ended up with much better data than one might have expected, and the data told an amazing story: Under easily accessible conditions of solute, concentration, molecular weight, and solvent, one can *in a single experiment* of increasing shear rate observe normal forces that first increase to a maximum, then suddenly become negative, then reach a negative maximum, then suddenly become positive again. We essentially had no explanation; we merely observed this phenomenon.

By complete coincidence, Roger Porter and I recently collaborated on a review chapter[8] to be published early in 1997. While researching this chapter we found that in the intervening years, the essential observation has been thoroughly confirmed, both for m-cresol solutions of poly-γ-benzyl-glutamate (PBG) and for aqueous solutions of hydroxypropylcellulose (HPC). The confirmation is gratifying, but not unexpected, since we had satisfied ourselves that the phenomenon was genuine. Even more gratifying, and much less expected, is the fact that subtle and insightful theoretical work has been accomplished by several sets of researchers which has

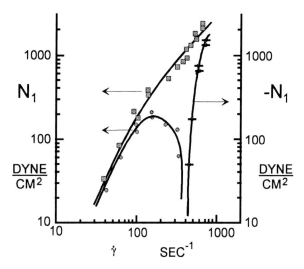

Figure 1. First normal stress difference vs. shear rate for 2% aqueous solution of polyacrylamide, with and without inertial correction.

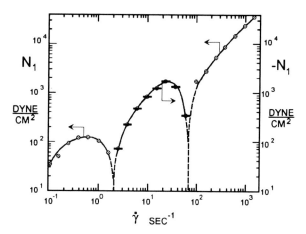

Figure 2. First normal stress difference vs. shear rate for 16.4 wt. % PBG/m-cresol.

completely explained the normal force observations. Compare the features of Figure 2, taken from our 1978 experimental paper, and the theoretical result shown in Figure 3, published in 1990 by Marrucci and Maffettone.[9] Indeed, the theory also predicts a similar series of transitions in N2, the second normal stress coefficient, which has been borne out experimentally.[10]

At the same time that we were confronted with the negative normal force phenomenon, we also noticed an odd morphological effect, that of a shear-induced band structure. To our knowledge, the formation of striations or bands was first mentioned in passing by Toth and Tobolsky,[11] who did not publish images. The first of the now familiar band images was published by Aharoni[12] who did not specify the shearing direction. The first solid report of shear-induced perpendicular band structures was.[13] Many polymer liquid crystal systems, both lyotropic and thermotropic, have now been seen to exhibit striations perpendicular to the shearing direction upon cessation of shear and the phenomenon has now been satisfactorily explained, as discussed in the review chapter.[8]

In 1980[13] we published polarized light micrographs taken during shear with stroboscopic illumination which appeared to show perpendicular striations *during* shear, as opposed to developing after the cessation of shear. The occurrence of these striations during shear seemed to correspond to the shear rate at which negative normal force occurred. This observation has not been confirmed and in fact

some evidence suggests that it could not be correct.[14] We suspect now that this was in fact an artifact due to insufficient parallelism of the rotating plates. This too is discussed in the review chapter.[8]

The two phenomena of Negative Normal Force and bands perpendicular to shear are both manifestation of liquid crystalline order, how this order is affected by forces imposed on molecules and domains during shear, and how the stored energy relaxes after cessation of shear. The former effect seems to have been confirmed only in lyotropes, the latter in thermotropes as well. Neither has been reported in non-ordered fluids nor in low molecular weight liquid crystals. Band structures perpendicular to the flow direction do however pop up in most unexpected places. For example, compare the image of a dried sheared film of lyotropic PBLG/dioxane solutions

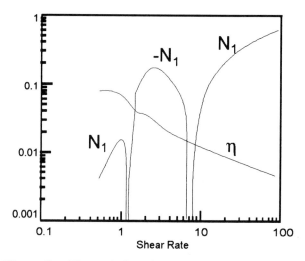

Figure 3. Theoretical prediction for viscosity and first normal stress difference vs. shear rate.

(Fig. 4) with that of ripples on a sand dune (Fig. 5). We suspect that an aerial photo of the irritating washboard surface on the outrun of any ski slope at the end of a busy day would look very similar. Despite the large body of work which has been performed on these two phenomena, a number of unaddressed questions occurred to us during the preparation of the review chapter.[8] We reproduce a few of these here using the same numbering as in the chapter.

1. The Negative N_1 Effect has been seen in very few systems. Oddly, the two most studied polymers, PBG and HPC are extremely different in flexibility. What is the factor or balance of factors which appears to be so difficult to achieve, and yet is readily achieved in these two extremely disparate systems? Why is it that two such different polymers exhibit this phenomenon, yet it is not general? PCBZL in m-cresol exhibits the Negative N_1 Effect[15] and should be intermediate in flexibility, yet has been little studied. Poly-benzyl-L-aspartate (PBA) does not show it.[16] What about other helical polypeptides?

2. Negative N_1 has been seen for PBG only in m-cresol. We found m-cresol to be convenient because its low vapor pressure prevented evaporation and concentration changes from becoming an issue. Is this merely a coincidence, or is there a connection? Why does HPC show a negative N_1 in aqueous solutions and in m-cresol, but not acetic acid?[17]

9. It appears likely that we observed band structures *during* shear due to an inadvertent

Figure 5. Death Valley sand dune (courtesy of G. C. Vogel).

superposition of oscillatory and steady shear. An attempt should be made to reproduce a quasi-steady state band structure and to explore this phenomenon. Either this will turn out to be relatively easy to do, or we were lucky to observe it. If this is accomplished, it would be interesting to insert a first order red plate at various angles to the shear direction.

10. Since there appears to be an upper shear rate for formation of band structures,[18] shearing beyond this critical rate should not produce recoil.

Many other interesting possibilities for future work would no doubt present themselves upon further reflection.

The introductory section of many investigations of Negative N_1 and band structures give as justification for the work the "technological importance"

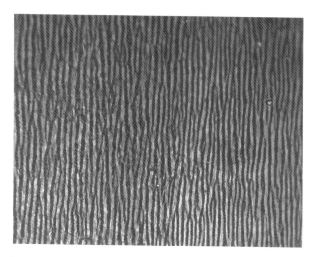

Figure 4. Dried sheared film of lyotropic PBLG/dioxane solutions.

of high strength and rigidity lyotropic (Kevlar) and thermotropic polymers. It is taken for granted that deeper understanding of orientation phenomena that occur during flow and deformation will somehow eventually translate into new products which would have been unobtainable in the absence of this understanding and knowledge. Or, at least, cheaper ways of making useful products.

Is this really so? Has this been simply an exercise in intellectual beauty, satisfying the curiosity, or can one hope for a practical result? Will the insights developed in explaining negative N_1 lead to an unexpectedly effective extrusion die design or processing technique? When I was a graduate student at the University of Massachusetts, friends would sometimes ask about my research. After a brief explanation, which always sounded rather tepid (". . . First the plates get pushed apart, then pulled back together, then pushed apart again!" "Yeah? Then what? What's it good for?"), I finally resorted, only half in jest, to the observation that "What it's good for is getting a doctorate."

In a review of the rheology of processing polymer liquid crystals, Muir and Porter[19] could give no indication that the insights gained in determining the origins of negative N_1 and band structures have had a pronounced impact on the control of orientation and the physical properties of thermotropes. There may be some link to the observation of near-zero or even negative die swell, which would be "very attractive to those interested in the design of dies and injection cavities, particularly for thin sections."[19] However the performance of thermotropic PLCs still lags that of lyotropes.

We hope that time will reveal that the first 20 years after our observation of the Negative N_1 Effect and band structures in lyotropic PLCs was simply the explanatory phase, interesting as it may be to a small niche of specialists. We hope that the second 20 years will prove to be the exploitation phase.

REFERENCES AND NOTES

1. Iizuka, E., (1974) *Mol. Cryst. Liq. Cryst.* **25**, 287.
2. Hermans, J. Jr., (1962) *J. Colloid Sci.* **17**, 638.
3. Flory, P. J., (1956) *Proc. Royal Soc. London* **A234**, 60.
4. Onsager, L., (1949) *Annals N. Y. Acad. Sci.* **51**, 627.
5. Porter, R. S., Barrall, E. M. II, and Johnson, J. F., (1966) *J. Chem. Phys.* **45**, 1452.
6. Porter, R. S., and Johnson, J. F., (1967) in *Rheology: Theory and Applications Vol. IV* (F. R. Eirich, Ed.), Academic Press, New York, 317.
7. Kulicke, W. M., Kiss, G., and Porter, R. S., (1977) *Rheol. Acta* **16**, 568.
8. Kiss, G. and Porter, R. S., Flow-Induced Phenomena of Lyotropic LCPs: The Negative Normal Force Effect and Bands Perpendicular to Shear in *Mechanical and Thermophysical Properties of Liquid Crystal Polymers*, (W. Brostow, Ed.), Chapman and Hall, London (1997)
9. Marrucci, G., and Maffettone, P. L., (1990b) *J. Rheol.* **34**, 1231.
10. Magda, J. J., Baek, S. G., Devries, K. L., and Larson, R. G., *Macromolecules* **24**, 4460 (1991).
11. Toth, W. J., and Tobolsky, A. V., (1970) *Polymer Lett.* **8**, 531.
12. Aharoni, S. M., (1979) *Macromolecules* **12**, 94.
13. Kiss, G., and Porter, R. S., (1980b) *Mol. Cryst. Liq. Cryst.* **60**, 267.
14. Picken, S. J., Moldenaers, P., Berghmans, S., and Mewis, J., (1992) *Macromolecules* **25**, 4759.
15. Kiss, G., and Porter, R. S., (1980a) *J. Polymer Sci. Phys.* **18**, 361.
16. Kiss, G., Orrell, T. S., and Porter, R. S., (1979) *Rheol. Acta* **18**, 657.
17. Navard, P., and Hauden, J. M., (1986) *J. Polymer Sci. Phys.* **24**, 189.
18. Vermant, J., Moldenauers, P., Mewis, J., and Pickens, S. J., (1994b) *J. Rheol.* **38**, 1571.
19. Muir, M. C., and Porter, R. S., (1989) *Mol. Cryst. Liq. Cryst.* **169**, 83.

PERSPECTIVE

Comments on "Rheology of Concentrated Solutions of poly(γ-benzyl-glutamate)," by Gabor Kiss and Roger S. Porter, *J. Polym. Sci., Polym. Symp.*, 65, 193 (1978)

R. G. LARSON

Bell Laboratories, Lucent Technologies, 700 Mountain Ave., Murray Hill, NJ 07974

In 1978, Kiss and Porter[1] published a paper in the Journal of Polymer Science showing a highly unusual rheological phenomenon: a *negative* sign of the first normal stress difference N_1 in shearing flow of liquid crystalline polymers, or LCP's. The particular LCP's studied were solutions of poly(γ-benzyl glutamate), or PBG. N_1 was found to be negative over an intermediate shear-rate range; at both high and low shear rates, N_1 had a positive sign, and so there were two sign changes in N_1 with increasing shear rate.

The first normal stress difference is the difference in normal stress components of the stress tensor, $N_1 \equiv \tau_{11} - \tau_{22}$, where "1" is the flow direction and "2" is the gradient direction in a shearing flow. A simple manifestation of a non-zero N_1 is that in a cone-and-plate rheometer, there is, in addition to the torque produced by viscous drag, a thrust along the axis of rotation of the cone or plate, tending to either force apart the cone and plate (in the case of positive N_1), or pull them together (in the case of negative N_1). For simple Newtonian fluids, N_1 is zero, while for viscoelastic fluids, N_1 was, prior to the work of Kiss and Porter, always supposed to be positive in sign.[2] The care shown by Kiss and Porter in ruling out experimental artifacts left no doubt that this supposition was incorrect.

The first normal stress difference N_1 is a basic rheological property of viscoelastic fluids, important in such phenomena as rod climbing (by which

N_1 was first ascertained to be non-zero by Weissenberg in 1947[3]), and in the swelling of an extrudate as it emerges from a capillary die.[2] As discussed by Ferry,[4] N_1 at low shear rates $\dot{\gamma}$ for *ordinary polymeric liquids* can be related to the elastic storage modulus G' at low frequencies ω by the relationship

$$\lim_{\dot{\gamma} \to 0} \left(\frac{N_1(\dot{\gamma})}{\dot{\gamma}^2} \right) = 2 \lim_{\omega \to 0} \left(\frac{G'(\omega)}{\omega^2} \right) \qquad (1)$$

Thus, N_1, like G', is ordinarily linked to the elastic character of the liquid. Hence, it was no small surprise to learn from work of Kiss and Porter that unlike $G'(\omega)$, $N_1(\dot{\gamma})$ is not necessarily a positive function.

The fluid studied by Kiss and Porter was, of course, an unusual one. Poly(γ-benzyl-glutamate) is a synthetic polypeptide, which in helicogenic solvents (such as metacresol) forms a rather rigid helix. In a gross sense, the molecule can be considered a slightly flexible rod; it forms a nematic or cholesteric liquid crystal when its concentration reaches around 8% or more by volume.[5] It was immediately clear from the work of Kiss and Porter that negative value of N_1 was linked to the liquid crystalline character of the solution; for, when the concentration of PBG was decreased below that necessary to form a liquid crystal, only positive values of N_1 were found.

The peculiarity of the Kiss and Porter result was recognized immediately by the rheology and polymer-physics communities. Interest in the negative-normal-stress phenomenon grew throughout the eighties, along with the commercial interest in liquid crystalline polymers (LCP's) as potential precursors of high modulus, high strength, materials. The rheological and flow properties of LCP's are crucial to their end use, since it is in their flow-induced alignment that high modulus and high strength are attained.

The lack of a credible explanation for the phenomenon of negative N_1 meant that no real trust could be placed in existing theories for the dynamics and flow properties of LCP's. Although rigid-rod molecular theories for LCP flow properties had been developed by Hess[6] and by Doi[7], only positive values of N_1 were predicted, *within the approximations used to solve the equations*, only positive values of N_1 were predicted. Hence, as Doi and Edwards stated in their 1986 book, *The Theory of Polymer Dynamics*[8]: "This effect [i.e., negative N_1] is not explained by the present constitutive equation and some other physical reason appears to be needed." Thus, while scientific study of the flow properties of LCP's was rapidly expanding, the lack of an explanation of the Kiss-Porter result was considered prima facie evidence that no deep understanding of LCP rheology had really been achieved.

Happily, this situation was soon to be turned on its head. In retrospect, it is now possible to recognize the first clues to a resolution of the negative N_1 mystery in a 1984 paper by Kuzuu and Doi.[9] They found by careful solution of the rigid-rod theory at low shear rates the prediction of *director tumbling*—i.e., the lack of a steady-state average alignment direction, and the tendency of the local preferred direction of alignment to rotate continuously in a shearing flow. This result was especially significant because it contradicted the prediction extracted from the approximate solutions of the rigid-rod equations that had earlier predicted only positive values of N_1. By the late eighties, Pino Marrucci at the University of Naples had began to suspect a connection between director tumbling and negative N_1. In 1989 he and Maffettone[10] published a seminal paper in which they solved more rigorously the rigid-rod equation within a two dimensional approximation and showed that it predicted not only director tumbling at low shear rates, but negative N_1 at higher shear rates! Confirmation of these predictions by a full three dimensional solution of the equations[11] soon followed, as well as direct experimental proof by Burghardt and Fuller[12] that PBG solutions do indeed exhibit director tum-

bling. Further detailed measurements of both N_1 and the second normal stress difference $N_2 \equiv \tau_{22} - \tau_{33}$ by Magda et al.[13] showed sign changes in both quantities that were in qualitative and nearly quantitative agreement with the theory.

The mechanism of negative N_1 was explained in the Marrucci-Maffetone paper. Ordinary polymeric liquids are isotropic at rest and hence shearing can only increase the orientation of the molecules. The elastic response is to try to reduce the orientation by pushing apart the cone and plate. Hence N_1 is always positive for ordinary polymeric liquids. Liquid crystalline polymers, however, have spontaneous orientation at rest. This orientation can, under some conditions, be reduced by a shearing flow; the elastic response of the liquid is to try to increase its orientation, which results in a pulling together of the shearing surfaces. Thus, while ordinary polymeric liquids can have only positive values of N_1, LCP's can have either positive or negative values.

The recognition of the role of director tumbling in the flow properties of PBG and other LCP's "broke the dam" and released a flood of experimental and theoretical discoveries related to the rheology of PBG and other LCP's. Many of the unusual flow properties of PBG are now quite well understood to be consequences of director tumbling.[14] The stimulus provided by the Kiss and Porter paper to these developments is evidenced by its widespread citation in the LCP literature.

The influence of the Kiss-Porter paper continues to this day. Although much is now known about the rheology of PBG and some other LCP's, negative values of N_1 appear to be absent in many other LCP's, particularly solvent-free, or thermotropic, ones.[15] The rheology of such thermotropic LCP's often differs from that of PBG in other ways as well. It may well be that an understanding for the lack of negative N_1 in these polymers will lead to a much better understanding of their other rheological properties. If so, then substantial credit for that progress will once again be attributable to that curious 1978 paper of Kiss and Porter.

REFERENCES AND NOTES

1. G. Kiss, R. S. Porter, *J. Polym. Sci., Polym. Symp.,* 65, 193 (1978).
2. R. I. Tanner, *Engineering Rheology,* Oxford University Press, New York, 1985.

3. K. Weissenberg, *Nature,* 159, 310 (1947).

4. J. D. Ferry, *Viscoelastic Properties of Polymers,* 3rd ed., Wiley, New York, 1980.

5. C. Robinson, J. C. Ward, R. B. Beevers, *Disc. Faraday Soc.,* 25, 29 (1958).

6. S. A. Hess, *Naturforsch,* 31A, 1034 (1976).

7. M. Doi, *Ferroelectrics,* 30, 247 (1980).

8. M. Doi and S. F. Edwards, *The Theory of Polymer Dynamics,* Oxford Press, New York, 1986.

9. N. Kuzuu and M. Doi, *J. Phys. Soc. Japan,* 53, 1031 (1984).

10. G. Marrucci and P. L. Mafettone, *Macromolecules,* 22, 4076 (1989).

11. R. G. Larson, *Macromolecules,* 23, 3983 (1990).

12. W. R. Burghardt and G. G. Fuller, 24, 2546 (1991).

13. J. J. Magda, S.-G. Baek, K. L. DeVries, and R. G. Larson, *Macromolecules,* 24, 4460 (1991).

14. G. Marrucci and F. Greco, in *Advances in Chemical Physics,* I. Prigogine and S. A. Rice, Eds., 86, 331, Wiley, New York, 1993.

15. S.-G. Baek, J. J. Magda, and R. G. Larson, *J. Rheol.,* **37,** 1201 (1993).

RHEOLOGY OF CONCENTRATED SOLUTIONS OF POLY(γ-BENZYL-GLUTAMATE)

GABOR KISS and ROGER S. PORTER

Materials Research Laboratory, Polymer Science and Engineering Department, University of Massachusetts, Amherst, Massachusetts 01003

SYNOPSIS

Steady shear viscosity, dynamic viscosity, dynamic modulus, and normal force were measured via rotational rheometry for concentrated solutions of racemic mixtures of poly(benzyl-L-glutamate) and poly(benzyl-D-glutamate) in m-cresol. A transition from the isotropic state to liquid–crystalline order with increase in concentration was indicated by optical anisotropy and maxima in all four material functions. This occurred at a critical concentration higher than the Flory prediction. Over a well-defined range of concentrations and shear stresses, some of the liquid–crystalline solutions exhibited negative first normal-stress differences that were not due to inertial effects.

INTRODUCTION

It has been known for some time that solutions of rigid rodlike particles form anisotropic, i.e., liquid–crystalline, solutions if the concentration is sufficiently high. Well-established theories [1–6] predict that as the concentration of a dilute (isotropic) solution is increased, a phase transition occurs to an anisotropic phase. As concentration is increased further, the fraction of anisotropic phase increases at the expense of the isotropic phase without, however, changing the concentrations in each phase until the solution becomes fully liquid–crystalline. This phenomenon has been demonstrated for systems of a variety of rodlike particles [7–10].

One class of rodlike molecules that has received considerable attention is the synthetic polypeptide poly(γ-benzyl-glutamate), usually the L enantiomer (PBLG). This polypeptide is obtainable in narrow molecular weight distributions and in appropriate solvents forms a helix that behaves in solution like a rigid rod [11]. When a single enantiomer is dissolved in an appropriate solvent, the helices all have the same optical rotary sense, resulting in a cholesteric liquid crystal. However, for equimolar mixtures of both L and D enantiomers, a nematic liquid crystal is obtained [12]. The early developments in this field have been reviewed [13]. The rheology of concentrated solutions of this polymer was studied by Hermans [14] and later by Iizuka [15]. Hermans studied several molecular weights of BPLG in m-cresol by capillary flow measurements. Iizuka used solutions of PBLG in CH_2Br_2 and in dioxane, and made steady-shear and oscillatory measurements using a cone-and-plate rheometer.

Journal of Polymer Science: Polymer Symposium 65, 193–211 (1978)
© 1978 John Wiley & Sons, Inc.   0360-8905/78/0065-0193$01.00

The present work is an extension of previous studies with distinctions in the following respects: (1) an equimolar mixture of PBLG and PBDG was used rather than a single enantiomer; (2) a wider range of homogeneous shear rates was accessible through the use of a variety of cones with different radii and cone angles for steady shear in the cone-plate geometry; (3) measurements of dynamic viscoelastic properties were made using the "eccentric rotating disk" geometry. This permits decoupling the viscous and elastic response of the system without the need for mechanical oscillation and the necessity of making precise phase angle measurements. This technique has been shown to give results equivalent to those from oscillatory shear for a variety of polymer solutions and melts [16].

EXPERIMENTAL

PBLG of molecular weight 350,000 and low dispersity was obtained from Biopolymer Corp., of Moreland Hills, Ohio, and PBDG of molecular weight 320,000 and low dispersity was obtained from Pilot Chemical Corp., of Watertown, Massachusetts. All solutions contained equal weights of both enantiomers; this will henceforth be indicated by the acronym PBG.

The (helicogenic) solvent selected was m-cresol in order to minimize concentration changes during measurement due to solvent evaporation. The m-cresol was distilled prior to use. Solutions of 3–10 wt % were made by successive evaporation of a dilute solution in a vacuum evaporator at 80°. Solutions of 25–11 wt % were made by successive dilutions of a 25 wt % solution that was obtained by introducing weighed amounts of PBG and m-cresol into a sealed container. A homogeneous solution was obtained in about 2 weeks.

Because of the high cost of the PBG and the high concentrations required, each solution was reused several times in the rheological measurements. Importantly, it has been shown that shear stresses higher than those achieved in this study neither disrupt the helical structure of PBG of comparable molecular weight dissolved in m-cresol nor cause mechanochemical degradation [17]. The solutions were observed to darken with time, but this was presumed to be due to oxidation of the solvent, which normally darkens on standing. In fact the ease with which m-cresol oxidizes may be a desirable property for inhibiting solute oxidation. After rheological measurements were completed, aliquots of each solution were retained, and the polymer was recovered by dissolving in methylene chloride followed by precipitation with methanol. The recovered PBG was then dissolved in dichloroacetic acid (a nonhelicogenic solvent) to 0.0467 ± 0.00002 wt %. Flow times through a Ubelhode viscometer were compared with that for a solution of fresh polymer. In all cases the flow times were found to be identical, indicating that no degradation had occurred.

Steady-shear viscosity and total normal-thrust measurements were made on a Rheometrics RMS-7200 mechanical spectrometer. Shear rates in the range 0.25–10,000 sec⁻¹ were accessible through the use of cone-and-plate geometries of the following radii (in cm) and cone angles (in rad) respectively: 5.00, 0.04; 2.50, 0.04; 2.50, 0.1; 1.25, 0.1; 1.25, 0.01. All data reported herein represented averages of many measurements. Precision in runs was about 5% for both viscosity and normal force; however, run-to-run reproducibility was no better than 20% in some cases. This scatter was not due to instrument instability but to some unknown phenomenon, occurring, possibly, at the platen surface.

An effort was made to obtain data at each shear rate using as wide a variation in geometry as possible. The data at intermediate shear rates were measured over the overlapping ranges for at least two different cones, with data at extremely low and high shear rates measured with one cone only. Data at shear rates above 1000 sec^{-1} do not refer to steady-state measurements due to exudation of sample from the gap at longer times. Reported values were extrapolated to time of initiation of shear. All data were taken with both clockwise and counterclockwise rotation, which gave identical results.

Dynamic viscosities and moduli were measured in the same instrument using the eccentric rotating-disk mode over a rotation speed range of 0.01–62.5 rad/ sec. Each reported point represents an average of many measurements generally made with plates of different radii and plate separation. To test whether results were in the linear viscoelastic region, each measurement was made at five different strains ranging from 0.15 to 0.75 at low rotation speeds and 0.03 to 0.15 at high rotation speeds. This procedure reduced the rather large scatter of data, which amounted to about 30% of the mean values reported.

All measurements were made at an ambient temperature of 24° ± 0.5°. Shear heating as measured by a thermocouple embedded in one platen was negligible.

RESULTS

Birefringence

Examination of the PBG solutions in m-cresol by polarized light microscopy indicated that solutions 8.1 wt % polymer and below were optically isotropic and solutions 9.9% and above were optically anisotropic. This result may be compared with conditions expected to produce anisotropy according to theory. For example, PBG of molecular weight 335,000 corresponds to 1506 residues. Since the diameter of the α-helix is 15 Å and the length of the helix is 1.5 Å per residue, the axial ratio would be 150 [18]. The equation given by Flory [3]

$$\phi_2^* = \frac{8}{p}\left(1 - \frac{2}{p}\right)$$

yields a critical volume fraction for formation of an anisotropic phase of 0.053. This corresponds to a concentration of 6.9 wt % for PBG in m-cresol. Thus the Flory theory appears to underestimate slightly the critical concentration for formation of the anisotropic phase. An error in this direction would be expected if the helices were not perfectly rigid [6].

Steady-Shear and Dynamic Viscosities

Steady-shear viscosity measurements were made over a wide range of shear rates for all concentrations. In all cases a low-shear limiting viscosity was obtained, but in no case could a high-shear limit be observed. Plotting the low-shear limiting viscosity reveals a rapid increase with concentration to a maximum at 11.0 wt % followed by an equally rapid decrease to a minimum at 22 wt % (see Fig. 1). At yet higher concentrations there is the suggestion of further gradual increase in viscosity. The shoulder that appears on the low concentration side of the viscosity maximum is unlikely to be due to experimental error. Moreover a comparable feature has been reported by Iizuka [15].

Steady-shear viscosity measurements for the three lowest concentrations did not extend to a shear rate high enough to obtain an unambiguous power law index, (i.e., the exponent n such that $\tau_{12} \propto \dot{\gamma}^n$), but an estimate of $n = 0.12$ can be made (see Fig. 2). The solutions of concentration 9.9 wt % and above all gave measurements in the power-law region; in some cases $\log \eta$ versus $\log \dot{\gamma}$ was linear for up to four decades of shear rate (see Fig. 3). The power-law indices obtained for these solutions were in the range $0.45 \le n \le 0.5$.

It can be seen in Figures 2 and 3 that the flow curves for the different solutions intersect, implying that the concentration at which the viscosity reaches a maximum depends on the shear rate. Indeed, it would be expected that the effect of shear orientation would be to ease the formation of the anisotropic phase. This behavior is seen explicitly in Figure 4, which shows that the concentration of maximum viscosity decreases with shear rate, in agreement with the observation of Hermans [14].

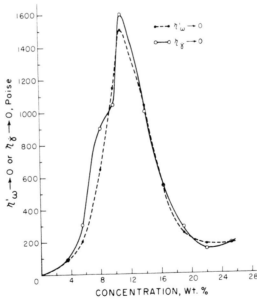

FIG. 1. Low-shear limit of steady-shear viscosity and low-frequency limit of dynamic viscosity versus concentration for PBG in m-cresol.

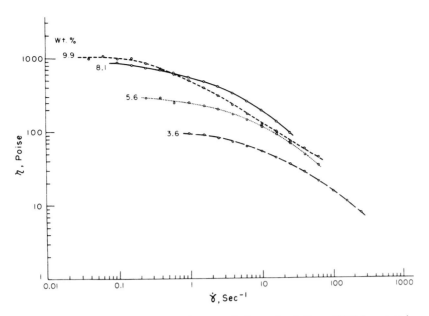

FIG. 2. Steady-shear viscosity versus shear rate for isotropic solutions of PBG in m-cresol.

Dynamic viscosity measurements were made for all concentrations over a frequency range of three decades in most cases (see Figs. 5 and 6). Clear low-frequency limiting values were indicated which, in most cases, agree well with the low-shear limit of η (see Fig. 1). The maximum value of $\eta'_{\omega \to 0}$ was observed at the same concentration as the maximum value of $\eta_{\dot\gamma \to 0}$. However, the shoulder on the low concentration side of the viscosity maximum was clearly absent. Also it is not clear whether the dynamic viscosity would have increased gradually with further concentration or had reached a limiting value.

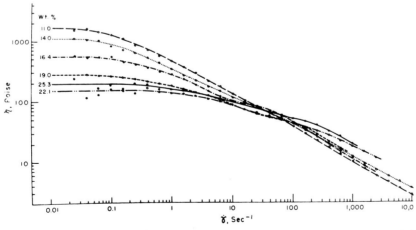

FIG. 3. Steady-shear viscosity versus shear rate for liquid–crystalline solutions of PBG in m-cresol.

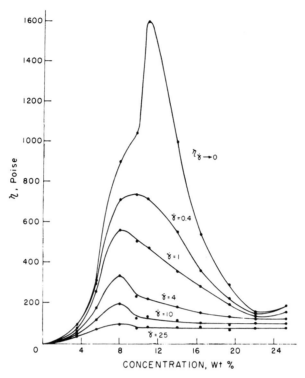

FIG. 4. Concentration dependence of steady-shear viscosity at several shear rates for PBG in *m*-cresol.

The curves of log η' versus log ω for various concentrations (see Figs. 5 and 6) generally did not cross, indicating that the peak in η' versus C would not be shifted to lower concentration by an increase in frequency. Thus, unlike steady shear, dynamic shear does not appear to drive the thermodynamic transition to the anisotropic phase to a lower concentration.

It is also suggested by Figure 6 that plots of log η' versus log ω converge to a common curve at high ω for all concentrations above that at which the viscosity peak was observed. This behavior is in contrast to that of the same solutions in steady shear for which curves of log η versus log $\dot\gamma$ all intersect at approximately the same point (see Fig. 3). It is of note that the curves of log η versus log $\dot\gamma$ change order at this intersection point; i.e., viscosity decreases with concentration for solutions between 11.0 and 25.3 wt % at lower shear rates and increases with concentration at higher shear rates.

The plateaus in curves of log η' versus log ω and to a lesser extent in log η versus log $\dot\gamma$ were experimentally significant and reproducible. They may have been a consequence of the use of a mixture of two polymers (PBLG and PBDG) of slightly different molecular weight.

It is worthwhile to contrast the behavior of solutions on either side of the viscosity-concentration maximum with increasing deformation rate, i.e., shear rate or frequency. The steady-shear viscosities of the isotropic solutions—those

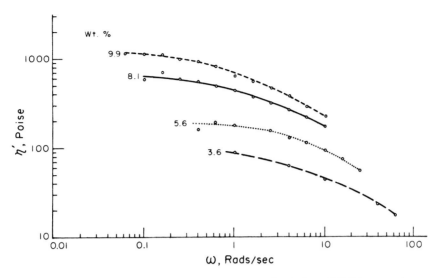

FIG. 5. Dynamic viscosity versus frequency for isotropic solutions of PBG in *m*-cresol.

below the viscosity maximum—was much more dependent on shear rate than those of the liquid-crystalline solutions—those above the viscosity maximum—as indicated by the difference in power-law exponents (see Figs. 2 and 3). The dependence of dynamic viscosity on frequency was also slightly greater for isotropic solutions, though this difference is less evident because of the curvature of the graphs and the restricted range of measurements (Figs. 5 and 6). A more easily seen difference in the dynamic viscosity behavior is that log η' versus log ω tends to converge to a common curve for the liquid-crystalline solutions (Fig. 6) but not for the isotropic solutions (Fig. 5).

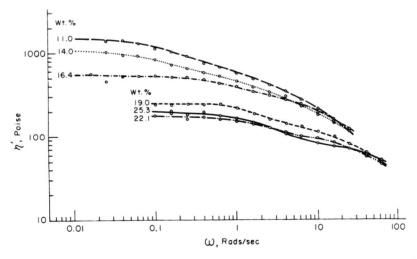

FIG. 6. Dynamic viscosity versus frequency for liquid-crystalline solutions of PBG in *m*-cresol.

200   KISS AND PORTER

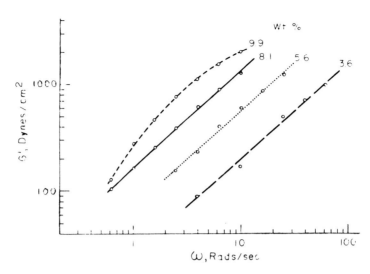

FIG. 7. Dynamic modulus versus frequency for isotropic solutions of PBG in *m*-cresol.

Dynamic modulus

Dynamic modulus G' measurements were made for all solutions over a range of about two decades of frequency ω (Figs. 7 and 8). Log G' versus log ω increases linearly with a slope of slightly less than unity for most concentrations. However, log G' versus log ω for the 9.9 wt % solution is curved and intersects the curve

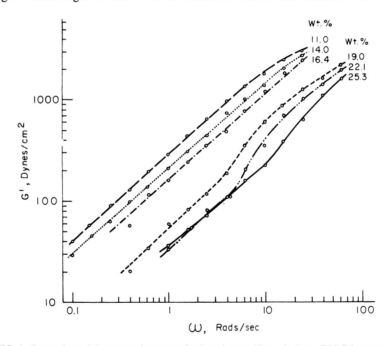

FIG. 8. Dynamic modulus versus frequency for liquid-crystalline solutions of PBG in *m*-cresol.

for 11.0 wt % at a frequency of 1 rad/sec. This means that for frequencies greater than 1 rad/sec the maximum modulus occurs at 9.9 wt %, whereas for frequencies less than 1 rad/sec the maximum modulus occurs at 11.0 wt %. In Figure 9, G' at a frequency of 4 rad/sec is plotted against concentration. It is quite similar in appearance to $\eta_{\dot\gamma\to 0}$ versus C and $\eta'_{\omega\to 0}$ versus C (Fig. 1). There is a suggestion of a shoulder on the high concentration side of the maximum, possibly due to the coexistence of an isotropic phase and an anisotropic phase at these concentrations. This would correspond to a concentration between Robinson's A point and B point [10]. This curve of G' versus C differs from those obtained by Iizuka [15] for PELG in dioxane or CH_2Br_2 in that G' does not appear to increase sharply again at the highest concentrations, where the solutions were fully liquid–crystalline.

Referring again to Figure 8, we observe that the curve of log G versus log ω deviates from linearity for the three highest concentrations at higher frequencies. The onset of these deviations appear to be at frequencies such that G' was approximately constant at 100–200 dyn/cm^2.

It is significant that the slopes of log G' versus log ω are similar for both liquid–crystalline and isotropic solutions, whereas log η versus log $\dot\gamma$ and log η versus log ω are quite different (see Figs. 2, 3, 5, and 6). The value of this slope, slightly less than unity, is in contrast to the Kirkwood-Auer [19] prediction of a slope of 2 at low frequencies, increasing to an asymptotic value at high frequency. Of course, the K-A theory was derived for dilute solutions of rigid rods and cannot be applied to concentrated solutions. Predictions of the dynamic elastic behavior of liquid–crystalline or concentrated solutions of rigid rods are lacking.

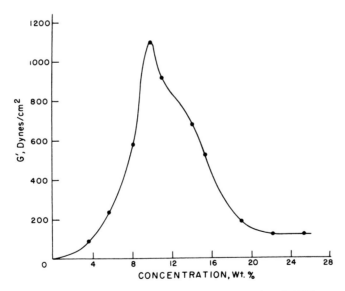

FIG. 9. Dynamic modulus at $\omega = 4$ rad/sec versus concentration of PBG in m-cresol.

First Normal-Stress Difference

The first normal-stress difference, $\tau_{11} - \tau_{22}$, was measured via total normal thrust for flow in cone-and-plate geometry for all solutions over a wide range of shear rates.

The four least concentrated solutions were investigated prior to improvements in instrumentation that greatly extended the accessible range of shear rate. All results are shown in Figure 10. The plots of log $\tau_{11} - \tau_{22}$ versus log $\dot{\gamma}$ are convex and parallel. The maximum value of $\tau_{11} - \tau_{22}$ was exhibited by the 8.1 wt % solution over the entire range of shear rate. Log $(\tau_{11} - \tau_{22})$ versus log $\dot{\gamma}$ for the 11.0 wt % solution is shown in Figure 11. This curve has a peculiar appearance, with an inflection point at 1 sec^{-1} followed by a convex region, another inflection point at 70 sec^{-1}, and finally a long linear region with slope 0.6. It is possible, but not likely, that the four least concentrated solutions would have behaved similarly if a greater range of shear rate had been accessible and that those measurements which could be made fell fortuitously in the convex region.

The curve for $\tau_{11} - \tau_{22}$ for the 14.0 wt % solution is presented in Figure 12. Its appearance is peculiar indeed, with $\tau_{11} - \tau_{22}$ increasing with $\dot{\gamma}$ for low shear rate, reaching a maximum at $\dot{\gamma} \approx 0.2$ sec^{-1}, abruptly becoming negative at $\dot{\gamma} \approx 0.6$ sec^{-1}, reaching a maximum negative value at $\dot{\gamma} \approx 3$ sec^{-1}, and abruptly becoming positive again at $\dot{\gamma} \approx 5$ sec^{-1}, then increasing to a linear region with slope of approximately 0.7. This remarkable behavior was also observed for solutions of concentrations 16.4, 19.0, and 22.1 wt % (see Figs. 13–15). The solution of concentration 25.3 wt % did not exhibit unusual (negative) normal stress behavior (see Fig. 16).

It can be observed in Figures 12–15 that the shear rates at which the reversals in sign of $\tau_{11} - \tau_{22}$ occurred were concentration-dependent. The shear stress

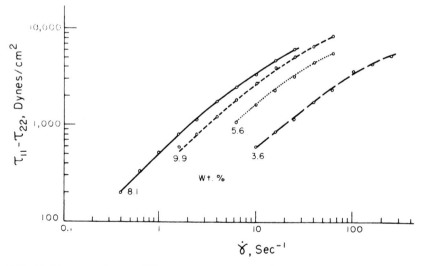

FIG. 10. First normal-stress difference versus shear rate for isotropic solutions of PBG in m-cresol.

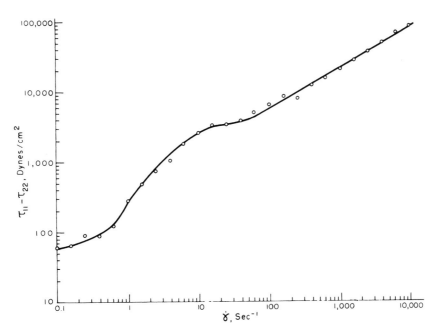

FIG. 11. First normal-stress difference versus shear rate for PBG in *m*-cresol (11.0 wt %).

to which the fluids were being subjected at these reversal points is shown in Figure 17 as a function of concentration. The experimental points frame an area in which negative first normal-stress differences were observed. The dashed extrapolations intersect at a concentration of 11 wt %, consistent with the positive $\tau_{11} - \tau_{22}$ measured at all shear rates for the 11.0 wt % solution.

An effort was made to exclude the possibility that these unusual (negative) normal stress observations were the result of artifact. It has been known for some time that inertial forces, which are neglected in the conventional analysis relating first normal-stress difference to total normal thrust in cone-and-plate flow, can make a negative contribution to the normal thrust [20]. The inertial contribution can in fact produce a spurious sign reversal in $\tau_{11} - \tau_{22}$ for certain polymer solutions that is eliminated when a correction is applied (see Fig. 18). This correction was applied to all normal-thrust data and found to be much smaller than the measured normal thrust in all our observations of PBG in *m*-cresol. Since the only material property that influences the magnitude of the inertial correction is density, inertial effects cannot be responsible for the negative first normal-stress difference observed here.

Another precaution that was taken to preclude artifacts was the use of a variety of cones of different radii and cone angles. It would be expected that any contribution due to edge effects or secondary flow would be of very different magnitudes for different cones and would have been immediately obvious.

Finally it should be noted that both total normal thrust, from which $\tau_{11} - \tau_{22}$ is calculated, and torque, from which η is calculated, were measured simultaneously, both on the bottom plate below the rotating cone. Therefore, it would

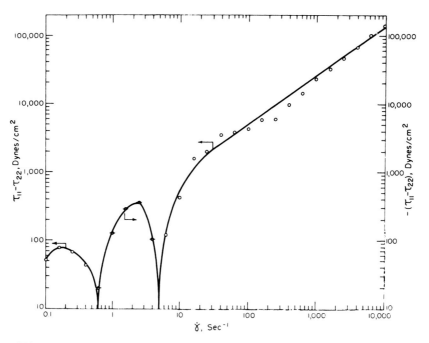

FIG. 12. First normal-stress difference versus shear rate for PBG in *m*-cresol (14.0 wt %).

be expected that any gross departure from the simple laminar flow field assumed by conventional analyses of cone-and-plate flow would also have been manifested in the apparent viscosity behavior as a discontinuity or change in slope.

A possible rationalization of a negative normal thrust can be found in postulating a dilatometric effect, i.e., a reduction of sample volume on shearing. Efforts to correlate the shape of the free surface at the edge of the cone with negative normal thrust observations were frustrated by changes in shape caused by incipient secondary flow or the fluid's being forced out by centrifugal force. Another pertinent observation is that normal thrust values, both positive and negative, were established almost immediately on initiation of shear and were time-independent for the duration of the measurement (up to 15 min). The same normal thrust values were obtained for both clockwise and counterclockwise rotation. Decay of the normal thrust on cessation of shear was very rapid. All these observations argue against the negative normal thrust's being a dilatometric effect.

DISCUSSION

In the course of the exposition of the experimental results we have had occasion to use the terms *liquid–crystalline* and *isotropic* to describe solutions of different concentrations. Strictly speaking, the term *liquid–crystalline* should be applied to any fluid that exhibits long-range order, e.g., a solution of rodlike molecules at the A point, at which the anisotropic phase first emerges. Previous workers

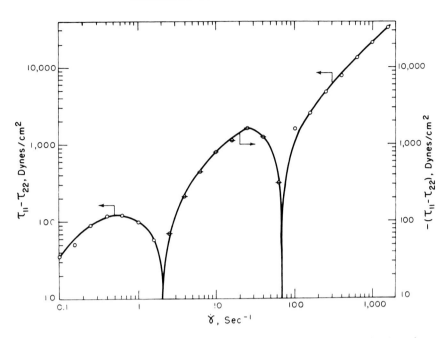

FIG. 13. First normal-stress difference versus shear rate for PBG in *m*-cresol (16.4 wt %).

have identified the concentration of maximum viscosity with the A point [21] and have accordingly labeled solutions of higher concentrations as liquid–crystalline and those of lower concentration as isotropic.

It is unlikely, however, that the formation of a small fraction of an anisotropic

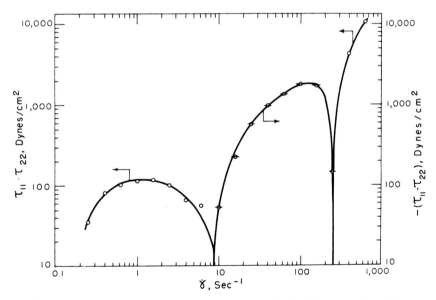

FIG. 14. First normal-stress difference versus shear rate for PBG in *m*-cresol (19.0 wt %).

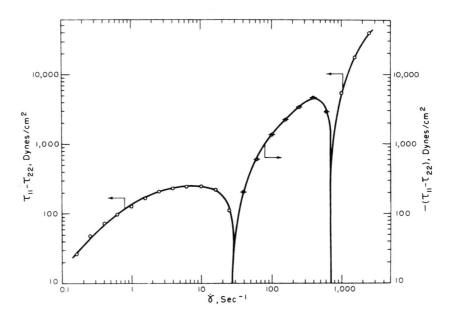

FIG. 15. First normal-stress difference versus shear rate for PBG in *m*-cresol (22.1 wt %).

phase at the A point would be reflected in an immediate and drastic change in the macroscopic properties of the solution as a whole. It is more likely that the anisotropic phase must comprise a substantial fraction of the solution; this would correspond to a concentration somewhere between the A point and the B point, that at which the solution becomes wholly anisotropic. It would not be unexpected that different material properties of the solution would be sensitive to the presence of an anisotropic phase to varying degrees. Thus depending on which property one were to observe, the solutions would appear to become liquid-crystalline at different concentrations. As shown in Table I, liquid-crystalline order was manifested at different concentrations by various properties, corresponding to different ratios of anisotropic phase to isotropic phase. The Flory prediction of the critical concentration for formation of an anisotropic phase is in fact lower than the indications of any of the observed properties. It is interesting that birefringence, which would be expected to be the most sensitive to the presence of a small amount of anisotropic phase coexisting with a large amount of isotropic phase, gave a higher estimate for the critical concentration than first normal-stress differences.

The steady-shear viscosity measurements presented here are in qualitative agreement with those of previous workers [14, 15] in that viscosity increased sharply with concentration to a maximum value after which it dropped sharply and also in that solutions beyond the viscosity maximum were partially or fully liquid-crystalline. We feel, however, that the viscosity peak does not uniquely mark the emergence of the anisotropic phase but that some of this phase is present before the maximum viscosity is achieved. The magnitudes of the viscosities observed were similar to those previously reported, but detailed

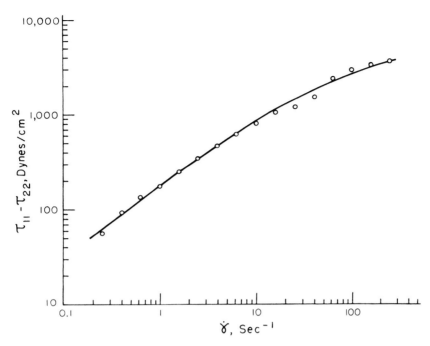

FIG. 16. First normal-stress difference versus shear rate for PBG in *m*-cresol (25.3 wt. %).

comparisons were impossible, since previous investigators used systems with slight differences, e.g., solvent, molecular weight of PBLG. In particular no previous reports of the rheological behavior of equimolar mixtures of PBLG and PBDG are available

It would be anticipated that concentrated solutions of an equimolar mixture of PBLG and PBDG, which form nematic liquid crystals [10], would exhibit different rheological properties from concentrated solutions of a single enantiomer, which form cholesteric liquid crystals. Cholesteric liquid crystals are generally considered to be special cases of nematics in which adjacent molecular layers are slightly displaced, leading to a helical superstructure with its screw axis perpendicular to the molecular layers [22]. Experiments on low-molecular-weight thermotropic liquid crystals indicate that the viscosity changes at the nematic → isotropic-liquid and cholesteric → isotropic-liquid transitions are markedly different [23]. Nematics generally have lower viscosities than their

TABLE I

Indication of liquid crystal order	Concentration (wt %)
Optical anisotropy (birefringence)	9.9
Maximum in $\eta_{\dot{\gamma} \to 0}$ vs c	11.0
Maximum in $\eta_{\omega \to 0}$ vs c	11.0
Maximum in G' vs c	9.9
Maximum in $\tau_{11} - \tau_{22}$ vs c	8.1
Flory theory	6.9

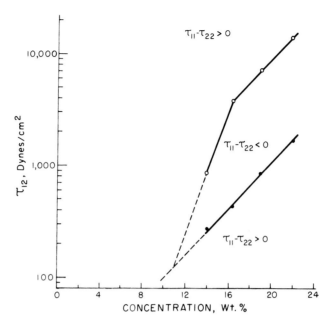

FIG. 17. Shear stress at points of change in sign of first normal-stress difference versus concentration for PBG in *m*-cresol.

isotropic liquids, since their molecules are readily oriented in the direction of flow measurements. Cholesterics generally have higher viscosities than their corresponding isotropic liquids. This behavior may be understood by reference

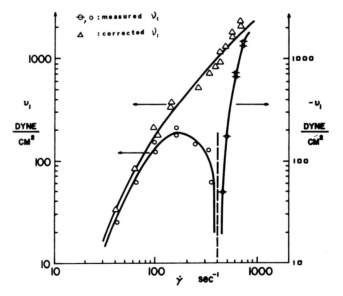

FIG. 18. First normal-stress difference versus shear rate for 2% aqueous polyacrylamide solution ($\overline{M}_n = 1.56 \times 10^6$).

to results obtained by Pochan and Marsh [24], who observed that mixtures of cholesteryl chloride and cholesteryl oleyl carbonate were found to exhibit the Grandjean texture at low shear rate (helical screw axes perpendicular to direction of shear). The effect of increased shear was to tilt the helical screw axes to a direction parellel to the direction of shear (dynamic focal conic texture), meaning that the molecules themselves would be oriented perpendicular to the direction of shear.

Another difference in the rheological behavior of thermotropic cholesterics and nematics is the observation that nematics are Newtonian whereas cholesterics are non-Newtonian [23]. Thus, the presence of cholesteric superstructure has a significant effect on the rheology of thermotropic liquid crystals.

For this reason it is interesting that both nematic and cholesteric solutions of helical polypeptides behave so similarly. The viscosity–concentration behavior of PBLG in m-cresol reported by Hermans [14] is very similar to that of our racemic mixtures and also for solutions of nonchiral rodlike macromolecules [25, 26]. A direct comparison of PELG in CH_2Br_2 with a mixture of PELG and PEDG in the same solvent revealed very similar viscosity behavior [15].

Our observation was that for concentrations up to 25.2 wt %, the liquid–crystalline solutions were of lower viscosity than the isotropic solutions at the highest concentration. It may be argued that since the cholesteric superstructure in solutions of a single enantiomer would be quite weak, probably much weaker than that in thermotropic cholesterics, and easily disrupted by shear, rheological measurements on such solutions were in fact made on nematic liquid crystals. Such a shear-induced cholesteric $\rightarrow$ nematic transition process would be analogous to the "unwinding" of a cholesteric in a magnetic field [27]. Such an argument would be strengthened by the observation of yield stresses in solutions of a single enantiomer (nominally cholesteric) and not in solutions of mixtures of both enantiomers (nematic).

The observation of normal stresses for liquid crystals has been infrequently reported. Erhardt, Pochan, and Richards [28] have reported normal-stress measurements on mixtures of cholesteryl chloride and cholesteryl oleyl carbonate. They observed behavior consistent with that of a second order fluid, i.e., that $\lim_{\omega \to 0} G'/\omega^2 = \lim_{\dot{\gamma} \to 0} [(\tau_{11} - \tau_{22})/2\dot{\gamma}^2]$ [29].

Normal-stress measurements on concentrated solutions of helical polypeptides, including PBLG, were conducted by Iizuka [15, 30]. However, the results were not reported but were used to calculate extinction angles from which the rotary diffusion constant was deduced and thence an apparent particle size for molecular clusters from tables given by Scheraga [31]. Therefore, our report of negative normal forces appears to be unprecedented in the liquid–crystal literature and in fact may be rare in the rheological literature. Coleman and Markovitz [29] demonstrated that for a second-order fluid in slow Couette flow the viscoelastic contribution to normal thrust must have a sign opposite to the inertial contribution on thermodynamic grounds. Walters [32] reports that measurements of first normal stress difference have invariably led to a positive quantity except for one which was later found to be in error. Adams and Lodge [33] reported the possible observation of a negative value of $\tau_{11} - \tau_{22}$ for solutions of polyisobutylene in decalin. This result was obtained by a combination of $\tau_{11} + \tau_{22} - 2\tau_{33}$, obtained from radial variation in normal stress in a cone-and-plate, with measurements of $\tau_{22} - \tau_{33}$ that were of uncertain accuracy. However, in a discussion of this point, Adams and Lodge concluded that negative values of $\tau_{11} - \tau_{22}$ would be curious but not impossible.

Other observations of negative normal stress have been reported. In a study of rheological properties of aqueous lecithin solutions Duke and Chapoy [34] observed a positive normal thrust in steady-shear cone-and-plate flow, but, on reversal of the direction of rotation, the normal thrust became negative and then increased to the positive steady-state value. This effect was attributed to incomplete normal-stress relaxation on cessation of flow.

Time-dependent, negative, normal thrust observations were reported by Huang [35] on melts of an SBS copolymer. In this case a negative normal thrust was generated on initiation of shear but decayed to zero after about 75 sec. This effect was thought to be due to a small volume decrease caused by shearing.

In a personal communication Iizuka [36] reported negative normal stress in solutions of PBLG in CH_2Br_2 of concentration 10 vol % or greater. This is in agreement with our observation of negative normal thrust only in liquid–crystalline solutions of PBG in m-cresol. This effect was ascribed by Iizuka to the adhesive force of the solution.

Our observations differ from those referred to above in that the negative normal thrust was not time-dependent, nor did it appear only on sudden reversal of the direction of rotation.

CONCLUSION

A theory for the rheological properties of liquid–crystalline solutions of rigid rodlike molecules is lacking. Progress has been made in the flow properties of thermotropic (bulk) liquid crystals from a continuum approach [37], and dilute solution theories of rigid rodlike molecule are very successful [19, 38]. However, neither of these could be expected to apply to this work.

Therefore, we have merely described our observations in the thought that they will stimulate progress toward an understanding of, and perhaps a theory for, the rheology of such fluids.

SYMBOLS AND ABBREVIATIONS

C	Concentration, wt %
G'	Dynamic modulus
n	Power-law index
P	Axial ratio
PBDG	Poly(γ-benzyl-D-glutamate)
PBLG	Poly(γ-benzyl-L-glutamate)
PBG	Equimolar mixture of BPLG and PBDG
$\dot{\gamma}$	Shear rate
η	Steady-shear viscosity

$\eta_{\dot{\gamma} \to 0}$	Low shear limit of steady-shear viscosity
η'	Dynamic viscosity
$\eta'_{\omega \to 0}$	Low-frequency limit of dynamic viscosity
ν_1	First normal-stress difference
τ_{ij}	Stress
$\tau_{11} - \tau_{22}$	First normal-stress difference
ϕ_2^*	Critical volume fraction of solute for phase separation

REFERENCES

[1] L. Onsager, *Ann. N.Y. Acad. Sci., 51,* 627 (1949).

[2] A. Ishihara, *J. Chem. Phys., 19,* 1142 (1951).

[3] P. J. Flory, *Proc. R. Soc. London, A234,* 60 (1956).

[4] E. A. DiMarzio, *J. Chem. Phys., 35,* 658 (1961).

[5] J. P. Straley, *Mol. Cryst. Liq. Cryst., 22,* 333 (1973).

[6] S. Ya. Frenkel, *J. Polym. Sci., C44,* 49 (1974).

[7] F. C. Bawden and N. W. Pirie, *Proc. R. Soc. London, B123,* 274 (1937).

[8] J. D. Bernal and I. Frankuchen, *J. Gen. Physiol., 25,* 111 (1941).

[9] S. Ya. Frenkel, V. G. Baranov, and T. I. Volkov, *J. Polym. Sci., C16,* 1655 (1967).

[10] C. Robinson, J. C. Ward, and R. B. Beevers, *Discuss. Faraday Soc., 25,* 29 (1958).

[11] P. Doty, J. H. Bradbury, and A. M. Holtzer, *J.A.C.S., 78,* 947 (1956).

[12] C. Robinson and J. C. Ward, *Nature, 180,* 1183 (1957).

[13] R. S. Porter and J. F. Johnson, "Rheology: Theory and Applications," Vol. IV, F. R. Eirich, Ed., Academic, New York, pp. 317–345.

[14] J. Hermans, Jr., *J. Colloid Sci., 17,* 638 (1962).

[15] E. Iizuka, *Mol. Cryst. Liq. Cryst., 25,* 287 (1974).

[16] C. W. Macosko and W. M. David, *Rheol. Acta, 13,* 814 (1974).

[17] J. T. Yang, *J.A.C.S., 81,* 3902 (1959).

[18] Y. Layec and C. Wolff, *Rheol. Acta, 13,* 696 (1974).

[19] J. G. Kirkwood and P. L. Auer, *J. Chem Phys., 19,* 281 (1951)

[20] W. M. Kulicke, G. Kiss, and R. S. Porter, *Rheol. Acta, 16,* 568 (1977).

[21] E. T. Samulski and A. V. Tobolsky in "Liquid Crystals and Plastic Crystals," Vol. 1, G. W. Gray and P. A. Winsor, Eds., Halsted Press, New York, 1974, pp. 175–199.

[22] F. D. Saeva, *Mol. Cryst. Liq. Cryst., 23,* 271 (1973).

[23] R. S. Porter, E. M. Barrall, II, J. F. Johnson, *J. Chem. Phys., 45,* 1452 (1966).

[24] J. M. Pochan and D. G. Marsh, *J. Chem. Phys., 57,* 1193 (1972).

[25] S. P. Papkov et al., *J. Polym. Sci. 12,* 1753 (1974).

[26] S. Kwolek, "Optically Anisotropic Aromatic Polyamide Dopes," U.S. Patent 3,671,542.

[27] P. G. DeGennes, *Solid State Commun., 6,* 163 (1965).

[28] P. F. Erhardt, J. M. Pochan, W. C. Richards, *J. Chem. Phys., 57,* 3596 (1972).

[29] B. D. Coleman and H. Markovitz, *J. Appl. Phys., 35,* 1 (1964).

[30] E. Eizuka, *J. Phys. Soc. Jpn., 35,* 1792 (1973).

[31] H. A. Scheraga, J. T. Edsall, J. D. Gadd, *J. Chem. Phys., 19,* 1101 (1951).

[32] K. Walters, "Rheometry," Halsted Press, New York, 1974, p. 88.

[33] N. Adams and A. S. Lodge, *Philos. Trans. R. Soc., 256,* 149 (1964).

[34] R. W. Duke and L. L. Chapoy, *Rheol. Acta, 15,* 548 (1976).

[35] T. A. Huang, Ph.D. Thesis, University of Wisconsin, Department of Engineering Mechanics, 1976, p. 93.

[36] E. Iizuka, personal communications, April 1977.

[37] F. M. Leslie, *Arch. Rat. Mech. Anal., 28,* 265 (1968).

[38] M. C. Williams, *A.I.C.E.J., 21,* 1 (1975).

COMMENTARY

Reflections on "Isobaric Volume and Enthalpy Recovery of Glasses. II. A Transparent Multiparameter Theory," by A. J. Kovacs, J. J. Aklonis, J. M. Hutchinson, and A. R. Ramos, *J. Polym. Sci., Polym. Phys. Ed.,* 17, 1097 (1979)

J. J. AKLONIS

c/o yacht "SILKE," Westhaven Marina, P.O. Box 1560, Auckland, New Zealand

In the past, the importance of polymer science was clearly recognized in industrial laboratories but was often overlooked in academia, especially by the most prestigious institutions. I believe that this situation is now rapidly changing. Broader recognition of our field could not have resulted without the existence of first rate scientific journals devoted to polymer science. It is indeed a pleasure to congratulate the editors, staff, and authors who have worked hard over the last half century on the *Journal of Polymer Science* as they continue to see the fruits of their labors ripen.

In the early 1970s, there was considerable debate about just what constituted the glass transition, or "Tee Gee" as it had become known. To be sure, this characteristic parameter could be easily measured for most polymers and many liquids in any laboratory using thermal analysis equipment and its practical importance was obvious. Still, the underlying physics of the "transition" was not clear.

Fundamental work aimed at understanding the glass transition separated into two broad categories. One, primarily led by Robert Simha and his associates[1] and Julian Gibbs and Ed DiMarzio,[2] had made considerable progress using thermodynamics and treating T_g within this framework.

The other approach concentrated on the kinetic aspects and here Andre Kovacs made his contributions. During his thesis work, Andre had started a monumental study of the kinetic aspects of the glass transition which culminated in his 1963 *Fortsch.* paper.[3] As was typical of Andre's work, the paper was very complete and thus quite long. In addition, Andre had a habit of working and reworking each word, sentence and paragraph to make the language as concise as possible. Being written in French in this style, the paper was quite difficult for most of the scientific community to fathom. It also didn't help that Figure 23, probably the most important figure in the paper, was mistakenly labeled. (I remember Andre's commenting that less than five people in the world had read and understood the paper after which he laughed heartily.) In retrospect, this situation was very unfortunate since there is a wealth of experimental information and insightful analysis in this paper and it is astounding that virtually all of the measurements were made using nothing more than simple mercury filled dilatometers and thermostated fish tanks!

I was lucky to spend a sabbatical with Andre and his group in Strasbourg. When I arrived anxious to work on the glass transition, Andre told me that there was no longer anything exciting happening in this area and I should help him with his new work on polymer crystals. John Hutchinson was on a post doctoral at the same time and was the only person actually working on amorphous glasses. Although I never did learn much about crystalline polymers, after several months we did come up with the model which is described in the paper.

In my opinion, the main contribution of this paper is the explanation of the complicated aspects of the kinetics of the glass transition phenomenon in terms of a very simple model. The fundamental asymmetry of the kinetics, its non-linearity as well as memory effects, are all satisfyingly explained in terms of nothing more than a distribution of relaxation times, like those so familiar to anyone acquainted with viscoelasticity, which moves along the relaxation time scale according to the current nonequilibrium state of the glass.

Approximately four years elapsed between the initial formulation of these ideas[4] and their formal publication. During this period, Andre continued to explore the model and, with John Hutchinson and A. R. Ramos, showed that much of the previously unexplained low temperature behavior of glasses also was in accord with this simple model.

Having been out of active research for several years now, I am flattered to have been asked to write this short commentary and am most happy that the editors have chosen our paper as one of those to be recognized. I hope that my memory of the events which occurred during this period is accurate and that my comments help to clarify the context in which the work was done.

REFERENCES AND NOTES

1. Robert Simha and R. F. Boyer, *J. Chem. Phys.,* **37,** 1003 (1962); J. Moacanin and Robert Simha, *J. Chem. Phys.,* **45,** 964 (1966).
2. Julian H. Gibbs and Edmund A. DiMarzio, *J. Chem. Phys.,* **28,** 373 (1958).
3. A. J. Kovacs, *Fortsch. Hochpolym. Forsch. (Adv. Polym. Sci.),* **3,** 394 (1963).
4. J. M. Hutchinson, J. J. Aklonis, and A. J. Kovacs, *Polym. Prepr.,* **16,** 94 (1975).

PERSPECTIVE

Comments on "Isobaric Volume and Enthalpy Recovery of Glasses. II. A Transparent Multiparameter Theory," by A. J. Kovacs, J. J. Aklonis, J. M. Hutchinson, and A. R. Ramos, *J. Polym. Sci., Polym. Phys. Ed.,* 17, 1097 (1979)

GREGORY B. McKENNA

Polymers Division, NIST, Gaithersburg, MD 20899 USA

The "transparent multiparameter model" developed in the paper by Kovacs, Aklonis, Hutchinson, and Ramos (KAHR)[1] has become a classic in the field of polymer glasses. Its impact can, perhaps, be seen by the fact that in 1995, 17 years after its original publication, it was cited 13 times.[2] Since its publication the paper has been cited approximately 300 times.[2] However, citations alone do not determine the worth of a paper; at best they reflect it. The KAHR paper brought to the field both the technical achievement of the model itself and an intellectual challenge by making clear where the known physics were incorporated into the model and where the model does not work. It is both of these aspects that make this paper a truly major contribution to both Polymer Physics and the Physics of Glasses.

The KAHR paper brought to the Polymer Physics community ideas that had been developed from a somewhat different perspective in the Inorganic Glass community, beginning with some of the pioneering work of Tool,[3,4] who worked at the National Bureau of Standards during the 1930s and 1940s. Tool was the first to recognize that the relaxation time of a glass depends on both the temperature and the fictive temperature, the latter being a measure of the non-equilibrium structure of the glass. Subsequently, in 1971 Narayanaswamy[5] published a paper which extended the ideas of Tool into a mathematical or constitutive equation framework that provided for a phenomenological description of many of the nonlinear events observed in temperature-jump and tem-

perature-scan experiments. In the KAHR model, these ideas were brought independently into the domain of polymer glasses and expanded upon to include concepts that are still current in polymers, e.g., the free volume description of the glass transition. In addition, rather than evoke a fictive temperature, KAHR postulated that the relaxation times depended upon a normalized departure δ [$=(v - v_\infty)/v_\infty$] from equilibrium—a similar, but not identical, measure of the structure of the nonequilibrium glass.

This paper became a classic because the concepts incorporated in the model were calculated quantitatively in the paper and compared qualitatively with the original results of André Kovacs' own classical volume dilatometry measurements.[6] The model really begins with a deep understanding of the prior experimental results of the kinetics of glass forming systems below the glass temperature. What one can think of as the "essential ingredients" are summarized in the following paragraphs.

The first important ingredient arises from the observation that if one jumps to the temperature T_0 from an equilibrium state at $T_0 - \Delta T$ (up-jump) or down to T_0 from $T_0 + \Delta T$, the responses are highly asymmetric. Hence, as seen in Figure 1 (compare to Figure 1 in the KAHR paper), the up-jump response is significantly different from the down-jump (or *intrinsic isotherm*) response to the same temperature. This asymmetry is a manifestation of a nonlinearity induced by the fact (recognized first by Tool) that the recovery time governing the evolution of the volume depends upon the instantaneous structure of the glass. In this regard, then, KAHR introduced the idea of a time-structure equivalence much as one does in time-temperature superposition. In the down-jump, the mobility progressively decreases as the volume decreases in its evolution towards equilibrium. In the up-jump the mobility increases as the volume increases towards equilibrium. Hence, the structural recovery times start fast and get progressively longer in the down-jump experiment while in the up-jump they do the opposite. Naturally, near equilibrium, they tend towards the same values. The resulting strong nonlinearity gives the asymmetry in the approach curves seen in Figure 1.

The second essential ingredient arises from the observations made in the so-called memory or cross-over experiment. Here, one performs experiments in which the glass is aged or annealed at different temperatures for times such that upon making a jump to the final temperature T_0 the departure from equilibrium is zero ($\delta = 0$). In this instance, as depicted in Figure 2 (compare with Figure 9 in the KAHR paper), the volume recovers from zero (or near to it) and goes through a maximum before the material comes back into equilibrium. This response, from prior work by Kovacs,[7] could not be described by a single relaxation time and led to the postulate that a nonexponential decay function is required for the fundamental volume retardation response. In the KAHR paper a sum of exponentials was used (Moynihan and co-workers[8] at about the same time introduced a stretched exponential into the Tool–Narayanaswamy equations). The final "essential ingredient" arose because the KAHR team had a strong background in viscoelasticity and was interested in making a constitutive equation that "looks like" linear viscoelasticity. Hence, one can write that the structural recovery response at some time t can be written as:

$$\delta(t) = -\Delta\alpha \int_0^z R(z - z') \frac{dT}{dz'} \, dz'$$

Where the variables are defined as:
δ = structural departure from equilibrium = $(v - v_\infty)/v_\infty$ for the volume departure
$\Delta\alpha$ = the change in coefficient of thermal expansion in going from the glass to the liquid
$R(t)$ = the retardation function describing structural recovery;

$$R(t) = \delta_0 \sum_1^N g_i e^{-t/\tau_i}$$

δ_0 = initial structure departure from equilibrium
g_i = weighting factors

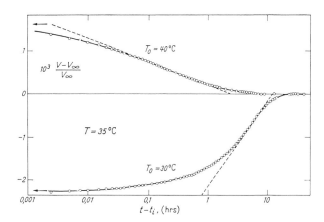

Figure 1. Asymmetry of approach in a polyvinyl acetate glass after down and up jumps of 5 K to 35°C. (After Kovacs, ref. 6, with permission).

τ_i = the retardation times
T = absolute temperature
z = reduced time:

$$z = \int_0^t \frac{d\xi}{a_T a_\delta}$$

a_T = time-temperature shift factor
a_δ = time-structure shift factor

The above set of equations, then, looks like linear viscoelasticity in the reduced time domain, but manifests a strongly nonlinear response in the time domain due to the fact that δ depends upon itself in the convolution integral through the reduced time z. By making physically realistic estimates of the model parameters, KAHR were able to solve the model numerically for the various thermal histories discussed. It was found that the equations manifested all of the nonlinear features seen in the experiments. This was a major development in the understanding of the physics of Polymer Glasses.

At the same time, KAHR also recognized that the model is lacking certain things and these are discussed in the paper. The observation, for example, that the τ-effective paradox and expansion gap in the asymmetry of approach experiments is not reproduced (see KAHR paper Figure 2 and Figure 23 in Reference 6) has led to efforts to expand upon the ideas in the KAHR paper using other reduced time concepts[9] or somewhat different constitutive equations.[10] The work also showed an equivalence between the total departure from equilibrium (rather than the normalized departure) and the fictive temperature and showed the essential equivalence of the stretched exponential and the sum of exponentials to describe the time dependence of the struc-

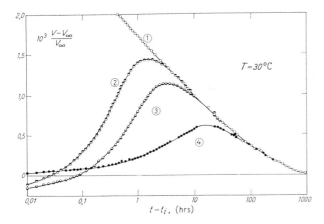

Figure 2. Memory response of a polyvinyl acetate glass at 30°C. Curve 1 is down jump from 40°C. Other curves are for annealing at (10°C, 160 h; 15°C 140 h and 25°C 90 h) prior to jumping up to 30°C. Note that the value of δ at the beginning of the curves is near to zero. (After Kovacs, ref. 6, with permission).

tural recovery function. In a set of appendices KAHR also proposed possible forms for the temperature and structure shift functions that form part of the calculational repertoire still in use today.

Finally, I expect because of the strong history in molecular and morphological aspects of polymers in the laboratories at the Centre de Recherches sur les Macromolecules in Strasbourg where the KAHR model was developed, the KAHR model attempts to include concepts of microstructure in the model. While such concepts are very difficult to validate and are nearly invariably disputable in amorphous systems, the idea that each local region has a local departure from equilibrium provides a physical picture of how the glassy structure is evolving. This picture has been tremendously elaborated in the works of, in particular Robertson,[11-13] and provide a reason, other than that provided by the phenomenological equations alone, for the glassy behavior in, e.g., the memory experiment. Furthermore, the concepts of the KAHR model for a distribution of environments permit greater nonlinearities than those actually calculated in their model since a simplifying assumption was made that the local mobility depends on the global departure from equilibrium to render the calculations more tractable. However, building models in which the local mobility depends on the local departure from equilibrium is a possibility that is now being explored and it shows promise for explaining some of the things that the KAHR model, in its simplified form, does not.[14]

REFERENCES AND NOTES

1. A. J. Kovacs, J. J. Aklonis, J. M. Hutchinson, and A. R. Ramos, *Journal of Polymer Science, Polymer Physics Edition,* **17,** 1097–1162 (1979).
2. Citation Index.
3. A. Q. Tool, *J. Am. Ceram. Soc.,* **29,** 240 (1946).
4. A. Q. Tool, *J. Res. Natl. Bur. Stand.* (U.S.), **37,** 73 (1946).
5. O. S. Narayanaswamy, *J. Am. Ceram. Soc.,* **54,** 491 (1971).
6. A. J. Kovacs, *Fortschritt. Hochpolym.-Forsch.,* **3,** 394 (1964).
7. J. M. Hutchinson and A. J. Kovacs, *J. Polym. Sci., Polym. Phys. Ed.,* **14,** 1575 (1975).
8. C. T. Moynihan, A. J. Esteal, and M. A. DeBolt, *J. Amer. Ceram. Soc.,* **59,** 12 (1976).
9. R. W. Rendell, J. J. Aklonis, K. L. Ngai, and G. R. Fong, *Macromolecules,* **20,** 1070 (1987).
10. S. R. Lustig, R. M. Shay, and J. M. Caruthers, *J. Rheology,* **V40,** 69 (1996).
11. R. E. Robertson, R. Simha, and J. G. Curro, *Macromolecules,* **17,** 911 (1984).
12. R. E. Robertson, R. Simha, and J. G. Curro, *Macromolecules,* **18,** 2239 (1985).
13. R. E. Robertson, R. Simha, and J. G. Curro, *Macromolecules,* **21,** 3216 (1988).
14. J. M. Caruthers, Purdue University, personal communication and A. B. Starry, Ph.D thesis, Purdue University, Dec. 1995.

Isobaric Volume and Enthalpy Recovery of Glasses. II. A Transparent Multiparameter Theory

A. J. KOVACS, J. J. AKLONIS,* J. M. HUTCHINSON,† and A. R. RAMOS,‡

Centre de Recherches sur les Macromolécules, Centre National de la Recherche Scientifique-Université Louis Pasteur, 67083 Strasbourg, France

Synopsis

A multiordering parameter model for glass-transition phenomena has been developed on the basis of nonequilibrium thermodynamics. In this treatment the state of the glass is determined by the values of N ordering parameters in addition to T and P; the departure from equilibrium is partitioned among the various ordering parameters, each of which is associated with a unique retardation time. These times are assumed to depend on T, P, and on the instantaneous state of the system characterized by its overall departure from equilibrium, giving rise to the well-known nonlinear effects observed in volume and enthalpy recovery. The contribution of each ordering parameter to the departure and the associated retardation times define the fundamental distribution function (the structural retardation spectrum) of the system or, equivalently, its fundamental material response function. These, together with a few experimentally measurable material constants, completely define the recovery behavior of the system when subjected to any thermal treatment. The behavior of the model is explored for various classes of thermal histories of increasing complexity, in order to simulate real experimental situations. The relevant calculations are based on a discrete retardation spectrum, extending over four time decades, and on reasonable values of the relevant material constants in order to imitate the behavior of polymer glasses. The model clearly separates the contribution of the retardation spectrum from the temperature-structure dependence of the retardation times which controls its shifts along the experimental time scale. This is achieved by using the natural time scale of the system which eliminates all the nonlinear effects, thus reducing the response function to the Boltzmann superposition equation, similar to that encountered in the linear viscoelasticity. As a consequence, the system obeys a rate (time) -temperature reduction rule which provides for generalization within each class of thermal treatment. Thus the model establishes a rational basis for comparing theory with experiment, and also various kinds of experiments between themselves. The analysis further predicts interesting features, some of which have often been overlooked. Among these are the impossibility of extraction of the spectrum (or response function) from experiments involving cooling from high temperatures at finite rate; and the appearance of two peaks in the expansion coefficient, or heat capacity, during the heating stage of three-step thermal cycles starting at high temperatures. Finally, the theory also provides a rationale for interpreting the time dependence of mechanical or other structure-sensitive properties of glasses as well as for predicting their long-range behavior.

INTRODUCTION

The glass transition of supercooled liquids, or rubbery polymers, occurs when the rate of molecular (structural, configurational) rearrangement due to thermal agitation becomes the same order of magnitude as the rate of cooling.[1-7] This

---

* Permanent address: Department of Chemistry, University of Southern California, University Park, Los Angeles, California 90007.

† Permanent address: Department of Metallurgy and Materials Science, The University of Nottingham, University Park, Nottingham N.G. 7 2rd, U. K.

‡ Present address: Alfa Division Industries, A. P. 5000 Monterrey N. L., Mexico.

Journal of Polymer Science: Polymer Physics Edition, Vol. 17, 1097–1162 (1979)
0098-1273/79/0017-1097$01.00

definition of the glass transition can be generalized to any experimental procedure which results in a gradual reduction of the rates of molecular motions, e.g., increase in pressure P or concentration c of the glass-forming component of the system, due to polymerization or solvent evaporation, etc. Whatever the cause, the effect of such treatment is a progressive departure of the system from (metastable) equilibrium as the temperature T becomes low enough (P or c high enough). Consequently, the free energy becomes larger than that of the equilibrium liquid at the same T, P, and c.

In the nonequilibrium glass, the random molecular motions are therefore biased by the excess free energy. This bias drives the density and energy fluctuation averages toward their equilibrium values uniquely determined by the intensive variables T, P, and c, thus giving rise to recovery of the equilibrium structure. Structural recovery can be detected easily by measuring the isothermal and/or isobaric variations of volume v, enthalpy H, or other structure sensitive parameters of the system. Glass-transition phenomena, originating from delayed structural recovery of liquids, are therefore controlled by the configurational mobility of molecular constituents and thus provide direct inspection of the overall (effective) rate of fluctuations and molecular motions in systems in which long-range order is absent.

Isobaric volume and enthalpy recovery generated by stepwise, or continuous temperature changes, at constant P, have been investigated for a great variety of glass forming systems, following the pioneering work of Tool.[2] Such investigations of isothermal recovery have shown three characteristic features: (i) *nonlinearity* with respect to the magnitude of the initial departure from equilibrium,[2] (ii) *asymmetry* with respect to its sign,[3,5–9] and (iii) *"memory effects,"* observed after two (or more) successive temperature steps.[6,7]

Both the nonlinearity and asymmetry of the approach to equilibrium have been attributed to the dependence of the retardation time(s) on the actual state (structure) of the glass[2,7] in addition to the dependence on the intensive variables: T, P, c. In fact, model calculations involving a *single* structure-dependent retardation time have been quite successful in describing both the nonlinearity and asymmetry of the recovery as observed in a limited class of experiments involving volume contraction,[5,7] whereas the agreement between such theories and experiments involving volume expansion is only qualitative.[7,8] One parameter theory, however, completely fails to describe memory effects. The latter must be attributed to the multiplicity of ordering parameters (or equivalent concepts) which are necessary for full characterization of the state of the glass. Since multiparameter models lead to linear behavior (provided that the departure from equilibrium is sufficiently small), here too one must assume that the retardation times (τ_i) involved depend on the instantaneous state of the system. These requirements for accurately describing the recovery behavior of glasses have been discussed in a recent review paper (Part I of this series),[9] which also introduces a new multiparameter treatment allowing the τ_i's to depend on structure.

In fact, in recent years, three similar multiparameter approaches have been proposed by Narayanaswamy,[10] Moynihan et al.,[11] and the present authors,[12] all involving the same basic assumptions but using different routes and approximations. The present paper deals with the development of the latter[12] and its theoretical predictions for a one-component glass ($c = 1$), subjected to a wide variety of thermal treatments at constant pressure.   Though this theory has already been tested in a semiquantitative manner using volume recovery data,[9,12] its full application to a variety of experimental data obtained with polymer glasses will be given elsewhere.[13]

PART I

Theory

The theoretical approach follows three consecutive steps, starting with the thermodynamic definition of the state of the system based on a set of ordering parameters.   This results in multiple retardation mechanisms, characterized, as in linear viscoelasticity, by a distribution of retardation times τ_i.   The corresponding spectrum is assumed to remain invariant; changes in temperature and structure merely shift the spectrum along the logarithmic time scale, by the same amount as τ_i, without altering its shape.   Thus the system is assumed to be "thermorheologically simple"[14] whatever the cause of the variation of τ_i.

In the second step, the dependence of τ_i on temperature and on the instantaneous state (structure) of the glass is introduced.   This results in a material response which is nonlinear and asymmetric with respect to the magnitude and the sign of the departure from equilibrium, respectively.

Finally, the response function is linearized by an appropriate reduction of the time scale.   This operation corresponds, in fact, to introducing the proper time scale of the system as determined by its spontaneous fluctuation rate in a reference equilibrium state.

The Thermodynamic State of the Glass

In the model adopted here, all states of the one-component system are determined by the intensive thermodynamic parameters, e.g., T and P, plus values of a set of ordering parameters ζ_i, where $1 \leq i \leq N$.   The actual number (N) of ordering parameters which is necessary to describe all states of the system is fixed, but its precise value is relatively unimportant in the development.   It is also unnecessary to define the physical nature of the ordering parameters; in thermodynamic equilibrium, however, one must have $\bar{\zeta}_i = \zeta_i(T,P)$, since the state of the system is uniquely defined by T and P alone.

The exact differential of any extensive property of the system, e.g., dv, is given as[15]

$$dv = \left(\frac{\partial v}{\partial T}\right)_{P,\zeta_1\cdots\zeta_N} dT + \left(\frac{\partial v}{\partial P}\right)_{T,\zeta_1\cdots\zeta_N} dP$$
$$+ \left(\frac{\partial v}{\partial \zeta_1}\right)_{T,P,\zeta_2\cdots\zeta_N} d\zeta_1 + \cdots + \left(\frac{\partial v}{\partial \zeta_N}\right)_{T,P,\zeta_1\cdots\zeta_{N-1}} d\zeta_N \quad (1)$$

Such an expression can be written and is useful whenever the control of the experimenter over the system ensures that the partial derivatives are meaningful.   In the derivation that follows, these partial derivatives will be related to the set of retardation times controlling the instantaneous rate of change of each ordering parameter.

Here, only constant pressure processes will be considered, so that $dP = 0$ and the subscript P will be omitted, although the partial derivatives may still depend on the actual values of both P and T. This simplification represents no limitation on the treatment which follows, but merely reflects the fact that most of the experiments considered here have been carried out under isobaric conditions.

The model thus implies the existence of an expansion coefficient:

$$\alpha_g = v^{-1}\left(\frac{\partial v}{\partial T}\right)_{\zeta_1\cdots\zeta_N} \tag{2}$$

which characterizes the thermal expansion of an ensemble of states (to be called the *glass* hereafter) for which the ordering parameters have any set of constant values, i.e., $d\zeta_i = 0$. On the other hand, the expansion of the system in liquid-like equilibrium is characterized by

$$\alpha_l = v_\infty^{-1}\left(\frac{dv_\infty}{dT}\right) \tag{3}$$

where v_∞, the equilibrium volume, is fully defined by T (and P) alone.

At equilibrium, the values of $\zeta_i = \bar{\zeta}_i$ are uniquely defined by T (and P), and in the present context, the only possible variations of v_∞ result from temperature changes. Therefore, applying eq. (1) to v_∞ yields

$$dv_\infty = \left(\frac{\partial v}{\partial T}\right)_{\bar{\zeta}_1\cdots\bar{\zeta}_N} dT + \sum_{i=1}^{i=N}\left(\frac{\partial v}{\partial\zeta_i}\right)_{T,\zeta_{j\neq i}}\left(\frac{d\bar{\zeta}_i}{dT}\right) dT \equiv v_\infty\alpha_l dT \tag{4}$$

which can be written

$$\frac{dv_\infty}{dT} = v_\infty\alpha_g + v_\infty\sum_{}^{N}\Delta\alpha_i \equiv v_\infty(\alpha_g + \Delta\alpha) \tag{5}$$

where the excess expansion coefficient of the liquid with respect to the glass

$$\Delta\alpha = \alpha_l - \alpha_g = \sum_{}^{N}\Delta\alpha_i \tag{6}$$

results from the thermal contributions of the equilibrium set of ordering parameters $\bar{\zeta}_i$ to v_∞, viz.,

$$\Delta\alpha_i = v_\infty^{-1}\left(\frac{\partial v}{\partial\zeta_i}\right)_{T,\zeta_{j\neq i}}\left(\frac{d\bar{\zeta}_i}{dT}\right) \tag{7}$$

Subtracting eq. (4) from eq. (1), and taking into account that, at fixed temperature, the equilibrium volume is independent of the state of the nonequilibrium glass, i.e.,

$$\left(\frac{\partial v_\infty}{\partial\zeta_i}\right)_{T,\zeta_{j\neq i}} \equiv 0 \tag{8}$$

one finally obtains

$$d(v - v_\infty) = (v - v_\infty)\alpha_g dT + \sum_{}^{N}\left(\frac{\partial(v-v_\infty)}{\partial\zeta_i}\right)_{T,\zeta_{j\neq i}} d\zeta_i - v_\infty\sum_{}^{N}\Delta\alpha_i dT \tag{9}$$

in which $v - v_\infty = \delta_v$ measures the volume departure of the system from equilibrium at T.

It is often more convenient to replace δ_v by a dimensionless variable $\delta = \delta_v/v_\infty$, viz.,

VOLUME AND ENTHALPY RECOVERY OF GLASSES 1101

$$\delta = (v - v_\infty)/v_\infty = (v/v_\infty) - 1 \tag{10}$$

which measures the relative departure from equilibrium, and can be identified[7] with the excess of the fractional free volume (f) of the system with respect to its equilibrium value f_T. In terms of δ, eq. (9) takes the form

$$d\delta = -\Delta\alpha\delta dT + \sum^N \left(\frac{\partial\delta}{\partial\zeta_i}\right)_{T,\zeta_{j\neq i}} d\zeta_i - \sum^N \Delta\alpha_i dT \tag{11}$$

in which $\Delta\alpha\delta$ can be neglected as compared to $\Delta\alpha = \sum^N \Delta\alpha_i$, since in the glass-transition range and below, $|\delta|$ cannot exceed[7,16] $f_g \simeq 0.025$. (Nevertheless, in the absence of such limitations on $|\delta|$, the term $\Delta\alpha\delta$ should be taken into account in the asymptotic glassy region, obtained on cooling to low temperatures and defined by $d\zeta_i = 0$.)

In this case, eq. (11) may be split into N individual contributions such that

$$\sum^N \delta_i = \delta \quad \text{and} \quad \sum^N d\delta_i = d\delta \tag{12}$$

where each individual δ_i corresponds to the fractional departure from equilibrium associated with the ordering parameter ζ_i. This is equivalent to assuming that

$$\left(\frac{\partial\delta}{\partial\zeta_i}\right)_{T,\zeta_{j\neq i}} = \left(\frac{\partial\delta_i}{\partial\zeta_i}\right)_{T,\zeta_{j\neq i}}$$

i.e., δ_i is independent of all ζ_j other than ζ_i, thus $(\partial\delta_i/\partial\zeta_j)_T = 0$. The variance of the system is therefore defined by N distinct differential equations

$$d\delta_i = \left(\frac{\partial\delta_i}{\partial\zeta_i}\right)_{T,\zeta_{j\neq i}} d\zeta_i - \Delta\alpha_i dT, \quad 1 \leq i \leq N \tag{13}$$

together with eq. (12).

To this point, the approach has been strictly thermodynamic; time dependence of the system is, however, of primary interest. By stipulating that all the states through which the system passes in time (t) are uniquely defined by the usual thermodynamic parameters and the set of ordering parameters, the rate of change in δ_i is controlled by that of ζ_i. Then eq. (13) is written

$$\frac{d\delta_i}{dt} = -\Delta\alpha_i q + \left(\frac{\partial\delta_i}{\partial\zeta_i}\right)_{T,\zeta_{j\neq i}} \frac{d\zeta_i}{dt}, \quad 1 \leq i \leq N \tag{14}$$

where $q = dT/dt$ is the experimental heating $(q > 0)$ or cooling $(q < 0)$ rate. The last term represents the rate of approach to zero (equilibrium) of δ_i under the conditions that T and P remain constant. Since $\bar\zeta_i$ is independent of time

$$\left(\frac{\partial\zeta_i}{\partial t}\right)_T = \left(\frac{\partial(\zeta_i - \bar\zeta_i)}{\partial t}\right)_T$$

one can reasonably assume proportional causality, and state that

$$\frac{d\delta_i}{dt}\bigg|_T = \left(\frac{\partial\delta_i}{\partial\zeta_i}\right)_{T,\zeta_{j\neq i}} \frac{d\zeta_i}{dt}\bigg|_T = -\frac{\delta_i}{\tau_i}, \quad 1 \leq i \leq N \tag{15}$$

i.e., the instantaneous rate of approach of each δ_i to equilibrium is proportional to the departure from equilibrium associated with the ith ordering parameter, the proportionality factor being $-\tau_i^{-1}$. Thus the model results in a distribution

1102 KOVACS ET AL.

of retardation mechanisms, each controlling a definite fraction δ_i of the total departure δ, and each involving a distinct retardation time τ_i associated with the rate of change of the ith ordering parameter and vice versa. This relationship between τ_i and ζ_i may be viewed as defining the ordering parameters in terms of their fractional contributions (δ_i/δ) to the total recovery of the system.

Equation (14) then becomes

$$-\frac{d\delta_i}{dt} = \Delta\alpha_i q + \frac{\delta_i}{\tau_i}, \quad 1 \le i \le N \tag{16}$$

This set of N differential equations together with eq. (12) and the initial conditions govern the time and temperature dependence of the system when subjected to any arbitrary thermal history. It should be noted that the same expression has been derived for one-parameter models ($N = 1$) and used by many authors[2,3,7,8] to describe the recovery behavior of glasses. Therefore, one could start the discussion of multiparameter models merely by postulating N independent retardation mechanisms, each obeying eq. (16). Nevertheless, the thermodynamic justification of this set of equations remains useful, since it clearly defines the various assumptions and approximations involved, and thus allows possible modifications of them if necessary.

The scope of eq. (16) can be extended to enthalpy recovery,[8] by substituting for δ_v the departure of the enthalpy $\delta_H = H - H_\infty$ from its equilibrium value in eq. (9), while replacing α_g and α_l by the specific heat of the glass $C_{p,g}$ and the liquid $C_{p,l}$, respectively. Furthermore, since in eq. (1), T and P are interchangeable, eq. (16) also applies to isothermal variations of δ_i with changing pressure, by defining q as dP/dt and substituting the negative excess isothermal compressibility of the liquid with respect to that of the glass $-\Delta\chi_i$ for $\Delta\alpha_i$. In fact, eq. (16) has been used (with $N = 1$) to describe bulk creep and recovery of glasses subjected to stepwise pressure changes.[17]

Structure and Temperature Dependence of the Retardation Times

Since retardation times are determined by the configurational mobility of the system and thus depend upon the time scale of molecular motions, it is clear that the τ_i's are functions of temperature. In addition to pure temperature dependence, however, it has been well established that molecular rate processes also depend on microscopic order, or structure of the system which, in the present model, is defined by the set of values of the ordering parameters. Since the departure of the structure from its equilibrium is measured by δ, the retardation times will depend upon δ, as well as on T.

One can conceive a model where each τ_i would depend on T and δ_i alone and would not be influenced by the values of the other ordering parameters $\zeta_{j\neq i}$. In such a situation, eq. (16) would consist of N independent differential equations. Although this approach is mathematically attractive, it appears physically unreasonable since the time scale of molecular motions must depend on the *overall* state of a system rather than, independently, on the particular value of each individual ordering parameter. Therefore we assume that it is the *total* departure δ from equilibrium which determines the structural effect on the rate of molecular motions, as reflected by the set of retardation times τ_i. Such a dependence mathematically couples the N expressions given in eq. (16) and also results in a generalized thermorheologically simple behavior,[14] i.e., structural, as well as thermal variations reduce *each* retardation time by the same factor.

The particular functionalities of τ_i can now be discussed. Many expressions have been used previously to express the temperature and structure dependence of retardation times based on activation processes,[10] free volume,[7,16] or configurational entropy.[18] In a recent review,[9] we have shown that all these expressions are related and, in a narrow temperature interval, they all reduce to that used earlier by Tool,[2] which can be written[9,11]

$$\tau_i(T,\delta) = \tau_{i,r} \exp[-\theta(T - T_r)] \exp[-(1 - x)\theta\delta/\Delta\alpha] \equiv \tau_{i,r}a_Ta_\delta \qquad (17)$$

where $\tau_{i,r}$ is the value of the ith retardation time at a reference temperature T_r in equilibrium ($\delta = 0$). Thus θ is a material constant, equal[9,11] to E_a/RT_r^2, which characterizes the temperature dependence of τ_i in equilibrium conditions (E_a being the apparent activation energy, and R the gas constant) and x is a partition parameter ($0 \le x \le 1$) which determines the relative contributions of temperature (T) and structure (δ) to τ_i. Clearly, $x = 1$ denotes pure temperature dependence while pure structural dependence arises when $x = 0$. Therefore, the shift factor $a_T = \exp[-\theta(T - T_r)]$ incorporates the temperature dependence of τ_i at equilibrium (or at $\delta = $ const), whereas the second shift factor $a_\delta = \exp[-(1 - x)\theta\delta/\Delta\alpha]$ represents the structure-dependent adjustment of the time scale, at constant T. In the theoretical treatment which follows, only the reduced form of $\tau_i(T,\delta) = \tau_{i,r}a_Ta_\delta$ will be used, for the sake of generality. The various expressions of the shift factors are discussed in Appendix A.

The fundamental mathematical statement of the present model is then obtained by combining eqs. (16) and (17) to give

$$\frac{d\delta_i}{dt} = -\Delta\alpha_i q - \frac{\delta_i}{\tau_{i,r}a_Ta_\delta}, \quad 1 \le i \le N. \qquad (18)$$

Alternatively, dividing both sides by $q = dT/dt$, one can write

$$\frac{d\delta_i}{dT} = -\Delta\alpha_i - \frac{\delta_i}{\tau_{i,r}a_Ta_\delta q}, \quad 1 \le i \le N. \qquad (18')$$

As anticipated in the Introduction, and fully discussed below, the nonlinearity and asymmetry of the recovery behavior of glasses originate in the dependence of τ_i on the instantaneous state of the system through a_δ, while "memory effects" result naturally from the multiplicity of retardation processes. Since both of these characteristics are incorporated in eq. (18), one can expect that the response of the model will closely approximate the behavior of real glasses when submitted to a wide variety of thermal treatments.

In Part II, the response functions of the model will be derived using the simplest thermal treatment involving a single instantaneous T-jump. These functions will then be discussed in terms of the retardation spectrum characterizing the material. The relevant numerical calculations are based on a hypothetical spectrum giving rise to a behavior which closely imitates that of polymer glasses. Finally, in the last part of the paper, the response of the model will be analyzed for various classes of complex thermal stimuli, while deriving the fundamental constitutive equation of the system applicable to any thermal history. Among these, thermal cycles involving cooling from high temperatures, subsequent isothermal recovery, and heating will be analyzed with particular interest given to specific features of the heating isobars.

PART II

Isothermal Recovery after a Single Instantaneous T-Jump

Consider a system, in equilibrium ($\delta_i = 0$; $1 \le i \le N$) at T_0, which is isobarically removed at $t = 0$ from this state by an *instantaneous* change of its temperature by ΔT, and subsequently kept at $T = T_0 + \Delta T$, for $t \ge 0^+$. According to the model, such treatment results in an instantaneous (Cauchy) strain:

$$[v(T;t = 0^+) - v_\infty(T_0)]/v_\infty(T_0) \simeq \alpha_g \Delta T$$

which represents the time-independent *elastic* deformation of the material when all the ζ_i are kept constant [cf. eqs. (1) and (2)]. Concomitantly, however, the new equilibrium volume becomes $v_\infty(T) \simeq v_\infty(T_0)(1 + \alpha_l \Delta T)$, and the initial departure at T is

$$\delta_0 = [v(T;t = 0^+) - v_\infty(T)]/v_\infty(T) \simeq -\Delta\alpha\Delta T$$

omitting the factor $v_\infty(T_0)/v_\infty(T)$. Therefore, δ_0 represents the initial value of the recoverable strain, although the system has been neither deformed nor submitted to any stress. (Note that thermal stresses originating from the temperature gradients inside the sample are neglected here although they can be taken into account[19] in more elaborate treatments.)

Both δ_0 and $\Delta\alpha$ are sums of the elementary contributions associated with each ordering parameter [eqs. (7) and (12)]. At $t = 0^+$, one has, according to eq. (18'), $|q| = \infty$:

$$\delta_{i,0} = -\Delta\alpha_i\Delta T, \quad \delta_0 = \sum_{i}^{N} \delta_{i,0} = -\Delta\alpha\Delta T, \quad 1 \le i \le N \tag{19}$$

whereas at $t \ge 0^+$, when T is kept constant, $q = 0$ and eq. (18) becomes

$$\left.\frac{d\delta_i}{dt}\right|_T = -\frac{\delta_i}{\tau_{i,r}a_T a_\delta}, \quad 1 \le i \le N \tag{20a}$$

$$\left.\frac{d\delta}{dt}\right|_T = -a_T^{-1} a_\delta^{-1} \sum_{i}^{N} \left(\frac{\delta_i}{\tau_{i,r}}\right) \tag{20b}$$

Clearly, in the present case all the $\delta_{i,0}$'s have the *same* sign, opposite to that of ΔT [eq. (19)]. Furthermore, $d\delta_i/dt_T$ maintains its sign until δ_i vanishes, since $\tau_{i,r}a_T a_\delta$ is strictly positive. Therefore, $d\delta/dt|_T$ is either negative ($\delta_0 > 0$; $\Delta T < 0$) or positive ($\delta_0 < 0$; $\Delta T > 0$) and consequently isothermal recovery occurs in a monotonic manner either by contraction or by expansion, respectively. Such experiments in which the system is removed from an initial equilibrium state (all $\delta_i = 0$) by a change of its temperature and is then kept at constant T until full recovery are called simple approach experiments, in contrast to those where the initial state is a nonequilibrium state (at least one $\delta_i \ne 0$).

The set of differential equations (20), together with the definition of τ_i [e.g., eq. (17)] can only be integrated numerically. Indications of the calculation procedure and the material constants used will be given below. Figure 1 represents calculated recovery isotherms at $T = T_r - 5$ K, in terms of δ-vs.-log$t$ plots, for various initial temperatures, involving different magnitudes and signs of ΔT, as indicated. Clearly, these isotherms have the same character as observed for real glasses under similar conditions (see for example Figs. 7, 8, 16, and 17 in ref.

VOLUME AND ENTHALPY RECOVERY OF GLASSES   1105

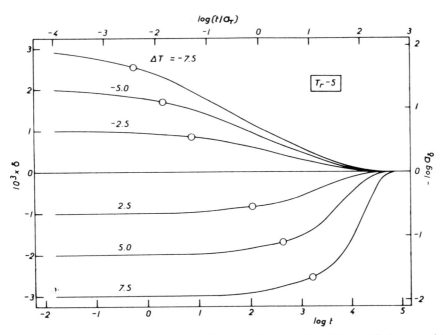

Fig. 1. Time dependence of the departure δ from equilibrium at $T_r - 5$ K, or at T_r (upper scale) after instantaneous T-jumps ($\Delta T = T - T_0$, as indicated). The circles indicate the time at which $\delta/\delta_0 = 0.85$. The right-hand scale shows the simultaneous change of the a_δ function assuming eq. (17) to hold, with $\theta = 1.0$ K^{-1} and $x = 0.4$.

7). In particular, the nonlinear and asymmetric character of the recovery is immediately apparent, since reduction of δ by $\delta_0 = -\Delta\alpha\Delta T$ would not result in a single master curve, as would be the case for a linear system. In fact the same value of the *fractional departure* $\rho = \delta/\delta_0$ is reached at rapidly increasing times for decreasing δ_0 as illustrated by the circles in Figure 1 for $\rho = 0.85$. Though other qualitative conclusions may also be drawn from such data (some of which will be discussed below), there is no direct method to extract from such recovery isotherms the effects of the individual ordering parameters and thus the retardation spectrum which is our present interest.

Reduced Time and the Retardation Spectrum

As mentioned above, the set of differential equations (20) cannot be integrated analytically. Nevertheless, they represent an "autonomous system"[20] subject to some simplification. Since at fixed T, a_T is invariant and a_δ has the same strictly positive value for every i, one can define a reduced time variable*

$$u \equiv \int_{0^+}^{t} \frac{dt'}{a_\delta}, \quad \frac{du}{dt} = a_\delta^{-1}. \tag{21}$$

Clearly, for a given $t \geq 0^+$, the value of u strongly depends on that of δ_0 (or ΔT) and also, to a lesser extent, on T, if a_δ is defined by a more elaborate expression (Appendix A) than that given in eq. (17).

---

* Such reduction of the time variable has previously been considered by Hopkins (ref. 21) in analyzing stress relaxation behavior during uniform heating or cooling of rubbery materials.

In terms of u, eq. (20) becomes

$$\frac{d\delta_i}{du} = -\frac{\delta_i}{\tau_{i,T}}, \quad 1 \le i \le N \tag{22}$$

where $\tau_{i,T} = a_T \tau_{i,r}$ is the ith retardation time in equilibrium at T. This can be integrated easily to yield

$$\delta_i(u) = \delta_{i,0} \exp(-u/\tau_{i,T}) \tag{23}$$

since at $t = 0^+$, $u = 0$ and $\delta_i(0) = \delta_{i,0} = -\Delta\alpha_i \Delta T$.

Therefore, by summing all the individual terms

$$\delta(u) = \sum_{i}^{N} \delta_{i,0} e^{-u/\tau_{i,T}} \equiv \delta_0 \sum_{i}^{N} g_i e^{-u/a_T \tau_{i,r}} \tag{24}$$

where

$$g_i = \delta_{i,0}/\delta_0 = \Delta\alpha_i/\Delta\alpha \tag{25}$$

is the fractional contribution of the ith retardation mechanism to δ_0 and $\Delta\alpha$, and satisfies the normalization condition $\Sigma^N g_i = 1$. Since each g_i is uniquely associated with the retardation time τ_i, the set of paired values $(g_i, \tau_{i,r})$ defines the normalized intensities associated with each retardation time at T_r, i.e., the retardation spectrum $G(\tau_{i,r})$ implied by the model. The set of τ_i considered here is discrete, thus $G(\tau_i)$ is a line spectrum. Alternatively, one could use a model based on a continuous spectrum $G(\log\tau)$, which measures the contribution of mechanisms involving retardation times lying between $\log\tau$ and $(\log\tau + d\log\tau)$. Assuming the set of g_i values [cf. eqs. (6) and (7)] independent of T and δ, it is obvious that the *spectrum is an invariant* of the system, since it is uniquely defined by the fixed set of values of $(g_i, \tau_{i,r})$. When T and/or δ vary, the spectrum merely shifts along the logarithmic time scale by an amount equal to $\log a_T$ and/or $\log a_\delta$.

The Isothermal Response Function

Equation (24), together with eqs. (21) and (25), uniquely describe the isothermal recovery of the system at T, after being removed from its initial equilibrium at T_0 by an instantaneous T-jump, equal to $\Delta T = T - T_0$. Accordingly, at fixed T the fractional departure $\rho = \delta/\delta_0$ is a *single-valued function* of the reduced time u and vice versa, whatever the magnitude and/or sign of the initial departure $\delta_0 = -\Delta\alpha\Delta T$. Therefore, in terms of u the system behaves *linearly*; in fact, nonlinearity and asymmetry have been removed through the reduction of the experimental time scale given in eq. (21), i.e., by introducing the proper time scale of the system as determined by the instantaneous rate of its structural fluctuations under equilibrium conditions at T.

Moreover, eq. (24) also shows that, at various fixed temperatures, the fractional departure ρ_T is a unique function of u/a_T, since g_i and $\tau_{i,r}$, which characterize the retardation spectrum at T_r, were assumed to be invariant. Consequently, the $\rho_T(u/a_T)$ isotherms exactly superimpose when plotted against u/a_T and coincide with that which would obtain at the reference temperature where $a_T = 1$, i.e., $\rho_T(u/a_T) = \rho_r(u)$. Thus, the $\rho_r(u)$ isotherm defined by

$$\rho_r(u) = \rho_T\left(\frac{u}{a_T}\right) = \sum_{i}^{N} g_i e^{-u/\tau_{i,r}a_T} \tag{24'}$$

is unique and represents the *linearized isothermal response function* of the system in simple approach experiments involving instantaneous T-jumps, whatever the value of T and/or ΔT. These important conclusions resulting from eqs. (21) and (24) will be widely used and further illustrated below.

If a_δ is independent of T, as in eq. (17), the $\delta_T(\delta_0, \log t)$ isotherms, such as are shown in Figure 1, can be superimposed even when determined at different temperatures, merely by shifting them along the logarithmic experimental time axis by $-\log a_T$ (cf. upper scale in Fig. 1), provided that the value of ΔT (or δ_0) is the same. In fact, according to eq. (20), $\tau_{i,r} a_\delta (d \ln \delta_i)_T$ is a single-valued function of $d(t/a_T)$ when the initial conditions [eq. (19)] are the same. Moreover, in this particular case [eq. (17)]

$$\delta = -2.303 \left[\Delta\alpha/(1-x)\theta \right] \log a_\delta.$$

Therefore, primary data, such as are shown in Figure 1, can immediately be scaled in terms of $-\log a_\delta$ (see the right-hand ordinate) by using the same reduction factor at any temperature, since the scaling parameters involved ($\Delta\alpha$, x, and θ) are material constants. Nevertheless, for more elaborate expressions of τ_i, listed in Appendix A, involving a slight temperature dependence of a_δ [to be distinguished hereafter by the notation $a_\delta(T)$], the above reduction of the experimental time scale by a_T would bring neither the $\delta_T[\delta_0, \log(t/a_T)]$, nor the $\log a_\delta(T)$ isotherms into strict coincidence, since their shapes depend slightly on T in addition to δ_0.

Furthermore, since $[\partial \log a_\delta(T)/\partial \delta]_T$ is always negative and finite, even at $\delta = 0$ [see eq. (17) and Appendix A], in a strict sense, there is no linear range of isothermal recovery except when $x = 1$, i.e., when τ_i depends on temperature alone and $a_\delta = 1$, an unrealistic case. Nevertheless, within experimental accuracy there will be a range of $|\delta|$ (of the order of 10^{-4}) in which recovery *appears* to be linear.[22] Here $a_\delta \simeq 1$, thus $u \simeq t$ and the isothermal bulk compliance $A_T(t)$ with respect to an instantaneous T-jump of the amount of ΔT can be approximated by

$$A_T(t) \equiv -\frac{\delta_T(t)}{\Delta T} = \Delta\alpha \sum_{}^{N} g_i e^{-t/\tau_{i,T}} = \Delta\alpha \rho_T(t) \tag{26}$$

from which the retardation spectrum $G(\tau_{i,T})$ can be extracted, using appropriate inversion techniques.[16] An extension of this procedure to the nonlinear case will be given below.

In fact, the bulk compliance can be given in terms of the reduced time u [eq. (21)]. For example, with a continuous retardation spectrum $G(\log\tau)$, subject to the normalization condition $\int_{-\infty}^{+\infty} G(\log\tau) \, d \log\tau = 1$, the bulk compliance $A_T(u)$ is defined [eq. (24)] by

$$A_T(u) \equiv -\frac{\delta_T(u)}{\Delta T} = \Delta\alpha \int_{-\infty}^{+\infty} G(\log\tau_T) e^{-u/\tau_T} d \log\tau_T = \Delta\alpha \rho_T(u) \tag{27}$$

which is similar to the expression of the *relaxation modulus* used in linear viscoelasticity, except that the kernel function is defined here in reduced time.

In spite of its apparent simplicity, this generalized function is not very useful when changes in the recoverable strain (δ) are produced by temperature variations which invalidate the applicability of the above equation; it only describes isothermal behavior of the system after being removed from its equilibrium by a single instantaneous T-jump.

In this respect recovery behavior of glasses is much more complex than linear viscoelastic behavior of rubberlike materials under sequential stresses or strains. It will be shown, however, that for more involved thermal treatments further simplifications are still possible which reduce the material's response to a single function, obeying Boltzmann's superposition principle.[16]

The Effective Rate

Instead of the response function $\rho_r(u)$, isothermal recovery of glasses can also be analyzed to good advantage[7] in terms of a very sensitive rate parameter, called τ *effective* and defined by

$$\tau_{\text{eff},T}^{-1} \equiv -\delta^{-1} \frac{d\delta}{dt}\bigg|_T > 0 \tag{28}$$

which can be determined easily by differential analysis of the primary data, such as are shown in Figure 1. Comparison of eqs. (20) and (28) reveals that τ_{eff} is the temperature-structure-dependent instantaneous retardation time that would obtain if only a *single* retardation mechanism, involving a unique ordering parameter ($N = 1$), were operative.

In Figure 2 some $\log \tau_{\text{eff}}^{-1}$-vs.-$\delta$ plots are shown, at three different temperatures, as calculated from eq. (28) using the present model with various ΔT values, as indicated. Details of the calculation and the material constants involved (the same as in Fig. 1) will be given below. At the present stage these plots can be considered as schematic although their relationship with the actual data will become clear later.

For $\delta_0 > 0$, thus $\Delta T < 0$, the τ_{eff} curves correspond to isothermal contraction, whereas for $\delta_0 < 0$, thus $\Delta T > 0$, they represent isothermal expansion data. A right-angle counterclockwise turn of Figure 2 reveals some resemblances to the primary data shown in Figure 1, the time scale here being defined by $\log \tau_{\text{eff}}$, rather than by $\log t$.

The considerable variation of τ_{eff} accompanying relatively minor changes in volume, as well as the asymmetry of the approach towards equilibrium, are obvious. Once again, these computed plots have a similar character to those obtained experimentally (e.g., Fig. 6 in ref. 9).

According to the present model, τ_{eff} can be expressed [eqs. (20) and (28)] as

$$\tau_{\text{eff},T}^{-1} = -\delta^{-1} \sum^N \frac{d\delta_i}{dt}\bigg|_T = (a_T a_\delta)^{-1} \sum^N \left(\frac{\delta_i}{\delta}\right) \tau_{i,r}^{-1} \tag{29}$$

Therefore, immediately after the T-jump, at $t = 0^+$, one has

$$a_{\delta_0} a_T (\tau_{\text{eff},T}^{-1})_0 = \sum^N \frac{\delta_{i,0}}{\delta_0} \tau_{i,r}^{-1} = \sum^N \frac{g_i}{\tau_{i,r}} \tag{30}$$

the value of which is invariant, since it depends solely on the parameters of the retardation spectrum $G(\tau_{i,r})$. On the other hand, when $\delta \to 0$ and $a_\delta \to 1$, one can show (see Appendix B) that $\sum^N (\delta_i/\delta)\tau_{i,T}^{-1}$ remains finite, its limiting value being $\tau_{N,T}^{-1}$, i.e., the reciprocal of the longest retardation time in equilibrium at T. Consequently, whatever the values of δ_0 and T, one has

$$\lim_{\delta \to 0} \tau_{\text{eff},T} = a_T \tau_{N,r} = \tau_{N,T} \tag{31}$$

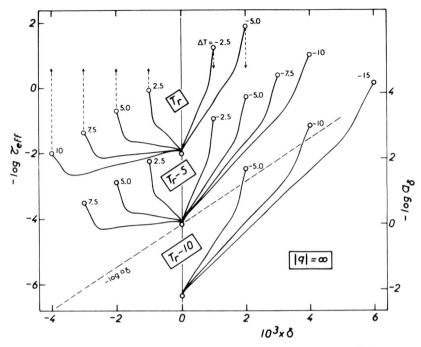

Fig. 2. τ_{eff} vs. δ isotherms at three temperatures (in frame) after instantaneous ($|q| = \infty$) T-jumps ΔT indicated by small numerals. Dashed line: variation of the a_δ function (right-hand scale) as defined by eq. (17).

In Figure 2, the initial and final values of τ_{eff}, as indicated by circles for each isotherm, result from the particular distribution adopted, involving finite values of $\tau_{1,r}$ and $\tau_{N,r}$, to be given below.

The Reduced Rate Function

For intermediate δ values, eq. (29) can be rearranged to give

$$a_\delta \tau_{\text{eff},T}^{-1} = a_T^{-1} \sum^N \frac{\delta_i}{\delta} \tau_{i,r}^{-1} \equiv \phi'_T(\delta,\delta_0) \equiv a_T^{-1} \phi'_r(\delta,\delta_0) \tag{32}$$

The rate function $\phi'_r(\delta,\delta_0) = \Sigma^N(\delta_i/\delta)\tau_{i,r}^{-1}$ has some interesting properties directly related to the shape of the retardation spectrum and to its extension along the logarithmic time scale. Furthermore, if the $a_\delta(T)$ function is known, $\phi'_r(\delta,\delta_0)$ can easily be extracted from experimental τ_{eff} data since, in a general manner, one has at T,

$$\log\phi'_T(\delta,\delta_0) = \log a_\delta(T) - \log\tau_{\text{eff},T} \tag{32a}$$

or alternately,

$$\log\phi'_r(\delta,\delta_0) = \log\alpha_\delta(T) - \log(\tau_{\text{eff},T}/a_T) \tag{32b}$$

even when one allows for a temperature dependence of a_δ (Appendix A).

Assuming eq. (17) to hold, $-\log a_\delta$ is represented in Figure 2 (right-hand scale) by the dashed straight line, the slope of which $-d\log a_\delta/d\delta = (1 - x)\theta/2.303\Delta\alpha$, is independent of both T and δ, involving the material constants θ, x, and $\Delta\alpha$

alone.* Such $-\log a_\delta(T)$ lines represent the variation of $\log(a_T \tau_{\text{eff},T}^{-1})$ with δ for a *single* retardation time model [cf, eq. (32) with $N = 1$, thus $\delta_i/\delta = 1$] and Figure 2 clearly illustrates the drastic differences in behavior of single and multiparameter models, as noted previously.[7] Note also that for a single parameter model, $\phi'_T(\delta,\delta_0) = \tau_T^{-1}$ is independent of both δ and δ_0.

The isothermal variation of $\phi'_T(\delta,\delta_0)$, at $T_r - 5$ K and that of $\phi'_r(\delta,\delta_0)$, at T_r, as derived for the present multiparameter model using eqs. (32a) and (32b) are both represented in Figure 3, for four ΔT values with symmetrical initial departures $|\delta_0|$. Here the simplification of the system is obvious, since the *tilt* of the $\log \tau_{\text{eff}}^{-1}$ isotherms, originating from the contribution of $-\log a_\delta$ and giving rise to the dominating asymmetry in Figure 2, has completely disappeared. In fact, according to eq. (32a), at each T the reduction of the initial $(\tau_{\text{eff},T})_0$ values by $a_{\delta_0}(T)$ defines a single value for $\phi'_T(\delta_0) = a_T^{-1} \sum^N (g_i/\tau_{i,r})$, which depends on T alone. The vertical dashed lines in Figure 2 illustrate this reduction of $(\tau_{\text{eff}}^{-1})_0$ at T_r, resulting in the constant value of $\phi'_r(\delta_0)$. Consequently, the envelope of $\log(\tau_{\text{eff},T}^{-1})_0$ values, associated with various initial departures δ_0, is parallel to the $-\log a_\delta(T)$ curve, which in the present case [eq. (17) and Fig. 2] is a straight line, the slope of which is independent of T, as shown above. In this simple case, the $\tau_{\text{eff},T}^{-1}$ vs. δ isotherms, involving the same value of δ_0, can thus be superimposed merely by reduction of the time scale by a_T [cf. eq. (32b) and Fig. 2].

In terms of the reduced time u [eq. (21)], $\phi'_T(\delta,\delta_0)$ can be written [cf. eqs. (23)–(25) and (32)]

$$\phi'_T(\delta,\delta_0) = \left(\frac{\delta_0}{\delta}\right)_T \sum^N \frac{\delta_i}{\delta_0} \tau_{i,T}^{-1} = \left(\frac{\delta_0}{\delta}\right)_T \sum^N \tau_{i,T}^{-1} g_i e^{-u/\tau_{i,T}} > 0 \tag{33}$$

Thus

$$\phi'_T(\delta,\delta_0) = \sum^N \tau_{i,T}^{-1} g_i e^{-u/\tau_{i,T}} \bigg/ \sum^N g_i e^{-u/\tau_{i,T}} \equiv \phi_T\left(\frac{\delta}{\delta_0}\right) \equiv \phi_T(\rho) \tag{33'}$$

which shows that ϕ'_T is a single-valued function of u, thus of $(\delta/\delta_0)_T = \rho_T(u)$, whatever the magnitude and/or sign of δ_0. In other words, $\phi'_T(\delta,\delta_0)$ data such as are shown in Figure 3, result in a single-valued function when plotted against ρ, rather than δ, as already shown above for $\delta = \delta_0$, i.e., for $\rho = 1$. The same obviously holds for

$$\phi'_r(\delta,\delta_0) = \phi_r(\rho) = a_T \phi_T(\rho) \equiv a_T \phi''_T(u/a_T) \tag{34}$$

which is represented by the solid line in Figure 4 (left-hand scale).

This *single* reduced rate function characterizes the contribution of the retardation spectrum to the effective rate of recovery in a *unique manner*, uncomplicated by nonlinear effects, which themselves depend, in a very sensitive way, on the exact conditions (T and ΔT) of the experimental procedure. This contribution is linear since it depends on the fractional departure alone. As anticipated above, nonlinearity and asymmetry in δ and τ_{eff} merely originate from the dependence of τ_i on the actual state of the system, as determined by the shift factor $a_\delta(T)$ alone. The effect of the latter [eqs. (32)] is to reduce the

---

* More elaborate expressions for $a_\delta(T)$ listed in Appendix A, result in lines with a slight downward curvature (refs. 7 and 9). In addition, the slope $- [d \log a_\delta(T)/d\delta]_T$, at fixed δ, slightly increases with decreasing T.

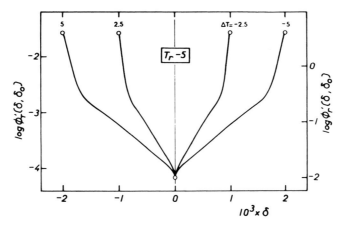

Fig. 3. $\phi'_T(\delta,\delta_0)$ isotherms at $T_r - 5$ K and at T_r (right-hand scale) derived from the data shown in Figures 1 and 2, using eqs. (32a) and (32b).

values of $\phi'_T(\delta,\delta_0)$ asymmetrically (compare Figs. 2 and 3), since for any value of $|\delta| \neq 0$, $a_{\delta>0} < a_{\delta<0}$, and if eq. (17) is applicable, $a_{\delta>0} = a_{\delta<0}^{-1}$.

Finally, as the system approaches equilibrium (t and $u \rightarrow \infty$ and all $\delta_i \rightarrow 0$), obviously $\phi'_T(\delta,\delta_0)$ tends to $\tau_{eff,T}^{-1}$, and thus to $\tau_{N,T}^{-1}$ [eq. (31) and Appendix B], since $a_\delta(T)$ approaches unity. One can thus define a dimensionless linear response

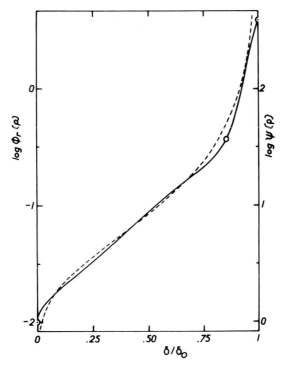

Fig. 4. Reduced rate function $\phi_r(\rho)$ and $\psi(\rho)$ as derived from Figure 3 or from the postulated retardation spectrum (Fig. 5). Circle corresponds to $p = 0.85$. Dashed line: $\overline{\phi}_r(\rho)$ as derived from an analytical expression [eq. (44b)] of the normalized response function shown by the dashed line in Figure 6 [eq. (44)].

1112   KOVACS ET AL.

function [eqs. (32) and (33)]

$$\psi(\rho) \equiv \tau_{N,T}\phi_T(\rho) = a_T a_\delta \tau_{N,r}\tau_{\text{eff},T}^{-1} = \tau_{N,r}\phi_r(\rho) \qquad (35)$$

the value of which ranges [eq. (30)] from $\Sigma^N(g_i\tau_{N,r}/\tau_{i,r})$ to unity when δ varies from δ_0 to zero, as shown in Figure 4 (right-hand scale).

In this section, on a purely analytical basis we have derived two linear response functions, $\rho_r(u)$ and $\phi_r(\rho)$, or $\psi(\rho)$, either of which, together with $a_\delta(T)$, fully characterize isothermal recovery of the present multiparameter model in simple approach experiments after instantaneous T-jumps. In the next section other properties of these functions and their interdependence will be further discussed and analyzed in terms of the retardation spectrum $G(\tau_{i,r})$, assumed here to be an invariant of the system.

Application of the Theory

The behavior of the model will now be explored in a quantitative manner and compared, qualitatively, with isothermal volume and enthalpy recovery of real glasses. To proceed in this fashion one must postulate a distribution function $G(\tau_i)$ and assume some reasonable values for the set of the material constants involved in the theoretical treatment given above.

Material Constants

In preliminary work[9,12] it has been shown that fair agreement between theory and experiment could be obtained by using a distribution function consisting of the two "boxes." The line spectrum $G(\tau_{i,r})$ adopted in the present calculation and shown in Figure 5 is inspired by this result. It consists of $N = 33$ evenly spaced lines on the $\log\tau_i$ scale, extending over four decades, with $\tau_{i+1}/\tau_i = 4/3$ and $\tau_{N,r} = 100$ time units (to be specified below). The first 16 lines ($-2 \le \log\tau_{i,r}$

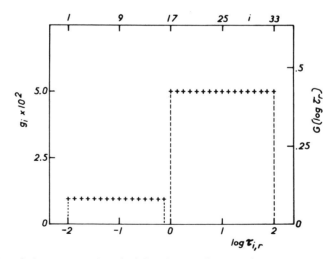

Fig. 5. Retardation spectrum (standard distribution) $G(\tau_{i,r})$ or $G(\log\tau_r)$, at T_r, used in the numerical calculations.

≤ -0.125, for $1 \leq i \leq 16$), each associated with a constant intensity $g_i = 9.375 \times 10^{-3}$, represent the short-time contribution to the recovery with a total weight $16 \times g_i = 0.15$. The complementary long-time contribution ($0 \leq \log\tau_{i,r} \leq 2$, for $17 \leq i \leq 33$) results from 17 lines of constant intensity $g_i' = 0.05$, with a total weight $17 \times g_i' = 0.85$.

There is no fundamental significance in the choice of this particular line spectrum (to be called hereafter the "standard distribution" of τ_i). Furthermore, it can be shown that for a given τ_N/τ_1 ratio, the separation of the distribution points does not appreciably affect the results whenever the density of equispaced lines is greater than four per decade. In fact, indistinguishable behavior would result from a continuous double-box distribution $G(\log\tau_r)$, as depicted in Figure 5 by the dashed lines, spanning the same time scale and having the same fractional contributions (0.15 and 0.85) as the discrete line spectrum. In this case the intensities would be 0.080 and 0.425 (right-hand scale), respectively, the ratio of which is nearly equal to g_i/g_i'.

The sharp cutoffs of this spectrum at $i = 1$ and 33, as well as the stepwise increase of g_i at $i = 17$, have been deliberately introduced to detect their specific effects on the characteristic functions introduced above.*

Clearly, the particular time unit (t.u.) associated with $\tau_{17,r}$ uniquely defines the absolute value of T_r and vice versa, whereas a_T depends on $T - T_r$ alone, whatever its expression (Appendix A). Therefore by scaling the temperature in terms of $X = T - T_r$, as done here, it is not necessary to specify the t.u. or T_r. More precisely, assuming eq. (17), a reduction of the t.u. by a factor c is equivalent to a shift of the temperature scale by $\Delta T = T' - T = T_{r'} - T_r = -\theta^{-1} \ln c$, with $c = \tau_{i,r'}/\tau_{i,r}$, while the value of $X = T - T_r = T' - T_{r'}$ remains invariant. Accordingly, the data shown in Figures 1–4 are independent of the t.u. They correspond to the practical time range of dilatometric experiments if $\tau_{17,r}$ is of the order of 1 min. It will be shown below that the adopted distribution (Fig. 5) implies a value of $T_r \simeq T_g(-1.0) + 1.1$ K, where $T_g(-1.0)$ is the conventional glass-transition temperature obtained on cooling the system from $T_0 \gg T_g$, at a constant rate of $q = -1.0$ K$/\tau_{17,r}$, whatever the value of $\tau_{17,r}$.

In the present calculations, eq. (17) was assumed to hold; thus the model involves only three material constants: $\Delta\alpha$ (or ΔC_p), θ, and x, the last two determining the relative contributions of temperature and structure to the retardation times. More elaborate expressions for τ_i listed in Appendix A also involve two parameters directly related to θ and x. Finally, if the results are to be expressed as specific volume, or in enthalpy units, rather than in δ, or ρ, one must specify an expansion coefficient, e.g., α_g [eq. (2)], or the specific heat $C_{p,g}$, together with a reference equilibrium volume, e.g., $v_\infty(T_r) = v_{r,\infty}$, or enthalpy $H_{r,\infty}$.

The appropriate values of all these material constants can easily be determined by independent experiments, except for x, which cannot be measured directly.[8] The following values, characteristic of polymer glasses, have been adopted:

---

* An obvious expansion of the model would involve allowing the short- and long-time contributions to shift differently along the time scale; thus each would have distinct a_T and/or a_δ functions. Such an assumption would be consistent with a secondary dispersion region below T_g involving a set of rapid segmental motions associated with a smaller activation energy than that controlling the main glass transition. For the sake of simplicity, however, this case will not be considered here.

1114 KOVACS ET AL.

$$\alpha_g = 2.0 \times 10^{-4} \text{ K}^{-1}$$

$$\Delta\alpha = 4.0 \times 10^{-4} \text{ K}^{-1}$$

$$\theta = 1.0 \text{ K}^{-1} \tag{36}$$

$$x = 0.4$$

$$v_{r,\infty} = 1.0000 \text{ cm}^3\text{g}^{-1}$$

which, together with the retardation spectrum shown in Figure 5, fully determine the behavior of the system when submitted to any arbitrary thermal history. This behavior will be examined below, starting with isothermal recovery after one, or several successive instantaneous T-jumps. Before this, however, an outline of the calculation procedure used will be described briefly.

Numerical Integration Procedure

Since the set of differential equations (20), combined with eq. (17) cannot be solved analytically, solutions have been obtained by numerical integration. The initial state of the system at $t = 0^+$ is determined by T and ΔT, or $\delta_{i,0}$ [eq. (19)]. The time is then increased by a small increment $\Delta t_0 = 0.1 \times a_{\delta_0}\tau_{1,T}$ and the corresponding $\Delta\delta_i$ increments are calculated according to eqs. (20) and summed to give $\Delta\delta$. The procedure is then repeated while taking into account the variation of τ_i with δ [eq. (17)], until the relevant portion of the $\delta(\delta_0, \log t)$ isotherm and other related data are generated.

Obviously, the rapid processes involving the smallest i values relax first so that their contributions to δ soon become negligible. In order to reduce the calculation time and to prevent "underflow" of the computer, values of δ_j for which δ_j/δ become less than 10^{-7}, were set at zero. Concomitantly, the time increment Δt is adjusted by setting it equal to $\frac{1}{10}$th of that value of $a_\delta\tau_{i,T}$ which is associated with the smallest nonzero δ_i value. In fact, the choice of the Δt value used in the integration procedure is crucial. If its magnitude is set comparable to $a_\delta\tau_{j,T}$, the time dependence of all processes for which $i < j$, is omitted, thus introducing an arbitrary truncation of the short time tail of the distribution (Fig. 5). On the other hand, setting Δt equal to $\frac{1}{100}$th of the smallest τ_i rather than to $\frac{1}{10}$th, results in alteration of only the fourth significant figure of δ values. Therefore, such a reduction of Δt does not significantly improve the accuracy of the results, but considerably increases the computation time.

All the calculations have been performed by using a Univac 1110 computer and classical library subroutines to ensure the reliability of the integration procedure. Calculation of a full recovery run takes less than 1 min of computer time, while listing about 20 values of δ, ρ, τ_{eff}, a_δ, ϕ_T, etc., per decade of the elapsed time t. In addition, the set of the 33 instantaneous values of δ_i, τ_i, and δ_i/δ_0, i.e., the profile of the actual intensities has been listed at eight fixed values of ρ, ranging between 1 and 0.05.

Figures 1 and 2 represent the isothermal recovery behavior of the system after instantaneous T-jumps have been generated in this manner, using various initial conditions (T and ΔT) as indicated. Since τ_i has been defined in terms of eq. (17), response curves at any temperature can be obtained merely by reduction of the experimental time scale by a_T, as shown above. In the following section some additional features of the characteristic functions $\phi_r(\rho)$ and $\rho_r(u)$ will be discussed with special emphasis on their dependence on the shape and extension of the retardation spectrum.

The Reduced Rate Function

The $\phi_r(\rho)$, or $\psi(\rho)$ curve shown in Figure 4 (solid line) displays some remarkable features which are directly related to the standard distribution (Fig. 5). Since the latter is sharply truncated at finite values of $\tau_{1,r}$ and $\tau_{N,r}$, the initial and final values of $\phi_r(\rho)$, at $\rho = 1$ and $\rho = 0$, also remain finite, being equal to $\sum^N g_i \tau_{i,r}^{-1}$ [cf. eqs. (30) and (32b)] and $\tau_{N,r}^{-1}$ [eq. (31)], respectively.

Furthermore, from eq. (35) one has

$$\frac{\phi_T(1)}{\phi_T(0)} = \frac{\phi_r(1)}{\phi_r(0)} = \psi(1) = \tau_{N,T} \sum_{}^{N} g_i \tau_{i,T}^{-1} \tag{37}$$

since $\psi(0) = 1$. Therefore, $\psi(1)$ defines the range of variation of the rate function $\phi'_T(\delta,\delta_0)$, whatever the value of δ_0 and/or T (Fig. 3). For the distribution adopted here, $\psi(1) = 390.2$. Clearly, for $N \geq 2$, $\psi(1)$ is always smaller than the ratio $\tau_{N,T}/\tau_{1,T}$. Note in this respect that the short-time contributions ($0.85 \leq \rho \leq 1$) occur in a logarithmic time interval of about one decade (Fig. 4), whereas the corresponding τ_i values ($1 \leq i \leq 16$, Fig. 5) cover almost two decades. On the other hand the long-time contributions ($0 \leq \rho \leq 0.85$) extend over about 1.6 decades in $\psi(\rho)$, whereas the corresponding portion of the spectrum covers two decades. Therefore, $\log \psi(1)$ defines the lower limit of the logarithmic time interval covered by the distribution of the retardation times, but its exact relationship with the latter depends critically on the whole set of $g_i/\tau_{i,r}$ ratios, thus on the detailed shape of the line spectrum.

For short times, when rapid processes dominate ($0.85 \leq \rho \leq 1$), the slope of $d \log \psi/d\rho$ is large and shows a first inflexion in the middle of this ρ interval (Fig. 4). If the distribution were smoothly tailed at short times, rather than being truncated, $\psi(1)$ would become infinite and the inflexion would disappear.

A second inflexion is apparent in the vicinity of $\rho \simeq 0.65$ and a third one at $\rho \simeq 0.45$, i.e., close to the middle of the interval: $0 \leq \rho \leq 0.85$, where recovery is controlled essentially by the slow processes ($17 \leq i \leq 33$, Fig. 5). The ratio of the first and this third inflexional slope is close to $g'_i/g_i \simeq 5.33$, i.e., to the ratio of the two intensities characterizing the standard distribution (Fig. 5). In a quite general (but only approximate) manner, one can conclude that the slope $d \log \psi/d\rho$ is reciprocally related to the intensities of the line spectrum. As a consequence, the final cutoff of the distribution at $i = 33$, results in a considerable increase of the slope of $\log \psi$ at very small values of $\rho(< 0.02)$, giving rise to a fourth inflexion in the vicinity of $\rho \simeq 0.2$. This downward drift of $\log \psi$ would extend to a larger range of ρ and ψ, if the long time end of the distribution were smoothly tailed rather than being sharply truncated.

This discussion shows that three of the four inflexional slopes of $\log \psi$ (or $\log \phi_r$) are directly associated with the discontinuities in the intensities of the standard distribution adopted, occurring at $i = 1, 17$, and 33, respectively, although their effect is considerably smeared out and would be barely apparent within the usual accuracy of the experimental data. The additional inflexion at $\rho \simeq 0.65$ merely ensures the continuity of the reduced rate function and would appear even in the absence of any discontinuity in g_i, e.g., for a bell-shaped distribution (dashed lines in Fig. 4, to be discussed below).

The shapes of the $\log\tau_{\text{eff},T}^{-1}$ isotherms shown in Figure 2 result from the combined effects of the dependence of $\log\phi'_T(\delta,\delta_0)$ and $-\log a_\delta(T)$ on δ [cf. eq. (32a)]. In fact,

$$\frac{d\,\log\tau_{\text{eff},T}^{-1}}{d\delta} = \frac{\delta_0^{-1}d\,\log\phi_r(\rho)}{d\rho} - \frac{d\,\log a_\delta(T)}{d\delta} \tag{38}$$

since

$$d\,\log\phi'_T(\delta,\delta_0)/d\delta = \delta_0^{-1}\,d\,\log\phi_r(\rho)/d\rho$$

[eqs. (33a) and (34)]. Taking into account that $-d\,\log a_\delta(T)/d\delta$ and $d\,\log\phi_r(\rho)/d\rho$ are both positive, eq. (38) shows that for isothermal contraction ($\delta_0 > 0$), the slope of $\log\tau_{\text{eff},T}^{-1}$ is always positive whatever the value of δ_0. For isothermal expansion ($\delta_0 < 0$), $d\,\log\tau_{\text{eff},T}^{-1}/d\delta$ is negative for small values of $-\delta_0$, as long as

$$-\delta_0 \leq \frac{[d\,\log\phi_r(\rho)/d\rho]_{\min}}{-\,d\,\log a_\delta(T)/d\delta} \tag{38a}$$

$[d\,\log\phi_r(\rho)/d\rho]_{\min}$ being the smallest (inflexional) slope of the reduced rate function. For larger values of $-\delta_0$, $d\,\log\tau_{\text{eff},T}^{-1}/d\delta$ will vanish at least twice in the whole range of ρ, giving rise to one minimum and one maximum in $\tau_{\text{eff},T}^{-1}$, occurring at relatively large ρ values and near equilibrium, respectively. Such reversions in the variation of the τ_{eff}^{-1} vs. δ isotherms are observed experimentally (e.g., Fig. 6 in ref. 9).

With the standard distribution (Fig. 5) and the material constants [eq. (36)] adopted here together with eq. (17), the critical value $-\delta_0^*$, above which a minimum in τ_{eff}^{-1} occurs, is independent of T since

$$-\delta_0^* = 2.303\Delta\alpha(1-x)^{-1}\theta^{-1}\left(\frac{d\,\log\phi_r(\rho)}{d\rho}\right)_{\min} \simeq 2.21 \times 10^{-3} \tag{38b}$$

and the minimum slope of $\log\phi_r(\rho)$ is equal to 1.44 (corresponding to the second inflexion, at $\rho \simeq 0.65$, in Fig. 4). Note that this value of δ_0^* is quite close to those obtained experimentally for polymer glasses.[7,9,13]

Unfortunately, the accuracy in determining τ_{eff} data from volume recovery isotherms (Fig. 1) becomes quite poor when δ approaches zero [cf. eq. (28)] and the downward drift of $\log\tau_{\text{eff}}^{-1}$ near equilibrium (Fig. 2) is barely apparent.[9,13] According to the above discussion, this suggests a rather sharp cutoff for the long time end of the retardation spectrum (a physically acceptable conclusion, since volume and enthalpy recovery are essentially controlled by the modifications in the short-range order of molecular segments occurring in a small volume element[16]).

For τ_{eff} plots derived from experimental expansion data, values of τ_{eff} extrapolated to $\delta = 0$ increase significantly as $-\delta_0$, the magnitude of the initial departure, increases.[7,9,13] In Figure 2, on the other hand, such extrapolation leads to essentially one value of τ_{eff} which is only slightly smaller than $\tau_{N,T}$. This discrepancy between theory and experiment is the only one which cannot be removed by modification of the shape of the retardation spectrum, provided the latter is assumed to be an invariant of the system, as in this treatment. This puzzling effect,[7,9] which has been overlooked previously by most authors, should

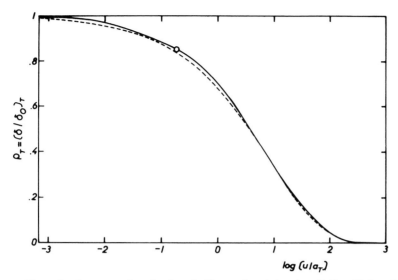

Fig. 6. Normalized response function [eq. (24′)] vs. reduced time [eq. (21)] at T_r, based on the postulated retardation spectrum (Fig. 5). Dashed line: The same based on the analytical expression (ref. 11) given by eq. (44), with $\beta = 0.4552$ and $\bar{\tau}_r = 8.32\tau_{17,r}$. Circle indicates $p = 0.85$.

be documented further by systematic experimental investigations in order to elucidate its origin.

The Isothermal Response Function

The normalized isothermal response function of the system, which is removed from its equilibrium by an instantaneous T-jump, is given by eq. (24) in terms of the reduced time variable u [eq. (21)]:

$$\rho_T(u) = \sum_{i}^{N} g_i \exp\left(-(a_T\tau_{i,r})^{-1} \int_{0^+}^{t} a_\delta(T)^{-1}\, dt'\right) \tag{39}$$

This function can easily be derived both from primary $\delta_T(\delta_0, \log t)$ and/or $\tau_{\text{eff},T}$ data, provided that the $a_\delta(T)$ function is known.

As shown above, the value of u, defined by the integral in eq. (39), depends on ρ alone, although both $a_\delta(T)$ and the time $t(\rho)$ necessary to reach a given value of the fractional departure δ/δ_0, depend on δ_0. In fact, for a fixed value of ρ (e.g., $\rho = 0.85$, as shown in Fig. 1), a_δ^{-1} increases with δ_0, whereas $t(\rho)$ decreases in such a way that they exactly compensate each other, so that the value of the integral in eq. (39) becomes independent of the magnitude and sign of δ_0. Furthermore, when plotted against u/a_T, the $\rho_T(u/a_T)$ isotherms obtained at various temperatures all reduce to a unique response function $\rho_r(u)$ which would be obtained at T_r [cf. eq. (24′)].

This function is represented in Figure 6 (solid line) as derived from any one of the $\delta_T(\delta_0, \log t)$ isotherms (such as those shown in Fig. 1) while computing the values of the reduced time

$$u(\rho) = \int_{0^+}^{t(\rho)} a_\delta^{-1}\, dt' = \int_{0^+}^{t(\rho)} \exp\left(\frac{(1-x)\theta\rho\delta_0}{\Delta\alpha}\right) dt' \tag{39a}$$

corresponding to decreasing values of ρ, reached at times $t(\rho)$, at T.

Alternatively, the reduced time function $u(\rho)$ can also be derived from the reduced rate function $\phi_r(\rho)$, discussed above (Fig. 4). In fact, by differentiating eq. (24) with respect to u, one has [cf. eq. (33a)]

$$\frac{d\rho_T}{du} = - \sum_{i}^{N} \tau_{i,T}^{-1} g_i e^{-u/\tau_{i,T}} = - \rho_T \phi_T(\rho). \tag{40}$$

Since the right-hand side of this equation is a unique function of ρ in the interval $0 \le \rho \le 1$, eq. (40) can be arranged to yield, at T,

$$a_T^{-1} u(\rho) = - \int_1^\rho [\rho' \phi_r(\rho')]^{-1} d\rho' \tag{41}$$

which satisfies the initial and final conditions: $u(1) = 0$ and $u(0) = \infty$, since $\phi_r(\rho)$ is finite over the whole range of ρ. Numerical integration of eq. (41) again leads to $\rho_T(u/a_T)$ shown in Figure 6 which coincides to within the fourth significant figure with that derived from eq. (39a). This coincidence provides a critical test for the internal consistency and the accuracy of the calculation program used.

The equivalence of eqs. (39a) and (41) can also be derived from eqs. (28) and (32), since

$$a_\delta(T) = \tau_{\mathrm{eff},T} \phi_T'/\delta, \delta_0 = - \delta \left(\frac{d\delta}{dt}\right)_T^{-1} \phi_T(\rho) \tag{42}$$

thus [eqs. (21) and (33′)]

$$du = a_\delta(T)^{-1} dt = -\rho^{-1} \phi_T(\rho)^{-1} d\rho. \tag{42a}$$

Therefore, at any temperature and time $t(\rho)$ one has, whatever the value of δ_0,

$$a_T^{-1} \int_{0^+}^{t(\rho)} a_\delta(T)^{-1} \, dt' = \int_\rho^1 \phi_r(\rho')^{-1} \, d \ln\rho' = \frac{u(\rho)}{a_T} \tag{43}$$

in which $a_\delta(T)$ is supposed to be a known function of the experimental time t as determined by any of the δ_T isotherms (cf. Fig. 1). Equation (43) thus relates the characteristic functions $\rho_r(u)$ and $\phi_r(\rho)$ of the model to the actual response $\delta_T(\delta_0, \log t)$, of the system, provided that the initial conditions (δ_0, T) and the a_δ and a_T functions are known.

The shape of the isothermal response function $\rho_T(u/a_T) = \rho_r(u)$ shown in Figure 6 (solid line) is similar to the $\delta(\delta_0, \log t)$ isotherms (Fig. 1). It appears, however, to have fewer marked features than the reduced rate function $\phi_r(\rho)$ represented in Figure 4 (solid line). In particular, the discontinuities of the standard distribution (Fig. 5) do not show up, since $\rho_r(u)$ has only one inflexional tangent in the vicinity of $\rho \simeq 0.45$.

Since $\rho_T(u/a_T)$ is independent of δ_0, it is identical with the $\rho_T(t/a_T)$ function one would obtain for an infinitesimal initial departure δ_0, for which $a_\delta(T) \simeq 1$. As noted above [eq. (26)], one can thus extract the retardation spectrum $G(\tau_{i,r})$ from $\rho_r(u)$, using appropriate inversion techniques,[16] simply by substituting for u the experimental time t. Nevertheless, owing to the smooth character of $\rho_r(u)$, i.e., the $\rho_r(t)$ function, in which the original features of the spectrum are almost completely smeared out, when inverting one should use approximations involving high-order derivatives (ref. 16) $d^n\rho/dt^n$. These, however, are difficult to determine accurately from the calculated response curve shown in Figure 6, and even more so from experimental data. This situation is somewhat improved if $G(\tau_{i,r})$ is derived from the $\phi_r(\rho)$ function, using eq. (33') with $u = t$. Such computational difficulties in the inversion of the response functions into spectra have been treated in pertinent work on linear viscoelasticity theory[16] and will not be discussed further. This may be the reason why Narayanaswamy[10] and Moynihan et al.[11,23] in a recent phenomenological approach describe the recovery behavior of glasses in terms of a fundamental *nonexponential decay function*, rather than using a retardation spectrum as in the present case. These two approaches are of course completely equivalent[24] provided that the postulated response function is sufficiently close to that derived from the assumed distribution of retardation times, or vice versa.

In this respect it is interesting to compare the present data with the analytical expression proposed by Moynihan et al.[11,23] which can be written

$$\bar{\rho}_T(u/a_T) = \exp[-(u/a_T\bar{\tau}_r)^\beta] \tag{44}$$

where $\bar{\tau}_r$ is a correlation time (at T_r) and β is an exponent less than unity, the value of which appears to be close to 0.5 for most glass-forming systems. This function is shown by dashed lines in Figure 6. The ratio $\bar{\tau}_r/\tau_{17,r}$ is taken as 8.32 with $\beta = 0.4552$, to bring the inflexional tangent [occurring at $\bar{\rho} = e^{-1} \simeq 0.37$ and involving a slope $(d\rho/d \log u) = -2.303\beta/e$] into close coincidence with that derived from the above analysis based on the retardation spectrum shown in Figure 5.

Figure 6 shows that the agreement between the response functions defined by eqs. (24) and (44) appears quite satisfactory in the whole range of ρ, except at short times, where the contribution of the standard distribution to ρ is slightly larger than that given by eq. (44). In fact, the latter involves an asymmetric bell-shaped spectrum[25] $G(\log\tau)$ which has a maximum at $\bar{\tau}_r$ and smoothly decreases to zero at both ends; the half-width extends over about two decades of the time scale. One can thus conclude that eq. (44), with the values of the characteristic parameters ($\bar{\tau}_r$ and β) given above, describes the isotherms such as those shown in Figure 1 in a successful manner, while using the material constants given in eq. (36).

Furthermore, from eq. (44) one can also derive an analytical expression for the reduced rate function $\phi_r(\rho)$. Differentiation with respect to u gives

$$\frac{d\bar{\rho}_T}{du} = -\frac{\beta}{a_T\bar{\tau}_r}\left(\frac{u}{a_T\bar{\tau}_r}\right)^{\beta-1} \exp\left[-\left(\frac{u}{a_T\bar{\tau}_r}\right)^\beta\right]$$

$$= -\bar{\rho}_T\left(\frac{\beta}{a_T\bar{\tau}_r}\right)(-\ln\bar{\rho}_T)^{(\beta-1)/\beta} \tag{44a}$$

and identification with eq. (40), leads to

$$a_T\bar{\phi}_T(\rho) \equiv (\beta/\bar{\tau}_r)(-\ln\bar{\rho}_T)^{(\beta-1)/\beta}. \tag{44b}$$

The $\bar{\phi}_r(\rho) \equiv a_T\bar{\phi}_T$ function is represented by dashed lines in Figure 4, and it can be seen that it is close to the $\phi_r(\rho)$ function derived from the standard distribution; the agreement is, however, poorer than for the $\rho_r(u)$ function, especially in the range of $\rho \sim 0.85$. In particular, $\log\bar{\phi}_r(\rho)$ has only one inflexional tangent [the slope of which is comparable to that of $\log\phi_r(\rho)$ at $\rho \sim 0.65$] and it extends from $-\infty$ to $+\infty$ for $\rho = 0$ and 1, respectively. This comparison shows again that the $\phi_r(\rho)$ function is more sensitive to the shape of the retardation spectrum than the material response function $\rho_r(u)$. Therefore, comparison of recovery behavior of glasses can be made in a more sensitive manner in terms of the characteristic reduced rate function, rather than in terms of $\rho_r(u)$.

Recovery of the Elementary Processes

In addition to the above reduced response and rate functions which determine the recovery of the *total* departure ($\delta = \Sigma^N \delta_i$) of the system from its equilibrium, it is interesting to analyze the recovery of the *individual* contributions δ_i to δ. This can be achieved by listing the set of 33 values of δ_i at constant ρ values, while plotting either $\delta_i(\rho)/\delta_0$, or $\delta_i(\rho)/\delta_{i,0}$ vs. $\tau_{i,r}$, i.e., the instantaneous profile of the partial departures from equilibrium. Such an analysis provides a deeper insight into the recovery phenomena, especially for more involved thermal histories than a single T-jump, and thus elicits the hidden features of the overall response.

For simple approach experiments involving a single instantaneous T-jump ($|q| = \infty$), such plots are independent of the initial conditions (δ_0 and T), since ρ_T is a single-valued function of u/a_T [eqs. (24) and (39) and Fig. 6], whatever the value of δ_0 and/or T.

(a) In Figure 7, $\delta_i(\rho)/\delta_0$ is plotted at the eight indicated ρ values as a function of $\tau_{i,r} = (a_T a_\delta)^{-1}\tau_i$. Instead of plotting the discrete values [cf. eqs. (23) and 25]

$$\delta_i(\rho)/\delta_0 = g_i \exp(-u/a_T\tau_{i,r}) \tag{45}$$

for each i, the data points have been connected by smooth curves except at $i = 1, 16, 17$, and 33 where major discontinuities occur. The boxes indicated by dashed lines represent the initial situation at $t = 0^+$, i.e., for $\rho = 1$, as depicted in Figure 5. This representation of the instantaneous profile of the elementary processes, at constant ρ values, will be used repeatedly in similar graphs.

The evolution of the actual profile $\delta_i(\rho)/\delta_0$ with decreasing values of ρ, represented in Figure 7, clearly displays the modifications occurring in the spectral distribution of g_i (Fig. 5), after the system is initially "excited" by an instantaneous T-jump: $\Delta T = -\delta_0/\Delta\alpha = -\delta_{i,0}/\Delta\alpha_i$ [eq. (19)], of any magnitude or sign.

(b) Alternatively, one can represent

$$\delta_i(\rho)/\delta_{i,0} = \exp(-u/a_T\tau_{i,r}) \tag{46}$$

as a function of $\tau_{i,r}$ (Fig. 8). This function is independent not only of δ_0 and T but also of g_i and thus of the particular shape of the retardation spectrum. Therefore the discontinuity occurring in the standard distribution between $i = 16$ and 17 vanishes.

For fixed values of ρ, $\delta_i(\rho)/\delta_{i,0}$ is a single-valued function of $\tau_{i,r}$, i.e., of i alone, which may be considered a continuous variable, as in Figure 8, where the individual points corresponding to discrete values of i have been connected by smooth curves within the interval $1 \leq i \leq 33$. For the standard distribution adopted one can write $\log\tau_{i,r} = 0.125 (i - 1) - 2$, which also holds for noninteger values of i, i.e., for a spectrum defined by a continuous double box distribution, $G(\log\tau_r)$, as shown in Figure 5.

VOLUME AND ENTHALPY RECOVERY OF GLASSES   1121

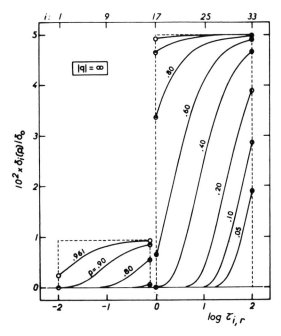

Fig. 7. Instantaneous distribution profiles (reduced to T_r), of the intensities during isothermal recovery (after an instantaneous T-jump) at various fixed values of ρ as indicated.

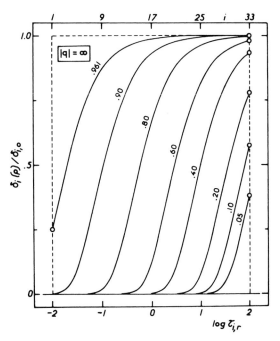

Fig. 8. Same as in Figure 7, but with intensities reduced by g_i [eq. (46)].

Therefore, for decreasing values of ρ, the $\delta_i(\rho)/\delta_{i,0}$ isotherms are merely shifted towards increasing i values and the curves shown in Figure 8 are indeed superimposable by horizontal shifts along the $\log \tau_{i,r}$ axis. Equation (46) shows that the shift involved for any pair of fixed ρ values (e.g., $0 < \rho_1$ and $\rho_2 < 1$) is equal to $\log[u(\rho_2)/u(\rho_1)]$, which is uniquely determined by the $\rho_T(u/a_T)$ function shown in Figure 6, at any value of T.

Concomitantly, at fixed i values, the reduction in $\delta_i(\rho)/\delta_{i,0}$ when ρ decreases from ρ_1 to ρ_2, is given by [eq. (46)]

$$\ln\left[\delta_i(\rho_1)/\delta_i(\rho_2)\right] = (a_T \tau_{i,r})^{-1}[u(\rho_2) - u(\rho_1)] \tag{46a}$$

which is again determined by the $\rho_T(u/a_T)$ function.

These properties of the $\delta_i(\rho)/\delta_{i,0}$ isotherms also apply to the $\delta_i(\rho)/\delta_0$ function [eq. (45)] in any particular range of the spectrum, where g_i is constant. Of course, the individual contributions $\delta_i(\rho)$ cannot be determined directly from experiment nor from the $\rho_r(u)$ function without specifying the distribution of $\tau_{i,r}$.

PART III

Recovery During and After Complex Thermal Treatments

The preceding discussion is restricted to simple approach experiments in which the system is removed from its initial equilibrium state ($\delta_i = 0$, for all i) by a *single* instantaneous T-jump. One can, however, easily extend the application of the model to any situation in which, *prior* to the instantaneous T-jump of interest, the specimen *is not* in equilibrium, i.e., at least one of the δ_i values is different from zero. In fact, the differential equation (20) is applicable to any isothermal ($q = 0$) recovery experiment and can always be integrated numerically when the initial conditions, T and $\delta_{i,0}$, are known. The latter are fully defined by the set of residual $\delta_{i,s}$ values at the temperature T_{n-1}, from which the last instantaneous T-jump to T_n has been performed. Under these conditions, one has initially at T_n,

$$\delta_{i,0} = \delta_{i,s} - \Delta\alpha_i \Delta T_n, \quad 1 \le i \le N \tag{47}$$

where $\Delta T_n = T_n - T_{n-1}$ and $n-1$ is the number of successive temperatures at which the system has been held for a finite time after having been removed from its *last equilibrium state*, reached at T_0. Clearly, eq. (19) represents a special case of eq. (47), for which all $\delta_{i,s} = 0$ and $n = 1$. The set of $\delta_{i,s}$ values thus depends on the past thermal history $T(t)$ of the specimen starting from its last equilibrium state.

The main difference between single and multiple T-jump experiments (including partial recovery at each T_j) concerns the sign of $\delta_{i,0}$ [eq. (47)] which in the latter is not uniquely determined by the sign of ΔT_n, as it was in the former case [eq (19)]. In fact, depending on the sign of ΔT_n and $\delta_{i,s}$ and on their respective magnitudes, $\delta_{i,0}$ can take positive or negative values and may even be equal to zero. The same holds for $\delta_0 = \Sigma^N \delta_{i,0}$ since, for a given set of $\delta_{i,s}$, the sign of $\delta_{i,0}$ may also depend on i, and in some particular situations, the positive and negative components of δ_0 may exactly cancel.

Moreover, the sign of each initial recovery rate at T_n, as defined by eqs. (20):

$$\left(\frac{d\delta_i}{dt}\right)_0 = -\frac{\delta_{i,0}}{a_{T_n}a_{\delta_0}\tau_{i,r}}, \quad 1 \le i \le N \tag{20c}$$

is also determined by the sign of $\delta_{i,0}$. Therefore, the overall initial rate $\Sigma^N(d\delta_i/dt)_0$ can be positive or negative or even equal to zero, when the partial rates of opposite sign exactly cancel each other. Nevertheless, one can easily show that the situation in which δ_0 and $(d\delta/dt)_0$ are *both* equal to zero after the ultimate T-jump, can never arise when $N \ge 2$. This leads to the important conclusion that equilibrium cannot be established instantaneously by any appropriate T-jump, if the state of the system is defined by more than one ordering parameter.

Consecutive T-Jumps

In this section an important class of thermal treatments will be analyzed. These consist of a succession of n instantaneous T-jumps, followed by isothermal periods of limited duration t_j, at each T_j, during which the specimen never reaches an equilibrium state. Starting from equilibrium at $t = 0$ and T_0, the thermal history $T(t)$ is thus a multistep function, with n discontinuities ΔT_j; $1 \le j \le n$. Consequently, δ is also discontinuous as well as a_T and a_δ, i.e., τ_i. For this reason it is convenient to introduce a new time variable, defined by[10]

$$z = \int_{0^+}^{t} (a_T a_\delta)^{-1} dt' \ge 0 \tag{48}$$

which is a continuous function of t whatever the thermal history of the system in spite of discontinuities in the product $a_T a_\delta$. This time scale reduces the instantaneous rates of structural modifications of the specimen, defined by the set of τ_i^{-1} values, to those which would be obtained in equilibrium at the reference temperature T_r.

After the first T-jump, from T_0 to T_1, one has at $t = 0^+$, $\delta_i(0^+, T_1) = -\Delta\alpha_i\Delta T_1$, as in the elementary case discussed above [eq. (19)], while after subsequent recovery, at time $t > 0$,

$$\delta_i(t, T_1) = -\Delta\alpha_i\Delta T_1 \exp\left(-\tau_{i,r}^{-1} \int_{0^+}^{t} (a_T a_\delta)^{-1} dt'\right)$$

$$= -\Delta\alpha_i\Delta T_1 e^{-z/\tau_{i,r}}, \quad 1 \le i \le N \tag{49}$$

similarly to eq. (23). Clearly, $z = u/a_{T_1}$ [eq. (21)], and thus during this first stage of the experiment, z is a unique function of $\rho_T = \delta_0^{-1} \Sigma^N \delta_i$, represented in Figure 6.

The situation becomes more involved when, at time t_1 ($<\tau_{N,T_1}$), the specimen is submitted to another instantaneous T-jump to T_2, at which the initial departure δ_i takes the value [cf. eq. (47)]

$$\delta_i(t_1^+; T_2) = \delta_i(t_1; T_1) - \Delta\alpha_i\Delta T_2, \quad 1 \le i \le N \tag{50}$$

where $\Delta T_2 = T_2 - T_1$. As in eq. (19), the term $-\Delta\alpha_i\Delta T_2$, originates from the difference between the instantaneous changes of the actual and the equilibrium volume (or enthalpy) during the second T-jump ($|q| = \infty$) as given by eq. (18′). The subsequent isothermal recovery at T_2 can be expressed again, for all i, similarly to eq. (23), as

$$\delta_i(t > t_1; T_2) = \delta_i(t_1^+; T_2) \exp\left(- \tau_{i,r}^{-1} \int_{t_1^+}^{t} (a_T a_\delta)^{-1} \, dt'\right)$$

$$= \delta_i(t_1^+; T_2) e^{-(z - z_1)/\tau_{i,r}} \tag{51}$$

which, together with eqs. (49) and (50), can be cast in a compact form

$$\delta_i(t > t_1; T_2) = -\Delta\alpha_i(\Delta T_1 e^{-z/\tau_{i,r}} + \Delta T_2 e^{-(z - z_1)/\tau_{i,r}}). \tag{52}$$

The use of the reduced time u [eq. (21)] would have resulted in a more complex expression.

Next at time t_2, the system is subjected to a T-jump to T_3, and one can easily derive the initial departure $\delta_i(t_2^+; T_3)$, as in eq. (50), and its subsequent recovery for $t > t_2$, similarly to eq. (51). By repeating the same procedure for further steps, one finally obtains at T_n, after the nth instantaneous T-jump occurring at t_{n-1}:

$$\delta_i(t > t_{n-1}; T_n) = -\Delta\alpha_i \sum_{j=1}^{n} \Delta T_j \exp\left(- \tau_{i,r}^{-1} \int_{t_{j-1}^+}^{t} (a_T a_\delta)^{-1} \, dt'\right)$$

$$= -g_i \Delta\alpha \sum_{j=1}^{n} \Delta T_j e^{-(z - z_{j-1})/\tau_{i,r}}, \quad 1 \leq i \leq N \tag{53}$$

with $t_0 = z_0 = 0$, for $j = 1$ and $\Delta\alpha_i = g_i \Delta\alpha$ [eq. (25)]. Each term of the right-hand side of eq. (53) represents, at $t > t_{n-1}$, the residual value of the partial departure $-\Delta\alpha_i \Delta T_j$ produced by the jth T-jump which occurred at t_{j-1}.

The overall departure at t is then obtained by summing over all the i values, viz.,

$$\delta(t > t_{n-1}; T_n) = \sum_{i=1}^{N} \sum_{j=1}^{n} \delta_i(t > t_{n-1}; T_n)$$

$$= -\Delta\alpha \sum_{j=1}^{n} \Delta T_j \sum_{i=1}^{N} g_i e^{-(z - z_{j-1})/\tau_{i,r}} \tag{54}$$

in which the summations have been commuted. The second sum in eq. (54) depends on z and j alone, and one can write

$$1 \geq \sum_{i=1}^{N} g_i e^{-(z - z_{j-1})/\tau_{i,r}} \equiv R(z - z_{j-1}) \geq 0 \tag{55}$$

where $R(z)$ is the *normalized response function* of the material, which is identical with the $\rho_r(u)$ function discussed above (Fig. 6), and proved to be an invariant of the system. The notation is merely changed to avoid confusion, since $R(z - z_{j-1})$ is no longer equal to δ/δ_0, except for $j = 1$.

Combining eqs. (54) and (55) yields

$$\delta(t > t_{n-1}; T_n) = -\Delta\alpha \sum_{j=1}^{n} \Delta T_j R(z - z_{j-1}) \tag{56}$$

which represents the solution of the present problem in terms of the reduced time z [eq. (48)], which can easily be determined using standard numerical integration procedures provided that the a_T and a_δ functions are known.

Equations (53), (54), and (56) are analogous to those used in linear viscoelasticity[16] when instantaneous stimuli are applied to the system at times t_j. They express the Boltzmann superposition principle for a succession of instantaneous

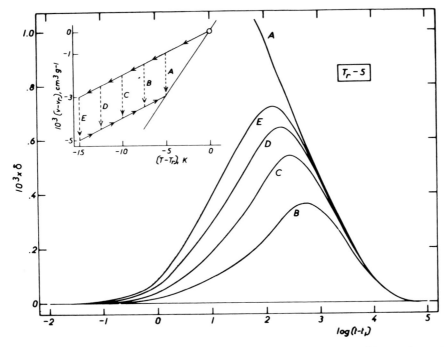

Fig. 9. Memory effects at $T_2 = T_r - 5$ K, after two consecutive instantaneous T-jumps of opposite sign. The partial recovery is the same (2×10^{-3}) at each T_1, as illustrated in the inset. (A) Direct quench from $T_0 = T_r$, (B) $T_1 = T_r - 7.5$ K, $\log t_1 = 3.386$; (C) $T_1 = T_r - 10$ K, $\log t_1 = 3.331$; (D) $T_1 = T_r - 12.5$ K, $\log t_1 = 3.430$; (E) $T_1 = T_r - 15$ K, $\log t_1 = 3.711$.

T-jumps stated in terms of the reduced time z which removes the inherent nonlinearity of structural recovery of glasses resulting from the dependence of the retardation times on the instantaneous state of the system.

Memory Effects

Examples of thermal treatments, involving two consecutive T-jumps of opposite sign, are depicted in the inset of Figure 9. The responses of the system to these treatments are compared through the recovery behavior at a fixed $T_2(= T_r - 5$ K), including that which would be obtained after direct quench ($q = -\infty$) from $T_0 = T_r$ (route A), discussed in the previous sections and shown in Figure 1 (with $\Delta T = -5$ K). The other routes B–E correspond to initial T-jumps from the same T_0 to various T_1 ($\leq T_r - 7.5$ K) values and subsequent recoveries for limited times t_1, during which the departures, initially equal to $\delta_0 = -\Delta\alpha(T_1 - T_0)$, decreased to δ_1, such that $\delta_0 - \delta_1 = 2 \times 10^{-3}$, the value of the initial departure $-\Delta\alpha(T_2 - T_0)$ involved in route A. Then at t_1, the specimen is instantaneously reheated ($q = \infty$) to T_2, where the isothermal recovery is observed as a function of the elapsed time $t - t_1$. The T_1 and t_1 values involved are listed in the legend of Figure 9.

The successive steps in such experiments will now be analyzed in terms of the present model. Clearly, in all cases the first two steps represent simple approach experiments, which obey eq. (24). Nevertheless, owing to the increasing δ_0 values as T_1 decreases, the relative departures $\rho_{T_1}(t_1) = \delta_1/\delta_0 = 1 - 2 \times 10^{-3}/\delta_0$ decrease with T_1 [$\rho(t_1) = 2/3, 1/2, 2/5, 1/3$ from B to E, respectively] and involve quite different

1126   KOVACS ET AL.

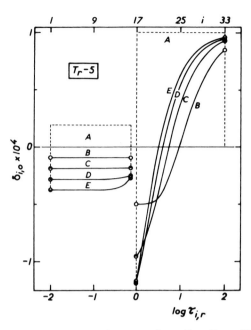

Fig. 10. Initial distribution of the partial departures $\delta_{i,0}$ at $T_2 = T_r - 5$ K immediately after instantaneous heating from T_1 for the routes B–E shown in the inset of Figure 9. (A), Standard distribution.

values of the residual $\delta_i(t_1;T_1)$ as calculated from eq. (49). Their dependence on i can be determined from Figure 7 using the $\rho(t_1)$ values given above. This distribution of δ_i is drastically modified by the second T-jump, according to eq. (50), and the initial states at T_2, defined by the set of $\delta_i(t_1^+;T_2) = \delta_{i,0}$ values, are plotted in Figure 10 for all routes, including A (for which $t_1 = 0$). This shows that an important fraction of the short-time contributions are negative and exactly compensate the long-time contributions. In fact, the parameters defining the thermal treatment B–E have been chosen such that the overall initial departures $\delta_0 = \Sigma \delta_{i,0}$ at T_2 are all equal to zero (see the insert in Fig. 9). In spite of this particular situation, the system does not remain in its *apparent* equilibrium state, since the individual values of $\delta_{i,0}$ are different from zero (with the possible exception of *one* of them, cf. Fig. 10) and recovery according to eq. (51) occurs with an initial rate defined by eq. (20a).

Since all the rapid processes involve negative $\delta_{i,0}$ values, thus positive rates, the system will first expand until the contribution of the slow processes (involving positive values of $\delta_{i,0}$) compensates this expansion and then finally dominates the recovery. Therefore, the system departs from its apparent equilibrium, passes through a maximum of δ, which increases in magnitude and occurs at shorter times $(t - t_1)$ as T_1 decreases, and then approaches true equilibrium (Fig. 9). These isotherms have been calculated using eq. (52), while summing over all values of i, as described by eq. (56), and adapted to the present experiments to yield at $t > t_1$, thus at $z > z_1$:

$$\delta(t > t_1;T_2) = -\Delta\alpha[\Delta T_1 R(z) + \Delta T_2 R(z - z_1)] \tag{57}$$

Since at t_1^+ in each case $\delta_0 = 0$, and $R(z - z_1) \equiv R(0) = 1$, one has $R(z_1) = -\Delta T_2/\Delta T_1$, whereas at long times, when z_1 becomes negligible as compared to

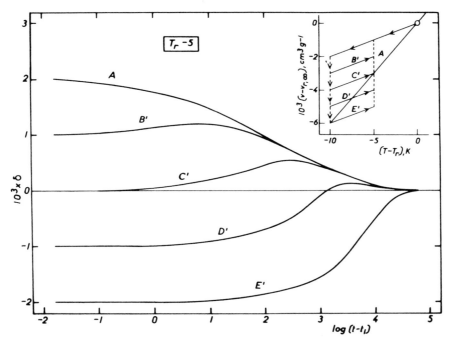

Fig. 11. Memory effects at $T_2 = T_r - 5$ K, after two consecutive instantaneous T-jumps of opposite sign and partial or full recovery at fixed $T_1 = T_r - 10$ K, as depicted in the inset. (A) Direct quench from $T_0 = T_r$, (B') $\delta_1 = 3 \times 10^{-3}$, $\log t_1 = 2.041$; (C') $\delta_1 = 2 \times 10^{-3}$, $\log t_1 = 3.331$; (D') $\delta_1 = 1 \times 10^{-3}$, $\log t_1 = 4.656$; (E') $\delta_1 = 0$, $t_1 = \infty$.

z, $\delta(t \gg t_1; T_2) \simeq -\Delta\alpha(T_2 - T_0)R(z)$. This explains why all the isotherms (B–E) asymptotically approach that obtained by quenching the specimen directly from T_0 to T_2 (route A).

Although surprising at first sight, such phenomena, called *memory*, or *crossover* effects, have often been observed experimentally[6,7,26,27] (compare Fig. 9 with Fig. 24 in ref. 7), and attributed to a multiplicity of retardation mechanisms. As anticipated above, one-parameter models ($N = 1$) cannot account for such effects, since $\delta_0 = 0$ implies necessarily $d\delta/dt = 0$ [eq. (20)].

Another series of such crossover experiments is shown in Figure 11, again involving two consecutive T-jumps of opposite sign and the same fixed values of $T_0(= T_r)$ and $T_2(= T_r - 5$ K) as above. Here, however, T_1 is also kept constant ($= T_r - 10$ K), but reheating ($q = \infty$) to T_2 occurs from evenly decreasing values of δ_1, such that $\rho(t_1) = \delta_1/\delta_0 = \delta(t_1; T_1)/\Delta\alpha(T_0 - T_1)$ is equal to $\frac{3}{4}(B')$, $\frac{1}{2}(C')$, $\frac{1}{4}(D')$, and $0(E')$, respectively. Clearly, the C' isotherm is the same as C in Figure 9, whereas A and E' correspond to simple approach experiments shown in Figure 1 involving the same initial departure $|\delta_0|$ but opposite signs. The B'–D' isotherms again exhibit a shallow maximum of δ, which occurs at increasing times $t - t_1$, as δ_1 decreases. These curves have been generated by eq. (57), taking into account the particular thermal treatments listed in the caption for Figure 11. Isotherms of such character have, in fact, been obtained experimentally by several authors[6,27] (cf. Figs. 3 and 4 in ref. 27) on subjecting glassy specimens to similar treatments.

Such calculations based on eq. (56) can be carried out for treatments involving any two, or more consecutive T-jumps. Clearly, when all ΔT_j have the same sign, isothermal recovery at T_n occurs solely by contraction ($\Delta T_j < 0$) or by expansion ($\Delta T_j > 0$), as in the case of a single T-jump, whereas each alternation in the sign of ΔT_j may give rise to a maximum or a minimum in δ,[7] as illustrated by the isotherms shown in Figures 9 and 11.

Finite Cooling or Heating Rate

The above treatments have been restricted to instantaneous T-jumps. The application of the model can, however, be extended easily to thermal histories involving finite cooling and/or heating rates, which correspond more closely to the experimental situation, since for specimens of finite size, thermal equilibrium cannot be reached without a time lag Δt. Nevertheless, the previous calculations are not only important and useful for a comprehensive analysis of the model's behavior, but they also provide a close approximation to real situations when Δt is of the same order of magnitude as, or smaller than the actual value of the shortest retardation time $a_T a_\delta \tau_{1,r}$, i.e., when structural modifications during quenching or heating are negligibly small.

Since the actual state of the system can only be defined when its thermal history $T(t)$ is known, we consider here a specimen which is removed at $t = 0$ from its initial equilibrium state, at T_0, by cooling or heating at a finite rate $q = dT/dt$. The latter may be kept constant, or vary with T, or t. Under these conditions one can immediately apply eq. (56), by decomposing the thermal treatment into infinitesimal T-jumps of magnitude dT, and isothermal periods of duration dt, thus by replacing the sum with an integral to yield at $T(t) = T_0 + \int_0^t q\, dt'$:

$$\delta(z) = -\Delta\alpha \int_0^z R(z - z') \frac{dT}{dz'}\, dz' \tag{58}$$

in which the reduced time z [eq. (48)] defined by

$$z = \int_{T_0}^{T} (q a_{T'} a_\delta)^{-1}\, dT' \geq 0 \tag{48a}$$

depends on T and q alone, provided that $0 < |q| < \infty$.

The convolution integral in eq. (58) is the fundamental constitutive equation of the system, based on Boltzmann's superposition principle, according to which the effects of sequential changes in stress (here temperature) are additive. Equation (58) was first proposed by Narayanaswamy[10] in analogy with linear viscoelasticity, in terms of the reduced time z. In fact, the $\Delta\alpha R(z)$ function may be thought of as the *isothermal bulk creep compliance* with respect to instantaneous temperature changes.

Recently, Moynihan and co-workers[11,23] have shown the applicability of eq. (58) for describing the isobaric behavior of glasses during uniform cooling and heating, while using the postulated normalized response function $\bar{\rho}_T(u/a_T) \equiv \bar{R}(z)$ given in eq. (44), and shown in Figure 6 (dashed lines). In the following sections some implications of the present model during similar treatments will be examined and discussed in terms of the characteristic features of the $\delta(T)$ and $v(T)$ isobars.

Cooling Isobars

The cooling isobars shown in Figure 12 have been generated by numerical

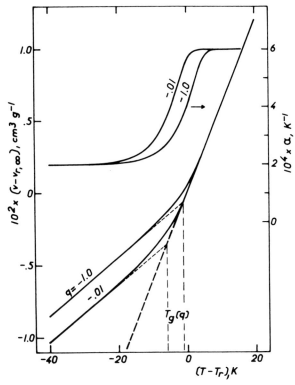

Fig. 12. Volume and $\alpha = v^{-1}(dv/dT)$ isobars obtained on cooling from $T_0 = T_r + 20$ K at a constant rate q, as indicated. Dashed lines represent the volumes as extrapolated from the asymptotic glassy regime reached at low temperatures [(cf. eq. (61)] and the corresponding values of the conventional glass-transition temperature $T_g(q)$.

integration of eq. (18′) to yield $\delta(T)$, e.g.,

$$v - v_{r,\infty} = v_\infty \delta(T,q) + v_{r,\infty}\alpha_l(T - T_r)$$

cf. eqs. (3) and (10). The temperature increments used in this procedure have been chosen equal to $0.1 \times q a_T a_\delta \tau_{1,r}$, similarly to the isothermal case, nevertheless, in order to account for the cumulative effect of the elementary T-jumps, none of the 33 values of δ_i were neglected even when δ_i/δ was less than 10^{-7}. Furthermore, for high starting temperatures T_0 when the departure from equilibrium is very gradual (thus $a_\delta \simeq 1$), an appropriate analytical expression[28] was used for each δ_i (as can be derived by assuming pure temperature dependence of $\tau_i = a_T \tau_{i,r}$), in order to avoid "underflow" of the computer. This procedure was followed until δ became larger than 5×10^{-7} [i.e., until $a_\delta < 1 \pm 7.5 \times 10^{-4}$, cf. eqs. (17) and (36)], while starting the numerical integration of eq. (18′) with the set of δ_i values reached in the analytical step.

Figure 12 represents the volume and $\alpha = v^{-1}(dv/dT)$ isobars at two different cooling rates, as indicated, with $T_0 = T_r + 20$ K. The hyperbolic shape of these volume isobars is similar to those obtained from experiment, showing in particular two asymptotic branches at high and at low temperatures, where the thermal expansion coefficient reaches its limiting values, equal to α_l and α_g, respectively.

The conventional glass-transition temperature $T_g(q)$, defined by the intersection of the asymptotes, increases with the cooling rate $|q|$. If a_δ depends on δ alone, as in eq. (17), one can write, according to the theoretical considerations given elsewhere[7,28,29] and developed below:

$$\frac{\Delta T_g(q)}{\Delta \log|q|}\bigg|_{T_0 \gg T_g(q)} = 2.303\theta^{-1} \tag{59}$$

provided that the starting temperature T_0 is high enough. As anticipated above, the reference temperature, as determined by the spectrum shown in Figure 5, is $T_r \simeq T_g(-1.0) + 1.1$ K, whatever the time unit, defined by the value of $\tau_{17,r}$. Therefore, the isobars shown in Figure 12 correspond to the practical range of rates which can be achieved in scanning calorimetry, or in dilatometry, if $\tau_{17,r}$ is assumed to be 1 sec, or 1 min, respectively.

In a more general context, the $\delta(T,q)$ isobars can be brought into exact coincidence merely by shifting them along the temperature scale. In fact, eq. (48a) shows that z depends on the experimentally controlled parameters (T, T_0, and q) alone, since δ, thus a_δ, is uniquely determined by z [eq. (58)]. Therefore, one should have $\delta(T,q) = \delta(T',q')$, if for all values of T and q one can associate, in a unique manner, a set of T',q' values such that $qa_T = q'a_{T'}$. If eq. (17) holds, this requirement is fulfilled both for cooling and heating isobars when

$$T_0 - T_0' = T - T' = \theta^{-1}\ln(q/q') \tag{60}$$

provided that q and q' have the same sign. This equation holds even for more involved expressions for a_T and a_δ, provided that θ is allowed to vary (slightly) with T, as implied by its various definitions given in Appendix A. In fact, for the nonlinear system, eq. (60) defines the time-temperature reduction method comparable to that used in linear viscoelasticity[16] and based on the principle of corresponding viscoelastic states.

Since $T_g(q)$ is uniquely defined by the temperature at which the low-temperature asymptote of $\delta(T,q)$ defined by

$$\bar{\delta}(T,q) = -\Delta\alpha[T - T_g(q)] \tag{61}$$

intercepts the ordinate $\delta = 0$, eq. (60) implies that

$$T_g(q) - T_g(q') = \theta^{-1}\ln(q/q') \tag{60'}$$

Therefore, eq. (59) represents a particular application of the method of reduced variables defined by eq. (60) and provides a convenient method for determining the material constant θ. One can thus conclude that all the $\delta(T,q)$ isobars obtained on cooling from $T_0 \gg T_g(q)$ at various rates q, will exactly coincide when plotted against $T - T_g(q)$, or $T - T_r(q)$, if $T_r(q)$ is allowed to vary with q as $T_g(q)$, i.e., keeping $T_g(q) - T_r(q)$ constant. Moreover, the validity of this statement can be extended to any value of T_0 at which cooling, or heating starts from an equilibrium state, at a rate q, provided that for a rate q' the starting temperature T_0' is related to T_0 as in eq. (60). This requirement, however, has no practical effect when $T_0 \gg T_g(q)$.

The same reduction of the temperature scale also brings the $\alpha(T,q)$ isobars into near coincidence, as can be seen in Figure 12, whereas superimposition of the volume isobars in addition requires a vertical shift, equal to $v_{r,\infty}\alpha_l[T_g(q) - T_g(q')]$, in order to maintain the equilibrium volumes invariant. The glass-

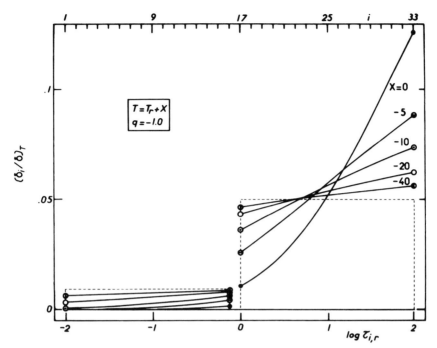

Fig. 13. Instantaneous distribution profiles of the intensities at various temperatures during uniform cooling from $T_0 > T_r + 10$ K at rate $q = -1.0$. Dashed line: the standard distribution as in Figure 5.

transition temperature $T_g(q)$ obtained *on cooling* from $T_0 \gg T_g(q)$ at a rate q is, therefore, the genuine reference temperature of the system in terms of which isobaric responses at any value of q can be reduced to a unique one. This important conclusion derived from eq. (60) will be further substantiated by the examples given below.

In addition to these isobars characterizing the experimentally measurable parameters of the system, it is interesting to analyze the thermal variation of the elementary departures δ_i during cooling through the glass-transition range. At temperatures high enough, all the δ_i values will be practically equal to zero and the system remains at equilibrium, determined by T and P alone. Appreciable departure from equilibrium occurs only when the longest retardation time $a_T \tau_{N,r}$ becomes of the same order of magnitude as the characteristic "cooling time" [28] $\theta^{-1}|q|^{-1}$. A similar situation prevails for the shorter retardation times $a_T \tau_{i,r}$ $(i < N)$ on further cooling, as the elementary processes progressively start to contribute to δ. Qualitatively, the evolution of δ_i is similar to that which would occur during a hypothetical *reversed* isothermal contraction experiment, in which the time would *decrease* from ∞ to zero while δ increases from zero to finite values (cf. Fig. 7, with increasing values of ρ).

The instantaneous profile of $(\delta_i/\delta)_T$, as derived from the present model, is depicted in Figure 13 at various temperatures, while cooling from $T_0 = T_r + 20$ K is monitored at a rate $q = -1.0$. Clearly, the contribution of the long-time processes to δ is enhanced, whereas that of the rapid processes is reduced, as compared to the standard distribution, shown by dashed lines. This skewing of the distribution slowly diminishes as T decreases, but is still appreciable even at $T_r - 40$ K, since the approach towards the standard distribution is asymptotic.

1132   KOVACS ET AL.

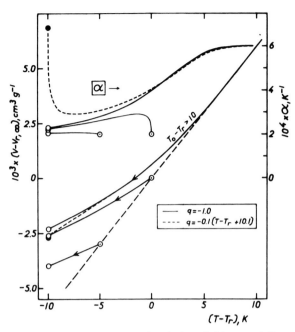

Fig. 14. Dependence of the volume and α cooling isobars (with $q = -1.0$) on the starting temperature T_0, indicated by $\odot$. The dashed lines correspond to an isobar obtained with an exponential temperature decay from $T_r + 20$ K to $T_r - 10$ K (see text).

If cooling were stopped at any temperature in the range considered in Figure 13, the subsequent isothermal recovery would involve an initial distribution of $(\delta_{i,0}/\delta_0)_T$ which would be quite different from that resulting from an instantaneous quench. The consequences of this modification of the initial distribution will be analyzed below.

One expects that the cooling isobars, and the corresponding distribution profile, also depend on the value of T_0, when cooling starts from an equilibrium state reached in the transition range, or below. Figure 14 shows this effect on some volume and α isobars for various values of $T_0 - T_r$ and $q = -1.0$ (solid lines). Clearly, the departure from equilibrium becomes more and more abrupt as T_0 decreases, whereas the expansion coefficient α shows a shallow maximum which is consistent with eq. (18'). For low values of T_0, the situation thus becomes comparable to that of an instantaneous quench. In fact when $a_{T_0}\tau_{1,r}$ is greater than $\theta^{-1}|q|^{-1}$, even the most rapid processes appear "frozen in" during cooling, or heating.*

The particular α and v isobars represented by dashed lines refer to *exponential cooling*, from $T_0 = T_r + 20$ K to $T_r - 10$ K, at a rate $q = -0.1(T - T_r + 10.1)$, thus $T - T_r + 10.1 = 30.1 \exp(-t/10)$. The value of the thermal retardation time (10 time units) has been chosen such that at $T_r - 10$ K approximate thermal equilibrium is reached about one minute after the quench, if the time unit is assumed to be one second. This treatment thus simulates the practical situation achieved by rapid quenching of a dilatometric specimen through its glass-transition range.[6]

---

* At equilibrium, with the adopted material constants [eq. (36)] and distribution (Fig. 5), the time scale is reduced by $10^4 = \tau_{N,r}/\tau_{1,r}$ when the temperature decreases by 9.2 K.

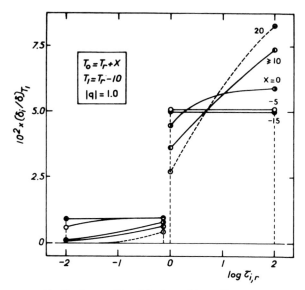

Fig. 15. Instantaneous distribution profiles of the intensities at $T_1 = T_r - 10$ K right after uniform cooling or heating from various initial temperatures T_0 (equilibrium) at a rate $|q| = 1.0$. Dashed line: the same but using the exponential temperature decay as in Figure 14.

Figure 14 shows that the effect of this treatment is similar to that obtained with uniform cooling, at a rate $q = -1.0$, starting at $T_0 > T_r + 10$ K. The α isobar, however, shows a dramatic increase in the vicinity of $T_r - 10$ K. This is due to the rapid reduction of the cooling rate in this temperature range resulting in a maximum of δ [cf. eq. (18′)], which implies a preceding inflexion of the v-T isobar, thus a change in the sign of $d\alpha/dT$.

The above conclusions are further substantiated by Figure 15, which shows the instantaneous profiles (solid lines) of the distribution $(\delta_i/\delta)_T$ at fixed $T = T_1 = T_r - 10$ K for the same values of T_0 and q as in Figure 14, and for $T_0 = T_r - 15$ K, corresponding to a heating process with $q = 1.0$. Clearly, when $T_0 < T_r - 5$ K, the initial distribution at T is practically the same as for an instantaneous quench. The distribution shown by dashed lines corresponds to the exponential cooling discussed above. Thus, this treatment considerably enhances the contribution of long time processes at the expense of the short time ones.

The situation depicted by Figures 13–15 can easily be extended to any other cooling rate by an appropriate shift of the temperature scale, defined by eq. (60). In particular, the distributions obtained at $T_r + X$ (Fig. 13) after cooling from $T_0 \gg T_g(q)$ at a rate $q = -1.0$, will be identical to that which would result for a rate q' at a temperature $T_r + X'$ such that $\theta(X' - X) = \ln|q'|$. On the other hand, when T_0 is located in the transition range, as in the case of Figures 14 and 15, T_0 and T should both be increased by $\theta^{-1} \ln|q'|$ in order to obtain the same situation as for $|q| = 1.0$. This is equivalent to a shift of the reference temperature by the same amount, as has been shown above.

Isothermal Recovery After Uniform Cooling or Heating

If q is made zero at any temperature T_1 and time t_1, the subsequent isothermal recovery of the individual processes is still controlled by eq. (20). At this point the departures $\delta_{i,1} \equiv \delta_i(t_1;T_1)$, resulting from the previous thermal history, are uniquely determined by T_0, T_1, and q, as shown in Figures 13 and 15 for $|q| = 1.0$. Therefore, one can rewrite eqs. (22)–(24) in terms of the reduced time z [eq. (48)] to yield at $t > t_1$:

$$\delta(t > t_1;T_1) = \sum^N \delta_{i,1}e^{-(z-z_1)/\tau_{i,r}} = \delta_1 \sum^N \gamma_i e^{-(z-z_1)/\tau_{i,r}} \equiv \delta_1 R^1(z - z_1) \quad (62)$$

where z_1 is defined by eq. (48a), with $T = T_1$, whatever the sign of q, while

$$z - z_1 = a_{T_1}^{-1} \int_{t_1^+}^{t} (a_\delta)^{-1} \, dt' \quad (48b)$$

and

$$\gamma_i \equiv \delta_i(t_1;T_1)/\sum^N \delta_i(t_1;T_1) \equiv \delta_{i,1}/\delta_1 \neq g_i \quad (63)$$

is the fractional contribution of the ith retardation mechanism to δ_1, satisfying the normalization condition $\sum^N \gamma_i = 1$. Further, δ_1 and z_1 are related through eq. (58), since

$$\delta(t_1;T_1) = -\Delta\alpha \int_0^{z_1} R(z_1 - z') \left(\frac{dT}{dz'}\right) dz'. \quad (64)$$

Equation (62) shows that $\rho_1 \equiv \delta(t > t_1;T_1)/\delta_1 \equiv R^1(z - z_1)$ is a single-valued function of $z - z_1$. Nevertheless, z_1, as well as the initial set of γ_i values, critically depend on the past thermal history. Therefore $R^1(z - z_1) \neq R(z - z_1)$ except when $t_1 = z_1 = 0$, which corresponds to an instantaneous T-jump, for which $\gamma_i = g_i$.

One can also express the isothermal recovery after cooling or heating at a finite rate from T_0 (equilibrium) in terms of the fundamental convolution integral given by eq. (58), and write

$$\delta(t > t_1;T_1) = -\Delta\alpha \int_0^{z_1} R(z - z') \left(\frac{dT}{dz'}\right) dz'$$

$$\equiv -\Delta\alpha \int_0^{z} R(z - z') \left(\frac{dT}{dz'}\right) dz' \quad (65)$$

That is, the integration [related to the changes in temperature, cf. eq. (56)] is carried out through the dummy variable z', from zero to z_1. The upper limit can be replaced by z, even though z increases during the isothermal period [cf. eq. (48b)] since $dT/dz' = 0$ in this situation. This expression is useful whenever the system is defined through the normalized response function $R(z)$ [cf. eq. (44) and Fig. 6], rather than by the retardation spectrum $G(\tau_{i,r})$. Combining eq. (65) with eqs. (62) and (64), one obtains for $z > z_1$:

$$\int_0^{z} R(z - z') \left(\frac{dT}{dz'}\right) dz' = R^1(z - z_1) \int_0^{z_1} R(z_1 - z') \left(\frac{dT}{dz'}\right) dz' \quad (66)$$

which relates the particular isothermal response function $R^1(z - z_1)$ to $R(z)$.

When the system is defined, as is done here, through its retardation spectrum (Fig. 5), integration of eq. (62) is straightforward provided that the set of γ_i values [eq. (63)], resulting from the previous thermal history, is known. As an example, Figure 16 represents the isothermal recovery of the system at various tempera-

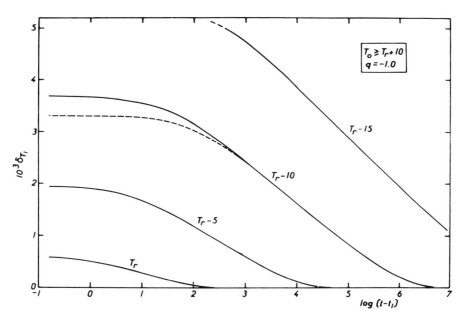

Fig. 16. Intrinsic δ isotherms at various indicated temperatures T_1, following cooling from T_0 > T_r + 10 K at a rate $q = -1.0$, or $q = -0.1$ $(T - T_r + 10.1)$ (dashed line).

tures T_1 after uniform cooling from $T_0 > T_r + 10$ K at a rate $q = -1.0$ (solid lines), or $q = -0.1$ $(T - T_r + 10.1)$ (at $T_r - 10$ K, dashed lines) discussed above. The relevant initial distributions $(\gamma_i, \tau_{i,r})$, which will be called $\Gamma_1(\tau_{i,r})$ from here on, are those shown in Figures 13 and 15.

Clearly, these isotherms are different from one involving an instantaneous T-jump which results in the same initial departure δ_1 at the same temperature T_1, such as shown in Figure 1. Isotherms obtained after cooling from temperatures $T_0 \gg T_g(q)$ to T_1 will be referred to as *intrinsic*,[6,7,9] though they depend on T_1 and q. Nevertheless, the intrinsic isotherm at any value of T_1, defines the *limiting approach curve* for all other isotherms, involving complex previous thermal histories (cf. memory effects, such as shown in Figs. 9 and 11) and/or lower values of T_0, provided that in the temperature range $T_0 > T > T_1$ the *first cooling* from T_0 is carried out at a rate $|q'| \le |q|$. The two isotherms shown in Figure 16, for $T_1 = T_r - 10$ K, obtained with different cooling rates (see legend) approximately fulfill this requirement and merge in their approach toward equilibrium even though they involve quite different sets of γ_i values (Fig. 15).

As mentioned above, however, the particular distribution $\Gamma_1(\tau_{i,r})$ reached at T_1 after cooling from $T_0 \gg T_g(q)$ at a rate -1.0 (Fig. 13), is identical to that which would result at T'_1 for a cooling rate q', provided that $T'_1 - T_1 = \theta^{-1} \ln |q'|$. The same holds when T_0 is located in the transition range, or below (cf. Figs. 14 and 15), if $T'_1 - T_1 = T'_0 - T_0$, as required by eq. (60). In fact, the set of values of $\delta_{i,1}$, thus $\Gamma_1(\tau_{i,r})$, is uniquely defined by the reduced time z_1, which depends on T_1 alone when $q a_T$ is a single-valued function of T [cf. eq. (48a)]. Furthermore, $\rho_1 = R^1(z - z_1)$ depends on $z - z_1$ alone. Therefore, all the $\delta(T'_0, T'_1, q')$ isotherms fulfilling the requirements of eq. (60), reduce to a single curve, when plotted against $(t - t'_1)/a_{T_1}$, as in Figure 1, provided that eq. (17) is applicable

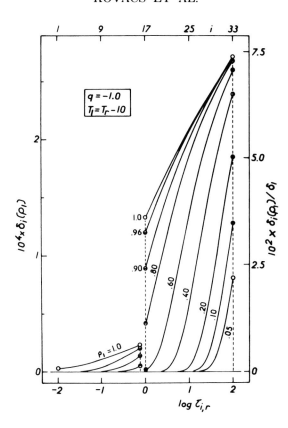

Fig. 17. Instantaneous δ_i values, or distribution profiles of the intensities (right-hand scale) during intrinsic isothermal recovery at $T_1 = T_r - 10$ K at fixed values of ρ_1, as indicated. Previous history: uniform cooling from $T_0 > T_r + 10$ K to T_1, at a rate $q = -1.0$, as in Figure 16.

in the temperature interval covered. Further consequences of this rate-temperature reduction will be discussed below.

Figure 17 represents the evolution of the distribution profile $\delta_i(\rho_1)$ or $\delta_i(\rho_1)/\delta_1$, during isothermal recovery for the intrinsic isotherm at $T_r - 10$ K, after cooling at a rate -1.0 (cf. Fig. 15). Like $\Gamma_1(\tau_{i,r})$, these plots depend on the previous thermal history and they may differ considerably from those involving instantaneous T-jumps (Fig. 7) since as in eq. (45):

$$\delta_i(\rho_1)/\delta_1 = \gamma_i e^{-(z-z_1)/\tau_{i,r}} \tag{45a}$$

In contrast, the curves representing $\delta_i(\rho_1)/\delta_{i,1}$ vs. $\log\tau_{i,r}$ (Fig. 18) are exactly superimposable on those shown in Figure 8 by appropriate shifts along the abscissa, since the shape of these curves is independent of γ_i [cf. eq. (46)]. The shift involved for a particular value of ρ is determined by the relationship between the corresponding $R^1(z - z_1)$ and the $\rho(u/a_T) \equiv R(z)$ function.

Concomitantly, the relevant reduced isothermal rate functions defined by eqs. (32) and (34) also depend on the past thermal history of the system. In fact, according to the definition of $\phi'_r(\delta, \delta_0)$ in eq. (32), at T_r, one has similarly to eq. (33):

VOLUME AND ENTHALPY RECOVERY OF GLASSES   1137

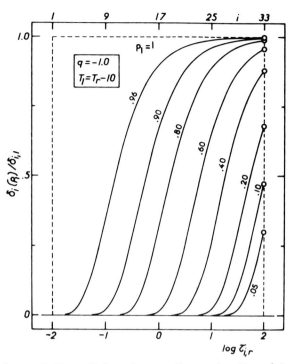

Fig. 18. Same as in Figure 17, but using an ordinate reduced by γ_i [cf. eq. (45a)].

$$\sum_{}^{N} \frac{\delta_i}{\delta} \tau_{i,r}^{-1} = [R^1(z - z_1)]^{-1} \sum_{}^{N} \frac{\gamma_i}{\tau_{i,r}} e^{-(z-z_1)/\tau_{i,r}} \equiv \phi_r^1(\rho_1) \qquad (67)*$$

where $\phi_r^1(\rho_1) \neq \phi_r(\rho)$, except when equilibrium is approached, since the limiting value of both of these functions is $\tau_{N,r}^{-1}$ [cf. eq. (31) and Appendix B].

Figure 19 represents the functions $\phi_r^1(\rho_1)$ and $\psi^1(\rho_1) \equiv \tau_{N,r}\phi_r^1(\rho_1)$ which characterize the intrinsic isotherms (Fig. 16) obtained at three different temperatures, with $q = -1.0$. Clearly, both the shape and the range of variation of these functions differ considerably from each other and from $\phi_r(\rho)$, shown by dashed lines, which characterizes isothermal recovery after any single instantaneous T-jump (Fig. 4). The $R^1(z - z_1)$ functions show a similar trend, since eq. (41) holds with the relevant substitutions. In the temperature range considered $\phi_r^1(1)$ increases, and approaches the value of $\phi_r(1)$ as T_1 decreases. Like $R^1(z - z_1)$, $\phi_r^1(\rho_1)$ reflects the distribution $\Gamma_1(\tau_{i,r})$ reached at the end of the cooling stage at T_1 (Fig. 13).[†] Furthermore, reduction of the cooling rate leaves

---

*From here on, the rate function will be denoted as $\phi_r^1(\rho_1)$ where the subscript refers to the temperature T_r [as for $\phi_r(\rho)$], while the superscript 1 indicates that it is related to the function R^1 obtained at T_1, i.e., to the truncated spectrum $\Gamma_1(\tau_{i,r})$.

[†]Owing to the skewed distribution and to the reduction of the time interval covered by $\Gamma_1(\tau_{i,r})$, intrinsic isotherms obtained in the transition range can be described with fair accuracy (refs. 7 and 9) in terms of a unique retardation mechanism, involving a single retardation time $\tau(T,\delta) = a_\delta a_T \tau_{N,r}$. Nevertheless, by referring to a one-parameter model, implying $\psi(\rho) = 1$ [cf. eqs. (33) and (35)] and $\tau_{\text{eff},T} = a_\delta a_T \tau_{N,r}$ [cf. eq. (29), with $i = 1$, and Fig. 2], one implicitly includes the contributions of ρ_1 and T_1 to $\phi_r^1(\rho_1)$ in the a_δ function. This leads to exaggerated values of $(1 - x)\theta/\Delta\alpha$ which are incompatible with the expansion isotherms obtained at the same temperature,[7] and to an underestimation of the value of x, i.e., of the pure contribution of T to τ.

1138 KOVACS ET AL.

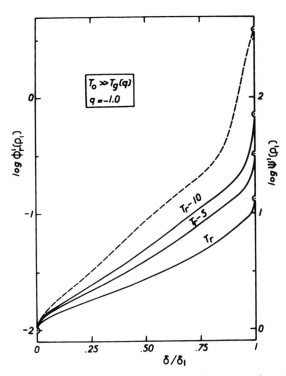

Fig. 19. Reduced rate function $\phi_r^1(\rho_1)$ of the intrinsic isotherms at various temperatures T_1, as indicated, reached after cooling from $T_0 > T_r + 10$ K at a rate $q = -1.0$. Dashed line: $\phi_r(\rho)$ characterizing recoveries after instantaneous T-jumps as in Figure 4; this curve is also the limit of $\phi_r^1(\rho_1)$ as T_1 approaches zero.

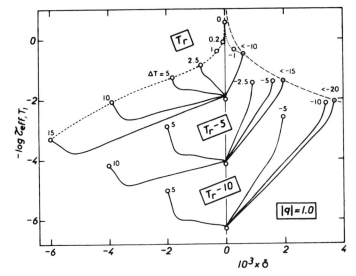

Fig. 20. τ_{eff} vs. δ isotherms during recovery at various fixed temperatures T_1 (framed) reached after uniform cooling, or heating from T_0 (equilibrium) at a rate $|q| = 1.0$; $\Delta T = T_1 - T_0$ values are indicated by small numerals on each curve. Dashed lines: upper bound defined by the initial τ_{eff} values of the intrinsic isotherms for varying values of T_1 (see the upper right-hand corner of the graph), and those of the expansion isotherms at fixed $T_1 = T_r$ with $T_0 \leqslant T_r$. Dotted line: the same for the contraction isotherms again at fixed $T_1 = T_r$, but $T_0 \geqslant T_r$.

the $a_T^{-1}\phi_r^1(\rho_1)$ function unaltered at T_1' so long as this temperature is related to T_1 by eq. (60).

In a more general context, the τ_{eff,T_1} isotherms [eqs. (28) and (29)] represented in Figure 20 (at three different temperatures T_1) clearly depict the characteristic modifications produced by the previous thermal history, in this case involving uniform cooling or heating from T_0 (equilibrium) to T_1, at a rate $|q| = 1.0$, rather than instantaneous T-jumps, as in Figure 2. In fact, during the isothermal period, eq. (29) is still applicable with the definition of $\Sigma^N(\delta_i/\delta)\tau_{1,r}^{-1}$ given by eq. (67), to yield at $t > t_1$:

$$\tau_{\text{eff},T_1}^{-1} = (a_{T_1}a_\delta)^{-1}\phi_r^1(\rho_1), \quad z \geq z_1 \tag{68}$$

Clearly, when δ approaches zero, the final value of τ_{eff,T_1} is uniquely determined by T_1, as given by eq. (31), since at equilibrium $a_\delta = 1$ and $\phi_r^1(0) = \phi_r(0) = \tau_{N,r}^{-1}$ (cf. Fig. 19). In Figure 20, the initial and final values of τ_{eff,T_1} are circled, and those of $\Delta T = (T_1 - T_0)$ are indicated on each isotherm.

Obviously, the initial $(\tau_{\text{eff},T_1})_0$ value depends on the past thermal history of the system, as do both δ_1 and $\phi_r^1(1)$, hence at $(t - t_1) = 0^+$, i.e., at $z = z_1$, one has

$$(\tau_{\text{eff},T_1}^{-1})_0 = (a_{T_1}a_{\delta_1})^{-1}\sum_{\tau_{i,r}}^{N}\frac{\gamma_i}{\tau_{i,r}} = (a_{\delta_1}a_{T_1})^{-1}\phi_r^1(1) \tag{69}$$

and $R^1(0) = \rho_1 = 1$. If $|q|$ is kept constant, as in Figure 20, the partial derivatives of $(\tau_{\text{eff},T_1}^{-1})_0$ either at fixed T_0, or T_1, reveal some interesting properties of the system.

(a) The effect of the truncation of the standard distribution $G(\tau_{i,r})$ at short times (cf. Fig. 13) on the τ_{eff,T_1} plots of the intrinsic isotherms is clearly apparent in Figure 20. Here cooling starts at $T_0 \gg T_g(q)$ and the initial $(\tau_{\text{eff},T_1}^{-1})_0$ values define an envelope (shown by dashed lines in the upper right-hand corner) which bounds an inaccessible region where values of τ_{eff,T_1} cannot be reached with the specific cooling rate whatever the value of T_0. This envelope asymptotically approaches two straight lines. The first asymptote, reached when T_0 and $T_1 \rightarrow \infty$ [practically when $T_0 > T_1 > (T_r + 10\text{ K})$] is obviously defined by $\bar{\delta}_1 = 0$, since at such high temperatures the system does not depart from equilibrium, yet $(\tau_{\text{eff},T_1}^{-1})_0$ increases with T_1 [cf. eq. (69)].

On the other hand, at low enough T_1 temperatures, the system asymptotically approaches the glassy regime [cf. eq. (61) and Fig. 12] defined by

$$\bar{\delta}_1 = -\Delta\alpha[T_1 - T_g(q)] \tag{61a}$$

and $(\tau_{\text{eff},T_1}^{-1})_0$ approaches another asymptote defined by

$$(\tau_{\text{eff},T_1}^{-1})_0 = (a_{T_1}a_{\delta_1})^{-1}\phi_r(1) = a_{\delta_1}^{-1}\phi_{T_1}(1), \quad T_1 \ll T_g(q) \tag{70}$$

since, in this temperature range, $\gamma_i \rightarrow g_i$ (cf. Fig. 13) and $\phi_r^1(1) \rightarrow \phi_r(1) = \Sigma^N(g_i/\tau_{i,r})$. In Figure 20, this low-temperature asymptote intersects the abscissa $\delta = 0$ at $\log\phi_{T_g(q)}(1)$, and its slope is defined by the partial derivative of $\log(\tau_{\text{eff},T_1}^{-1})_0$ with respect to δ_1 at fixed T_0 and q, to yield, with reference to eq. (17):

$$\left(\frac{\partial\log(\tau_{\text{eff},T_1}^{-1})_0}{\partial\delta_1}\right)_{q,T_0} = -x\theta/2.303\Delta\alpha \tag{70a}$$

which characterizes the pure contribution of temperature to τ_i along the asymptotic glassy line [eq. (61a)], where $(\partial\bar{\delta}_1/\partial T_1)_{q,T_0} = -\Delta\alpha$.

Equations (70) and (70a) hold for any fixed value of T_0, thus even when $T_0 < T_1$ (i.e., for $\delta_1 < 0$), provided that $(\partial \delta_1/\partial T_1)_{q,T_0} = -\Delta\alpha$. In fact, the asymptotic glassy regime is reached more rapidly as T_0 is decreased (cf. Figs. 14 and 15). Accordingly, the asymptotic variation of $(\tau_{\text{eff},T_1})_0$ with δ_1, at fixed q and T_0, provides a convenient method for determining the important material constant x, since θ and $\Delta\alpha$ can easily be derived from independent data {e.g., from the rate dependence of the cooling isobars, as shown in Figure 12 [cf. eq (59)], or from the temperature dependence of $\tau_{N,T}$, defined by the final values of $\tau_{\text{eff},T}$, at $\delta = 0$}.

(b) For fixed T_1, the variation of $(\tau_{\text{eff},T_1})_0$ with δ_1 is more involved due to the peculiar dependence of $\phi_r^1(1)$ on $(T_1 - T_0)$. In fact, $\phi_r^1(1)$ approaches its limiting value $\phi_r(1)$ both at large and at very small values of $|\delta_1|$, while passing through a minimum in between. As an example, Figure 20 shows (by dashed lines) the peculiar dependence of $(\tau_{\text{eff},T_1})_0$ on δ_1 for the expansion isotherms ($\delta_1 < 0$) obtained at T_r. An even more dramatic situation prevails on the contraction side ($T_0 > T_r$; dotted line), where the increase of δ_1 is limited by the relevant intrinsic isotherm (Fig. 16). At moderately large *negative* values of δ_1, e.g., for $(T_1 - T_0) > 2.5$ K, the slope

$$\left(\frac{\partial \log(\tau_{\text{eff},T_1}^{-1})_0}{\partial \delta_1}\right)_{q,T_1} = (1-x)\theta/2.303\,\Delta\alpha + \left(\frac{\partial \log\phi_r^1(1)}{\partial \delta_1}\right)_{q,T_1} \tag{70b}$$

is generally smaller than $-(\partial \log a_\delta/\partial\delta)_{T_1}$, since in this range $\phi_r^1(1)$ increases as δ_1 decreases, while approaching its limiting value $\phi_r(1)$ and this effect is accentuated as T_1 decreases. Accordingly, the slope defined by eq. (70b) approaches $(1-x)\theta/2.303\Delta\alpha$, whereas the corresponding asymptote intercepts the ordinate $\delta = 0$ at $\phi_{T_1}(1)$, as in Figure 2. In fact, as T_0 decreases, $\gamma_i \to g_i$ and the distribution $\Gamma_1(\tau_{i,r})$ approaches $G(\tau_{i,r})$, (cf. Fig. 15 with $X = -15$ K). If this is the case, one should have $\delta_1 = -\Delta\alpha(T_1 - T_0)$ as for an instantaneous T-jump [eq. (19)]. Thus such selected isotherms can be used to determine both the normalized response function (Fig. 6) and the reduced rate function (Fig. 4) which characterize the contribution of the retardation spectrum $G(\tau_{i,r})$ to the isobaric behavior of the system.

Finally, the validity of the τ_{eff,T_1} map shown in Figure 20 can easily be extended to rates other than $|q| = 1.0$, using the rate-temperature reduction method defined by eq. (60), i.e., by referring to a temperature $T_r(q') = T_g(q') + C$, where C is a constant temperature difference (in the present case 1.10 K; cf. Fig. 12), while reducing the value of the time unit $\tau_{17,r}$ (Fig. 5) by a factor $|q'|^{-1} = \exp\{-\theta[T_r(q') - T_r]\}$.

The above conclusions, appropriately modified, hold even when $\tau_i(\delta,T)$ is defined by expressions more involved than eq. (17). In most cases, however, the latter provides an excellent approximation within the glass-transition region, where equilibrium can be reached in a time period of reasonable length. In fact, the τ_{eff} plots shown in Figure 20 have the same character as those obtained with polymeric glasses,[7,13] except for the $\tau_{N,T}$ discrepancy mentioned above. In particular, the drastic truncation of the intrinsic τ_{eff} isotherms is quite similar to what is observed experimentally (cf. Fig. 23 in ref. 7) and should be contrasted with the situation involved in instantaneous quenching (see Fig. 2) which cannot be approached by cooling, whatever the rate, when $T_0 \gg T_g(q)$. Consequently, the intrinsic isotherms alone do not provide a means of accurately determining the retardation spectrum, or the normalized response function $R(z)$ of the system, since, especially in the transition region, they reflect a distribution strongly truncated at short times (cf. Fig. 13). This limitation of the use of the intrinsic isotherms has been overlooked by most authors. In contrast, the τ_{eff} plots of other, appropriately selected isotherms, in conjunction with the asymptotic relationships of $(\tau_{eff,T_1}^{-1})_0$, provide all the necessary information for determining the material constants [eq. (36)] and $G(\tau_{i,r})$, or $R(z)$, characterizing the isobaric recovery behavior of glasses subjected to any arbitrary, but fully specified thermal treatment.

Heating Isobars

Though this class of treatment has already been mentioned above, it deserves a special consideration when heating is pursued to temperatures high enough to reestablish equilibrium. In fact, such heating isobars often display an *auto-catalytic* increase of δ with T [similar to that of the expansion isotherms involving large initial $|\delta_0|$ values; cf. eq. (38a)], resulting in a rather sharp peak in α, or C_p, which has occasionally been attributed, erroneously in our opinion, to a first-order transition. It has been shown, indeed, that such effects can be ascribed to the recovery phenomena of glass-forming liquids[2,3,7,8,11,29] and that they can be simulated, at least qualitatively, by the simplest one-parameter models.[2,8,28] One can thus reasonably expect that the present multiparameter model will reach quantitative agreement with experimental data, as claimed in the recent work of Moynihan and co-workers[11,23] using a similar approach.

Since the material response on heating depends on thermal history, the analysis which follows necessarily considers that the heating is preceded by at least one cooling stage and possibly an isothermal recovery period; further generalization is straightforward. Hence, we deal with thermal cycles in which *cooling*, starting at $t = 0$, from T_0 (equilibrium) at a rate q_1, is followed by an *isothermal period* of limited duration $(t_2 - t_1) \geq 0$ at T_1, and finally *heating*, at a rate q_2, starts at t_2; the departure δ_2 prior to heating is positive or zero. Although such a thermal cycle is the simplest one, it still involves five independent parameters: T_0, q_1, T_1 (or t_1), t_2, and q_2. Though some simplifications are possible, e.g., $T_0 \gg T_g(q_1)$, the characteristic features of such heating isobars depend on the particular set of experimental parameters, in addition to the material constants [eq. (36)] and the retardation spectrum, or the $R(z)$ function of the system. This results in a rather complex material response, which cannot be analyzed in a meaningful way without determining the fundamental parameters of the system from a set of independent and suitably less involved experiments, such as those described in the previous section. Nevertheless, this consideration has been overlooked by most authors dealing exclusively with C_p heating isobars, as determined by differential scanning calorimetry (DSC). Hence, some of the conclusions based on such experiments and the discrepancies between C_p isobars and isothermal data[11] should be considered with caution.

If the system is fully specified, as in the present case, one can easily calculate its response during the heating stage of such cycles. In fact, eq. (62) or (65), together with eqs. (48a) and (48b), describe the state of the system at t_2, thus at $z = z_2$, prior to heating. The heating period involves an additional term defined by eq. (48), viz.:

$$\delta(t > t_2; T) = -\Delta\alpha \int_0^{z_1} R(z - z') \left(\frac{dT}{dz'}\right) dz' - \Delta\alpha \int_{z_2}^{z} R(z - z') \left(\frac{dT}{dz'}\right) dz'$$

$$\equiv -\Delta\alpha \int_0^{z} R(z - z') \left(\frac{dT}{dz'}\right) dz' \qquad (71)$$

where z is defined by the thermal history as

$$z(T_0, T_1, q_1, t_2, q_2) = \int_{T_0}^{T_1} (q_1 a_T a_\delta)^{-1} \, dT'$$

$$+ a_{T_1}^{-1} \int_{t_1}^{t_2} a_\delta^{-1} \, dt' + \int_{T_1}^{T} (q_2 a_T a_\delta)^{-1} \, dT' \qquad (72)$$

in which the consecutive integrals can be recognized as z_1, $(z_2 - z_1)$, and $(z - z_2)$, respectively [cf. eqs. (48a) and (48b)]; t_1 is related to T_0, T_1, and q_1 by $t_1 = \int_{T_0}^{T_1} q_1^{-1} \, dT'$, and $dT/dz' = 0$ during the isothermal annealing at T_1.

Equations (71) and (72) can be generalized to any thermal treatment involving n cooling and/or heating steps (characterized by the starting temperature and the rate) and m intermittent isothermal periods (each defined by its duration) during which $dT/dz' = 0$. Hence, the thermal history involves $(2n + m)$ experimental parameters which fully determine the reduced time z and the particular $z_j (j < n + m)$ values, at which the thermal program is modified.

For the three-step cycles considered here, the first integral in eq. (71) [which may also be expressed by the particular $R^1(z - z_1)$ function, as in eq. (66)] is always positive, since $q_1 < 0$, whereas the second one is negative, since $q_2 > 0$. These two integrals generally counteract each other so that, at a temperature $T_\times > T_1$, $\delta = 0$ even though the individual values of δ_i are different from zero (with the possible exception of one of them) which is like the case in the memory effects discussed above (cf. Fig. 10). In fact, the latter also involve a reversal in the variation of temperature ($T_0 > T_1 < T_2$) and partial recovery at T_1. Cooling and heating at finite rates, however, involve the convolution integral of the response function, rather than the function itself as in eq. (57).

For the present thermal cycles, two limiting situations must be considered: (a) The first arises when heating starts immediately after cooling, i.e., when $z_1 = z_2$. Such isobars will be referred to as *intrinsic* $(q_1 q_2)$ or *intrinsic* $|q|$ (when $|q_1| = q_2$). (b) The other occurs when $z_2 \gg a_{T_1} \tau_{N,r}$ such that the system reaches equilibrium at T_1 and the first integral in eq. (71) vanishes. One can thus define z_2 as the origin of time and the lower limit of the second integral becomes zero, while z is given by eq. (48a) with $T_0 = T_1$ and δ by eq. (58). This case has been discussed above, and used prior to the isothermal-expansion experiments ($\delta_1 < 0$) shown in Figure 20 through their z_{eff} plots.

In both of these limiting situations the rate-temperature reduction method [eq. (60)] applies if

$$T_0 - T_0' = T_1 - T_1' = \theta^{-1} \ln(q_1/q_1') = \theta^{-1} \ln(q_2/q_2'). \qquad (60b)$$

In fact, with these conditions the values of z and z_1 are uniquely defined by the relevant set of experimental parameters: T_0, T_1, q_1, and q_2. In intermediate situations, when $(t_2 - t_1)$ is finite, the reduction method is still applicable, provided that $(z_2 - z_1)$ is held constant [i.e., $(t_2 - t_1)/a_{T_1} = (t_2' - t_1')/a_{T_2'}$]. This requirement is realized whenever heating starts at the same value of δ_2, and thus

VOLUME AND ENTHALPY RECOVERY OF GLASSES   1143

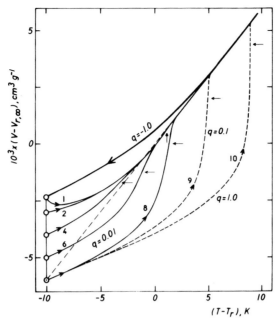

Fig. 21. Volume isobars obtained for the thermal cycles specified in eq. (73) and Table I. Arrows locate the temperatures T_p at which α displays a maximum (see Fig. 22).

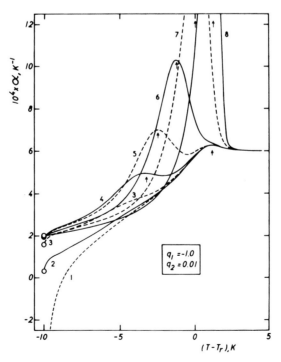

Fig. 22. Isobars of α obtained during the heating stage of the thermal cycles listed in Table I (runs 1–8). Arrows indicate the temperatures at which the α peaks occur.

TABLE I

Characteristics of the Heating Isobars Obtained with $T_0 > T_r + 10$ K, $q_1 = -1.0$ (K/t.u.) and $T_1 = T_r - 10$ K

	Previous history (cf. Fig. 16)			Heating stage (cf. Figs. 21 and 22)					
Curve	$\log(t_2 - t_1)$	$10^3 \times \delta_2$	$\delta_2/\Delta\alpha$ (K)[a]	q_2 (K/t.u.)	$10^4 \times \alpha_0$ (K⁻¹)	$T_\times - T_r$ (K)	$\hat{T} - T_r$ (K)	$10^4 \times \alpha_{max}$ (K⁻¹)	$\hat{T} - T_r$ (K)
1	$-\infty$	3.690	9.23	0.01	−28.58	−0.76			1.16
2	2.245	3.000	7.50	0.01	0.313	−0.77		b	1.16
3	2.904	2.503	6.25	0.01	1.569	−0.78		b	1.16
4	3.528	2.001	5.00	0.01	1.896	−0.86	−3.31	4.94	1.16
5	4.151	1.500	3.75	0.01	1.976	−5.32	−2.47	7.01	1.09
6	4.788	1.003	2.50	0.01	1.995	−7.25	−1.22	10.32	
7	5.495	0.503	1.26	0.01	1.999	−8.71	0.04	16.41	
8	∞	0	0	0.01	2.000	−10.0	1.29	27.02	
9	∞	0	0	0.10	2.000	−10.0	4.91	78.26	
10	∞	0	0	1.00	2.000	−10.0	8.91	360.17	

[a] This column defines $T^* - T_1$, T^* being the fictive temperature (ref. 2) of the system prior to heating (cf. Appendix A).
[b] Shoulder.

at the same $(\delta_2/\delta_1) = \rho_1 = R^1(z_2 - z_1)$ values [cf. eq. (62)], since δ_1 is constant when eq. (60b) is satisfied.

Typical volume isobars associated with such cycles are shown in Figure 21 and were generated with the following parameters:

$$T_0 > T_r + 10 \text{ K}$$

$$q_1 = -1.0 \text{ (K/time unit)}$$

$$T_1 = T_r - 10 \text{ K} \tag{73}$$

$$q_2 = 0.01 \text{ (K/time unit)(for curves 1–8)}$$

Values of $\log(t_2 - t_1)$ and δ_2 are listed in Table I, together with other parameters characterizing the heating isobars. The relevant $\alpha = v^{-1}(\partial v/\partial T)_{q_2}$ isobars are shown in Figure 22 for eight values of δ_2, ranging from δ_1 (intrinsic isobar) to zero (equilibrium isobar) some of which are not shown in Figure 21. The characteristics of two heating isobars starting from equilibrium at $T_1 \equiv T_0 = T_r - 10$ K obtained with larger values of q_2 (cf., runs 8 and 10 in Fig. 21) are also included in Table I.

As expected, all the heating isobars of v cross (overshoot) the equilibrium line ($\delta = 0$) at temperatures $T_\times$, whereas those of α display at least one maximum (at T_p indicated by arrows in Figs. 21 and 22). These characteristic temperatures, as well as the values of $\alpha_{\max}$ and α_0 (defined by the initial slope at T_1) are listed in Table I. Unexpectedly, however, some α isobars display either a shoulder *and* a maximum (curves 2 and 3), or two maxima (4 and 5), at $\hat{T}$ and $\hat{T}$,* the second of which is essentially invariant always occurring at virtually the same temperature $\hat{T}$ ($= T_r + 1.16$ K) and involving the same values of δ ($\simeq -3.0 \times 10^{-4}$) and $\alpha_{\max}$ ($\simeq 6.27 \times 10^4$ K^{-1}). Furthermore, when $3 \geq 10^3 \delta_2 \geq 2$, the first maximum (or shoulder) of α occurs below $T_\times$, where δ is still positive (curves 2–4) and the corresponding values of $\alpha_{\max} < \alpha_l$. The v and α isobars obtained for lower values of T_1, everything else being the same, show similar features; see Table II, with $T_1 = T_r - 20$ K. For this particular T_1 they display two well-separated α peaks (even for the intrinsic α isobar) when $\delta_2 \geq 3.7 \times 10^{-3}$. Again, the second one is essentially invariant and occurs at the same temperature $\hat{T}$, as for $T_1 = T_r - 10$ K, matching the peak of run 1 in Figure 22.

Though surprising at the first sight, such features have recently been observed with a polystyrene glass (see Figs. 3 and 4, of ref. 8) during slow heating ($q_2 = 0.017$ K sec^{-1}) after being quenched from $T_0 = 105°$C and moderately annealed ($t_2 - t_1 = 2.1$ hr) at $80°$C, similar to the treatment under discussion [eq. (73)]. In addition, the C_p isobars of poly(vinyl acetate) reported by Sharonov and Volkenstein[30] also show a "misplaced" peak for the intrinsic ($|q_1| \gg q_2$) isobar, similar to that of Figure 22. These results are in qualitative agreement with the present model and thus lend some measure of experimental support to it.

Finally, there is a narrow range of δ_2 in which the v isobars actually cross the equilibrium line three times, at $T_{\times_j}$ ($1 \leq j \leq 3$), rather than just once (cf. Table II for $\delta_2 = 4.3 \times 10^{-3}$). Under these particular circumstances, the first α peak is then reached between $T_{\times_1}$ and $T_{\times_2}$, and the second above $T_{\times_3}$ (in both cases

---

* The temperatures T_p have been relabeled $\hat{T}$ and $\hat{T}$, to distinguish between the two separate maxima which arise under certain circumstances.

TABLE II

Characteristics of the Heating Isobars Obtained with $T_0 > T_r + 10$ K, $q_1 = -1.0$ (K/t.u.), $T_1 = T_r - 20$ K, and $q_2 = 0.01$ (K/t.u.)

	Previous history			Heating stage				
Run	$\log(t_2 - t_1)$	$10^3 \times \delta_2$	$\delta_2/\Delta\alpha$ (K)	$10^4 \times \alpha_0$ (K^{-1})	$T_\times - T_r$ (K)	$\hat{T} - T_r$ (K)	$10^4 \times \alpha_{max}$ (K^{-1})	$\hat{T} - T_r$ (K)
1	$-\infty$	7.564	18.91	−0.401	−0.76	−14.70	1.47	1.16
2	3.080	7.002	17.51	1.812	−0.76	−16.94	1.94	1.16
3	3.839	6.499	16.25	1.949	−0.76	−14.24	2.30	1.16
4	4.972	5.403	13.51	1.995	−0.71	−5.04	4.78	1.18
5	5.445	4.897	12.24	1.998	−0.63	−4.03	7.21	1.20
6	6.006	4.300	10.75	1.999	−8.55[a]	−2.60	12.60	1.26
7	6.578	3.700	9.25	2.000	−10.61	−1.11	24.3	1.42
8	7.161	3.097	7.74	2.000	−12.20	0.53	53.0	[b]
9	7.744	2.503	6.26	2.000	−13.74	2.12	135.0	
10	8.250	1.999	5.00	2.000	−15.00	3.53	352.5	
11	8.877	1.398	3.50	2.000	−16.51	5.26	1341	
12	9.559	0.800	2.00	2.000	−18.00	7.03	6023	
13	∞	0	0	2.000	−20.00	9.48	52,544	

[a] Value of $T_{\times 1} - T_r$, while: $T_{\times 2} - T_r = -2.0$ K and $T_{\times 3} - T_r = -0.5$ K.   Note the sudden decrease of $T_\times$.
[b] Shoulder at ≃ 1.2.

$\delta < 0$), whereas a rather sharp minimum ($\alpha_{\min} = 5.06 \times 10^{-4}\,\mathrm{K}^{-1}$) occurs between $T_{\times_2}$ and $T_{\times_3}$, where $\delta > 0$. Obviously the system ignores the false equilibrium situations ($\delta = 0$ but $\delta_i \neq 0$), and at high temperature always approaches its real equilibrium through negative values of δ. In fact, all the v isobars shown in Figure 21 approach the intrinsic (q_1, q_2) isobar and they merge with the latter at increasing temperatures as δ_2 decreases. The shapes of the isobars vary in a smooth and systematic manner so that they do not cross each other and remain between the intrinsic and equilibrium (curve 8) isobars which appear to be upper and lower bounds of all such curves. Again, this situation is analogous to the memory effects shown in Figure 11.

Table I and Figure 22 show that, at the start of heating at T_1, the value of the expansion coefficient α_0 is always smaller than or equal to $\alpha_g (= 2 \times 10^{-4}\,\mathrm{K}^{-1})$, and that it may even become negative if δ_2 is large enough. This effect clearly originates from the predominance of the first integral in eq. (71), which still involves rapid processes at T_1 (cf. Fig. 17) as compared to the heating rate q_2. In fact, α_0 increases with q_2, everything else being the same. At higher temperatures, after a rapid increase, the α isobars show considerable crisscrossing in contrast to the v isobars. When the latter cross the equilibrium volume v_∞ for the first time at $T_\times$, the value of the expansion coefficient $\alpha_\times$ is larger than α_g (except when $\delta_2 = 0$), but smaller than α_l. Hence $(d\delta/dT)_\times = \alpha_\times - \alpha_l > -\Delta\alpha$, whereas for a one-parameter model $(d\delta/dT)_\times = -\Delta\alpha$ [cf. eq. (18′) with $i = 1$ and $\delta_i = \delta = 0$]. As noted by Moynihan et al.,[23] this behavior, like the memory effects, also reflects the fact that recovery is controlled by more than one retardation mechanism.

Furthermore, there are two critical heating isobars, corresponding to the upper and lower limits of the narrow δ_2 range, in which the v isobars cross the equilibrium line three times (cf. Run 6 in Table II). These critical isobars appear either when $(T_{\times_2} - T_{\times_1}) \to 0$, or when $(T_{\times_3} - T_{\times_2}) \to 0$ and at those limits they are tangent to the equilibrium line. In other words, one has simultaneously $\delta = 0$ and $d\delta/dT = 0$, hence $\alpha_\times = \alpha_l$ and $\Sigma^N(\delta_i/\tau_{i,r}a_T a_\delta) = \Delta\alpha$ [cf. eq. (18′)]. This critical behavior is comparable to the isotherm C', shown in Figure 11, though for the latter at $\delta_0 = 0$, $(d\delta/dt)_0 \neq 0$, since, in the isothermal case, recovery is controlled by eq. (20) in which the $\Delta\alpha q$ term is absent.

The above features can be understood, at least qualitatively, by considering the competition between the two integrals in eq. (71), originating from the positive and negative contributions of the individual processes to δ. As an example, Figure 23 shows the evolution of the δ_i profile during uniform heating for isobar 5 in Figure 22 which displays two distinct α peaks. Unlike the previous plots of this kind, the relevant $\delta_i(T)$ values are now plotted against the actual values of the dimensionless product $\theta q_2 \tau_i = \theta q_2 a_T a_\delta \tau_{i,r}$, and for clarity, only the long-time contributions are represented, while indicating the individual values of δ_{33} and δ_{17}. In the present example, in fact, the short-time contributions ($i < 17$) are negligible near T_1 and above $T_1 + 7\,\mathrm{K}(= T_r - 3\,\mathrm{K})$ and remain smaller than 1×10^{-5} in between.

The distributions depicted in Figure 23 result, in part, from the recovery of the positive δ_i contributions reached at T_1 and t_2 (close to those shown in Fig. 17, for $\rho_1 = 0.40 \simeq \delta_2/\delta_1 = 0.406$) which essentially involve the slowest processes ($i > 20$). In addition to these, subsequent heating by dT generates a series of negative δ_i contributions [through the second integral in eq. (71)], each

1148   KOVACS ET AL.

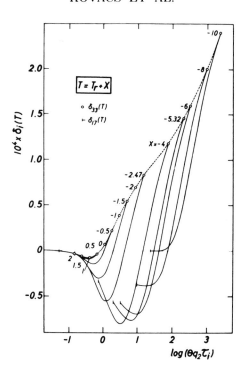

Fig. 23. Instantaneous profiles of $\delta_i(T)$ values reached at various temperatures (as indicated) during the heating stage of isobaric run 5 in Table I, plotted here as a function of the actual values of the product $\theta q_2 \tau_i = 0.01\tau_i$, for i values ranging from 17 to 33 (cf. Fig. 5). For this particular isobar $T_\times = T_r - 5.32$ K, $\hat{T} = T_r - 2.47$ K and $\hat{\hat{T}} = T_r + 1.09$ K.

amounting to $-g_i \Delta\alpha dT$ ($= 2 \times 10^{-5} dT$), which is equivalent to a uniform decrease of *all* values of δ_i by the same amount. Simultaneously, however, the latter also relax, according to eq. (18'), at an increasing rate for decreasing i values, giving rise to a minimum in δ_i, its associated τ_i value being designated hereafter as $\check{\tau}_i$. Clearly, the relevant value of i increases with T, whereas $\check{\tau}_i = a_T a_\delta \check{\tau}_{i,r}$ decreases.

Furthermore, the recovery rate of all the contributions, controlled by $\delta_i \tau_i^{-1}$ [cf. eq. (18')] compete with the heating rate q_2, resulting in the rather complex evolution of the distribution profile shown in Figure 23. Within a fair approximation, the *first* α peak occurs at $\hat{T} = T_r - 2.47$ K when $\theta q_2 \check{\tau}_i \simeq 1$, whereas the *second* appears when δ_N reaches its minimum (i.e., its largest negative) value, thus involving $d\delta_N/dT = 0^*$ ($\hat{\hat{T}} = T_r + 1.16$ K). The second "misplaced" α peak is therefore related to the retardation mechanisms involving the longest retardation times.

As δ_2 decreases from its maximum value δ_1, everything else being the same, the first criterion ($\theta \check{\tau}_i q_2 \simeq 1$) is met at increasing temperatures, thus $\hat{T}$ increases, whereas the second one, defining $\hat{\hat{T}}$, remains practically independent of δ_2 and

---

* The exact definition of the occurrence of the α peaks involves a much more complex relationship, to be discussed elsewhere (refs. 29 and 31). The criteria given above should thus be considered as empirical, applying only to box distributions, which is the present case since the short-time contributions are negligible.

T_1 (see below). Consequently, there is a critical δ_2^* value, for which these independent events arise at the same temperature ($\simeq \hat{T}$) and the two α peaks overlap (cf. curves 6–8, in Fig. 22). The critical δ_2^* value, of course, increases when T_1 decreases (e.g., for $T_1 = T_r - 10$ K, $\delta_2^* \simeq 5 \times 10^{-5}$, whereas for $T_1 = T_r - 20$ K, $\delta_2^* \simeq 2.9 \times 10^{-3}$). Whatever the value T_1, the intrinsic α isobar appears as the lower envelope of all the others as long as $\delta_2 \geq \delta_2^*$ (cf. Fig. 22).

In situations where the second condition ($d\delta_N/dT = 0$) is met before the first one, the "invariant" α peak is overshadowed by the main one, still defined by $\theta \check{\tau}_i q_2 \simeq 1$. Therefore, the appearance of the secondary α peak is related to the order in the succession of the critical conditions defined above. These considerations therefore provide a criterion for identification of the nature of the α peaks, the importance of which will become clear below.

Figures 21 and 22 show only a few examples of the simplest thermal cycles in which either δ_2, or q_2 is varied in a systematic manner, while keeping the three other experimental parameters (T_0, T_1, and q_1, or δ_1) constant. As mentioned, however, the heating isobars obtained with lower values of T_1 are similar to those discussed above (cf. Table II), though in this case the range of (δ_2/δ_1) which can be covered in practice will be smaller, due to the increasing length of time required to reach equilibrium. A few other situations, in which the ratio $|q_1| q_2^{-1}$ is varied, will be considered in the next section dealing with the variations of $\hat{T}$ and $\check{T}$.

Finally, independently of the present theoretical analysis, it should be pointed out that the value of the circular integral between $T_0 \gg T_g(q_1)$ and T_1 of *any* pair of α isobars (whatever the magnitudes and signs of q), amounts to the difference in the relevant values of δ at T_1, provided that T_0 is chosen high enough to reach true equilibrium in both cases (i.e., $\alpha = \alpha_l$, $d\delta/dT = 0$ and $d^2\delta/dT^2 = 0$). As an example, in comparing any of the heating isobars of α to the intrinsic one (curve 1 in Fig. 22), or to that corresponding to the cooling step (shown in Fig. 14, by the solid line), since both involve the same value of $\delta(= \delta_1)$ at T_1, one has [cf. eq. (10)] for $T_0 \gg T_g(q)$:

$$\int_{T_0}^{T_1} v\alpha' \, dT' + \int_{T_1}^{T_0} v\alpha'' \, dT' = \oint_{T_0}^{T_1} v\alpha \, dT'$$
$$= v_1 - v_2 = v_\infty(\delta_1 - \delta_2) \quad (74)$$

where α' and α'' are defined by the respective thermal treatments considered, involving the volumes v_1 and v_2, respectively, at T_1.

Similarly, by referring to any one of the equilibrium heating isobars of α (e.g., curves 8–10 in Fig. 21), one can write

$$\left| \oint_{T_0}^{T_1} v\alpha \, dT \right| = v_2 - v_\infty \simeq v_\infty \delta_2 \quad (74')$$

which uniquely determines the initial departure from equilibrium at T_1, for the relevant nonequilibrium situation prior to heating. These relationships are useful, especially in determining the time dependence of the enthalpy for fixed thermal history (T_0, T_1, q_1, δ_2, and q_2) as derived from C_p data obtained by DSC for various values of t_2. The α isobars represented in Figure 22 show, however, that one must take into account the full range of variation of C_p between T_1 and $T_0 \gg T_p$ rather than merely the peak region; an obvious requirement which has not always been respected.[32]

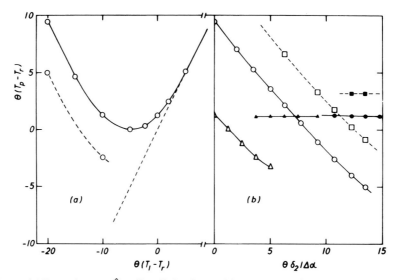

Fig. 24. (a) Dependence of $\hat{T}$ on $T_1 = T_0$ for the equilibrium isobars ($\delta_2 = 0$) obtained on heating at a rate $q_2 = 0.01$. Dashed line: connects values of $\hat{T}$ obtained in thermal cycles starting at $T_0 > T_r + 10$ K, with $q_1 = -1.0$, $q_2 = 0.01$, and involving a fixed value of $\delta_2 = 1.5 \times 10^{-3}$. (b) Dependence of T_p on δ_2 for $T_1 = T_r - 10$ K (triangles) and $T_1 = T_r - 20$ K (circles) obtained in thermal cycles starting at $T_0 > T_r + 10$ K, with $q_1 = -1.0$ and $q_2 = 0.01$. Open and closed signs denote values of $\hat{T}$ and $\hat{\hat{T}}$, respectively. Dashed lines connect T_p values (squares) obtained in similar cycles involving $T_1 = T_r - 20$ K and $q_2 = 0.1$, everything else being the same.

Dependence of T_p on Thermal History

The above analysis clearly shows that, for a given system, the temperatures T_p at which the α peaks occur depend critically upon the thermal treatment. Hence, for the simplest cycles considered, the variations of $\hat{T}$ and $\hat{\hat{T}}$ each involve five partial derivatives of T_p, one with respect to each of the experimental parameters (T_0, T_1, q_1, δ_2, and q_2), viz.:

$$dT_p = s(T_0)dT_0 + s(T_1)dT_1 + s(\delta_2)d\delta_2 + s(q_1)d\ln|q_1| + s(q_2)d\ln q_2 \quad (75)$$

where $s(x_i) = (\partial T_p/\partial x_i)_{x_{j\neq i}}$, i.e., the variable x_i in parentheses implies that the four other experimental parameters $(x_{j\neq i})$ are kept constant. Furthermore, since the partial derivatives of $\hat{T}$ and $\hat{\hat{T}}$ are not the same, they will be designated as $\hat{s}(x_i)$, and $\hat{\hat{s}}(x_i)$, respectively. *None* of these ten derivatives can be expressed in an analytical form involving the material constants of the system and the experimental parameters. In the practical range of the latter, however, their variations are generally small.

The values of $\theta(T_p - T_r)$, obtained from the isobars shown in Figures 21 and 22 (Tables I and II) as well as other cycles, are plotted in Figures 24(a) and 24(b) as a function of $\theta(T_1 - T_r)$ and $\theta\delta_2/\Delta\alpha$, respectively. The normalizing factors θ and $\theta/\Delta\alpha$ for T and δ have been introduced here for sake of generality. In fact, it can easily be shown[29] that, in terms of the dimensionless variables θT and $\theta\delta/\Delta\alpha$, eqs. (18) and (18') become independent of the material constants θ and $\Delta\alpha$. The same is true for the appropriately modified partial derivatives, i.e., for $\theta s(q)$, and $\Delta\alpha s(\delta)$.

VOLUME AND ENTHALPY RECOVERY OF GLASSES   1151

(a) In the simplest case, heating starts from equilibrium ($T_1 = T_0$ and $\delta_2 = 0$) and the isobars depend on T_1 and q_2 alone, so eq. (75) yields

$$d\hat{T} = \hat{s}(T_1)dT_1 + \hat{s}(q_2)d\,\ln q_2 \qquad (76)$$

In fact, the relevant α peak is of the first kind, since δ_N is always negative and its minimum is reached in the range of, or below $\hat{T}$, at which $\theta q_2 \check{\tau}_i = \theta q_2 \tau_N$ becomes of the order of unity (see the preceding section).

Figure 24(a) shows (solid line) the variation of $\hat{T}$ vs. T_1, for $q_2 = 0.01$. The hyperbolic curve approaches two asymptotes, one at high and one at low temperatures, the slopes of which are equal to 1 and -1.4, respectively. [The latter value is reached in the temperature range of $T_r - 40$ K, not shown in Fig. 24(a)]. In the transition range, therefore, $\hat{s}(T_1)$ increases with T_1 between these two limiting values, while passing through zero near $T_r - 5$ K, at which $\hat{T} \simeq T_r$. A similar situation prevails in cyclic experiments when $\delta_2 > 0$ (dashed lines, with $\delta_2 = 1.5 \times 10^{-3}$) which will be discussed below. In such cycles, a second α peak may also occur at $\hat{T}$ [not shown in Fig. 24(a)] the value of which is practically independent of T_1 and δ_2 (cf. Tables I and II) if q_1 and q_2 are kept constant. Nevertheless, for $\delta_2 > 0$, the variation of $\hat{T}$ is limited to the low-temperature range of T_1 (and more so as δ_2 is increased), since for a fixed cooling rate $\delta_2 \leq \delta_1$ (cf. Fig. 12). Therefore, in such cycles $\hat{s}(T_1)$ is generally negative and its value again approaches -1.4 as T_1 decreases, everything else being the same.

The value of $\hat{s}(q_2)$, the second term characterizing the equilibrium isobars [eq. (76)] can be derived from the data given in Table I (curves 8–10, in Fig. 21). In the range of $0.01 \leq q_2 \leq 1.0$ and for $T_1 = T_r - 10$ K, these yield $(\partial \hat{T}/\partial \log q_2) = 2.303 \hat{s}(q_2) = (3.8 \pm 0.4)\theta^{-1}$. This value increases slightly with q_2, and also as T_1 decreases. The situation is again similar in cyclic experiments when $\delta_2 > 0$. For example, from isobars 1 and 2 in Table III, and from isobar 5 in Table I, all involving the same values of q_1 and δ_2 ($= 1.5 \times 10^3$), one can derive $2.303 \hat{s}(q_2) = (3.7 \pm 0.3)\theta^{-1}$, for $T_1 = T_r - 10$ K (note, that the second α peak does not show up when $q_2 \geq 0.1$). One can thus conclude that $\hat{s}(T_1)$ and $\hat{s}(q_2)$ do not critically depend on the intermediate state δ_2 obtained at fixed T_1 and q_1.

It is interesting to compare these partial derivatives with those of T_g, which depend, as do the equilibrium heating isobars, on only two experimental parameters T_0 and q_1.* To do so one can write

$$dT_g = s_g(T_0)dT_0 + s_g(q_1)d\,\ln|q_1| \qquad (77)$$

using the same notation as for T_p. In fact, many authors[23,32] dealing with C_p isobars actually associate T_p with the glass-transition temperature. If $T_0 \gg T_g(q_1)$, T_g is independent of T_0, thus $s_g(T_0) = 0$, whereas $s_g(q_1) = dT_g/d\,\ln|q_1| = \theta^{-1}$ is a fundamental invariant of the system [cf. eq. (59)]. Using the values of $\hat{s}(q_2)$ given above, the ratio $\hat{s}(q_2)/s_g(q_1) = (1.65 \pm 0.2)$ turns out to be larger than 1.0 as observed experimentally.[8] Furthermore, as T_0 decreases, $s_g(T_0)$ increases from zero[29] to 1 (cf. Fig. 14), whereas $\hat{s}(T_1)$ approaches the value of -1.4 (Fig. 24a). This comparison clearly shows, consistent with experiment, that T_g and $\hat{T}$, obtained on cooling and heating from equilibrium at the same rate $|q|$,

---

* $T_g(T_0,q)$ is defined as above [cf. eq. (61)] by the temperature at which the low-temperature asymptote of δ, obtained on cooling from any value of T_0 (equilibrium) at a rate q, intercepts the ordinate $\delta = 0$.

TABLE III

Characteristics of Some Heating Isobars Obtained with $T_0 > T_r + 10$ K

Run	Previous history					Heating stage			
	$-q_1$ (K/t.u.)	$T_1 - T_r$ (K)	$10^3\delta_2$	q_2 (K/t.u.)	$10^4\alpha_0$ (K^{-1})	$T_\times - T_r$ (K)	$\hat{\tau} - T_r$ (K)	$10^4\alpha_{max}$ (K^{-1})	$\hat{T} - T_r$ (K)
1	1.0	−10	1.500	0.1	1.998	−6.02	1.11	12.56	⋯
2	1.0	−10	1.500	1.0	2.000	−6.21	5.01	28.38	⋯
3	1.0	−20	2.500	0.1	2.000	−13.75	6.62	704.3	⋯
4	1.0	−20	4.300	0.1	2.000	−9.20	1.71	29.16	⋯
5	1.0	−20	5.403	0.1	1.999	−5.01	−0.87	8.48	3.20
6	0.1	−20	4.903	0.01	1.998	−4.53	−3.62	6.46	0.87

not only do not coincide, but they shift quite differently with $|q|$ and T_0. The same conclusion holds also for $\hat{T}$, as will be shown below.

(b) Returning to the variation of T_p in cyclic experiments starting at $T_0 \gg T_g(q_1)$, now consider the other partial derivatives entering into eq. (75). In Fig. 24(b), the values of T_p, at fixed values of q_1 and q_2, are represented as a function of δ_2 for two values of T_1 as indicated in the legend. Clearly, the second α peak occurs at practically the same temperature $\hat{T}$ when δ_2 is large enough, irrespective of the value of T_1. Therefore, $\hat{T}$ depends on q_1 and q_2 alone and one can write to good approximation

$$\hat{s}(\delta_2) = 0 \quad \text{and} \quad \hat{s}(T_1) = 0 \tag{78}$$

On the other hand $\hat{T}$ decreases rapidly as δ_2 increases, involving a value of $\hat{s}(\delta_2) \simeq -(1.10 \pm 0.15)\Delta\alpha^{-1}$ as T_1 ranges from $T_r - 10$ K to $T_r - 20$ K, while $\partial \hat{s}(\delta_2)/\partial \delta_2 > 0$. Nevertheless, when δ_2 increases and approaches its maximum value δ_1, this α peak either vanishes (cf. Fig. 22), or $\hat{T}$ decreases dramatically in a nonuniform manner (cf. runs 1–3 in Table II).

Figure 24(b) also shows (dashed lines) a few values of T_p obtained with $q_2 = 0.1$ and $T_1 = T_r - 20$ K (see Table III, runs 3–5). Comparison of these isobars with those obtained with $q_2 = 0.01$ (cf. runs 4–9 in Table II) leads to the value of $(\partial \hat{T}/\partial \log q_2) = 2.3 \hat{s}(q_2) \simeq (4.3 \pm 0.2)\theta^{-1}$ and to $2.3 \hat{s}(q_2) = 2.0\theta^{-1}$ which is about half as large as $\hat{s}(q_2)$. Note that the value of $\hat{s}(q_2)$ is slightly larger than that derived above with $T_1 = T_r - 10$ K, which shows that this derivative increases with decreasing T_1.

The value of $s(q_1)$ can be estimated by using the rate-temperature reduction relationship [eq. (60b)], implying $dT_1 = dT_p = \theta^{-1} d \ln|q_1| = \theta^{-1}d \ln q_2$, and δ_2 = constant. Since in the present cycles $s(T_0) = 0$, eq. (75) can be written

$$s(q_1) = \frac{dT_p}{d \ln|q_1|} - s(q_2) - \frac{s(T_1)dT_1}{d \ln|q_1|} = \theta^{-1} - s(q_2) - s(T_1)\theta^{-1}. \tag{79}$$

Substituting the relevant values of $\hat{s}(q_2)$ and $\hat{s}(T_1)$ with their ranges of variation, one obtains for $T_1 = T_r - 10$ K, $2.3 \hat{s}(q_1) \simeq -(1 \pm 0.4)\theta^{-1}$, whereas for $T_1 = T_r - 20$ K, $2.3 \hat{s}(q_1) = -(0.5 \pm 0.4)\theta^{-1}$. The relatively large variation range originates from the $\hat{s}(q_2)$ term in eq. (79).

Application of eq. (79) to $\hat{T}$ yields

$$\frac{d\hat{T}}{d \ln|q_1|} = \frac{d\hat{T}}{d \ln q_2} = \hat{s}(q_1) + \hat{s}(q_2) = \theta^{-1} \tag{80}$$

since $\hat{s}(T_1) = 0$ [cf. Fig. 24(b) and eq. (78)]. With the value of $\hat{s}(q_2)$ given above this leads to $2.3 \hat{s}(q_1) \simeq 0.3\theta^{-1}$, consistently with run 6 in Table III, and run 5 in Table II. Further comparison of these two runs gives a value of $2.3 \hat{s}(q_1) \simeq -0.4\theta^{-1}$ which is consistent with that derived from eq. (79), at $T_1 = T_r - 20$ K.

(c) Alternately, one can define the derivatives $s_2(q_1)$ and $s_2(q_2)$ for which the variable $\delta_1 - \delta_2$ (rather than δ_2) is held constant.[31] The values of $\hat{s}_2(q_1)$ and $\hat{s}_2(q_2)$, as well as their ranges of variation can be determined in a direct way from the intrinsic isobars obtained with various $|q_1|/q_2$ ratios. As mentioned in the previous section, such isobars always display an α peak satisfying the criterion $d\delta_N/dT = 0$. The relevant variation of $\hat{T}$ is plotted in Figure 25 as a function of $\log|q_1|q_2^{-1}$ for nine combinations of $|q_1|$ and q_2; $|q_1|$ and q_2 ranging between

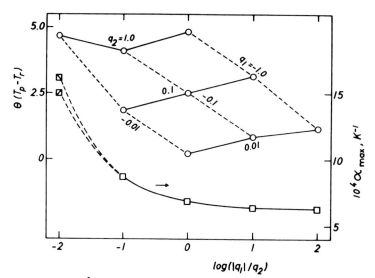

Fig. 25. Dependence of $\hat{T}$ and α_{max} on the ratio $|q_1|q_2^{-1}$ for a family of intrinsic heating isobars, as well as for the heating stage of similar cycles, all starting at $T_0 > T_r + 10$ K and involving relatively large δ_2 values (cf. runs 1–5 in Table I, 1–7 in Table II, and 5–6 in Table III). Solid and dashed lines connect T_p values obtained with fixed q_2 and q_1, respectively. Note that these lines merely shift vertically when q_1 and q_2 are multiplied by the same factor [cf. eq (81)]. [At $|q_1|q_2^{-1} = 0.01$ the value of α_{max} (squares) is larger for $T_1 = T_r - 20$ K, than the one for $T_1 = T_r - 10$ K.]

0.01 and 1.0, as indicated. Furthermore, the set of T_p values represented in Figure 25 is practically independent of T_1 and δ_2 [cf. Fig. 24(b)], implying that eq. (78) holds. Likewise, the corresponding values of α_{max} (see the right-hand scale) only depend on $|q_1 q_2^{-1}|$, except for $|q_1 q_2^{-1}| = 0.01$, at which α_{max} varies slightly with T_1 (see below). Therefore, the data shown in Figure 25 apply to all isobars showing a distinct secondary α peak occurring at $\hat{T}$, at which $d\delta_N/dT = 0$. Thus for thermal cycles giving rise to such isobars, eq. (75) reduces, *at fixed values of* $|q_1 q_2^{-1}|$, to

$$\frac{d\hat{T}}{d \ln|q_1|} = \frac{d\hat{T}}{d \ln q_2} = \hat{s}_2(q_1) + \hat{s}_2(q_2) = \theta^{-1} = s_g(q_1) \tag{81}$$

since in this case $d \ln|q_1| = d \ln q_2$ as in eq. (80), and $s_2(T_0) = \hat{s}_2(T_1) = 0$. That is, the shift of $\hat{T}$ resulting from a simultaneous reduction of $|q_1|$ and q_2 by the same factor is identical with the shift of $T_g(q_1)$, though neither $\hat{s}_2(q_1)$, nor $\hat{s}_2(q_2)$ is equal to $s_g(q_1)$, as defined by eq. (59).* In fact, the individual values of $\hat{s}_2(q_1)$ and $\hat{s}_2(q_2)$ vary, as can be seen in Figure 25 by considering the slopes of the lines along which either q_1 or q_2 is invariant. In the range $|q_1 q_2^{-1}| \geq 0.1$, i.e., with the exception of the point at $|q_1 q_2^{-1}| = 0.01$, one has

$$0.32 \leq 2.3\theta\,\hat{s}_2(q_1) \leq 0.82; \quad 2.0 \geq 2.3\theta\,\hat{s}_2(q_2) \geq 1.5 \tag{81a}$$

which shows that the variation of $s(q_1)$ exactly compensates that of $\hat{s}(q_2)$, their sum being constant, as postulated by eq. (81). For the "irregular" point at $|q_1 q_2^{-1}| = 0.01$, involving a negative value of $\hat{s}_2(q_1) = -0.58/2.3\theta$ and a much larger

---

* Such cycles thus provide for an independent determination of θ through the variation of $\hat{T}$ with the heating and cooling rates, while keeping their ratio invariant.

value of $\hat{s}_2(q_2) = 2.84/2.3\theta$, the two criteria defined in the previous section are satisfied at nearly the same temperature. Since the latter depends (as $\alpha_{\max}$) slightly on T_1, the nature of this α peak is somewhat ambiguous.

More practical criteria for identifying the nature of the α peaks can be established on the basis of the shifts of $\hat{T}$, as defined by eqs. (78) and (81), which are independent of the shape of the retardation spectrum. In fact, the above values of the partial derivatives do not critically depend on its shape,[31] whereas they do depend on the value of x in a very sensitive manner.

The dependence of these derivatives on x has been discussed extensively for a one-parameter model,[29] involving only one α peak which always occurs at $T_p > T_\times$ (see above). For this model, the heating isobars are fully defined by three experimental parameters: δ_2, T_1, and q_2, whatever the thermal history prior to heating. The relevant values of $-\Delta\alpha s(\delta_2)$, $s(T_1)$, and $\theta s(q_2)$, however, do not differ considerably from those derived above for the variations of $\hat{T}$ provided that x is assumed to be near to 0.4, as here [cf. eq. (36)]. Nevertheless, these values dramatically increase as x approaches zero (involving pure structural dependence of τ_i), and there is evidence that a similar situation prevails for the present multiparameter model. The variations of these partial derivatives, with x, as well as those of $s(T_0)$ and $\theta s(q_1)$ involved by such models, will be reported elsewhere.[31]

The above discussion shows the complexity of the thermal behavior of glasses on heating. Clearly, the various characteristics of the α and/or C_p isobars have no simple physical meaning; rather, they reflect involved interactions between the different retardation mechanisms and the thermal history. Thus the deconvolution of these processes requires a sound theoretical basis, which, in turn, will lead to the proper choice of experiments necessary to obtain the fundamental material parameters of the system.

CONCLUSION

A transparent multiparameter model for isobaric volume and enthalpy recovery of glasses has been developed using a thermodynamic approach for irreversible processes. In this treatment, the state of the system (glass) is determined by the values of N ordering parameters ζ_i, in addition to T and P. Accordingly, the *total* departure δ (or δ_H) from equilibrium is partitioned among the ζ_i, each of which is associated with a unique retardation time τ_i, the latter are assumed to depend on δ (or δ_H) in addition to T and P. The analysis further implies that the contribution of any one ζ_i to the total departure is independent of changes occurring in any of the other ordering parameters. This partitioning is similar to the familiar normal coordinate analysis and leads to N distinct differential equations. These are, however, coupled through the dependence of the individual τ_i on the total departure δ, reflecting the effect of the environment on the segmental mobility controlling the effective rates of molecular rearrangements. A similar theoretical frame has also been put forward recently by Struik.[33]

It should be pointed out that this model is the simplest one conceivable to account for *all* the characteristic features of structural recovery of glasses, mentioned in the introduction and in various sections of this paper. In fact the model involves only three material constants (two of them merely being scaling factors which can be determined from independent experiments) in addition to the fundamental distribution function (the retardation spectrum), or alternatively to the material response function.[10,11] These material parameters fully define the isobaric behavior of the system when subjected to any thermal treatment.

The present formulation of the model is merely phenomenological in the sense that the material constants and the shape of the spectrum are determined from experiment rather than from the molecular characteristics of the particular glass. (Attempts to relate transition probabilities of isomeric states of short chain segments to structural recovery of glasses have recently been made by Robertson[34] and the results are in qualitative agreement with the present model.) Thus, the importance of this work is that it separates the intrinsic molecular information (in the form of a retardation spectrum) from the observed response which critically depends on the particular experimental conditions. This situation is analogous to that encountered in linear viscoelasticity in shear, where a single distribution function (e.g., the relaxation spectrum) fully determines the behavior of the material subjected to any variety of dynamic or transient experimental stimuli under isothermal conditions. Unlike shear viscoelasticity, however, volume or enthalpy recovery involves the dependence of the effective rates of molecular motions on the actual state (structure) of the glass. This results in a genuine nonlinear and asymmetric response depending on the magnitude and on the sign of the departure from equilibrium, respectively; this specific feature of structural recovery of glasses, has been deconvoluted following Narayanaswamy's approach[10] by using an appropriately reduced time scale, as determined by the rate of fluctuations in a reference equilibrium state.

Although many consequences of the theory have been developed a few of them are worth restating, since they have been overlooked by most authors, including those using a similar approach.[10,11,23]

(a) The present analysis provides a rational foundation for comparing the theory with experiment, or for comparing various kinds of experiments between themselves. As an example, it has been shown that on, or after, cooling (whatever the rate) the material response always reflects a truncated situation, in which the contributions of the rapid processes have partially or fully vanished, and more so when cooling starts at relatively high temperatures which is the procedure adopted by most authors. Thus such data are unsuitable for determining the fundamental material response or reduced rate functions.

(b) The model implies a rate (time) -temperature reduction method as in linear viscoelasticity, an interesting achievement for a nonlinear system. Owing to the nonlinear response, however, the application of the reduction method is here more restricted, as has been discussed for various types of thermal treatments, including three-step thermal cycles. A trivial example of inappropriate application of the reduction method is the determination of the apparent activation energy from the shifts, with q, of the α or C_p peaks or the related peak values of the mechanical or electrical loss factors observed during heating (see below).

(c) In thermal cycles involving cooling, partial recovery and relatively slow heating, the calculated volume or enthalpy isobars display some unexpected features during the heating stage which, to our knowledge, have not yet been reported. In this respect, the present theory not only accounts for most of the observed effects, but also predicts some new ones which appear to be worth further investigation.

Application of the present model can easily be extended to nonisobaric situations, for instance, when glass formation is achieved by hydrostatic compression of the liquid. Such an extension to combined pressure and temperature effects is currently in progress in this laboratory.[35]

Finally, the theory can also be applied to changes in mechanical, electrical or other structure-sensitive properties of glasses which depend on the rate of segmental motions. Since the latter are controlled by the $a_T a_\delta$ reduction factors similar to those involved in structural recovery (cf. Appendix A), the time scale of the mechanical (electrical, etc.) relaxation spectrum is determined by the instantaneous state of the glass, i.e., by T and δ. These spectra thus shift simultaneously with the spontaneous structural recovery of the glass, giving rise to important changes in the material properties, as already reported by several authors.[33,36]

Therefore combination of mechanical (electrical, etc.) measurement with structural recovery is the only consistent way of investigating and predicting long-range behavior of polymeric glasses subjected to independent thermal and mechanical (electrical, etc.) stimuli. The practical importance of this type of "physical aging" has only been recognized recently, and the present treatment provides an appropriate theroetical framework for carrying out such investigations in a consistent manner.

The authors are very indebted to Mrs. L. Dziewulski for writing the computer program and for carrying out many calculations, only the most illustrative of which are reported in this paper. They also wish to acknowledge the aid of Professor B. Schmitt (Institute of Mathematics of the Université Louis Pasteur) for pointing out the properties of autonomous systems, as well as the help of Dr. J. Thomann (Centre de Calcul of the Centre National de la Recherche Scientifique, Cronenbourg) for using the potentialities of the Univac 1110 computer in the most efficient way.

APPENDIX A

Comparison Between Various Expressions for the Retardation Times

Basic Assumptions and Definitions

In this paper all the retardation times τ_i are assumed to depend on temperature (T) and on the instantaneous state (structure) of the glass as defined by the *total* departure $\delta = \Sigma \delta_i$ (or δ_H) from equilibrium at T and P, the pressure being kept constant. Furthermore, all values of τ_i are reduced by the same factor when T and δ vary, hence one can write, omitting the subscript i,

$$\tau(T,\delta) = \frac{\tau(T,\delta)}{\tau(T,0)} \frac{\tau(T,0)}{\tau(T_r,0)} \tau(T_r,0) = a_\delta a_T \tau_r \tag{82}$$

where the *first* factor $a_\delta = \tau(T,\delta)/\tau(T,0)$ denotes the reduction of τ due to the departure δ from equilibrium ($\delta = 0$) at T, whereas the *second* $a_T = \tau(T,0)/\tau(T_r,0)$ is the reduction factor of τ due to the temperature difference $T - T_r$ under equilibrium conditions. The rather inconsistent but widely used convention for subscripts of a thus indicate the span covered by *one* of the variables while keeping the other constant; in addition two different reference (equilibrium) states: $(T_r,0)$ and $(T,0)$, are implied but not indicated.

In order to calculate the pure contribution of T to τ at fixed structure it is convenient to refer to the temperature T^* at which the line passing through the coordinates (T,δ) with a slope $-\Delta\alpha$, thus defined by

$$\bar{\delta} = -\Delta\alpha(T - T^*), \tag{83}$$

intersects the equilibrium line, along which $\delta = 0$ (Fig. 26). Note that T^* is the *fictive temperature* characterizing the instantaneous structure of the glass at (T,δ) according to Tool's definition[2,10,11]

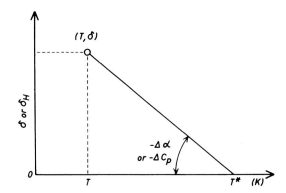

Fig. 26. Operational definition of the fictive temperature T^*, for a nonequilibrium state at T, characterized by the total departure δ.

which can be used safely since the values of τ_i are assumed here to depend on δ, rather than on the individual values of δ_i.

For the present model, eq. (83) defines the changes in δ resulting from any instantaneous T-jump ($|q| = \infty$) starting at (T,δ), during which the structure of the system remains invariant [cf. eq. (19)], since along such a line, involving a thermal expansion coefficient α_g [cf. eq. (2)] the values of all the ordering parameters ζ_i are fixed. Accordingly, invariant structure means constant T^*, while δ varies as in eq. (83) and vice versa. In equilibrium, of course, $T = T^*$. Therefore, along the line shown in Figure 26 (which is not necessarily straight if $\Delta\alpha$, or ΔC_p vary with T) changes in τ (if any) result from temperature alone and the relevant reduction factor of τ is [cf. eq. (82) with $T^* = T_r$]

$$b_T^* = \tau(T,\delta)/\tau(T^*,0) = a_\delta a_T^* \tag{84}$$

where both $a_T^* = \tau(T,0)/\tau(T^*,0)$ and b_T^* refer to the fictive T^*. Clearly, in the absence of structural dependence $a_\delta = 1$, whereas if τ depends on structure alone $b_T^* = 1$. Furthermore, by referring as above to a conventional equilibrium state at (fixed) T_r, located in the glass-transition range, one has

$$a_T = \tau(T,0)/\tau(T_r,0) = a_T^* \tau(T^*,0)/\tau(T_r,0) \tag{85}$$

in which the ratio $\tau(T^*,0)/\tau(T_r,0) = a_{T^*}$, consistently with the notation adopted for a_T. Thus, $b_T^* = (a_T/a_{T^*})a_\delta$.

Comparison of the Shift Factors

Expressions for a_T, a_δ, and b_T^* will now be derived using five formulations of $\tau(T,\delta)$, or $\tau(T,T^*)$ which have been proposed by different authors and discussed in a previous paper.[9] These expressions will then be compared with those given in eq. (17), viz.,

$$a_T = \exp[-\theta(T - T_r)], \quad T_r = \text{const}$$

$$a_\delta = \exp[-(1 - x)\theta\delta/\Delta\alpha] \tag{86}$$

in order to relate the relevant material constants involved to θ and x used in the present calculations [cf. eq. (36)]. Since the algebra is trivial, intermediate calculation steps (see ref. 9) will be omitted.

(i) If τ is defined by the expression introduced by Tool[2,5]

$$\tau(T,T^*) = A \exp(-K_1 T - K_2 T^*) \tag{87}$$

where A, K_1, and K_2 are constants, one has according to the above definitions [cf. eqs. (82)–(84)],

$$a_T = \exp[-(K_1 + K_2)(T - T_r)]$$

$$a_\delta = \exp[K_2(T - T^*)] = \exp(-K_2\delta/\Delta\alpha)$$

VOLUME AND ENTHALPY RECOVERY OF GLASSES 1159

$$b_T^* = \exp[-K_1(T - T^*)] = \exp(K_1\delta/\Delta\alpha) \tag{88}$$

Comparing Eqs. (88) with Eqs. (86), one obtains

$$\theta = K_1 + K_2, \quad x = K_1(K_1 + K_2)^{-1}$$

which shows that eq. (87) is strictly equivalent to eq. (17), and

$$b_T^* = \exp(x\theta\delta/\Delta\alpha) \tag{89}$$

(ii) If τ is defined by an Arrhenius equation,[10,11] one can write, omitting the preexponential factor A,

$$\tau(T,T^*) \propto \exp[x'E_a/RT + (1 - x')E_a/RT^*] \tag{90}$$

where E_a is the apparent activation energy in equilibrium conditions ($T = T^*$) and $x'E_a$ its fractional value ($0 \leq x' \leq 1$) at constant T^*, i.e., at invariant structure, one has

$$a_T = \exp[(E_a/R)(T^{-1} - T_r^{-1})]$$

$$a_\delta = \exp[-(1 - x')(E_a/R)(T^{-1} - T^{*-1})] = \exp[-(1 - x')(E_a/RTT^*)(\delta/\Delta\alpha)]$$

$$b_T^* = \exp[(x'E_a/R)(T^{-1} - T^{*-1})] = \exp[(x'E_a/RTT^*)(\delta/\Delta\alpha)] \tag{91}$$

Identification with eqs. (86) leads with a fair approximation to

$$\theta \simeq E_a/RT^2, \quad x \equiv x' \tag{92}$$

provided that the temperature range ($T - T^*$) covered is small enough (as compared to T), which is generally the case. This comparison defines the partition coefficient x used in eq. (17) in a physically meaningful manner.

(iii) In the Doolittle equation[7,9,16,37]

$$\tau(T,\delta) \propto \exp(b/f) \tag{93}$$

in which b is a constant of the order of *one* and f is the fractional free volume of the system defined by

$$(v - v_0)/v_\infty \equiv f = f_T + \delta$$

$$(v_\infty - v_0)/v_\infty \equiv f_T = f_r + \alpha_f(T - T_r) \tag{94}$$

where v_0 is the "occupied" volume and $\alpha_f = df_T/dT$, the expansion coefficient of f in equilibrium conditions, or at constant δ. Accordingly,

$$\ln a_T = -\frac{(b/f_r)(T - T_r)}{(f_r/\alpha_f) + T - T_r}$$

$$\ln a_\delta = \frac{(b/f_T)(T - T^*)}{(f_T/\Delta\alpha) - (T - T^*)} = -\frac{(b/f_T)\delta}{f_T + \delta}$$

$$\ln b_T^* = -\frac{(b/f_{T^*})(T - T^*)}{f_{T^*}(\alpha_f - \Delta\alpha)^{-1} + T - T^*} = \frac{(b/f_{T^*})(\delta/\Delta\alpha)}{f_{T^*}(\alpha_f - \Delta\alpha)^{-1} + (\delta/\Delta\alpha)} \tag{95}$$

Note that all the three reduction factors of τ are of the Williams-Landel-Ferry (WLF) type,[16] but they involve different expansion coefficients.

In a narrow temperature interval in which one can assume $f \simeq f_{T^*} \simeq f_r$, and neglect $T - T_r$ as compared to (f_r/α_f) identification with eqs. (86) results in[9]

$$\theta \simeq b\alpha_f/f^2 \simeq E_a/RT^2, \quad x \simeq b(\alpha_f - \Delta\alpha)/\alpha_f \tag{96}$$

This shows that, in case of pure free volume dependence of τ ($x = 0$, $b_T^* = 1$), one should have $\alpha_f = \Delta\alpha$ which is generally not the case.[16]

(iv) A "hybrid" form of the free volume dependence of τ, proposed by Litovitz and Macedo[38] can be written

$$\tau(T,\delta) \propto \exp(E/RT + b/\bar{f}) \tag{97}$$

in which E is the activation energy controlling the pure temperature dependence of τ (i.e., at constant T^*), and

1160   KOVACS ET AL.

$$(v - v_0)v_\infty^{-1} \equiv \bar{f} = \bar{f}_T + \delta$$

$$(v_\infty - v)v_\infty^{-1} \equiv \bar{f}_T = f_r + \Delta\alpha(T - T_r) \tag{98}$$

are defined similarly to eqs. (94). Here the expansion coefficient of $\bar{f}_T$, however, is $\Delta\alpha$ implying that the fractional free volume during instantaneous T-jumps remains invariant [cf. eq. (83)]. Accordingly, structural modifications are uniquely associated with the changes in free volume and one has

$$\ln a_T = -\frac{E(T - T_r)}{RTT_r} - \frac{(b/\bar{f}_r)(T - T_r)}{(\bar{f}_r/\Delta\alpha) + T - T_r}$$

$$\ln a_\delta = -(b\delta/\bar{f}_T)(\bar{f}_T + \delta)^{-1}$$

$$\ln b_T^* = -(E/RTT^*)(T - T^*) = (E/RTT^*)(\delta/\Delta\alpha) \tag{99}$$

Again, if the temperature interval is small

$$\theta \simeq (E/RT^2) + (b\Delta\alpha/\bar{f}^2), \quad x \simeq \theta^{-1}(E/RT^2) \simeq E/E_a \tag{100}$$

which is consistent with the definition of x as introduced in eq. (90).

(v) Finally, let τ be defined through the Adam-Gibbs equation[18] as

$$\tau(T,\delta_H) \propto \exp(\epsilon/TS_c) \tag{101}$$

where ϵ is a constant energy barrier (per monomer unit) and S_c the configurational entropy of the glass at T^*, which here is associated with the excess enthalpy $\delta_H = -\Delta C_p(T - T^*)$ with respect to its equilibrium value at T. In fact, extending the definition of the authors[18] to nonequilibrium situations S_c can be defined in a self-consistent manner as

$$S_c \equiv S_{cT^*} = \int_{T_2}^{T^*} \Delta C_p \left(\frac{dT}{T}\right) = \int_{T_2}^{T} \Delta C_p d\ln T + \Delta C_p \ln(T^*/T) \tag{102}$$

provided that the excess heat capacity ΔC_p is constant in the temperature interval $T^* - T$. The lower limit T_2 of the integrals in eq. (102) is the critical temperature at which $S_{c,T}$ (the configurational entropy in equilibrium at T) vanishes.[39] Limited expansion of the last term in eq. (102) yields

$$S_c \simeq S_{c,T} + \delta_H/T \tag{103}$$

which is similar to the definition of $\bar{f}$ in eq. (98). Hence along the lines where S_c, i.e., the structure, is invariant (and where $\delta_H \simeq -\Delta C_p(T - T^*)$ with a fair approximation) the pure contribution of temperature to τ is provided by the factor T associated with S_c in eq. (101), thus involving an activation energy $E = \epsilon R/S_c$, which slightly increases when T^* decreases.

According to these definitions one has

$$\ln a_T = \epsilon \frac{T_r S_{c,r} - T S_{c,T}}{T_r T S_{c,r} S_{c,T}} \simeq -\frac{(\epsilon/T_r S_{c,r})(T - T_r)}{(T_r S_{c,r}/\Delta C_p) + T - T_r}$$

$$\ln a_{\delta_H} = -\frac{\epsilon \delta_H/T S_{c,T}}{T S_{c,T} + \delta_H} \tag{104}$$

which again are similar to the WLF equation,[16] whereas

$$\ln b_T^* = (\epsilon/S_c \dot{T}_*)(T^{-1} - T^{*-1}) \tag{105}$$

is analogous to the relevant expression given above in eqs. (91) and (99), based on activated rate processes, with $(\epsilon/S_c) = xE_a/R = E/R$.

Identification with eqs. (86) and (92) yields

$$\theta \simeq \epsilon\Delta C_p/T^2 S_{c,T}^2 \simeq E_a/RT^2 \tag{106}$$

and comparison of b_T^*, as expressed by eqs. (91) and (105) results in

$$x \simeq \theta^{-1}\epsilon/T^2 S_c \simeq S_c/\Delta C_p \simeq \ln(T^*/T_2) \tag{107}$$

[cf. eq. (102)], provided that ΔC_p is invariant in the *whole* temperature interval $(T^* - T_2)$. Moreover, since T^* is always located in the glass-transition range and $(T_g/T_2) \simeq (1.30 \pm 0.12)$ for a great variety of glassy polymers,[18] eq. (107) predicts an average value for $\bar{x} \simeq 0.26$, the magnitude of which appears reasonable for these systems. Deviations from this value should mainly be attributed to the temperature dependence of ΔC_p which generally decreases with T.[11,23]

Clearly, the definition of τ by eq. (101) is well suited for enthalpy recovery, though eqs. (104)–(107) may also be applied to volume recovery if for any nonequilibrium state the fictive temperatures, as determined from volume and enthalpy measurements, are the same. In this case obviously $\delta/\Delta\alpha = \delta_H/\Delta C_p = T^* - T$ (cf. Fig. 26) and δ_H can be replaced in the above equations by $(\Delta C_p/\Delta\alpha)\delta$.

These calculation clearly show that whatever the definition of the $\tau(T,\delta)$ function, the various expressions for the shift factors reduce to eqs. (86) and (89) in a narrow temperature interval. For larger temperature changes a_T, a_δ, and b_T^* depend slightly on both T and δ (or δ_H), nevertheless eq. (17) still remains a fair approximation in most practical cases.

APPENDIX B

Limiting Value of τ_{eff} near Equilibrium

Equation (29), together with eqs. (32) and (33) can be expressed as

$$\tau^{-1}_{eff,T} = a_\delta^{-1}\,\Sigma(\delta_i/\delta)(a_T\tau_{i,r})^{-1} = a_\delta^{-1}\phi_T(\rho) \tag{108}$$

with

$$\phi_T(\rho) = \Sigma\tau^{-1}_{i,T}g_i e^{-u/\tau_{i,T}}/\Sigma g_i e^{-u/\tau_{i,T}} \tag{109}$$

where $\tau_{i,T} = a_T\tau_{i,r}$. When t and $u \to \infty$, $\delta \to 0$, and $a_\delta \to 1$, thus the limiting values of τ^{-1}_{eff} and $\phi_T(\rho) = \phi_T(0)$ are identical.

If $\tau_{N,T}$ is the longest retardation time, at T, eq. (109) can be rearranged as

$$\phi_T(\rho) = \frac{\dfrac{g_N}{\tau_{N,T}}e^{-u/\tau_{N,T}} + \sum^{N-1}\dfrac{g_i}{\tau_{i,T}}e^{-u/\tau_{i,T}}}{g_N e^{-u/\tau_{N,T}} + \sum^{N-1}\dfrac{g_i}{\tau_{i,\hat{T}}}e^{-u/\tau_{i,T}}} \tag{110}$$

in which the last term involving $\tau_{N,T}$ has been separated from the other $(N-1)$ terms of the sum. Dividing both the numerator and the denominator by $e^{-u/\tau_{N,T}}$ yields

$$\phi_T(\rho) = \frac{\dfrac{g_N}{\tau_{N,T}} + \sum^{N-1}\dfrac{g_i}{\tau_{i,T}}\exp[-u(\tau^{-1}_{i,T} - \tau^{-1}_{N,T})]}{g_N + \sum^{N-1}g_i\exp[-u(\tau_{i,T} - \tau_{N,T})]} \tag{111}$$

Now, since $\tau_{N,T} > \tau_{i,T}(1 \leqslant i < N)$, all the differences $\tau^{-1}_{N,T} - \tau^{-1}_{N,T}$ are positive and the sums in both the numerator and denominator approach zero as u becomes large enough and δ approaches zero. Thus,

$$\lim_{\delta\to 0}\tau^{-1}_{eff,T} = \lim_{\rho\to 0}\phi_T(\rho) = \tau^{-1}_{N,T} \tag{112}$$

Q.E.D.

References

1. F. Simon, *Ergeb. Exakten Naturwiss.*, **9**, 222 (1930); and *Z. Anorg. Allg. Chem.*, **203**, 220 (1931).
2. A. M. Tool, *J. Am. Ceram. Soc.*, **29**, 240 (1946).
3. R. O. Davies and G. O. Jones, *Adv. Phys. (Philos. Mag. Suppl.)*, **2**, 370 (1953).
4. W. Kauzmann, *Chem. Rev.*, **43**, 219 (1948).
5. H. N. Ritland, *J. Am. Ceram. Soc.*, **37**, 370 (1954); **39**, 403 (1956).
6. A. J. Kovacs, *J. Polym. Sci.*, **30**, 131 (1958).
7. A. J. Kovacs, *Fortschr. Hochpolym. Forsch. (Adv. Polym. Sci.)*, **3**, 394 (1963).
8. J. M. Hutchinson and A. J. Kovacs, *J. Polym. Sci. Polym. Phys. Ed.*, **14**, 1575 (1975).
9. A. J. Kovacs, J. M. Hutchinson, and J. J. Aklonis, *The Structure of Non-Crystalline Materials*, P. H. Gaskell, Ed., Taylor & Francis, 1977, p. 153.
10. O. S. Narayanaswamy, *J. Am. Ceram. Soc.*, **54**, 491 (1971).

11. M. A. De Bolt, A. J. Easteal, P. B. Macedo, and C. T. Moynihan, *J. Am. Ceram. Soc.,* **59,** 16 (1976).

12. J. M. Hutchinson, J. J. Aklonis, and A. J. Kovacs, *Polym. Prepr.,* **16**(2), 94 (1975); J. J. Aklonis and A. J. Kovacs, *J. Polym. Sci. Polym. Symp.,* to appear.

13. J. M. Hutchinson, A. J. Kovacs, and J. J. Aklonis, to be submitted.

14. F. Schwarzl and A. J. Staverman, *J. Appl. Phys.,* **23,** 838 (1952).

15. J. Meixner, *Z. Naturforsch.,* **9a,** 654 (1954).

16. J. D. Ferry, *Viscoelastic Properties of Polymers,* 2nd ed., Wiley, New York, 1971.

17. A. J. Kovacs, *Trans. Rheol. Soc.,* **5,** 285 (1961).

18. G. Adams and J. H. Gibbs, *J. Chem. Phys.,* **43,** 139 (1965).

19. R. Gardon and O. S. Narayanaswamy, *J. Am. Ceram. Soc.,* **53,** 380 (1970).

20. G. Sausone and R. Conti, *Non Linear Differential Equations,* Pergamon, New York, 1964.

21. I. L. Hopkins, *J. Polym. Sci.,* **28,** 631 (1958).

22. G. Goldbach and G. Rehage, *J. Polym. Sci., Part C,* **16**(4), 2289 (1965); and *Rheol. Acta,* **6,** 30 (1967).

23. C. T. Moynihan, A. J. Easteal, and M. A. De Bolt, *J. Am. Ceram. Soc.,* **59,** 12 (1976).

24. M. Goldstein, *Modern Aspect of Vitreous State,* Vol. 3, J. D. Mackenzie, Ed., Butterworth, London, 1964.

25. G. Williams and D. C. Watts, *Trans. Faraday Soc.,* **66,** 80 (1970).

26. S. Spinner and A. Napolitano, *J. Res. Nat. Bur. Stand.,* **A70,** 147 (1966).

27. S. Hozumi, T. Wakabayashi, and K. Sugihara, *Polym. J.,* **2,** 632 (1970).

28. J. M. Hutchinson and A. J. Kovacs, *The Structure of Non-Crystalline Materials,* P. H. Gaskell, Ed., Taylor & Francis, 1977, p. 167.

29. A. J. Kovacs and J. M. Hutchinson, *J. Polym. Sci. Polym. Phys. Ed.,* to appear.

30. M. V. Volkenstein and Yu. Sharonov, *Vysokomol. Soedin.,* **3,** 1739 (1961); **4,** 917 (1962).

31. A. R. Ramos, J. M. Hutchinson, and A. J. Kovacs, to be submitted.

32. S. E. B. Petrie, *J. Polym. Sci., A-2,* **10,** 1255 (1972).

33. L. C. E. Struik, *Physical Aging in Amorphous Polymers and Other Materials,* Elsevier, Amsterdam, 1978.

34. R. E. Robertson, *J. Polym. Sci., Part C,* **63,** 173 (1978); and *J. Polym. Sci. Polym. Phys. Ed.,* to appear.

35. A. R. Ramos and A. J. Kovacs, in preparation.

36. A. J. Kovacs, R. S. Stratton, and J. D. Ferry, *J. Phys. Chem.,* **67,** 152 (1963).

37. A. K. Doolittle, *J. Appl. Phys.,* **22,** 1471 (1951).

38. T. A. Litovitz and P. B. Macedo, *Proceedings of the International Conference on the Physics of Non-Crystalline Solids,* J. A. Prins, Ed., North-Holland, Amsterdam, 1965, p. 220.

39. J. H. Gibbs and E. A. Di Marzio, *J. Chem. Phys.,* **28,** 373 (1958); **28,** 807 (1958).

Received July 12, 1978
Revised January 20, 1979

COMMENTARY

Reflections on "Some Phenomenological Consequences of the Doi-Edwards Theory of Viscoelasticity," by William W. Graessley, *J. Polym. Sci., Polym. Phys. Ed.,* 18, 27 (1980)

WILLIAM W. GRAESSLEY

School of Engineering and Applied Science, Department of Chemical Engineering, Princeton University, Princeton, NJ 08544

The work described in this paper began almost from the moment I received preprints of the tube model papers[1-3] from Masao Doi. A few years before I had reviewed molecular rheology with particular emphasis on the chain entanglement concept[4] and was struck then by the mismatch between experiment and theory. Neutron scattering had recently cleared away all the lingering doubts about the chain dimensions of polymers at high concentrations. Anionic polymerization was rapidly replacing fractionation as the prime source of molecularly well-defined polymers. With those materials and new instrumentation, systematic studies were being made on many polymer species and chain architectures. This growing body of data had already revealed that the flow properties of undiluted and highly concentrated polymers were related to architecture by physical laws of great generality. In many cases the relationships were tantalizingly simple, and a variety of evidence pointed toward chain entanglement as the root of them. Yet, the supporting molecular theory had hardly progressed beyond Bueche's pioneering efforts in the 1950s.[5] Here then this new theory appears, dealing in detail with exactly that experimentally rich but theoretically starved topic, so I got busy to see how well it worked.

I have overstated the theory shortage during that era. Actually, there were many theories about chain entanglement. Coined by Busse in 1932,[6] the term itself had come to mean an interaction that arises because molecular chains cannot literally cross through one another. Relative to the phantom chains of theory, it retards diffusion and relaxation in liquids—the chains must move around one another's contours—and it raises the modulus of elasticity in networks by introducing topological constraints. Essentially a truism but lacking in specifics—What does an entanglement look like? Is it a knot or loop or something more elaborate?—the idea led to a proliferation of theories aimed at explaining specifics rather than providing a general framework.

One exception to that pattern was what came to be called the modified Rouse model,[5,7] which did not assume a specific picture at all but supposed merely that entanglement did nothing more than shift the slowest modes of motion to longer times. The cut-off mode for shifting was chosen to fit the plateau modulus G_N^0; the shift magnitude itself was chosen to fit the viscosity η, through adjusting the Rouse friction coefficient. How well it worked was judged by comparisons with the other properties. As it happened, the limited data then available on diffusion coefficient D agreed rather well, but the prediction for recoverable compliance J_e^0 was clearly wrong. Nontheless, modified Rouse provided a highly effective organizing principle for data, so it seemed quite natural to adapt the same approach to the new theory.

The Doi–Edwards theory uses three elements to approximate the various entanglement necessities— the Edwards tube, representing "uncrossability" by a confining field surrounding each contour, the deGennes reptation idea for the dynamics, and the Doi stress law to connect with the mechanical response. The result is a two-parameter theory for the dynamics of Rouse chains moving independently in a meshwork of spatial constraints. Both parameters are species-dependent but independent of chain length, the monomeric friction coefficient and the mesh size. Each could be evaluated independently so as to allow unambiguous comparisons, magnitudes as well as chain length dependence, with entangled polymer data for η, J_e^0, and D. The results were mixed, but even outright failures like the viscosity and polydispersity predictions were instructive. I remember thinking at the time that now entangled polymer rheology finally has its proper theoretical starting point. We still didn't know what an entanglement looks like, but that somehow seems less important now.

REFERENCES AND NOTES

1. M. Doi and S. F. Edwards, *J. Chem. Soc. Faraday Trans. 2*, **74,** 1789–1801 (1978).
2. M. Doi and S. F. Edwards, *J. Chem. Soc. Faraday Trans. 2*, **74,** 1802–1817 (1978).
3. M. Doi and S. F. Edwards, *J. Chem. Soc. Faraday Trans. 2*, **74,** 1818–1832 (1978).
4. W. W. Graessley, *Adv. Polym. Sci.,* **16,** 1 (1974).
5. F. Bueche, "Physical Properties of Polymers," Wiley, New York, 1962.
6. W. F. Busse, *J. Phys. Chem.,* **36,** 2862 (1932).
7. J. D. Ferry, "Viscoelastic Properties of Polymers," 2nd ed., Wiley, New York, 1970.

PERSPECTIVE

Comments on "Some Phenomenological Consequences of the Doi-Edwards Theory of Viscoelasticity," by William W. Graessley, *J. Polym. Sci., Polym. Phys. Ed.,* 18, 27 (1980)

RALPH H. COLBY

Department of Materials Science and Engineering, The Pennsylvania State University, University Park, PA 16802

In 1978 Doi and Edwards[1-3] published three papers that derived a constitutive relation for linear polymers from the reptation idea of de Gennes.[4] Graessley[5] extracted the essential physics regarding chain diffusion and linear viscoelasticity from the Doi and Edwards papers, and compared them with experiments. At the time, this paper played a central role in alerting scientists interested in experimental polymer physics and rheology about the new theory.

Graessley showed that the Doi–Edwards theory gave remarkable predictions of both the molecular weight exponent and the magnitude of chain diffusion. The linear viscoelastic response was not predicted quite as well by the theory. The predicted breadth of the relaxation time distribution is too narrow, as reflected in the product of recoverable compliance J_e^0 and the plateau modulus G_N^0, with $J_e^0 G_N^0 = 6/5$ predicted and $2.5 < J_e^0 G_N^0 < 3$ typically observed for near-monodisperse melts of linear chains. The viscosity η was predicted to be considerably larger than the experimental values, with an exponent that was not quite correct ($\eta \sim M^3$ was predicted, while $\eta \sim M^{3.5}$ is observed). Graessley correctly surmised that these two facts were related, and that other high frequency relaxation processes would both broaden the relaxation time distribution and lower the predicted viscosity.

This insight proved quite valuable for the development of improved tube model ideas. It is now well-appreciated that the Doi–Edwards (and de Gennes) reptation is an oversimplification of the correct physics—Rouse motion of a chain confined in a tube. The original reptation solution does not allow the contour length of the tube to change with time. Doi[6,7] subsequently published a better representation of the Rouse motion of a chain in a tube, which both broadened the spectrum of relaxation times and lowered the viscosity, with $\eta \sim M^{3.5}$ predicted in the typical experimental range. An exact analytical solution to the problem of a Rouse chain confined in a tube is still not known, but numerous other approximate and numerical solutions have been suggested,[8-13] and the agreement with experiment is quite impressive.[11,14-16]

REFERENCES AND NOTES

1. M. Doi and S. F. Edwards, *J. Chem. Soc. Faraday Trans. 2,* **74,** 1789 (1978).
2. M. Doi and S. F. Edwards, *J. Chem. Soc. Faraday Trans. 2,* **74,** 1802 (1978).
3. M. Doi and S. F. Edwards, *J. Chem. Soc. Faraday Trans. 2,* **74,** 1818 (1978).
4. P. G. DeGennes, *J. Chem. Phys.,* **55,** 572 (1971).
5. W. W. Graessley, *J. Polym. Sci., Polym. Phys. Ed.,* **18,** 27 (1980).
6. M. Doi, *J. Polym. Sci. Polym. Lett. Ed.,* **19,** 265 (1981).
7. M. Doi, *J. Polym. Sci., Polym. Phys. Ed.,* **21,** 667 (1983).
8. M. Rubinstein, *Phys. Rev. Lett.,* **59,** 1946 (1987).
9. J. M. Deutsch and T. L. Madden, *J. Chem. Phys.,* **91,** 3252 (1989).
10. R. Ketzmerick and H. C. Öttinger, *Continuum Mech. Thermodyn.,* **1,** 113 (1989).
11. J. DesCloizeaux, *Macromolecules,* **23,** 4678 (1990).
12. J. Reiter, *J. Chem. Phys.,* **94,** 3222 (1991).

13. N. P. T. O'Connor and R. C. Ball, *Macromolecules,* **25,** 5677 (1992).

14. M. Rubinstein and R. H. Colby, *J. Chem. Phys.,* **89,** 5291 (1988).

15. R. Kimmich, M. Kopf, and P. Callaghan, *J. Polym. Sci., Polym. Phys. Ed.,* **29,** 1025 (1991).

16. T. T. Perkins, D. E. Smith, and S. Chu, *Science,* **264,** 819 (1994).

Some Phenomenological Consequences of the Doi–Edwards Theory of Viscoelasticity

WILLIAM W. GRAESSLEY, *Chemical Engineering and Materials Science Departments, Northwestern University, Evanston, Illinois 60201*

Synopsis

Doi and Edwards have recently proposed a molecular theory for the dynamics of entangled polymer liquids based on a tube model to represent the mutual constraints on configurational rearrangement of the chains. Expressions for diffusion coefficient, plateau modulus, zero-shear viscosity, steady-state recoverable compliance, and terminal relaxation time can be devloped, and relations among these properties that depend only upon observable quantities can be obtained. Several such relations are derived and are compared with experimental observations.

INTRODUCTION

Doi and Edwards have recently developed a molecular theory of viscoelasticity for entangled linear polymer liquids[1–3] based on the reptating chain model of de Gennes.[4] Their results can be cast in the form of relations among various dynamic properties. This paper presents several such relations having to do with diffusion and linear viscoelasticity and compares them with experiment.

DEVELOPMENT OF EQUATIONS

An essential idea in the Doi–Edwards theory is that the transverse segmental motions of each chain are impeded by the meshwork of strands from other chains in its neighborhood. Large-scale configurational rearrangement and diffusion are assumed to proceed mainly by reptation, i.e., by random snakelike motions of each chain parallel to its own contour. Each chain continually disengages itself from its current cage of strands, creating new cage and eventually a new configuration as its emerging end wanders randomly through the mesh. Since the strands of the mesh are parts of chains which diffuse similarly, the lifetime of each cage constraint is itself comparable to the disengagement time for the entire chain.[5] Thus, for chains which are long compared to the mesh size, the frequency of transverse jumps is small enough to make plausible the idea of a semipermanent cage and predominantly snakelike motion.

If the system is deformed the cages are distorted and the chains are carried into new configurations. Stress relaxation proceeds first by a relatively rapid equilibration of chain configurations within the distorted cages, then by a relatively slow diffusion of chains out of the distorted cages into random configurations.

Doi and Edwards represent the cage by a tube of diameter a and contour length L enclosing each chain. The chains are random coils with molecular weight M and mean-square end-to-end distance R^2. There are ν chains per unit volume.

Journal of Polymer Science: Polymer Physics Edition, Vol. 18, 27–34 (1980)
© 1980 John Wiley & Sons, Inc.   0098-1273/80/0018-0027$01.00

The tube diameter is assumed to be independent of M and the tube length directly proportional to M. Each chain and the path of its associated tube have random walk configurations with the same end-to-end vector. The distance a represents the mesh size and is assumed to correspond also to the step length of the tube path. Therefore, we have

$$aL = R^2 \tag{1}$$

Alternatively, the tube path is a random walk of N steps, each of length a, and

$$Na^2 = R^2 \tag{2}$$

The self-diffusion coefficient for the chains is given by[1]

$$D = kT/3N\zeta_r \tag{3}$$

where k is the Boltzmann constant, T is the absolute temperature, and ζ_r is the molecular friction coefficient of a Rouse chain[6] $(D_r = kT/\zeta_r; \zeta_r \propto M)$.

Doi and Edwards calculate the stress following an instantaneous deformation. Immediately after deformation* the stress is that for an affinely deformed Gaussian network with νN strands/volume[2]:

$$\sigma_{ij}(0) = 3\nu NkT\langle (\mathbf{F} \cdot \mathbf{u})_i (\mathbf{F} \cdot \mathbf{u})_j \rangle \tag{4}$$

in which σ is the extra stress tensor, $\mathbf{F}$ is the deformation gradient tensor, and $\langle\ \rangle$ denotes an average taken over all directions of the unit vector $\mathbf{u}$. Equilibration within the distorted tubes requires a time of the order of the Rouse relaxation time $[\tau_r = (1/6\pi^2)\ \zeta_r R^2/kT; \tau_r \propto M^2]$. After this process is complete the stress is[2]

$$\sigma_{ij}(\tau_r) = 3\nu NkT\left\langle \frac{(\mathbf{F} \cdot \mathbf{u})_i (\mathbf{F} \cdot \mathbf{u})_j}{|\mathbf{F} \cdot \mathbf{u}|} \right\rangle \frac{1}{\langle |\mathbf{F} \cdot \mathbf{u}| \rangle} \tag{5}$$

where the first $|\mathbf{F} \cdot \mathbf{u}|$ accounts for the equilibration of local stretches within the tube and the second for the assumed retraction of the tube path to its equilibrium length L. For longer times $(t \gtrsim \tau_r)$ the stress decays according to[2]

$$\sigma_{ij}(t) = \sigma_{ij}(\tau_r)\frac{8}{\pi^2}\sum_{\text{odd } n}\frac{1}{n^2}\exp\left(-\frac{n^2 t}{\tau_d}\right) \tag{6}$$

in which τ_d is the tube disengagement time given by

$$\tau_d = L^2\zeta_r/\pi^2 kT \tag{7}$$

The long-time process [eq. (6)] corresponds to diffusion of chains out of their original cages. Since $\tau_d \propto M^3$, the equilibration and disengagement processes will be widely separated in time scale if the chains are long enough.

For an instantaneous simple shear deformation γ in the x direction the components of $\mathbf{F}$ are

$$\mathbf{F} = \begin{vmatrix} 1 & \gamma & 0 \\ 0 & 1 & 0 \\ 0 & 0 & 1 \end{vmatrix} \tag{8}$$

---

* Actually a brief time τ^*, independent of M, must pass to allow equilibration over distances of order a, corresponding to the transition region in linear viscoelasticity (ref. 7).

DOI–EDWARDS THEORY 29

If γ is sufficiently small the shear stress $\sigma_{yx} = \sigma$ at short times is given by [from eq. (4)]

$$\sigma(0) = \nu NkT\gamma \tag{9}$$

and, after a time lapse of the order of τ_r, by [from eq. (5)]

$$\sigma(\tau_r) = 4\nu NkT\gamma/5 \tag{10}$$

It seems reasonable to assume that the difference between these values arises from a linear viscoelastic relaxation of stress in the plateau region and that the subsequent relaxation, described by eq. (6), corresponds to the terminal region.[7] Thus, the stress-relaxation modulus in the terminal region for long chains is given by

$$G(t) = \frac{4}{5}\nu NkT\frac{8}{\pi^2}\sum_{\text{odd } n}\frac{1}{n^2}\exp\left(-\frac{n^2 t}{\tau_d}\right) \tag{11}$$

The experimental values obtained for the plateau modulus G_N^0 by the usual methods[7] probably do not include contributions from plateau relaxation processes. Thus, the appropriate expression for plateau modulus is that suggested by eq. (10):

$$G_N^0 = 4\nu NkT/5 \tag{12}$$

Expressions for the zero shear viscosity η_0 and the recoverable compliance J_e^0 of sufficiently long chains ($\tau_d \gg \tau_r$) can be obtained from eq. (11) using[7]

$$\eta_0 = \int_0^\infty G(t)\,dt \tag{13}$$

$$J_e^0 = \frac{1}{\eta_0^2}\int_0^\infty tG(t)\,dt \tag{14}$$

which yield

$$\eta_0 = \pi^2\nu NkT\tau_d/15 \tag{15}$$

$$J_e^0 = 3/2\nu NkT \tag{16}$$

All these results have been obtained without the independent alignment approximation, a simplifying assumption used to develop a constitutive equation (3). Slightly different results are obtained when the approximation is used. The $4/5$ factor in eq. (12), for example, becomes $3/5$.

Three parameters of the tube model a, L, and N appear in the expressions for several dynamic properties: the diffusion coefficient [eq. (3)], a terminal zone time constant [Eq. (7)], the plateau modulus [eq. (12)], the viscosity [eq. (15)], and the recoverable compliance [eq. (16)]. The three parameters are related to each other and observable quantities by two equations [eqs. (1) and (2)], so only one is independent. We can therefore use one of the dynamic properties G_N^0 to eliminate the last tube model parameter in expressions for all other dynamic properties. The Rouse molecular friction coefficient ζ_r also appears in several expressions. This is the frictional coefficient which controls the viscosity of short-chain polymers ($M < M_c$).[7,8] The Rouse expression for viscosity is

$$\eta_r = \zeta_r\nu R^2/36 \tag{17}$$

so ζ_r can be estimated from observable quantities by

$$\zeta_r = 36M\eta(M_c)/\nu R^2 M_c \tag{18}$$

where $\eta(M_c)$ is the viscosity for chains of molecular weight M_c under the conditions (concentration, temperature, etc.) of the experiment.

The results of these substitutions are the following equations:

$$D = \frac{G_N^0}{135} \left(\frac{cR_GT}{G_N^0} \right)^2 \left(\frac{R^2}{M} \right) \frac{M_c}{M^2\eta_0(M_c)} \tag{19}$$

$$\tau_d = \frac{45}{\pi^2} \left(\frac{G_N^0}{cR_GT} \right)^2 \frac{M^3\eta_0(M_c)}{G_N^0 M_c} \tag{20}$$

$$\eta_0 = \frac{15}{4} \left(\frac{G_N^0}{cR_GT} \right)^2 \frac{M^3\eta_0(M_c)}{M_c} \tag{21}$$

$$J_e^0 = 6/5G_N^0 \tag{22}$$

in which c is the polymer concentration (wt/vol) and R_G is the universal gas constant. The tube diameter can be expressed in terms of observables

$$a^2 = \frac{4}{5} \left(\frac{R^2}{M} \right) \left(\frac{cR_GT}{G_N^0} \right) \tag{23}$$

and of course the other tube parameters L and N can be similarly expressed with eqs. (1) and (2) if desired.

COMPARISONS WITH EXPERIMENT

Diffusion Coefficient

Klein has measured diffusion coefficients of five deuterated polyethylene fractions ($3600 < M < 23,000$) in a commercial polyethylene melt ($\overline{M}_w = 160,000$; $\overline{M}_w/\overline{M}_n \approx 15$) at 176°C, obtaining[9]

$$D = 0.26/M^2 \text{ (cm}^2/\text{sec)} \tag{24}$$

With eq. (19) and data on polyethylene melts at 176°C [$G_N^0 = 2 \times 10^7$ dyn/cm^2; $\rho = 0.767$ g/ml; $R^2 = 1.0 \times 10^{-16}M$ cm^2, $M_c = 3800$, and $\eta_0(M_c) = 0.32$ poise][8,10,11]:

$$D = 0.34/M^2 \text{ (cm}^2/\text{sec)} \tag{25}$$

Thus, the tube model predicts both the correct molecular weight dependence and the correct magnitude (within a factor of 1.3) of the diffusion coefficient.

It is instructive to compare the tube model predictions with those of other theories. An equation for D can be developed based on the idea of a uniformly effective friction coefficient.[7] If the Rouse expressions for η_0 and D are combined and ζ/M is adjusted to fit the observed viscosity behavior, the relation between diffusion coefficient and molecular weight becomes

$$D^* = \frac{cR_GT}{36\eta_0(\overline{M}_w)} \left(\frac{R^2}{M} \right) \frac{\overline{M}_w}{M} \tag{26}$$

where $\overline{M}_w$ and $\eta_0(\overline{M}_w)$ are the weight-average molecular weight and viscosity of the medium. The ratio of D^* to D in eq (19), using $\eta_0(\overline{M}_w) = \eta_0(M_c)(\overline{M}_w/M_c)^{3.5}$, then becomes

$$\frac{D^*}{D} = \frac{4}{15} \left(\frac{cR_GT}{G_N^0} \right) \left(\frac{\overline{M}_w}{M_c} \right)^{2.5} \frac{1}{M} \tag{27}$$

This ratio, calculated for Klein's polyethylene system[9] with a midrange molecular weight $M = 10^4$ for the diffusing species, has a value of approximately 0.0023. Thus the uniformly effective friction model not only predicts the incorrect dependence on molecular weight ($D^* \propto M^{-1}$) but also yields values of D which are smaller than observed by more than two orders of magnitude.

Some early measurements of D for polystyrene in concentrated solutions[12] ($c = 0.3$–0.6 g/ml, $M = 30{,}000$–$200{,}000$) were found to correlate with η_0 through

$$\eta_0 D = c R_G T (R^2/M)/36 \tag{28}$$

which is simply eq. (26) with $M = \overline{M}_w$. From eqs. (19) and (21) the corresponding expression for the tube model is

$$\eta_0 D = G_N^0 R^2/36 \tag{29}$$

The ratio of these expressions is

$$\eta_0 D/(\eta_0 D)_{\text{tube}} = (G_N^0/c R_G T)M \tag{30}$$

The values of this ratio [using $G_N^0 = 2.0 \times 10^6$ dyn/cm^2 for undiluted polystyrene and assuming $G_N^0 \propto c^2$ (ref. 10)] range from 1 to 4 for the solutions studied. Thus, the somewhat scattered experimental results are in this case consistent with both theories.

Relaxation Time

Kraus and Rollman have measured the dynamic moduli associated with free polybutadiene molecules ($1 \times 10^4 < M < 6 \times 10^5$) in the polybutadiene phase of butadiene-styrene block copolymers at 25°C.[13] The volume fraction of free chains was approximately 0.3 in all cases. For polybutadiene (ca. 10% vinyl, 50% *trans*, 40% *cis*) $G_N^0 = 1.25 \times 10^7$ dyn/cm^2, $M_c = 5000$, $\rho = 0.90$ g/ml, and $\eta_0(M_c) = 20$ poise.[10,14] From eq. (20) for the disengagement time, we have

$$\tau_d = 4.3 \times 10^{-16} M^3 \text{ sec} \tag{31}$$

which gives values that are about four times larger than experimental values estimated as $1/\omega_m$, where ω_m is the frequency at the maximum of loss modulus associated with the free chains, and only about 50% larger than those estimated similarly from the frequency at the maximum in loss tangent. The observed molecular weight dependence, $\tau \propto M^{3.1}$, agrees well with eq. (31).

The uniformly effective friction theory cannot provide an alternative comparison here because the viscosity of the medium is infinite. However, it is possible to estimate frictional coefficients for free chains in a network from a theory based on pair-wise interactions[15] in which one component (the network component) is taken to have infinite molecular weight. The value of ω_m for free strands (volume fraction 0.3) is predicted to be less by about a factor of 2 than $(\omega_m)_0$, the value for the undiluted free strands alone. For undiluted polybutadiene of similar microstructure at 25°C (ref 14):

$$(\tau_m)_0 = 1/(\omega_m)_0 = 1.2 \times 10^{-19} M^{3.5} \tag{32}$$

so for free strands in the network

$$\tau_m \approx 2.4 \times 10^{-19} M^{3.5} \tag{33}$$

Although the dependence on molecular weight is slightly stronger than observed the numerical values of ω_m from eq. (33) straddle those obtained experimentally. Thus, both theories predict relaxation times in the observed range.

Viscosity–Molecular-Weight Relation

The quantity $M\eta_0(M_c)/M_c$ is the viscosity predicted by the Rouse model at molecular weights above M_c. Thus, in eq. (21) the quantity $15(G_N^0/cR_GT)^2M^2/4$ is the predicted enhancement factor for viscosity associated with entanglement. Experimentally, $\eta_0 \propto M^{3.4-3.6}$ above M_c,[8] so the observed enhancement factor has a somewhat stronger molecular weight dependence. On the other hand, the magnitude of the predicted enhancement is much greater than that observed over the accessible range of molecular weights. Note that cR_GT/G_N^0 is M_e, a characteristic molecular weight that is roughly $M_c/2$.[8,10] With this, eq (21) can be rewritten as

$$\eta_0 = 15\eta_0(M_c)(M/M_c)^3 \tag{34}$$

$(M \gg M_c)$, whereas the observed behavior can be expressed as

$$\eta_0 = \eta_0(M_c)(M/M_c)^{3.5} \tag{35}$$

Predicted viscosities [eq. (34)] thus remain above observed values [eq. (35)] for $M \lesssim 200M_c$, and $200M_c$ is well beyond the range of available data ($200M_c$ is 6×10^6 for undiluted polystyrene and 1×10^6 for undiluted polybutadiene).

With its total suppression of sustained transverse motions the tube model may in fact provide only an upper bound on disengagement time and therefore on the viscosity. Klein's analysis[5] appears to rule out significant contributions to the configurational rearrangement rate from transverse jumping (tube constraint "evaporation"). However, other transverse motions through the meshwork are certainly possible. For example, a chain could "leak" from its tube anywhere along its length by projecting loops out through the surroundings. Sustained motions of this sort have not yet been considered and may well be important for chains of intermediate length. Thus it seems possible that the stronger experimental dependence of η_0 on M in the observable range merely reflects a slow approach to the limit provided by reptation processes alone.

Recoverable Compliance

Equation (22) predicts that J_e^0, like G_N^0, is independent of molecular weight in highly entangled systems. Such behavior is observed above a characteristic molecular weight M_c',[10] although the product $G_N^0J_e^0$ is 2.5–3.0 in narrow distribution polymers rather than the predicted value of $6/5$ for monodisperse chains. On the other hand, polydispersity can increase J_e^0 substantially, so the disagreement may not be as bad as appears.

Plateau Modulus

Experimentally the plateau modulus is independent of M and approximately proportional to c^2 in concentrated systems.[7,10] The characteristic ratio R^2/M should be practically independent of c in concentrated systems. Thus, from eq.

(23) the tube diameter a apparently varies as $c^{-1/2}$. One would expect $a \propto c^{-1/3}$ if the mesh scales like the mean distance between chain units in the liquid. If tunnel diameter is regarded as a correlation distance then one expects $a \propto c^{-3/5}$ (mean field) or $a \propto c^{-3/4}$ (scaling),[1] but in each of these cases R^2 will presumably also vary with concentration.

In any case the values of tube diameter calculated from eq. (23) are considerably larger than the distance between neighboring chain segments. With the data for undiluted polystyrene ($T = 200°C$, $\rho = 0.96$ g/ml, $G_N^0 = 2.0 \times 10^6$ dyn/cm^2, $R^2 = 0.46 \times 10^{-16} M$ cm^2) the calculated a is 83 Å; for undiluted polyethylene at 176°C, $a = 34$ Å.

EFFECTS OF MOLECULAR WEIGHT DISTRIBUTION

The effects of dispersity on η_0 and J_e^0 can be worked out with equations given by Doi and Edwards.[3] It turns out that η_0 is a function of the product $\overline{M}_w \overline{M}_z \overline{M}_{z+1}$, requiring a much stronger dependence on high-molecular-weight components than the observed dependence on $\overline{M}_w$.[8] For J_e^0 the predicted dispersity factor is $(\overline{M}_{z+4}\overline{M}_{z+3}\overline{M}_{z+2})/(\overline{M}_{z+1}\overline{M}_z\overline{M}_w)$. Although J_e^0 is quite sensitive to the presence of high-molecular-weight components, this prediction is almost certainly too strong. The Rouse polydispersity factor $\overline{M}_z\overline{M}_{z+1}/\overline{M}_w^2$ is roughly correct for blends and for polymers with the most probable distribution.[15–17]

This problem with dispersity effects may again reflect the upper limit character of the tube model. Chains of all sizes are compelled to diffuse only by reptation, and yet the cage lifetimes, a communal property of the system, may be very small compared to the disengagement time of the largest chains in the system. It would perhaps be better to consider diffusion strictly by reptation for chains much smaller than the average, by randomly directed segmental jumps (Rouse-like motions) for chains much larger than the average, and a smooth transition between these extremes for intermediate sizes. Daoud and de Gennes have recently considered diffusion and configurational relaxation in binary mixtures of chains of different size.[18]

This work was supported by the National Science Foundation Grant No. ENG 75-15683A01. Helpful comments by Dr. D. S. Pearson and Professor J. D. Ferry are acknowledged with gratitude.

References

1. M. Doi and S. F. Edwards, *J. Chem. Soc. Faraday Trans. 2*, **74**, 1789 (1978).
2. M. Doi and S. F. Edwards, *J. Chem. Soc. Faraday Trans. 2*, **74**, 1802 (1978).
3. M. Doi and S. F. Edwards, *J. Chem. Soc. Faraday Trans. 2*, **74**, 818 (1978).
4. P. G. de Gennes, *J. Chem. Phys.*, **55**, 572 (1971).
5. J. Klein, *Macromolecules*, **11**, 852 (1978).
6. F. Rouse, *J. Chem. Phys.*, **21**, 1272 (1953).
7. J. D. Ferry, *Viscoelastic Properties of Polymers*, 2nd Ed., Wiley, New York, 1970.
8. G. C. Berry and T. G Fox, *Adv. Polym. Sci.*, **5**, 261 (1967).
9. J. Klein, *Nature*, **271**, 143 (1978); private communication.
10. W. W. Graessley, *Adv. Polym. Sci.*, **16**, 1 (1974).
11. V. R. Raju, G. G. Smith, G. Marin, J. R. Knox, and W. W. Graessley, *J. Polym. Sci. Polym. Phys. Ed.*, **17**, 1183 (1979).
12. F. Bueche, W. M. Cashin, and P. Debye, *J. Chem. Phys.*, **20**, 1956 (1952).
13. G. Kraus and K. W. Rollman, *J. Polym. Sci. Polym. Symp.*, **48**, 87 (1974).

14. W. E. Rochefort, G. G. Smith, H. Rachapudy, V. R. Raju, and W. W. Graessley, *J. Polym. Sci. Polym. Phys. Ed.*, **17,** 1197 (1979).

15. W. W. Graessley, *J. Chem. Phys.*, **54,** 5143 (1971).

16. K. Ninomiya, J. D. Ferry, and Y. Oyanagi, *J. Phys. Chem.*, **67,** 2297 (1963).

17. W. W. Graessley and H. W. Pennline, *J. Polym. Sci. Polym. Phys. Ed.*, **12,** 2347 (1974).

18. M. Daoud and P. G. de Gennes, private communication.

Received April 2, 1979
Accepted May 2, 1979

COMMENTARY

Reflections on "Crazing and Shear Deformation in Crosslinked Polystyrene," by Chris S. Henkee and Edward J. Kramer, *J. Polym. Sci., Polym. Phys. Ed.,* 22, 721 (1984)

EDWARD J. KRAMER

Department of Materials Science and Engineering and the Materials Science Center, Cornell University, Ithaca, NY 14853

I am very pleased that the Editors have selected this paper for the 50th anniversary celebration of the *Journal of Polymer Science*. This work was made possible by a transmission electron microscopy (TEM) method that allowed us to measure the extension ratio λ of craze fibrils. The method was developed simultaneously and independently by Hugh Brown,[1] then a faculty member in Australia, and Bruce Lauterwasser,[2] a graduate student in our group at Cornell. The coincidence was discovered before publication of either paper when Hugh wrote to me, asking if it would be possible to spend part of his sabbatical at Cornell and describing his new TEM method. After spending half of his sabbatical at Cornell, Hugh went on to invent another important TEM technique[3] for the study of craze microstructure, low-angle electron diffraction, in the second half of his sabbatical (with Andrew Keller at Bristol).

At Cornell our initial emphasis was to discover the mechanisms of craze widening (fibril drawing[2]) and craze tip growth (meniscus instability[4,5]), but with the arrival of a new postdoctoral student, Athene Donald, who like Hugh Brown and myself had been trained as a metal physicist rather than as a polymer scientist, the emphasis gradually shifted. Athene discovered that some glassy polymers, e.g., polycarbonate, did not craze in thin films but formed two dimensional necks we called plane stress deformation zones, or DZs, for short.[6] She also observed that the λs of either crazes or DZs for a given glassy polymer were correlated with the theoretical extension ratio λ_{max} of a single strand of the entanglement molecular weight M_e.[7,8] Even today the idea that a linear viscoelastic measurement of the rubbery plateau modulus in the melt could elucidate the large strain plastic deformation behavior of the glass seems nothing short of miraculous; but there were certainly many hints in previous work that entanglements were important for both deformation and fracture.

Athene also realized that there was a strong correlation between M_e and which polymers crazed and which formed DZs in response to tensile stress. Those with large M_e formed crazes while those with small M_e formed DZs.[9] It was thus natural to propose that an entanglement network, inherited from the melt, existed in the polymer glass and that if crazes were formed, strands in this network would have to be broken to form the fibril surfaces. With typical fibril dimensions of $\sim$ 10 nm, the number of such strands that must break to form the fibrils in such a highly entangled polymer as polycarbonate is very large and the work done in such chain scission could increase the crazing stress above that for DZ formation.[10]

Naturally this proposal initially met with some skepticism and it is certainly true that other factors, such as cohesive energy density, can also play a role in inhibiting craze fibril formation.[11] As a crucial test Chris Henkee began experiments to look at the effects of a true crosslinked network in the glass. He chose polystyrene, which normally deforms by crazing, and crosslinked the thin films with the electron beam of an electron microprobe. Chris's success shows the im-

675

portance of persistence; earlier, Nigel Farrar, a post-doctoral student in the group, tried a similar experiment using gamma radiation to crosslink the films. Alas, due to the high temperature in our ^{60}Co gamma cell, chain scission rather than crosslinking resulted, leading to very brittle behavior of the film.

Finally it is important to say something about the support provided by the National Science Foundation through the Materials Science Center (MSC) at Cornell. Not only did the MSC provide funding for Chris and for the central facilities for electron microscopy that play such a prominent part in this paper, but in addition its director at that time, the late Herb Johnson, recognized the importance of polymers early on and strongly supported the effort to reemphasize the study of polymeric materials at Cornell. At a time when research budgets on the national scene are becoming tightly constrained, it is worth emphasizing that the NSF Materials Research Laboratories and their present equivalents, the Materials Research Science and Engineering Centers, have been crucial to the success of much research on polymers.

REFERENCES AND NOTES

1. H. R. Brown, *J. Materials Sci.,* **14,** 237 (1979).
2. B. D. Lauterwasser and E. J. Kramer, *Philos. Mag.,* **39A,** 469 (1979).
3. H. R. Brown, *J. Polym. Sci., Polym. Phys. Ed.,* **21,** 483 (1983).
4. A. S. Argon and M. M. Salama, *Materials Science and Engineering,* **23,** 219 (1977).
5. A. M. Donald and E. J. Kramer, *Philos. Mag.,* **43A,** 857 (1981).
6. A. M. Donald and E. J. Kramer, *J. Materials Science,* **16,** 2967 (1981).
7. A. M. Donald and E. J. Kramer, *J. Polym. Sci., Polym. Phys. Ed.,* **20,** 899 (1982).
8. A. M. Donald and E. J. Kramer, *Polymer,* **23,** 1183 (1982).
9. A. M. Donald and E. J. Kramer, *J. Materials Sci.,* **17,** 1739 (1982).
10. E. J. Kramer, in Crazing in Polymers, H. H. Kausch, Ed., *Adv. in Polym. Sci.,* **52/53,** 1 (1983).
11. R. P. Kambour, *Polym. Comm.,* **24,** 292 (1983).

PERSPECTIVE

Comments on "Crazing and Shear Deformation in Crosslinked Polystyrene," by Chris S. Henkee and Edward J. Kramer, *J. Polym. Sci., Polym. Phys. Ed.,* 22, 721 (1984)

HUGH R. BROWN

BHP Steel Institute, Wollongong University, Northfields Ave. Wollongong, NSW2522, Australia

In this paper Chris Henkee and Ed Kramer produced convincing evidence that the mode of deformation (craze or yield zones) of thin films of glassy polymers is controlled by the entanglement or crosslink network within the material. The idea that the entanglement network controlled both the craze fibril extensions ratios and the balance between craze and yield deformation had been developed by Athene Donald and Ed Kramer a couple of years earlier.[1,2] However, as Donald and Kramer had used a range of different polymers to obtain the different entanglement densities, it had been possible to argue that the effects they observed could have been caused by changes in local mobility between the polymers rather than changes in the entanglement density. By sticking to just one polymer, polystyrene, and changing the crosslink density, Henkee and Kramer clearly showed that it is indeed the strand network density that controlled the deformation mode. They also demonstrated the equivalence of a crosslink to an entanglement when considering glassy deformation.

The idea that the entanglement network may have a profound influence on deformation and fracture of glassy polymers had been around for a number of years at the time of this work.[3,4] For example, Ian Ward and coworkers in the 1960s studied the deformation of a number of polymers including poly(ethylene terephthalate) and related the natural draw ratio to the extensibility of the entanglement network. They also ascribed the al-most complete recovery of glassy polymers that are stretched at a temperature below T_g and then heated above T_g, to the existence of the entanglement network. Another hint that the entanglement network may be important had come from studies of the effects of molecular weight on fracture of glassy polymers. It was known that the strength and toughness of a glassy polymer tended to increase with molecular weight and then saturate at a molecular weight above 2–4 times M_e. However, it was Kramer with his coworkers who realized that the entanglement network would control crazing and, perhaps more importantly, devised an elegant series of experiments to provide proof.

The entanglement network does more than just control the craze or deformation zone extension. Henkee and Kramer pointed out that, in a crosslinked polymer, the process of crazing has to involve considerable chain scission to form the new fibril surfaces. This scission both increases the effective fibril surface energy and decreases the network density in the fibrils, thereby increasing their extension ratio over that of deformation zones. The agreement between the results for crosslinked materials with previous work on uncrosslinked polymers is excellent evidence that this scission process must be important in uncrosslinked materials. I for one was not entirely convinced at the time by this evidence of the importance of chain scission in crazing. However, perhaps following Henkee and Kramer's suggestion that "disentanglement may play a more significant role at

677

temperatures approaching T_g," some years later Chris Plummer and Athene Donald studied the effects of temperature on crazing. They demonstrated the existence of scission crazes at low temperatures and disentanglement crazes close to T_g.[5] The disentanglement crazing was suppressed by crosslinking.[6] It is still true that we do not really know just how much scission is involved in crazing, as the geometry of craze fibrils is complicated by the existence of mechanically important crossties that may contain many entangled strands.

REFERENCES AND NOTES

1. A. M. Donald and E. J. Kramer, *J. Polym. Sci., Polym. Phys. Ed.,* **20,** 899 (1982).
2. A. M. Donald and E. J. Kramer, *Polymer,* **23,** 1183 (1982).
3. W. Whitney and R. D. Andrews, *J. Polym. Sci. C,* **16,** 2981 (1967).
4. S. W. Allison, P. R. Pinnock, and I. M. Ward, *Polymer,* **7,** 66 (1966).
5. C. J. G. Plummer and A. M. Donald, *J. Polym. Sci., Polym. Phys. Ed.,* **27,** 325 (1989).
6. C. J. G. Plummer and A. M. Donald, *J. Mater. Sci.,* **26,** 1165 (1991).

Crazing and Shear Deformation in Crosslinked Polystyrene

CHRIS S. HENKEE and EDWARD J. KRAMER, *Department of Materials Science and Engineering, and the Materials Science Center, Cornell University, Ithaca, New York 14853*

Synopsis

Thin films of polystyrene (PS) are bonded to copper grids and crosslinked with electron irradiation. When the films are strained in tension regions of local plastic deformation, either crazes or plane stress deformation zones (DZs), nucleate and grow from dust particles. The nature of the local deformation, as well as the local extension ratio λ, is determined by transmission electron microscopy. The behavior of the PS glass is consistent with its being a network of molecular strands of total density $v = v_E + v_X$, where v_E is the entangled strand density inferred from melt elasticity measurements of uncrosslinked PS and v_X is the density of crosslinked strands determined from the ratio of the applied electron dose to the electron dose for gelation. When v is less than 4×10^{25} m^{-3} ($<1.3v_E$), only crazes are observed whose microstructure is similar to those in uncrosslinked PS. As v increases from 4×10^{25} to 8×10^{25} m^{-3} (from $1.3 v_E$ to $2.5v_E$) shear deformation begins to compete with crazing. As v increases above 8×10^{25} m^{-3}, only shear DZs are observed, the strain in which becomes progressively more diffuse as v increases. The λ in the crazes and DZs correlate well with λ_{max}, the maximum extension ratio of a strand in a network of density v computed using the Porod–Kratky model. For crazes $\ln(\lambda) \simeq 0.9 \ln(\lambda_{max})$ and for DZs $\ln(\lambda) \simeq 0.55 \ln(\lambda_{max})$. The strain at which crack nucleation is first observed increases as v increases from $<5\%$ in uncrosslinked PS with $v = 3.3 \times 10^{25}$ m^{-3} to $>20\%$ in PS with $v = 33 \times 10^{25}$ m^{-3} ($v = 10v_E$); crosslinking to still higher crosslink densities, e.g., $v = 14v_E$, results in cracks which propagate in a catastrophic manner at low applied strains. An optimum v thus exists, one not too high to suppress local shear ductility but high enough to suppress crazes which can act as crack nucleation sites. These results are compared with previous results on a variety of linear homopolymers, copolymers, and polymer blends that are characterized by a wide range of v ($v = v_E$). The transitions from crazing to crazing plus shear and from crazing plus shear to shear only take place at almost identical values of v. In addition the correlation between λ in the crazes and DZs and λ_{max} for a single network strand is the same for both classes of polymers. This agreement implies that chain scission is the major mechanism by which strands in the entanglement network are removed in forming fibril surfaces. Craze suppression, by either increasing v in the crosslinked polymer or v_E in the uncrosslinked ones, is due to the extra energy required to break more main-chain bonds to form these surfaces.

INTRODUCTION

While polymer glasses are specified increasingly for load bearing applications, they still suffer from a tendency to fail in a macroscopically brittle manner. Ironically, although these polymers crack with little global plastic deformation, large-strain plastic deformation on a small scale leads to microscopic crazes which can serve as sites for crack nucleation.

Crazing involves the production of fine polymers fibrils, with diameters 5–30 nm, which are drawn across from one craze interface to the other.[1-8] These fibrils give the craze a load bearing character. At long times, or at high loads, these fibrils can gradually break down to form large voids which grow slowly to become

Journal of Polymer Science: Polymer Physics Edition, Vol. 22, 721–737 (1984)

subcritical cracks. When one of these cracks reaches critical size, rapid fracture follows.

Some polymer glasses may also deform by shear deformation, without forming the fibril/void structure of the craze.[9–17] This mode may be observed during tensile deformation of thin sheets or films of these polymers as plane stress shear deformation zones, or DZs.[18–22] These zones consist of polymer drawn to a uniform extension ratio λ_{DZ} at "shoulders" which bound the zone, much as "shoulders" delineate a neck drawn in a textile fiber. No voiding and no premature crack formation are observed in these zones.

Recent experimental observations[23–26] of crazes and DZs in thin films of various linear homopolymers, copolymers, and polymer blends suggest that the entanglement network of the polymer plays an important role in determining both the micromechanical properties of crazes and DZs and the competition between them. To analyze these effects an oversimplified model of this network was constructed, one which has a density of network chains v_E between localized points of entanglement. We are well aware that such a model cannot fully describe the viscoelastic properties of the melt (the tube models, for example, seem superior in that case[27–30]). Nevertheless, to describe the deformation properties of the glassy state, in which reptation times must be long, the network model may still be appropriate. The entanglement density is given by

$$v_E = \rho N_A / M_e \tag{1}$$

where ρ is the density of the polymer, N_A is Avogadro's number, and M_e is the entanglement molecular weight as determined from measurements of the shear modulus of the melt in the rubbery plateau region just above the glass transition temperature T_g.

The contour length l_e of each chain is

$$l_e = (l_0/M_0)M_e = \rho N_A l_0 / M_0 v_E \tag{2}$$

where l_0 is the average projected length of stiff units along the chain and M_0 is the average molecular weight of these units. Each chain may also be described by an end-to-end vector $\mathbf{d}$. The magnitude of d, or the entanglement mesh size, is the root-mean-square end-to-end distance of a chain of molecular weight M_e. From the Porod–Kratky model[31,32] of a wormlike chain, d is found to be

$$d = (2al_e\{1 - (a/l_e)[1 - \exp(-l_e/a)]\})^{1/2} \tag{3}$$

where a is the persistence length of the chain. Since for long chains d is given asymptotically by

$$d = k(M_e)^{1/2} = k(\rho N_A/v_E)^{1/2} \tag{4}$$

where k may be determined directly from measurements of the molecular coil size in the melt or in a theta solvent, the persistence length is

$$a = k^2 M_0 / 2l_0 \tag{5}$$

The maximum extension ratio of a single chain stretched along d is simply

$$\lambda_{\max} = l_e/d \tag{6}$$

For thin films of a series of homopolymers, copolymers, and polymer blends in which v_E varied between 1×10^{25} chains/m³ and 30×10^{25} chains/m³, the crazing and shear deformation behavior was found to vary systematically with

v_E. For polymers with $v_E < 4 \times 10^{25}$ chains/m^3 only crazing was observed and the craze extension ratio λ_{craze} decreased with $\lambda_{\max}$ as v_E increased [$\lambda_{\text{craze}} \simeq (1.0\text{--}0.8)\lambda_{\max}$]. For polymers with $4 \times 10^{25} < v_E < 8 \times 10^{25}$ chains/m^3 both crazes and DZs were normally observed in competition, although some polymers in this range [e.g., poly(methyl methacrylate) (PMMA)] only craze. Finally, for polymers with $v_E > 8 \times 10^{25}$ chains/m^3 only shear deformation zones are observed, although some of these polymers can be caused to craze by physical aging (annealing the film at temperatures just below T_g). The extension ratios in these DZs also correlate with $\lambda_{\max}$ but are always lower then those of crazes in the same polymer ($\lambda_{\text{DZ}} \simeq 0.6\lambda_{\max}$).

The transition from crazing (which involves fine fibril formation) to shear deformation (which does not) may be rationalized[33] as due to an extra energy required to break chains during formation of fibril surfaces. If all chains in the entanglement network which originally cross the surface of a fibril are broken, the required surface work Γ is given approximately by

$$\Gamma = \lambda + \tfrac{1}{4}dv_E U_b \tag{7}$$

where γ is the van der Waals surface energy (of intermolecular separation) and U_b is the energy to break a main-chain bond. Since $dv_E \propto v_E^{1/2}$, Γ increases as v_E increases, thus making fibril formation and crazing unfavorable relative to shear deformation. The fact that chains in the entanglement network must be either broken or disentangled in forming the fibril surfaces also implies that the entanglement density v_E' left in the fibrils is less than the v_E in the DZs, providing a reason why $\lambda_{\text{craze}} > \lambda_{\text{DZ}}$.

If these ideas are correct it should be possible to start with a polymer with a low v_E which only crazes, e.g., polystyrene (PS), and convert it to a polymer which is craze resistant simply by crosslinking. Although some previous experiments indicate that such a transformation may be possible,[34,35] there has never been a quantitative study of the effects of crosslinking on crazing. If v_X represents the density of crosslinked chains, the total chain or strand (a strand is defined as a chain segment bounded by crosslinks or entanglement points) density after crosslinking would be

$$v = v_E + v_X \tag{8}$$

The strand contour length l_e of the new network can be obtained by substituting v for v_E in eq. (2); d can then be found by substituting this value for l_e into eq. (3). If the model holds, λ_{craze} and λ_{DZ} for the crosslinked polymer should have the same correlation with $\lambda_{\max}$ as observed for the uncrosslinked polymer series. Crazing should be observed in the crosslinked polymer (well beyond the gel dose) as long as $v < 4 \times 10^{25}$ chains/m^3. Competition between crazing and shear should be observed for $4 \times 10^{25} < v < 8 \times 10^{25}$ chains/m^3 followed by a transition to shear deformation only above 8×10^{25} chains/m^3. We demonstrate below that all these features may be observed in a series of PS thin films, crosslinked by electron irradiation.

EXPERIMENTAL

Monodisperse PS with a weight-average molecular weight of $M_w = 390{,}000$ and $M_w/M_n < 1.10$ was dissolved in toluene. Thin films of PS ($\simeq 0.9\ \mu$m thick) were produced by drawing a glass slide from the solution at a constant rate. After drying, a film could be floated off the glass slide onto the surface of a water bath. It could then be picked up on an annealed copper grid, the bars of which had previously been coated with a film of the same polymer solution. Bonding of the film to the grid was achieved by a short exposure to toluene vapor.

After drying, the specimens were exposed to electron irradiation using a JEOL 733 electron microprobe. The specimen (which was in contact with a liquid-nitrogen cold finger) was exposed using a beam current of 1×10^{-7} A and an accelerating voltage of 40 kV. The electron beam was defocused to a diameter of 0.4 mm and rastered across the sample to expose an area of 8.1 mm^2. The scanned area was measured directly, as the video output reveals only a portion of the actual area scanned. Each scan had 1000 lines and each line required 0.5 ms to complete. Thus the entire 8.1 mm^2 was irradiated in 0.5 s.

Crosslink densities were determined by measuring the gel point of the PS under identical microprobe irradiation conditions. PS thin films were irradiated for various lengths of time. After being aged for 48 h, the irradiated samples were immersed in toluene for several hours to remove the sol fraction. In order to produce samples of uniform thickness with which to measure the gel fraction, the samples could not simply be removed from the toluene bath, but had to be precipitated by adding first acetone, and then methanol. After drying, the films of gel were mounted on glass slides and the gel thickness was measured using an interference microscope.

From the thickness measurements a Charlesby–Pinner plot[36] was constructed and extrapolated to find the gel point of 65 ± 10 s. To ensure linearity of the crosslinking rate up to long exposure times, monodisperse samples of molecular weight 37,000 and 2,000 were also irradiated. It was observed that gelling of these samples did take place within the time calculated by using the crosslinking rate determined from the 390,000-molecular-weight samples. The largest source of variation in the crosslinking rate was the current instability of the electron beam, which had to be monitored constantly.

From the gel time of 65 s for a molecular weight of 390,000, one can calculate the G value for crosslinking, $G(x)$. [$G(x)$ is defined as the number of crosslinks formed per 100 eV absorbed in the material.] Depending on the exact value for the rate of energy transfer from the electron beam to the polymer film,[37] one obtains a value of $G(x) \simeq 0.019$. Experimental values for $G(x)$ using γ irradiation fall in the range[38] of 0.027–0.051. Thus the value obtained in this experiment seems reasonable in view of the uncertainty in the local temperature rise of the polymer under the electron beam, which would decrease[39,40] the effective value of $G(x)$ observed. From the crosslinking rate determined from these measurements, the density of crosslinked strands v_X can be found for any irradiation time.

After exposure, the grids were set aside for 48 h, after which they were mounted in a strain frame and strained in tension until observation with an optical microscope indicated crazes or DZs had nucleated. The copper grid deforms plastically in the procedure, so that the strain in the polymer film is maintained even when the grid is removed from the strain frame. Grid squares of interest were then cut from the copper grid for examination by transmission electron microscopy using a JEOL 200CX microscope operating at 200 kV.

To characterize the local extension ratio of a craze/DZ, the method developed by Lauterwasser and Kramer[41] and by Brown[42] may be used. For a craze, v_f, the volume fraction of fibrils in the craze, is found from densitometry of the electron image plate to give values of the optical densities of the craze (ϕ_{craze}), the film (ϕ_{film}), and a hole through the film (ϕ_{hole}). The value of v_f is then given by

$$v_f = 1 - \frac{\ln(\phi_{craze}/\phi_{film})}{\ln(\phi_{hole}/\phi_{film})} \qquad (9)$$

For the case of unfibrillated regions, v_f is simply the ratio of the thickness within the zone to the undeformed film thickness; the expression for v_f is identical to eq. (9) but with ϕ_{craze} replaced by ϕ_{DZ}, the optical density of the DZ image. Since both crazing and DZ formation are plastic deformation processes which occur at a constant polymer volume, the extension ratio λ of the craze or DZ is related to v_f by

$$\lambda = 1/v_f \qquad (10)$$

RESULTS

As the crosslink density is increased, optical microscopy reveals a change in deformation mode from crazing to shear, a change confirmed by more detailed TEM observations. Figure 1 shows optical micrographs of films with various crosslink densities that have been strained in tension. Figure 1(a) shows the craze morphology in an unirradiated PS film strained to 1.2%. When the total strand density $v = 1.3v_E$ the film still crazes as seen in Figure 1(b) but the crazes are shorter and the strain to achieve a given craze density higher than in the unirradiated film. Unirradiated specimens have about ten times the craze density shown in Figure 1(b) at a comparable strain level. One also observes the formation of several very short crazelike structures. Figures 1(c), 1(d), and 1(e) show that as the strand density increases ($v = 1.5, 2.0,$ and 3.0 times v_E, respectively), the formation of long crazes which span the grid bars is suppressed, while numerous short regions of local deformation are observed to form. It is impossible to say whether these are crazes or shear DZs from the optical evidence alone, but TEM reveals they have at least some shear character in addition to fibrillation. The tensile strain level required to observe significant densities of these regions rises from 2.6% in Figure 1(c) to 4.4% in 1(d) to 5.3% in 1(e).

The strain required to observe localized regions of deformation continues to increase as v increases. At $v = 10v_E$ Figure 1(f) shows that little, if any, localized deformation has occurred even at a strain level of 23%. A short crack has initiated at a dust particle but has not propagated. At the crack tip there is a very diffuse region of local shear deformation thinning the film as evidenced by the gradual change in interference color (darkness of the film image) in Figure 1(g). At the highest v level, $v = 14v_E$, even diffuse shear deformation is suppressed and cracks propagate (in a truly brittle fashion). Figure 1(h) reveals the brittle "catastrophic" failure that occurred in such a film which had been strained to 11.5%.

The microstructure within the regions of local plastic deformation may be revealed using TEM. In unirradiated PS these regions are crazes with the typical craze fibril structure shown in Figure 2(a). As v is increased to $1.5v_E$, the spec-

726 HENKEE AND KRAMER

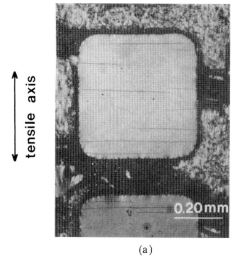

tensile axis

0.20 mm

(a)

(b)

imens still craze but the craze fibrils, Figure 2(b), do not appear to be as straight as in unirradiated PS. This increase disorientation of the craze fibrils implies the existence of more connecting or "cross-tie" fibrils running between the main fibrils as the strand density increases. Along with this change in craze fibril structure, shear deformation begins to compete with crazing at this level of crosslink density. About one-half the crazes did not taper to a sharp tip, as is typical in unirradiated PS, but were blunted by shear deformation as illustrated in Figure 2(c). Sometimes one surface of the film continued to craze but the other began to deform by shear, giving rise to the microstructures seen in Figure 2(d). These hybrid crazes are typical of those previously observed in thin films

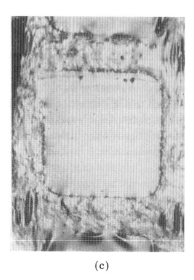

(c)

(d)

Fig. 1. Craze or deformation zone morphology in PS films for various total strand densities at a tensile strain ϵ: (a)$v = v_E$, uncrosslinked, $\epsilon = 1.2\%$; (b)$v = 1.3v_E$, $\epsilon = 2.0\%$; (c)$v = 1.5v_E$, $\epsilon = 2.6\%$; (d)$v = 2.0v_E$, $\epsilon = 4.4\%$; (e)$v = 3.0v_E$, $\epsilon = 5.3\%$; (f)$v = 10v_E$, $\epsilon = 23\%$; (g) higher-magnification view of crack in film (f); (h)$v = 14v_E$, $\epsilon = 11.5\%$.

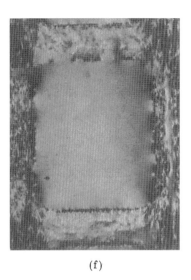

| (e) | (f) |

of linear polymers, e.g., poly(styrene-co-26% acrylonitrile),[43] which have intermediate values of v.

As v is increased still further, shear deformation begins to be more important than fibrillation in these hybrid structures, as can be seen in Figures 3(a), 3(b), and 3(c). These micrographs are representative of the local deformation in PS films where $v = 1.7$, 2.0, and $2.4v_E$, respectively. These regions nucleate at dust particles which have debonded from the PS under the applied tension. The first region to form is fibrillated, but it rapidly dissolves into a DZ at the craze tip. The region of fibrillation within the DZ becomes shorter and shorter as v is increased. Further deformation as the strain is increased occurs by widening and lengthening of this DZ. Under large strain conditions the fibrillated region may break down to form a crack.

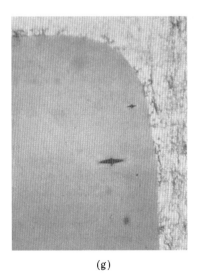

| (g) | (h) |

Fig. 1 *(Continued)*

728   HENKEE AND KRAMER

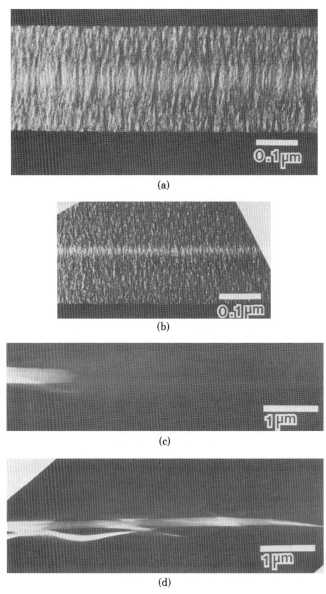

Fig. 2. Transmission electron micrographs showing craze microstructure in PS films for various total strand densities v: (a)$v = v_E$, uncrosslinked; (b)$v = 1.5v_E$; (c)$v = 1.5v_E$, craze tip blunted by shear deformation; (d)$v = 1.5v_E$, craze and shear competition.

As v is increased above $3v_E$ only pure DZs are observed, with no evidence of fibrillation around the dust particle nucleus. At $v = 4v_E$ the DZs are well defined and localized as shown in Figure 4(a). The extension ratio within the zone is approximately constant over the zone and there is an abrupt "shoulder" at the edge of the zone where the extension ratio decreases to its value in the elastically strained film. As v is increased further, the DZs become more diffuse, as shown in Figure 4(b) for a film with $v = 6v_E$. Now the thickness of the film, and thus the extension ratio of the zone, varies smoothly and gradually from the debonded

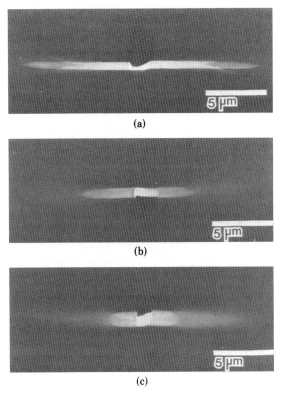

Fig. 3. Transmission electron micrographs of hybrid craze/deformation zone structures in crosslinked PS films at various values of total strand density v: (a)$v = 1.7v_E$, (b) $2.0v_E$, (c) $2.4v_E$.

dust particle to the surrounding film. There is no abrupt shoulder at which drawing takes place.

The mass thickness contrast of the TEM micrographs was analyzed as described in eq. (9) to yield local values of λ_{craze} and λ_{DZ}. For all crazes the local extension ratio was found to be constant along the craze length, except at the craze tip and in the craze midrib, where λ_{craze} increases. These results coincide with those on linear polymers and indicate that the crazes thicken by drawing more polymer into the craze fibrils from the craze–bulk polymer interfaces. The well-defined DZs also have a constant extension ratio and thicken by drawing. No such statement holds for the diffuse DZs. Nevertheless, it was observed that λ_{DZ} measured in the thinnest region of the film, i.e., near the debonded dust particle, was constant for a particular v. It seems reasonable to assume that this value corresponds to fully strain-hardened polymer and it is this value that is reported here. The measured values of λ_{craze} and λ_{DZ} are plotted as a function of v in Figure 5.

DISCUSSION

It seems clear from Figure 5 that λ_{craze} and λ_{DZ} decrease as v increases, in much the same way as they do when linear polymers with increasing v_E are examined.

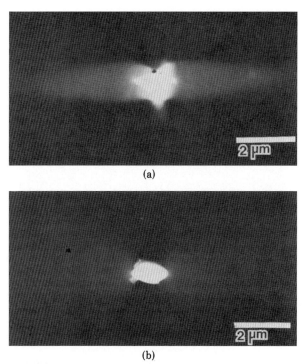

(a)

(b)

Fig. 4.   Transmission electron micrographs of deformation zones in crosslinked PS films at various values of total strand density v: (a)$v = 4v_E$, (b) $6v_E$.   The void in the center of each zone is where a dust particle debonded from the PS; such debonds are frequent nuclei of crazes and deformation zones.

If the network model is correct, one expects λ_{craze} and λ_{DZ} to be correlated with λ_{max}, the maximum extension ratio of an individual strand in the network.   The λ at which rapid strain hardening of the highly oriented polymer in either the

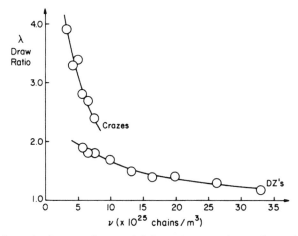

Fig. 5.   Extension ratios in crazes λ_{craze} and deformation zones λ_{DZ} as determined by the mass thickness contrast technique as a function of the total strand density v in uncrosslinked and crosslinked PS films.

CRAZING AND SHEAR DEFORMATION 731

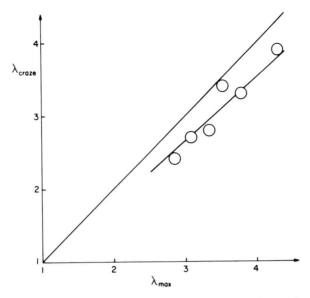

Fig. 6. Extension ratios in crazes in uncrosslinked and crosslinked PS films plotted against λ_{max}, the maximum extension ratio of a single strand in the network.

fibrils or the DZs sets in should scale with λ_{max}. This expectation is borne out by viewing Figures 6 and 7, which are plots of λ_{craze} vs. λ_{max} and λ_{DZ} vs. λ_{max}, respectively, where λ_{max} has been computed using the Porod–Kratky model [eq. (2)]. Use of the Porod–Kratky model for λ_{max} is important for the highly crosslinked PS since l_e in these networks approaches the Kuhn step length at

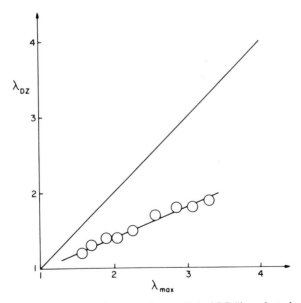

Fig. 7. Extension ratios in deformation zones in crosslinked PS films plotted against λ_{max}, the maximum extension ratio of a single strand in the network.

high v; using eq. (4) would give rise to $\lambda_{\max}$ falling significantly below the correct values.

Comparison of these data with those on linear homopolymers, copolymers, and polymer blends is more easily accomplished using a plot of the ratio ϵ_R of $\ln\lambda$ to $\ln\lambda_{\max}$ vs. v shown in Figure 8. In this ratio λ is either λ_{craze} or λ_{DZ}. The ordinate ϵ_R is simply the ratio of the true strain in a craze or deformation zone to the maximum true strain obtained by stretching a single chain in the network. The filled symbols, triangles and circles, represent the data on crazes and DZs, respectively, in crosslinked PS. The open squares represent data on crazes and DZs in linear homopolymers and copolymers. The open triangles and open hexagons represent data on polystyrene–poly(phenylene oxide) blends. All the data are for polymers that have not been physically aged by heating to temperatures just below T_g.

Note that the true strain ratio ϵ_R for crazes in the crosslinked PS overlap almost exactly the values for crazes in the uncrosslinked polymers. The same thing is true for the values of ϵ_R for the deformation zones. The overlap observed strongly supports the use of the entanglement network model in estimating the large-strain behavior of the uncrosslinked polymer glasses.

The extension ratio in the crazes of both the crosslinked and uncrosslinked polymers lies much closer to $\lambda_{\max}$ ($\epsilon_R \simeq 1.0$–0.8) than that in the deformation zones, in which $\epsilon_R \simeq 0.6$–0.4. It seems logical that the λ_{DZ} represents the natural draw ratio of the entanglement or crosslinked network in which strands are neither broken nor disentangled but that λ_{craze} represents the natural draw ratio of the entanglement or crosslinked network in which some strands must be either broken or disentangled to form fibril surfaces. It has recently been shown[44] that only a fraction q of the strands in a random network can survive the fibrillation that accompanies crazing. The fraction q is a function only of the ratio d/D_0. The "phantom" fibril diameter D_0 is the diameter of the cylinder of unoriented polymer that is subsequently drawn into the fibril; i.e.,

$$D_0 = D\lambda_{\mathrm{craze}}^{1/2} \tag{11}$$

where D is the final fibril diameter. The fraction q can be computed numerically for various d/D_0, but a reasonable empirical expression for $d/D_0 < 1$ which agrees within ± 0.02 with the correct result is given by[33]

$$q = 1 - (d/D_0) + B\,(d/D_0)^m \tag{12}$$

where $B = 0.155$ and $m = 3.23$. An earlier upper-bound formula for q has been derived[44] in which $B = 0.333$ and $m = 2$, but this seriously overestimates q near $d/D_0 = 1$. Since

$$q = v'/v \simeq (\lambda_{\mathrm{DZ}}/\lambda_{\mathrm{craze}})^2 \tag{13}$$

for large l_e, where v' is the strand density of the network in the fibrils, measurements of the ratio of λ_{DZ} to λ_{craze} can give an estimate of q. Unfortunately, uncrosslinked PS does not form DZs in thick films whereas uncrosslinked PS is the only polymer in which the fibril diameter D of room-temperature crazes has been measured. The value of D is necessary to compute D_0 for the theoretical estimate of q. Nevertheless the ϵ_R for a DZ in uncrosslinked PS can be deter-

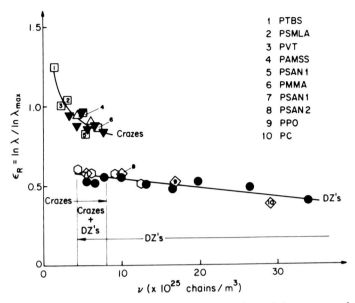

Fig. 8. True strain ratio $\epsilon_R = \ln\lambda/\ln\lambda_{max}$, where λ is the extension ratio in a craze or a deformation zone, plotted against total polymer strand density v: ($\blacktriangledown$) crazes in crosslinked PS, $v = v_E + v_X$; ($\bullet$) DZs in crosslinked PS, $v = v_E + v_X$; ($\square$) crazes in various linear homopolymers and copolymers from refs. 24 and 44, $v = v_E$; ($\diamond$) DZs in various linear homopolymers and copolymers from ref. 24, $v = v_E$; ($\triangle$) crazes in PS-PPO blends from ref. 25, $v = v_E$; ($\bigcirc$) DZs in various PS-PPO blends from ref. 26, $v = v_E$.

mined by extrapolation on Figure 8, which yields $\epsilon_R = 0.60 \pm 0.05$. Since λ_{max} of PS is known to be 4.3, $\lambda_{DZ} \simeq 2.3 \pm 0.1$. This value agrees closely with the measured λ in the "perforated-sheet" crazes observed in very thin ($<$150 nm thick) films of PS.[45] These crazes contain very coarse fibrils. From eq. (13), the experimental estimate of q is 0.35 ± 0.08, in reasonable agreement with the theoretical value of $q = 0.275$.

Not only are the ϵ_R of crazes and DZs the same for the crosslinked PS as for the linear polymers of the same v, but also the range of v over which first crazing and then shear deformation predominate is identical for the crosslinked and uncrosslinked polymers. For $v < 4 \times 10^{25}$ strands/m^3 crosslinked PS only crazes, whereas for $v > 8 \times 10^{25}$ strands/m^3 DZs are observed predominantly. For v between 4×10^{25} and 8×10^{25} strands/m^3 both crazes and shear deformation are observed. As shown in Figure 8 the same ranges of v for crazing and shear deformation are found for the series of linear homopolymers, copolymers, and polymer blends. Not only are limits the same but the morphology of the competing shear deformation and crazing regions in the overlap regime are very similar in the two classes of polymer. For example, the microstructures of shear-blunted craze tips in films of PSAN1 have been published previously[43] which are virtually identical to the microstructures shown in Figures 2(c) and 2(d) for crosslinked PS with $v = 5 \times 10^{25}$ strands/m^3; the entanglement density of PSAN1 is 5.6×10^{25} strands/m^3.

According to the hypothesis outlined above the increasing competition of shear deformation with crazing as v is increased is due to an increase in the stress for

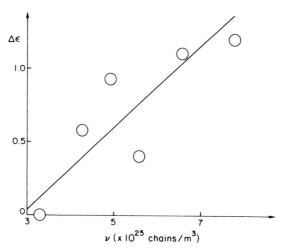

Fig. 9. Increase in strain for first craze nucleation versus total strand density in crosslinked PS films.

crazing relative to the stress for shear deformation. This increase is due to the extra surface work required to create the fibril surfaces if most strands must be broken. That the stress for crazing increases can be demonstrated by measuring the increase in strain $\Delta\epsilon$ required to observe the first craze in a crosslinked film spanning one grid square over that required to see the first craze in a neighboring but unirradiated grid square. This $\Delta\epsilon$ is plotted versus v, the total strand density, in Figure 9. There is no doubt that the critical strain (and stress) for crazing increases with total strand density.

In creating craze fibril surfaces in crosslinked PS strands in the molecular network must be broken since disentanglement is not possible. The correspondence between results for the crosslinked PS and the uncrosslinked polymers implies that at room temperature (which is at least 70 degrees below T_g for all these polymers) chain scission is the most important mechanism of strand loss in the entanglement network when craze fibril surfaces are formed. If disentanglement were the main mechanism one would observe a strong dependence of craze kinetics on polymer molecular weight which is not found in practice. While disentanglement may play a more significant role at temperatures approaching T_g, the identical effect of increasing chemical crosslink density or increasing entanglement density on craze suppression means that chain disentanglement cannot play the most significant role in craze fibril creation at low temperatures.

Finally, the remarkable increase in crack nucleation strain of the PS thin films by irradiation crosslinking deserves comment. Whereas uncrosslinked PS will craze at strains above 0.5% and the crazes so produced will begin to break down to form cracks at strains greater than 3%, the irradiated PS film with a total strand density $v = 10v_E$ can be pulled in tension to a strain greater than 20% without crazing or crack propagation. Figure 10 shows a different comparison of two PS films overlying two separate grid squares in the same grid sheet. One of these film squares was electron crosslinked until the total strand density was about 2.5 v_E; the other film square was uncrosslinked. The grid was then

(a)   (b)

Fig. 10. Optical micrographs of two PS film squares in the same grid pulled in tension to a strain of 5%: (a) uncrosslinked PS, $v = v_E$; (b) crosslinked PS, $v = 2.5v_E$.

stretched to a strain of 5%. As can be seen in Figure 10, the unirradiated film square has crazed and the crazes have been converted into cracks by fibril breakdown. On the other hand the crosslinked film square has neither crazed nor cracked at this strain.

Thus craze suppression by crosslinking represents a potentially important method for improving fracture properties. Two important caveats should be noted, however. First, too high a crosslink density leads to a very low "natural" extension ratio after shear deformation and thus a loss of local ductility. For example, using Figure 9 one can estimate ϵ_R for $v = 14v_E$ to be about 0.4 and thus from the $\lambda_{max} = 1.4$ computed for this strand density, the maximum strain in a deformation zone should be only 0.14. Even though such deformation may be spread over a relatively larger area ahead of a crack tip than for the case of crazing, the total plastic work done in propagating the crack, and thus the fracture toughness, is severely reduced. Cracks once started can propagate easily with little discernible plastic deformation. Our results show that an optimum v exists which is not too high to severely suppress local ductility but high enough to suppress crazes which can act as crack nucleation sites. The other caveat is that these results are for thin polymer films. Shear deformation zone sizes, even at the tips of initially sharp cracks, are large compared to the film thickness. In sheets where the plastic zone size is small compared to the sheet thickness, however, the plastic deformation at a crack or notch tip is localized on the surface and a triaxial state of tension builds up at depths greater than the zone size, tending to suppress further shear deformation there. In this case crazing will occur below the surface even if it is suppressed at the surface or in thin films. Brown[46] has analyzed this situation in detail for conventional impact test geometries and has shown that the critical specimen thickness for "brittle" crack propagation depends on σ_c^2/σ_y^4, where σ_c is the craze breakdown stress and σ_y is the shear yield stress. While σ_c is clearly increased by crosslinking PS, σ_y is also increased if only due to the increase in T_g on crosslinking. In addition, σ_y is rather large to begin with in uncrosslinked PS. Thus it is not obvious that the toughening induced in PS thin films by crosslinking will be translated into an observable increase in toughness for thicker specimens. What the present results do signify, however, is that one may be able to toughen glassy polymers by

736   HENKEE AND KRAMER

crosslinking if a means to relieve the triaxial stresses induced by local plastic deformation is found.   There is evidence[47,48] in the higher-entanglement-density polymers that second-phase rubber particles can play such a role by cavitating within shear bands or zones.

In the past thermosetting or crosslinked polymers have always been treated as a different class of materials from the thermoplastic or linear polymers. Certainly from the standpoint of rheology at temperatures above T_g such a distinction is amply justified.   But what we have shown in this article is that for plastic deformation well below T_g, these two classes of materials apparently behave identically.   In the thermoplastics the large strain behavior is governed by a network of entangled chains (in fact the same network that can be inferred from melt elasticity measurements).   In the crosslinked polymers the large strain behavior is governed by a network of entangled and crosslinked chains.   In both cases it appears that chain scission must accompany crazing.   It now seems plausible that for plastic deformation and fracture properties a similar treatment can be used for both classes of polymers.

The financial support of this work by the National Science Foundation through the Cornell Materials Science Center is gratefully acknowledged.   The previous experiments of Dr. Nigel Farrar on crosslinked PS at Cornell were important in developing the electron irradiation technique used for these studies.

References

1. P. Beahan, M. Bevis, and D. Hull, *Philos. Mag.*, **24**, 1267 (1971).
2. S. Rabinowitz and P. Beardmore, *Crit. Rev. Macromol. Sci.*, **1**, 1 (1972).
3. R. P. Kambour, *J. Polym. Sci. Macromol. Rev.*, **7**, 1 (1973).
4. T. E. Brady and G. S. Y. Yeh, *J. Mater. Sci.*, **8**, 1083 (1973).
5. P. Beahan, M. Bevis, and D. Hull, *J. Mater. Sci.*, **8**, 162 (1974).
6. S. T. Wellinghoff and E. Baer, *J. Macromol. Sci. Phys.*, **B11**, 367 (1975).
7. P. Beahan, M. Bevis, and D. Hull, *Proc. R. Soc. London Ser. A*, **A343**, 525 (1975).
8. D. L. G. Lainchbury and M. Bevis, *J. Mater. Sci.*, **11**, 2222 (1976).
9. A. S. Argon, R. D. Andrews, J. A. Godrick, and W. Whitney, *J. Appl. Phys.*, **39**, 1899 (1968).
10. P. B. Bowden and S. Raha, *Philos. Mag.*, **22**, 463 (1970).
11. E. J. Kramer, *J. Macromol. Sci. Phys.*, **B10**, 191 (1974).
12. G. A. Adam, A. Cross, and R. N. Haward, *J. Mater. Sci.*, **10**, 1582 (1975).
13. J. B. C. Wu and J. C. M. Li, *J. Mater. Sci.*, **11**, 434 (1976).
14. J. B. C. Wu and J. C. M. Li, *J. Mater. Sci.*, **11**, 445 (1976).
15. S. T. Wellinghoff and E. Baer, *J. Appl. Polym. Sci.*, **22**, 2025 (1978).
16. C. C. Chau and J. C. M. Li, *J. Mater. Sci.*, **14**, 1593 (1979).
17. C. C. Chau and J. C. M. Li, *J. Mater. Sci.*, **14**, 2172 (1979).
18. N. J. Mills, *Eng. Fract. Mech.*, **6**, 537 (1974).
19. I. Narisawa, M. Ishikawa, and H. Ogawa, *Polym. J.*, **8**, 181 (1976).
20. M. Iskikawa, I. Narisawa, and H. Ogawa, *Polym. J.*, **8**, 391 (1976).
21. A. M. Donald and E. J. Kramer, *J. Mater. Sci.*, **16**, 2967 (1981).
22. A. M. Donald and E. J. Kramer, *J. Mater. Sci.*, **16**, 2977 (1981).
23. A. M. Donald, E. J. Kramer, and R. A. Bubeck, *J. Polym. Sci. Polym. Phys. Ed.*, **20**, 1129 (1982).
24. A. M. Donald and E. J. Kramer, *J. Polym. Sci. Polym. Phys. Ed.*, **20**, 899 (1982).
25. A. M. Donald and E. J. Kramer, *Polymer*, **23**, 461 (1982).
26. A. M. Donald and E. J. Kramer, *Polymer*, **23**, 1183 (1982).
27. P. G. de Gennes, *J. Chem. Phys.*, **55**, 572 (1971).
28. M. Doi and S. F. Edwards, *J. Chem. Soc. Faraday Trans. 2*, **74**, 918 (1978).
29. M. Doi and S. F. Edwards, *J. Chem. Soc. Faraday Trans. 2*, **74**, 1789 (1978).

30. M. Doi and S. F. Edwards, *J. Chem. Soc. Faraday Trans. 2*, **74**, 1802 (1978).

31. G. Porod, *Monatsh. Chem.*, **80**, 351 (1949).

32. O. Kratky and G. Porod, *Rec. Trav. Chim.*, **68**, 1106 (1949).

33. E. J. Kramer, Cornell University Materials Science Center Report No. 5038, submitted for publication.

34. D. E. Kline, U.S. Pat. 3,137,633 (1964).

35. A. Van der Boogaart, in *Physical Basis of Yield and Fracture, Conf. Proc.* Institute of Physics and Physical Society, London, (1966), p. 167.

36. A. Charlesby and S. H. Pinner, *Proc. R. Soc. London Ser. A*, **A249**, 367 (1959).

37. F. A. Makhlis, *Radiation Physics and Chemistry of Polymers*, Wiley, New York, 1975, Chap. 1.

38. W. W. Parkinson and R. M. Keyser, in *The Radiation Chemistry of Macromolecules*, *Vol. 2*, M. Dole, Ed., Academic, New York, 1973, Chap. V.

39. Y. Shimizu and H. Mitsui, *J. Polym. Sci. Polym. Chem. Ed.*, **17**, 2307 (1979).

40. T. N. Bowmer, J. H. O'Donnell, and D. J. Winzor, *J. Polym. Sci. Polym. Chem. Ed.*, **19**, 1167 (1981).

41. B. D. Lauterwasser and E. J. Kramer, *Philos. Mag.*, **39A**, 469 (1979).

42. H. R. Brown, *J. Mater. Sci.*, **14**, 237 (1979).

43. A. M. Donald and E. J. Kramer, *J. Mater. Sci.*, **17**, 1871 (1982).

44. E. J. Kramer, *Adv. Polym. Sci.*, **53/54**, 1 (1983).

45. T. Chan, A. M. Donald, and E. J. Kramer, *J. Mater. Sci.*, **16**, 676 (1981).

46. H. R. Brown, *J. Mater. Sci.*, **17**, 467 (1982).

47. A. M. Donald and E. J. Kramer, *J. Appl. Polym. Sci.*, **27**, 3729 (1982).

48. A. M. Donald and E. J. Kramer, *J. Mater. Sci.*, **17**, 2351 (1982).

Received August 26, 1983
Accepted November 4, 1983